Fundamentals of
Nuclear Physics

Fundamentals of Nuclear Physics

Jagdish Varma
Professor of Physics (Retd.),
M.L. Sukhadia University,
Udaipur (India)

Roop Chand Bhandari
Professor of Physics (Retd.),
Rajasthan University,
Jaipur (India)

D.R.S. Somayajulu
Professor of Physics,
M.S. University of Baroda,
Vadodara (India)

CBS

CBS Publishers & Distributors Pvt. Ltd.
New Delhi • Bengaluru • Chennai • Kochi • Kolkata • Mumbai
Hyderabad • Nagpur • Patna • Pune • Vijayawada

ISBN: 81-239-1159-9 (PB)
ISBN: 81-239-1194-7 (HB)

First Edition: 2005
Reprint: 2010, 2013, 2017

Published by **Satish Kumar Jain** and produced by **Varun Jain** for
CBS Publishers & Distributors Pvt. Ltd.,
4819/XI Prahlad Street, 24 Ansari Road, Daryaganj, New Delhi - 110002
delhi@cbspd.com, cbspubs@airtelmail.in • www.cbspd.com
Ph.: 23289259, 23266861, 23266867 • Fax: 011-23243014

Corporate Office: 204 FIE, Industrial Area, Patparganj, Delhi - 110 092
Ph: 49344934 • Fax: 011-49344935
E-mail: publishing@cbspd.com • publicity@cbspd.com

Branches:
- ***Bengaluru:*** 2975, 17th Cross, K.R. Road, Bansankari 2nd Stage, Bengaluru - 70 • Ph: +91-80-26771678/79 • Fax: +91-80-26771680 E-mail: cbsbng@gmail.com, bangalore@cbspd.com
- ***Chennai:*** No. 7, Subbaraya Street, Shenoy Nagar, Chennai - 600030 Ph: +91-44-26681266, 26680620 • Fax: +91-44-42032115 E-mail: chennai@cbspd.com
- ***Kochi:*** Ashana House, 39/1904, A.M. Thomas Road, Valanjambalam, Ernakulum, Kochi • Ph: +91-484-4059061-65 Fax: +91-484-4059065 • E-mail: cochin@cbspd.com
- ***Kolkata:*** 6-B, Ground Floor, Rameshwar Shaw Road, Kolkata - 700014 Ph: +91-33-22891126/7/8 • E-mail: kolkata@cbspd.com
- ***Mumbai:*** 83-C, Dr. E. Moses Road, Worli, Mumbai - 400018 Ph: +91-9833017933, 022-24902340/41 • E-mail: mumbai@cbspd.com

Representatives:
- Hyderabad: 0-9885175004
- Patna: 0-9334159340
- Vijayawada: 0-9000660880
- Nagpur: 0-9021734563
- Pune: 0-9623451994

Printed at:
J.S. Offset Printers, Delhi (India)

PREFACE

Teaching postgraduate classes for over three decades, our experience has been that consultation of several books is essential for doing justice to a nuclear physics course. Many of these books are rare and in a large number of universities and postgraduate colleges they are not easily available to the students and teachers. The present book is an effort to provide relevant physics at one place. The book takes care of the postgraduate nuclear physics course in most of the universities. It is our hope that the book will revive interest in the study of nuclear physics.

This book has been written under the auspices of University Grants Commission. The authors appreciate the help of Dr. K. Venugopalan, Dr. (Mrs.) N. Laxmi, Dr. K. C. Sebastian and Mr. Mukesh Chavde in the preparation of this manuscript. We wish to acknowledge the cooperation of M. L. Sukhadia University in our effort.

Authors

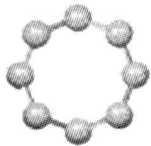

CONTENTS

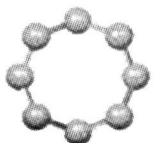

1

PROPERTIES OF STABLE NUCLEI

1.1 NUCLEAR CONSTITUENTS

J. J. Thomson in 1898 proposed that the nucleus of an atom consists of protons and electrons and its size is almost as big as that of the atom. In 1911 Geiger and Marsden performed an experiment on the scattering of α-particles by gold foil under the guidance of Lord Rutherford. The experiment showed that the gold nucleus has a positive charge of 79 e and a radius of about 5×10^{-15} meters. On the basis of such experiments, Rutherford proposed that the nucleus of an atom is heavy and positively charged—the size of the nucleus is $\approx 10 \times 10^{-15}$ m and electrons, whose number is equal to the atomic number of the atom, revolve round the nucleus to form a neutral atom. The atomic mass of gold nucleus is about 197 times the mass of a proton. The large mass of the nucleus remained a mystery till 1932 when Chadwick discovered neutrons. The atomic nucleus is thus proposed to consist of protons and neutrons. At the energies available in a nucleus, the uncertainty principle of Heisenberg could be employed to show that an electron cannot be confined in the nuclear volume. If Z is the atomic number, A the atomic mass number, then a nucleus consists of Z protons and $A - Z = N$ number of neutrons, bound together in a volume of radius about 5×10^{-15} m. The size of an atom is about 10^{-10} m which is about 10^4 times larger than that of the nucleus.

Thomson's positive ray analysis showed the presence of two types of neon atoms, one having a mass of 20 and the other mass 22 times heavier than a proton. Each of these two types of neon atoms has the same number of protons (equal to Z), hence they must differ in the number of neutrons. ^{20}Ne should have 10 protons and 10 neutrons while ^{22}Ne should have ten protons and twelve neutrons. Ten electrons revolve round the nucleus to form a neutral neon atom in both ^{20}Ne and ^{22}Ne.

Nuclei, which have the same number of protons Z but different number N of neutrons, are known as Isotopes. The element neon, therefore, has two isotopes

^{20}Ne and ^{22}Ne. Nuclei, which have the same number of neutrons but differ in their proton number are known as Isotones. The nuclei in which the proton number and the neutron number are interchanged are known as mirror nuclei e.g. ^{25}Al and ^{25}Mg. Nuclei with same mass number A but different proton number are called Isobars. The word nucleon is often used to represent a neutron or a proton without distinguishing between the two. The mass number of a nucleus is thus the number of nucleons ($N + Z$) present in it. A widely used term nuclide refers to a particular species of nuclei. Each isotope of an element is a nuclide. Its atomic number Z, mass number A and the chemical symbol of the element identify a nuclide. The ^{13}C isotope of carbon is depicted as $^{13}_{6}C_{7}$. Similarly the ^{238}U isotope of uranium is depicted as $^{238}_{92}U_{146}$.

The chemical properties of an element are determined by the number of electrons in the outermost orbit of the atom. These properties are not affected by the number of neutrons present in the nucleus and are, therefore, same for all the isotopes of an element. The atomic weight of an element is the weighted average of the atomic masses of all the isotopes present. In chlorine for example, two isotopes are present $^{37}_{17}Cl$ and $^{35}_{17}Cl$ in the ratio of 1 : 3. This gives the atomic weight of chlorine as 35.46. The abundance of ^{20}Ne and ^{22}Ne isotopes is 10 : 1 giving the atomic weight of neon as 20.18.

If one plots the number of neutrons as a function of protons present in all the known stable as well as radioactive nuclides, one obtains a curve, rather a band as shown in Fig. 1.1. It is observed that in stable nuclides upto $Z = 20$, the number of neutrons and protons is equal. As the number of protons increases, the number of neutrons increases at a faster rate. ^{56}Fe has $Z = 26$ and neutron number $N = 30$ while ^{197}Au has $Z = 79$ and $N = 118$. In uranium 238, $Z = 92$ and $N = 146$. It may be argued that as Z increases, the Coulomb repulsion between the protons becomes very strong and a larger number of neutrons are required to keep the nucleus intact.

The stable nuclei lie approximately in the middle of the band in Fig. 1.1 The zig zag line joining the stable nuclei is known as the line of β-stability. The nuclei lying above this line have an excess of neutrons; hence they decay with the emission of β-particles, changing a neutron into a proton. The nuclei, which lie below the line of β-stability, have an excess of protons and decay by β^+ emission. The heavy nuclei which decay by emission of α-particles all lie above $Z = 82$.

1.2 NUCLEAR MASS

After the pioneer experiment of Thomson, showing the presence of isotopes $A = 20$ and 22 in neon, great efforts were devoted in developing mass spectrometers, also known as mass spectroscopes. Mass spectrometers were extensively used for measuring atomic masses and isotopic abundance ratio in elements. Presently they find application in identifcation of products of nuclear reactions, chemical analysis of complicated vapour mixtures and routine testing of gas leaks in enclosures and vacuum chambers. A mass spectrometer basically,

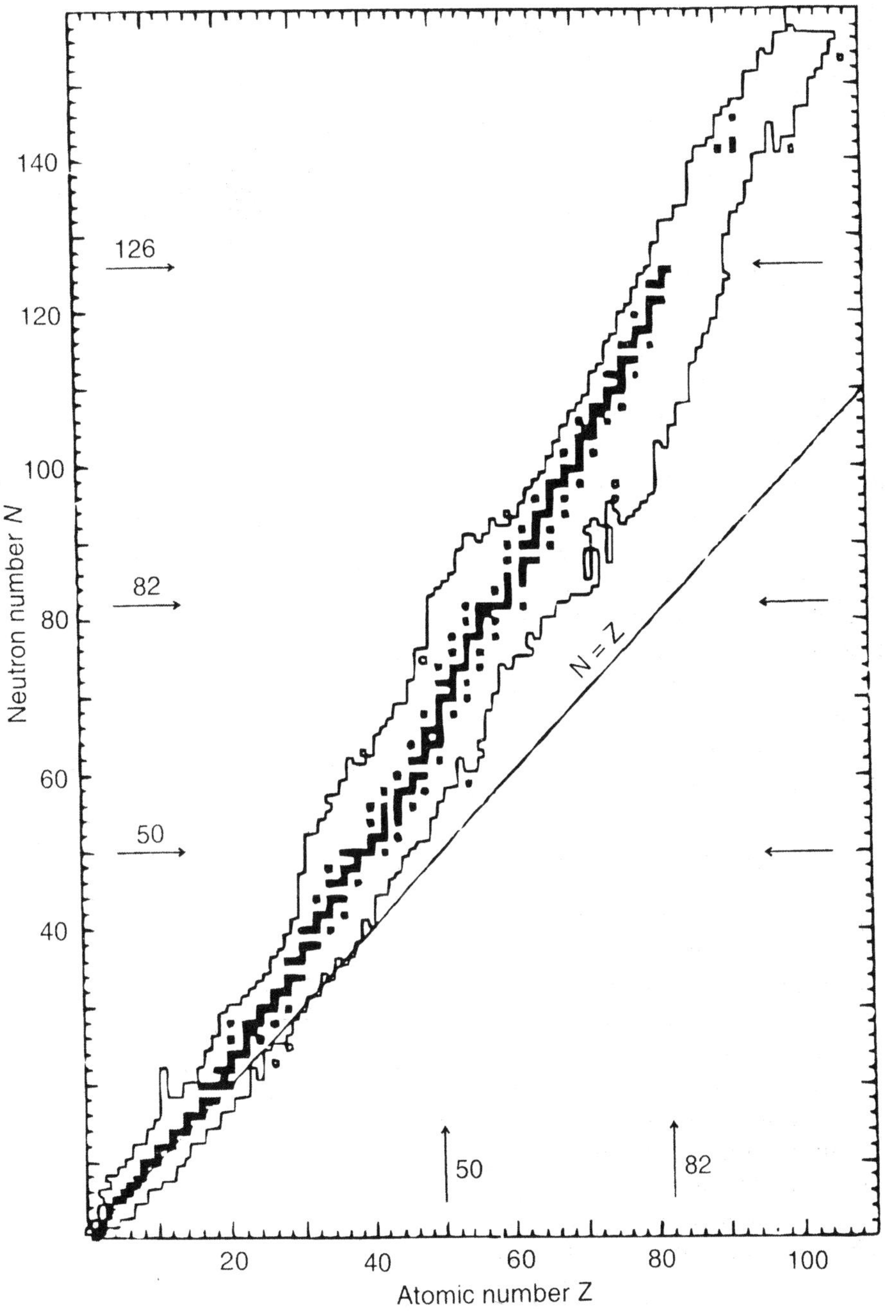

Fig. 1.1: A plot of neutron number N as a function of atomic number Z of nuclei. Filled squares denote the stable nuclei. Radioactive nuclei fill the area bounded by the lines. The nuclei falling outside the lines decay by nucleon emission (Chart of Nuclides, 1977, General Electric Co. Schenectady, USA).

consists of an ion source which produces positively charged ions of the atoms under study, and an arrangement of electrostatic and magnetic fields to measure the energy and momentum of the ions. The positively charged ions usually have charge $+e$ or some times a charge Q which is multiple of $+e$ so that $Q = ne$ where n is an integer. The electric and the magnetic fields bring to a focus, ions of the same mass (assuming all of them to carry the same charge) and varying speeds and directions within a small range. An ion has associated with it the mass M, charge Q, momentum p and kinetic energy W_k. The accuracy with which the mass m can be determined depends upon the accuracy in the measurement of p and W_k.

The point at which the ion beam coming from the ion source is focussed has a finite size and we can define the resolving power of a mass spectrometer as $R_M = M/\Delta M$ where ΔM is the spread in the focal point on the mass scale. The main factors which determine the mass resolving power are (1) the width of the slit at the exit of the ion source, (2) departure from perfect focussing by the electric and the magnetic fields (3) scattering of ions by the residual gas in the spectrometer and (4) imperfections in the construction of the apparatus. If the size of the ion source slit is reduced to increase the resolving power, the intensity of the transmitted ion beam decreases correspondingly. A mass spectrometer should have a high ion gathering power which depends upon the range of speeds and directions which the spectrometer can focus. The ion gathering power or transmission of the spectrometer is the fractional solid angle in which the ions are focussed in the spectrometer.

Ion sources have been developed with different designs. It may be a high voltage low current electric discharge in the rarefied gas, composed of atoms under study. The ions are extracted from the discharge tube through a narrow canal in the cathode. The emerging ions have usually a large spread in their speed. In the ion source in which a copious beam of electrons bombards the neutral gas or vapour particles produces ions having a narrow range of speeds. If the substance to be studied is in liquid or solid form, it can be first vaporised in an oven and then bombarded by the electron beam. In another type of ion source a low voltage arc is struck between two electrodes, with the anode having a cavity containing the source material. The ions extracted from an ion source through a canal in the cathode are accelerated through a voltage of 4000 to 40000 volts and collimated by a very fine slit of width about 0.1 mm. The ion beam coming out from the collimating slit is subjected to different configurations of electric and magnetic fields.

If an ion beam is passed through crossed electric field E and magnetic field B, perpendicular to the beam, the velocity v of the ions which remain undeflected is given as

$$v = c\frac{E}{B}$$

where c is the velocity of light. In such a velocity selection, the ions diverging from the beam direction are lost.

A more elegant velocity selector which has a directional focussing property, is made of two closely spaced concentric, coaxial, cylindrical plates, forming a cylindrical condenser. If the outer plate is at a potential of $+V'$ with respect to the inner plate, the electric field E points radialy inwards. If E_0 is the intensity of the electric field at the mean radius r_0 of the plates, the ions moving in a circular path along the mean radius with velocity v_0 experience the centripetal force QE_0 so that

$$QE_o = \frac{Mv_o^2}{r_o} \qquad \ldots(1.2.1)$$

If after extraction from the ion source, the ions were accelerated through a potential V then $QV = \frac{1}{2}Mv_o^2$ which combined with equation (1.2.1) gives

$$QV = \frac{1}{2}E_o r_o Q \qquad \ldots(1.2.2)$$

From the knowledge of E_0 and r_o the energy of the ions coming out of the cylindrical condenser can be calculated. The intensity of electric field E between the plates at a distance r from the axis of the plates is given as

$$E = E_o \frac{r_o}{r}$$

The electric field outside the mean radius is weaker than the field inside it. The cylindrical condenser sector has optimum directional focussing property when the sector angle is $\pi / \sqrt{2}$ radians = 127°17'. The ions having same kinetic energy but diverging in direction by about 1°, as they enter the cylindrical condenser, are brought to a focus at the exit slit.

If the object (ion source slit) and the image point (where the ions are focussed) are kept at certain critical positions outside the electric field, directional focussing of ions can be obtained by a "sector field" using cylindrical condenser subtending an angle of less than $\pi / \sqrt{2}$ radians. If the plates subtend an angle α at the axis, directional focussing is obtained when the object, and image points are symmetrically situated at a distance $(r/\sqrt{2}) \cot(\alpha/2)$ from the centre.

A magnetic field acting perpendicular to the motion of the ions, exerts a force normal to the ion velocity. If the magnetic field acts over the whole trajectory, the ions with slight difference in velocity are focussed after they have deflected through an angle of 180°. A wedge shaped sector magnetic field in a region bound by two radii subtending an angle α, produces both directional and velocity focussing of ions. The conditions for focussing are (1) the object point, the image point and the vertex of the wedge are colinear and (2) the angle, the incident and the outgoing beams make with the line is, $\alpha/2$. For a given geometry, the magnetic field can be adjusted to fulfil the above conditions. The above mentioned arrangements of electric and magnetic fields have been utilised by various workers to design mass spectrometers.

Bainbridge and Jordon constructed a mass spectrometer in which the cylindrical condenser subtended an angle of 127° and the sector magnetic field subtended an angle of 60°. Fig. 1.2 shows a schematic diagram of their mass spectrometer. The ions were accelerated through a potential of about 4000 volts in the ion source itself. The mean radius of the ion path in both electric and magnetic fields was about 25 cm. The ions get focussd at a photographic plate. where they produce black spots. The mass scale on the photographic plate is linear. It may be noticed that the exit slit of the condenser, the apex of the magnet and the photographic plate are colinear. Bainbridge and Jordon were able to obtain a resolving power $M/\Delta M$ of about 10000. As the blackness produced in the photographic plate is not linear with the number of ions focussed on it, the spectrometer can not be fruitfully used for measuring isotopic abundance ratios. It is used primarily for determination of nuclear masses.

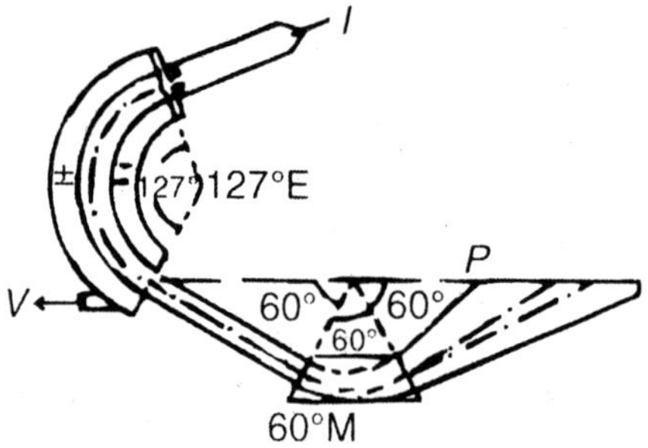

Fig. 1.2: Schematic diagram of the mass spectrometer of Bainbridge and Jordan. *I* is the ion source, *E* the 127° electrostatic velocity analyzer, *M* the 60° sector magnet and *P* the photographic plate The whole system is evacuated by a vacuum pump at *V*.

Nier and Roberts developed a mass spectrometer (Fig. 1.3) in which the energy analyser (cylindrical condenser) subtended an angle of 90° with a mean radius of 19 cm. The magnetic analyser subtended an angle of 60° with a mean radius of 15 cm. The ions after extraction from the ion source were accelerated by a potential of 20000 volts. After the focal slit of the magnetic analyser, the ions were collected in an electrometer in which the ion current of about 10^{-12} amperes was detected. In this spectrometer the mass spectrum is scanned by changing the electrostatic field in the condenser and the voltage accelerating the ions.

With advances in the technology of ion sources and detection of ions, it is now routinely possible to get a mass resolving power of the 10^6. The masses of almost all stable and long lived radioactive nuclei have been measured with these mass spectrometers.

While kilogram is the basic unit of mass, nuclear masses are measured in units of atomic mass unit (amu or μ). The atomic mass unit is defined as 1/12 the mass of the most abundant carbon atom ^{12}C. The value of the atomic mass unit is as follows:

$$1 \text{ amu} = 1.6604 \times 10^{-27} \text{ kg}$$
$$= 931.48 \text{ MeV}$$

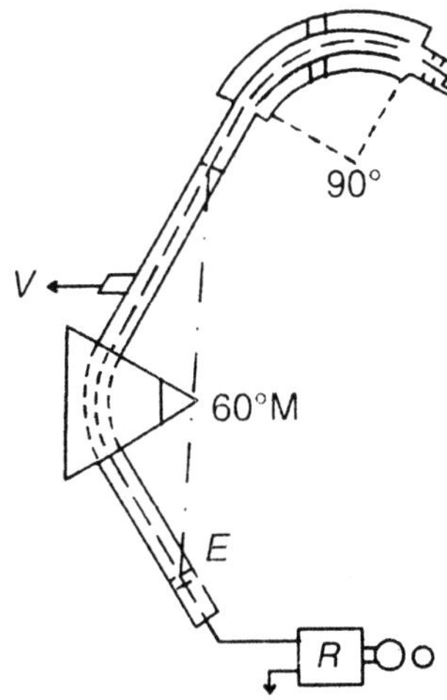

Fig. 1.3: Schematic diagram of the mass spectrometer of Nier and Roberts. The ions are produced in the ion source *I* and accelerated to an energy of about 20000 eV by the electrode *E*. After passing through the 90° electrostatic velocity selector the ions pass through a 60° sector magnet *M*. The ions are detected by an electrometer *E* whose deflection is recorded in *R*.

The equivalence of one atomic mass unit to 931.48 MeV follows from the Einstein's mass energy equivalence equation ($E = mc^2$). The masses of some common particles and their energy equivalent is given below.

Mass of electron, $M_e = 0.000548\ \mu$

$= 0.511$ MeV

Mass of proton, $M_p = 1.0072766\ \mu$

$= 938.25$ MeV

Mass of hydrogen atom $= 1.007825\ \mu$

$(M_p + M_e)$

Mass of neutron $= 1.008665\ \mu$

$= 939.55$ MeV

Mass of deuterium atom $= 2.014102\ \mu = 1876.09$ MeV

The deuterium atom, an isotope of hydrogen, consists of a proton and a neutron bound together to form the nucleus and an electron going round. It is seen from the masses given above, that the deuterium or heavy hydrogen atom is lighter than the combined masses of a neutron and a hydrogen atom. The loss in mass is attributed to the binding energy between the proton and the neutron in the nucleus. From the above mass data it is seen that $M_H + M_n - M_d$ (1.007825 + 1.008665 − 2.014102) = 0.002393 which is equivalent to 2.22 MeV, the binding energy of deuteron. In general one can write the mass $M(Z, A)$ of a nucleus of atomic number Z and neutron number $A - Z = N$ as

$$M(Z, A) = ZM_H + NM_n - \Delta M$$

where ΔM is the mass defect and is related to binding energy of the nucleus as

$$B.E. = (ZM_H + NM_n - M(Z, A)\,c^2 \quad ...(1.2.3)$$

The binding energy of a nucleus is thus the energy required to break a nucleus into its constituent particles. It may be observed that the binding energy of the two-nucleon nucleus—the deuteron—is about 10^5 times the binding energy of an electron in the atom. Hence in calculating the binding energy of a nucleus (which is same as binding energy of the atom), the binding energy of the electron in the atom is completely ignored. In the case of an atom—say hydrogen atom—when a free electron is bound to a proton, the binding energy of the electron in the hydrogen atom is emitted in the form of electromagnetic radiations. In a similar fashion, when a neutron and a proton join together to form a deuteron, the binding energy of the system which is 2.22 MeV is emitted in the form of electromagnetic radiations *i.e.* gamma rays of energy 2.2 MeV.

From the measured masses of the stable nuclides, their binding energy (B.E.) can be calculated. Dividing this binding energy by the number of nucleons, A in the nucleus, one can obtain the average binding energy of a nucleon in the nucleus. This average binding energy of a nucleon is only approximately equal to the energy required to remove a nucleon from the nucleus.

If one plots binding energy per nucleon (BE/A) for all the known stable nuclei one obtains a curve as shown in Fig. (1.4). The binding energy per nucleon for deuteron is 1.1 MeV and rises to 7 MeV for ^{4}He and to 8.6 MeV for $A \approx 50$. Beyond $A \approx 60$ the value of BE/A decreases gradually and becomes about 7.8 MeV for uranium. Generally speaking the binding energy per nucleon is almost constant at about 8 MeV, in all the nuclei ($A > 4$). A few kinks can be observed in the curve, the most prominent being for $A = 4$ *i.e.* $^4_2He_2$. Other kinks appear at A = 8, 12 and 16. These kinks are attributed to a shell structure in nuclei (Chapter 4).

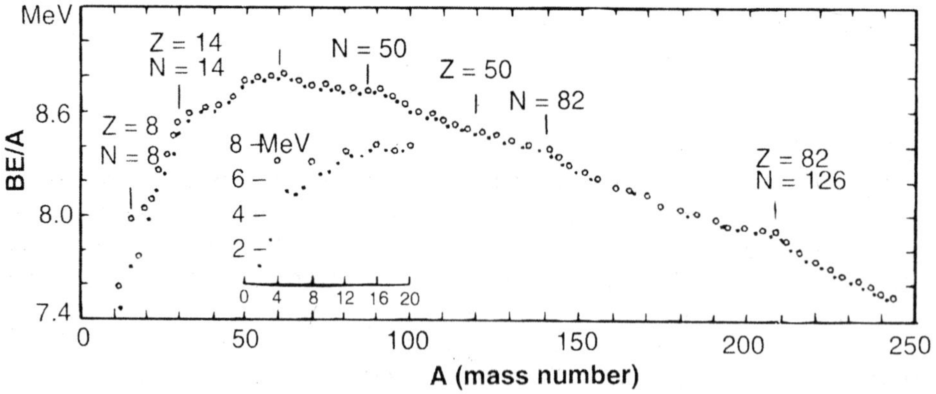

Fig. 1.4: Binding energy per nucleon as a function of atomic mass A of nuclei. The positions of the magic numbers where the curve shows kinks are indicated. In the insert the binding energy per nucleon is plotted for the lightest nuclei.

The binding energy per nucleon in a nucleus is much larger than the average binding energy of an electron in an atom (almost a factor of 10^5). This implies that the nuclear forces, which bind the neutrons and protons inside the nucleus, are much stronger than the electromagnetic forces binding the electrons. So the nuclear forces are "strong" (Chap. 2).

If in a nucleus every nucleon attracts every other nucleon with the same strong force, the number of bonds in a nucleus of atomic mass A will be $A(A-1)$ and the binding energy would also be proportional to $A(A-1) \approx A^2$. This gives BE/A to be proportional to A. The observed value of BE/A is almost independent of A, signifying that a particular nucleon in a nucleus attracts only few other nucleons in its neighbourhood. This implies that the nuclear forces are saturable (Chapter 2).

The measurement of nuclear masses thus gives important information about the nuclear forces, that they are strong and saturable.

1.3 NUCLEAR SIZE

A nucleus of an atom except hydrogen, consists of neutrons and protons. The protons being positively charged, give rise to an electrostatic Coulomb potential about the nucleus. In addition both neutrons and protons give rise to the nuclear potential due to the strong nuclear forces. The different types of radii that can be attributed to a nucleus are:

(*a*) radius of neutron distribution R_n.
(*b*) radius of proton or charge distribution R_p
(*c*) Coulomb potential radius R_c
(*d*) nuclear potential radius R_N

The particular radius measured, is determined by the experiment employed. Analysis of direct nuclear reactions e.g. stripping reactions, indicates that the radius of the neutron distribution in a nucleus is very nearly the same as that of proton distribution. This is in conformity with the strong nuclear forces between a proton and a neutron. It is known that the Coulomb potential is maximum at the surface of a uniformly charged sphere, beyond which it decreases as $1/r$. If the density of protons inside the nucleus is uniform throughout, the Coulomb potential radius R_c and the proton distribution radius R_p would be same. If the proton density is not uniform then the two radii R_p and R_c would be different. The nuclear potential does not follow a simple decay law as the Coulomb potential. Experiments show that the nuclear potential falls off to zero after a certain distance r_o, which is, called the range of nuclear potential. The effect of the nuclear potential becomes significant at a distance less than r_o from the nuclear surface. This means that R_N is approximately equal to $R_p + r_o$.

Some of the experiments performed to estimate the size of the nucleus are discussed in subsequent sections.

1.3.1 Rutherford Scattering of α-particles

The experiment on the Coulomb scattering of α-particles by gold nucleus was performed in 1911 by Geiger and Marsden. In this experiment a collimated beam of α-particles emitted from a radon tube was incident on a thin foil of gold. The number of α-particles scattered by the foil at different angles was detected by the scintillations they produced in a ZnS screen observed through a microscope.

The gold foil scatterer could be replaced by foils of other metals. The energy of the α-particles could be changed by placing very thin metal foils of known thickness in front of the α-particle source. The arrangement of the α-particle source, the scatterer and the ZnS screen was placed in high vacuum. As the light produced by the α-particles on the ZnS screen was very faint, the experiment was performed in a photographic dark room.

The expression for the cross section for Coulomb scattering of α-particles can be derived on the principle of conservation of energy and angular momentum. Let N be a nucleus with charge Ze and mass M. An α-particle of charge ze and mass m is incident along the direction AB (Fig. 1.5). As the α-particle approaches the nucleus, it experiences a Coulomb repulsion F which changes the path of the particle. The scattered particle goes out in the direction CD, while the nucleus recoils. If the nucleus is much heavier than the α-particle, the nuclear recoil can be neglected.

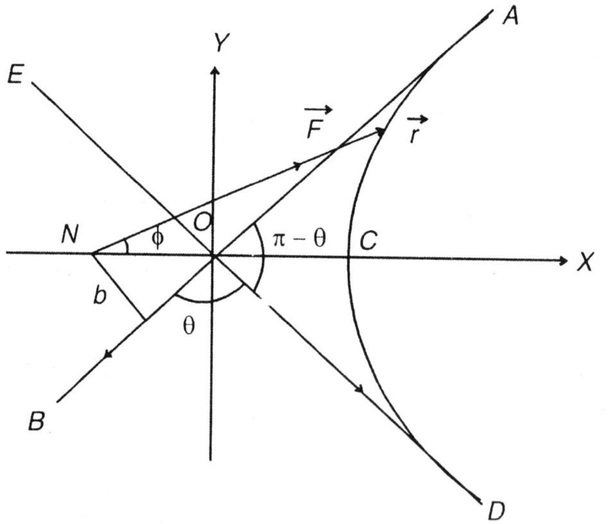

Fig. 1.5: Coulomb scattering of α-particle by a nucleus N. AB would be the trajectory of the α-particle in the absence of the Coulomb field. Due to Coulomb repulsion $\vec{F}$ the trajectory becomes ACD.

Taking position of the nucleus as one pole, let (r, ϕ) be the co-ordinates of the α-particle at any instant of time during its transit. The electrostatic repulsive force acting on the α-particle at this instant is given as

$$F = \frac{Ze\,ze}{4\pi\varepsilon_o} \cdot \frac{\vec{r}}{r^3} \qquad ...(1.3.1)$$

The radial acceleration due to this force is

$$a_r = \frac{d^2r}{dt^2} - r\left(\frac{d\phi}{dt}\right)^2 = \frac{Ze\,ze}{4\pi\varepsilon_o mr^2} \qquad ...(1.3.2)$$

The acceleration a_ϕ perpendicular to the radius vector is zero because F has no component perpendicular to r so that

$$a_\phi = \frac{1}{r}\frac{d}{dt}\left(r^2\frac{d\phi}{dt}\right) = 0 \qquad ...(1.3.3)$$

Equation 1.3.3 on integration gives

$$r^2\frac{d\phi}{dt} = C\text{, a constant} \qquad ...(1.3.4)$$

The angular momentum of the α-particle with respect to the nucleus is mv_0b where v_0 is its velocity and b the perpendicular distance of the line representing velocity v_0, from the nucleus. b is called the impact parameter and represents the distance from the nucleus at which the α-particle would have passed if it were not deflected by the Coulomb force. As the force acting at the α-particle is always along the radius vector, the angular momentum remains constant throughout the trajectory. This follows from equation (1.3.4) also, so that

$$mv_0b = mr^2\frac{d\phi}{dt} = mC \qquad ...(1.3.5)$$

mr^2 in equation (1.3.5) represents the moment of inertia of the α-particle about an axis passing from the nucleus N and $d\phi/dt$ is the instantaneous angular speed. To obtain the trajectory of the α-particle from equation (1.3.2) and (1.3.5) we change the variable r to u such that $u = \frac{1}{r}$

so that $$\frac{dr}{d\phi} = -\frac{1}{u^2}\frac{du}{d\phi} \qquad ...(1.3.6)$$

and $$\frac{dr}{dt} = \frac{dr}{d\phi}\cdot\frac{d\phi}{dt} \qquad ...(1.3.7)$$

as $$\frac{d\phi}{dt} = \frac{C}{r^2} = Cu^2$$

we have

$$\frac{d^2r}{dt^2} = \frac{d}{dt}\left(\frac{dr}{dt}\right) = \frac{d}{dt}\left(-C\frac{du}{d\phi}\right)$$

$$\frac{d^2r}{dt^2} = -C\frac{d}{d\phi}\left(\frac{du}{d\phi}\right)\cdot\frac{d\phi}{dt}$$

$$= -C\frac{d^2u}{d\phi^2}\cdot\left(\frac{C}{r^2}\right)$$

$$= -C^2u^2\frac{d^2u}{d\phi^2} \quad ...(1.3.8)$$

Substituting the value of d^2r/dt^2 from equation (1.3.8) in equation (1.3.2) gives

$$-C^2u^2\frac{d^2u}{d\phi^2} - r\left(\frac{C}{r^2}\right)^2 = \frac{Zze^2}{4\pi\varepsilon_0\, mr^2} \quad ...(1.3.9)$$

Putting $u = 1/r$ in (1.3.9)

$$\frac{d^2u}{d\phi^2} + u = -\frac{Zze^2}{4\pi\varepsilon_0 mC^2} \quad ...(1.3.10)$$

as the right hand side is a constant one can write

$$\frac{d^2}{d\phi^2}\left(u + \frac{Zze^2}{4\pi\varepsilon_0 C^2m}\right) = -\left(u + \frac{Zze^2}{4\pi\varepsilon_0 C^2m}\right) \quad ...(1.3.11)$$

The solution of the equation gives

$$u + \frac{Zze^2}{4\pi\varepsilon_0 C^2m} = A\cos(\phi - \alpha) \quad ...(1.3.12)$$

where A and α are constants of integration. By a proper choice of the x-axis one can have $\alpha = 0$ giving

$$u = \frac{1}{r} = -\frac{Zze^2}{4\pi\varepsilon_0 C^2m}\left(1 - \frac{4\pi\varepsilon_0 AmC^2}{Zze^2}\cos\phi\right) \quad ...(1.3.13)$$

Putting $\varepsilon = \dfrac{4\pi\varepsilon_0 AmC^2}{Zze^2}$ gives

$$\frac{1}{r} = -\frac{Zze^2}{4\pi\varepsilon_0 mC^2}(1 - \varepsilon\cos\phi) \quad ...(1.3.14)$$

This equation connects the radius vector r with the polar angle ϕ and therefore, gives the equation of the trajectory of the α-particle. It is a conic section with eccentricity ε.

At the point (r, ϕ) the conservation of energy gives

$$\frac{1}{2}mv_0^2 = \frac{1}{2}mv^2 + \frac{Zze^2}{4\pi\varepsilon_0 r} \quad ...(1.3.15)$$

v is the instantaneous speed of the α-particle and the second term gives its potential energy.

Now

$$v^2 = \left(\frac{dr}{dt}\right)^2 + \left(r\frac{d\phi}{dt}\right)^2$$

$$= \left(\frac{d\phi}{dt}\right)^2\left[\left(\frac{dr}{d\phi}\right)^2 + r^2\right]$$

$$= \frac{C^2}{r^4}\left[\left(\frac{dr}{d\phi}\right)^2 + r^2\right] \qquad \text{...(1.3.16)}$$

Equation (1.3.14) gives

$$-\frac{1}{r^2}\frac{dr}{d\phi} = -\frac{Zze^2}{4\pi\varepsilon_0 mC^2} \in \sin\phi$$

or $$\frac{dr}{d\phi} = \frac{Zze^2}{4\pi\varepsilon_0 m}\cdot\frac{r^2}{C^2} \in \sin\phi \qquad \text{.... (1.3.17)}$$

Substituting the value of $dr/d\phi$ in equation (1.3.16) and using equation (1.3.14) gives

$$v^2 = \left(\frac{Zze^2}{4\pi\varepsilon_0 mC}\right)^2\left[\varepsilon^2\sin^2\phi + (1-\varepsilon\cos\phi)^2\right] \qquad \text{... (1.3.18)}$$

Putting the value of v^2 in equation (1.3.15) gives

$$v_0^2 = \left(\frac{Zze^2}{4\pi\varepsilon_0 mC}\right)^2(\varepsilon^2 + 1 - 2\varepsilon\cos\phi) + \frac{2}{m}\frac{Zze^2}{4\pi\varepsilon_0 r}$$

Substituting $\frac{1}{r} = -\frac{Zze^2}{4\pi\varepsilon_0 mC^2}(1-\varepsilon\cos\phi)$ gives

$$v_0^2 = \left(\frac{Zze^2}{4\pi\varepsilon_0 mC}\right)^2(\varepsilon^2 + 1 - 2\varepsilon\cos\phi) + \frac{Zze^2}{4\pi\varepsilon_0 m}2\cdot\left[-\frac{Zze^2}{4\pi\varepsilon_0 mC^2}(1-\varepsilon\cos\phi\right] \qquad \text{... (1.3.19)}$$

$$= \left(\frac{Zze^2}{4\pi\varepsilon_0 mC}\right)^2(\varepsilon^2 + 1 - 2\varepsilon\cos\phi - 2 + 2\varepsilon\cos\phi)$$

$$= \left(\frac{Zze^2}{4\pi\varepsilon_0 mC}\right)^2(\varepsilon^2 - 1)$$

$$\therefore \qquad (\varepsilon^2 - 1) = \left(\frac{4\pi\varepsilon_0 m v_0 C}{Zze^2}\right)^2 \qquad \text{...(1.3.20)}$$

It is evident that $\varepsilon > 1$, hence the trajectory of the α-particle is a hyperbola with the nucleus at the external focus. The initial direction *AB* is an asymptote of the hyperbola while the other asymptote *CD* gives the direction of the scattered α-particle. As the nucleus is assumed to be very heavy, the velocity of the scattered α-particle is same as the incoming velocity. The angle of scattering is given by the angle between the lines *AB* and *CD*, the two asymptotes to the hyperbola.

In Cartesian co-ordinates the equation of the hyperbola is

$$\frac{x^2}{a_0^2} - \frac{y^2}{b_0^2} = 1, \qquad \text{...(1.3.21)}$$

where $$a_0 = \frac{NO}{\varepsilon}$$

and $$\frac{b_0^2}{a_0^2} = \varepsilon^2 - 1 \qquad \text{...(1.3.22)}$$

The equations of the asymptotes are obtained from

$$\frac{x^2}{a_0^2} - \frac{y^2}{b_0^2} = 0$$

or $$\frac{b_0^2}{a_0^2} = \frac{y^2}{x^2} = \tan^2\left(\frac{\pi - \theta}{2}\right) = \cot^2\left(\frac{\theta}{2}\right)$$

Thus

$$\varepsilon^2 - 1 = \cot^2(\theta/2) \qquad \text{...(1.3.23)}$$

Combining equations (1.3.23), (1.3.20) and (1.3.5) gives

$$\cot\left(\frac{\theta}{2}\right) = \frac{4\pi\varepsilon_0 m v_0^2}{Zze^2} \cdot b \qquad \text{...(1.3.24)}$$

The angle of scattering of the α-particle is related to the impact parameter *b*. For a head on collision the impact parameter *b* is zero and the α-particle is scattered in the backward direction ($\cot \theta/2 = 0$,. or $\theta/2 = \pi/2$).

To obtain the expression for the cross-section for α-particle scattering consider N_0 number of α-particles of some fixed energy falling normally on an area *S* of a thin foil of thickness *t* containing *n* atoms per unit volume. It is further assumed that there is no multiple scattering of the α-particle and a single encounter scatters the particle by angle θ. The thickness of the foil is assumed to be small so that there is no change in the energy of the α-particle. It is seen from equation 1.3.24 that all α-particles whose impact parameter *b* lies between the value 0 and *b* are scattered by an angle greater than θ. Thus an α-particle incident on an area πb^2 around a nucleus is scattered by angle θ or more. The number of nuclei in the foil is nSt so the total area of the foil which contributes to α-particle scattering by angle greater than θ is $nSt\pi b^2$. The fraction *f* of the α-particles which fall on this area is

$$f = \frac{nSt\pi b^2}{S} = \pi ntb^2 \quad ...(1.3.25)$$

$$f = \pi nt\left(\frac{Zze^2}{4\pi\varepsilon mv_0^2}\right)^2 \cot^2(\theta / 2) \quad ...(1.3.26)$$

The probability that an α-particle will have an impact parameter between b and $b + db$ and will be scattered by angle between θ and $\theta + d\theta$ is obtained by differentiating the above equation as

$$df = -\pi nt \cdot \left(\frac{Zze^2}{4\pi\varepsilon_0 mv_0^2}\right)^2 \cot(\theta/2)\text{cosec}^2(\theta/2) \cdot d\theta \quad ...(1.3.27)$$

The particles scattered in the angle range θ and $\theta + d\theta$ move in a solid angle $d\Omega = 2\pi \sin\theta \, d\theta$. Hence the fraction of particles scattered at angle θ in unit solid angle is given by

$$\frac{df}{d\Omega} = \frac{\pi nt \cdot \left(\frac{Zze^2}{4\pi\varepsilon_0 mv_0^2}\right)^2 \cot(\theta / 2)\text{cosec}^2(\theta / 2) \cdot d\theta}{2\pi \sin\theta d\theta}$$

$$= \left(\frac{Zze^2}{4\pi\varepsilon_0 mv_0^2}\right)^2 \cdot \frac{nt \cdot \cot(\theta/2)\text{cosec}^2(\theta/2)d\theta}{2 \times 2\sin(\theta/2)\cos(\theta/2)d\theta}$$

$$= \frac{nt}{4}\left(\frac{Zze^2}{4\pi\varepsilon_0 mv_0^2}\right)^2 \text{cosec}^4(\theta/2) \quad ...(1.3.28)$$

The cross-section σ(θ) is defined as the probability that an α-particle is scattered by one scattering centre ($nt = 1$) at angle θ in a solid angle $d\Omega$, so that

$$\sigma(\theta) = \left(\frac{Zze^2}{8\pi\varepsilon_0 mv_0^2}\right)^2 \text{cosec}^4(\theta / 2) \quad ...(1.3.29)$$

Putting $z = 2$ for α-particle

$$\sigma(\theta) = \left(\frac{Ze^2}{4\pi\varepsilon_0 mv_0^2}\right)^2 \text{cosec}^4(\theta / 2) \quad ...(1.3.30)$$

The kinetic energy of the α-particle $E_\alpha = \frac{1}{2}mv_0^2$ so that

$$\sigma(\theta) = \left(\frac{Ze^2}{8\pi\varepsilon_0 E_\alpha}\right)^2 \frac{1}{\sin^4(\theta / 2)} \quad ...(1.3.31)$$

The dependence of α-particle scattering on the atomic number Z of the scattering material, energy E_α of the α-particles and the scattering angle θ, were verified in a series of experiments by *H.*Geiger and *E.* Marsden. Using α-particles of 7.7 *MeV* energy, Geiger and Marsden could not observe any deviation from the Rutherford's formula. They showed that the size of the gold nucleus should be smaller than the smallest distance of approach of the α-particle from the nucleus. This would happen when the α-particle approaches the nucleus head on (impact parameter $b = 0$) and gets scattered by 180°. At the instant of closest approach r_o the initial kinetic energy E_α of the particle is completely converted into its potential energy due to the positively charged nucleus; so that

$$E_\alpha = \frac{1}{4\pi\varepsilon_0}\frac{2Ze^2}{r_0} \qquad \text{...(1.3.32)}$$

For 7.7 MeV α-particles and gold nucleus, the smallest distance of approach r_0 is

$$r_o = \frac{(9\times10^9)\cdot 2.79(1.6\times10^{-19})^2}{7.7\times(1.6\times10^{-13})}$$

$$= 3\times10^{-14}\ \text{m}$$

The results of the experiments of Geiger and Marsden showed that the size of the gold nucleus is less than 3×10^{-14}m. If α-particles of energy 40 MeV are scattered by a gold foil, particles with small impact parameter could overcome the Coulomb barrier and reach so close to the nucleus as to experience the nuclear potential. Such α-particles are absorbed by the nucleus. The scattering cross-section drops sharply at the scattering angle corresponding to the impact parameter equal to $R_N + R_\alpha$ where R_α is the radius of the α-particle. The value of R_N or the distance at which the nuclear potential becomes effective depends upon the form of nuclear potential used. Kerlee *et al.* studied the scattering of α-particles of energies varying from 12 MeV to 44 MeV from a number of nuclei ranging from Ni to Pu. Using a sharp cut off model (square well potential) the analysis of their data gave the nuclear radius as $R_N = R_O A^{1/3}$ where $R_O = 1.414 \times 10^{-15}$m and the radius of the α-particle $R_\alpha = 2.19 \pm 0.2F$. Igo *et al.* studied the scattering of 40 MeV α-particles by various nuclei. Assuming a diffused nuclear potential as given by Woods and Saxon optical model they found that the nuclear radius is given as

$$R_N = (1.35\,A^{1/3} + 1.3) \times 10^{-15}\ \text{m}.$$

1.3.2 α-Decay Systematics

In very heavy nuclei ($A > 200$) the number of protons is large ($Z > 82$) and the Coulomb energy which varies as Z^2 reduces the binding energy of the nuclei substantially. Many heavy nuclei, therefore, decay with the emission of α-particles, which tends to increase their stability. A proton can not come out of the nucleus because it has a binding energy of 6–7 MeV. An α-particle due to its large binding energy of 28 MeV can come out of a heavy nucleus with some kinetic energy (4-8 MeV). An α-particle to be able to escape a nucleus must overcome the potential energy barrier due to the Coulomb field of the positive charge of the daughter nucleus. At the Coulomb radius R_C of the daughter nucleus having atomic number Z, the height of the potential barrier is V_O (Fig. 1.6), where

$$V_O = \frac{Zze^2}{4\pi \epsilon_0 R_C} \quad ...(1.3.33)$$

(z = charge of α-particle = 2)

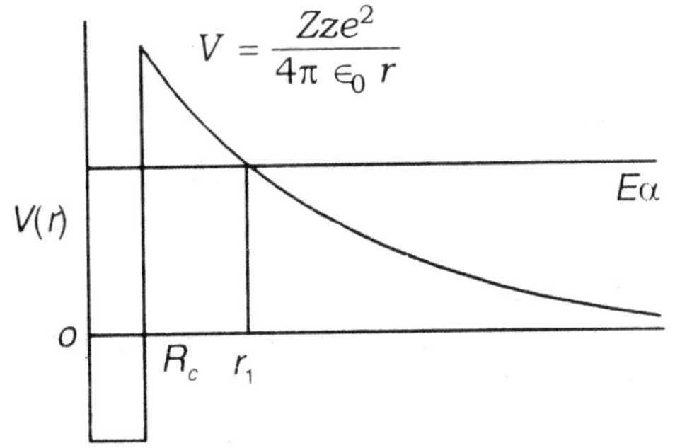

Fig. 1.6: Coulomb potential due to a nucleus of radius R_c, experienced by an α-particle emerging with a kinetic energy E_α

For a uranium nucleus the height of the Coulomb potential barrier is 35.6 MeV. The energy of the α-particle coming out is only 4-8 MeV and classically it can not escape the nucleus. Quantum mechanically, however, there is a certain probability that the α-particle can tunnel through the potential barrier. Condon developed the theory of α-decay on the following assumptions.

(*i*) An α-particle may exist as an entity within a heavy nucleus. The probability of the formation of an α-particle inside a nucleus is dependent on the model of nucleus assumed.

(*ii*) The α-particle inside the nucleus is in constant motion with kinetic energy E_α, which is equal to the kinetic energy of the escaping α-particle. The α-particle keeps striking the potential barrier and is reflected back. The α-particle is thus prevented from escaping from the nucleus by the Coulomb potential barrier.

(*iii*) There is a definite probability—though small—that the α-particle hitting the potential barrier with kinetic energy E_α tunnels through the barrier and escapes the nucleus.

The probability λ that a nucleus will decay with the emission of the α-particle is thus a product of the probability f of finding an α-particle at the potential barrier with kinetic energy E_α and the probability P of its tunnelling through the barrier. Thus,

$$\lambda = f \cdot P \quad ...(1.3.34)$$

Assuming that an α-particle moving with kinetic energy E_α moves back and forth along the nuclear diameter, then

$$f = \frac{v_\alpha}{2R_C} \quad \text{where} \quad \frac{1}{2}mv_\alpha^2 = E_\alpha \qquad \text{...(1.3.35)}$$

Typical values of v_α and R_C might be 2×10^7 m/sec and 10^{-14} m respectively, thus

$$f \approx 10^{21}\ \text{sec}^{-1}$$

The α-particle thus knocks at the potential barrier about 10^{21} times per second but it waits upto 10^{10} years to escape the nucleus.

The probability P for the α-particle to tunnel through the Coulomb potential barrier can be calculated by *WKB* approximation as

$$\ln P = -\frac{2}{\hbar}\int_{R_C}^{R_e}\left\{2m\left(\frac{Zze^2}{4\pi\varepsilon_0 r} - E_\alpha\right)\right\}^{\frac{1}{2}} dr \qquad \text{...(1.3.36)}$$

where R_e is the distance from the nuclear centre at which the α-particle can escape (Fig. 1.6). At this distance the kinetic energy of the α-particle is equal to its potential energy so that

$$E_\alpha = \frac{Zze^2}{4\pi\varepsilon_0 R_e}$$

$$R_e = \frac{Zze^2}{4\pi\varepsilon_0 E_\alpha}$$

Under the assumption that $R_e >> R_c$ equation 1.3.36 can be solved to give

$$\ln P = \frac{4e}{\hbar}\left(\frac{m}{\pi\varepsilon_0}\right)^{\frac{1}{2}} Z^{\frac{1}{2}} R_c^{\frac{1}{2}} - \frac{e^2}{\hbar\varepsilon_0}\left(\frac{m}{2}\right)^{\frac{1}{2}} Z E_\alpha^{-\frac{1}{2}},$$

where m is the mass of the α-particle.

Substituting the values of various constants and taking the energy E_α in MeV and radius R_c in Fermi ($1F = 10^{-15}$ m) one gets

$$\ln P = 2.97\, Z^{\frac{1}{2}} R_c^{\frac{1}{2}} - 3.95 Z E_\alpha^{-\frac{1}{2}} \qquad \text{...(1.3.37)}$$

Substituting equation (1.3.35) and (1.3.37) in equation (1.3.34) gives

$$\ln \lambda = \ln \frac{v_\alpha}{2R_c} + 2.97\, Z^{\frac{1}{2}} R_c^{\frac{1}{2}} - 3.95\, Z E_\alpha^{-\frac{1}{2}}$$

$$\log_{10} \lambda = \log_{10}\left(\frac{v_\alpha}{2R_c}\right) + 1.29 Z^{\frac{1}{2}} R_c^{1/2} - 1.72 Z E_\alpha^{-\frac{1}{2}} \qquad \text{...(1.3.38)}$$

The above equation has been verified experimentally for a large number of α-decays of nuclei. A systematic analysis has been made for different nuclides to give the variation of R_c with the atomic mass. It is found that

$$R_c = R_o A^{1/3} \text{ where } R_o = 1.4 \times 10^{-15}\ \text{m} = 1.4\ \text{F}$$

1.3.3 Electron Scattering

In the scattering of α-particles, protons or neutrons by a nucleus, the nuclear potential plays a dominant role. The nuclear potential is not well known hence there is some uncertainty in the interpretation of the experimental data. The electrons are not capable of interacting with nuclear potential. The only way an electron can interact with a nucleus is through the electromagnetic interaction, which is well known. Measurement of nuclear radius employing electron scattering is, therefore, ideal. If one employs electrons of deBroglie wavelength much less than the nuclear size, one can even obtain the shape of the charge distribution in a nucleus. An electron of energy 10^9 eV has a deBroglie wavelength $\lambda = \lambda/2\pi \sim 2 \times 10^{-16}$ m and can be fruitfully used to study the nuclear charge distribution.

A beam of high energy electrons is incident on a target and the angular distribution of the scattered electrons is studied. Considering only the elastic Coulomb scattering of the electrons, the differential cross-section for a point nucleus of charge Z is given by Mott's relation.

$$\frac{d\sigma_M}{d\Omega} = \sigma_M(\theta) = \left(\frac{Ze^2}{2E_0}\right)^2 \frac{1}{\sin^4(\theta/2)}\left[\frac{\cos^2(\theta/2)}{1+(2E_0/Mc^2)\sin^2(\theta/2)}\right], \quad \text{...(1.3.39)}$$

where θ is the laboratory scattering angle, E_0 the electron energy in laboratory system and M the mass of the nucleus. The finite size of the nucleus modifies the scattering (Fig. 1.7) and the scattering cross-section can be written as

$$\sigma(\theta) = \sigma_M(\theta) \cdot [F(q)]^2, \quad \text{...(1.3.40)}$$

where $F(q)$ is the form factor and

$$q = \frac{|\vec{p}-\vec{p}'|}{\hbar} = |\vec{k}-\vec{k}'| = \frac{2}{\hbar} p\sin(\theta/2), \ [|\vec{p}| = |\vec{p}'|] \quad \text{...(1.3.41)}$$

Here q gives the change in momentum (direction only) of the scattered electron. In Born approximation the form factor is given as

$$F(q) = \frac{1}{Ze}\cdot\frac{4\pi}{q}\int_0^\infty \rho(r)\sin(qr)r\cdot dr \quad \text{...(1.3.42)}$$

The density of nuclear charge ρ is normalised

as $$\int_0^\infty \rho(r)4\pi r^2 dr = Ze \quad \text{...(1.3.43)}$$

For low energy electrons $\lambda >> R$ and $qR << 1$ so that

$$F(q) = \frac{1}{Ze}\cdot\frac{4\pi}{q}\int_0^\infty \rho(r)\left[qr - \frac{1}{6}q^3r^3 + \ldots\right]\cdot rdr \quad \text{...(1.3.44)}$$

$$= \frac{1}{Ze}\cdot\int_0^\infty 4\pi r^2\rho(r)dr - \frac{1}{6}\frac{q^2}{Ze}\int_0^\infty 4\pi r^2\rho(r)r^2 dr$$

$$F(q) = 1 - \frac{1}{6}(qa)^2,$$

where $$a^2 = \frac{1}{Ze}\int_0^\infty r^2\rho(r)4\pi r^2 dr$$

a^2 is the mean square radius of the charge distribution. For high values of momentum transfer q one can take an inverse transform of equation for $F(q)$ and have

$$\rho(r) = \frac{1}{2\pi^2 r}\int_0^\infty F(q)\sin(qr)\,q\,dq \qquad \text{...(1.3.45)}$$

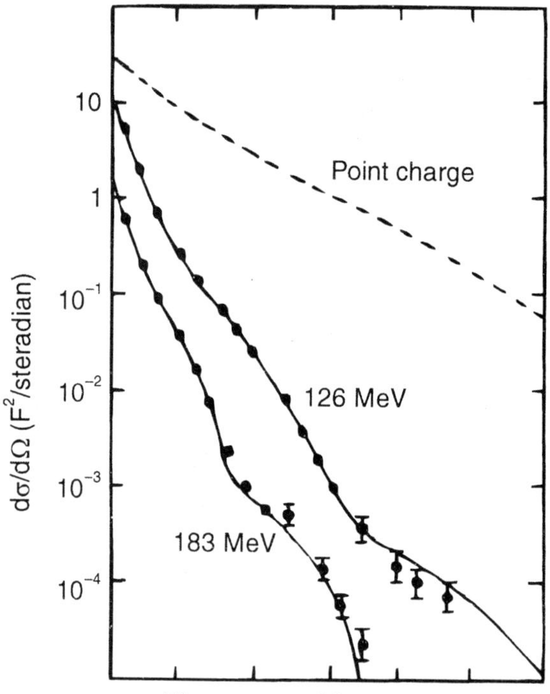

Fig. 1.7: Elastic scattering differential cross-section for electrons, from gold at energies 126 MeV and 183 MeV. The experimental points are fitted with a Fermi function with R = 6.63 fm and a = 0.45 fm. The dashed curve is expected differential cross-section from a point charge nucleus [R. Hofstadter (1963). "*Electron Scattering and Nuclear and Nucleon Structure*". Benjamin Press, New York.]

To make use of the above equation one can assume a suitable form for $\rho(r)$ and compare the calculated angular distribution with the observed one. It is found that the function $\rho(r)$ is fairly complicated but a good fit with the experimental data is obtained if $\rho(r)$ is assumed as a Fermi function *i.e.*

$$\rho_F(r) = \frac{\rho_o}{1 + \exp\left(\frac{r - c}{a_o}\right)} \qquad ...(1.3.46)$$

The charge density of the nucleus as given by equation 1.3.46 is shown in Fig. 1.8. The best fit for medium heavy nuclei is obtained with $c = 1.07\ A^{1/3}F$ and $a_o = 0.545F$. It is seen that c is the half density radius of the nucleus. The charge density has a skin whose thickness as defined by the distance in which the charge density drops from 0.9 ρ_o to 0.1 ρ_o is given as

$$4.40\, a_o = 2.4F \qquad ...(1.3.47)$$

This is independent of A.

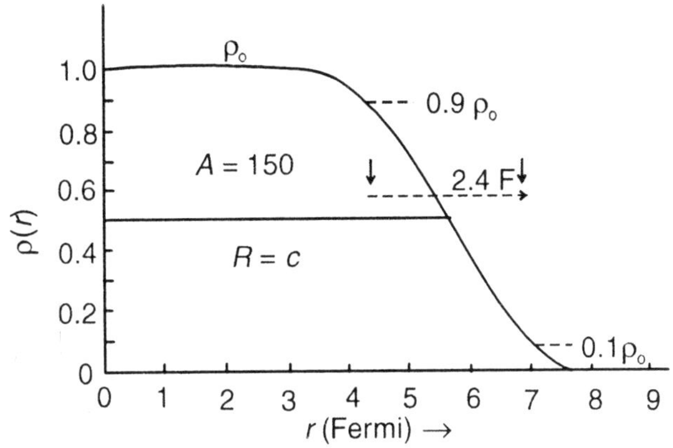

Fig. 1.8: Charge density $\rho(r)$ of a nucleus as a function of distance r from its centre, as given by Fermi function. $R = c$ is the half density radius. The skin thickness corresponding to $\rho = 0.1\ \rho_0$ to $0.9\ \rho_0$ is $4.4\ a_0 = 2.4$ F.

The charge distribution in a number of nuclei is shown in Fig. 1.9. It is observed that in the case of heavy nuclei the charge density is constant near the centre. The charge density given in equation (1.3.46) is not suitable for light nuclei. In the case of ^{12}C the experimental results indicate a charge distribution as

$$\rho(r) = \rho_0\left(1 + \frac{wr^2}{a_0^2}\right)e^{-\frac{r^2}{a_0^2}} \qquad ...(1.3.48)$$

where w is a constant. For ^9Be, ^6Li and ^7Li the form of charge distribution is obtained as

$$\rho(r) = \rho_0\left(1 + \frac{r}{a_0}\right)e^{-\frac{r}{a_0}} \qquad ...(1.3.49)$$

while for ^4He the charge distribution is given as

$$\rho(r) = \rho_0 e^{-\frac{r^2}{a_0^2}} \qquad ...(1.3.50)$$

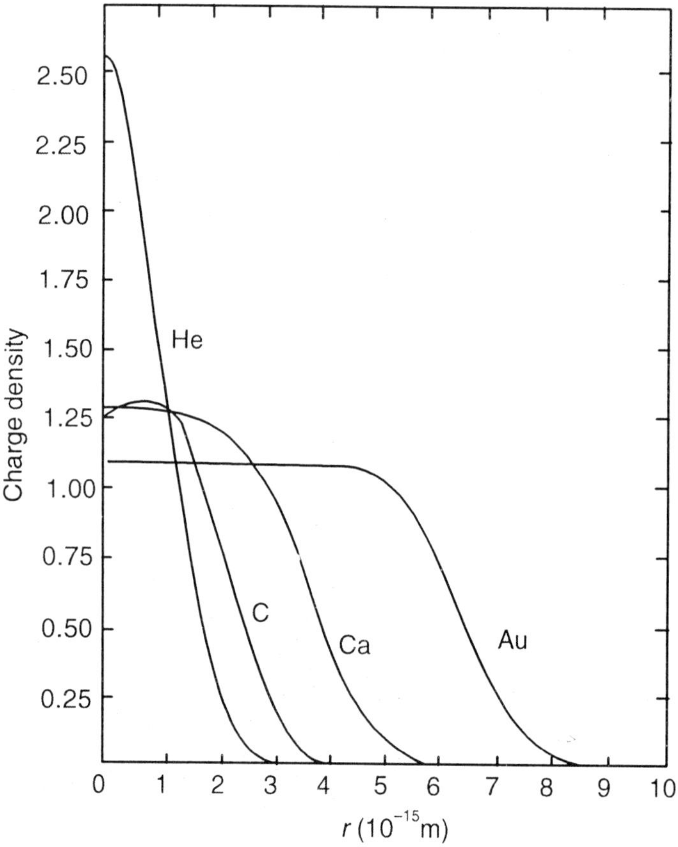

Fig. 1.9: Nuclear charge density as a function of distance from centre for some nuclei [R. Hofstadter (1957) *Ann. Rev. Nucl. Sc*, **7**; 231].

The Fermi form of nuclear charge density (eqn 1.3.46) indicates a skin of thickness of 2.4 F in all the medium and heavy nuclei. A nucleus thus does not have a well-defined radius. It can be shown that in the lead nucleus almost half the number of protons is present in the skin region.

The electron scattering gives the distribution of the positive charge, protons—inside the nucleus. It does not give the neutron distribution. Due to the strong binding between the neutrons and protons it is safe to assume that the neutron distribution follows the proton distribution. Thus equation (1.3.46) may be regarded to express the mass or nucleon distribution in a nucleus.

It may be mentioned that the analysis of stripping reaction data indicates a slight difference between the radii of neutron and proton distributions.

The scattering of electrons at very high energy has been employed to obtain the charge distribution in a proton and a neutron. In the interpretation of the data, one has to take into consideration the spin and anomalous magnetic moments of the electron, proton and neutron. It has been shown that as a first approximation, the electric charge density of a proton can be represented as

$$\rho(r) = \rho(o)\, e^{-r/a} \qquad ...(1.3.51)$$

where $a = 0.23$ F. The root mean square radius for this distribution is $a\sqrt{12}$ = 0.8×10^{-15} m.

An analysis of more refined experiments on scattering of electrons by protons and deuterons gives the charge distribution in a proton and a neutron as shown in Fig. 1.10.

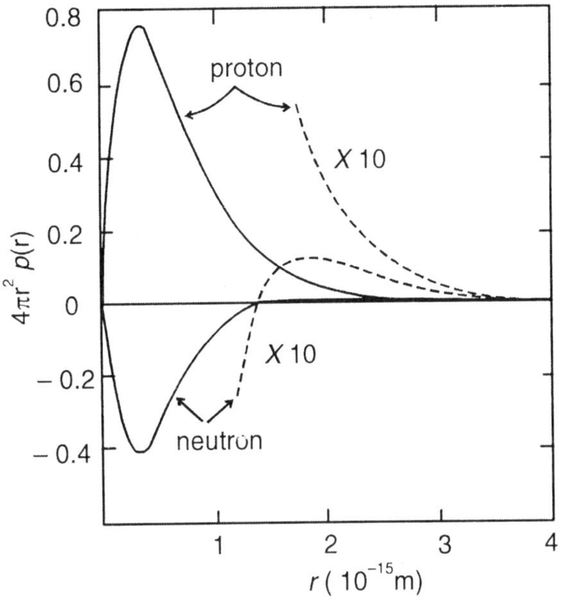

Fig. 1.10: The charge density distribution of proton and neutron as given by electron scattering. At the origin there is a very sharp peak of positive charge for the proton, containing a charge of +0.3e. In the case of the neutron the positive charge content of the central peak neutralises the negative charge distribution. These narrow positive charge peaks have not been shown. [R. Hofstadter and R. Herman (1961), *Phys. Rev. Lett*, **6**; 293].

1.3.4 Hyperfine Splitting of Spectral Lines

The energy levels of atomic electrons can be calculated for a point nucleus. When the nucleus has a finite volume, there is some probability that an atomic electron is inside the nuclear volume. This probability is maximum for the s-electrons. During the period the electron is inside the nuclear volume (radius R) the Coulomb potential experienced by the electron at a distance r from the centre of the nucleus is given as

$$V(r) = -\frac{Ze^2}{R}\left(\frac{3}{2} - \frac{1}{2}\frac{r}{R}\right),\ r < R \qquad ...(1.3.52)$$

Consequently the *s* electron has a shift in its energy which depends upon the nuclear radius R. Different isotopes of an element thus show a shift in their spectral lines. The hyperfine splitting in the spectral lines associated with transitions to the S orbit in heavy elements is attributed to this isotope effect. The nuclear radius has been estimated from this splitting. The radius is found as $R = R_0 A^{1/3}$ where A is the atomic mass of the isotope and $R_0 = 1.3 \times 10^{-15}$ m.

The spectral lines of very light elements show a hyperfine splitting which is independent of the orbitals of the levels involved in the transition. This splitting is due to the centre of mass effect and is also known as isotope effect. The isotope ^2H (deuterium) was first discovered from the splitting of the hydrogen lines.

Similar to the isotope effect due to finite size of the nucleus, in spectral lines, the X-ray emission lines from different isotopes of an element show a shift in their wavelength. Due to the low resolution available in X-ray spectroscopy, reliable calculation of the nuclear radius is rather difficult. The X-rays emitted from μ-mesic atoms show a very large energy shift from the point nucleus calculations, and can be fruitfully employed to give nuclear radius. A μ-meson is very similar to an electron and has mass of 207 m_e (m_e is mass of an electron) and half-life of 2.2×10^{-6} seconds. The μ-meson is not affected by nuclear force and behaves like a heavy electron. μ-mesons can be produced in large numbers in a high-energy cyclotron. When the μ-mesons are brought to rest in an absorber, some mesons can be captured in an atom in the outer most orbit. The meson so captured makes transition to the inner shells, emitting Auger electrons and electromagnetic radiations. When it makes the final transition from the 2p orbit to ls orbit it emits K-X-rays whose energy can be measured. The μ-meson having a mass 207 m_e has the radius of the ls orbit 207 times smaller than the corresponding electron's radius. The energy of the K-X-rays is expected to be 4×10^4 times larger. In a μ–mesic lead atom the radius of the K–shell is 3.1×10^{-15} m and that of the L–shell 12×10^{-15} m. For a point lead nucleus the K-X-ray energy of the μ–mesic atom is 16 MeV while for a nucleus of 6.5×10^{-15}m size, the energy of the K-X-rays is only 5.5 MeV. The change in the energy of the K-X-rays is thus very large and can be measured with good accuracy.

The energy of the ls state of a μ–mesic atom depends sensitively on the nuclear charge distribution. The energy shift due to finite size of the nucleus in light nuclei is very small (2% for aluminium). For heavy atoms where the energy shift is appreciable, the theoretical calculations using Dirac theory are very approximate. Fitch and Rainwater studied the energy shift in the X-ray energies of μ-mesonic atoms with $Z = 13$ to $Z = 83$. They obtained the nuclear radius to be given by–

$$R = 1.17 \times 10^{-15} A^{1/3} \text{ m} \qquad ...(1.3.53)$$

The value of $R_0 = 1.17 \times 10^{-15}$ m is in agreement with the results of electron scattering experiments.

1.3.5 Mass Difference between Mirror Nuclei

The Coulomb repulsion between the protons in a nucleus reduces its binding energy. In the case of mirror nuclei the number of neutrons and protons are interchanged, keeping the mass number same. While the size of two mirror nuclei is the same (same A) their energies are different. The Coulomb energy of a nucleus with radius R and atomic number Z is given as

$$E_c = \frac{3}{5}\frac{(Ze)^2}{R} \quad \text{...(1.3.54)}$$

The energy difference between two mirror nuclei with atomic number Z and $(Z + 1)$ is

$$\Delta E_c = \frac{3}{5R}\left((Z+1)^2 - Z^2\right)e^2$$

$$= \frac{3}{5R}(2Z+1)e^2 \quad \text{...(1.3.55)}$$

The energy difference between the mirror nuclei $({}^7_3\text{Li}, {}^7_4\text{Bi})$ $({}^{11}_5\text{B}, {}^{11}_6\text{C})$ $({}^{15}_7\text{N}, {}^{15}_8\text{O})$ $({}^{17}_8\text{O}, {}^{17}_9\text{F})$ and $({}^{19}_9\text{F}, {}^{19}_{10}\text{Ni})$ can be obtained from the end point energy of the β-transitions between the nuclei. The value of R_0 calculated from these measurements is given as $R_0 = 1.5 \times 10^{-15}$ m. The calculation of the Coulomb energy is dependent upon model of the charge distribution of the nucleus. The Coulomb energy given in equation (1.3.54) holds for a uniform charge distribution in the nucleus. Applying corrections to this model Kofoed and Hausen estimated the nuclear radius parameter from mirror nuclei Coulomb energy to be

$$R_0 = 1.28 \pm 0.05\,\text{F} \quad \text{...(1.3.56)}$$

1.3.6 Scattering of Neutrons by Nuclei

The nuclear radius can be obtained from experiments on the scattering of neutrons and protons. Though the measurements on proton scattering are relatively simple, the analysis of the data becomes complicated due to the Coulomb effect. In the scattering of neutrons the Coulomb effect is absent and the analysis of the results becomes a little simpler. When a neutron collides with a target nucleus it could be absorbed or scattered either elastically or inelastically. The total cross-section for the interaction of neutrons with a target can be measured by the method of attenuation of a beam of neutrons. The neutron beam of initial intensity I_0 on passing through a target of thickness x and having n nuclei per unit area is attenuated to intensity I where

$$I = I_0\, e^{-n\sigma x} \text{ (absorption and scattering)} \quad \text{...(1.3.57)}$$

where σ is the total interaction cross-section. For neutron energies in the MeV range, the total interaction cross-section for neutrons with nuclei of radius R is given as

$$\sigma = 2\pi R^2 \quad \text{...(1.3.58)}$$

Experiments have been performed using neutrons of energy 14 MeV. Fig. 1.11 shows the calculated and the experimental results of σ for nuclear radius R, where

$$R = R_0 A^{1/3}, \; R_0 = 1.4 \times 10^{-15} \text{ m}$$

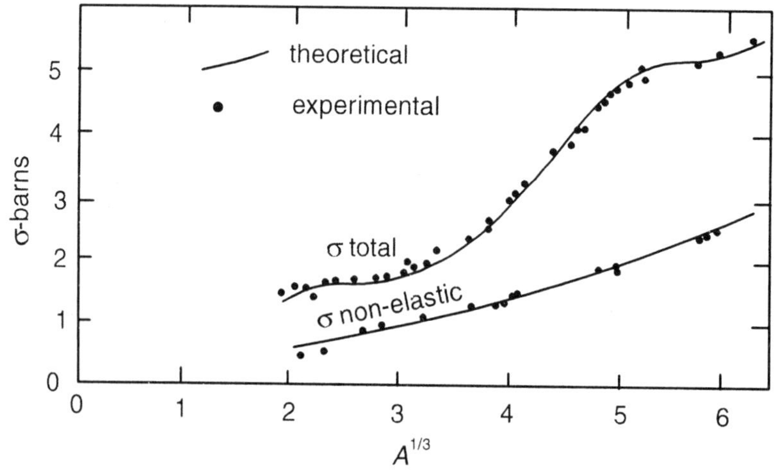

Fig. 1.11: Theoretical fit of the experimental values of total and non-elastic scattering cross-sections of 14 MeV neutrons for different nuclei [S Fernback (1958) *Rev. Mod. Phys*; ***30***, 414].

The intensity of neutrons elastically scattered by a target shows an angular distribution, similar to the diffraction pattern of light scattered by an object. Assuming the nucleus to be completely opaque to neutrons-behaving as a black disc of radius R, diffracting the de Broglie waves of the incident neutrons, the differential scattering cross sections in a solid angle $d\omega$ is given as

$$\frac{d\sigma}{d\omega} = \frac{4\pi^2 R^2}{\sin^2\theta} J_1^2(R\sin\theta / \lambda), \quad \text{...(1.3.59)}$$

where θ is the angle of scattering, J_1 the Bessel function of order unity and $\lambda = \dfrac{h}{mv}$ is the de Broglie wavelength of the incident particle. The function $J_1(x)$ has the first zero at

$$x = 0.610 \times 2\pi \quad \text{...(1.3.60)}$$

substituting $x = R\sin\theta/\lambda$ one gets the angle of the first minima as

$$\sin\theta = 0.610 \times 2\pi\lambda/R \quad \text{...(1.3.61).}$$

The results of the measurements of the angular distribution of 14 MeV neutrons elastically scattered by some elements is shown in Fig. 1.12. The data gives the nuclear radius as

$$R = R_0 A^{1/3}$$

where $\quad R_0 = 1.2 \times 10^{-15}$ m

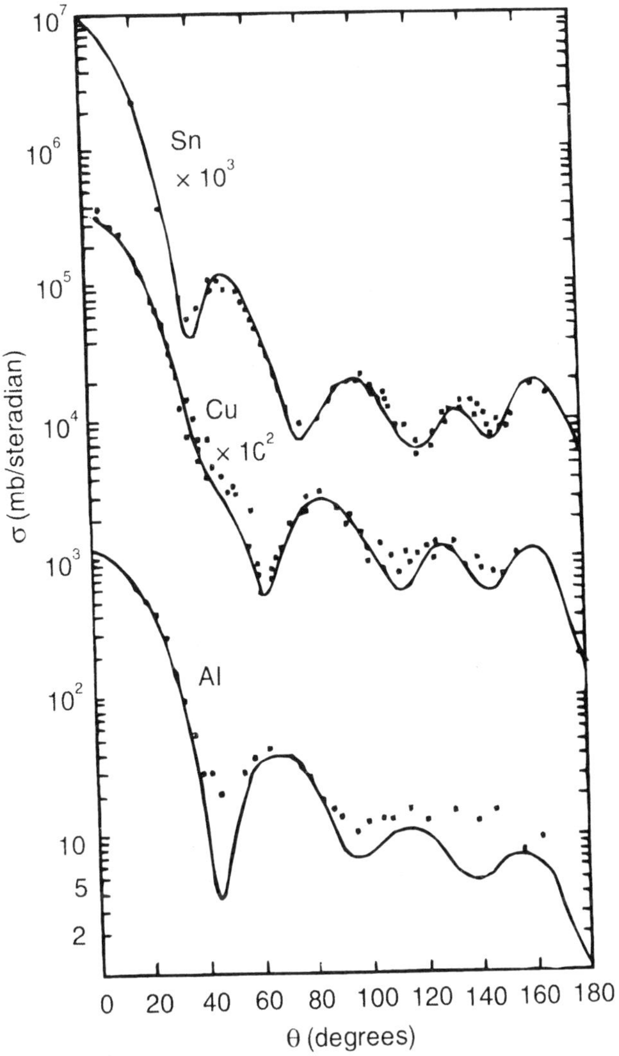

Fig. 1.12: Differential cross-sections for 14 MeV neutrons elastically scattered from Sn, Cu, and Al [S Fernback (1958) *Rev. Mod. Phys. 30*; 414].

All the methods mentioned above conclusively show that the nuclear radius varies as $A^{1/3}$. The values of R_0 obtained by different methods differ from each other. The most accurate measurements are made by the electron scattering, which shows the variation of the nuclear matter density as a function of distance from the centre. It may be mentioned that not all nuclei have a spherical shape as given by electron scattering experiments. Measurements of electric quadrupole moments of nuclei show a large number of them to have a spheroidal shape.

1.4 LIQUID DROP MODEL AND WEIZSACKER'S MASS FORMULA FOR NUCLEI

The radius of a nucleus with atomic mass A has been shown to be $R = R_0 A^{1/3}$. The volume of the nucleus is thus proportional to the number of nucleons in it. This implies that the nucleons fill up the nuclear volume like small balls filling up a jar or water molecules accumulating to form a drop. The density of a nucleus, similar to liquid drop, is independent of its size. Taking the mass of a nucleon as 1.67×10^{-27} kg and $R_0 = 1.2 \times 10^{-15}$ m the density of the nuclear matter is obtained as 2.3×10^{17} kg/m^3. The total binding energy of a nucleus varies as A. The nucleus thus behaves very much like a liquid drop.

Weizsacker on the basis of liquid drop model gave the mass of an atom $M(A, Z)$ as $M(A, Z) = (A - Z)\,M_n + ZM_p + ZM_e - \Delta M$

or $$M(A, Z) = (A - Z)M_n + ZM_H - \Delta M \qquad ...(1.4.1)$$

where mass defect ΔM corresponds to the binding energy of the nucleus. M_n, M_p and M_H are the masses of a neutron, a proton and a hydrogen atom, respectively.

In terms of energy

$$\Delta Mc^2 = BE = a_1 A - a_2 A^{2/3} - a_3 \frac{\left(\frac{A}{2} - Z\right)^2}{A} - a_4 \frac{Z^2}{A^{1/3}} + \delta \qquad ...(1.4.2)$$

In the above formula the first term—some times called the volume term follows from the experimental observation that to a first approximation the binding energy per nucleon in a nucleus (BE/A) is constant

The second term called the surface term arises due to the fact that, as in a liquid, the particles at the surface of a nucleus do not experience the same force of attraction as those in the interior. The number of such nucleons is proportional to the surface area of the nucleus which in turn is proportional to $A^{2/3}$. This term is negative and reduces the total binding energy. The third term called the asymmetry term arises due to the difference in the number of neutrons and protons in a nucleus. This term can be derived on the basis of Fermi-Dirac statistics as follows:

Neutrons and protons follow F.D. statistics as both the particles have spin angular momentum $s = \frac{1}{2}\hbar$. The Fermi energy of the N neutrons in a nucleus is

$$\epsilon = C_0 \left(\frac{N}{A}\right)^{2/3} \qquad ...(1.4.3)$$

The total Fermi energy of all the N neutrons in the nucleus is

$$E_N = C_1 \frac{N^{5/3}}{A^{2/3}} \qquad ...(1.4.4)$$

where C_0 and C_1 are constants.

As the electrostatic energy of the protons is taken care of in the fourth term of the BE formula, the total Fermi energy of the protons is

$$E_Z = C_1 \frac{Z^{5/3}}{A^{2/3}} \qquad ...(1.4.5)$$

The total energy of the nucleus is

$$E = E_N + E_Z = E_{N+Z} = C_1\left(\frac{N^{5/3} + Z^{5/3}}{A^{2/3}}\right) \qquad ...(1.4.6)$$

The minimum of the total energy is E_{N+Z}(min) when $N = Z = A/2$

The correction term in the BE will be proportional to

$$E_{N+Z} - E_{N+Z}(\text{min}) = C_1 A^{-2/3}\left(N^{5/3} + Z^{5/3} - 2\left(\frac{A}{2}\right)^{5/3}\right)$$

Putting $\qquad x = \dfrac{N-Z}{2} = \dfrac{A}{2} - Z$

or $\qquad Z = \dfrac{A}{2} - x$ and $N = \dfrac{A}{2} + x$

one gets

$$E_{N+Z} - E_{N+Z}(\text{min}) = C_1 A^{-2/3}\left[\left(\frac{A}{2}+x\right)^{5/3} + \left(\frac{A}{2}-x\right)^{5/3} - 2\left(\frac{A}{2}\right)^{5/3}\right]$$

Expanding the first two terms by Taylor series one gets

$$E_{N+Z} - E_{N+Z}(\text{min}) = C_1 A^{-2/3}\left[2\cdot\frac{\frac{5}{3}\cdot\frac{2}{3}}{2}\left(\frac{A}{2}\right)^{-\frac{1}{3}} x^2\right]$$

$$= C_2\frac{\left(\frac{A}{2} - Z\right)^2}{A}, \qquad ...(1.4.7)$$

where, C_2 is a constant.

This corresponds to the third term in the BE formula.

The fourth term in the BE formula corresponds to the electrostatic energy due to the mutual repulsion between the protons in the nucleus. The final correction term known as pairing energy δ depends upon the stability of nuclei with respect to whether the number of protons and neutrons is even or odd. When A is odd, the value of δ is zero. When both N and Z are even (even-even nuclei), the value of δ is positive and for odd N odd Z nuclei (odd-odd nuclei) the value of δ is negative. The constants a_1, a_2, a_3 and a_4 are determined by least square fit of the known masses of a large number of nuclides. The values of these constants in millimass units and energy units are as given below (A.E.S. Green, 1958 *Rev. Mod. Phys.* Vol. 30, p. 569).

a_1 = 16.710 mμ = 15.565 MeV

a_2 = 18.500 mμ = 17.232 MeV

$a_3 = 100 \text{ m}\mu = 93.148 \text{ MeV}$

$a_4 = 0.750 \text{ m}\mu = 0.699 \text{ MeV}$

$$\delta = \frac{36}{A^{3/4}} \text{m}\mu = \frac{33.533}{A^{3/4}} \text{ MeV}$$

where, $1\text{m}\mu = 931.478$ keV

The binding energy of a nucleus (A, Z) can be written as

$$BE(A, Z) = 15.565A - 17.232\, A^{2/3} - 93.148\, \frac{(A/2 - Z)^2}{A}$$

$$-0.699 \frac{Z^2}{A^{1/3}} \pm \frac{33.533}{A^{3/4}} \text{ MeV} \qquad \ldots(1.4.8)$$

The mass of an atom $M(A, Z)$ can be written as

$$M(A, Z) = (A - Z)\, M_n + ZM_H - 16.710A + 18.500A^{2/3}$$

$$+100 \frac{\left(\frac{A}{2} - Z\right)^2}{A} + 0.750 \frac{Z^2}{A^{1/3}} \pm \frac{36}{A^{3/4}} \text{ m}\mu \qquad \ldots(1.4.9)$$

The binding energy of a nucleus with proton number Z, neutron number N and mass number A can be calculated using relation (1.4.8). The binding energies of some nuclei, giving the contribution of each term, is given in Table 1.1 (all energies are in MeV). The contribution of various terms of the Weizsacker's formula in the value of binding energy per nucleon (BE/A) is shown in Fig. 1.13.

TABLE 1.1

Nucleus	a_1A	$a_2A^{2/3}$	$a_3\frac{\left(\frac{A}{2}-Z\right)^2}{A}$	$a_4\frac{Z^2}{A^{1/3}}$	δ	*Total*	BE/*A*
$^{16}O_8$	249	−19.4	0	−17.8	+4.19	126	7.87
$^{56}Fe_{30}$	871.6	−252.2	−6.65	−123.5	+1.64	491	8.77
$^{238}U_{146}$	3704.5	−661.8	−285.3	−954.7	+0.55	1803.3	7.57
$^{235}U_{143}$	3657.8	−656.2	−257.7	−958.7	0	1785.2	7.60

It may be observed from the above table that the surface energy term, the asymmetry term and the Coulomb term increase sharply with the atomic number of the nucleus.

If one assumes that Z takes continuous values (actually values of Z are discrete), it can be shown that eqn. (1.4.9) represents a parabola for mass of a nuclide versus Z.

The most stable isotope for a given value of A, the mass number, corresponds to the minimum value of $M(Z, A)$ and corresponds to the value of Z given by

$$Z = \frac{A}{2 + 0.015A^{2/3}} \qquad \text{...(1.4.10)}$$

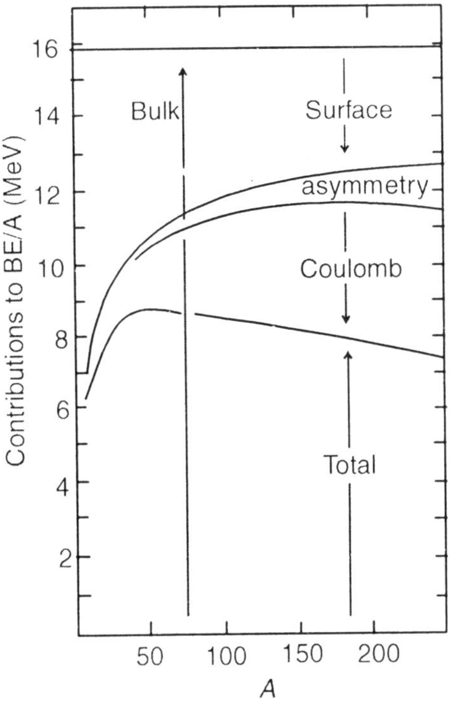

Fig. 1.13: Contribution of various terms of the Weiszacker's mass formula to the binding energy per nucleon as a function of atomic mass

For light elements the stable nuclides have $A = 2Z$. *i.e.* the number of neutrons and protons are equal. As the mass number increases the number of neutrons is larger than that of protons and $Z < A/2$.

The parabola represented by eqn. (1.4.9) for an odd value of A (e.g. $A = 101$) for which the pairing energy term δ is zero is as shown in Fig. 1.14.

As Z takes only discrete values, it is not necessary that a nuclide would have Z exactly at the minimum of the parabola. In Fig. 1.14, the $Z = 44$ point lies close to the minimum and corresponds to the most stable isotope. Other nuclides corresponding to $Z = 42, 43, 45, 46$ have masses greater than that of nuclide with $Z = 44$. The nuclide with greater mass decays to nuclide with smaller mass. In the case of $A = 101$ as shown in Fig. 1.14, the nuclide (101, 42) which is ^{101}Mo decays by β- emission to ^{101}Tc which in turn decays to ^{101}Ru by β-emission. The probability that ^{101}Mo decays to ^{101}Ru directly by double β-decay i.e. emission of two β-particles simultaneously is extremely low. The lifetime of double Beta-decay has been shown experimentally to be greater than 10^{20} years. Similarly $^{101}_{46}$Pd decays by positron emission to $^{101}_{45}$Rh which again decays to $^{101}_{44}$Ru by β^+ emission.

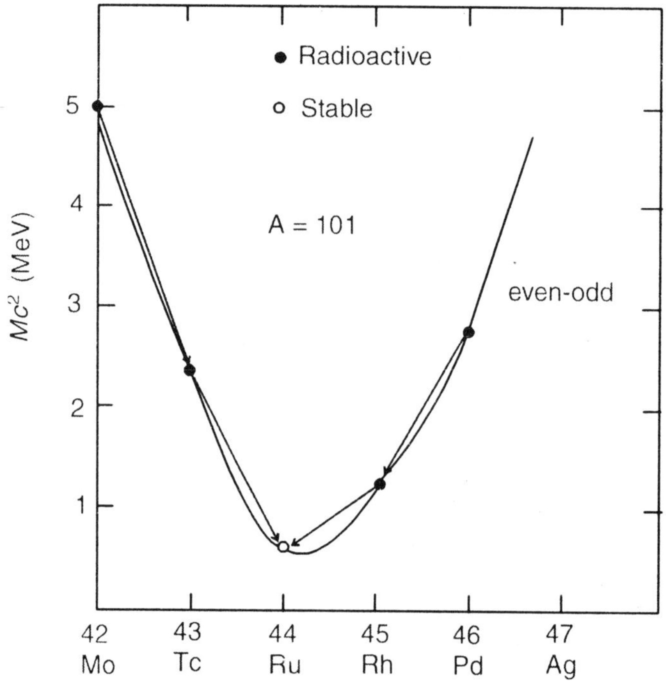

Fig. 1.14: Difference in the energies of nuclei of mass number 101. The parabola corresponds to the atomic mass energy parabola. The zero point on the energy scale is arbitrary.

For even *A*, even *Z* (*N*-even) nuclei the last term (pairing energy) in eqn. 1.4.9 is negative and shifts the mass parabola to lower mass value. For the same value of *A* and odd *Z*, the pairing energy term is positive and shifts the parabola to higher mass value. The two parabolas for even *A* = 106 are shown in Fig. 1.15.

It may be seen from Fig. 1.15 that $^{106}_{49}$In decays by β^+ emission to $^{106}_{48}$Cd. The nuclide $^{106}_{48}$Cd is lower in mass to $^{106}_{47}$Ag and can not decay to it. Double β-decay of $^{106}_{48}$Ca to $^{106}_{46}$Pd is not possible (very low probability) so that $^{106}_{48}$Cd is a stable nuclide. The isotope $^{106}_{47}$Ag can decay by β^+ emission to $^{106}_{46}$Pd or by β^- emission to $^{106}_{48}$Cd. The nuclide $^{106}_{44}$Ru decays by β-emission to $^{106}_{45}$Rh which in turn decays to $^{106}_{46}$Pd. Thus in this case of *A* = 106 there should be two stable nuclides corresponding to atomic number *Z* equal to 46 and 48. In some cases there could be more than two stable nuclides.

The Weizsacker's mass formula thus predicts that:

(*i*) For odd mass number there can be only one stable isotope (one value of *Z*)
(*ii*) For even mass number and even *Z* nuclei there can be more than one stable isobars.
(*iii*) All odd *A*, odd *Z* isotopes are unstable.

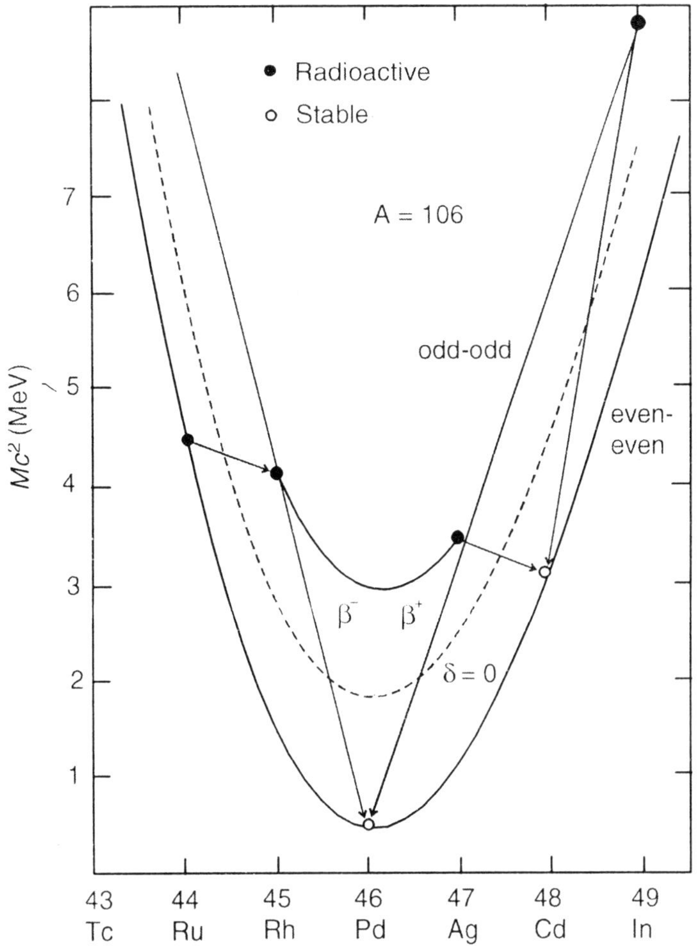

Fig. 1.15: Difference in the energies of nuclei of mass number 106. The atomic mass energy parabolas correspond to odd-odd and even-even nuclei. The zero point on the energy scale is arbitrary. The energy difference between the two is 2δ.

The only exceptions to the above predictions are the nuclides ^{14}N, ^{10}B and ^{6}Li. These are very light nuclei and it is not surprising that the liquid drop model is not applicable to them.

The Weizsacker's mass formula can be employed to calculate the energy release in any nuclear transformation e.g. β-decay, α-decay and nuclear reactions. In β-decay an atom ${}_Z^AX$ transforms into atom ${}_{Z+1}^{A}Y$ with emission of an electron and an antineutrino as–

$$ {}_Z^AX \rightarrow {}_{Z+1}^{A}Y + e^- + \bar{\nu} + \text{energy} $$

The energy released in this transformation can be found by writing the mass equation. The energy released E_β is

$$E_\beta = \{m(A, Z) - m(A, Z+1) - M_e - M_\nu\}\, c^2 \qquad ...(1.4.11)$$

M_e is mass of electron and M_ν is mass of neutrino which is found to be zero. $m(A, Z)$ and $m(A, Z+1)$ are the nuclear masses of ${}^{A}_{Z}X$ and ${}_{Z+1}{}^{A}Y$ nuclei respectively. One can write equation (1.4.11) as

$$E_\beta = \{m(A, Z) + ZM_e - ZM_e - m(A, Z+1) - M_e\}\, c^2$$

$$= \{M(A, Z) - M(A, Z+1)\}\, c^2 \qquad ...(1.4.12)$$

where $M(A, Z)$ and $M(A, Z+1)$ are the masses of the neutral atoms of X and Y respectively. It is seen from equation (1.4.12) that the difference of the atomic masses of the atoms X and Y (multiplied by c^2) gives the maximum energy available in β^--decay. This energy appears as the kinetic energy of the emitted particles (e^- and ν) and the recoil nucleus. In a similar manner one can find the energy released in e^+ decay where the atom ${}^{A}_{Z}X$ decays to atom ${}_{Z-1}{}^{A}Y$ as

$${}^{A}_{Z}X \rightarrow {}_{Z-1}{}^{A}Y + e^+ + \nu$$

In terms of the nuclear masses the energy released can be written as before

$$E_{\beta^+} = \{m(A, Z) - m(A, Z-1) - M_e - M_\nu\}\, c^2$$

$$M_{e^+} = M_{e^-}, \text{ and } M_\nu = 0$$

$$= \{m(A, Z) + ZM_e - ZM_e - m(A, Z-1) - M_e\}\, c^2$$

$$= [\{m(A, Z) + ZM_e\} - \{m(A, Z-1) + (Z-1)M_e\} - 2M_e]\, c^2$$

$$= \{M(A, Z) - M(A, Z-1) - 2M_e\}c^2$$

where, as before $M(A, Z)$, $M(A, Z-1)$ are the masses of the neutral atoms of X and Y respectively. Thus we see that in terms of the atomic masses the energy released in β^+ decay is the difference in the masses of the two atoms, multiplied by c^2 minus $2M_ec^2$. It can be shown that in the electron capture decay ${}^{A}X$ to ${}^{A}Y$ as shown below.

$${}^{A}_{Z}X \rightarrow {}_{Z-1}{}^{A}Y + \nu$$

The decay energy is given by the mass difference of the neutral X and Y atoms.

In the case of α-decay of an atom ${}^{A}_{Z}X$ to ${}^{A-4}_{Z-2}Y$ the energy released is

$$E_\alpha = \{m(A, Z) - m(A-4, Z-2) - M_\alpha\}c^2$$

$$= \{M(A, Z) - M(A-4, Z-2) - M_{He}\}\, c^2$$

It can be shown that for $A > 210$ the energy released E_α is positive. The probability of the α-transformation depends upon the energy released E_α. For small value of E_α the lifetime can be greater than the life of the universe in which case the atom can be regarded as stable.

The liquid drop model is a simplified picture of a nucleus. It does not take into consideration the shell effects in the nucleus which can contribute significantly to the energy released in β-decay or α-decay.

In the fission process a heavy nucleus breaks up into two lighter nuclei whose masses are nearly equal. When the atom of $^{235}_{92}U$ absorbs a neutron it forms $^{236}_{92}U$, which undergoes fission. The energy released in the symmetric fission of an atom $^{A}_{Z}X$ into two atoms of $^{A/2}_{Z/2}Y$ can be calculated as:

$$E_f = \{M(A, Z) - 2M(A/2, Z/2)\}c^2$$

Substituting from equation (1.4.9) one gets

$$E_f = c^2\left\{18.5A^{2/3}(1 - 2^{1/3}) + 0.75\frac{Z^2}{A^{1/3}}(1 - 2^{-2/3})\right\}$$

$$= c^2\left\{-18.5 \times 0.26A^{2/3} + 0.75 \times 0.37\frac{Z^2}{A^{1/3}}\right\}$$

$$E_f = c^2\left\{-4.8A^{2/3} + 0.2775\frac{Z^2}{A^{1/3}}\right\}$$

For the symmetric fission of ^{236}U one gets.

$$E_f = c^2 \cdot 197 \text{ m}\mu$$

$$= 183 \text{ MeV}$$

One finds that the energy released in fission of uranium is very large. Normally one does not observe a symmetric fission. Invariably one fission fragment is heavier than the other. A few neutrons are also emitted in the fission process. The above estimate gives an idea of the enormous energy released in the fission process.

1.5 SPIN AND MAGNETIC MOMENT OF NUCLEI

1.5.1 Total Angular Momentum or Spin of a Nucleus

Stern and Gerlach experiment showed that an electron has a spin angular momentum whose measured value is $\pm(1/2)\hbar$. Similar experiments confirmed that protons and neutrons also have spin angular momentum whose measured value is $\pm(1/2)\hbar$. Quantum mechanically each of these particles is assigned a spin angular momentum $\vec{S}$ (in units of $\hbar$) whose absolute magnitude is $|S| = \sqrt{s(s+1)}\hbar$ where s is called the spin quantum number and has a value 1/2. When a direction is defined—say z direction, the expectation value or the observable value of $\vec{S}$ is its component S_z or $m_s\hbar$ along the z direction. The value of m_s is either +1/2 or −1/2. The vector $\vec{S}$ is not a constant of motion. It precesses round the z direction, defined by the experiment. The quantities that are constant of motion are S_z and $|S|^2$. This is shown in Fig. 1.16.

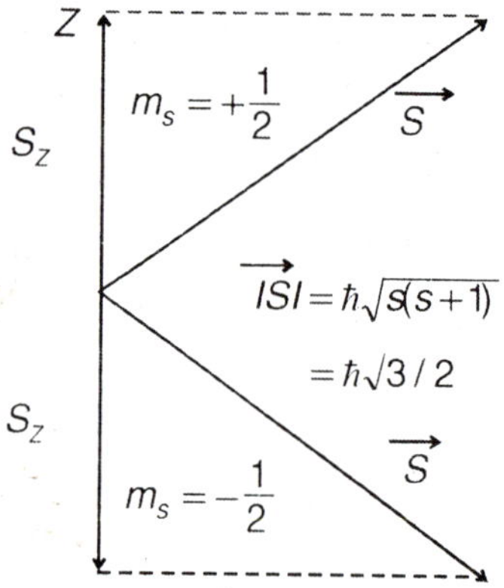

Fig. 1.16: The spin angular momentum vector $\vec{S}$ precesses around the Z direction with components $S_z = \pm 1/2\hbar$. As the direction of $\vec{S}$ is changing due to precession, it is not constant of motion. Only its magnitude $|\vec{S}|$ and S_z are constant of motion

A nucleus is composed of a number of neutrons and protons whose spin angular momenta add vectorially to give the spin angular momentum of the nucleus as a whole. The spin angular momentum $\vec{S}$ of the nucleus is thus

$$\vec{S} = \sum_{1}^{A} \vec{S}_k \qquad ...(1.5.1)$$

where, $\vec{S}_k$ is the spin angular momentum of the kth particle and the summation is over all the A particles in the nucleus. For a pair of particles the individual spin angular momenta can align only in such a way that the z components are either parallel (↑↑) or antiparallel (↑↓) so that S_z = 1 or 0. (in units of $\hbar$)

A nucleon inside a nucleus has its intrinsic spin angular momentum $\vec{S}$ as also an orbital angular momentum $\vec{L}$. The nucleons in a nucleus are packed closely and it is hard to imagine that they could have orbits of different orbital angular momentum as in the case of atoms. When the Schrodinger equation for a nucleon inside a nucleus, is solved one obtains a quantum number l which is associated with the orbital angular momentum $\vec{L}$. The quantity $\sqrt{l(l+1)}\hbar$ has same behaviour as the magnitude of $\vec{L}$. The orbital angular momentum $\vec{L}$ is not constant of motion while its magnitude $|L| = \sqrt{l(l+1)}\hbar$ and its z component L_z are constants of motion. The orbital angular momentum quantum number l can take only integral values as l = 0, 1, 2, 3, ... Along the z direction (defined by the experiment) the components L_z (or m_l) can have values as

$$m_l = +l, (l-1), (l-2) \ldots -l \qquad ...(1.5.2)$$

m_l is known as magnetic quantum number. For a given value of l, the magnetic quantum number m_l can have $(2l + 1)$ possible values. Fig. 1.17 shows the five possible values of m_l for $l = 2$.

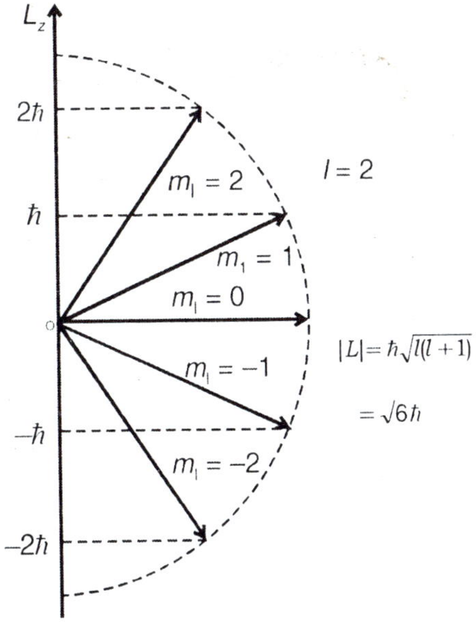

Fig. 1.17: Orbital angular momentum $\vec{L}$ has magnitude $|L| = \sqrt{l(l+1)}\hbar$ and components $L_z(m_l\hbar)$ along the z-direction for different orientations. The figure shows the different orientations for $l = 2$.

As in the case of spin angular momentum, the total orbital angular momentum of a nucleus is the vector sum of the orbital angular momentum of each of the A nucleons. Thus

$$\vec{L} = \sum_{k=1}^{A} \vec{L}_k \qquad \text{...(1.5.3)}$$

$\vec{L}_k$ is the orbital angular momentum of the kth nucleon.

As discussed above, a nucleus has a spin angular momentum $\vec{S}$ and an orbital angular momentum $\vec{L}$. The total angular momentum of a nucleus, sometimes called the "spin", is the vector sum of its spin and orbital angular momenta. So that

$$\vec{I} = \vec{L} + \vec{S}$$

$$= \sum_{k=1}^{A} \vec{L}_k + \sum_{k=1}^{A} \vec{S}_k \qquad \text{...(1.5.4)}$$

The magnitude of the vector $\vec{I}$ is $\sqrt{I(I+1)}\hbar$ and its components $I_z = m_I\hbar)$ can have any of the $(2I + 1)$ possible values as

$$m_I = I, (I-1), (I-2) \ldots -I \qquad \text{...(1.5.5)}$$

I is the quantum number associated with the total angular momentum $\vec{I}$.

Classically it may be said that vectors $\vec{L}$ and $\vec{S}$ precess round their resultant vector $\vec{I}$ which in turn precesses round the z direction in different orientations to give the components $m_I\hbar$ as shown in Fig. 1.18. As the spin of a nucleon is half integral and its orbital angular momentum is integral the total angular momentum of a nucleus with odd number of particles (A odd) is half-integral while for even A, it is integral or zero. Fig. 1.19 shows the possible values of m_I for $I = 3/2$

As nucleons obey Pauli exclusion principle, each substate m_1 can accommodate two particles so that for spin I, $2(2I + 1)$ protons and same number of neutrons can be accommodated.

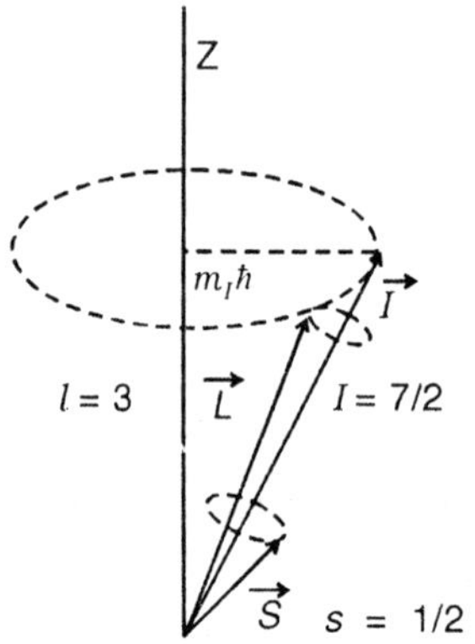

Fig. 1.18: Spin and orbital angular momenta couple to give total angular momentum $\vec{I}$ which precesses about the z direction, giving components I_z. The vectors $\vec{S}$ and $\vec{L}$ precess arond the vector $\vec{I}$.

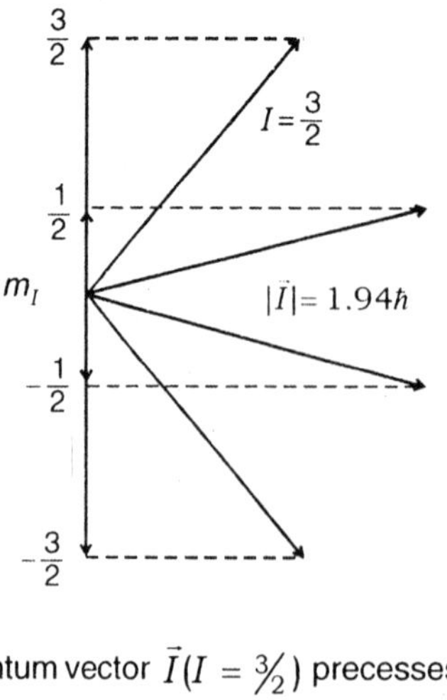

Fig. 1.19: The total angular momentum vector $\vec{I}(I = 3/2)$ precesses around z-axis to give components along z direction with $m_I = \pm 1/2, \pm 3/2$.

1.5.2 Gyromagnetic Ratio

If a current i flows through a circular loop of area S then the magnetic moment of the loop according to Biot-Savert law is

$$\mu = iS$$

When a charged particle of mass m and charge e moves in a circular orbit of radius R, with a constant speed v the current in the loop is

$$i = \frac{ev}{2\pi R}$$

and the magnetic moment is

$$\mu = \frac{ev}{2\pi R} \cdot \pi R^2 = \frac{evR}{2}$$

The orbital angular momentum of the particle $P_\phi = mvR$ so that

$$\frac{\mu}{P_\phi} = \frac{evR}{2} \cdot \frac{1}{mvR} = \frac{e}{2m}$$

The ratio of the magnetic moment to the orbital angular momentum of a charged particle is called its gyromagnetic ratio. Both μ and P_ϕ are vector quantities and one can write

$$\vec{\mu} = \frac{e}{2m} \vec{P}_\phi$$

For an electron the charge e is negative hence the two vectors $\vec{\mu}$ and $\vec{P}_\phi$ are opposite in direction.

The maximum observable orbital angular momentum of a proton in a nucleus is $\vec{L}$ in units of $\hbar$, the magnetic moment is thus

$$\mu = \frac{e\hbar}{2m_p} l$$

$$= \mu_N l \qquad \text{...(1.5.6)}$$

where, m_p is mass of the proton, $\mu_N = e\hbar/2m_p$ is called the nuclear magneton and

$$\mu_N = 5.0505 \times 10^{-27} \text{ Joules per Tesla or Jm}^2/\text{Wb}$$

For an electron the Bohr magneton μ_B is given as

$$\mu_B = \frac{e\hbar}{2m_e} = 9.27 \times 10^{-24} \text{ Joule/Tesla or (Jm}^2/\text{Wb)}$$

where, m_e is mass of an electron. It can be seen that

$$\mu_B = \frac{m_p}{m_e} \mu_N \approx 1837 \mu_N \qquad \text{...(1.5.7)}$$

A spinning charged particle also has a magnetic moment which is related to its spin angular momentum $\vec{S}$. For an electron the intrinsic magnetic moment is given as

$$\mu_s = g\mu_B s \qquad \text{...(1.5.8)}$$

The factor g is called the Linde g factor and has been calculated for an electron from Dirac theory. The calculated value of g agrees with the experimental value $g = 2.00290716$.

On analogy with the electron, the spin magnetic moment of a nucleon is written as

$$\mu_s = g\mu_N s$$

The values of the "g factor" for a proton and a neutron measured experimentally are

$$g_p \text{ (proton)} = 5.58550$$

$$g_n \text{ (neutron)} = -3.8270 \qquad \text{...(1.5.9)}$$

As the spin angular momentum quantum number for both the particles is $\frac{1}{2}$, their intrinsic magnetic moment is

$$\mu_p \text{ (proton)} = 2.79275\ \mu_N$$

$$\mu_n \text{ (neutron)} = -1.9135\ \mu_N \qquad \text{...(1.5.10)}$$

The observed values of the magnetic moment of neutron and proton can be explained on the basis of their non-uniform charge distribution as shown in Fig. 1.10. It can also be explained by the pion exchange theory of n-p forces, proposed by Yukawa. According to this theory the strong nuclear force between two nucleons is due to exchange of π mesons. A proton emits a π^+ meson, which is absorbed by the neutron. In this process the proton is converted into a neutron and the neutron, into a proton. The process goes on, giving rise to a force between a proton and a neutron. It can also happen that a neutron emits a π^- meson changing itself into a proton. The π^- meson is absorbed by a proton, changing it into a neutron. The two processes can be depicted as follows

$$n + p \rightarrow n + (\pi^+ + n) \rightarrow (n + \pi^+) + n \rightarrow p + n$$

$$n + p \rightarrow (p + \pi^-) + p \rightarrow p + (\pi^- + p) \rightarrow p + n$$

The measured mass of $273m_e$ for π^+ and π^- mesons explains the strength of the n-p force correctly. The Yukawa theory implies that for part of the time, a proton exists as a neutron surrounded by a cloud of π^+ mesons and a neutron as a proton with a cloud of π^- mesons. The spin quantum number for π mesons has a value 1. The magnetic moment of π meson is, therefore,

$$\mu_\pi = \mu_N \frac{m_p}{m_\pi} = 6.73\ \mu_N$$

also $\qquad \mu_{\pi^-} = -\mu_{\pi+}$

Taking the "g factor" of bare proton, on the analogy of an electron to be 2 and that of a μ meson to be 1. The magnetic momenta of the various particles is

$$\mu_p \text{ (bare proton)} = e\hbar / 2m_p = \mu_N$$

μ_n (bare neutron) $= 0$

$$\mu_{\pi^+} = 6.7\ \mu_N$$

$$\mu_{\pi^-} = -6.7\ \mu_N$$

Assuming that for x fraction of time a neutron exists as a bare proton and a π^- meson the observed magnetic moment of the neutron would be

$$\mu_{n_{obs}} = x(\mu_P + \mu_{\pi^-}) + (1 - x)\ \mu_n \quad \text{...(1.5.10)}$$

as $\mu_n = 0$

$$\mu_{n_{obs}} = x(\mu_N - 6.7\mu_N)$$

$$= -5.7\ x\ \mu_N$$

As $\mu_{n_{obs}} = -1.91\ \mu_N$

one gets

$$x = \frac{1.91\mu_N}{5.7\ \mu_N} \approx 0.33$$

Likewise if a proton is assumed to exist as a bare neutron and a π^+ meson for x part of the time then

$$\mu_{p_{obs}} = x(\mu_n + \mu_{\pi^+}) + (1 - x)\ \mu_p$$

as $\mu_n = 0$

$$\mu_{p_{obs}} = x(6.7\ \mu_N - \mu_N) + \mu_N$$

$$= -5.7\ \mu_N \cdot x + \mu_N$$

As $\mu_{p_{obs}} = 2.79\ \mu_N$

$$x = \frac{2.79\mu_N - \mu_N}{5.7\mu_N}$$

$$x = \frac{1.79}{5.7} \approx 0.31$$

Above calculations show that for one third of the time both neutron and proton are surrounded by the meson (π^- or π^+) cloud. The meson cloud concept helps to understand the possible origin of the enhanced magnetic moment of a proton and a negative magnetic moment of a neutron.

1.5.3 Magnetic Moment of Nuclei and Schmidt Lines

The magnetic moment of a nucleus, consisting of Z protons and N neutrons ($A = N + Z$), is the vector sum of the magnetic momenta associated with the orbital and spin angular momenta of all the particles, so that

$$\vec{\mu} = \vec{\mu}_l + \vec{\mu}_s \quad \text{...(1.5.11)}$$

The orbital angular momentum of a neutron, because of its neutral charge does not contribute to magnetic moment. The magnetic moment μ_L, associated with orbital angular momentum is only due to protons so that

$$\vec{\mu}_l = \sum_{k=1}^{Z} \mu_N \vec{L}_k \qquad ...(1.5.12)$$

The summation here is on all the Z protons. The spin angular momenta of both protons and neutrons give rise to magnetic moment, so that

$$\vec{\mu}_s = \sum_{k=1}^{Z} \mu_N \cdot g_p \cdot \vec{S}_k + \sum_{k=Z+1}^{A} \mu_N \cdot g_n \cdot \vec{S}_k \qquad ...(1.5.13)$$

g_p and g_n are the g factors for a proton and a neutron respectively as given in equation (1.5.9). The summation from $k = z + 1$ to A is over all the neutrons in the nucleus.

The magnetic moment of the nucleus is thus.

$$\vec{\mu} = \left[\sum_{k=1}^{Z} \left(\vec{L}_k + g_p \vec{S}_k \right) + \sum_{k=z+1}^{A} g_n \vec{S}_k \right] \mu_N \qquad ...(1.5.14)$$

The total angular momentum $\vec{I}$ of the nucleus is

$$\vec{I} = \sum_{k=1}^{A} \left(\vec{L}_k + \vec{S}_k \right) \qquad ...(1.5.15)$$

The vectors $\vec{\mu}$ and $\vec{I}$ are both composed of the same vectors $\vec{L}_k$ and $\vec{S}_k$, but their weightage in the two are different. The directions of vectors $\vec{\mu}$ and $\vec{I}$ are, therefore, different. The component of vector $\vec{\mu}$ along the vector $\vec{I}$ is the effective magnetic moment of the nucleus.
Thus

$$\vec{\mu}_{eff} = \vec{\mu} \cos(\vec{\mu}, \vec{I}) = \frac{\vec{\mu} \cdot \vec{I}}{I^2} \cdot \vec{I}$$

$$= \mu_N \cdot g_I \cdot I \qquad ...(1.5.16)$$

The observable value of the magnetic moment μ_Z of the nucleus is its component along the z-direction defined by the experiment, so that

$$\mu_z = \frac{\vec{\mu} \cdot \vec{I}}{I(I+1)} \cdot m_I$$

$$= \mu_N \cdot g_I \cdot m_I \qquad ...(1.5.17)$$

The maximum value of observable magnetic moment is

$$\mu_I = \mu_N \cdot g_I \cdot I \qquad ...(1.5.18)$$

The "g factor" for the nucleus is given as

$$g_I = \frac{\vec{\mu} \cdot \vec{I}}{I(I+1)}$$

$$= \frac{1}{I(I+1)} \vec{I} \cdot \left[\sum_{k=1}^{Z} (\vec{L}_k + g_p \vec{S}_k) + \sum_{k=z+1}^{A} g_n \vec{S}_k \right] \qquad \text{...(1.5.19)}$$

It is evident that unless the spin and orbital angular momenta for each nucleon are known, the g-factor can not be calculated. To calculate the magnetic moment, Schmidt assumed a model in which, all the even number of nucleons align in pairs with opposite spin and orbital angular momenta. The total angular momentum I and the magnetic moment μ of the nucleus is determined by the spin and orbital angular momentum of the last odd nucleon only. According to this model, the total angular momentum I of an even Z even N nucleus should be zero. This has actually been verified in a large number of nuclei. In the light of the above assumption, equation (1.5.19) is modified in that the summation is deleted and the vectors $\vec{L}_k$ and $\vec{S}_k$ refer to only the last odd nucleon. The last odd nucleon could be a proton or a neutron. One can then write a general expression for the g-factor as

$$g_I = g_l a_l + g_S a_S \qquad \text{...(1.5.20)}$$

If the last odd nucleon is a proton then $g_l = 1$ and $g_S = g_p = 5.585$ and if it is a neutron then $g_l = 0$ and $g_S = g_n = -3.827$. In equation (1.5.20)

$$a_l = \frac{\vec{I} \cdot \vec{L}}{I(I+1)} \qquad \text{...(1.5.21)}$$

and

$$a_S = \frac{\vec{I} \cdot \vec{S}}{I(I+1)} \qquad \text{...(1.5.22)}$$

Since the spin I of the nucleus is given as

$$\vec{I} = \vec{l} + \vec{s}$$

where $\vec{l}$ and $\vec{s}$ represent the orbital and spin angular momenta of the last odd nucleon one can write

$$\vec{I} - \vec{l} = \vec{s}$$

Squaring both sides gives

$$I^2 + l^2 - 2\vec{I} \cdot \vec{l} = s^2$$

$$\vec{I} \cdot \vec{l} = \frac{I^2 + l^2 - s^2}{2}$$

$$= \frac{I(I+1) + l(l+1) - s(s+1)}{2}$$

Similarly (using the relation $\vec{I} - \vec{s} = \vec{l}$) one gets

$$\vec{I} \cdot \vec{s} = \frac{I(I+1) - l(l+1) + s(s+1)}{2}$$

Substituting values of $\vec{I} \cdot \vec{l}$ and $\vec{I} \cdot \vec{s}$ in equations (1.5.21 and 1.5.22) respectively gives

$$a_l = \frac{I(I+1) + l(l+1) - s(s+1)}{2I(I+1)} \qquad ...(1.5.23)$$

$$a_s = \frac{I(I+1) - l(l+1) + s(s+1)}{2I(I+1)} \qquad ...(1.5.24)$$

The last odd nucleon has spin $s = 1/2$ so the total angular momentum I of the nucleus could have a value $I = l + 1/2$ or $I = l - 1/2$.

For $\qquad I = l + 1/2 \qquad (l = I - 1/2)$

$$a_l = \frac{I(I+1) + \left(I - \frac{1}{2}\right)\left(I + \frac{1}{2}\right) - \frac{1}{2} \cdot \frac{3}{2}}{2I(I+1)}$$

$$= \frac{I - 1/2}{I}$$

and

$$a_s = \frac{I(I+1) - \left(I - \frac{1}{2}\right)\left(I + \frac{1}{2}\right) + \frac{1}{2} \cdot \frac{3}{2}}{2I(I+1)}$$

$$= \frac{1}{2I}$$

Substituting the values of a_l and a_s in equation (1.5.20) one gets for $I = l + 1/2$

$$g_I = \frac{1}{I}\left\{\frac{1}{2} g_s + \left(I - \frac{1}{2}\right) g_l\right\} \qquad ...(1.5.25)$$

Similarly for $I = l - 1/2$ it can be shown that

$$a_l = \frac{I + 3/2}{I+1}$$

and

$$a_s = -\frac{1}{2(I+1)}$$

and

$$g_I = \frac{1}{I+1}\left\{-\frac{1}{2} g_s + \left(I + \frac{3}{2}\right) g_l\right\} \qquad ...(1.5.26)$$

If the last odd nucleon is a proton for which $g_l = 1$ and $g_s = 5.5855$ the magnetic moment of the odd Z nucleus of spin I is

$$\mu_I = \frac{1}{I}\left\{2.7927 + \left(I - \frac{1}{2}\right)\right\}\mu_N \text{ for } I = l + \frac{1}{2} \quad \text{...(1.5.27)}$$

and

$$\mu_I = \frac{1}{I+1}\left\{-2.7927 + \left(I + \frac{3}{2}\right)\right\}\mu_N \text{ for } I = l - \frac{1}{2} \quad \text{...(1.5.28)}$$

Similarly if the last odd nucleon is a neutron for which, $g_l = 0$, and $g_s = -3.8270$ the magnetic moment of the odd N nucleus of spin I is

$$\mu_I = -\frac{1.9135}{I} \cdot I\mu_N \text{ for } I = l + 1/2 \quad \text{...(1.5.29)}$$

$$= -1.9135\ \mu_N$$

and

$$\mu_I = -\frac{1.9135}{I+1} \cdot I\mu_N \text{ for } I = l - 1/2 \quad \text{...(1.5.30)}$$

The values of the effective magnetic moment for nuclei with last odd nucleon being a proton or a neutron and different total angular momentum I are tabulated in Table 1.2.

TABLE 1.2: VALUES OF EFFECTIVE MAGNETIC MOMENT OF NUCLEI IN UNITS OF μ_N WHEN LAST ODD PARTICLE IS A PROTON OR A NEUTRON AND $I = l + 1/2$ AND $I = l - 1/2$

	Last odd nucleon a proton		*Last odd nucleon a neutron*	
Nuclear spin	$I = l + 1/2$ $\mu_{eff}(\mu_N)$	$I = l - 1/2$ $\mu_{eff}(\mu_N)$	$I = l + 1/2$ $\mu_{eff}(\mu_N)$	$I = l - 1/2$ $\mu_{eff}(\mu_N)$
1/2	2.79	− 0.26	− 1.91	0.64
3/2	3.79	0.12	− 1.91	1.15
5/2	4.79	0.85	− 1.91	1.36
7/2	5.79	1.71	− 1.91	1.49
9/2	6.79	2.65	− 1.91	1.56

The calculated and the experimental values of the magnetic moment of odd Z and odd N nuclei for different spin I are plotted in Fig. 1.20 and 1.21 respectively. The lines joining the calculated values of magnetic moment are called the Schmidt lines.

It is observed from Fig. 1.20 and 1.21 that the measured values of magnetic moment of nuclei lie between the two Schmidt lines corresponding to $I = l + 1/2$ and $I = l - 1/2$. This implies that more than one nucleon contribute to the magnetic moment and the total angular momentum I of a nucleus. The single particle model of a nucleus needs refinements which will be discussed later.

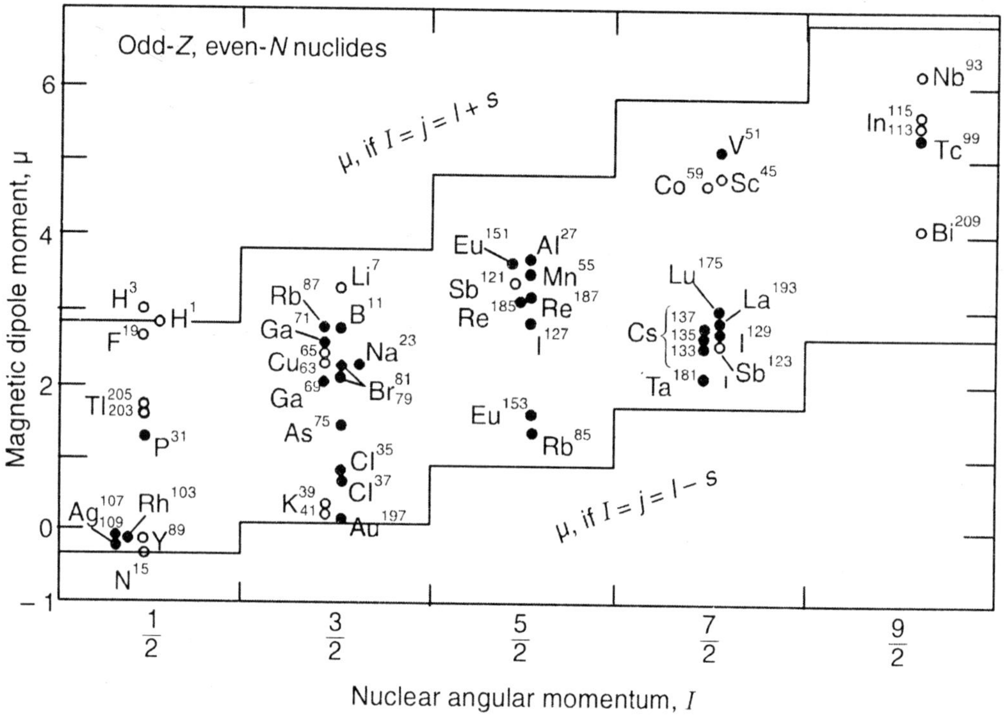

Fig. 1.20: Schmidt diagram showing magnetic moment μ of odd Z even N nuclides for different spin I. The measured values of magnetic momenta of various nuclei are also shown

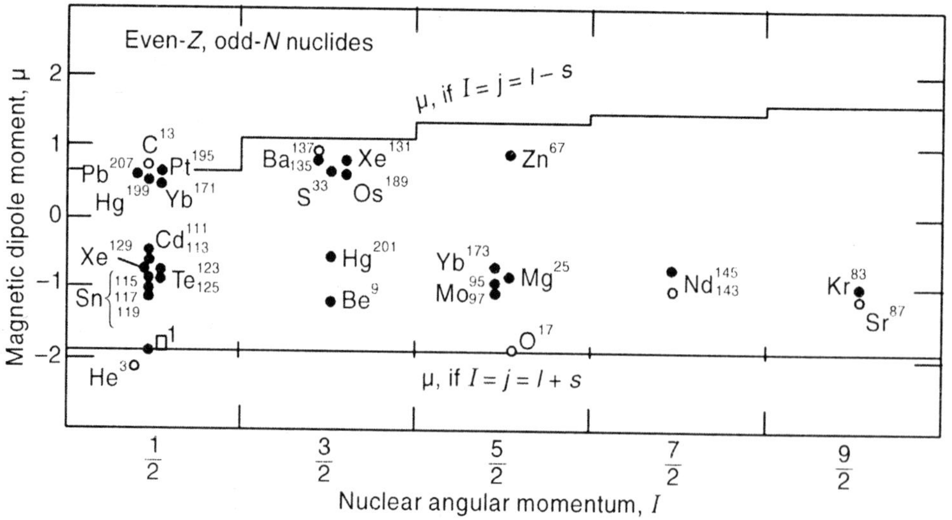

Fig. 1.21: Schmidt diagram showing magnetic moment μ of odd N even Z nuclides for different nuclear spin I. The measured values are also shown

1.5.4 Nuclear Statistics

When a system is composed of two identical particles its properties are governed by qantum mechanics in a different way. The Schrodinger equation for a system of two identical particles can be written as

$$H(1, 2)\,\Psi(r_1, r_2) = E\,\Psi(r_1, r_2) \quad ...(1.5.31)$$

or $$\frac{\hbar^2}{2m}(\nabla_1^2 + \nabla_2^2)\,\psi(r_1, r_2) + [E - V(r_1, r_2)]\psi(r_1, r_2) = 0 \quad ...(1.5.32)$$

Here r_1 or r_2 and 1 or 2 indicate all the coordinates including spins of particles 1 and 2 respectively. Because of indistinguishability of two particles $H(1, 2) = H(2, 1)$ hence if $\Psi(r_1, r_2)$ is a solution of the Schrodinger equation, $\Psi(r_2, r_1)$ is also a solution with the same eigen value. A linear combination of $\Psi(r_1, r_2)$ and $\Psi(r_2, r_1)$ is also a solution of the Schrodinger equation 1.5.32. The total wave function Ψ can be written as

$$\Psi(r_1, r_2) = a\,\psi(r_1, r_2) + b\,\psi(r_2, r_1) \quad ...(1.5.33)$$

If $a = b$ $$\psi(r_1, r_2) = \psi(r_2, r_1) \quad ...(1.5.34)$$

If $a = -b$ $$\psi(r_1, r_2) = -\psi(r_2, r_1) \quad ...(1.5.35)$$

The first case, when $a = b$ refers to a system being symmetrical in the exchange of the two particles. The second case refers to an anti-symmetric wave function. As the Hamiltonian operating on the wave function is symmetrical, the symmetric or antisymmetric property of the wave function of the system remains unchanged. Pauli has shown, using relativistic quantum mechanics that particles having zero or integral spin have symmetric wave functions and particles having half integral spin have antisymmetric wave functions. Thus particles with zero or integral spin (symmetric wave functions) follow Bose-Eienstein statistics and are called bosons. Particles with half integral spins (antisymmetric wave function) follow Fermi-Dirac statistics and are known as fermions.

Consider two non-interacting identical particles in a system. The Hamiltonian of the system separates as the sum of two terms

$$H(1, 2) = H(2, 1) = H(1) + H(2) \quad ...(1.5.36)$$

If $\psi_1(1), \psi_2(1) \ldots \psi_n(1) \ldots$ are eigen functions of $H(1)$ with eigen values $E_1(1), E_2(1), \ldots E_n(1)$ and $\psi_1(2), \psi_2(2), \ldots \psi_m(2)$ are eigen function of $H(2)$ with eigen values $E_1(2), E_2(2) \ldots E_m(2) \ldots$ then the eigen function of $H(1, 2)$ belonging to the eigen value $E_n(1) + E_m(2)$ is degenerate and has the form

$$\Psi_{nm}(1,2) = \frac{1}{\sqrt{2}}[\psi_n(1)\psi_m(2) \pm \psi_m(1)\Psi_n(2)] \quad ...(1.5.37)$$

where the plus sign is for symmetric wave functions (ψ_n or ψ_m) and minus sign for antisymmetric wave functions. If the final wave function $\Psi(1, 2)$ has to be antisymmetric, then it vanishes when $n = m$. This means that two identical particles with antisymmetric wave function (e.g. electrons, protons, neutrons) can not occupy the same state ($n = m$). This is the original formulation of Pauli's exclusion principle.

For a diatomic molecule having two identical atoms the wave function can be written as

$$\psi = \psi_{el}\, \psi_{vib}\, \psi_{rot}\, \psi_{spin} \quad ...(1.5.38)$$

Where ψ_{el} is the electronic wave function, which we assume to be symmetric. The wave function ψ_{vib} depends only upon the relative distance between the two nuclei in the molecule and is symmetric. The third term ψ_{rot} can be written as

$$\psi_{rot} \sim P_l^m (\cos\theta)\, e^{im\phi}$$

Where θ and ϕ are co-altitude and longitude of the axis forming the two nuclei. Interchange of two nuclei corresponds to change in variables of the wave function; θ goes to $\pi - \theta$ and ϕ goes to $\pi + \phi$. If expressed in x, y, z, system, it corresponds to a change in the space coordinates. $P_l^m(\cos\theta)e^{im\phi}$ is a spherical harmonic of order l. If l is even the wave function ψ_{rot} remains unchanged on interchange of particles while it changes sign if l is odd.

If each nucleus in the diatomic molecule has spin ½, its wave funcntion could have only two values corresponding to spin up or spin down. These wave functions may be indicated as α = spin up and β = spin down. Thus wave function $\alpha(2)$ corresponds to the second particle with spin up and $\beta(2)$ corresponds to the second particle with spin down. For two nuclei with spin $\frac{1}{2}$ the spin wave function ψ_{spin} has one of the following forms corresponding to the triplet state (parallel spins).

$$\alpha(1)\,\alpha(2),\ \frac{1}{\sqrt{2}}[\alpha(1)\,\beta(2) + \alpha(2)\,\beta(1)],\ \beta(1)\,\beta(2)$$

If the spins of the two nuclei are antiparallel, the eigen function for the singlet state has the form

$$\frac{1}{\sqrt{2}}[\alpha(1)\,\beta(2) - \alpha(2)\,\beta(1)]$$

The triplet state is symmetric and the singlet state is antisymmetric with respect to change of spin only. If the total wave function Ψ has to be antisymmetric with respect to change of nuclei, it is clear that the eigen function ψ_{rot} with odd values of l must be associated with the triplet state. Conversly, states with even values of l are associated with singlet spin states. The weight factor—probability of occurrence—for triplet state is 3 and for singlet state it is 1.

In the emission or absorption of electromagnetic radiations, transitions between different levels of the molecule take place. As the nature of a state does not change in a transition, these transitions could take place between symmetric states or between anti-symmetric states with respect to exchange of ordinary coordinates (not spins). The intensity of a line depends upon the statistical weight of the state from which transition takes place. In the diatomic molecule considered above, the spectral lines corresponding to l odd would be three times as strong as those corresponding to l even. If the nuclei in the diatomic molecule are bosons, the lines with even l would be stronger. If the spin of each of the nuclei in the homonuclear diatomic molecule is I the statistical weights of the neighbouring rotational states are in ratio $(I + 1)/I$.

1.6 TOTAL ANGULAR MOMENTUM OF AN ATOM

In an atom, the nucleus has the total angular momentum $\vec{I}$ and the electrons have a total angular momentum $\vec{J}$ (both in units of $\hbar$). The two angular momenta couple vectorially to give the total angular momentum $\vec{F}$ to the atom, so that

$$\vec{F} = \vec{I} + \vec{J} \qquad ...(1.6.1)$$

The magnitude of the vector F is $\sqrt{F(F+1)}\hbar$, where F is the quantum number associated with the total angular momentum of the atom.

Along the z-axis defined by the experiment, the vector $\vec{F}$ has components $m_F\hbar$ where,

$$m_F = F, F-1, F-2 \ldots -F \qquad ...(1.6.2)$$

There are $(2F + 1)$ possible values of the atomic magnetic quantum number m_F. For given values of I and J, the total angular momentum quantum number F can take the values as

$$F = (I + J), (I + J - 1) \ldots |I - J| \qquad ...(1.6.3)$$

There are thus $(2I + 1)$ values that F can take if $I < J$ and $(2J + 1)$ values if $J < I$.

Like L-S coupling, classically the vectors $\vec{I}$ and $\vec{J}$ for the atom precess around their resultant $\vec{F}$ which in turn precesses around the z-axis in such orientations as to give different values of m_F [Fig. 1.22].

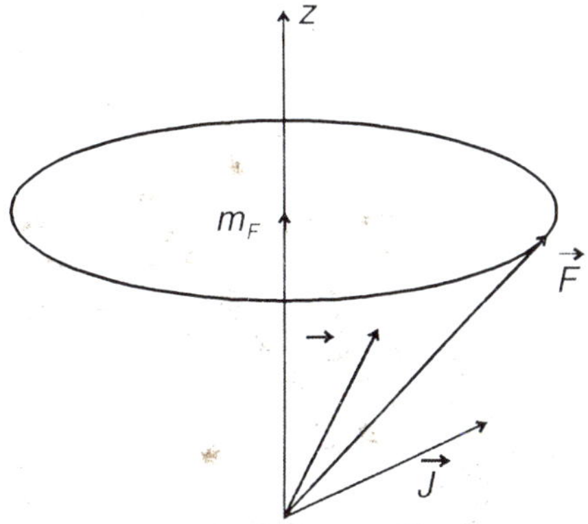

Fig. 1.22: Total angular momentum $\vec{I}$ of the nucleus couples with total angular momentum $\vec{J}$ of the atomic electrons to give the resultant total angular momentum $\vec{F}$ of the atom, The vector $\vec{F}$ precesses around z direction to give different components m_F.

1.7 MEASUREMENT OF SPIN AND MAGNETIC MOMENT OF NUCLEI

In the early days, optical spectroscopy methods were extensively employed to measure the spin and magnetic moment of nuclei. Light emission from atomic beams allowed much higher spectroscopic resolution than could be obtained for emission of light from discharge tubes.

All the spectroscopic methods were superseded later on by the radiofrequency resonance methods which gave a higher accuracy in the measurement of magnetic moment of nuclei. Various methods are discussed briefly.

1.7.1 Molecular Rotational Spectrum

The spin or total angular momentum of a nucleus can be obtained from the intensity ratio of the alternate lines of the band spectrum of a diatomic homonuclear molecule. The relative intensity of each of the spectral lines in a rotational band spectrum of a homonuclear diatomic molecule (e.g. $^{1}H^{1}H$, $^{12}C^{12}C$, $^{14}N^{14}N$ $^{16}O^{16}O$) is determined by the statistical weight of the states involved in the transition. If I is the total angular momentum quantum number of each nucleus of a diatomic molecule, then the total nuclear angular momentum quantum number T of the molecule can have any of the values $T = 2I, 2I - 1 \ldots 2I - 3 \ldots 0$. It can be seen that the states with even values of T (*i.e.* $2I, 2I - 2 \ldots 0$) belong to one of the two types of rotational states--symmetric or antisymmetric in the space co-ordinates of the nuclei (even or odd parity). The states with odd values of T would correspond to thc symmetry opposite to that of the even T states. Each state with total angular momentum quantum number T consists of $2T + 1$ magnetic substates which coincide, in the absence of an external magnetic field. Each of the magnetic substates have an equal chance of occurrence. The statistical weight of state T is thus $2T + 1$ times the statistical weight of the $T = 0$ state $(2T + 1 = 1)$.

Transitions between states of opposite symmetry is completely forbidden. Transitions between states of similar symmetry can occur only when accompanied by an electronic transition. The homonuclear diatomic molecules, therefore, do not have a pure rotational or rotational vibrational spectrum. Alternate lines in the rotational fine structure of the electronic spectra arise from transitions between states belonging to one of the symmetry types, depending upon the electronic states involved. For example, the first, third and the fifth lines could be from antisymmetric rotational states and second, fourth and sixth lines could be from transitions among symmetric states. For any rotational state, all the permitted values of T are possible. The statistical weight of the symmetric or the antisymmetric states is the sum of the statistical weights of all the T states of the corresponding symmetry. The ratio of the statistical weights of the symmetric states to the antisymmetric states is $(I + 1)/I$. The successive lines, therefore, have an intensity ratio of $(I + 1)/I$. This is the ratio of the intensity of the symmetric lines to that of the antisymmetric lines for nuclei obeying Bose-Einstein statistics $(I = 0, 1, 2, 3, 4 \ldots)$. The ratio of the intensities of the symmetric lines to that of antisymmetric lines for nuclei obeying Fermi-Dirac statistics ($I = 1/2, 3/2, 5/2, ...$) is $I/(I+1)$. Thus regardless of the type of statistics obeyed by the nuclei in a homonuclear diatomic molecule, the average ratio of the intensities of the more intense to the less intense lines is always equal to $(I+1)/I$. Measurement of the intensities of the rotational lines thus gives the total angular momentum quantum number I for the nuclei forming the homonuclear di-atomic molecule. It is clear from the intensity ratio that if $I = 0$ the ratio of the intensities of the alternate lines is infinity, which means that the alternate lines are missing from the spectrum. The total angular momenta or spins of the nuclei ^1H, ^2H, ^4He, ^7Li, ^{12}C, ^{13}O, ^{14}N, ^{15}N, ^{16}O, ^{19}F, ^{23}Na, ^{31}P, ^{32}S and ^{35}Cl have been measured by this method.

1.7.2 Hyperfine Structure of Atomic Spectral Lines

The study of the hyperfine structure of atomic spectral lines has been used to measure the spin and the magnetic moment of nuclei. It is known that the yellow line of sodium is composed of two lines of wavelengths 5890Å and 5896Å arising from the l - s coupling for the electron in the p-orbit. This is known as the fine structure of the yellow line of sodium. If one examines the 5890Å line, which arises from the transition from the $^2p_{3/2}$ state to $^2s_{1/2}$ ground state of sodium atom, under a very high resolution, one finds that the line is composed of two lines. This is known as the hyperfine structure of the line.

In an atom, the electrons produce a magnetic field at the nucleus. The field is strongest when there is a single electron in the s-orbit (e.g. in alkali metals). The intensity of the field is proportional to the probability of finding the electron in the nuclear volume, which is largest for the s-electron. The magnitude of the magnetic field produced at the nucleus by a single electron in the s and p orbits of atoms of alkali metals is tabulated in Table 1.3.

The interaction of the magnetic moment of the nucleus in different orientations, with the magnetic field produced by the atomic electrons gives rise to a splitting in the energy levels of the atom. Consequently, the spectral lines show the hyperfine splitting which is much smaller than the fine structure splitting.

TABLE 1.3

Atom	*Principal quantum number* n	*H(o) Gauss* $^2s_{1/2}$	*Principal quantum number* n	*H(o) Gauss* $^2p_{1/2}$	*H(o) Gauss* $^2p_{3/2}$
H	1	1.74×10^5	–	–	–
Li	2	1.3×10^5	–	–	–
Na	3	4.4×10^5	3	4.2×10^4	2.5×10^4
K	4	6.3×10^5	4	7.9×10^4	4.6×10^4
Rb	5	1.3×10^6	5	1.6×10^5	8.4×10^4
Cs	6	2.1×10^6	6	2.8×10^5	1.3×10^5

Consider a nucleus with total angular momentum quantum number I and magnetic moment μ_I placed in a magnetic field H(0). The energy E of the nucleus due to magnetic interaction is

$$E = -\vec{\mu}_I \cdot \vec{H}(0) \qquad ...(1.7.1)$$

$$\vec{\mu}_I = \mu_I \cdot \frac{\vec{I}}{I} \qquad ...(1.7.2)$$

μ_I is the maximum observable magnetic moment of the nucleus. Eqns. (1.7.1) and (1.7.2) give:

$$E = -\mu_I \frac{\vec{I} \cdot \vec{H}(0)}{I}$$

The magnetic field H(0) produced by the electron at the nucleus is proportional to its total angular momentum, J. The field $\vec{H}$ (0) can be written as

$$\vec{H}(0) = H(0)\frac{\vec{J}}{J} \qquad ...(1.7.3)$$

which gives

$$E = -\mu_I H(0)\frac{\vec{I} \cdot \vec{J}}{IJ}$$

$$= A_J(\vec{I} \cdot \vec{J}), \qquad ...(1.7.4)$$

where $$A(J) = \frac{-\mu_I H(0)}{IJ} \qquad ...(1.7.5)$$

Because of the negative charge of the electron, the direction of $\vec{H}$ (0) is opposite to that of $\vec{J}$. If the magnetic moment of the nucleus is positive the energy of the atom is lowest when μ_I and $\vec{H}$ (0) point in the same direction (*i.e.* $\vec{\mu}$ and $\vec{J}$ are anti-parallel).

The total angular momentum $\vec{F}$ of the atom is

$$\vec{F} = \vec{I} + \vec{J}$$

which gives,

$$\vec{I} \cdot \vec{J} = \frac{F^2 - I^2 - J^2}{2}$$

$$= \frac{F(F+1) - I(I+1) - J(J+1)}{2}$$

$$\vec{I} \cdot \vec{J} = C/2, \qquad \text{...(1.7.6)}$$

where $$C = F(F+1) - I(I+1) - J(J+1) \qquad \text{...(1.7.7)}$$

Equations (1.7.4), (1.7.5) and (1.7.6) give

$$E = A_J \cdot \frac{C}{2} \qquad \text{...(1.7.8)}$$

The hyperfine splitting of an atomic level for which $I = 3/2$, $J = 2$ and a positive magnetic moment is shown in Fig. 1.23.

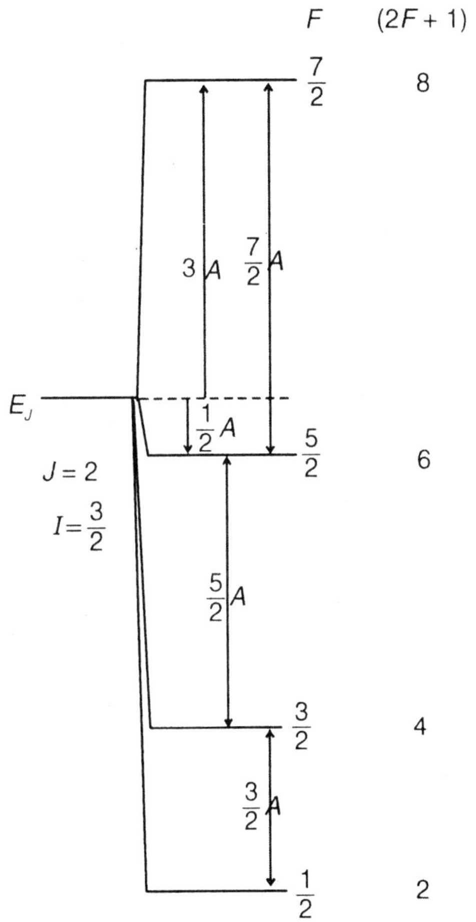

Fig. 1.23: An atom with $I = 3/2$ and $J = 2$ has total angular momentum quantum number F having values 7/2, 5/2, 3/2 and 1/2. Due to the internal magnetic field, the energy level of the atom splits into levels of different values of F. Each F level is $2F + 1$ fold degenerate. The magnetic moment of nucleus is assumed to be positive. If the nuclear magnetic moment is negative, the order of the level is inverted.

The energy difference between the atomic levels having total angular momentum quantum number F and $F - 1$ respectively is given by $E_F - E_{F-1}$.

Using equations (1.7.7) and (1.7.8) it can be shown that

$$E_F - E_{F-1} = A_J \cdot F \qquad ...(1.7.9)$$

A_J is called the "interval factor" of hyperfine structure. The energy interval between the hyperfine split levels is in the ratio

$$I+J : I+J-1 : I+J-2...$$

This is known as the interval rule for hyperfine splitting. If J is known, measurement of the interval ratio gives the value of I, the nuclear spin.

The nuclear magnetic moment is taken to be positive if it is parallel to the nuclear spin I and negative if it is antiparallel to $\vec{I}$. As $\vec{H}$ (0) is antiparallel to $\vec{J}$ the hyperfine state with highest value of F ($\vec{I}$ and $\vec{J}$ parallel) lies highest in the energy level diagram provided the nuclear magnetic moment is positive.

According to equation (1.7.9) the hyperfine splitting of an atomic level depends upon A_j which depends upon H(0) as given in equation (1.7.5). If the magnetic field H(0) for a particular electron state is weak, the hyperfine splitting is too small to be observed. In another electron state the magnetic field may be appreciable. If transition takes place from an unsplit state to hyperfine split state the emitted spectral line is split in number of lines. The number of such lines is $(2J+1)$ if $J < I$ and $(2I+1)$ if $I < J$.

The 5890 Å line of sodium arises from the transition of an electron from $^2p_{3/2}$ state to the $^2s_{1/2}$ ground state. The magnetic field H(0) at the nucleus in the $^2p_{3/2}$ state is very weak (Table 1.3) hence the hyperfine splitting of the state is negligible. The total angular moment F of this state ($I = 3/2$) could have the values $F = 3, 2, 1$ and 0, hence the state is degenerate. In the $^2s_{1/2}$ ground state ($I = 3/2$, $J = 1/2$) the magnetic field H(0) is appreciable and the state splits in $2J + 1 = 2$ energy levels having the quantum number $F = 2$ and 1 respectively. The $F = 2$ state is higher in energy as the magnetic moment of the sodium nucleus is positive. The hyperfine splitting of the energy levels and the transitions between them is shown in Fig. 1.24.

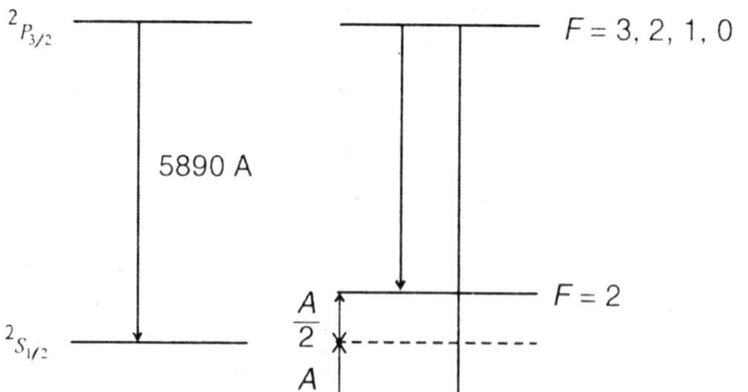

Fig. 1.24: Hyperfine splitting of the 5890 Å line of sodium. (The transition $^2p_{3/2} \rightarrow {}^2s_{1/2}$ is also shown)

The selection rules for transition between the hyperfine multiplets is

$$\Delta F = \pm 1, 0$$

Under a high resolution, the 5890 Å line of sodium is found to consist of a doublet. From the separation of these two lines of the doublet, the value of the interval factor A_j can be obtained. As the magnitude of the electronic field H(0) can be calculated, the magnetic moment of the nucleus can be determined.

A quantum mechanical derivation of the interval factor, using Dirac equation, has been given by Fermi. For a single s-electron the interval factor A_s is obtained as

$$A_s = \frac{8\pi}{3} \cdot 2\mu_B^2 g_I |\psi_{n,0}(0)|^2 \qquad \text{...(1.7.10)}$$

where $|\psi_{n,0}(0)|^2$ is the square of the value of the eigen function of the electron at the centre of the nucleus and is given as

$$|\psi_{n,0}(0)|^2 = \frac{1}{\pi a_H^3} \cdot \frac{Z^3}{n^3} \qquad \text{...(1.7.11)}$$

where a_H is the radius of the first Bohr orbit in hydrogen atom, n the principal quantum number of the electron orbit and Z the atomic number of the nucleus. Equations (1.7.9) and (1.7.10) directly give the value of g_I from the separation of the hyperfine multiplets.

It may be pointed out that, for an electron state with $J > 1/2$ and nuclear spin $I > 1/2$, the interval between the hyperfine multiplets is affected by the electric quadrupole moment of the nucleus. In many cases the departure of the interval between hyperfine multiplets from the interval rule is employed to calculate the quadrupole moment of the nucleus.

1.7.3 Zeeman and Paschen Back Splitting of Hyperfine Levels

When an atom with a non zero spin (total angular momentum) is placed in a weak external magnetic field H_e, the total angular momentum of the atom $\vec{F}$ precesses round the field $\vec{H}_e$ with components m_F. As the external field is increased, the coupling between vectors $\vec{I}$ and $\vec{J}$ breaks down so that F is no longer a good quantum number. Vectors $\vec{I}$ and $\vec{J}$ orient themselves independently about the field and the system is thus described by the magnetic quantum numbers m_I and m_J.

In the external field H_e the magnetic energy of the electron in state J is

$$E_J = -\mu_B g_J \vec{J} \cdot \vec{H}_e \qquad \text{...(1.7.12)}$$

while the energy of the nucleus with spin I is

$$E_I = -\mu_N g_I \vec{I} \cdot (\vec{H}_e + \vec{H}(0)) \qquad \text{...(1.7.13)}$$

The external magnetic field is said to be very weak if

$$\mu_N g_I \vec{I} \cdot \vec{H}(0) >> g_J \mu_B \vec{J} \cdot \vec{H}_e \qquad ...(1.7.14)$$

As $\mu_B \approx 2000\ \mu_N$ an external field H_e of magnitude of few hundred Gauss can be regarded as a very weak field. In this situation the hyperfine splitting of the atomic level is much larger than the magnetic splitting. The magnetic interaction energy of the atom for a given value of F in a very weak field is

$$E = -\mu_B\, g_J\, J\, H_e \cos(\vec{J} \cdot \vec{F}) \cos(\vec{F} \cdot \vec{H}_e)$$

$$+ \mu_N\, g_I\, I\, H_e \cos(\vec{I} \cdot \vec{F}) \cos(\vec{F} \cdot \vec{H}_e), \qquad ...(1.7.15)$$

where $J \cos (J \cdot F)$ is the projection of $\vec{J}$ along $\vec{F}$ and $I \cos(I \cdot F)$ is the projection of $\vec{I}$ along $\vec{F}$ .cos $(F \cdot H_e)$ is the cosine of angle between vectors $\vec{F}$ and $\vec{H}_e$.

As $\mu_B >> \mu_N$ the second term is very small and can be neglected. Further

$$\cos(\vec{F} \cdot \vec{H}_e) = \frac{m_F}{F}$$

and $$\cos(\vec{J} \cdot \vec{F}) = \frac{\vec{J} \cdot \vec{F}}{JF}$$

so that

$$E = -\mu_B g_J J H_e \cdot \frac{m_F}{F} \cdot \frac{\vec{J} \cdot \vec{F}}{JF}$$

$$= -\mu_B g_J H_e \cdot m_F \frac{\vec{J} \cdot \vec{F}}{F^2}$$

$$= -\mu_B \cdot g_F \cdot m_F H_e \qquad ...(1.7.16)$$

where, $$g_F = g_J \frac{\vec{J} \cdot \vec{F}}{F^2}$$

As $$\vec{F} = \vec{I} + \vec{J}$$

so that $$\vec{F} - \vec{J} = \vec{I}$$

squaring both sides and substituting values of F^2, J^2 and I^2 gives

$$\vec{J} \cdot \vec{F} = \frac{F(F+1) + J(J+1) - I(I+1)}{2}$$

$$g_F = g_J \frac{F(F+1) + J(J+1) - I(I+1)}{2F(F+1)} \qquad ...(1.7.17)$$

Equation (1.7.16) shows that the atomic hyperfine level with total angular momentum quantum number F is split in $(2F + 1)$ levels. This gives rise to the Zeeman splitting of the hyperfine levels. As the field is increased such that condition (1.7.14) is no longer satisfied, the $\vec{I}\vec{J}$ coupling breaks down and the splitting of levels is described by the magnetic quantum numbers m_I and m_J (Paschen Back Effect). Figure 1.25 shows the Zeeman and Paschen Back splitting of the levels with $I = 1$ and $J = ½$.

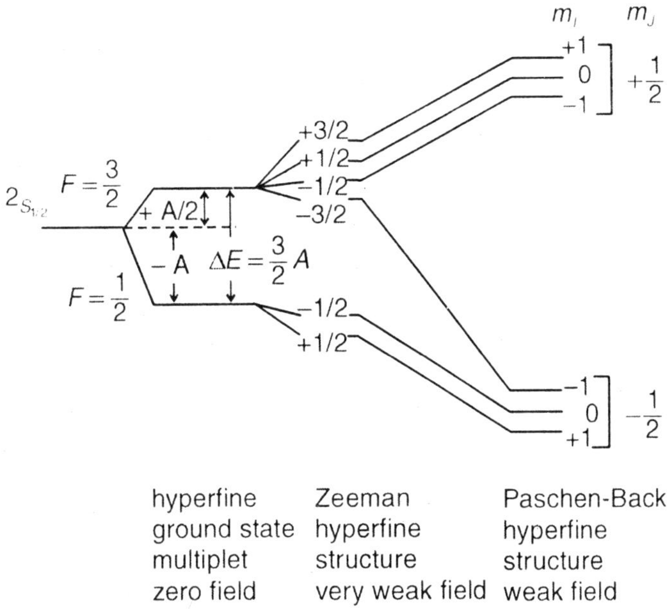

Fig. 1.25: Hyperfine, Zeeman and Paschen-Back splitting of the atomic state with $I = 1$ and $J = 1/2$ and a positive magnetic dipole moment of the nucleus

When the external magnetic field H_e is such that the magnetic splitting and hyperfine splitting of atomic levels is comparable, (field $H_e \approx$ 1000 Gauss) the energy of an atom is given as

$$E = -(\mu_N g_I m_I + \mu_B g_J m_J)H_e + \mu_I H(0)\frac{\vec{I}\cdot\vec{J}}{IJ} \qquad ...(1.7.18)$$

The first term is negligible as $\mu_N << \mu_B$. The average value of $I \cdot J = m_I m_J$ so that putting $A = \mu_I \dfrac{H(0)}{IJ}$

$$E = -\mu_B g_J m_J H_e + A m_I m_J \qquad ...(1.7.19)$$

It is observed that the atomic level is split into $(2J + 1)$ levels corresponding to the different values of m_J. Each m_J level is further split into $(2I + 1)$ levels corresponding to different m_I values. This is the region of Paschen Back effect and is shown in Fig. 1.25. The variation of the energies of the sub levels of an atom as the external field is increased from zero to the Paschen Back region is shown in Fig. 1.26. The Zeeman region corresponds to a magnetic field where $x < 0.5$. The region for $x > 1.5$ corresponds to the Paschen Back effect region. In the Zeeman region where F is a good quantum number, the selection rule for transition between the different substates is

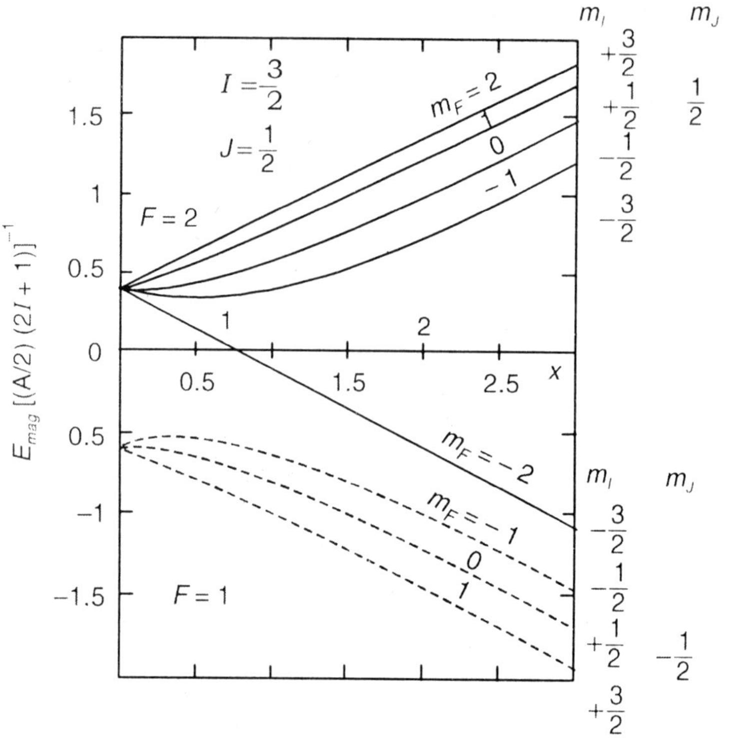

Fig. 1.26: Variation of the energies of the Zeeman split atomic levels with $J = 1/2$ and $I = 3/2$ as a function of the applied magnetic field described by the parameter x.

$$\Delta F = \pm 1, 0$$

$$\Delta m_F = 0 \text{ for } \pi \text{ component}$$

$$\Delta m_F = \pm 1 \text{ for } \sigma \text{ component}$$

In the region of Paschen Back effect where the IJ coupling has broken down, the selection rule for transition is

$$\Delta m_I = 0$$

$$\Delta m_J = 0 \text{ for } \pi \text{ component}$$

$$\Delta m_J = \pm 1 \text{ for } \sigma \text{ component}$$

The transition corresponding to the π component emits light which is polarised in the direction of the applied magnetic field. The light emitted in transition for σ component is polarised perpendicular to the external field. When observed perpendicular to the applied field one observes only π component, while along the field one observes only σ component.

The Zeeman and Paschen Back effects in the hyperfine multiplets of sodium were observed by Jackson and Kuhn. A beam of sodium atoms was passed through a magnetic field of a few hundred Gauss. Light was passed through the beam of sodium atoms perpendicular to the magnetic field and also the direction of the beam. Under very high resolution the 5890 Å absorption line of sodium was found to split into eight components corresponding to the π transitions. The observed absorption spectrum can be understood from Fig. 1.27 where the hyperfine multiplets and their Zeeman splitting is shown.

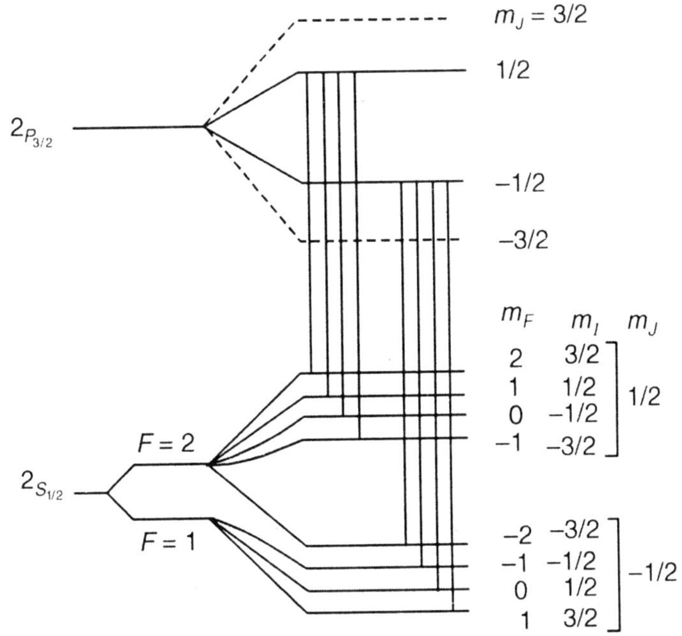

Fig. 1.27: Zeeman splitting of the hyperfine doublet of the 5890 Å line of sodium. The transitions have been observed in the absorption spectrum [D.A, Jackson and H. Kuhn (1938) *Proc. Roy. Soc. London*, **167**; 210].

In the excited $p_{3/2}$ state of the atom, the magnetic field $H(0)$ at the nucleus due to the p electron is very small, hence the second term in equation (1.7.19) is negligible. The $p_{3/2}$ level therefore splits only in the various m_J levels. The $s_{1/2}$ ground state is composed of a hyperfine doublet with $F = 2$ and $F = 1$. The $F = 2$ level in the external magnetic field splits in five $(2F + 1)$ substates with m_I and m_J values as shown in Fig. 1.27. Similarly the $F = 1$ state splits in three substates.

The π transitions require $\Delta m_J = 0$ restricting to the transitions shown in Fig. 1.27.

The major problem in the spectroscopic investigations of hyperfine structure is the production of spectral lines with sufficiently small half width. The most important effect causing the broadening of spectral lines is the Doppler effect, brought about by the random motion of the emitting atoms. This line broadening can be much larger than the resolution required for studying hyperfine structure. The halfwidth of a spectral line emitted by atoms of a vapour with molecular weight M and at absolute temperature T is

$$\Delta\nu = 2(2R \log 2)^{1/2} (\nu/c)\, (T/M)^{1/2}\ \text{cm}^{-1} \qquad \ldots(1.7.20)$$

where, R is the universal gas constant, c the speed of light and ν the wave number of the line in question. For small half width, the temperature of the vapour should be as low as possible. A hollow cathode discharge source is often employed. The material under study is placed in a cylindrical hole in a metal block, which forms the cathode. An electrical discharge in an inert gas concentrates in the hollow cathode and sputters the material under study into the gas. The atoms of the material are excited and emit light which can be studied spectroscopically. Such a type of source can be cooled by liquid nitrogen or liquid hydrogen. The hollow cathode source employs a very small sample under study.

The hollow cathode source can be cooled to liquid helium temperature by direct refrigeration technique. It is possible to attain temperatures of about 3 K in the atomic beam source where the atoms are collimated in a narrow beam. If the source is used in the study of emission spectra, the atomic beam is excited by electrons moving perpendicular to the beam and the radiation is observed in a direction perpendicular to both the beam and the electron stream. When radiations are observed in this way, the Doppler effect due to the velocity of the atoms in the beam is minimal. A considerable simplification of the apparatus is achieved if the atomic beam is used to absorb radiations from an auxiliary source. In this way one studies the absorption spectrum. An atomic beam source can produce spectral lines with half widths approaching the natural line widths of the radiation.

For large hyperfine splitting a grating spectrograph can be employed. For small hyperfine splitting, the high resolving power required is usually obtained with a Fabry-Perot interferometer. The maximum resolving power of the Fabry-Perot etalon is matched with, the line width obtainable from an atomic beam source. The Fabry-Perot etalon is used in conjunction with a spectrograph, which isolates the lines to be examined, from other wavelengths present.

1.7.4 Molecular and Atomic Beam Method (Stern and Gerlach's Experiment)

In a magnetic field H_e in which I and J coupling remains intact, the magnetic moment of an atom, according to equation (1.7.16) is

$$\mu = \frac{dE}{dH_e}$$

$$= \mu_B g_F m_F \quad \ldots(1.7.21)$$

When, the I, J coupling breaks down the magnetic moment from equation (1.7.16) is

$$\mu = \mu_B g_J m_J + \mu_N g_I m_I \quad \ldots(1.7.22)$$

The variation of the magnetic moment of an atom of sodium ($J = 1/2$, $I = 3/2$) as a function of the applied magnetic field is shown in Fig. 1.28. The curves in this figure represent the slopes of various curves in Fig. 1.26.

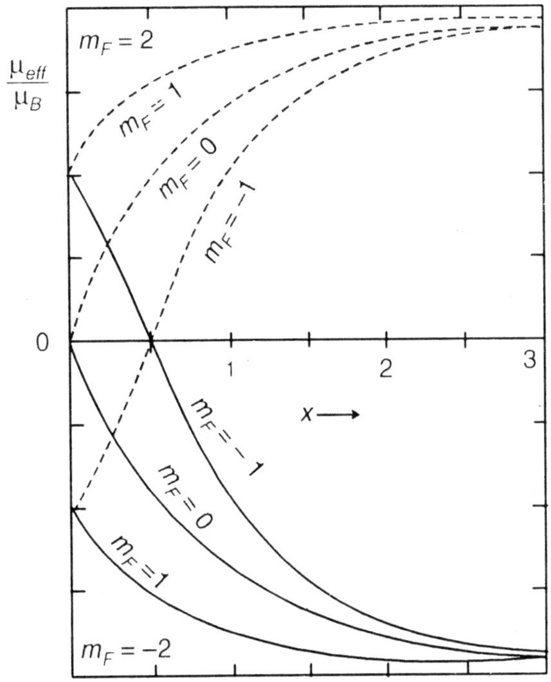

Fig. 1.28: Variation of the magnetic moment of a sodium atom as a function of the applied magnetic field defined by the parameter x

An atom having magnetic moment and passing through a homogeneous magnetic field does not experience any force. In an inhomogeneous magnetic field—having a field gradient in the z direction, the force acting on the atom in the z direction is

$$F = \mu_{\text{eff}} \frac{dH_e}{dz} \qquad \text{...(1.7.23)}$$

If M is the mass of the atom, the acceleration produced due to the force is

$$\ddot{z} = \mu_{\text{eff}} \frac{dH_e}{dz} \frac{1}{M}$$

If the atom traverses a distance l in the inhomogeneous magnetic field with velocity v, its displacement in the z direction from its original path is

$$z = \frac{1}{2} \mu_{\text{eff}} \frac{dH_e}{dz} \frac{1}{M} \left(\frac{l}{v}\right)^2$$

$$= \frac{1}{4} \mu_{\text{eff}} \frac{dH_e}{dz} \frac{l^2}{E_{\text{kin}}} \qquad \text{...(1.7.24)}$$

where, $E_{\text{kin}} = \frac{1}{2} Mv^2$ is the kinetic energy of the atom.

Originally Stern and Gerlach used this principle to measure the spin and magnetic moment of electron. Later they applied the method to measure the magnetic moment of proton using a beam of orthohydrogen. Kellog, Rabi and Zacharias in 1936 measured the magnetic moment of proton using a beam of atomic hydrogen. The method was employed to measure the spin of sodium nucleus. When passed through an inhomogeneous magnetic field the beam of sodium atoms was found to divide into eight $\{(2I + 1)(2J + 1)\}$ components corresponding to nuclei spin $I = 3/2$. The separation between the various components gives the value of the magnetic moment of the atom under study. Rabi and co-workers were able to measure the nuclear magnetic moment of a number of alkali atoms.

The source of atomic beam, in the case of metals, is an oven heated above the melting point of the metal. The atoms in the vapour of the metal move out through a narrow slit in the form of a beam which is collimated. The atomic beams of diatomic gases are obtained from a discharge tube. The displacement of the beam produced by a strong inhomogeneous magnetic field is only a few hundredth of a millimetre. The collimating slits for the beam are usually about 0.01 mm wide. The atomic beam after collimation passed through the inhomogeneous magnetic field produced by a wedge shaped magnet (Fig. 1.29) in a vacuum of about 10^{-7} mm of Hg. The high vacuum is necessary so that there are no collisions between the atoms of the beam and the air molecules in the transit tube.

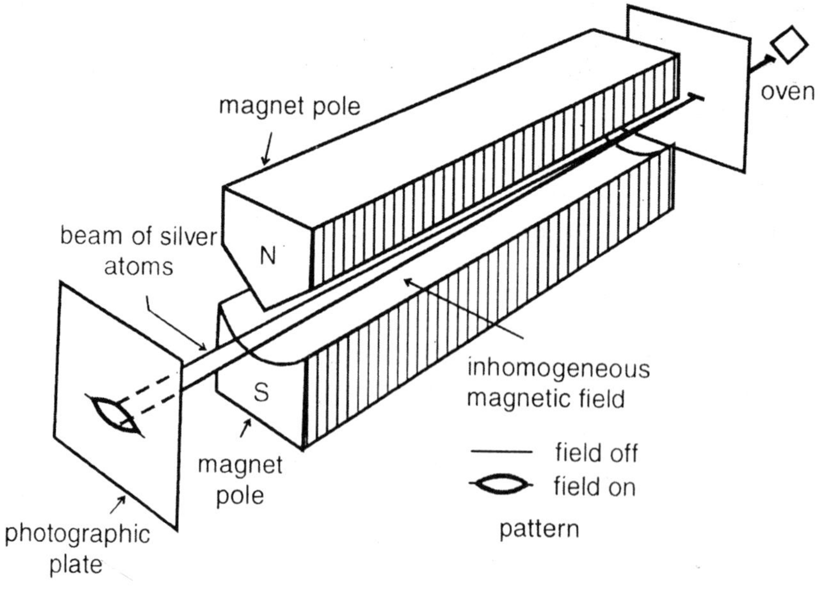

Fig. 1.29: Wedge shaped magnet giving an intense inhomogeneous magnetic field

A Pirani vacuum gauge has been used to detect beams of hydrogen and other light gases. An ionization gauge has been used for the detection of heavy elements. Both these instruments measure the change in pressure in a chamber which is closed except for a narrow entrance slit for the beam. These detectors are rather insensitive and have a very long response time. A more sensitive detector with a rapid response time is based on the production of positive ions when atoms are evaporated from the surface of a heated metallic filament. When an atomic beam strikes a heated tungsten filament, it is evaporated as a positive ion provided the ionization potential of the atoms is close to or less than the work function of the surface. This ion current which is proportional to the intensity of the atomic beam is collected on an electrode, amplified and measured. Another method which is sometimes employed is to ionize the atomic beam by electron bombardment and to detect the ionized atoms of the beam in a mass spectrograph. This method extends the number of elements which can be detected. The production of ionized atoms by electron bombardment is a very inefficient process and requires a strong source of the atomic beam.

1.7.5 Molecular Beam Magnetic Resonance Method

The term "molecular beam" includes beams of neutral atoms as well as neutral molecules. As mentioned above, the deflection of the atomic beam passing through an inhomogeneous magnetic field is very small and it is difficult to measure it accurately. Rabi developed the method of magnetic resonance employing the molecular beams.

In a strong magnetic field the atomic angular momentum $\vec{J}$ and the nuclear angular momentum $\vec{I}$ decouple and the energy of the nucleus is given by

$$E = g_I \mu_N H_e m_I \qquad \text{...(1.7.25)}$$

The magnetic energy levels of the nucleus for different values of m_I are equally spaced and the energy difference between any two consecutive levels is

$$\Delta E = g_I \mu_N H_e \qquad \text{...(1.7.26)}$$

The population of different energy levels is given by Boltzman distribution and at room temperature the difference between the population of two consecutive levels is only about 7×10^{-6}. When energy ΔE is supplied to the atom situated in the magnetic field, there is a transition from one magnetic substate to a higher state. There is an absorption of energy, which is re-emitted by the atom by transition back to the lower state. The energy ΔE is supplied to the atom as an electromagnetic quanta in a radiofrequency electromagnetic field of frequency ν. Resonance absorption of energy takes place when

$$h\nu = \Delta E = g_I \mu_N H_e \qquad \text{...(1.7.27)}$$

To detect the transitions among various m_I states, due to resonance absorption of energy, Rabi and co-workers employed two wedge shaped magnets of equal length and equal but opposite magnetic field gradients. A schematic diagram of the apparatus used by Rabi and co-workers is shown in Fig. 1.30.

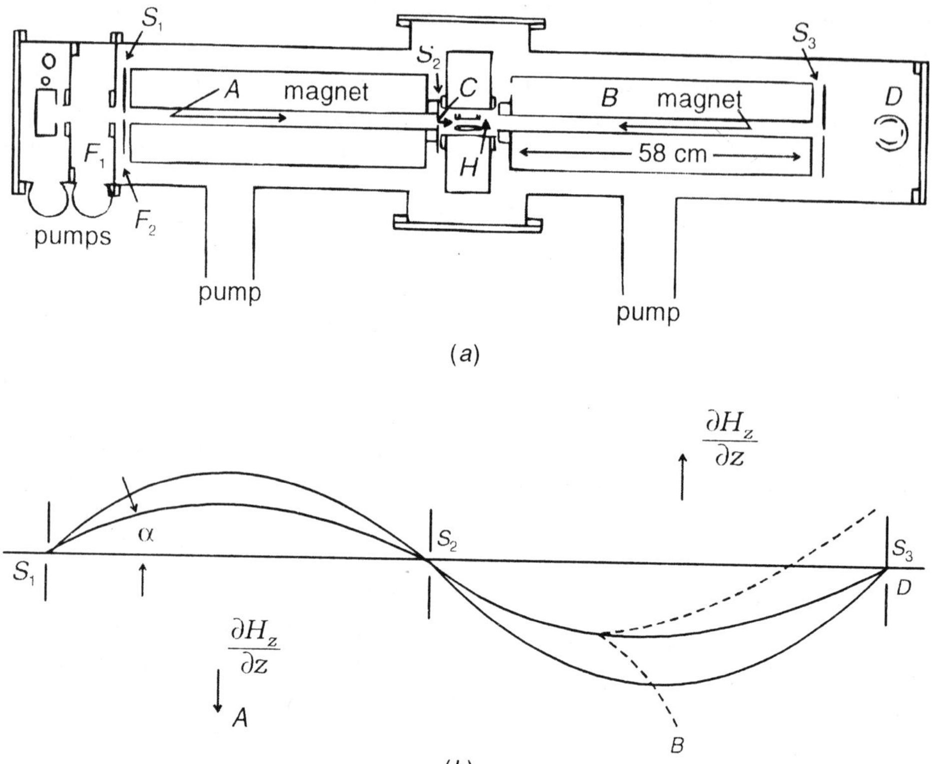

Fig. 1.30: Schematic diagram of the apparatus (Fig. *a*) used for measuring magnetic moment of nuclei using molecular beam resonance method. Fig. *b* shows the trajectory of the molecular beam [J.B.M. Kellogg and S. Millman (1948) *Rev. Mod. Phys.* **18:** 323].

The beam of atoms or molecules is produced in an oven O (Fig. 1.30 *a*). The beam passes through a slit S_1 having a width of about 0.01 mm. The beam then passes through region A where the field gradient points downwards. The beam crosses slit S_2 and then a small region C where there is a homogeneous magnetic field of a few thousand Gauss. The beam then passes through region B where the field gradient points upward. Crossing slit S_3, it enters the detector D. The magnitude of the magnetic field gradient is about 10^5 Gauss/cm and the length of each of the wedge shaped magnets is 20 cm. In region C the homogeneous magnetic field has a strength of about 1000 Gauss. There is a hairpin shaped coil H in region C through which radio frequency current is passed, producing a varying magnetic field perpendicular to both—the velocity of the atoms and the homogeneous magnetic field H_e. The frequency of the radio frequency current can be varied, so as to produce the resonance absorption of energy.

The atoms or molecules in the beam have a distribution of velocity and magnetic moment. These atoms emerge from slit S_1 in the form of a cone. These atoms, depending upon their effective magnetic moment, are acted upon by a force due to the field gradient. The direction of the force is towards the axis or away from it depending upon the sign of the magnetic moment of the atoms. Those atoms which have the right velocity and magnetic moment (depending upon their m_I and m_J) are focussed at slit S_2. These atoms emerge from slit S_2 again in a cone. If the magnetic moment distribution is not disturbed by the radiofrequency field in region *C*, the atoms emerging from slit S_2 experience equal and opposite force in region *B*. These atoms are again focussed at slit S_3 and after emerging from it are detected in detector *D*.

In the absence of a radiofrequency magnetic field, the neutral atoms or molecules passing through the homogeneous magnetic field H_e in region *C* are not affected at all. When the radiofrequency field, of just the right frequency, is applied, there are transitions among the various m_I states of the atoms, thus changing their effective magnetic moment. Such atoms passing through region *B* are either deflected in opposite direction or deflected too much, hence they are not focussed at slit S_3. The number of atoms detected in the detector goes down. Thus a reduction in the detector counting rate signals the resonance absorption of energy by the atoms in the beam. From the knowledge of the field H_e and the resonance frequency, equation (1.7.27) is used to obtain the value of g_I.

If one substitutes approximate values of the constants in equation (1.7.27) the frequency ν of the r–f current is obtained as

$$\nu \approx 10^3 \text{ Hz/Gauss} \qquad ...(1.7.28)$$

Thus in a homogeneous magnetic field of about 1000 Gauss, the frequency of the varying current is about 10^6 Hz. This current can be easily produced using an electronic oscillator with variable frequency.

Davis modified the above apparatus, in that the lengths and the field gradients of the magnets *A* and *B* were different. The length of magnet *A* was 8 cm while that of *B* was 20.8 cm. The homogeneous field magnet was 8.9 cm long. When no radiofrequency current in the hair pin coil was present, the atoms focussed at slit S_2 could not be focussed at slit S_3 and hence were not counted. When the radiofrequency magnetic field of proper frequency was present, the refocusing field of magnet *B* was arranged to focus only those atoms at slit S_3 which had undergone a Zeeman transition while passing through the homogeneous field at *C*. The absorption of energy from the radiofrequency field is signalled by a sudden increase in the counting rate of the detector placed after the slit S_3. In this method, at resonance, the detector counting rate jumps from zero to some finite value. The method is therefore more sensitive than that of Rabi and the quantity of the material required is extremely small. The method can therefore be employed on radioactive atoms also. Davis measured the spin and magnetic moment of the nucleus ^{22}Na using only 4×10^{-10} mole of the radioactive metal.

1.7.6 Nuclear Paramagnetic Resonance Absorption Method

Purcel, Torrey and Pound developed the radiofrequency resonance absorption method for measuring the nuclear g factors in bulk solid, liquid or gaseous samples. In this method a sample of volume about 1 cc is placed in a homogenous magnetic field H_e of a few thousand Gauss. A coil of about ten turns is wound round the sample such that when radiofrequency current is passed through it, the magnetic field produced is perpendicular to the field H_e. At the frequency ν of the radiofrequency current when the resonance absorption of energy takes place in the sample, the Q of the coil suddenly goes down. This reduction of the Q-value of the coil is detected electronically by employing a balanced bridge circuit. The experimental arrangement employed is shown in Fig. 1.31 *a*.

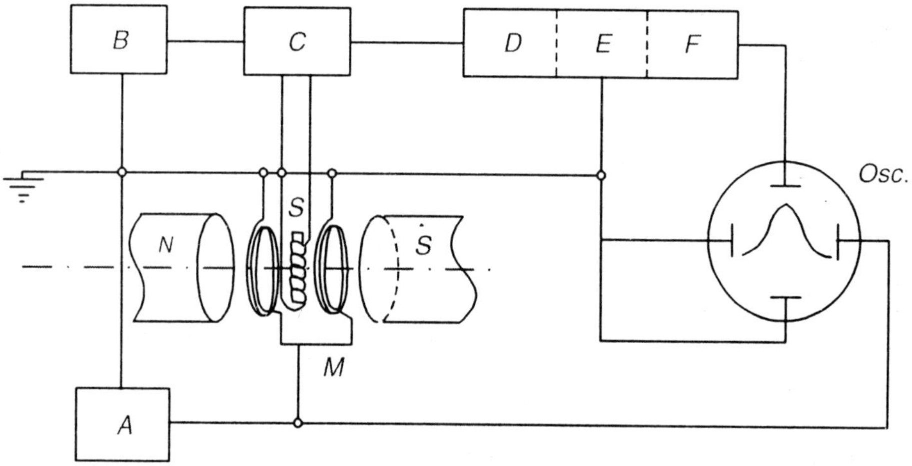

Fig. 1.31: (a) Schematic diagram of the apparatus for measuring magnetic moment of nuclei using paramagnetic resonance absorption method. N-S is an electromagnet and *M* the coil for modulating the magnetic field to which 30-100 cy/sec current is supplied by power supply *A*. *B* is the high frequency generator. A coil of about 10 turns is wound around the sample *S*. The coil is connected to the bridge circuit *C*. The signal generated in the bridge is amplified in a high frequency amplifier *D*, rectified in *E* and the low frequency rectified signal is amplified in *F*. The signal is then fed to the *Y* plates of an oscilloscope. The *X* plates one connected to the low frequency power supply

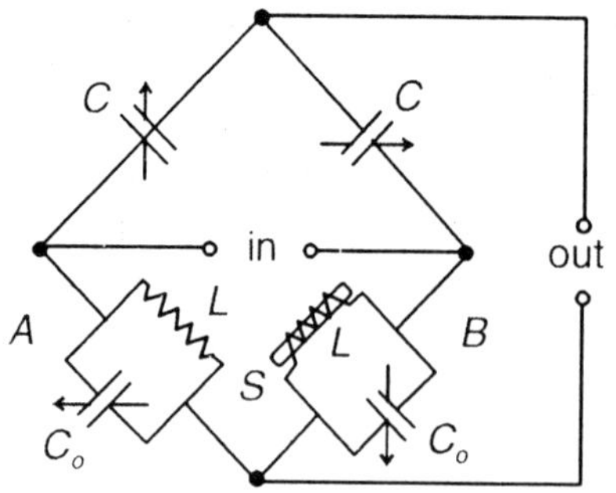

Fig. 1.31: (*b*) The r.f. bridge circuit

N-S are the pole pieces of an electromagnet which produce a highly homogeneous magnetic field of 5000 to 20000 Gauss at the sample S and M is a coil of about 100 turns wound on the pole pieces. When 30-100 cycles/sec alternating current of a few amperes is passed through the coil M an alternating field H_m of a few Gauss modulates the strong field H_e. A coil of about ten turns is wound round the sample S and is connected to the balancing circuit as shown in Fig. 1.31 b. In the balancing circuit L, C, R, C_0 are identical components. The output of a variable frequency oscillator is passed through the two arms A and B of the bridge with opposite phase. Normally the two arms are adjusted such that the signal reaching the r.f. detector is zero. When at resonance, the Q of the coil wound round the sample decreases, the balance of the two arms of the bridge is disturbed and there is a signal in the detector, which gives an output voltage to the Y plates of the oscilloscope. The voltage producing the modulating current after suitable amplification is applied to the X-deflection plates of the oscilloscope.

As the modulating field changes, the magnetic field at the sample with resonance occurring only at the point when $H_e + H_m$ has just the right value. At resonance there is a pulse appearing at the oscilloscope screen. The duration of the pulse is for a time for which total field at the sample corresponds to resonance field. The mid point of the oscilloscope trace corresponds to $H_M = 0$. By changing the oscillator frequency, the resonance pulse on the oscilloscope can be brought to the centre of the trace. The oscillator frequency ν at this point corresponds to the magnetic field H_e so that

$$h\nu = g_I \cdot \mu_N \cdot H_e \qquad \ldots(1.7.29)$$

The field H_e is measured in a separate experiment either by a Hall effect magnetometer or by a search coil or by observing the nuclear magnetic resonance in a proton. From the knowledge of the field H_e and resonance frequency ν one obtains the value of g_I for the nucleus.

It may be mentioned that in a strong magnetic field, even in a compound, the nuclear and the atomic angular momenta are decoupled and one observes the Paschen-Back effect for the ground state of the atomic nucleus.

The details of the experiment can be found in any book of NMR.

1.7.7 Nuclear Resonance Induction Method

This method was developed by Bloch and co-workers. As in the resonance absorption method, the sample is placed in a homogeneous magnetic field, H_e and a variable frequency radiofrequency current is passed through a coil wound round the sample. When the resonance absorption of energy takes place, the absorbed energy is re-emitted in all directions. If a small pick up coil is placed near the sample perpendicular to the r.f. coil this emitted energy—which is in the form of e.m.waves, induces electromagnetic oscillations in the coil. This signal is amplified and rectified in a radiofrequency receiver. The output of the receiver is connected to the Y deflecting plates of an oscilloscope. The X-deflecting plates of the oscilloscope are connected to the coil modulating the magnetic field as described in the resonance absorption method.

1.7.8 Perturbed Angular Correlation of Gamma-rays

The methods discussed so far are employed to measure the magnetic moment of nuclei in their ground state. The method of perturbed angular correlation (PAC) of cascade gamma rays is employed to measure the g factor (magnetic moment) of nuclei in their excited states, which have a lifetime greater than 10^{-11} sec. The method can be employed for nuclear excited states formed in nuclear reactions or α or β transitions. Only the method of PAC of cascade gamma rays is discussed here.

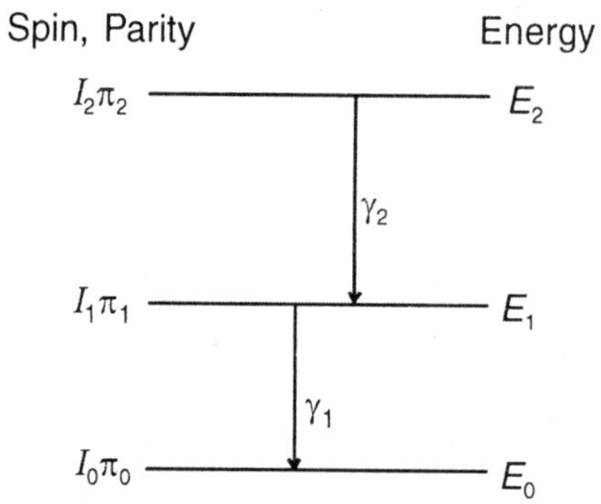

Fig. 1.32: The nucleus is formed in state E_2 which decays to state of energy E_1 emitting a photon γ_2. State E_1 makes transition to state E_0 with emission of photon γ_1

When a nucleus after its formation is in excited state, emits two gamma-ray photons in cascade as shown in Fig. 1.32, there is a definite correlation between the directions of emission of the two photons. The probability of emission of the two photons γ_2 and γ_1 at an angle between θ and $\theta + d\theta$ with respect to each other is given by the angular correlation function $W(\theta)$ where

$$\begin{aligned} W(\theta)d\theta &= 1 + A_2P_2(\cos\theta) + A_4P_4(\cos\theta) \\ &= 1 + a_2\cos^2\theta + a_4\cos^4\theta \\ &= 1 + b_2\cos 2\theta + b_4\cos 4\theta \end{aligned} \quad \text{...(1.7.30)}$$

where, P_2, P_4 are Legendre polynomials and the coefficients A_2, A_4, a_2, a_4, b_2 and b_4 depend upon the spin and parity of the three levels involved and the multipolarities of the transitions. Generally the coefficient A_4, a_4 and b_4 have a small value and can be neglected as a first approximation. The angular correlation between the two gamma-rays can be explained by the fact that after the emission of the first photon γ_2 the nucleus is left with its spin in a direction which is correlated with the direction of emission of the second photon γ_1. The angular correlation between the two gamma-rays remains unperturbed as long as the direction of the spin of the nucleus in the intermediate state E_1 remains unchanged. If a magnetic field is acting at the nucleus, the magnetic moment in the intermediate state interacts with the field. Consequently the spin I_1 precesses around the magnetic field. If the lifetime of this state is large, the precession may become large enough to perturb the angular correlation between the two gamma-rays. A similar situation arises when the nuclear quadrupole moment in the intermediate state interacts with an electric field gradient acting at the nucleus. The degree of perturbation of the angular correlation would depend upon the amount of precession of the spin I_1, before the second photon γ_1 is emitted. This naturally would depend upon the time t elapsed between the emission of γ_2 and the emission of γ_1. The time dependent perturbed angular correlation is given as

$$W(\theta) = 1 + G_2(t)\, A_2P_2\,(\cos\theta) + G_4(t)A_4P_4\,(\cos\theta)$$
$$= 1 + G_2(t)\, a_2\cos^2\theta + G_4(t)a_4\cos^4\theta$$
$$= 1 + G_2(t)b_2\cos(2\theta) + G_4(t)b_4\cos(4\theta) \qquad \text{...(1.7.31)}$$

As pointed out above, the last term in this equation is small and can be neglected.

In the case of magnetic hyperfine interaction the perturbation factor $G_2(t)$ has a simple form, when the magnetic field is applied perpendicular to the plane of emission of the two photons γ_2 and γ_1 as shown in Fig. 1.33. In such a case

$$G_2(t) \approx \cos 2\,\omega_L \qquad \text{...(1.7.32)}$$

and
$$W_\perp\,(\theta, t, H) = 1 + A_2P_2\,(\cos\theta - \omega_L t)$$
$$= W_\perp\,(\theta - \omega_L t) \qquad \text{...(1.7.33)}$$

where, ω_L is the Larmor precession frequency of I_1 around the magnetic field.

$$\omega_L = g_I\,\mu_N\,H_e/\,\hbar \qquad \text{...(1.7.34)}$$

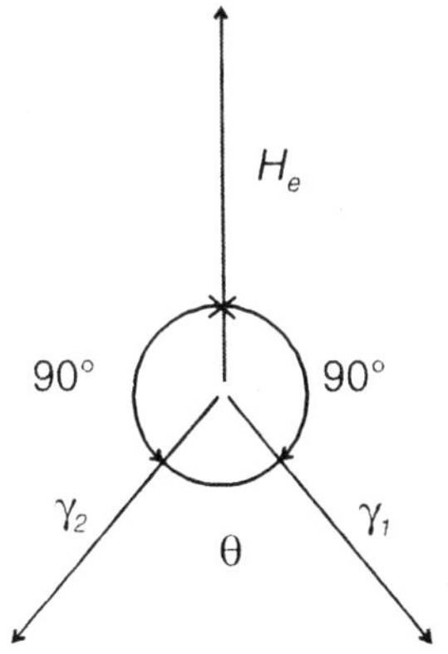

Fig. 1.33: The applied magnetic field H_e is perpendicular to the plane of γ_1 and γ_2

It is clear that the angular correlation is rotated by an angle $\omega_L t$. The direction of rotation depends upon the sign of the magnetic moment and the direction of the field. If the direction of the field is reversed, the rotation is also reversed.

Because of the great advances in electronics and the detectors, the experimental arrangement used in the study of the perturbed angular correlation has under gone many changes. The following discussion gives only the basic principle employed in these measurements.

The gamma-rays, emitted from a radioactive source S are detected in the two detectors C_1 and C_2 which subtend equal solid angles at the source in Fig. 1.34. The detectors C_1 and C_2 detect gamma-rays ν_1 and ν_2 respectively. The output pulses of the two detectors after amplification (A) and delay (D) are fed into a coincidence circuit C of resolving time $2\tau_0$. The variable delays D_1 and D_2 are introduced to correct for the delay difference in the two detectors. The smallest resolving time $2\tau_0$ of the coincidence circuit depends upon the detectors and the electronics employed. A resolving time of $2\tau_0 \approx 1$ to 3×10^{-9} sec is desirable. The output of the coincidence circuit signifies that the photons γ_1 and γ_2 giving rise to the coincidence event are emitted from the same nucleus. The coincidence output may be counted in a scaler circuit (S_{12}). The coincidence counting rate is determined at different angle θ between the two detectors. The coincidence counting rate at any angle is normalised to the counting rate at 90°. ($\cos\theta = 0$) to give the angular correlation function. The angular correlation function of the cascade gamma-rays emitted from a ^{60}Co source is shown in Fig. 1.35.

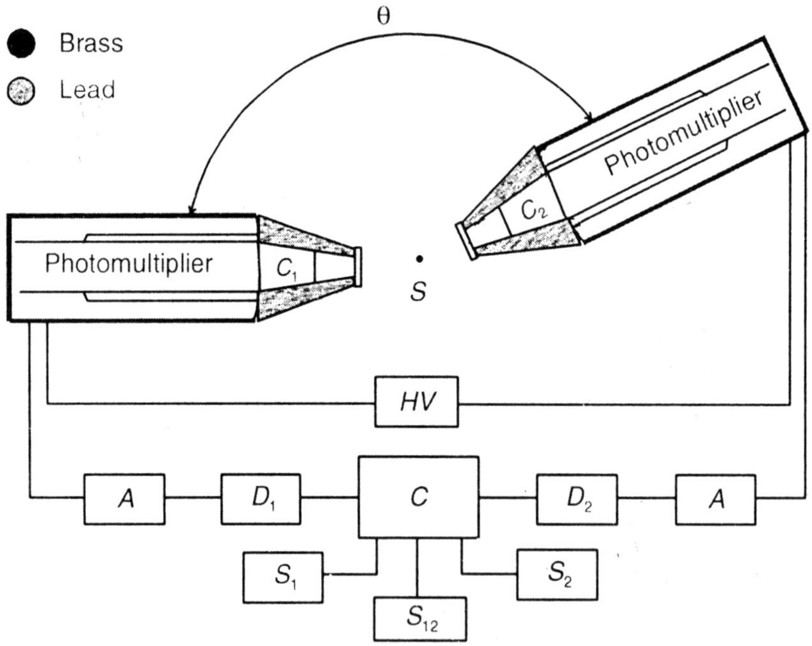

Fig. 1.34: Apparatus to study the angular correlation of γ-rays

If the A_4 terms of the angular correlation function is zero or negligible, the coefficient A_2 (equation 1.7.30) can be determined by measuring the coincidence counting rates $N(90)$ and $N(180)$ at $\theta = 90°$ and $\theta = 180°$ respectively. It can be shown that

$$2\frac{N(180) - N(90)}{N(180) + N(90)} = \frac{\frac{3}{2}A_2}{1 + \frac{A_2}{4}} \approx \frac{3}{2}A_2 \quad \text{... (1.7.35)}$$

and $$\frac{N(180) - N(90)}{N(90)} = a_2 \qquad ...(1.7.36)$$

$$\frac{N(180) - N(90)}{N(180) + N(90)} = b_2 \qquad ...(1.7.37)$$

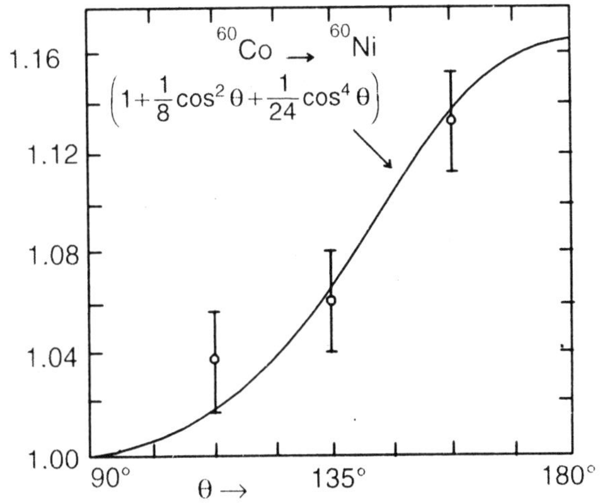

Fig. 1.35: Angular correlation function $W(\theta)$ for ^{60}Co γ-ray cascade. [E.L. Brady and M. Deutsch (1950) *Phys. Rev*, **78**; 558]

For the nuclear excited state having life time τ_N much smaller than the resolving time τ_o of the coincidence circuit, the method of observing the total angular correlation function for applied field up and down is most suitable. In such circumstances the rotation of the angular correlation function is integrated over the life time of the nuclear state so that

$$W_{\perp}(\theta, \omega_L \tau_N) = \frac{1}{\tau_N} \int_0^{\infty} e^{t/\tau_N} W(\theta, \omega_L t) dt$$

$$= 1 + b_2 \cos 2(\theta - \omega_L \tau_N)$$

where, $\omega_L \tau_N << 1$ ...(1.7.38)

The rotation of the angular correlation of the 276-161 keV gamma-ray cascade emitted in the decay of ^{133}Ba to ^{133}Cs is shown in Fig. 1.36. The magnetic field of 9.5 T from a super conducting magnet was applied perpendicular to the plane of the two detectors. The 161 keV level of ^{133}Cs has a life time of 0.19 nsec. The experiment gave the magnetic moment of the nucleus in this state as μ_I(161 keV) = 2.0 ± 0.2 μ_N.

One can measure the rotation of the angular correlation by measuring the coincidence counting rate at θ = 135° with the magnetic field pointing upwards (with respect to the plane of the detectors) and then pointing downwards. It can be shown that the ratio of the two counting rates is given by:

$$R = \frac{W(135° H\uparrow) - W(135° H\downarrow)}{W(135° H\uparrow) + W(135° H\downarrow)} = b_2 \cos 2\omega_L \tau_N \qquad ...(1.7.39)$$

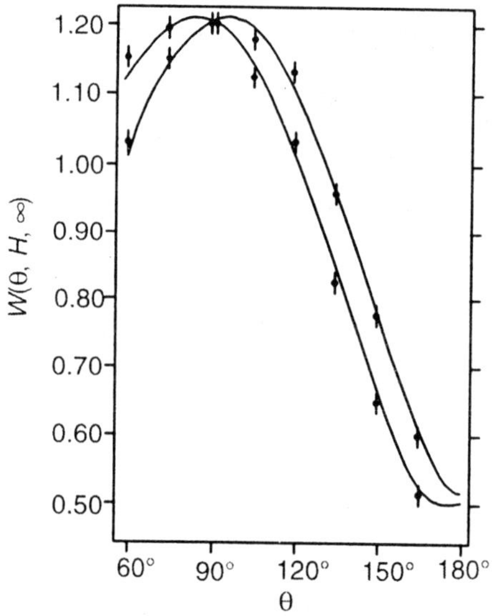

Fig. 1.36: Rotation of the angular correlation of 276-161 keV γ-rays cascade of ^{133}Cs in an external magnetic field of 9.5 Tesla. [Bodenstedt *et al.* (1959) *Nucl. Phys.* **11**: 584]

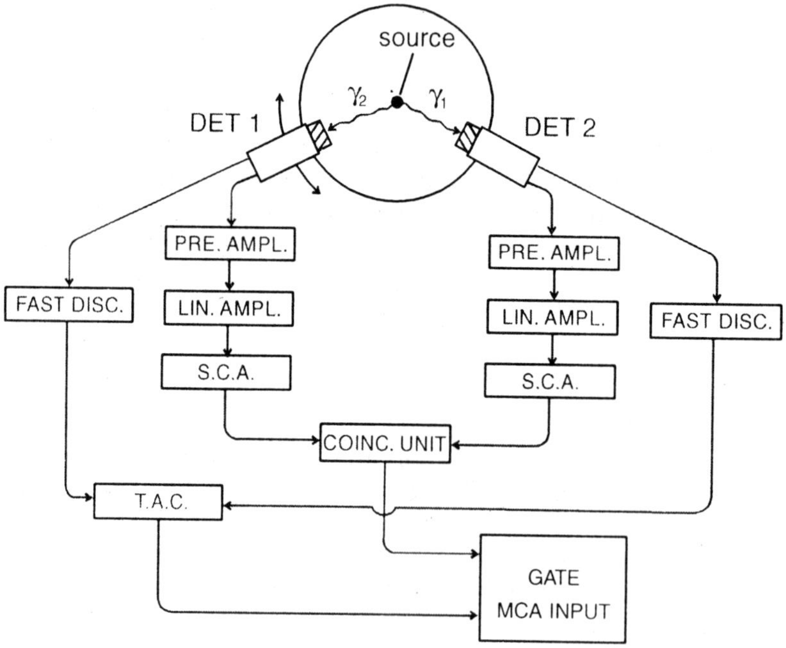

Fig. 1.37: Slow-fast coincidence arrangement for the study of time dependent perturbed angular correlation of γ-rays. The energy of γ-rays is selected in the slow channels, consisting of a pre-amplifier, a linear amplifier and a single channel analyser (SCA). The output of the slow concidence circuit in the slow channel, gates the input of the MCA. The fast pulses from the detectors are shaped in fast discriminators and fed to time to amplitude converter (TAC) whose output is fed to the input of the MCA.

The coefficient b_2 and the life time of the nuclear state τ_N are measured in separate experiments. From the ratio R the value of ω_L can be obtained. The value of the magnetic moment is obtained from the knowledge of the magnetic field. In this type of experiment magnetic moment of nuclear state with life time of ~ 10^{-10} sec can be measured using a magnetic field of about 20000 Gauss. For reasonable accuracy about 10^7 coincidences have to be collected for each direction of the magnetic field.

When the life time of the nuclear state is more than 10^{-8} sec., time differential perturbed angular correlation measurement can be employed to measure the g factor of the state. In this method, the coincidence counting rates for the two gamma-ray photons for $\theta = 90°$ and $\theta = 180°$ is recorded as a function of the time delay between the emission (and detection) of the two photons. In this experiment the coincidence circuit of Fig. 1.34 is replaced by a time to amplitude converter (TAC). The output of the TAC circuit is a coincidence pulse whose height is proportional to the delay between the start pulse (from γ_2 detector) and stop pulse from the detector detecting γ_1. The scaler circuit is replaced by multichannel pulse height analyser (MCA) which records the pulse height spectrum of the TAC pulses (representing the time delay spectrum) simultaneously (Fig. 1.37). Each channel of the MCA represents a certain fixed time delay between the pulses from the two gamma-ray detectors. The variable delay D_1 and D_2 are adjusted to give zero time delay between the two pulses from C_1 and C_2 at the zeroth channel of the MCA. Delay line D_1 can be employed to calibrate the MCA also. The spin rotation pattern obtained plotting $R(t)$ as a function of time (Eqn. 1.7.35) is a sine curve with time period $T = 1/\omega_L$ (Eqn. 1.7.32).

Time dependent PAC can also be studied by placing the two detectors at 135° and recording the pulse height spectrum of the TAC output with magnetic field H_e acting at the source S first in one direction and then in the reversed direction. If N^+ and N^- are the coincidence counting rates for magnetic field up and down respectively at any channel in the MCA one can plot the factor (for time difference t in the decay of γ_1 and γ_2) $R(t)$ defined below as function of time.

$$R(t) = 2\frac{N^+ - N^-}{N^+ - N^-}$$

It can be shown that

$$R(t) = B\sin 2\omega_L t$$

where, B is a constant independent of time. From the sine wave pattern obtained, one can obtain the value of ω_L and hence the nuclear g factor. The radioisotope 2.8 day ^{111}In decays to ^{111}Cd by electron capture and subsequent emission of 173 keV and 247 keV photons in cascade. The life time of the 247 keV level of ^{111}Cd is 84 nsec. The radioisotope ^{111}In is for this reason is highly suitable for angular correlation experiments. The perturbed angular correlation of the 173-247 keV γ cascade of ^{111}Cd was studied employing an external magenatic field of 32600 Gauss and θ = 135. The spin rotation pattern was obtained by plotting $R(t)$ as a function of time, as shown in Fig. 1.38. From these measurments the g factor of the 247 keV state of ^{111}Cd is determined as $g = -0.318 + 0.007$ which is in good agreement with the earlier results, $g = -0.313 + 0.009$. (R. M. Steffan and W. Zobel, (1955) *Phys. Rev.* **97**: 832 and (1956) *Phys. Rev.* **103**: 126.)

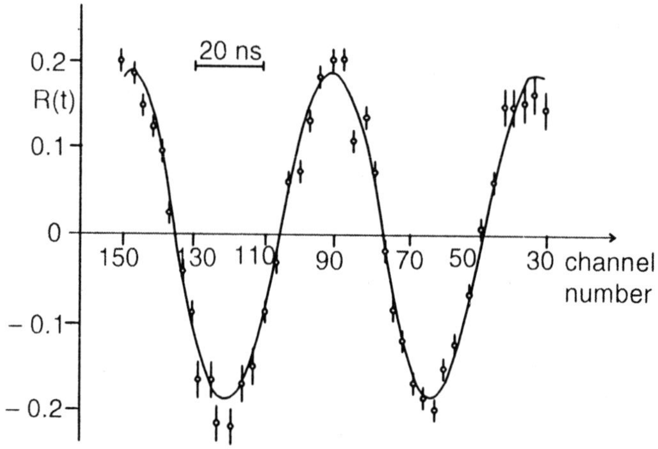

Fig. 1.38: Spin rotation pattern for the 173-247 keV γ-rays cascade of ^{111}Cd in an external magnetic field of 32600 G. [E. Matthias, L. Bostrom A Maciel, M. Solomon and T. Lindquist (1963). *Nucl. Phys*; **40**: 656]

1.8 QUADRUPOLE MOMENT OF NUCLEI

There is evidence that some nuclei do not have a spherical shape. The departure of the nuclear charge distribution from a spherical shape gives rise to electric moments. A nucleus as will be seen later, can not have an electric dipole moment. If the shape of a nucleus is spheroidal, it shall have an electric quadrupole moment. A more complicated shape would give rise to higher electric moment. Such complicated shapes of nuclei are rare and will not be discussed here. To calculate the quadrupole moment of a nucleus one considers its potential energy in an electrostatic potential $\phi(r)$ produced by an external charge at P (Fig. 1.39). The potential varies at every point of the nucleus. Taking the co-ordinate origin at the centre of the nucleus, the energy of the nucleus in the potential can be written as:

$$E = \int \rho(\vec{r})\phi(\vec{r})dV \qquad \text{...(1.8.1)}$$

where, $\rho(r)$ is the charge density of the nucleus at a point $\vec{r}$. The integration is carried over the whole volume of the nucleus. The charge density of the nucleus satisfies the condition that the total change on the nucleus is Ze. So that,

$$\int \rho(\vec{r})dV = Ze \qquad \text{...(1.8.2)}$$

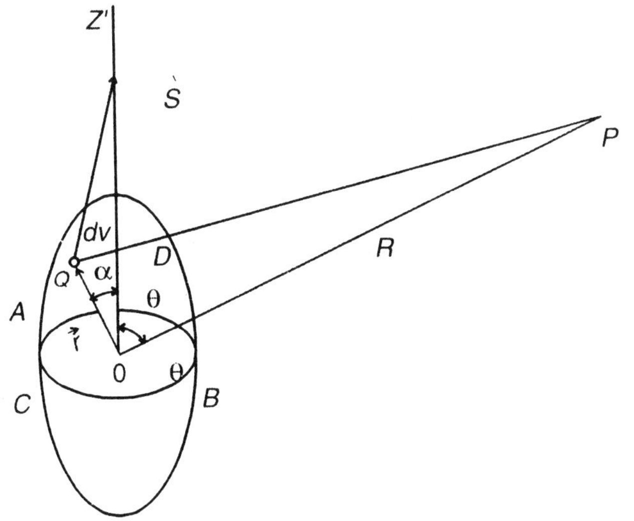

Fig. 1.39: A deformed nucleus *ABC* in the field of an external charge placed at *P*

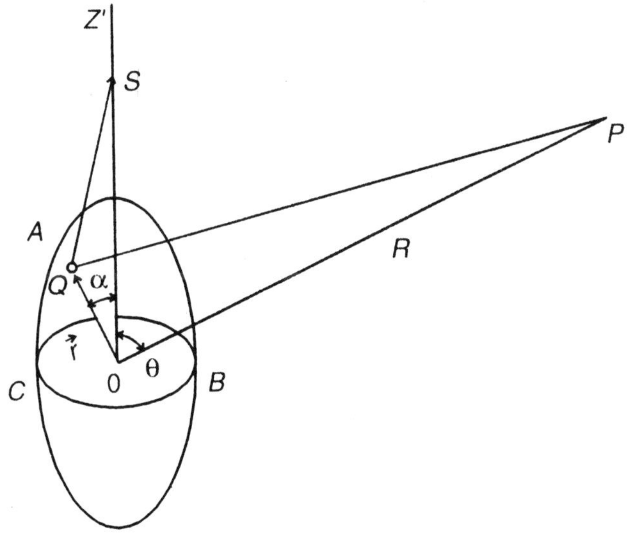

Fig. 1.40: Potential at point *P* due to a deformed nucleus *ABC*

The potential $\phi(r)$ can be expanded as a Taylor series as

$$\phi(\vec{r}) = \phi(0) + \vec{r} \cdot \vec{\nabla}\phi(0) + \frac{1}{2}\sum_{ij} x_i x_j \left[\frac{\partial^2 \phi}{\partial x_i \partial x_j}\right]_{r=0} + \ldots \qquad \ldots(1.8.3)$$

$i = 1, 2, 3$ and $j = 1, 2, 3$

Here i, j refer to the three axis x, y, z of the co-ordinate system respectively. Equation (1.8.3) can be written as,

$$\phi(r) = \phi(0) + r\,\nabla\phi(0) + \frac{1}{2}\Sigma\, x_i x_j \left(\frac{\partial^2\phi}{\partial x_i \partial x_j}\right)$$

$$+\frac{1}{6}\Sigma\, r^2\delta_{ij}\left(\frac{\partial^2\phi}{\partial x_i \partial x_j}\right)_{r=0} - \frac{1}{6}\Sigma\, r^2\delta_{ij}\left(\frac{\partial\phi}{\partial x_i \partial x_j}\right)_{r=0} + \ldots$$

$$= \phi(0) + r\nabla\phi(0) + \frac{1}{6}\sum_{ij}\delta_{ij} r^2\left(\frac{\partial^2\phi}{\partial x_i \partial x_j}\right)_{r=0}$$

$$+\frac{1}{6}\sum_{ij} q_{ij}\frac{\partial^2\phi(0)}{\partial x_i \partial x_j}\ldots \qquad \ldots(1.8.4)$$

where, $\delta_{ij} = 1$ if $i = j$

$\delta_{ij} = 0$ if $i \neq j$

and

$$q_{ij} = 3x_i y_j - r^2\delta_{ij} \qquad \ldots(1.8.4a)$$

q_{ij} = is a 3 × 3 tensor.

The third term of equation (1.8.4) can be written as

$$\frac{1}{6}\sum_{ij}\delta_{ij} r^2\left(\frac{\partial^2\phi}{\partial x_i \partial x_j}\right)_{r=0} = \frac{1}{6}\sum_{i} r^2\frac{\partial^2\phi(0)}{\partial x_i^2}$$

$$= \frac{1}{6} r^2\,\nabla^2\phi(0) \qquad \ldots(1.8.5)$$

Combining equation (1.8.1), (1.8.4) and (1.8.5) gives the potential energy of the nucleus as

$$E = \int\rho(r)\phi(0)dV + \int\rho(r)\vec{r}\,\vec{\nabla}\phi(0)dV + \frac{1}{6}\int\rho(r)r^2\,\nabla^2\phi(0)dV$$

$$+\frac{1}{6}\int\rho(r)q_{ij}\frac{\partial^2\phi(0)}{\partial x_i \partial x_j}dv + \ldots \qquad \ldots(1.8.6)$$

In the above equation the first term represents the electrostatic energy for a spherically symmetric charge distribution. The second term is the energy of the electric dipole moment which is zero due to the fact that there are equal and opposite terms in the integration, corresponding to the positive and negative values of r. The third term is called the isotopic energy and has a finite value only if the charge producing the electric potential ϕ is situated inside the nuclear volume. If the charge producing the potential is outside the nuclear volume, then according to Laplace's equation $\nabla^2\phi = 0$. In the fourth term, the quantity

$\partial^2\phi(0)/\partial x_i \partial x_j$ is the gradient of the electric field produced by the external charge. This term, thus represents the energy of interaction of the electric quadrupole moment of the nucleus with the electric field gradient. The electric quadrupole tensor of the nucleus is eQ_{ij} defined as,

$$eQ_{ij} = \int \rho(r)(3x_i x_j \sim \delta_{ij} r^2) dV \qquad \text{...(1.8.7)}$$

where, e is the charge on the proton. If the nucleus is of spheroidal shape and the Z' axis is taken along the axis of symmetry Fig. 1.39. The Z' component of the quadrupole tensor, which is called the intrinsic quadrupole moment of the nucleus, is given as

$$eQ_0 = eQ_{z'z'} = \rho(r)\,(3z'^2 - r'^2)\,dV \qquad \text{...(1.8.8)}$$

If the nucleus has a shape elongated along the symmetry axis (cigar shape) the term $(3z'^2 - r'^2)$ is positive and the quadrupole moment of the nucleus is positive. For an orange shape of the nucleus $3z'^2 < r'^2$ and the quadrupole moment is negative.

It can be shown classically that a spheroidal charge distribution of uniform density with semi axis c along z' direction and semi axis a perpendicular to it, has a quadrupole moment.

$$eQ = 2/5\,(c^2 - a^2) \cdot Ze \qquad \text{...(1.8.9)}$$

where, Ze is the total charge of the distribution. The mean square radius of the distribution is:

$$R^2 = 1/2(c^2 + a^2) \qquad \text{...(1.8.10)}$$

which on substitution in equation (1.8.9) gives

$$eQ = 4/5\,\eta R^2 Ze \qquad \text{...(1.8.11)}$$

where the asymmetry parameter η is defined as

$$\eta = \frac{c^2 - a^2}{c^2 + a^2} \qquad \text{...(1.8.12)}$$

If the mean radius is defined as

$$R = \frac{c + a}{2} \qquad \text{...(1.8.13)}$$

then also the quadrupole moment is given by equation (1.8.11) but the asymmetry parameter is then given by,

$$\eta = 2\frac{c - a}{c + a} \qquad \text{...(1.8.14)}$$

The quadrupole moment of a nucleus can also be obtained from the potential it produces at a point far from its centre. As the point P (Fig. 1.40) is outside the nuclear charge distribution the Laplace's equation gives,

$$\nabla^2\phi(P) = 0$$

Taking the origin at the centre of the nucleus, the Laplace's equation can be written in polar co-ordinates as

$$\nabla^2\phi = \frac{1}{R^2}\frac{\partial}{\partial R}\left(R^2\frac{\partial\phi}{\partial R}\right) + \frac{1}{R^2\sin\theta}\frac{\partial}{\partial\theta}\left(\sin\theta\frac{\partial\phi}{\partial\theta}\right) = 0 \quad \text{...(1.8.15)}$$

Due to the circular symmetry about z' axis the azimuthal part of the equation does not contribute. The solution of this equation is

$$\phi(P) = \frac{1}{R}\sum_{n=0}^{\infty}\frac{a_n}{R_n}P_n(\cos\theta) \quad \text{...(1.8.16)}$$

P_n is the Legendre polynomial of order n and a_n is a constant which depends upon the shape of the nucleus. To calculate the coefficient a_n consider the potential at a point S on the z' axis such that $OS = R$. As $\theta = 0$ one has

$$P_n\,(\cos\theta)_{\theta=0} = 1$$

and
$$\phi\,(S) = \frac{1}{R}\sum\frac{a_n}{R_n} \quad \text{...(1.8.17)}$$

The potential at S can also be calculated by taking a charge element of volume dV at position $\vec{r}$. The potential at S due to this charge element is

$$d\phi = \frac{\rho(r)dV}{|\vec{R}-\vec{r}|}$$

Now
$$\frac{1}{|\vec{R}-\vec{r}|} = \frac{1}{R}\sum_{n=0}^{\infty}\left(\frac{r}{R}\right)^n P_n(\cos\alpha) \quad \text{...(1.8.18)}$$

where α is the angle the vector $\vec{r}$ makes with the z' axis. The potential at S due to the charge element is

$$d\phi = \frac{\rho(r)dV}{R}\sum\left(\frac{r}{R}\right)^n P_n(\cos\alpha)$$

The potential due to the nucleus is obtained by integrating over the volume of the nucleus, so that

$$\phi\,(S) = \int\frac{\rho(r)}{R}\left(\frac{r}{R}\right)^n P_n(\cos\alpha)dV \quad \text{...(1.8.19)}$$

Comparing the expressions (1.8.17) and (1.8.19) for the potential at point S one gets,

$$a_n = \int\rho(r)r^n P_n(\cos\alpha)dV \quad \text{...(1.8.20)}$$

The integration is carried over the whole volume of the nucleus as in equation (1.8.6). The coefficients a_n give the different electric multipole momenta of the nucleus. It is seen that

$$a_0 = \int\rho(r)dr^3 = Ze = \text{Total nuclear charge} \quad \text{...(1.8.21)}$$

$$a_1 = \int \rho(r) r \cdot P_1(\cos\alpha) dV$$

$$= \int \rho(r) r \cdot \cos\alpha dV \quad \text{Dipole moment} \qquad ...(1.8.22)$$

$$a_2 = \int \rho(r) r^2 \cdot P_2(\cos\alpha) dV$$

$$= \int \rho(r) r^2 \cdot \frac{1}{2}(3\cos^2\alpha - 1) dV$$

$$= \frac{1}{2}\int \rho(r)(3z'^2 - r'^2) dV \quad \text{quadrupole moment} \qquad ...(1.8.23)$$

Classically the nucleus is regarded as a continuous charge distribution with total charge Ze. The charge on the nucleus is infact due to discrete particles, the protons. At a given volume element a proton has only some probability to be present. Let $\psi(r_1, r_2, \ldots r_z, r_z + 1, r_A)$ be the wave function of the nucleus consisting of Z protons and A-Z neutrons. The first Z co-ordinates $(r_1, \ldots, r_z)$ refer to the protons and the remaining ones refer to the neutrons. The probability to find the ith proton in the volume element dv_1 at the position $\vec{r}$ is given by $P_i\, dv_1$.

where,

$$P_i(\vec{r}) = \int |\psi(r_1, r_2, \ldots r_i, r_{i+1}, r_A)|^2 dv_1\, dv_2 \ldots dv_{i-1}\, dv_{i+1}\, dv_A \qquad ...(1.8.24)$$

The integration extends over the co-ordinates of all the particles except the ith particle which is at position $\vec{r}$ $(r_i = r)$. This implies that the ith proton is at position $\vec{r}$ while the remaining particles can be any where in the nuclear volume. The charge density of the nucleus, which is only due to the protons is given, as

$$\rho(r) = \sum_{i=1}^{z} eP_i(r) \qquad ...(1.8.25)$$

On substituting the charge density, one gets.

$$a_o = \int \rho(r) dV = \sum_{i=1}^{z} \int eP_i(r) dV$$

$$= \sum_{i=1}^{z} \int e|\psi(r_1 \ldots r_A)|^2 dv \qquad ...(1.8.26)$$

Similarly, $$a_1 = \sum \int er|\psi(r_1 \ldots r_A)|^2 dV \qquad ...(1.8.27)$$

As the nucleus has a definite parity which could be positive, if

$$\psi(-r_1, -r_2, \ldots -r_A) = \psi(r_1, r_2, \ldots r_A) \qquad ...(1.8.28)$$

or it could be negative, if

$$\psi(-r_1, -r_2, \ldots -r_A) = -\psi(r_1, r_2, \ldots r_A) \qquad ...(1.8.29)$$

The value of $|\psi(r_1 \ldots r_A)|^2$ is in both the cases positive. In the integration in equation (1.8.27) there are terms corresponding to $+r$ as well as $-r$ which will cancel to give zero. The electric dipole moment a_1 of a nucleus is always zero.

On substituting value of $\rho(r)$, the third term (equation 1.8.23) gives

$$a_2 = \int e\psi^* (r_1 \ldots r_A)\, r^2 P_2 (\cos\alpha)\psi (r_1 \ldots r_A)\, dV \quad \ldots(1.8.30)$$

As the nucleus has a certain total angular momentum I, the wave function, taking this fact into account, is written as $\psi^I (r_1 \ldots r_A)$. In quantum mechanics $r^2P_2(\cos\alpha)$ is an operator which is operating on the nuclear wave function. This operator corresponds to an angular momentum $I = 2$ and one can write

$$r^2P_2\psi^I (r_1 \ldots r_A) = \Sigma a^{IJ}\psi^J(r_1 \ldots r_A) \quad \ldots(1.8.31)$$

where the eigen function $\psi^J(r_1 \ldots r_A)$ corresponds to the resultant angular momentum J with amplitude a^{IJ}. The addition of angular momenta gives the possible values of J varying from $|2 + I|$ to $|2 - I|$. The quadrupole moment of the nucleus is thus

$$eQ = \Sigma \int ea_{ij}^{IJ}\psi^{*I}(r_1 \ldots r_A)\psi^J(r_1 \ldots r_A)dV \quad \ldots(1.8.32)$$

From the orthogonality condition of the wave functions, one gets a non-zero value of Q only if $I = J$. It may be observed that if $I = 0$ then $J = 2$ hence $eQ = 0$. So that all nuclei with total angular moment $I = 0$ have zero quadrupole moment. If $I = \frac{1}{2}$ the possible values of J are $J = 2 + \frac{1}{2} = 5/2$ or $J = 2 - \frac{1}{2} = 3/2$. As the values of J are not equal to that of I, the expectation value of eQ is zero. As a general rule the observable quadrupole moment of a nucleus with total angular momentum $I = 0$ or $\frac{1}{2}$ is always zero. This statement does not rule out the possibility that the shape of the nucleus could still be spheroidal.

The quadrupole moment of a nucleus as given in equation 1.8.7 is in any arbitrary system of co-ordinates. Choosing the z' axis along the axis of symmetry of the spheroid (Fig. 1.39) x', y' axes perpendicular to it, the components of the quadrupole moment are

$$eQ_{z'z'} = \int\rho(r)\,(3z'^2 - r^2)\, dV$$

$$eQ_{x'x'} = \int\rho(r)\,(3x'^2 - r^2)\, dV$$

$$eQ_{y'y'} = \int\rho(r)\,(3y'^2 - r^2)\, dV$$

Adding the above three equation one gets

$$e(Q_{z'z'} + Q_{y'y'} + Q_{x'x'}) = \int\rho(r)\,\{3(x'^2 + y'^2 + z'^2) - 3r^2\}dV = 0$$

So that

$$Q_{z'z'} = -(Q_{x'x'} + Q_{y'y'})$$

The cross-section of a spheroidal nucleus in the xy plane is a circle. The x and y direction are therefore equivalent and $Q_{x'x'} = Q_{y'y'}$. Thus the entire quadrupole moment is specified by a single constant quantity $Q_{z'z'}$. The unit of quadrupole moment as is clear from equation (1.8.9) is that of area.

The quantity $eQ_{z'z'}$ is known as the intrinsic quadrupole moment of a nucleus and is represented by eQ_o,

$$eQ_o = eQ_{z'z'}$$

In any system of co-ordinates the quadrupole moment tensor Q_{ij} (equation

1.8.7) has components along the x, y and z direction. The total angular momentum $\vec{I}$ of the nucleus also has component I_x, I_y and I_z along three direction. The quadrupole moment tensor can therefore be expressed in terms of the components of vector $\vec{I}$ so that

$$Q_{ij} = C\left(I_i I_j + I_j I_i - \frac{2}{3} < I^2 > \delta_{ij}\right) \qquad \text{...(1.8.35)}$$

The term I_i and I_j are taken separately as they may not commute in some cases. C is a constant, which has to be evaluated. The vector $\vec{I}$ is not a constant of motion. When an electric field gradient is applied at a nucleus in z direction, the vector $\vec{I}$ precesses around the z axis in different orientations giving rise to the components m_I (or I_z). One can calculate the quadrupole moment about the z axis, in that orientation of the vector I where m_I is maximum i.e. $m_I = (I_z) = I$. In this case equation 1.8.35 gives,

$$Q = C\left(I_z I_z + I_z I_z - \frac{2}{3} < I^2 >\right)$$

$$= \frac{2}{3} C[3I^2 - I(I+1)]$$

$$= \frac{2}{3} CI(2I-1)$$

or

$$C = \frac{3Q}{2I(2I-1)}$$

Substituting the value of C in equation 1.8.35 gives

$$eQ_{ij} = \frac{3eQ}{2I(2I-1)}\left[I_i I_j + I_j I_i - \frac{2}{3} I(I+1)_{\delta_{ij}}\right] \qquad \text{...(1.8.36)}$$

eQ is the component of the quadrupole moment about the space axis defined by the electric field gradient $\partial^2 V/\partial z^2$.

The component of quadrupole moment eQ_{ij} in the direction of nuclear orientation, such that the spin I has compound m_I along the z direction is given as,

$$eQ_m = \frac{3m_I^2 - I(I+1)}{4I(2I-1)} eQ \qquad \text{...(1.8.37)}$$

If $\partial^2\phi/\partial z^2$ is the electric field gradient acting at the nucleus, the energy of quadrupole interaction is

$$E = eQ \frac{3m_I^2 - I(I+1)}{4I(2I-1)} \frac{d^2\phi}{dz^2} \qquad \text{...(1.8.38)}$$

The energy of the nucleus varies with its orientation with respect to the z-direction. A nucleus having spin 5/2 can have orientation with $m_I = \pm 5/2, \pm 3/2$ and $\pm 1/2$. The quadrupole interaction energy depends upon m^2 so that it is same for positive and negative values of m. The energy levels of the nucleus are shown in Fig. 1.41.

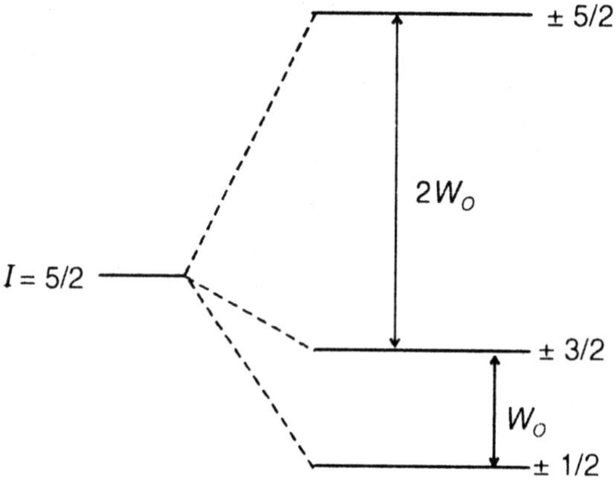

Fig. 1.41: The splitting of the energy level I = 5/2 in an electric field gradient

The difference in the energies between the levels $\pm 5/2$ and $\pm 3/2$ and $\pm 3/2$ and $\pm 1/2$ is W_o and 2 W_o respectively where,

$$W_o = eQ\frac{\partial^2\phi}{\partial z^2}$$

In the experiments where the quadrupole interaction energy is measured, one observes the quadrupole moment eQ, which is along the space axis. The intrinsic quadrupole moment eQ_o of a nucleus is observed in gamma transitions, Coulomb excitation, electron scattering and muonic atoms. The value of quadrupole moment eQ is always less than the intrinsic quadrupole moment eQ_o because the space axis never coincides the body axis (axis of symmetry of the spheroid).

In the ground state of a deformed nucleus, the intrinsic quadrupole moment eQ_o and the space quadrupole moment Q are related as

$$Q = \frac{I(2I-1)}{(I+1)(2I+3)}Q_o$$

The relation shows that even if Q_o is finite $Q = 0$ for $I = 0$ or $1/2$.

The electric field gradient at an atomic nucleus can be produced by the electrons in the atomic orbit with $J > 1/2$. In a molecule, due to asymmetric charge distribution around an atomic nucleus, there is generally a strong electric field gradient. In non cubic metallic lattice there is a strong electric field gradient which can be calculated from the crystal structure. The electric field gradients encountered in the above cases are very large (about 10^{21} volts/m^2) which can not be produced in the laboratory.

1.9 MEASUREMENT OF THE ELECTRIC QUADRUPOLE MOMENT OF NUCLEI

The electric quadrupole moment of a nucleus is generally measured by studying its interaction energy with a known electric field gradient (*efg*). In case of free atoms and molecules the electric field gradient acting on a nucleus of a constituent atom can be calculated. When the atom is embedded in a crystal lattice, the calculation of electric field gradient acting at its nucleus is affected to a great extent by the impurities and the defects in the lattice. Some of the methods employed for measuring the electric quadrupole moment of an atomic nucleus are discussed below.

1.9.1 Hyperfine Splitting of Atomic Spectral Lines

The hyperfine splitting of the atomic spectral lines, as discussed earlier, can arise due to the interaction of the nuclear magnetic moment with the magnetic field acting on it. The magnetic field is maximum for atomic orbital $J = 1/2$. The atomic electrons produce an electric field gradient at the nucleus, which has an expectation value different from zero only for atomic orbital $J > 1$. Thus when magnetic field acting at the nucleus is maximum (for $J = 1/2$ and 1) the electric field gradient is zero. In cases where $J > 1$ the hyperfine splitting of atomic levels due to electric quadrupole interaction is comparable to magnetic hyperfine splitting and can be measured. Consider an atom with $J = {}^{11}\!/_2$ and $I = {}^5\!/_2$. If both magnetic dipole moment μ and the electric quadrupole moment Q of the nucleus were zero, the state will be unsplit—a sharp level. The hyperfine splitting of the level when $Q = 0$, $\mu \neq 0$ and when $Q \neq 0$, $\mu = 0$ is shown in Fig. 1.42. The actual magnitude of the splitting depends upon the values of Q, μ, magnetic field and the electric field gradient.

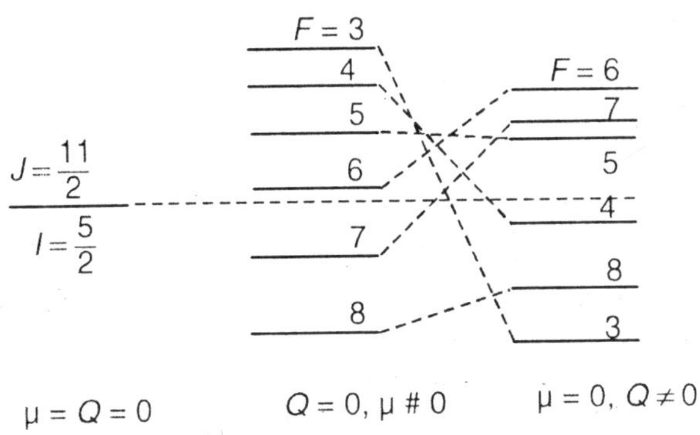

Fig. 1.42: Hyperfine splitting of an atomic level with $J = {}^{11}\!/_2$ and $I = {}^5\!/_2$ when the quadrupole moment of the nucleus $Q = 0$ and $\mu \neq 0$ and when its magnetic moment $\mu = 0$ and $Q \neq 0$. When both $Q = \mu = 0$, the atomic level is unsplit [J.R. McNally (1952). *Jr. Am. J. Phys*, **20**: 152]

In the presence of both magnetic dipole and electric quadrupole interactions, the energy splitting of an atomic level is given by the relation

$$E = \frac{A_J C}{2} + \frac{B}{8}\frac{3C(C+1) - 4IJ(I+1)(J+1)}{IJ(2I-1)(2J-1)} \quad ...(1.9.1)$$

where the first term gives the magnetic splitting as given in equation 1.7.8 according to which

$$A_J = -\frac{\mu_I H(0)}{IJ}$$

while, $B = eQ\,\phi_{zz}$,

and $C = F(F+1) - I(I+1) - J(J+1)$

The term B is proportional to the electric quadrupole moment of the nucleus. The hyperfine splitting of the 4260 Å line of singly ionized osmium atom is shown in Fig. 1.43. From the spacing of the hyperfine split lines the term A and B of equation (1.9.1) can be determined. From the calculation of the magnetic field and the electric field gradient acting at the nucleus, the magnetic moment μ and the quadrupole moment Q can be determined.

From similar measurements, the quadrupole moment of the deuteron was determined as

$$Q = +0.273 \times 10^{-26}\ \text{cm}^2$$

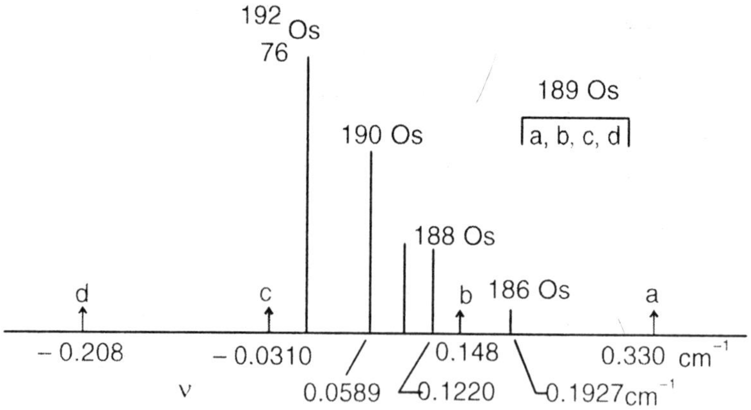

Fig. 1.43: Hyperfine structure and isotope shift in the 4260 Å line of singly ionized osmium, as measured by Fabry-Parot etalon. The lines due to even Z even N isotopes. ^{186}Os, ^{188}Os, ^{190}Os and ^{192}Os are single corresponding to nuclear spin $I = 0$. The line due to ^{189}Os is split into four hyperfine components marked a, b, c, d. The relative spacings of these lines show a deviation from the interval rule. From the above measurements $I = 3/2$, $\mu = 0.7 \pm 0.1\ \mu_N$ and $Q = +(2.0 + 0.8) \times 10^{-24}$ cm^2. [K. Murakawa and S. Suwa (1952) *Phys. Rev*, **87**: 1048].

1.9.2 Hyperfine Structure of Molecular Rotational Lines

In a polyatomic molecule the molecular electric field produces an electric field gradient at the nucleus of an atom. The interaction of the quadrupole moment of the nucleus with the electric field gradient gives rise to a hyperfine structure to the molecular rotational lines. The energy of interaction depends upon the relative orientation of the nuclear spin I, the angular momentum of the molecular rotation and the quadrupole moment of the nucleus.

For heavy molecules, the energy difference between the rotational levels of a molecule lie in the microwave frequency (10^{10} Hz) region. The rotational levels can be excited by the absorption of microwaves of the right frequency. The hyperfine splitting energy of the rotational lines is in the region of 10 MHz to 100 MHz. With the development in microwave techniques it is possible to study the absorption of microwaves as a function of frequency. The structure of the absorption line can be studied with fairly high resolution. The hyperfine structure of the absorption lines has been employed to obtain the nuclear spin I and the quadrupole moment of ^{14}N, ^{18}O, ^{33}S, ^{34}S, ^{35}Cl, ^{37}Cl, ^{79}Br, ^{81}Br and ^{127}I.

In a few cases it has been possible to place the sample under study in an external magnetic field of a few thousand Gauss and study the magnetic hyperfine splitting of the microwave absorption line along with the quadrupole splitting. This experiment gives both the magnetic moment and the quadrupole moment of the nucleus under study.

1.9.3 Nuclear Quadrupole Resonance Method

The radiofrequency resonance method to measure nuclear magnetic moment has been discussed earlier. The same method, with slight modification, is used to measure the quadrupole moment of nuclei in a molecular beam or in a bulk material. In the NMR method the radiofrequency of the oscillator is fixed suitably and the external magnetic field varied slightly to observe the resonance absorption. In the nuclear quadrupole resonance (NQR) method, as the electric field gradient at the nucleus can not be varied, the frequency of the oscillator is varied to observe the absorption.

When nuclear quadrupole resonance is observed in a molecular beam the quadrupole moment of the nucleus can be obtained from the knowledge of the electric field gradient acting at the nucleus. From the structure of the molecule, this electric field gradient can be calculated. In NQR experiments, using bulk matter, the knowledge of the electric field gradient acting at a nucleus depends strongly upon the charge distribution surrounding it. For a symmetric charge distribution as in the case of cubic lattice, the electric field grdient is zero. If however there are lattice defects or impurity atoms in the neighbourhood of the nucleus under study, the value of the electric field gradient could be appreciable. In tetragonal and hexagonal lattices, free from impurities or defects, the electric field gradient acting at a lattice site is well defined.

1.9.4 Gamma-ray Transition Energy and Life Time

When a nucleus is formed in an excited state, as a result of a nuclear reaction or decay of the parent atom, it decays to lower excited states with the emission of gamma ray quanta. In the nonspherical nuclei (having quadrupole moment), the lowest excited states can be considered to be due to the rotation of the nucleus similar to the rigid rotator. This can happen if the nucleons in the nucleus are bound tightly. The energy of excitation due to the different modes of rotation is only a few hundred keV. For these rotational states, the component Ω of the total angular momentum of the nucleus along the axis of deformation remains constant (Fig. 1.44). The energy of excitation of the nucleus in the rotational states is given by the relation,

$$E_{\text{rot}} = \frac{h^2}{2\mathcal{I}} I(I+1)$$

where, I is the total angular momentum of the nucleus in the excited state and $\mathcal{I}$ the effective moment of inertia of the nucleus. In the even Z even N nuclei the spin sequence of the levels generally is 2, 4, 6 ... while in odd A nuclei the spin sequences increases by one unit (Fig. 1.45 and Fig. 1.46.)

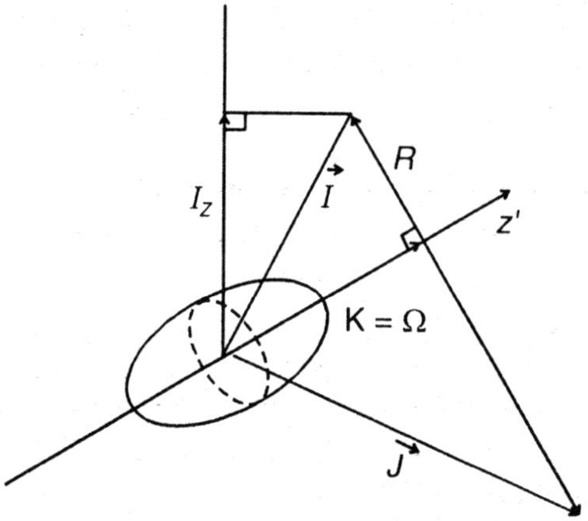

Fig. 1.44: For a deformed nucleus, the total angular momentum $\vec{I}$ is the resultant of rotational angular momentum $\vec{R}$ (which is perpendicular to the symmetry axis z') and the angular momentum $\vec{J}$ of all the nucleons, so that $\vec{I} = \vec{J} + \vec{R}$. The components of $\vec{I}$ and $\vec{J}$ along the symmetry axis are K and Ω respectively. For the rotational band of excited states corresponding to the ground state, $K = \Omega$. For higher rotational bands the value of K increases.

Rotational energy levels have been observed in a large number of medium and heavy nuclei. The effective moment of inertia $\mathcal{I}$ is related to quadrupole moment of the nucleus. Assuming a spheroidal nucleus of constant mass and charge density the moment of inertia to a first approximation is given as

$$\mathcal{I} = \frac{5}{8} \frac{m_A}{R^2} \frac{Q_0^2}{Z^2} \qquad ...(1.9.3)$$

m_A is the total nuclear mass, R the mean radius and Q_0 the intrinsic quadrupole moment of the nucleus as defined in equation (1.8.8).

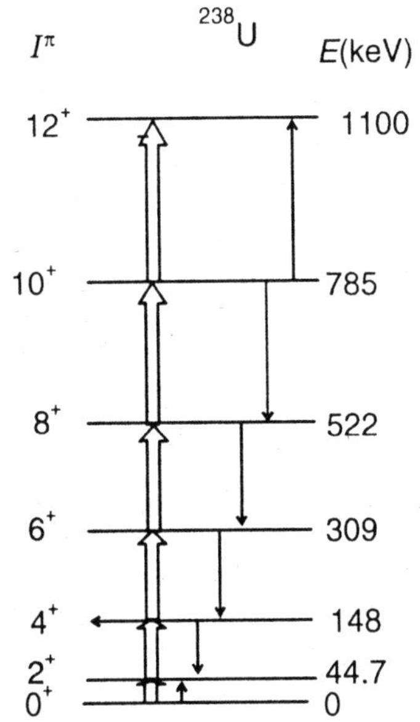

Fig. 1.45: Rotational energy levels of ^{238}U

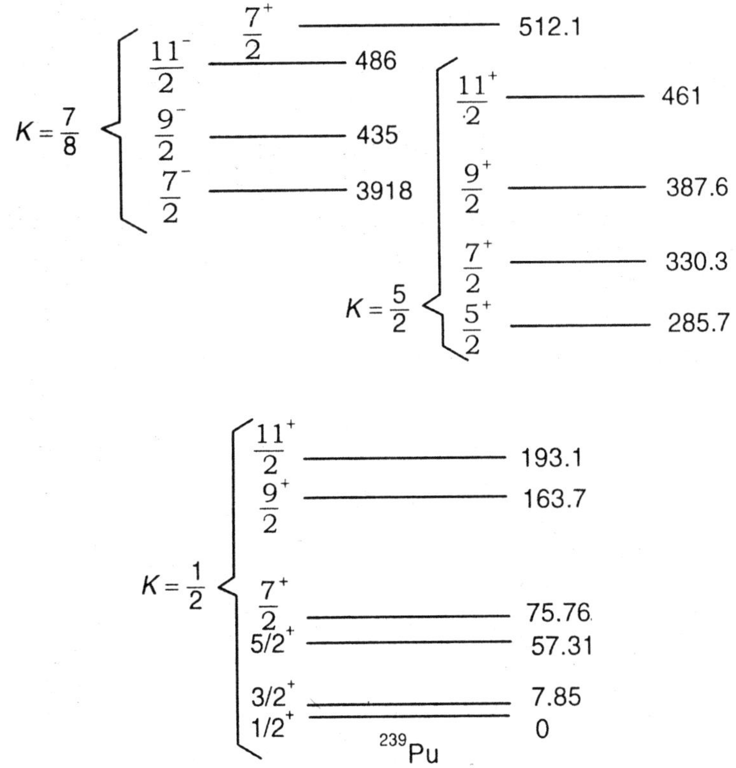

Fig. 1.46: Rotational energy levels of ^{239}Pu. The ground state rotational band has $K = 1/2$. The higher energy bands have $K = 5/2$ and $7/2$

If a gamma transition takes place between rotational energy states with spin I and $I+2$, the energy of the gamma ray gives the intrinsic quadrupole moment Q_0 as

$$Q_0^2 = \frac{8}{5}\frac{h^2(2I+3)R^2Z^2}{m_A \Delta E} \qquad \text{...(1.9.4)}$$

Thus from the energy interval between the successive rotational levels, the quadrupole moment Q_0 can be obtained. It has been found that the value of Q_0 obtained from eqn. (1.9.4) is too large by a factor of two, in comparison to the values obtained by other methods. This implies that the moment of inertia of equation (1.9.3) is too small by a factor of four. It is now postulated that the nucleus does not rotate as a rigid body. It is only the deformation of the nucleus which rotates. This may increase the effective moment of inertia of the nucleus.

Apart from the energy, the life time or the transition probability of a gamma-transition between two rotational states gives the quadrupole moment of a nucleus. The transition probability per second, for a gamma-transition is related to its decay constant λ and half life $T_{1/2}$ as

$$P = \lambda = \frac{0.693}{T_{1/2}(\text{sec.})} \qquad \text{...(1.9.5)}$$

The probability of gamma transition taking place due to the interaction of the quadrupole moment of a nucleus with the electric quadrupole electromagnetic field is given as,

$$P(E2) = \frac{4\pi}{75}\frac{1}{h}\left(\frac{\Delta E}{hc}\right)\cdot B(E2) \qquad \text{...(1.9.6)}$$

where, ΔE is the energy of the gamma transition and $B(E2)$ the reduced transition probability which is closely related to quadrupole matrix element. For an even Z even N nucleus exhibiting rotational energy levels, given by equation (1.9.2), the transition probability for transition from level $I+2$ to level with spin I is given as,

$$B(E2) = \frac{15}{32\pi}e^2Q_0^2\frac{(I+1)(I+2)}{(2I+3)(2I+5)} \qquad \text{...(1.9.7)}$$

For a odd A nucleus the decay probability for the electric quadrupole transition from level $(I+1)$ to level I is given as

$$B(E2) = \frac{15}{16\pi}e^2Q_0^2\frac{\Omega^2(I+1-\Omega)(I+1+\Omega)}{I(I+1)(2I+3)(2I+5)} \qquad \text{...(1.9.8)}$$

where, Ω as defined earlier is the component of nuclear spin I along the axis of symmetry.

In odd A nuclei the gamma-transitions are generally from level $I+1$ to level I and the transition is mixed magnetic dipole ($M1$) and electric quadrupole ($E2$). The magnetic and electric multipole gamma transitions will be discussed in Chapter 3. If the nucleus is strongly deformed, one may observe pure $E2$

transition from level $I+2$ to level I. For such transitions

$$B(E2) = \frac{15}{32\pi} e^2 Q_0^2 \frac{(I+1-\Omega)(I+1+\Omega)(I+2-\Omega)(I+2+\Omega)}{(I+1)(2I+3)(2I+5)(I+2)} \quad ...(1.9.9)$$

This expression holds good for $\Omega = 1/2$

1.9.5 Coulomb Excitation

When a charged particle e.g. an alpha particle is scattered by a nucleus, its distance of closest approach depends on its charge and kinetic energy. If the distance of closest approach is greater than the nuclear radius, the incoming charged particle is only Coulomb scattered. This holds, if the kinetic energy E is less than the Coulomb barrier energy.

$$E < \frac{z_1 Z_2 e^2}{4\pi E_0 R} \qquad (= E_c \text{ Coulomb barrier})$$

where, $z_1 e$ and $Z_2 e$ are the charges on the incoming particle and the nucleus respectively. When the incoming charged particle is close to the nucleus ($E < E_c$) it produces a very strong electromagnetic field and the electric quadrupole moment of the nucleus may interact with it. In the process, the nucleus absorbs energy from the kinetic energy of the incoming particle and gets excited to one of its excited states. The probability for the excitation by Coulomb interaction is large if the nucleus has a large quadrupole moment and shows rotational energy levels as given by eqn. (1.9.2). The total cross section for Coulomb excitation of a nucleus from level I to level $I+2$ is related to the gamma transition probability $B(E2)$. From the measured cross-section, the value of $B(E2)$ can be evaluated, which in turn gives the quadrupole moment of the nucleus. The cross-section for the Coulomb excitation can be measured with an accuracy of only about 20 per cent which puts the limit on the accuracy of the quadrupole moment measured by this method.

1.9.6 Perturbed Angular Correlation of Gamma Rays

Measurement of the perturbed angular correlation of the cascade gamma rays is some times employed to measure the quadrupole moment of a nucleus, in its excited state, having a life time greater than 10^{-8} seconds. As discussed in article 1.7 the perturbed angular correlation function is given as in Eqn. 1.7.31

$$W(\theta, t) = 1 + A_2 G_2(t) P_2(\cos\theta) + A_4 G_4(t) P_4(\cos\theta)$$

$$\approx 1 + A_2 G_2(t) P_2(\cos\theta) \quad ...(1.9.10)$$

The energy of a nucleus with quadrupole moment Q interacting with an axially symmetric electric field gradient (efg) in z direction $\partial^2 V/\partial z^2$ is given as

$$E_m - \frac{3m^2 - I(I+1)}{4I(2I-1)} eQ \frac{\partial^2 V}{\partial z^2} \quad ...(1.9.11)$$

An axially symmetric electric field gradient implies that

$$\frac{\partial^2 V}{\partial x^2} = \frac{\partial^2 V}{\partial y^2} = -\frac{1}{2}\frac{\partial^2 V}{\partial z^2}$$

It is convenient to introduce the quadrupole frequency ω_Q as

$$\omega_Q = \frac{1}{\hbar}\frac{eQ\dfrac{\partial^2 V}{\partial z^2}}{4I(2I-1)} \qquad \text{...(1.9.12)}$$

The angular frequency ω_0 corresponding to the smallest non-vanishing energy difference between the hyperfine split levels is

$$\omega_0 = 3\omega_Q \quad \text{for integral values of } I$$

$$\omega_0 = 6\omega_Q \quad \text{for half integral values of } I$$

For an axially symmetric electric field gradient oriented randomly with respect to the emission of gamma-rays (as in the case of a crystalline powder) the perturbation factor $G_k(t)$ is given as

$$G_k(t) = \sum_{n=0}^{n} S_{kn} \cos n\omega_0 t \qquad \text{...(1.9.13)}$$

where,

$$n = |m^2 - m'^2| \quad \text{for integral values of } I$$

$$n = \frac{1}{2}|m^2 - m'^2| \quad \text{for half integral values of } I$$

We shall be dealing with only the $k = 2$ term. The coefficients S_{kn} are tabulated in the literature.

Unlike magnetic field, it is not possible to apply an external electric field gradient of as high a value as required ($\approx 10^{21}$ volts/m^2). The simplest way is to embed the radioactive atom in a noncubic metallic lattice. The electric field gradient produced at the lattice site acts at the nuclear excited state E_1 when it is formed, giving rise to electric hyperfine interaction. The radioactive atom is introduced in the lattice by thermal heating or ion implantation. From the structure of the metallic lattice the electric field gradient can be calculated. The metallic crystal must be of a high chemical purity. Impurity atoms in the lattice change the electric field gradient at a lattice point to a great extent. The effective electric field gradient acting at the desired nucleus has to take into account the shielding due to the polarisation of the electron sea and the effect of the polarisation of the electron orbits in the atom of the nucleus under study. This effect is taken care of by Sternheimer antishielding factor, which has been tabulated in literature for different elements.

The perturbed angular correlation is observed by measuring the pulse height spectrum of the TAC output in a MCA at $\theta = 90°$ and $\theta = 180°$ as discussed in section 1.7.8. One plots the asymmetry ratio $R(t)$ as defined below as a function of time.

$$R(t) = 2\frac{N(180,t) - N(90,t)}{N(180,t) + N(90,t)} \approx \frac{3}{2} G_2(t) A_2 \qquad \text{...(1.9.14)}$$

The spin rotation pattern for the spin sequence is compared with the pattern expected from equation (1.9.13) and the value of ω_0 is determined. From the knowledge of ω_0 and the electric field gradient acting at the nucleus, the quadrupole moment in the excited state can be calculated.

Figure 1.47 shows the spin rotation pattern for the 173-247 keV gamma-ray cascade emitted in the decay of 2.8 day ^{111}In to ^{111}Cd. The 247 keV state of ^{111}Cd has a life time $T_{1/2}$ = 84n sec. The radioactive ^{111}In was embedded in cadmium matrix by thermal diffusion.

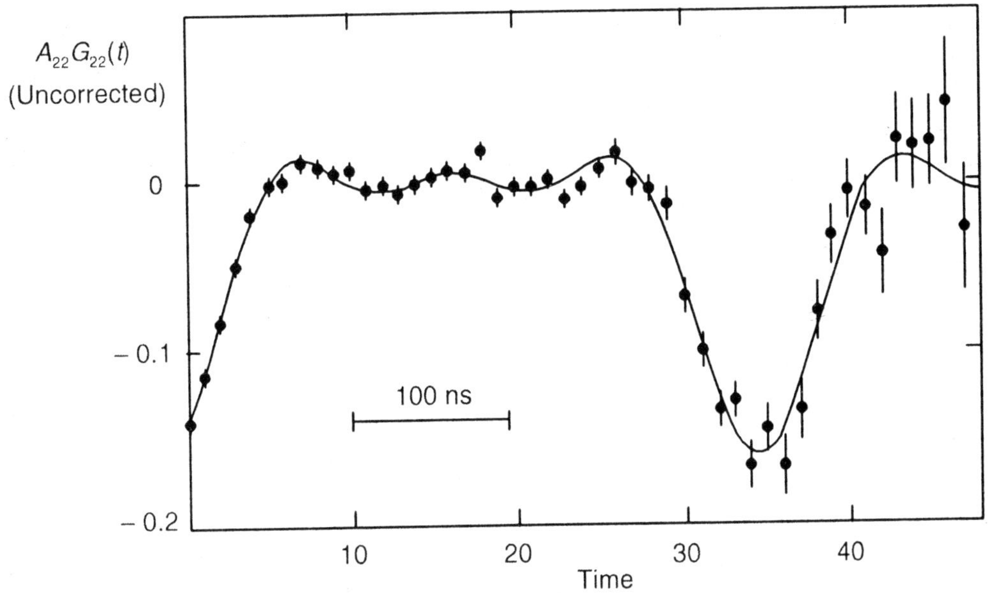

Fig. 1.47: Spin rotation pattern for the 173-247 keV γ-rays cascade of ^{111}Cd embedded in cadmium lattice. The time at which the first dip in the pattern ($T \approx$ 350 ns) appears corresponds to $\omega_0 T = 2\pi$. [P. Lehmanm and J. Miller (1956) *J. Phys. Radium*, **17**: 526].

1.10 ISOTOPIC SPIN OF NUCLEI

The experiment on low energy scattering of neutrons and protons have shown that apart from the Coulomb interaction—the nuclear forces between a neutron and a proton and between two neutrons and two protons are identical. This charge independence of nuclear forces manifests itself in similar properties of mirror nuclei. Mirror nuclei are obtained by transforming all neutrons into protons and all protons into neutrons. Except for a small difference due to Coulomb effect, the excited states of mirror nuclei (^{7}Li, ^{7}Be), (^{9}Be, ^{9}B) show a remarkable similarity. Even nuclei in which a neutron is replaced by a proton e.g. (^{6}He, ^{6}Li,^{6}Be), (^{8}Li, ^{8}Be, ^{8}B), (^{10}C, ^{10}Be, ^{10}B) show similar structure of excited states. The mass difference between mirror nuclei is attributed to the Coulomb energy, which is employed to estimate their radius (see, 1.2).

Neutrons and protons are ordinarily treated as separate entity. Charge independence of nuclear forces suggests that neutron and proton are perhaps two states of a nucleon. Apart from the degrees of freedom of co-ordinates and spin, a nucleon has an internal degree of freedom called isotopic spin, or isospin or *i*-spin. For a nucleon, the isotopic spin can take only two values $+1/2$ corresponding to a proton and $-1/2$ corresponding to a neutron. For a nucleus, the isotopic spin can be defined as $T = (Z - N)/2$, where Z and N are the proton and neutron numbers respectively. The isotopic spin for the nucleus ^{5}Li is $+1/2$ while that of ^{7}Li is $-1/2$.

The isospin operators are found to obey the same commutation relations as the Pauli spin operators. The two are in fact a perfect analogue. Like spin S, the isospin T has components and its third component T_3 determines the charge of the particle. For a nucleon $T = 1/2$ and its third component T_3 can have values $+1/2$ corresponding to a proton and $-1/2$ corresponding to a neutron. The π mesons which interact with nuclei by strong interaction (nuclear force) have isotopic spin $T = 1$ with its third component $T_3 = +1, 0, -1$. The component $T_3 = +1$ corresponds to $\pi+$ meson, $T_3 = 0$ corresponds to π° meson and $T_3 = -1$ corresponds to π^- meson.

The isotopic spin formalism is of great significance in the theory of β decay in which a neutron changes into a proton (β^- decay) or vice-versa (β^+ decay). The concept of isospin helps in the classification of charged particles encountered in the high energy particle physics.

For an isolated system or a nucleus the isospin, is a constant of motion. This conservation law, however, does not apply to heavy nuclei where the Coulomb interaction is quite strong. Though the foundation of isotopic spin formalism is empirical, it proves to be very important in pion physics.

1.11 PARITY OF A NUCLEUS AND ITS CONSERVATION

The wave function of a system of particles or a nucleus describes completely its behaviour as a function of time and space. The behaviour of the wave function $\Psi(r)$ under reflection of space co-ordinates (r changes to $-r$) determines the parity of the system. Under space reflection, the wave function may change sign or may remain unchanged. The change of sign, which requires an odd power wave function, represents the negative parity. The even power wave function which does not change sign represents an even parity so that

$$\psi(-\vec{r}) = -\psi(\vec{r}) \qquad \text{Odd parity}$$

$$\psi(-\vec{r}) = +\psi(\vec{r}) \qquad \text{Even parity}$$

The product of two odd wave functions results in an even parity function, while the product of an even and an odd wave function results in an odd parity wave function.

The parity of an isolated neutron or proton wave function is even while that of a π meson is odd. For a system of interacting nucleons the parity-denoted by π, is given by the orbital angular momentum of the system so that

$$\pi = (-1)^l$$

The parity of the deuteron (actually of its wave function) which is in $l = 0$ state is even.

For an odd parity function the integral over all space is always zero, because the contribution to the integral of elements of volume at $\vec{r}$ and $-\vec{r}$ cancel each other. It has been shown that the electric dipole moment of a nucleus for this reason is always zero.

The law of conservation of parity states that the parity of a system of nuclear particles remains constant under any reaction or transformation. In all electromagnetic and strong interactions (due to nuclear forces) the conservation of parity has been verified experimentally. The law of conservation of parity is regarded as a fundamental law of physics where strong and electromagnetic interactions are involved. This law as discussed in Chapter three (3) has been found to be violated in transformations involving weak interactions e.g. β-decay.

A consequence of the law of conservation of parity is that a nucleus or a system of nuclear particles can not co-exist in states of different parities. The total wave function ψ is a sum of wave function of the different possible states in which a nucleus or system of particles can exist, so that

$$\psi = \Sigma a_i \Psi_i$$

where, a_i is the amplitude of the wave function ψ_i for the ith state. The quantity a_i^2 gives the probability of finding the system at any time in the ith state. According to the conservation of parity all the wave functions ψ and ψ_i should have the same parity (same even or odd angular momentum states). A deuteron can co-exist in $l = 0$ (s state) and $l = 2$ (d state) but not in $l = 0$ and $l = 1$ states. Parity is thus a fundamental property of a nuclear state.

1.12 EXCITED STATES OF NUCLEI

A nucleus can exist in its lowest energy state called the ground state or in any of its excited states. An excited state of a nucleus has a discrete energy which is specific to the nucleus, No two nuclides have been found to have the same excitation energy. An excited state of a nucleus is specified by its excitation energy E, orbital angular momentum L, total angular momentum I, parity π, magnetic moment μ and electric quadrupole moment Q. A nucleus in its excited state tends to make transition to any other excited state of lower energy or to its ground state with the emission of gamma rays or internal conversion electrons. The probability of the gamma transition from one state to another depends upon the changes in energy, total angular momentum and parity involved in the transition. If the transition probability λ from an excited state is very small ($T = 1/\lambda > 10^{-8}$ sec) the state is called an isomeric state. The life times of nuclear excited states may vary from thousands of years to 10^{-18} seconds.

A nucleus can absorb energy in various modes and form an excited state. The main modes of excitation of low energy (< 2 MeV) are as follows:

(*i*) Single particle excitation in which only the last nucleon is excited to higher energy states leaving the other nucleons undisturbed.

(ii) Multiparticle excitation in which two or more nucleons are excited to higher energy states. The excited state so formed has properties derived from the mixing of the wave functions of all the excited particles.

(iii) Vibrational excitation in which the nucleus performs small oscillations about its equilibrium shape. The vibrational mode of excitation occurs not only among spherical even even nuclei but also among deformed nuclei. The spin sequence of the ground, first and second excited states is generally O^+, 2^+, 2^+ or 4^+. The superscript giving the parity of the state. The ratio of the energy of the second excited state to that of first state is approximately two.

(iv) Rotational excitation in which a deformed nucleus may rotate with its axis perpendicular to the symmetry axis. The energy of excitation is in a few hundred keV region. The energies of the excited states follow the pattern as:

$$E_I = \frac{h^2}{2\mathcal{J}} I(I+1)$$

In odd A nuclei the spin sequence is I, $I+1$, $I+2$ ..., and in even-even nuclei the spin sequence is 0, 2, 4, 6 and 8. In highly deformed nuclei, the low energy excited states can be interpreted as purely due to the rotational motion.

Generally it has been observed that the excited states of nuclei do not follow a single mode of excitation. The measurement of magnetic moment and quadrupole moment of a nucleus in its excited state gives an indication of the mixture of the various modes of excitation in forming the state.

EXERCISES

1. Upto what kinetic energy, the scattering of α-particles by hydrogen follows the Rutherford's scattering law.
2. What is the density of a nucleus ?
3. What relation will describe a linear fall off of the density of a nucleus?
4. The energy released in fission of ^{235}U is about 180 MeV. Approximately what fraction of the original mass (^{235}U + n) is converted into energy.
5. How are the Q values in laboratory system related for the following two reactions? $^1H(t, n)\,^3He$ and $^3H(p, n)\,^3He$
6. Calculate the binding energy of deuteron from the Q values of the following reactions.

   $^2H(d, p)\,^3H$ (Q_1 = 4.036 ± 0.012 MeV)
   $^2H(n, \gamma)\,^3H$ (Q_2 = 6.251 ± 0.008 MeV)
7. Calculate the atomic mass difference between ^{28}A and ^{28}Si from the following data on nclear reactions.

   $^{27}Al(d, p)\,^{28}Al$, Q = 5.494 MeV
   $^{28}Si(d, p)\,^{29}Si$, Q = 6.246 MeV
   $^{29}Si\ (d, \alpha)\,^{27}Al$, Q = 5.994 MeV

Also, $2H - {}^2H = 1.549 \times 10^{-3}$ amu
$2\,{}^2H - {}^4He = 25.596 \times 10^{-3}$ amu
$n - {}^1H = 0.782$ MeV

$2^2H - {}^4He = 25.596 \times 10^{-3}$ amu
$n - {}^1H = 0.782$ MeV

8. What is the change in binding energy of a nucleus when a neutron is removed from it?
9. Explain the reason why
Mass of ^3H > Mass of ^3He

$^{14}_{6}C > ^{14}_{7}N$

$^{41}_{19}K < ^{41}_{20}Ca$

$^{42}_{19}K < ^{42}_{20}Ca$

10. The Q values of the following reactions are given
$^{14}N + {}^4He \rightarrow {}^{17}O + {}^1H - 1.26$ MeV
$^{16}O + {}^2H \rightarrow {}^{14}N + {}^4He + 3.13$ MeV
Calculate the Q value of the reaction $^{16}O(^2H, {}^1H)\ {}^{17}O$
11. ^{13}N decays to ^{13}C by emission of positrons of maximum energy 1.2 MeV. The masses of ^{14}N and ^{13}C are known as
M (^{14}N) = 14.00307 amu
M (^{13}C) = 13.00335 amu
Find the threshold energy for the γ-n reaction in ^{14}N?
12. For what atomic number of a nucleus the first Bohr orbit of a Mu-meson is equal to the nuclear radius?
13. It is known that
and $^{63}Zn \rightarrow {}^{63}Cu + \beta^+ + \nu, E_{\beta^+} = 2.3$ MeV
and $n - {}^1H = 0.782$ MeV
Find the threshold energy for the (p, n) reaction
$^{63}Cu + {}^1H \rightarrow {}^{63}Zn + n$
14. Assume a proton to be a sphere of uniform density and radius of 1.5×10^{-13} cm. It has the intrinsic angular momentum $\hbar\sqrt{s(s+1)}$ and $s = 1/2$.
   (a) Find the angular velocity of the proton
   (b) what is the speed of the surface layer of the proton
   (c) what is rotational kinetic energy
   (d) how many amperes of current is going round the axis of rotation.
15. Assume a nucleon moving in a circular orbit of radius 4×10^{-13} cm. and having angular momentum quantum number $l = 1$. Find the velocity of the nucleon.
16. Calculate the magnetic field due to the intrinsic magnetic dipole moment of a proton at the equator of the proton, and at a distance of one Bohr radius.
17. Calculate the magnetic field at the nucleus of (a) hydrogen atom and (b) a cesium atom.
18. Consider two particles of masses m_1 and m_2 revolving round their centre of mass. If their charges are e_1 and e_2 respectively, calculate the gyromagnetic ratio of the system. Apply the result to obtain the orbital g-factor for a neutron and a proton in deuterium and μ^- meson bound to a proton.
19. Show that the electrostatic energy of a uniformly charged sphere of radius R and having charge Q is $\frac{3}{5}\ Q^2/R$.
20. Calculate the difference in the binding energies of the isobaric pairs ^{113}(Cd-In), ^{187}(Os – Re) and ^{123}(Sb – Te).
21. Calculate the hyperfine splitting Δν expected for ground state of hydrogen atom in units of (a) cycles/sec. (b) cm^{-1} (c) and electron volts ($h\Delta\nu$).

2

NUCLEAR FORCE AND TWO BODY PROBLEM

2.1 INTRODUCTION

In the preceding chapter on the static properties of nuclei, it has been discussed that there exists a strong, short range, attractive force inside a nucleus. For a strong binding between nucleons inside a static nucleus, there has to be mutual interaction between neutrons and protons. There could be three types of such interactions namely an interaction between a neutron and a proton (*n-p*), between a proton and a proton (*p-p*) and between a neutron and a neutron (*n-n*). In order to study these forces, if we consider nuclei containing many nucleons, the problem becomes more complex, because the law of nuclear force between a pair of nucleons is still unknown. Even if the internucleon forces were known, it does not follow that the force on a particular nucleon in a 3-body system is merely a sum of two body forces. There may be many body forces also. However, so far, no systematic use of such complicated many body forces has been made. It is, therefore, of great importance to study the nuclear force between a pair of nucleons. A step in this direction is to consider a system where only two particles exist together. Nature has provided such a two particle system. The nucleus of a deuterium (heavy hydrogen) atom consists of one neutron and one proton only. No physical system consisting of two protons or two neutrons exists. The (*p-p*) and (*n-n*) forces are studied through scattering experiments. Such experiments are limited to (*n-p*) scattering and (*p p*) scattering. It is not possible to prepare a neutron target. The force between two neutrons (*n-n* force) is, therefore, studied indirectly either through nuclear reactions or through scattering of neutrons by deuteron.

2.2 THE GROUND STATE OF DEUTERON

The experimental information about the deuteron nucleus is as follows:

(*i*) **Binding Energy:** The binding energy of deuteron is E_B = 2.225 MeV. The binding energy is obtained using the radiative capture of a neutron by hydrogen nucleus forming a deuteron ($n + p \rightarrow d + \gamma$). The energy of emitted gamma-ray in the reaction gives the binding energy of the deuteron. Likewise, by measuring the gamma ray energy sufficient to break the bond between proton and neutron inside the deuteron ($d + \gamma \rightarrow n + p$), or through the measurement of masses of neutron, proton and deuteron, the binding energy of deuteron is obtained.

(ii) **The total angular momentum or spin I of the deuteron is 1 (in units of $\hbar$):** The parity of the ground state of deuteron is even (+).

Spin and Parity of Deuteron

For the deuteron, it is known that the parity of the ground state wave function is even (positive). To understand this, the wave function of the state is separated into a product of three wave functions, namely the intrinsic wave function each of proton and neutron and the orbital wave function for the relative motion between the two nucleons. The intrinsic wave function of neutron and proton has same parity, because they are just two states of a nucleon. As a result the product of their intrinsic wave function has even parity. The parity of deuteron is thus, to be determined by the relative motion between the neutron and the proton. The parity of angular momentum wave function is given by $(-1)^L$ where L is the angular momentum for the state. Since the parity of the deuteron is even (positive), the value of L must be even.

The total angular momentum of the ground state of the deuteron is $I = 1$ where $I = L + S$. The possible values of S are 0 and 1 $\left(S = \left|\frac{1}{2} + \frac{1}{2}\right| = 0 \text{ or } 1\right)$ since it is not possible to combine $S = 0$ with any of the allowed even values of L to form $I = 1$ state, we can eliminate $S = 0$. Likewise even values of L greater than 2 are also ruled out. The only possible values of L and S are $L = 0$, $S = 1$ and $L = 2$, $S = 1$, to produce $I = 1$ state. The dominant part of the ground state wave function is the $L = 0$, $S = 1$ component. We shall see that there is a small but significant admixture of $L = 2$ state also. Thus for the ground state of deuteron $L = 0$, $I = L + S$ and $I = S = 1$.

(iii) Magnetic moment of deuteron as measured by magnetic resonance method is $\mu_d = 0.8574\ \mu_N$.

(iv) Electrical quadrupole moment of deuteron is measured as

$$Q = +2.82 \times 10^{-31}\ \text{m}^2.$$

(v) **Radius of Deuteron:** High energy electron scattering experiments yield the radius r_d of the deuteron of about 4.2F. This corresponds to root mean square distance $\left(\sqrt{< r^2 >}\right)$ between the neutron and proton

$$r_d = \sqrt{\langle r^2 \rangle} = 4.2\text{F} \ (1\ F = 10^{-15}\ \text{m})$$

2.3 SCHRODINGER WAVE EQUATION FOR DEUTERON AND ITS SOLUTION

The deuteron, from the point of view of quantum mechanics, is regarded as a stationary state formed by the mutual attraction between a neutron and a proton. The Schrodinger wave equation for a two body problem in the centre of mass system is

$$-\frac{\hbar^2}{2\mu}\nabla^2\Psi + V\Psi = E\Psi \qquad ...(2.3.1)$$

where μ is reduced mass defined as

$$\mu = \frac{m_1 m_2}{m_1 + m_2}$$

m_1 and m_2 being the masses of neutron and proton respectively. The difference between the masses being very small, we can put $m_1 = m_2 = M$, the mass of a nucleon, so that

$$\mu = \frac{M}{2}$$

E is total energy of the system and is equal to the sum of kinetic energy (T) and potential energy (V).

$\Psi = \Psi(r, \theta, \phi)$ where r, θ, ϕ are the co-ordinates of one particle relative to the other.

$V = V(r)$ is the potential describing the attractive force between neutron and proton inside the deuteron. It is a short range force and depends on the separation distance r between the particles. The potential decreases as the distance increases and vanishes for practical purposes for $r > b$, where b describes the range of nuclear force.

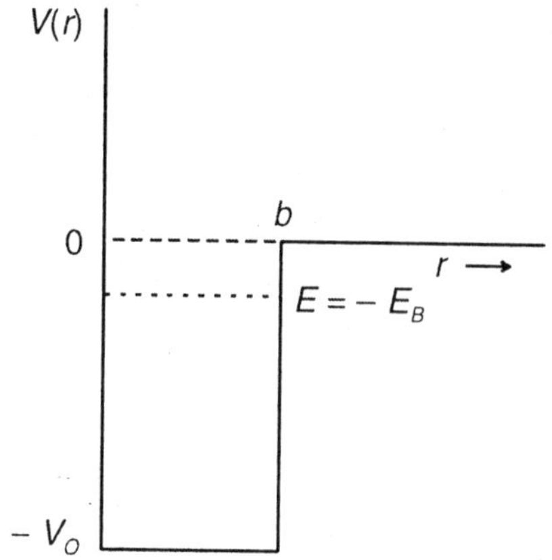

Fig. 2.1: Square well potential $V(r)$ for deuteron. E_B denotes binding energy of deuteron. 'b' is width of the potential

The actual form of the nuclear potential is still unknown. It is, therefore, arbitrarily assumed that the potential $V(r)$ is a square well type with depth V_o

and width 'b'. The Fig. 2.2 shows the square well potential which can be described as $V(r) = -V_o$ for $r \leq b$ and $V(r) = 0$ for $r > b$.

There is no reason to believe that the attractive part of the potential is a square well type. The only justification is its simplicity. The Schrodinger equation with such a potential can be solved exactly.

It may be mentioned that few other types of potential given below (Fig. 2.2) have been used, each giving almost similar results.

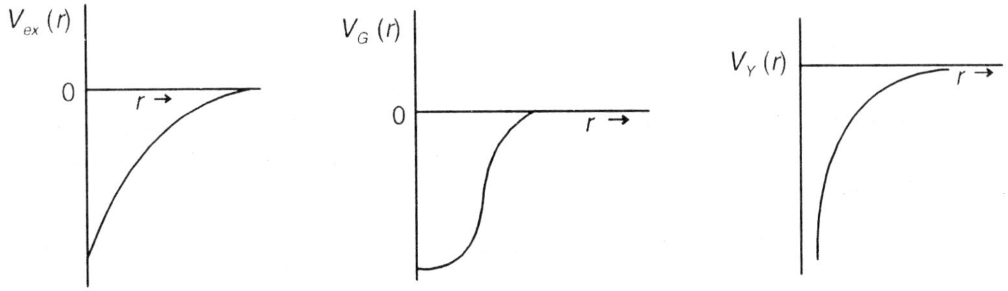

(*a*) Exponential (*b*) Gaussian (*c*) Yukawa

Fig. (2.2): Two nucleon potentials

(*a*) Exponential Potential

$$V_{ex}(r) = -V_o e^{\frac{-r}{b}}$$

(*b*) Gaussian Potential

$$V_G(r) = -V_o e^{\frac{-r^2}{b^2}}$$

(c) Yukawa Potential

$$V_Y(r) = -\frac{V_o e^{\frac{-r}{b}}}{\frac{r}{b}}$$

The solution of the Schrodinger equation (2.3.1) for a square well type of potential is given by

$$\Psi(\vec{r}) = \psi(r)(\Theta)\Phi(\phi) \qquad \ldots(2.3.2)$$

and for $\ell = 0, \Theta(\theta)$ and $\Phi(\phi)$ are constant. The potential $V(r)$ is spherically symmetric being independent of θ and ϕ. With this description and using

$$\psi(r) = \frac{u(r)}{r}$$

the Schrodinger equation (2.3.1) reduces to the radial equation.

$$\frac{d^2u(r)}{dr^2} + \frac{M}{\hbar^2}(E - V)u(r) = 0 \qquad \text{...(2.3.3)}$$

The function $u(r)$ satisfies the boundary conditions satisfied by $\psi(r)$. As an example, the probability for the particle to lie between r and $r + dr$ is

$$p(r)dr \propto \int_{\theta,\,\phi} \psi^2 r^2 \sin\theta\, dr\, d\theta\, d\phi$$

$$\propto \int_r u^2 dr$$

For the square well potential, (Fig. 2.1) $V = -V_o$ and the total energy E is given by the binding energy of deuteron *i.e.* $E = -E_B$. With these values the equation (2.3.3) then becomes

$$\frac{d^2u}{dr^2} + \frac{M}{\hbar^2}(V_o - E_B)u = 0 \qquad \text{...(2.3.4)}$$

Now consider the situations:

(*i*) For $r \le b$ the equation (2.3.4) becomes

$$\frac{d^2u_I}{dr^2} + k^2u_I = 0 \qquad \text{...(2.3.5)}$$

where u_I is wave function inside the potential well and

$$k^2 = \frac{M}{\hbar^2}(V_o - E_B) \qquad \text{...(2.3.6)}$$

(*ii*) For $r > b$, eqn. (2.3.4) becomes

$$\frac{d^2u_o}{dr^2} - \gamma^2u_o = 0 \qquad \text{...(2.3.7)}$$

where u_o is wave function outside the potential well and

$$\gamma^2 = \frac{M}{\hbar^2}E_B \qquad \text{...(2.3.8)}$$

The general solution of eqn. (2.3.5) is given by

$$u_I = A\sin kr + B\cos kr \qquad \text{...(2.3.9)}$$

Since at $r = 0$, $\psi(r) = \dfrac{u_I(r)}{r}$ is zero or finite, $u_I(0)$ is zero. This requires that the constant B in (2.3.9) is zero. (For $B \neq 0$, $u_I \to \infty$ as $r \to 0$).

Hence

$$u_I = A\sin kr \qquad \text{...(2.3.10)}$$

A being the normalisation constant.

Similarly, the solution of eqn. (2.3.7) is given by

$$u_o = C_e^{-\gamma r} + D\,e^{\gamma r} \qquad \text{...(2.3.11)}$$

At infinity $(r \to \infty), \psi(r) = \frac{u_o(r)}{r} \to 0$

or $u_o(\infty)$ is zero or finite. This requires $D = 0$ in eqn. (2.3.11)
(For $D \neq 0$, $u_o \to \infty$ as $r \to \infty$)
Hence

$$u_o = Ce^{-\gamma r} \qquad ...(2.3.12)$$

C being the normalisation constant.

Since u and $\frac{du}{dr}$ must be continuous at $r = b$ the equations (2.3.10) and (2.3.12) give

$$A \sin kb = Ce^{-\gamma b} \qquad ...(2.3.13)$$

$$Ak \cos kb = -C\gamma e^{-\gamma b} \qquad ...(2.3.14)$$

These two equations, (2.3.13) an (2.3.14) yield

$$k \cot kb = -\gamma$$

or
$$\cot kb = -\frac{\gamma}{k} \qquad ...(2.3.15)$$

The value of $\gamma = \sqrt{\frac{M}{\hbar^2} E_B}$ can be calculated from the values of M, $\hbar$ and known value of E_B. Equation (2.3.15) is a relation between the depth 'V_o' and width 'b' of the assumed square well potential. The transcendental equation (2.3.15) can be solved numerically and pairs of values of V_o and b could be obtained to fit the observed value of the binding energy E_B. Such pairs of values of V_o and b are given in Table (2.1).

TABLE 2.1: PAIRS OF VALUES OF V_o AND b

Range 'b' 10^{-15} m	*Depth V_o* MeV
1.0	120
1.5	53
2.0	36
2.5	25
∞	2.83

Many of the experiments give an approximate range of the nuclear force to be of the order of 10^{-15} m. Taking the approximate value for the range $b \simeq 2 \times 10$ m^{-15} the depth V_o should be around 36 MeV (Table 2.1). The depth-range relation (2.3.15) is

$$\cot kb = -\frac{\gamma}{k} = -\left(\frac{E_B}{V_o - E_B}\right)^{\frac{1}{2}}$$

Since $V_o >> E_B$

$$\cot kb \approx 0$$

and this demands that

$$kb \simeq \frac{\pi}{2},\ 3\frac{\pi}{2},\ \ldots \qquad \ldots(2.3.16)$$

or
$$\sqrt{\frac{M}{\hbar^2}(V_o - E_B)}.b \simeq \sqrt{\frac{M}{\hbar^2}V_o}b \simeq \frac{\pi}{2},\ \frac{3\pi}{2},\ \ldots$$

or
$$V_o b^2 \simeq \frac{\pi^2\hbar^2}{4M} \text{ or } \frac{9\pi^2\hbar^2}{4M} \text{ and so on } \ldots \qquad \ldots(2.3.17)$$

For a given V_o, the product kb should be such as to be consistent with the ground state of the deuteron. This condition be satisfied, if only the first term ($kb = \pi/2$) of the multiple values for kb is retained

$$kb = \sqrt{\frac{M}{\hbar^2}(V_o - E_B)} \cdot b \simeq \sqrt{\frac{M}{\hbar^2}V_o} \cdot b = \frac{\pi}{2}$$

or
$$V_o b^2 = \frac{\hbar^2\pi^2}{4M} \qquad \ldots(2.3.18)$$

This also agrees with the fact that the ground state wave function $u(r)$ cannot have a node inside the square well potential. It should converge towards zero at the boundary, b, of the potential. The higher terms $kb = 3\pi/2$ will correspond to a node in the wave function. From equation (2.3.18), for $b = 2 \times 10^{-15}$ m the value of V_o comes out around 30 Mev. (Note that even if one includes E_B, though of negligible magnitude compared to V_o, the order of magnitude of V_o does not change significantly).

The kinetic energy of the particle in the bound state ($b = 2 \times 10^{-15}$ m) is around 28 MeV ($E = T + V$ or $T = E - V \simeq 2 + 30 \simeq 28$ Mev) and this is essentially the energy indicated by the uncertainty principle for a nucleon to be confined within this distance.

2.4 SHAPE OF THE GROUND STATE WAVE FUNCTION

Consider the equation (2.3.15)

$$\cot kb = -\frac{\gamma}{k} = -\left(\frac{E_B}{V_O - E_B}\right)^{\frac{1}{2}}$$

In order that this equation is satisfied, the left hand side must be negative which requires that kb must be greater than $\pi/2$ and less than π. Therefore, the sinusoidal wave function u_I within the square well potential should be slightly greater than one quarter of a wave length. u_I is zero at the origin and joins smoothly the exponential function u_o at $r = b$. The complete wave function for the ground state of deuteron is shown in Fig. 2.3.

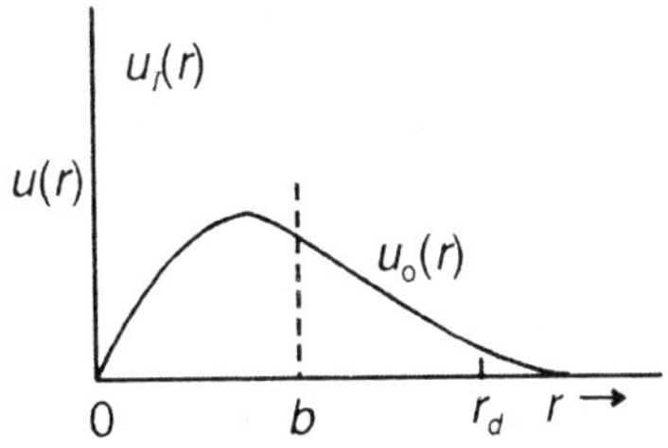

Fig. 2.3: Wave function of deuteron

2.5 NORMALISATION OF DEUTERON WAVE FUNCTION

The normalisation constant A and C in equations (2.3.10) and (2.3.12) are obtained from the requirement that the integral of $|\psi|^2$ over all space must be equal to unity

i.e. $$4\pi\int_o^\infty |\psi|^2 r^2 dr = 4\pi\int_o^\infty u^2 dr = 1 \qquad \ldots(2.5.1)$$

or $$\int_o^\infty u_I^2 dr + \int_o^\infty u_0^2 dr = \frac{1}{4\pi} \qquad \ldots(2.5.2)$$

Using equations (2.3.10) and (2.3.12) for u_I and u_o, one gets

$$A^2\int_o^b \sin^{-2} kr\, dr + C^2\int_b^\infty e^{-2\gamma r} dr = \frac{1}{4\pi} \qquad \ldots(2.5.3)$$

Using $C = A\sin kb e^{\gamma b}$ from eqn. (2.3.13)

$$A^2\int_o^b \sin^2 kr\, dr + A^2 \sin^2 kb\, e^{2\gamma b}\int_b^\infty e^{-2\gamma r} dr = \frac{1}{4\pi}$$

or $$A^2\left[\frac{1}{2}\left\{b - \frac{\sin^2 kb}{2k}\right\}\right] + \frac{A^2 \sin^2 kb\, e^{2\gamma b} e^{-2\gamma b}}{2\gamma} = \frac{1}{4\pi}$$

Taking $kb \simeq \frac{\pi}{2}$, so that $\sin kb \simeq 1$

$$A^2\left\{\frac{b}{2} + \frac{1}{2\gamma}\right\} = \frac{1}{4\pi}$$

$$A^2 = \frac{1}{4\pi}\frac{2\gamma}{1 + b\gamma} \qquad \ldots(2.5.4)$$

$$A = \left[\frac{1}{2\pi}\cdot\frac{\gamma}{(1 + b\gamma)}\right]^{\frac{1}{2}} \qquad \ldots(2.5.5)$$

Since $C = A \sin kb e^{\gamma b} = Ae^{\gamma b}$ (using $kb = \frac{\pi}{2}$)

Putting the value of A from eqn. (2.5.5) one gets

$$C = \left(\frac{1}{2\pi} \cdot \frac{\gamma}{(1+\gamma b)}\right)^{\frac{1}{2}} e^{\gamma b} \quad \text{...(2.5.6)}$$

Hence the radial wave functions for the deuteron are

$$u_I = \left[\frac{\gamma}{2\pi(1+\gamma b)}\right]^{\frac{1}{2}} . \sin kr \quad \text{...(2.5.7)}$$

$$u_o = \left[\frac{\gamma}{2\pi(1+\gamma b)}\right]^{\frac{1}{2}} e^{\gamma(b-r)} \quad \text{...(2.5.8)}$$

The probability of finding the neutron and proton within the range of the nuclear potential is given by

$$P = 4\pi \int u_I^2 dr = 4\pi \frac{\gamma}{2\pi(1+\gamma b)} \int_o^b \sin^2 kr \cdot dr$$

$$= \frac{2\gamma}{(1+\gamma b)} \cdot \frac{b}{2}$$

$$P = \frac{b\gamma}{1+\gamma b} \quad \text{...(2.5.9)}$$

where $\gamma = \sqrt{\frac{M}{\hbar^2} E_B}$

using $M = 1.67 \times 10^{-27}$ kg. $E_B = 2.225$ MeV

$\hbar = 1.054 \times 10^{-34}$ Js and $b = 2 \times 10^{-15}$ m

the probability P will be

$P = 0.315$

or $P = 31.5\%$s ...(2.5.10)

Thus one finds that the probability of finding the neutron and proton within the range of nuclear force is only about 30%. The probability for the neutron and proton to stay beyond the range of nuclear force is more than twice (around 70%) that of finding them within the range of nuclear force. This indicates that the deuteron is a loosely bound structure (low binding energy, 1.113 MeV per nucleon). From classical point of view, however, the nucleons cannot stay bound outside the potential well. But quantum mechanically we find that they do stay outside the well. This quantum mechanical phenomena is to be associated with the wave penetration or the tunnelling through the potential barrier.

2.6 SIZE OR RADIUS OF THE DEUTERON

The distance r_d at which the exponential function u_o falls to $\frac{1}{e}$ of its value, say at $r = r_d$ is taken as a measure of the radius of the deuteron. This gives

$$\gamma r_d = 1 \text{ or } r_d = \frac{1}{\gamma}, \text{ but } \gamma = \sqrt{\frac{M}{\hbar^2}E_B} = 0.232 \times 10^{15} \text{ m}^{-1}$$

$$\therefore \qquad r_d = 4.3 \times 10^{-15} \text{ m} = 4.3 \text{ F} \qquad \text{...(2.6.1)}$$

One finds that the value of r_d is about twice the range of nuclear potential. This explains the fact that the deuteron is a loosely bound structure. The measured value for the size of the deuteron is about 4.2 F.

2.7 MIXING OF ORBITALS IN DEUTERON

It is seen from the above discussion that the observed size and binding energy of deuteron are consistent with the quantum mechanical description of the nucleus, based on a strong attractive central potential of specified depth and width. However, the assumption of central potential failed to explain the electric quadrupole moment and magnetic dipole moment of the deuteron.

It is shown in section (2.1) that the ground state of the deuteron is predominantly an s-state ($l = 0$). The total intrinsic spin of the deuteron is $S = 1$. This indicates that the spin vectors of neutron and proton are parallel ($\uparrow\uparrow$) and in spectroscopic notation, the state of the deuteron is a triplet s-state $^3S_1$. If, however, the nuclear forces were independent of spin directions, we should expect a singlet state $^1S_0$ also with the same energy. In this case the two spin vectors will be antiparallel ($\uparrow\downarrow$). But no bound state with a neutron and a proton having $I = 0$ ($L = 0$, $S = 0$) has been observed. One, therefore, finds that the mutual force between the neutron and a proton inside the nucleus is strong enough when their spin vectors are parallel compared to when spins are antiparallel. Nevertheless, this does not mean that nuclear forces do not exist for antiparallel spin vectors. (This point has been discussed in detail in a later article in this chapter on scattering of slow neutrons by a proton.)

In a simple picture of a pure $^3S_1$ state, the charge distribution is spherically symmetric, suggesting thereby that the electric quadrupole moment, Q, should be zero. Also the magnetic dipole moment, μ_d, should be the algebraic sum of the magnetic moments, μ_n, μ_p of neutron and proton respectively. In fact, the deuteron does have a finite, though small electric quadrupole moment

$Q = +2.88 \times 10^{-31}$ m^2. Likewise the observed magnetic moment μ_d is not equal to the sum of the magnetic momenta of neutron and proton ($\mu_d \neq \mu_n + \mu_p$). The observed value of $\mu_d = 0.8573\ \mu_N$ whereas $\mu_n + \mu_p = -1.9132 + 2.7928 = 0.87\ \mu_N$. The small diference ($0.0223\ \mu_N$) between the expected value and observed value of μ_d and the non-zero quadrupole moment suggest that the deuteron ground state is not a pure $^3S_1$ state.

These two disagreements of experimentally known results with the corresponding predictions from the central potential led Rarita and Schwinger (1941) to suggest that the nuclear force between a neutron and a proton is not purely central. It was found necessary to add a term to the central potential which produces a non-central or a tensor force. The nuclear force between nucleons not only depends on the radial distance between them (central force), it also depends on the relative orientation of their spin directions (non-central). The non-central or the tensor component depends on the angle which their spin axes make with the line joining the two nucleons. The tensor force, in vector notation is, therefore, a function of ($\vec{s} \cdot \vec{r}$). For the same separation distance, the tensor force is different for different relative orientation of spins.

The orbital motion of the nucleons under the effect of non-central force, is such that the charge distribution acquires a spheroidal shape. (For spherically symmetric charge distribution, the orbital angular momentum of their relative motion around the centre of mass is zero). It is prolate or cigar shaped when the probability density of particles is more along the spin axis (axis of symmetry). The quadrupole moment Q is positive and the mutual interaction between the neutron and proton is attractive. When the probability density for the charge distribution is more along the line perpendicular to the spin axis, the shape is oblate, the mutual interaction is repulsive and Q is negative.* The Figs. (2.4 *a*) and (2.4 *b*) show these two situations.

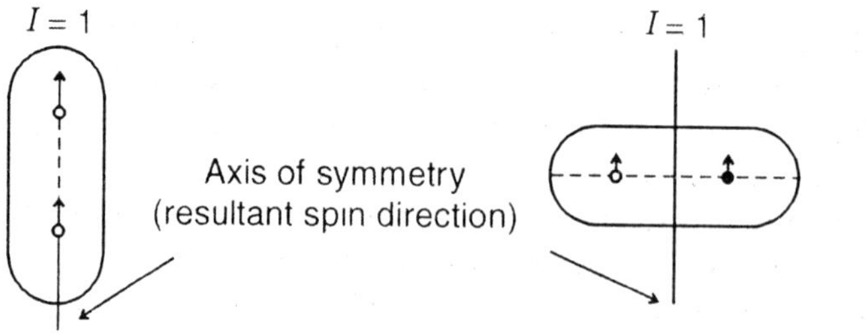

(*a*) Prolate, Q is positive (*b*) Oblate, Q is negative

*The attractive and repulsive nature of the mutual interaction between neutron and proton in the two situations of spin orientation resembles similar behaviour of magnetic dipoles when the two lie along the line parallel to their magnetic axis, resulting into attraction and the two repel each other when the magnetic axis is perpendicular to the line joining them.

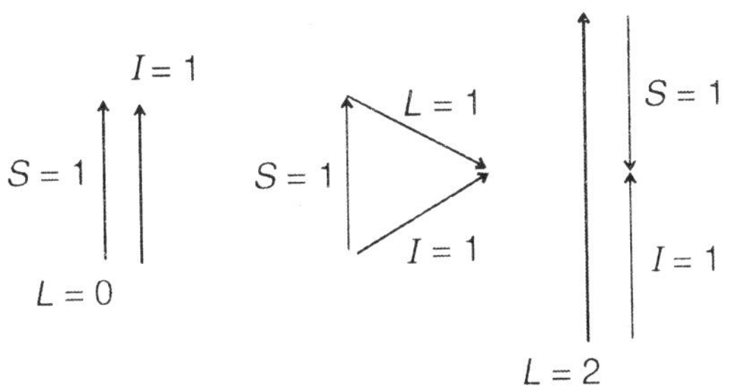

(c)

Fig. 2.4: Effect of tensor force on charge distribution in deuteron in the ground state

The orbital motion contributes towards the magnetic moment which must be combined with the intrinsic magnetic moment of neutron and proton. This accounts for the difference between the observed μ_d and $(\mu_n + \mu_d)$.

Thus one finds that besides a predominant s-state $(l = 0)$, there has to be another state $(l \neq 0)$ with $I = 1$ when the deuteron stays in the ground state. When the wave functions of such states are to be combined the resultant wave function must have definite parity and $I = 1$. It is shown (Section 2.1) that states with $(1, S)$ values (0, 1) and (2, 1) each have even parity, and on combination can give $I = 1$, while $l = 1$ and $S = 0$ and 1 can also be combined vectorially to give $I = 1$, but these P-states ($^1P_1$ and $^3P_1$) have odd parity (Fig. 2.4 c). Conservation of parity requires that only states of same parity can co-exist as such the wave-function could be a linear combination of same parity wave functions. The ground state of deuteron due to tensor force is therefore not a pure $^3S_1$ state but a mixture of $^3S_1$ and $^3D_1$ states. The non-central force giving rise to the mixture of states ($^3S_1 +$ $^3D_1$) explain satisfactorily the magnetic moment and the quadrupole moment of the deuteron.

2.8 MAGNETIC MOMENT OF DEUTERON

The magnetic moment of deuteron is due to the orbital motion of the proton having positive electric charge and due to the intrinsic spin of each nucleon. The orbital motion of neutron, being electrically neutral, does not contribute towards the magnetic moment.

The magnetic moment μ_l arising from orbital motion in units of nuclear magneton μ_N

$$\mu_l = g_l l \qquad \text{...(2.8.1)}$$

where g_l is gyromagnetic ratio and l is orbital angular momentum quantam number of a nucleon,

$$g_l = 1 \text{ for proton} \qquad \text{...(2.8.2)}$$

$$= 0 \text{ for neutron}$$

Therefore,

$$\mu_l(p) = l_p \qquad \text{...(2.8.3)}$$

Similarly from spin motion of the nucleons, the contribution towards magnetic moment is

$$\vec{\mu}_s = g_s\vec{s}_p + g_n\vec{s}_n \quad ...(2.8.4)$$

$\vec{s}_p$ and $\vec{s}_n$ are spin angular momenta of proton and neutron respectively in units of $\hbar$. That is

$$s_p = s_n = \frac{1}{2}$$

$$g_p = 2\mu_p = 2 \times 2.7928 = 5.5856\ \mu_N$$

$$g_n = 2\mu_n = -2 \times 1.9132 = -3.8264\ \mu_N$$

The magnetic moment of deuteron is, therefore, given by

$$\mu_d = g_p\vec{s}_p + g_n\vec{s}_n + \vec{l}_p \quad ...(2.8.5)$$

The masses of the nucleons are nearly equal, it may be assumed that each one carries one half of the orbital angular momentum associated with the relative motion between them. If L is the orbital angular momentum of deuteron, then

$$\vec{\mu}_d = g_p\vec{s}_p + g_n\vec{s}_n + \frac{1}{2}\vec{L}$$

$$= \frac{1}{2}(g_p + g_n)(\vec{s}_p + \vec{s}_n) + \frac{1}{2}(g_p + q_n)(\vec{s}_p - \vec{s}_n) + \frac{1}{2}\vec{L} \quad \left(s_p = s_n = \frac{1}{2}\right) \quad ...(2.8.6)$$

For a triplet state, the eigen value of $(\vec{s}_p - \vec{s}_n)$ is zero. $\vec{s}_p - \vec{s}_n = \frac{1}{2}(\vec{\sigma}_p - \vec{\sigma}_n)$, where σ are Pauli matrices and $\vec{s}_p + \vec{s}_n = \vec{S}$, the total angular momentum.

$$\therefore \quad \mu_d = \frac{1}{2}(g_p + g_n)\vec{S} + \frac{1}{2}\vec{L} \quad ...(2.8.7)$$

The expectation value of the operator $\vec{\mu}_d$ depends on the value of I_z, given by the projection of spin I of the state on z-axis. By convention, one calculates the expectation value of the z-component of the magnetic moment.

Taking the projection of μ_d along z-axis, the expectation value of the magnetic moment is given by (in Dirac notation)

$$\mu_d\langle\mu_z\rangle = \langle I, I_z = I|\mu_z|I, I_z = I\rangle \quad ...(2.8.8)$$

Using the formalism of spherical tensor and Rotation matrix for I, I_z and μ_z, along with C.G. coefficients and Wigner-Eckart theorem it is possible to obtain the expectation value of μ_z which is given by (The details of mathematical calculations can be seen in any book on advanced quantum mechanics).

$$\langle I, I_z|\mu_z|I, I_z\rangle = \frac{I_z}{I(I+1)}\langle I, I_z|\vec{\mu}_d \cdot \vec{I}|I, I_z\rangle \quad ...(2.8.9)$$

$$\therefore \quad \mu_d = \langle \mu_z \rangle = \frac{I_z}{I(I+1)} \left\langle I, I_z \left| \frac{1}{2}(g_p + g_n)\vec{S}, \vec{I} + \frac{1}{2}\vec{L}, \vec{I} \right| I, I_z \right\rangle \quad \ldots(2.8.10)$$

Now $\vec{I} = \vec{L} + \vec{S}$ or $I^2 = L^2 + S^2 + 2\vec{L} \cdot \vec{S}$

$$\therefore \quad L \cdot S = \frac{1}{2}(I^2 - L^2 - S^2) \quad \ldots(2.8.11)$$

also $\qquad \vec{I} = \vec{L} + \vec{S}$

$$\therefore \quad \vec{S} \cdot \vec{I} = \vec{S} \cdot \vec{L} + \vec{S} \cdot \vec{S} = S^2 + \frac{1}{2}(I^2 - L^2 - S^2)$$

$$= \frac{1}{2}(I^2 - L^2 + S^2) \quad \ldots(2.8.12)$$

and, similarly

$$\vec{L} \cdot \vec{I} = L^2 + \vec{L} \cdot \vec{S} = L^2 + \frac{1}{2}(I^2 - L^2 - S^2)$$

$$= \frac{1}{2}(I^2 + L^2 - S^2) \quad \ldots(2.8.13)$$

Using equation (2.8.12) and (2.8.13) in equation (2.8.10)

$$\mu_d = \langle u_z \rangle = \frac{I_z}{I(I+1)} \left[\left\langle I, I_z \left| \frac{1}{2}(g_p + g_n), \frac{1}{2}(I^2 - L^2 + S^2) \right| I, I_z \right\rangle \right.$$

$$\left. + \left\langle I, I_z \left| \frac{1}{2} \cdot \frac{1}{2}(I^2 + L^2 - S^2) \right| I, I_z \right\rangle \right] \quad \ldots(2.8.14)$$

and the value of the magnetic moment in a state *I*, *L*, *S* and $I = I_z$ reduces to

$$\mu_d = \frac{1}{4(I+1)} \left[(g_p + g_n)\{I(I+1) - 1(L+1) + S(S+1)\} + \{I(I+1) + L(L+1) - S(S+1)\} \right] \quad \ldots(2.8.15)$$

For a triplet *s*-state, ${}^3S_1$-state

$L = 0$, $I = 1$, and $S = 1$ we have,

$$\mu_d({}^3S_1) = \frac{1}{2}(g_p + g_n) = \mu_p + \mu_n = 0.8796\ \mu_N$$

For a ${}^3D_1$-state, $L = 2$, $I = 1$, $S = 1$ the magnetic moment is given as

$$\mu_d({}^3D_1) = \frac{1}{8}\left[(g_p + g_n)(-2) + 6\right] = 0.3102\ \mu_N$$

Since $\mu_d({}^3D_1)$ is smaller than $\mu_d({}^3S_1)$ a suitable mixture of the two can be obtained. Using the experimental value of μ_d and calculated values of $\mu_d(^3S_1)$ and $\mu_d(^3D_1)$, the respective proportion of the mixture could be obtained.

Taking the ground state wave function of deuteron, as

$$\psi = a\psi_S + b\psi_D \quad ...(2.8.16)$$

where ψ_S and ψ_D are wave functions for $^3S_1$ and $^3D_1$ states respectively. a and b are the corresponding wave amplitudes. The normalisation condition being

$$a^2 + b^2 = 1 \quad ...(2.8.17)$$

Since there are no cross-terms for the magnetic moment between the states $^3S_1$ and $^3D_1$. The deuteron magnetic dipole moment is, therefore, given by

$$\mu_d = a^2\mu_d(^3S_1) + b^2\mu_d(^3D_1) \quad ...(2.8.18)$$

$$0.8573 = a^2(0.8796) + b^2(0.3102) \quad ...(2.8.19)$$

where we have used the experimentally known value for μ_d in equation (2.8.18).

From the two simultaneous equations (2.8.17) and (2.8.19), one gets $b^2 = 0.04$ and $a^2 = 0.96$. This suggests that there is a 4% mixture of $^3D_1$-state in the ground state. In other words, the deuteron spends about 4% of its time in $^3D_1$ state.

2.9 QUADRUPOLE MOMENT OF DEUTERON

The quadrupole moment has been defined in Chapter 1, as departure from the spherical charge distribution in a nucleus. It is expressed as

$$Q = (3z^2 - r^2)$$

$$= r^2(3\cos^2\theta - 1)$$

In the centre of mass system, the distance of proton and neutron from the centre of mass is $\frac{r}{2}$ and only the proton contributes to the quadrupole moment, neutron being chargeless.

$$\therefore \quad Q = \frac{r^2}{4}(3\cos^2\theta - 1)$$

$$Q = \frac{r^2}{2}P_{2,0}(\cos\theta) \quad ...(2.9.1)$$

where $P_{2,0}(\cos\theta)$ is Legendre Polynomial.

The expectation value of the quadrupole moment is given by

$$Q_d = \int \psi_d^* Q \psi_d d^3r = \int \psi_d^* \frac{r^2}{2} P_{2,0}(\cos\theta)\psi_d d^3r \quad ...(2.9.2)$$

where the ground state wave function, ψ_d, of the deuteron is regarded as an admixture of $^3S_1$ and $^3D_1$ states, as done in the case of magnetic moment. It is described by equation (2.8.16)

$$\psi_d = a\psi_S + b\psi_D$$

where ψ_S and ψ_D are wave functions of ${}^3S_1$ and ${}^3D_1$ states respectively and a and b are normalisation constants.

$$\therefore \qquad Q_d = \int \frac{r^2}{2} P_{2,\,0}(\cos\theta)\,|\psi_d|^2\, d^3r$$

$$= \int \frac{r^2}{2} P_{2,\,0}(\cos\theta)\,|a\psi_S + b\psi_D|^2 \cdot r\sin\theta d\theta\, d\phi\, dr. \qquad \text{...(2.9.3)}$$

$$Q_d = \int \frac{r^2}{2} P_{2,\,0}(\cos\theta)\left\{|a\psi_S|^2 + |b\psi_D|^2 + 2ab\,\psi_S\psi_D\right\} d^3r \qquad \text{...(2.9.4)}$$

Since the pure ${}^3S_1$-state cannot contribute to quadrupole moment, as such

$$\int \frac{r^2}{2} \cdot P_{2,\,0}(\cos\theta)\, a^2|\psi_S|^2\, d^3r = 0$$

The second term corresponds to the contribution to the quadrupole moment from ${}^3D_1$-state.

Using $$\psi_D = \frac{u_2(r)}{r}\left[Y_{2,\,2}\chi_{1,\,1} - \sqrt{\frac{3}{10}}Y_{2,\,1}\chi_{1,\,0} + \frac{1}{10}Y_{2,\,0}\chi_{1,\,1}\right]$$

where Y's are spherical Harmonics and χ's are spin wave functions.

The quadrupole moment is a function of space part only, it does not depend on spin. Also the spin wave functions drop out because of orthogonality of spin wave function.

Expressing the spherical harmonics in terms of Associated Legendre Polynomial and using orthogonality of the Polynomials the second term on integration gives

$$-\frac{b^2}{20}\int_o^\infty r^2 u_2^2(r)dr \qquad \text{...(2.9.5)}$$

as the contribution from ${}^3D_1$ state.

Similarly, the third term can be written as

$$2ab\int_o^r r_2^2 P_{2,0}(\cos\theta)\psi_S\psi_D d^3r$$

$$= \frac{2ab}{4\pi}\int_0^\infty \frac{1}{2} r^2\, u_o(r)u_2(r)dr \int_o^\pi P_{2,\,0}(\cos\theta)\frac{1}{\sqrt{2}}P_{2,\,0}(\cos\theta)\sin\theta\, d\theta \int_o^{2\pi} d\phi$$

$$= \frac{1}{\sqrt{50}}\int_o^\infty r^2 \cdot u_o(r)\, u_2(r)dr \qquad \text{...(2.9.6)}$$

In order to estimate the magnitude of Q_d, it is essential to know the radial wave functions, which depends on the form and nature of nuclear potential.

Magnetic moment and positive quadrupole moment establishes the fact that the ground state of deuteron is an admixture of predominately ${}^3S_1$-state (about 96%) and a small fraction of ${}^3D_1$ state (about 0.04%).

2.10 EXCITED STATE OF DEUTERON

The depth of the assumed square well potential for the nuclear force between the nucleons inside the deuteron has been estimated from the known value of its binding energy. The ground state with lowest energy corresponds to a state with $l = 0$. Experimentally no excited state for the deuteron has been observed so far.

In order to understand that there cannot be bound states with higher angular momenta ($l \neq 0$), we first consider the general solution (2.3.2) for the Schrodinger equation (2.3.1). The radial equation for any angular momentum is given by

$$\frac{d^2u_l}{dr^2} + \frac{M}{\hbar^2}(E - V)\,u_l - \frac{l(l+1)}{r^2}u_l = 0 \qquad \text{...(2.10.1)}$$

When equation (2.10.1) is compared with equation (2.3.3) one finds that it is equivalent to an S-wave radial equation with the potential

$$V_{\text{eff}} = V(r) + \frac{\hbar^2 l(l+1)}{Mr^2} \qquad \text{...(2.10.2)}$$

The second term on the right hand side of eqn. (2.10.2) arises because of the orbital motion. That is, the motion in the θ-direction gives an effective force in the r-direction, namely the centrifugal force which gives this potential. It is positive, therefore, repulsive and as it increases with l, the binding energy or the total energy of the lowest state for a given l decreases as l increases. This fact is used to show that no bound P-state ($l = 1$) exists for deuteron.

The Schrodinger equation for $l = 1$ in the two regions are obtained by putting $l = 1$ in equation (2.10.1)

$$\frac{d^2u_{1i}}{dr^2} + k^2u_{1i} - \frac{2}{r^2}u_{1i} = 0 \qquad \text{for } r < b \qquad \text{...(2.10.3)}$$

$$\frac{d^2u_{10}}{dr^2} + \gamma^2u_{10} - \frac{2}{r^2}u_{10} = 0 \ \text{ for } r > b \qquad \text{...(2.10.4)}$$

where k^2 and γ^2 are defined by equation (2.3.6) and (2.3.8) respectively. u_{1i} and u_{10} are respectively the wave functions inside and outside the potential well for $l = 1$ state. The solution of these equations in the limit of $E_B \to 0$, can be obtained through the use of Neumann and Bessel equations. While solving these equations in terms of the Bessel and Neumann functions one has to match the wave functions and their derivatives at the boundary b of the potential well. This leads to the condition that kb should be equal to π.

But $$kb = \frac{M}{\hbar^2}(V_o - E_B)^{\frac{1}{2}} \cdot b$$

which for $E_B \to 0$ reduces to $kb = \left(\frac{M}{\hbar^2}V_o\right)^{\frac{1}{2}} b.$

According to the condition of $kb = \pi$, we get

$$\frac{M}{\hbar^2} V_o b^2 = \pi^2$$

$$\therefore \qquad Vo = \frac{\pi^2 \hbar^2}{Mb^2} \qquad \text{...(2.10.5)}$$

Taking $b = 2 \times 10^{-15}$ m, the value of the depth of potential will be around 144 MeV, which is almost four times the value of potential in the ground state. For large l values, the potential needed to produce a bound state, would be even deeper and deeper. Hence no bound state exists for $l > 0$.

Even no excited s-state exists. If we assume that there is a bound first excited state, then the value of $kb = 3\pi/2$, which would give the corresponding value of potential. However, when $kb = 3\pi/2$ then the wave function produces a node inside the potential well at $kb = \pi$ and as described earlier, the wave function for a bound state should not produce any node. Thus for $l = 0$ also there can be no excited state.

Therefore, in deuteron no excited bound state exists. This agrees with the experimental observations.

2.11 NUCLEON-NUCLEON SCATTERING

We have seen that experimentally known value of the binding energy of the deuteron has been useful in establishing the mutual attractive, short range nature of the forces between the proton and the neutron inside the deuteron. The spins of the neutron and proton in deuteron are parallel and this does not give information about the mutual interaction between neutron and proton when their spins are antiparallel. One would also like to know the interaction between two protons and two neutrons. One can learn more about two body unbound systems by shooting one nucleon towards another. For a brief moment the two come within the range of the nuclear force and then separate again. In the process the projectile is scattered from its original path. By studying such scattering we get an idea of the force that caused them to scatter. In practice one allows a beam of nuclear particles, like neutrons or protons, to hit a target which is made up of a suitable material. For the study of scattering of neutrons by protons one allows a beam of neutrons to hit a target of a hydrogeneous material like wax or paraffin. Neutron-neutron interaction can be studied by the scattering of neutrons from a deuterium target. In this experiment, the effect of neutron-proton scattering (due to proton of deuteron) is taken into account suitably.

A simple theoretical description of scattering of nuclear particles is obtained in a co-ordinate system in which the centre of mass of the system (incident particle plus the target particle) is at rest, both before and after collision. Such a system of co-ordinates is known as centre of mass co-ordinate system (CM-system). In this system the total momentum is always zero. Because of this the

CM system is also known as zero-momentum system. The experimental observations on scattering of particles are made in the laboratory where the target particle is initially at rest. The co-ordinate system, where the observations are made, is known as laboratory system, or L-system, It is interesting to consider the collision of two particles in these two systems and transformation of certain relevant physical parameters from one system into another.

L-System

Let a particle of mass m_1 move with an initial velocity u towards a stationary target particle of mass m_2. As the particle m_1 moves towards m_2, the centre of mass acquires a velocity v_{cm} in laboratory system and moves in the direction of the incident particle.

After elastic collision or interaction, the incident particle moves with a velocity u_1 making an angle θ_L with its initial direction and the target particle moves with a velocity u_2 making an angle θ_L' with the initial direction of the incident particle. Fig. 2.5 (a) and 2.5 (b) show the particles before and after collision in Lab-system.

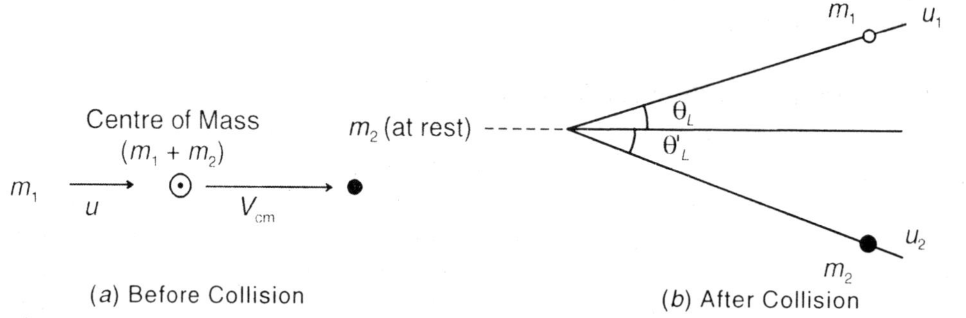

Fig. 2.5: Collision of two particles in lab-system

If v_{cm} is the velocity of centre of mass in the laboratory system then the condition of conservation of momentum gives,

$$(m_1 + m_2)v_{cm} = m_1 u$$

$$v_{cm} = \frac{m_1 u}{(m_1 + m_2)} \quad ...(2.11.1)$$

From the Fig. (2.5 *b*) one can write the momentum conservation conditions

$$m_1 u = m_1 u_1 \cos\theta_L + m_2 u_2 \cos\theta'_L \quad ...(i)$$

$$0 = m_1 u_1 \sin\theta_L - m_2 u_2 \sin\theta'_L \quad ...(ii)$$

and the energy conservation condition is (collision being elastic)

$$\frac{1}{2} m_1 u^2 = \frac{1}{2} m_1 u_1^2 + \frac{1}{2} m_2 u_2^2 \quad ...(iii)$$

These three conditions, (i), (ii) and (iii) relate four quantities u_1, u_2, θ_L and θ'_L as such one needs one more equation for further analysis and determination of these quantities. However, if any one is known, the rest can be determined.

Centre of Mass-system (C-system)

In the centre of mass system, the centre of mass of the system is to remain at rest. This can be achieved by providing the target particle a velocity v_2 sufficient enough to bring the centre of mass at rest. The two particles then approach the centre of mass before collision and after interaction recede from the centre of mass in opposite direction that is the centre of mass always lie on the line joining the two particles. The collision process in C.M. system is shown in Fig. 2.6, where θ_c is the angle of scattering of the incident particle.

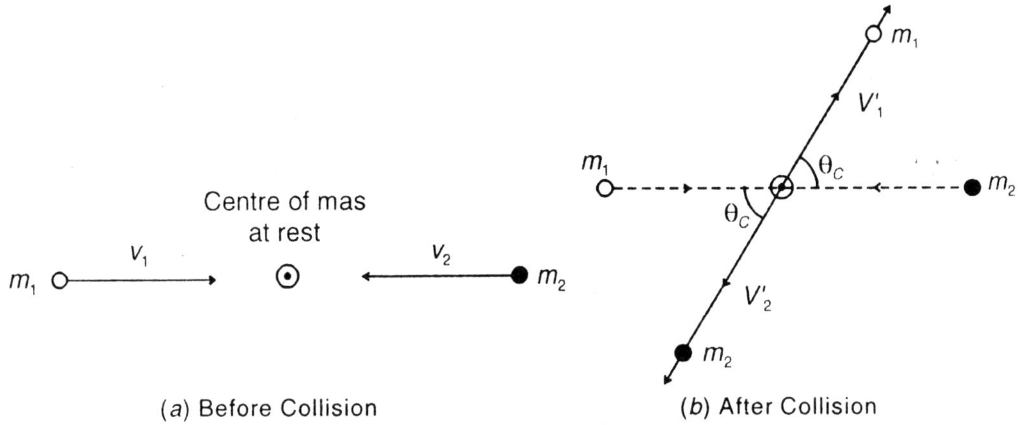

Fig. 2.6: Collision of two particles in centre of mass system

If u is the velocity of incident particle in Lab system and v_{cm} is the velocity of centre of mass in lab system then the velocity of the incident particle in C.M. system is given by,

$$v_1 = u - v_{cm} = u - \frac{m_1 u}{m_1 + m_2}$$

$$v_1 = \frac{m_2 u}{m_1 + m_2} \qquad ...(2.11.2)$$

and the velocity provided to m_2 so that centre of mass comes to rest, is equal to the negative of the velocity v_{cm} of the centre of mass in Lab-system,

$$\therefore \qquad v_2 = -\frac{m_1 u_1}{m_1 + m_2} \qquad ...(2.11.3)$$

For elastic scattering, the particles are each scattered through the same angle θ_c. The momentum of each particle before collision is,

$$P_1 = m_1 v_1 = \frac{m_1 m_2 u}{m_1 + m_2} \qquad ...(2.11.4)$$

and

$$P_2 = m_2 v_2 = -\frac{m_2 m_1 u}{m_1 + m_2}$$

If μ is the reduced mass of the particles i.e.

$$\mu = \frac{m_1 m_2}{m_1 + m_2}$$

then

$$P_1 = \mu u \text{ and } P_2 = -\mu u$$

The two particles have equal and opposite momentum before collision, such that the total momentum is zero

$$P_1 + P_2 = \mu u - \mu u = 0.$$

The conservation of momentum requires that the total momentum of the two particles, after collision must remain zero. If P_1' and P_2' are the momenta after collision, then

$$P_1' + P_2' = 0$$

The conservation of energy gives

$$E_1 + E_2 = E_1' + E_2'$$

where E_1' and E_2' are the kinetic energies, after scattering but

$$E_1 = \frac{p_1^2}{2m_1}, \; E_2 = \frac{p_2^2}{2m} \quad E_1' = \frac{p_1'^2}{2m_1}, \; E_2' = \frac{p_2'^2}{2m_2}$$

$$\therefore \quad \frac{p_1^2}{2m_1} + \frac{p_2^2}{2m_2} = \frac{p_1'^2}{2m_1} + \frac{p_2'^2}{2m_2}$$

But $\quad P_1 = -P_2$

and $\quad P_1' = -P_2'$

$$\frac{P_1^2}{2m_1} + \frac{P_1^2}{2m_2} = \frac{P_1'^2}{2m_1} + \frac{P_1'^2}{2m_2}$$

or

$$\frac{P_1^2}{2\mu} = \frac{P_1'^2}{2\mu} \qquad ...(2.11.5)$$

$$\therefore \quad P_1 = P_1' \qquad \text{or } v_1 = v'_1 \qquad ...(2.11.6)$$

Likewise $\quad P_2 = P_2' \qquad \text{or } v_2 = v'_2 \qquad ...(2.11.7)$

In elastic scattering, thus, the velocities of the particles before and after scattering then remain unchanged in the centre of mass system.

Relationship between θ_L and θ_C

The velocity u_1 of particle m_1 after scattering in the Lab-system is obtained by adding vectorially the velocity V_{cm} of centre of mass in Lab-system to v_1' of m_1 in centre of mass-system after scattering. This vector relationship is shown in Fig. 2.7.

$$\vec{u}_1 = \vec{v}_1' + \vec{v}_{cm}$$

From the figure we get

$$u_1 \cos\theta_L = v_1' \cos\theta_c + v_{cm} \quad \text{...(2.11.8)}$$

$$u_1 \sin\theta_L = v_1' \sin\theta_c \quad \text{...(2.11.9)}$$

Scattering being in the same plane, the azimuthal angle in the two systems remain the same.

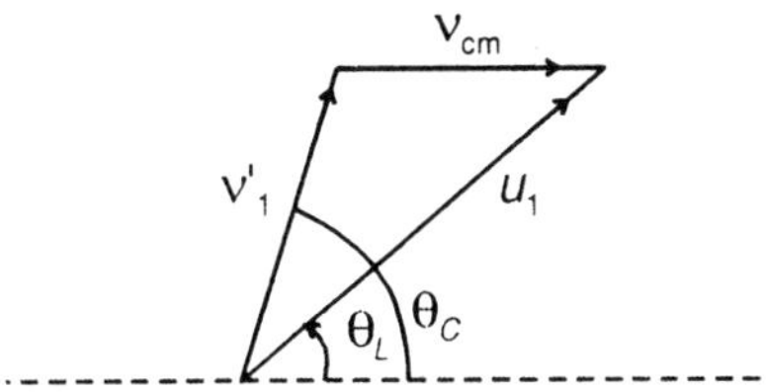

Fig. 2.7: Relationship between velocity vectors in the lab-system and centre of mass system

Hence by dividing, we find

$$\tan\theta_L = \frac{v_1' \sin\theta_c}{v_1' \cos\theta_c + v_{cm}} \quad \text{...(2.11.10)}$$

$$= \frac{\sin\theta_c}{\cos\theta_c + \dfrac{v_{cm}}{v_1'}}$$

Defining $\gamma = \dfrac{v_{cm}}{v_1'}$, $\dfrac{\text{speed of centre of mass in lab system}}{\text{speed of the scattered particle in C.M. system}}$

$$\tan\theta_L = \frac{\sin\theta_c}{\cos\theta_c + \gamma} \quad \text{...(2.11.11)}$$

The equation (2.11.11) gives a relation between the angle θ_L at which m_1 is scattered in the Lab-system as a function of θ_c in the C.M. system.

Since $v_{cm} = \dfrac{m_1 u}{m_1 + m_2}$ and $v_1' = v_1$ from (2.11.6)

But from eqn. (2.11.2)

$$v_1 = \frac{m_2 u}{m_1 + m_2}$$

$$\therefore \quad \frac{v_{cm}}{v_1} = \frac{m_1 u / m_1 + m_2}{m_2 u / m_1 + m_2} = \frac{m_1}{m_2}$$

$$\therefore \quad \gamma = \frac{v_{cm}}{v_1} = \frac{m_1}{m_2} \qquad ...(2.11.12)$$

Special Cases for Elastic Collisions

1. If the mass m_2 of target particle is very much greater than the mass m_1 of the incident particle, then γ is very small. In equation (2.11.11) γ becomes negligible and $\tan\theta_L \approx \tan\theta_C$ or $\theta_L \approx \theta_C$. This means that in such cases, the angles of scattering in the two systems are nearly equal.
2. if $m_1 > m_2$, then γ is greater than unity, the denominator in equation (2.11.11) is never zero. The maximum value of θ_L will always be less than $\frac{\pi}{2}$.
3. If $m_1 = m_2$, then $\gamma = 1$, the equation (2.11.11) in that case becomes

$$\tan\theta_L = \frac{\sin\theta_C}{1+\cos\theta_C} = \tan\frac{\theta_C}{2}$$

or
$$\theta_L = \frac{\theta_C}{2} \qquad ...(2.11.12)$$

The angle of scattering in Lab-system is just half the corresponding angle in centre of mass system. Furthermore, since the angle of scattering of target particle is $\pi - \theta_C$ in the centre of mass system, as is clear from the Fig. 2.6 (*b*), then the same angle in Lab-system is $(\pi - \theta_C)/2$. Therefore, when the masses of the incident and target particles are equal, the two particles leave the point of impact at right angles to each other as seen in the laboratory system. The kinetic energies of the incident and target particles in the centre of mass system, before scattering, are,

$$E_1 = \frac{1}{2}m_1 v^2 \text{ and } E_2 = \frac{1}{2}m_2 v_2^2$$

Total energy is

$$E_c = \frac{1}{2}m_1 v_1^2 + \frac{1}{2}m_2 v_2^2 \qquad ...(2.11.13)$$

Using equations (2.11.2) and (2.11.3) for v_1 and v_2 we get,

$$E_c = \frac{1}{2}m_1\left(\frac{m_2 u}{m_1 + m_2}\right)^2 + \frac{1}{2}m_2\left(-\frac{m_1 u}{m_1 + m_2}\right)^2$$

$$= \frac{1}{2}\frac{m_1 m_2}{m_1 + m_2}\cdot u^2 = \frac{1}{2}\mu u^2 \qquad ...(2.11.14)$$

But the kinetic energy in the Lab-system is

$$E_L = \frac{1}{2} m_1 u^2$$

$$\therefore \qquad E_c = \frac{m_2}{m_1 + m_2} \cdot E_L \qquad \text{...(2.11.15)}$$

When $m_1 = m_2$, then

$$E_c = \frac{1}{2} E_L \qquad \text{...(2.11.16)}$$

Thus the energy of the colliding system of two equal mass particles in the centre of mass system is half of the corresponding energy in the laboratory system.

The kinetic energy of the centre of mass in Lab-system is,

$$E_{cm} = \frac{1}{2}(m_1 + m_2) v_{cm}^2 = \frac{1}{2} \frac{(m_1 + m_2) m_1^2 u^2}{(m_1 + m_2)^2}$$

$$= \frac{1}{2} m_1 \cdot \frac{m_1 u^2}{(m_1 + m_2)}$$

$$= \frac{m_1 E_L}{(m_1 + m_2)} \qquad \left(E_L = \frac{1}{2} m_1 u^2 \right) \qquad \text{...(2.11.17)}$$

The kinetic energy of the centre of mass (eqn. 2.11.17) is not available for producing any inelastic effect like nuclear excitation of nuclear reaction. The energy available for the scattering is, therefore,

$$E_L - E_{cm} = \frac{1}{2} m_1 u^2 - \frac{1}{2}(m_1 + m_2) \frac{m_1^2 u^2}{(m_1 + m_2)^2} = \frac{1}{2} \frac{m_1 m_2}{(m_1 + m_2)} u^2$$

$$= \frac{1}{2} \mu u^2 \qquad \text{...(2.11.18)}$$

This is same as given by equation (2.11.14), which shows that the behaviour of the initial system in centre of mass system may be described in terms of a particle of mass μ moving with incident velocity u in the laboratory system. The energy of m_1 after scattering can be obtained by considering the vector relationship diagram (Fig. 2.8). The velocity of m_1 in lab-system is, from the Fig. 2.8,

$$u_1 = v_1' + v_{cm}$$

$$u_1^2 - v_1'^2 + v_{cm}^2 + 2v_1' \, v_{cm} \cos \theta_c$$

But $v_1' = v_1$

$\therefore$ $u_1^2 = v_1^2 + v_{cm}^2 + 2v_1 v_{cm} \cos\theta_c$

using $v_{cm} = \dfrac{m_1 u}{(m_1 + m_2)}$ and $v_1 = \dfrac{m_2 u}{(m_1 + m_2)}$ (eqn. 2.11.2)

$$u_1^2 = \frac{m_2^2 u^2}{(m_1 + m_2)^2} + \frac{m_1^2 u^2}{(m_1 + m_2)^2} + 2 \cdot \frac{m_2 u}{(m_1 + m_2)} \cdot \frac{m_1 u}{(m_1 + m_2)} \cos\theta_c$$

$$u_1^2 = u^2 \frac{m_2^2 + m_1^2 + 2m_1 m_2 \cos\theta_c}{(m_1 + m_2)^2}$$

$$\frac{u_1^2}{u^2} = \frac{1 + A^2 + 2A\cos\theta_c}{(1 + A)^2} \qquad \text{...(2.11.19)}$$

where $A = \dfrac{m_2}{m_1}$ ...(2.11.20)

If E_1' is the kinetic energy of m_1 after scattering and E_1 before scattering in lab-system respectively then,

$$\frac{E_1'}{E_1} = \frac{(1 + A^2 + 2A\cos\theta_c)}{(1 + A)^2}$$

or $$E_1' = E_1 \frac{(1 + A^2 + 2A\cos\theta_c)}{(1 + A)^2} \qquad \text{...(2.11.21)}$$

Inelastic Collision

If the scattering is such that some internal energy of the system is converted into kinetic energy of the scattered particles, then the scattering is termed as inelastic scattering or inelastic collision.

The momentum conservation condition in the centre of mass system remains the same, both before and after scattering i.e.

$$P_1 + P_2 = P_1' + P_2' = 0 \qquad \text{...(2.11.22)}$$

The energy conservation condition, now is

$$\frac{p_1^2}{2m_1} + \frac{p_2^2}{2m_2} = \frac{p_1'^2}{2m_1} + \frac{p_2'^2}{2m_2} + Q \qquad \text{...(2.11.23)}$$

Q represents the amount of energy that is lost or gained as a result of impact. In nuclear reactions, Q is of fundamental importance. If Q is positive, the nuclear reaction is said to be exothermic and for Q negative, it is endothermic. In this case also the equation (2.11.8), (2.11.9) and (2.11.11) are valid. γ is still defined as the ratio of speed of the centre of mass in Lab system to the speed of the scattered particle in the centre of mass system. However γ is no longer equal to $\frac{m_1}{m_2}$ but is expressed as

$$\gamma = \frac{m_1}{m_2}\left[1 - \frac{Q}{E}\left(1 + \frac{m_1}{m_2}\right)\right]^{-\frac{1}{2}} \qquad \text{...(2.11.24)}$$

where E is kinetic energy of the incident particle as measured in the laboratory system.

If, however, the emergent particles are different than the initial particles, then assigning masses, say m_3 and m_4 to the outgoing particles after scattering, then $m_1 + m_2 = m_3 + m_4$. Such phenomena is possible in nuclear reactions where the impact is such that outgoing particles are different from the incoming particles. The energy conservation condition (eqn. 2.11.23) now becomes

$$\frac{p_1^2}{2m_1} + \frac{p_2^2}{2m_2} + \frac{p_3^2}{2m_3} + \frac{p_4^2}{2m_4} + Q \qquad \text{...(2.11.25)}$$

Now γ is expressed as

$$\gamma = \left(\frac{m_1 m_3}{m_2 m_4} \cdot \frac{E_o}{E_C + Q}\right)^{1/2} \qquad \text{...(2.11.26)}$$

where E_C is the energy initially associated with the system in the centre of mass

i.e.
$$E_C = \frac{1}{2}\mu u^2 = \frac{1}{2}\frac{m_1 m_2}{(m_1 + m_2)}u^2$$

2.12 SCATTERING CROSS-SECTION

In performing scattering experiments, one of the useful information obtained is the scattering cross-section. The meaning of scattering cross-section can be understood by saying that it gives the probability of scattering of a projectile particle from its original path, on interaction with the target. It also gives how many particles are scattered from the initial beam of projectile particles in passing through a target of given material. However, the angular distribution of particles scattered by a target particle is determined by differential scattering cross-section.

If a monoenergetic parallel beam of projectile/incident particles, having flux n impinge on N scattering centres, then the number of particles that are scattered in unit time in an element of solid angle $d\Omega$ along a direction, θ with respect to the incident direction, is proportional to n, N and $d\Omega$. Here it is assumed that (*i*) the incident flux n is small so that there is no interaction between the incident particles. (*ii*) There is no appreciable decrease in the number of target particles as a result of recoil out of the target (*iii*) the target particles are far enough so that only one scattering centre takes parts in scattering.

If $N_s(\theta)$ is the number of particles scattered in a solid angle $d\Omega$ per sec along a direction θ, then

$$N_s(\theta) \propto N n d\Omega$$

$$= \sigma(\theta) N n d\Omega \quad ...(2.12.1)$$

The proportionality factor, $\sigma(\theta)$ is called differential scattering cross-section.

$$\therefore \quad \sigma(\theta) = \frac{N_s(\theta)}{N n d\Omega} \quad ...(2.12.2)$$

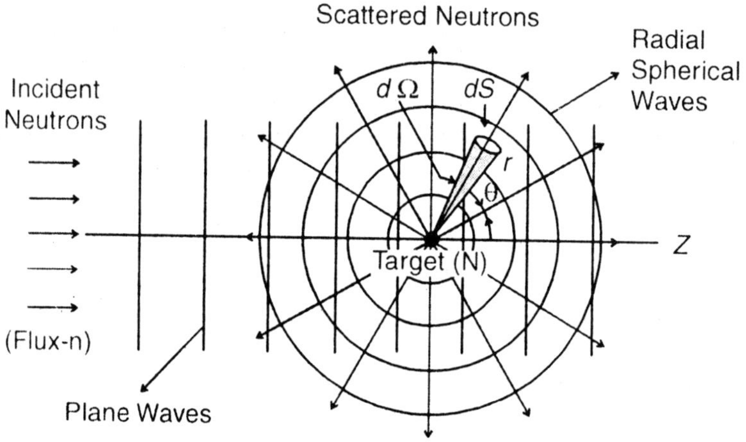

Fig. 2.8: Scattering of a parallel beam of incident particles (neutrons) by a target containing *N*-particles

The differential scattering cross-section is thus defined as the ratio of number of particles scattered by a single target particle into an element of solid angle $d\Omega$ along a direction making an angle θ with the incident direction per second per unit solid angle to the number of incident particles crossing unit area per sec.

Here it is assumed that the scattering is independent of azimuthal angle (ϕ) with respect to the incident direction. For an unpolarised beam of incident particles and spherically symmetric scattering centre (spherically symmetric potential). The scattering is independent of the azimuthal angle.

The number of particles scattered into the solid angle $d\Omega$ per unit incident flux by a target of unit surface density is given by $\sigma(\theta)d\Omega$.

The total scattering cross-section is given by integrating differential scattering cross section over all directions.

$$\sigma = \int \sigma(\theta) d\Omega = \int_o^{2\pi} d\phi \int_o^{\pi} \sigma(\theta) \sin\theta d\theta$$

Since $\sigma(\theta)$ is independent of $d\phi$

$$\sigma = 2\pi \int_o^{\pi} \sigma(\theta) \sin\theta \, d\theta \quad ...(2.12.3)$$

The total scattering cross-section is thus defined by the total number of particles deflected (scattered) in any direction whatsoever. Classically speaking, it represents the cross-sectional area which the target particle presents to the incoming particle.

The total scattering cross-section is expressed in barns where

$$1 \text{ barn} = 10^{-28} \text{ m}^2$$

The differential scattering cross-section is expressed in barns/steradian.

From equation (2.12.1), it is clear that dimensions of $\sigma(\theta)$ are that of an area.

If the fixed scatterer is heavy enough, the centre of mass of the system coincides with the scattering centre and then the centre of mass always remains at rest. The equation (2.12.1) is valid in the laboratory system as well as in centre of mass-co-ordinate system. Consequently $\sigma(\theta)$ given by (2.12.2) remains same in both the coordinate systems. When the target particle is light, having a finite mass then (2.12.2) in general, applies only to laboratory system, because in that case the centre of mass also moves. Since calculations are done with the centre of mass coordinate system while experiments are performed in the laboratory, it is, therefore, required that we should be able to transform the cross-section from one system to other system of coordinates.

Since by definition of the scattering cross-section it is required that the number of particles scattered into an element of solid angle $d\Omega_L$ about θ_L in Laboratory System should be same as scattered into an element of solid angle $d\Omega_c$ about θ_c in centre of mass system.

i.e. $$Nn\sigma_L(\theta_L)d\Omega_L = Nn\sigma_c(\theta_c)\,d\Omega_c \qquad \text{...(2.12.4)}$$

Since scattering is independent of ϕ, $\phi_L = \phi_c$

$$d\phi_L = d\phi_c$$

$\therefore$ $$Nn\sigma_L(\theta_L)\cdot 2\pi \sin\theta_L d\theta_L = Nn\sigma_c(\theta_c)\cdot 2\pi\sin\theta_c\,d\theta_c$$

or $$\sigma_L(\theta_L)\sin\theta_L d\theta_L = \sigma_c(\theta_c)\sin\theta_c d\theta_c$$

$$\sigma_L(\theta_L) = \frac{\sin\theta_c}{\sin\theta_L}\frac{d\theta_c}{d\theta_L}\sigma_c(\theta_c) \qquad \text{...(2.12.5)}$$

From equation (2.11.11)

$$\frac{d\theta_c}{d\theta_L} = \frac{(\gamma+\cos\theta_c)^2}{(1+\gamma\cos\theta_c)}\frac{1}{\cos^2\theta_L} \qquad \text{...}(i)$$

and

$$\sin\theta_L = \frac{\sin\theta_c}{(1+\gamma^2+2\gamma\cos\theta_c)^{\frac{1}{2}}} \qquad \text{...}(ii)$$

$$\cos^2\theta_L = \frac{(\gamma+\cos\theta_c)^2}{1+\gamma^2+2\gamma\cos\theta_c} \qquad \text{...}(iii)$$

Using (*i*), (*ii*) and (*iii*) in (2.12.5), we get,

$$\sigma_L(\theta_L) = \frac{(1+\gamma^2+2\gamma\cos\theta_c)^{3/2}}{1+\gamma\cos\theta_c}\sigma_c(\theta_c) \qquad ...(2.12.6)$$

where $\gamma = \frac{m_1}{m_2}$, for elastic scattering and for inelastic scattering, γ is given by (2.11.24) and by (2.11.26) when the masses m_1 and m_2 are transformed into m_3 and m_4.

It should be noted that the total scattering cross-section, σ, remains same in both the systems of coordinates.

If $m_1 = m_2$ and $\gamma = 1$, then (2.12.6) reduces to

$$\sigma_L(\theta_L) = 4\cos\theta_L \cdot \sigma_c(\theta_c) \qquad (\text{since } \theta_c = 2\theta_L) \qquad ...(2.12.7)$$

2.13 NEUTRON PROTON SCATTERING BELOW 10 MeV (Scattering of Neutrons by Free Protons)

In an actual laboratory experiment, a beam of neutrons from a nuclear reactor or some other neutron source is projected towards a target, containing hydrogeneous material having mostly protons. Electrons have practically no effect on incoming neutrons. The target is usually thin enough so that no multiple scattering takes place. The scattered particle is the result of a single event. The energy of the neutrons in the incident beam is kept much higher than the chemical binding energy of the protons in the molecule. The proton, thus, appears relatively free to the incoming neutrons.

To investigate theoretically, the scattering process cannot be entirely regarded in terms of sharp trajectories of nuclear particles involved in the scattering, because of their smallness in size. At energies of interest, the particles have to be treated as waves and the scattering process is the mutual interaction between such waves. Nevertheless, to have certain amount of understanding, the particle picture is retained in the treatment of the interaction process.

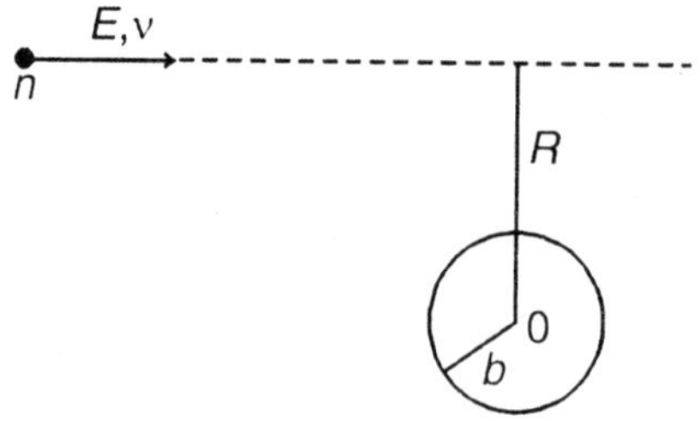

Fig. 2.9: A neutron n moving with velocity v, approaches a scattering centre (proton) 0. The impact parameter and the range of nuclear force are denoted by R and b.

Let us consider a neutron of energy E and velocity v moving towards a proton, which has a sphere of influence around it, by way of range of nuclear forces. Let b be the radius of this sphere of influence or the range of the nuclear forces. In

Fig. (2.9), 0 is the proton or the scattering centre, R is the perpendicular distance between direction of neutron and centre of scattering. It is the impact parameter. If p is linear momentum of neutron, then classically the angular momentum is $p \cdot R$.

Quantum mechanically, this angular momentum should be quantised having only integral values,

$$L = pR = l\hbar \qquad ...(2.13.1)$$

where l takes up values 0, 1, 2, ...

$$l = \frac{p.R}{\hbar} \qquad ...(2.13.2)$$

As is clear from the figure, classically speaking, there will be an interaction between neutron and proton, so long as the impact parameter is less than the range of the nuclear force. Even if $R \sim b$, then only those momentum states will contribute towards scattering which satisfy the condition,

$$l < \frac{bp}{\hbar} \qquad ...(2.13.3)$$

From de-Broglie relation $\lambda = \hbar / p$, where λ is the de-Broglie wavelength, then (2.13.3) becomes,

$$l < \frac{b}{\lambda} \qquad ...(2.13.4)$$

If the energy of the incoming neutrons is such that the de-Broglie waves associated with them have wavelength greater than that of range of nuclear force i.e. if $\lambda \gtrsim b$, then,

$$l < 1$$

It, therefore, follows, that essentially momentum state $l = 0$ will contribute to scattering for those neutrons whose wavelength is of the order of range of nuclear force. For such neutrons, it is said to be s-wave scattering.

Since $$\lambda = \frac{\hbar}{(2mE)^{\frac{1}{2}}}$$

using $\lambda = b$, the energy of the incoming neutrons comes around 13 MeV in laboratory system. Hence for s-wave scattering ($l = 0$) the energy of neutrons should be less than this value. To be on safe side, the neutron energy should be below 10 MeV in lab system. For such waves ($l = 0$), a spherically symmetric potential, can be considered to discuss the n-p scattering quantum mechanically. For such energies (below 10 MeV), the proton in the molecule is effectively free for the neutron collision because the binding energy of a proton in a molecule is less than 10 eV. The zero value for the angular momentum in classical terms means that the distance of closest approach of the two nucleons is zero, as if the particle passes right through the target particle. This is hard to imagine, and hence the wave picture..

In the scattering arrangement, a collimated beam of neutrons travelling with velocity v along the positive z-axis is incident on a target (proton) situated at the origin. The beam can be represented by a plane wave,

$$\psi_{inc} = e^{(ikz - i\omega t)} \quad ...(2.13.5)$$

where the wave number k is the reciprocal of the de-Broglie wavelength for the particle of reduced mass and corresponding energy in the C.M. system. The wave amplitude is taken to be unity so that the wave represents a particle density of one particle per unit volume and thus a flow of υ particles takes place through unit cross-sectional area in unit time. In other words, the incident flux of particles is just υ.

In the event of continuous sequence of interactions the problem becomes that of steady state and it is then possible to do away with the time factor in the incident plane wave given by (2.13.5). The incident plane wave then is,

$$\psi_{in} = e^{ikz} \quad ...(2.13.6)$$

The probability current, as defined quantum mechanically is

$$S(\vec{r},\ t) = \frac{\hbar}{2i\mu}\left[\psi^*\nabla\psi - (\nabla\psi^*)\psi\right] \quad ...(2.13.7)$$

and using equation (2.13.6) for ψ, we get

$$S(\vec{r},\ t) = \frac{\hbar}{2i\mu}\left[e^{-ikz}\frac{d}{dz}e^{ikz} - \left(\frac{d}{dz}e^{-ikz}\right)e^{ikz}\right] \quad ...(2.13.8)$$

$$= \frac{\hbar}{2i\mu}(ik - (-ik)) = \frac{\hbar \cdot 2ik}{2i\mu} = \frac{\hbar k}{\mu} = \frac{\mu\upsilon}{\mu}$$

$$= \upsilon \quad ...(2.13.8)$$

Thus, at far away distances from the interaction centre, the incident flux is υ, where υ is velocity of the particle at these distances.

The scattered particle outside the centre of interaction is described by a spherical wave having axial symmetry radiating outward from the centre of interaction (Fig. 2.8). At large distances from the origin, therefore, the wave function is a linear combination of the incident beam not scattered by the potential and a spherical wave representing scattered particles. This may then be expressed as,

$$\psi(r) = \psi_{in} + \psi_{sc}$$

or

$$\psi(r) \xrightarrow[r\to\infty]{} e^{ikz} + f(\theta)\frac{e^{ikr}}{r} \quad ...(2.13.9)$$

where $f(\theta)$ is the scattering amplitude for elastic scattering. The particle density in the incident beam is usually small and as such no interference takes place between incident and scattered particles. As discussed earlier, the scattering amplitude is independent of azimuthal angle $\phi \cdot f(\theta)$ is a measure of scattered wave in the direction having polar angle θ. Since the amplitude of the scattered spherical wave must fall off as $1/r$ and the radial flux as $1/r^2$, we shall neglect terms higher than $1/r^2$ in estimating the probability current density for the scattered spherical wave.

$$S_{sc} = \frac{\hbar}{2i\mu}\left[\psi^*\nabla\psi - (\nabla\psi^*)\psi\right]$$

$$= \frac{\hbar}{2i\mu}\left[\left(f(\theta)\frac{e^{ikr}}{r}\right)^* \cdot \frac{d}{dr}\left(f(\theta)\frac{e^{ikr}}{r}\right) - \frac{d}{dr}\left(f(\theta)\frac{e^{-ikr}}{r}\right) f(\theta)\frac{e^{ikr}}{r}\right]$$

$$= \frac{\hbar}{2iu}|f(\theta)|^2 \frac{2ik}{r^2}$$

$$= |f(\theta)|^2 \frac{\hbar k}{\mu r^2} = \frac{v}{r^2}|f(\theta)|^2 \qquad \text{...(2.13.10)}$$

The number of particles in the scattered wave which crosses an element of area dA perpendicular to the radius vector in unit time is,

$$S_{sc} \cdot dA = \frac{v}{r^2}|f(\theta)|^2 \cdot dA \qquad \text{...(2.13.11)}$$

If the area dA is at a distance r from the scattering centre than the solid angle substended by the area at the origin is,

$$d\Omega = \frac{dA}{r^2}$$

∴ The scattered particles reaching dA per unit time at large r is,

$$v|f(\theta)|^2 \cdot d\Omega \qquad \text{...(2.13.12)}$$

By definition of differential scattering cross section, $\sigma(\theta)$ is given as the number of particles scattered into a solid angle $d\Omega$ at an angle θ divided by incident flux

$$\sigma(\theta) = \frac{d\sigma}{d\Omega} = \frac{v|f(\theta)|^2}{v} = |f(\theta)|^2 \qquad \text{...(2.13.13)}$$

It may be noted that differential scattering cross-section is also represented by $d\sigma/d\Omega$.

The total cross-section is obtained by integrating eqn. (2.13.13) over all angles

$$\sigma = \int \sigma(\theta)\, d\Omega = \int |f(\theta)|^2 \sin\theta\, d\theta\, d\phi$$

$$= 2\pi \int |f(\theta)|^2 \cdot \sin\theta\, d\theta \qquad \text{...(2.13.14)}$$

Since the angular momentum of the incoming neutron with respect to proton can take up many values, it is, therefore convenient to express the incoming plane wave as a sum of spherical harmonic waves. Each wave then may be associated with a particle having a definite angular momentum about the target proton, with no component along z-direction (direction of motion of the neutron). The incident wave or the wave function of the incident wave is expressed as a sum of partial waves, each defined by the corresponding wave function *i.e.*

$$\psi_{\text{inc}} = \Sigma\psi_l \qquad \text{...(2.13.15)}$$

where ψ_l represents a partial wave for a given angular momentum quantum number l. Such a partial wave method is most applicable for low energies and a spherically symmetric potential of short range. The incident plane wave is (eqn. 2.13.6)

$$\psi_{inc} = e^{ikz} = e^{ikr\cos\theta}$$

It can be expanded in terms of spherical harmonic functions

$$\psi_{\text{inc}} = e^{ikr\cos\theta} = \sum_{l=0}^{\infty}(2l+1)i^l j_l(kr)P_l(\cos\theta) = \Sigma\psi_l \qquad \text{...(2.13.16)}$$

$P_l(\cos\theta)$ is Legendre polynomial and $j_l(kr)$ is Bessel function. Since the particles are detected at large r, far away from the scattering centre, the asymptotic value of the Bessel function being

$$j_l(kr) \xrightarrow[r\to\infty]{} \frac{\sin(kr - l\pi/2)}{kr} \qquad \text{...(2.13.17)}$$

The wave function of the incident wave asymptotically at large r would then be,

$$\psi_{\text{inc}}(r) = \underset{r\to\infty}{e^{ikr\cos\theta}} = \sum_{l=0}^{\infty}(2l+1)i^l P_l(\cos\theta)\frac{\sin(kr - l\pi/2)}{kr} \qquad \text{...(2.13.18)}$$

or $$\psi_{\text{inc}} = e^{ikr\cos\theta} = \frac{1}{kr}\Sigma(2l+1)i^l P_l(\cos\theta)\frac{e^{i(kr-l\pi/2)} - e^{-i(kr-l\pi/2)}}{2i} \qquad \text{...(2.13.19)}$$

The asymptotic form of the incident plane wave, expressed in eqn. (2.13.19), can be regarded as a series of outgoing spherical wave $e^{i(kr-l\pi/2)}$ alongwith a series of incoming waves $e^{-i(kr-l\pi/2)}$ of equal amplitudes. The presence of scattering potential affects the outgoing waves both in amplitude and in phase. The incoming waves remain unaffected. Thus the effect of the potential is to change the outgoing waves. The asymptotic form of the total wave function then can be expressed as

$$\psi = \frac{1}{kr}\Sigma(2l+1)i^l P_l(\cos\theta)\frac{\eta_l e^{i(kr-l\pi/2)} - e^{-i(kr-l\pi/2)}}{2i} \qquad \text{...(2.13.20)}$$

The factor η_l in (2.13.20) represents the effect of interaction potential on the outgoing waves. It is a complex number such that, if η_l is real or $|\eta_l| = 1$, then the phase of the outgoing waves is changed, which corresponds to elastic scattering. When η_l is imaginary or $|\eta_l|<1$, both phase and amplitude are affected, which then corresponds to absorption of particles. Since we are interested only in elastic scattering (no loss of incident particles on scattering). We shall consider $|\eta_l| = 1$ only.

The total wave function given by (2.13.20) can be regarded as superposition of incident and scattered waves,

$$\psi_{r\to\infty} = \psi_{\text{inc}} + \psi_{\text{sc}}$$

where according to (2.13.9)

$$\psi_{\text{sc}} \underset{r\to\infty}{\to} f(\theta)\frac{e^{ikr}}{r}$$

where $f(\theta)$ is scattering amplitude for elastic scattering

Now $$\psi_{\text{sc}} = \psi - \psi_{\text{inc}}$$

$$= \frac{1}{kr}\sum_{l=0}^{\infty}(2l+1)i^l P_l(\cos\theta)\frac{\eta_l e^{i(kr-l\pi/2)} - e^{-i(kr-l\pi/2)}}{2i}$$

$$-\frac{1}{kr}\sum_{l=0}^{\infty}(2l+l)i^1 P_l(\cos\theta)\frac{e^{i(kr-l\pi/2)} - e^{-i(kr-l\pi/2)}}{2i}$$

$$\psi_{\text{sc}} = \frac{1}{kr}\sum_{l=0}^{\infty}(2l+1)i^l P_l(\cos\theta)\frac{e^{i(kr-l\pi/2)}}{2i}(\eta_l - 1)$$

$$= \frac{1}{2ik}\sum_{l=0}^{\infty}(2l+1)P_l(\cos\theta)\frac{e^{ikr}}{r}i^l \cdot e^{-il\pi/2}(\eta_l - 1) \qquad \ldots(2.13.21)$$

Comparing (2.13.21) with (2.13.9) and using $i^l = e^{il\pi/2}$ we get for the scattering amplitude,

$$f(\theta) = \frac{1}{2ik}\sum_{l=0}^{\infty}(\eta_l - 1)(2l+1)P_l(\cos\theta) \qquad \ldots(23.13.22)$$

Since for elastic scattering, there is almost no loss of incident particles, (the loss is negligible as $f(\theta)$ is small)

|Incoming wave| $\simeq$ |outgoing wave|

In (2.13.20) we have,

$$|\eta_l|^2 = 1$$

so that

$$\eta_l = e^{2i\delta_l}$$

where δ_l is a real quantity. This gives,

$$f(\theta) = \frac{1}{2ik}\sum_{l=0}^{\infty}(2l+1)(e^{2i\delta_l} - 1)P_l(\cos\theta)$$

$$f(\theta) = \frac{1}{k}\sum_{l=0}^{\infty}(2l+1)e^{i\delta_l}\left(\frac{e^{i\delta_l} - e^{-i\delta_l}}{2i}\right)P_l(\cos\theta)$$

$$= \frac{1}{k}\Sigma(2l+1)e^{i\delta_l}\sin\delta_l P_l(\cos\theta) \qquad ...(2.13.23)$$

Now using $\eta_l = e^{2i\delta_l}$ in (2.13.20) and expressing in sinusoidal form we get for the total wave function

$$\psi(r)_{r\to\infty} = \frac{1}{kr}\sum_{l=0}^{\infty}(2l+l)i^l P_l(\cos\theta)e^{i\delta_l}\cdot\sin(kr - l\pi/2 + \delta_l) \qquad ...(2.13.24)$$

We find that the last term is $e^{i\delta_l}\sin(kr - l\pi/2 + \delta_l)$ in place of sin $(kr - l\pi/2)$. The effect of scattering potential is to introduce a phase shift. Thus, δ_l has a physical significance of phase shift in the asymptotic form (large r) of the partial l-wave.

The differential scattering cross-section is defined as (eqn. 2.13.13)

$$\sigma(\theta) = \frac{d\sigma}{d\Omega} = |f(\theta)|^2$$

which from eqn. (2.13.23) for $f(\theta)$, becomes,

$$\sigma(\theta) = \frac{d\sigma}{d\Omega} = \frac{1}{k^2}\left|\sum_{l=0}^{\infty}(2l+1)e^{i\delta_l}\sin\delta_l\cdot P_l(\cos\theta)\right|^2 \qquad ...(2.13.25)$$

The total elastic scattering cross section is

$$\sigma = \int\sigma(\theta)d\Omega = \int\sigma(\theta)\cdot\sin\theta d\theta dQ = 2\pi\int\sigma(\theta)\cdot\sin\theta d\theta$$

$$= 2\pi\frac{1}{k^2}\int\left|\sum_{l=0}^{\infty}(2l+1)e^{i\delta_l}\sin\delta_l\cdot P_l(\cos\theta)\right|^2\sin\theta d\theta$$

using $$\int_0^{\pi}|P_l(\cos\theta)|^2\sin\theta d\theta = \frac{2}{(2l+1)}$$

$$\sigma = \frac{4\pi}{k^2}\sum_{l=0}^{\infty}(2l+1)\sin^2\delta_l \qquad ...(2.13.26)$$

For low energy neutron where s-wave scattering is predominate $l = 0$ and the scattering amplitude for s-wave scattering is, putting $l = 0$ in (2.13.23), and $P_o(\cos\theta) = 1$.

$$f(\theta)_{l=0} = \frac{1}{k}e^{2\delta_o}\sin\delta_o \qquad ...(2.13.27)$$

Putting $l = 0$ in equation (2.13.25), we get the differential scattering cross-section

$$\sigma(\theta)_{l=0} = \left(\frac{d\sigma}{d\Omega}\right)_{l=0} = \frac{\sin^2\delta_0}{k^2} \qquad ...(2.13.28)$$

and equation (2.13.26) gives the total scattering cross-section by putting $l = 0$, as

$$\sigma_o = \frac{4\pi}{k^2}\sin^2\delta_o \qquad \text{...(2.13.29)}$$

The equations (2.13.28) and (2.13.29) show that differential scattering cross-section and total scattering cross-section for s-waves, are independent of angle of scattering as is clear from Fig. 2.10. The scattering is isotropic in centre of mass system. The isotropic nature of differential scattering cross-section indicates that equal number of particles is scattered at every angle per unit solid angle.

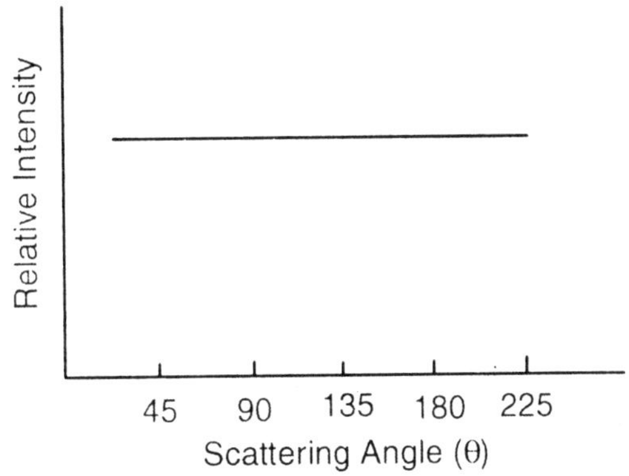

Fig. 2.10: Isotropic scattering of S-waves in the centre of mass system

Phase Shift

In order to know the cross section one must know the phase shift δ_o for s-waves ($l = 0$). We have discussed the deuteron problem quantum mechanically which is a s-state. It is, therefore, possible to adopt the same procedure considering the radial equation in the region of the interaction potential and outside of it. Assuming the potential to be of square well type of depth $-V_o$ and width b (range of nuclear force), but the energy E is now the total energy of unbound system and is therefore positive.

The radial part of the Schrodinger equation for $l = 0$ in the two regions is given below:

(*i*) For $r \leq b$, the radial equation is

$$\frac{d^2u_I}{dr^2} + \frac{M}{\hbar^2}(E + V_o)u_I = 0 \qquad \text{...(2.13.30)}$$

where $u_I = r\psi(r)$, the radial wave function inside the potential well (interaction region). The solution of (2.13.30) is

$$u_I(r) = A\sin Kr \qquad \text{...(2.13.31)}$$

where

$$K^2 = \frac{M(E + V_o)}{\hbar^2} \qquad \text{...(2.13.32)}$$

and A is a constant.

(*ii*) For $r > b$, the potential $V = 0$ and the radial equation is

$$\frac{d^2u_o}{dr^2} + \frac{M}{\hbar^2}Eu_o = 0 \qquad ...(2.13.33)$$

where u_o is wave function outside the potential well. The solution of (2.13.33) is

$$u_o = B\sin(kr + \delta_o) \qquad ...(2.13.34)$$

where B is a constant and δ_o is a constant phase shift as introduced by the scattering potential in the incoming waves as observed at large distances, and

$$k^2 = \frac{ME}{\hbar^2} \qquad ...(2.13.35)$$

The condition of continuity for the wave function in the two regions requires that the quantity $\frac{1}{u}\frac{du}{dr}$ of both u_I and u_o must be equal at the boundary of the potential $r = b$. This gives, from the equation (2.13.31) and (2.13.34)

$$K\cot Kb = k\cot(kb + \delta_o) \qquad ...(2.13.36)$$

For low energy neutrons, the wave function of a free neutron can be taken to be same as that for neutron inside the deuteron. Thus we can approximate the quantity $\frac{1}{u}\frac{du}{dr}$ for the region $r < b$ by the corresponding quantity for the deuteron. In the case of deuteron

$$K\cot Kb = -\gamma \qquad ...(2.13.37)$$

where $\gamma = \frac{1}{\hbar}\sqrt{ME_B}$, E_B being the binding energy of deuteron.

∴ The equation (2.13.36) becomes

$$k\cot(kb + \delta_o) = -\gamma$$

or $$\cot(kb + \delta_o) = -\gamma/k$$

$$kb + \delta_o = \cot^{-1}(-\gamma/k)$$

or $$\delta_o = \cot^{-1}(-\gamma/k) - kb \qquad ...(2.13.38)$$

Since for low energies (of the order of few electron volts) of the neutron $kb << 1$, we can write,

$$\delta_o \simeq \cot^{-1}(-\gamma/k)$$

or $$\cot\delta_o = -\gamma/k \qquad ...(2.13.39)$$

and $$1 + \cot^2\delta_o = 1 + \frac{\gamma^2}{k_2} = \frac{k^2 + \gamma^2}{k^2}$$

∴ $$\sin^2\delta_o = \frac{k^2}{k^2 + \gamma^2}$$

The total scattering cross-section (eqn. 2.13.29), therefore, becomes

$$\sigma_o = \frac{4\pi}{k^2}\sin^2\sigma_o = \frac{4\pi}{k^2}\cdot\frac{k^2}{k^2+\gamma^2}$$

$$\therefore \qquad \sigma_o = \frac{4\pi}{k^2+\gamma^2} \qquad \text{...(2.13.40)}$$

Since $$k^2 = \frac{1}{\hbar^2}ME$$

and $$\gamma^2 = \frac{1}{\hbar^2}ME_B$$

Using these expressions for k^2 and γ^2 in equation (2.13.40), the elastic scattering cross-section for $l = 0$ as a function of energy becomes,

$$\sigma_o = \frac{4\pi\hbar^2}{M}\frac{1}{(E+E_B)} \qquad \text{...(2.13.41)}$$

where E is neutron energy in C.M. system and E_B is binding energy of the deuteron. The expression (2.13.41) gives the variation of the total scattering cross-section with energy and Fig. 2.11 shows this variation upto about 10 MeV by dotted curve. The continuous curve corresponds to the experimental values obtained from transmission experiments. The experimental values lie much above the theoretical values for energies comparable to the binding energy of the deuteron. For low energies the calculated value of the cross section is about 5 barns while the experimental value is around 20 barns for the same energy values.

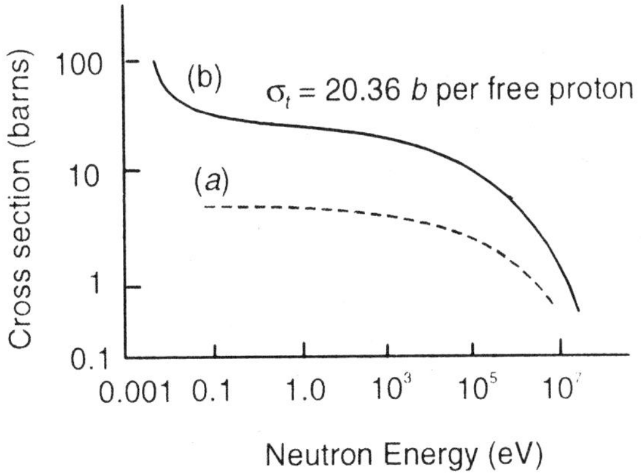

Fig. 2.11: Total scattering cross section as a function of neutron energy curve (a) calculated and (b) experimental

In the calculations, we have used the binding energy and other parameters pertaining to deuteron which is a triplet state. In an actual experiment, the spins of neutrons in a beam are oriented at random and as such the collisions can take where the spins of proton and neutron are, both, parallel and antiparallel. This may be the possible cause of discrepancy occurring between the experimental and theoretically calculated values of total scattering cross-section. In 1935, Winger, as a step to resolve the discrepancy, pointed out that though singlet state is unbound in deuteron, but it should be included in the calculation of scattering cross-sections. In a neutron-proton scattering process, the probability of a triplet collision (spins parallel, $I = 1$) is three times the probability of a singlet collision (spins antiparallel, $I = 0$). This is on account of the fact that the statistical factor or the number of spin states, is given by $(2I + 1)$. This gives three states for $I = 1$ and one state for $I = 0$. These three triplet states and one singlet state are all equally likely to exist. The total scattering cross section for low energies, should then be

$$\sigma = \frac{3}{4}\sigma_t + \frac{1}{4}\sigma_s \qquad \text{...(2.13.42)}$$

where σ_t and σ_s are scattering cross-section for triplet and singlet states respectively and the corresponding weight factors are 3/4 and 1/4 for these states. It is possible to estimate σ_s so as to get an agreement with the experimental value of total scattering cross-section. Using the experimental value of σ to be around 20 barns, we must have,

$$20 = \frac{3}{4}\sigma_t + \frac{1}{4}\sigma_s$$

and for low energy neutrons (0-5 MeV) σ_t is around 5 barns then σ_s should be around 65 barns. This shows that during the interaction the singlet states though exists only for one fourth of the time, the scattering due to it is still quite large.

If the binding energy for the singlet state is assumed to be E_s and $\delta_{o,\,s}$ is the corresponding phase shift for $l = 0$ in singlet state then

$$\sigma_{os} = \frac{4\pi}{k^2}\sin^2\delta_{o,\,s} \qquad \text{...(2.13.43)}$$

and

$$\sigma = \frac{4\pi}{k^2}\left(\frac{3}{4}\sin^2\delta_{o,\,t} + \frac{1}{4}\sin^2\delta_{o,\,s}\right) \qquad \text{...(2.13.44)}$$

The total scattering cross-section as a function of neutron energy (below 10 MeV) will then be given as (using eqn. 2.13.41)

$$\sigma = \frac{\pi\hbar^2}{M}\left(\frac{3}{E + E_B} + \frac{1}{E + |E_s|}\right) \qquad \text{...(2.13.45)}$$

and for $E \to 0$

$$\sigma_{E\to 0} = \frac{\pi\hbar^2}{M}\left(\frac{3}{E_B} + \frac{1}{|E_S|}\right) \qquad \text{...(2.13.46)}$$

Using the measured value of the zero energy neutron-proton scattering cross-section ($\sigma_{E\to 0}$) to be around 20 barns and numerical value of E_B = 2.22 MeV, we can estimate the binding energy E_S for the singlet state. The value of E_S for agreement with known value of σ comes around 60 keV. This establishes the fact that *n-p* force is spin dependent and the well depth should be different for spin parallel and spin antiparallel. It is found that the potential (rectangular) for singlet state is of low depth, around 14 MeV as compared to the large depth, around 30 MeV for the triplet state. However, the neutron-proton scattering cross-section, by itself, does not prove that the binding energy E_S is positive showing that S_O state is unbound. But this may be inferred from the coherent scattering of slow neutrons by hydrogen molecule (discussed later)..

Scattering Length

For scattering of very low energy neutrons by free protons, the wavelength in the asymptotic region, beyond the scatterer is very large ($\lambda \to \infty$ and $k \to 0$) as such the sine curve representing the radial wave function (eqn. 2.13.34) degenerates into a straight line i.e.

$$u_o = B\sin(kr + \delta_o) \approx B(kr + \delta_o) \qquad \ldots(2.13.47)$$

At low enough energies of neutrons, one finds from equation (2.13.27) that as $k \to 0$, δ_o must also approach zero so that $f(\theta)$ remains finite. In the limit of low energy and small δ_o, the scattering factor becomes,

$$f(\theta) \to \frac{e^{i\delta_o}}{k}\sin\delta_o \to \frac{\delta_o}{k} \qquad \ldots(2.13.48)$$

The equation (2.13.47) is an equation of a straight line. We find that the joining of the wave function inside the potential well with the outer wave function (straight line) leads to a negative slope. The straight line outside the range of the potential intersects the distance axis at a certain distance from the origin. This distance is called as scattering length and is represented by 'a_t'. If the potential is such that the slope is positive at the boundary of the potential well, the intersection of the straight line is on the negative slide of the *r*-axis. This scattering length is denoted by 'a_s'. These two cases are shown in Fig. 2.12.

Since equation (2.13.47) gives the wave function for a bound state, the scattering length a_t is for bound state which is a triplet state (hence the index '*t*'). a_s, represents scattering length for an unbound state which is a singlet state.

Now, for $r = a$, $u \to 0$

and then $(ka + \delta_o) = 0$

or
$$a = -\frac{\delta_o}{k}$$

or $$\frac{\delta_o}{k} = -a$$

The total scattering cross-section for low energy neutrons ($k \to 0$) is given as

$$\sigma = \frac{4\pi}{k^2}\sin^2\delta_o \approx \frac{4\pi\delta_0^2}{k^2} = 4\pi a^2 \qquad \ldots(2.13.49)$$

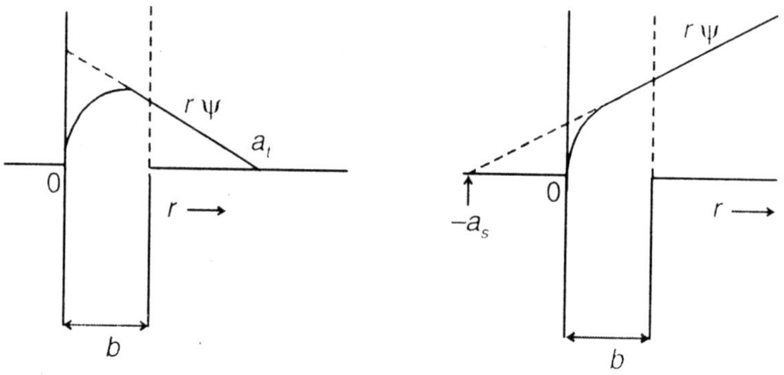

(i) 'a_t' For a Bound State (ii) 'a_s' For an Unbound State

Fig. 2.12: Wave function and scattering length

Thus the scattering length 'a' has geometrical significance of the radius of a hard sphere from which a point neutron (of almost zero energy) is scattered. The scattering cross-section, thus gives an information about the magnitude of the scattering length but not its sign. In accordance with (2.13.49), if a_t is scattering length for triplet state and a_s is scattering length for singlet state, then the total scattering cross-section will be

$$\sigma_{\text{total}} = \frac{3}{4}\sigma_t + \frac{1}{4}\sigma_s$$

$$= \frac{3}{4}\cdot 4\pi a_t^2 + \frac{1}{4}\cdot 4\pi a_s^2$$

$$= 4\pi\left(\frac{3}{4}a_t^2 + \frac{1}{4}a_s^2\right) \qquad \ldots(2.13.50)$$

Thus

$$a_t = \lim_{k\to 0}\left(-\frac{\delta_o}{k}\right)$$

$$a_s = \lim_{k\to 0}\frac{\delta_o}{k} \qquad \ldots(2.13.51)$$

Since $$\sigma = \frac{4\pi}{k^2}\sin^2\delta_o$$

or $$\sigma = \frac{4\pi}{k^2(1+\cot^2\delta_o)} = \frac{4\pi}{k^2 + k^2\cot^2\delta_o}$$

As k^2 decreases, k also decreases which means both σ and $\sigma(\theta)$ increases and in the limit of $k^2 \to 0$, $(k \to 0)$, both will tend to be infinite, meaning thereby that the number of particles scattered become infinitely large which is physically not possible. Even as $k^2 \to 0$ the number of scattered particles has to be finite. To achieve this, we put

$$k\cot\delta_o = -\frac{1}{a_k}$$

so that $$\sigma = \frac{4\pi}{k^2 + 1/a_k^2}$$

where a_k is scattering length for energy E. In order that σ should remain finite as $k \to 0$ we define

$$\lim_{k\to 0} k\cot\delta_o = \lim_{k\to 0}\left(-\frac{1}{a_k}\right) = -\frac{1}{a}$$

where 'a' is finite and is known as scattering length. In the limit of zero energy, the scattering cross-section is

$$\sigma = 4\pi a^2$$

2.14 SCATTERING OF NEUTRONS BY ORTHO AND PARA HYDROGEN

We have seen that the scattering cross-section (Eqn. 2.13.49) gives information about the scattering length. It is positive for a bound state (tiplet state) and negative for an unbound state (singlet state) formed by neutron-proton system respectively (Fig. 2.12). The singlet state or a virtual state can be physically pictured as if the n–p forces are not strong enough to bend the incoming wave past 90°. The presence of negative scattering length is confirmed from the scattering of low energy neutrons (cold neutrons at about 20 K) by hydrogen molecule. Such studies also confirmed that nuclear interactions are spin dependent.

So far we have described the scattering of the low energy neutrons by free protons. Such calculations can be extended for protons bound in molecules or crystals. A hydrogen molecule is a homonuclear diatomic molecule. The distance between the two atoms is large enough ($\sim 10^{-10}$ m) compared to the range of nuclear force ($\sim 10^{-15}$ m) as such one can ignore the nuclear interaction between the two protons in the molecule.

A hydrogen molecule can exist in two possible states in its ground state. One in which the spins of the two protons are parallel ($S = 1$) is known as Ortho hydrogen and the other is known as Para hydrogen, where the spins of the protons are antiparallel ($S = 0$). The statistical weight for ortho hydrogen is 3

$(2S + 1 = 3)$ and for para hydrogen it is 1 $(2S + 1 = 1)$. At normal temperatures, say around room temperatures, the ordinary hydrogen gas consists of an equilibrium mixture of ortho and para hydrogen in the ratio of 3 : 1. Though the energy state for para hydrogen is lower than that for ortho hydrogen, the transition from ortho to para state is very slow. At low temperatures, $(T \sim 20$ K), however, most of the ortho hydrogen molecules go over to para state and one gets almost pure para hydrogen molecules.

In a hydrogen molecule, the scattering of low energy neutrons is coherent and interference effects could be observed in the scattered beam because of large wavelength of incident neutrons compared to the distance between the two protons. The process of neutron scattering is analogous to the diffraction of light from a double slit where the width is much smaller than the wave length of light.

The incident neutron having arbitrary spin direction looks ortho hydrogen molecule and para hydrogen molecule differently. In ortho hydrogen, the *n-p* system gives rise to either triplet scattering. For para hydrogen on the other hand, if there is triplet scattering for one proton other there will be singlet scattering for the other proton. These situations for scattering is shown in Fig. 2.15.

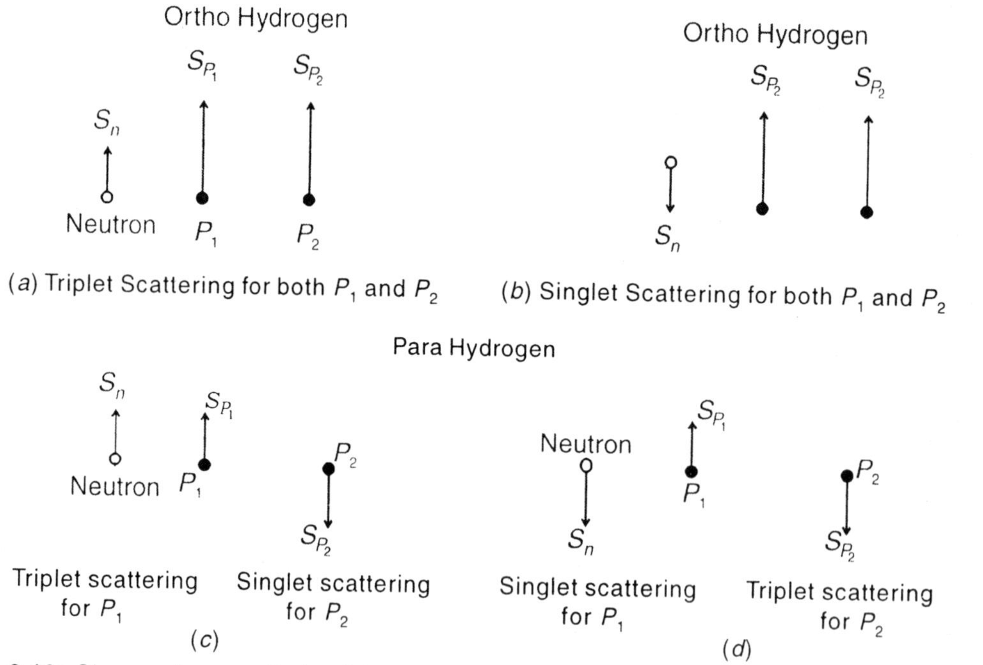

Fig. 2.13: Slow neutron scattering (*a*) Triplet scattering from both the protons (p_1, p_2) in ortho hydrogen (*b*) Singlet scattering from both the protons (p_1, p_2) in ortho hydrogen. (*c*) Triplet scattering from proton p_1 and singlet from proton p_2 in para hydrogen (*d*) Singlet scattering from before p_1 and triplet scattering from proton p_2 in para hydrogen.

The experiments on scattering of slow neutrons by pure para hydrogen and also by an equilibrium mixture of ortho and para hydrogen in the ratio of 3 : 1

would give scattering cross-sections for ortho and para hydrogen, from which the corresponding scattering lengths can be calculated.

High energy neutrons obtained, either from atomic reactors or by bombarding beryllium target with deuterons from a cyclotron are slowed down in paraffin cooled to liquid air temperature. The hydrogen, used as target for scattering experiment is enclosed in a tube cooled to liquid hydrogen temperature (at a pressure of about 60 cm of Hg). The neutrons attain energies in the range of 10 K to 30 K. The de-Broglie wavelength of a neutron at 20 K is about 7 Å which is nearly ten times the separation (0.78 Å) between the protons in a hydrogen molecule. The scattering of such neutrons by the two protons is therefore, coherent. The scattering cross-sections for ortho and para hydrogen are obtained, from which the scattering lengths for triplet and singlet state, respectively, are deduced.

The scattering length for scattering of a low energy neutron from the two protons in hydrogen molecule can be expressed as,

$$A = a_1 + a_2 \qquad \text{...(2.14.1)}$$

where a_1 and a_2 are scattering lengths for scattering from proton 1 and proton 2 respectively. The scattering from the two protons being coherent, we have added the scattering lengths. From Fig. 2.13, it is clear that a_1 and a_2 will be a_{t1} and a_{t2} in ortho hydrogen when neutron spin s_n is parallel and when it is antiparallel, they will a_{s_1} and a_{s_2}. While for para hydrogen, these will be a_{t_1} and a_{s_2}, because, if s_n is parallel with s_{p_1}, it will be antiparallel to s_{p_2}. If s_n is antiparallel with s_{p_1} and parallel to s_{p_2}, then a_1 and a_2 will be a_{s_1} and a_{t_2} respectively.

The triplet and singlet scattering lengths can be combined in a single expression in terms of projection operation as

$$a = a_t p_t + a_s p_s \qquad \text{...(2.14.2)}$$

where $$p_t = \frac{3}{4} + \vec{s}_n \cdot \vec{s}_p \qquad \text{...(2.14.3)}$$

$$p_s = \frac{1}{4} - \vec{s}_n \cdot \vec{s}_p \qquad \text{...(2.14.4)}$$

For a n-p system, the total spin is $\vec{s} = \vec{s}_n + \vec{s}_p$ (Algebraic sum).

Now $$S^2 = s_n^2 + s_p^2 + 2\vec{s}_n \cdot \vec{s}_p \qquad \text{...(2.14.5)}$$

Since $S = 1$ for tiplet and $S = 0$ for singlet state. The eigen-values or the expectation values of S^2 are

$$\langle S^2 \rangle = S(S+1) = 2 \text{ for triplet}$$

$$= 0 \text{ for singlet}$$

Similarly,

$$\langle s_n^2 \rangle = \langle s_p^2 \rangle = \frac{1}{2}\left(\frac{1}{2}+1\right) = \frac{3}{4}$$

From equation (2.14.5) we get

$$\langle \vec{s}_n \cdot \vec{s}_p \rangle = \frac{1}{2}\{\langle S^2 \rangle - \langle s_n^2 \rangle - \langle s_p^2 \rangle\}$$

$$= \frac{1}{2} S(S+1) - \frac{3}{4}$$

$$= \frac{1}{4} \text{ for triplet}$$

$$= -\frac{3}{4} \text{ for singlet} \qquad ...(2.14.6)$$

The projection operators now become

$$P_t = 1 \text{ for triplet}$$

$$= 0 \text{ for singlet} \qquad ...(2.14.7)$$

$$P_s = 1 \text{ for singlet}$$

$$= 0 \text{ for triplet}$$

With eqn. (2.14.7) the equation (2.14.2) becomes

$$a = \frac{1}{4}(3a_t + a_s) + (a_t - a_s)\vec{s}_n \cdot \vec{s}_p \qquad ...(2.14.8)$$

Now it is possible to write the scattering length a_1 for proton 1 and a_2 for proton 2.

$$a_1 = \frac{1}{4}(3a_t + a_s) + (a_t - a_s)\vec{s}_n \cdot \vec{s}_{p_1}$$

$$a_2 = \frac{1}{4}(3a_t + a_s) + (a_t - a_s)\vec{s}_n \cdot \vec{s}_{p_2}$$

and

$$A = a_1 + a_2 = \frac{1}{2}(3a_t + a_s) + (a_t - a_s)\, \vec{s}_n \cdot (\vec{s}_{p_1} + \vec{s}_{p_2})$$

$$= \frac{3a_t + a_s}{2} + (a_t - a_s)\vec{s}_n \cdot \vec{s}_H \qquad ...(2.14.9)$$

where $\vec{s}_H = \vec{s}_{p_2} + \vec{s}_{p_2}$

For para hydrogen, the spins of the two protons are antiparallel

$$\vec{s}_H = \vec{s}_{p_1} + \vec{s}_{p_2} = 0$$

$$\therefore \qquad A_{\text{para}} = \frac{3a_t + a_s}{2} \qquad ...(2.14.10)$$

The scattering cross-section for para hydrogen is given as,

$$\sigma_{\text{para}} = 4\pi A^2 = \frac{4\pi(3a_t + a_s)^2}{4} = \pi(3a_t + a_s)^2 \qquad ...(2.14.11)$$

The scattering cross-section for ortho hydrogen molecule is

$$\sigma_{\text{ortho}} = 4\pi A^2 = 4\pi\left[\frac{3a_t + a_s}{2} + (a_t - a_s)(\vec{s}_n \cdot \vec{s}_H)\right]^2$$

$$= 4\pi\left\{\left(\frac{3a_t + a_s}{2}\right)^2 + (a_t - a_s)^2(\vec{s}_n \cdot \vec{s}_H)^2 + (3a_t + a_s)(a_t - a_s)(\vec{s}_n \cdot \vec{s}_H)\right\}$$

...(2.14.12)

The incident neutron beam being unpolarised, the neutron spin must be averaged over all possible polarisation of the incident beam. Averaging over all possible orientations of the intrinsic spin of the neutron gives (properties of Pauli matrices)

$$(\vec{s}_n \cdot \vec{s}_H)_{av} = 0$$

Now

$$(s_n . s_H)^2_{av} = \{(s_{nx} + s_{ny} + s_{nz})(s_{Hx} + s_{Hy} + s_{Hz})\}^2_{av}$$

$$= \{s^2_{nx}s^2_{Hx} + s^2_{ny}s^2_{Hy} + s^2_{nz}s^2_{Hz}\} + \text{cross terms}$$

The cross, terms, according to the properties of Pauli matrices each will average to zero. e.g., terms of the type.

$$(s_{nx}\ s_{ny}\ s_{Hx}\ s_{Hy})_{av} = 0$$

and $$(s^2_{nx})_{av} = (s^2_{ny})_{av} = (s^2_{nz})_{av} = \frac{1}{4}$$

$$(s_n . s_H)^2_{av} = \frac{1}{4}(s^2_{Hx} + s^2_{Hy} + s^2_{Hz})$$

$$= \frac{1}{4}(s^2_H)_{av} = \frac{1}{4} s_H\ (s_H + 1)$$

$$= \frac{1}{4} \cdot 2 = \frac{1}{2}$$

...(2.14.13)

(For ortho-hydrogen, $s_H = s_{p_1} + s_{p_2} = 1$)

Using eqn. (2.13.13) and $(s_n \cdot s_H)_{av} = 0$, the equation (2.14.12) becomes

$$\sigma_{\text{ortho}} = 4\pi\left\{\frac{1}{4}(3a_t + a_s)^2 + \frac{1}{2}(a_t - a_s)^2\right\}$$

...(2.14.14)

where $$A^2_{\text{ortho}} = \frac{1}{4}(3a_t + a_s)^2 + \frac{1}{2}(a_t - a_s)^2$$

...(2.14.15)

Using equation (2.14.11) in equation (2.14.14), we get,

$$\sigma_{\text{ortho}} = \sigma_{\text{para}} + 2\pi(a_t - a_s)^2$$

...(2.14.16)

The second term in σ_{ortho} arises because of interference. From the expressions for σ_{ortho} and σ_{para}, we get

$$\frac{\sigma_{ortho}}{\sigma_{para}} = 1 + \frac{2(a_t - a_s)^2}{(3a_t + a_s)^2} \qquad ...(2.14.17)$$

The experimental values of the scattering cross-sections are,

σ_{ortho} = 125 barns and σ_{para}= 4 barns.

The second term on the right hand side of eqn. (2.14.17) will be very small, if $a_s > 0$ and then $\sigma_{ortho} \approx \sigma_{para}$.

Even if $a_s = a_t$, $\sigma_{ortho} = \sigma_{para}$. In either situation, the results obtained are contradictory to experimental values. If $a_s < 0$ i.e. if singlet scattering length is negative, $|3a_t + a_s|$ will be much smaller than $(a_t - a_s)$, since a_t is positive. σ_{ortho} will then be much greater than σ_{para}, as required by experimental values as well. Hence singlet scattering length a_s should be negative. The small experimental value of σ_{para} also indicates that a_t and a_s are of opposite signs. Thus the negative scattering length a_s confirms the Wigner's hypothesis that *n-p* interactions are spin dependent. The singlet state of deuteron is a virtual state (unbound state). The experimentally obtained values of the scattering lengths are

$$a_t = 5.37 \text{ F and}$$

$$a_s = -23.73 \text{ F}$$

With these values of a_t and a_s, the scattering lengths for ortho and para hydrogen molecules will be

$$A_{ortho} = 20.92 \text{ F} \qquad \text{(eqn. 2.14.15)}$$

$$A_{para} = -3.81 \text{ F} \qquad \text{(eqn. 2.14.10)}$$

The calculations of the scattering lengths of para and ortho hydrogen are based on the assumption that the protons are free. In the actual molecule, on the other hand, the protons are bound as such for close comparison with the experimental results, the calculations must be modified, because of the changed reduced mass. For the *n–p* system, the reduced mass when the protons are bound is twice than when the protons are free. Also the thermal motion of the molecules and the presence of certain amount of the ortho component introduces inaccuracy in the experimental results. More accurate experiments for the scattering lengths for singlet are (*i*) the coherent scattering of slow neutrons by hydrogen atoms in crystals and (*ii*) the reflection of slow neutrons by liquid hydrocarbon mirrors. The value of A_{para} obtained from the reflexion method is

$$A_{para} = -3.78 \text{ F}$$

2.15 EFFECTIVE RANGE THEORY FOR *n–p* SCATTERING (Shape Independence of Nuclear Potential)

The description of *s*-wave scattering of low energy neutrons by protons has been based on the assumption that the interaction potential between neutron and proton is square well type. The only justification for this shape has been its mathematical simplicity. It was, however, mentioned in the description that the accuracy of the results was retained even with any other shape of potential. Nevertheless, it has ben found possible to express the energy dependence of scattering cross-section in terms of a parameter, termed as effective range. It has dimensions of length. We now show that the introduction of this parameter does not depend on the shape of the potential between the nucleons.

Let us consider wave equations for neutrons with $E = \frac{\hbar^2 k^2}{M}$ and $E = 0$

$$\frac{d^2u}{dr^2} + \left(k^2 + \frac{MV(r)}{\hbar^2}\right) u = 0 \quad \text{...(2.15.1)}$$

$$\frac{d^2u_o}{dr^2} + \frac{MV(r)}{\hbar^2} u_o = 0 \quad \text{...(2.15.2)}$$

u and u_o are the radial wave functions corresponding to energy E and $E = 0$ and $V(r)$ is the nuclear potential. Multiplying equation (2.15.1) by u_o and (2.15.2) by u and subtracting we get,

$$\left(u_o \frac{d^2u}{dr^2} - u\frac{d^2u_o}{dr^2}\right) + k^2uu_o = 0 \quad \text{...(2.15.3)}$$

or $$\frac{d}{dr}(u_ou' - uu_o') + k^2uu_o = 0$$

or $$\frac{d}{dr}(uu_o' - u_ou') = k^2uu_o \quad \text{...(2.15.4)}$$

where the primes indicate the first derivative.

Integrating equation (2.15.4) from 0 to gives ∞ gives,

$$[uu_o' - u_ou']_o^\infty = k^2\int_o^\infty uu_o\, dr \quad \text{...(2.15.5)}$$

This is true for any arbitrary potential including $V(r) = 0$. Next we introduce comparison functions v and v_o with the equations,

$$\frac{d^2v}{dr^2} + k^2v = 0 \quad \text{...(2.15.6)}$$

$$\frac{d^2v_o}{dr^2} = 0 \quad \text{...(2.15.7)}$$

where $k^2 = \frac{ME}{\hbar^2}$ and $V(r) = 0$. Like u_o the radial wave function v_o corresponds to $E = 0$, but $V(r) = 0$. It is clear that asymptotically, for large distances beyond the range of nuclear potential where $V(r)$ is zero, the wave function u and u_o go over to v and v_o respectively. Proceeding as before, we multiply (2.15.6) by v_o and (2.15.7) by v, subtract and integrate from 0 to ∞. We then get,

$$[vv_o' - v_ov']_o^\infty = k^2\int_o^\infty vv_o dr \quad \text{...(2.15.8)}$$

Now subtracting eqn. (2.15.8) from eqn. (2.15.5) gives,

$$\left[uu_o'-u_ou'-vv_o'+v_ov\right]_o^{\infty} = k^2\int_o^{\infty}(uu_o - vv_o)dr \quad ...(2.15.9)$$

As mentioned above, asymptotically at large distances where $V = 0$, $u(r) = v(r)$ and $u_o(r) = v_o(r)$. At $r = 0$ at the origin, the wave functions must vanish. $u(0) = 0$, $u_o(0) = 0$. With this, the equation (2.15.9) then reduces to

$$(v_ov'-vv_o')_{r=0} = k^2\int_o^{\infty}(uu_o - vv_o)\, dr \quad ...(2.15.10)$$

Since the functions u and v are identical asymptotically at large distances, their form should be same beyond the range of nuclear potential where $V = 0$.

i.e. $$u(r) = v(r) = C\sin(kr + \delta) \quad ...(2.15.11)$$

We choose the normalisation constant C for the sake of convenience such that $v(0) = 1$.

$\therefore$ $$v(0) = C\sin\delta = 1$$

or $$C = 1/\sin\delta$$

and thus

$$u(r)_{r\to\infty} = v(r) = \frac{\sin(kr+\delta)}{\sin\delta} \quad ...(2.15.12)$$

Outside the range of potential well $V(r) = 0$ and the equation (2.15.7) has a straight line solution. Also asymptotically $u_0(r)$ goes over to $v_0(r)$ at $r \to \infty$, since $V(r) = 0$. Hence the asymptotic solution for the function $u_0(r)$ is a straight line. Thus,

$$u_o(r)_{r\to\infty} = v_o(r) = A(r - a) \quad ...(2.15.13)$$

The normalisation constant is such that $v_o(0) = 1$

This gives

$$A = -\frac{1}{a}$$

$$v_o(r) = -\frac{1}{a}(r - a) = 1 - \frac{r}{a} \quad ...(2.15.14)$$

where 'a' is scattering length.

From equation (2.15.12), $v'(r) = \dfrac{k\cos(kr+\delta)}{\sin\delta}$ and at $r = 0$

$$v'(0) = k\cot\delta$$

and from eqn. (2.15.14), $v_o'(o) = -\dfrac{1}{a}$

Using these results in eqn. (2.15.10) we get

$$-\left(1\times k\cot\delta + 1\times\frac{1}{a}\right) = k^2\int_o^{\infty}(uu_o - vv_o)\, dr$$

(The negative sign outside the bracket on the left hand side is because $r = 0$ is at the lower limit of the integration)

$$-\frac{1}{a} - k\cot\delta = k^2\int_0^\infty (uu_o - vv_o)\,dr \qquad ...(2.15,15)$$

On the right hand side of equation (2.15.15) the integrand vanishes at the upper limit of integration, because the functions u and u_o go over to v and v_o at infinity where the potential is zero. These functions however differ inside the region of nuclear potential. Within the range of nuclear potential u and v are almost independent of the energy, because for the energies that we consider, the depth of potential is much larger than the energy of the system. In the zero energy approximation therefore, it is possible to take

$$u \approx u_o \text{ and } v \approx v_o$$

In the limit of zero energy, let us define a quantity r_o as

$$2\int_0^\infty \left(v_o^2 - u_o^2\right) dr = r_o \qquad ...(2.15.16)$$

With this representation, we finally get,

$$-\frac{1}{a} - k\cot \simeq \frac{k^2 r_o}{2}$$

or

$$k\cot\delta \simeq -\frac{1}{a} + \frac{k^2 r_o}{2} \qquad ...(2.15.17)$$

The factor 2 in (2.15.16) is introduced, because r_o is to define the range of the potential i.e. r_o may then be approximately equal to the range of the potential. The quantity r_o is a constant and is called the effective range of nuclear potential, In terms of the effective range the scattering cross-section is given as

$$\sigma = \frac{4\pi}{k^2(1 + \cot^2\delta)} = \frac{4\pi}{k^2 + \left(\frac{1}{a} - \frac{k^2 r_o}{2}\right)^2} \qquad ...(2.15.18)$$

We thus find that for potential of any shape, the scattering cross-section is determined by the two parameters, the scattering length 'a' and the effective range r_o.

2.16 PROTON PROTON SCATTERING AT LOW ENERGIES

The stability of nuclei containing more than one neutron and one proton shows that like strong attractive force between a neutron and a proton, there must be strong short range attractive force between protons within a nucleus. Any information about nuclear force between two protons is possible only through scattering of a proton by a proton, because no bound state of two protons (diproton, ${}_2^3He$) has been observed. The proton-proton scattering experiments can be performed with great ease and accuracy. These days well collimated monoenergetic beams of protons are available from various accelerators. The protons can be easily detected by the ionisation they produce and their energies can be measured easily. However, theoretical analysis of experimental results is difficult and complicated, because protons being electrically charged particles, undergo Coulomb scattering in addition to nuclear scattering. Also allowance have to be made for the indisinguishability of the two particles i.e. one cannot distinguish between the scattering proton and the scattered proton. Pauli's exclusion principle excludes certain states of motion for identical particles, so that there well be either singlet or triplet states but not both the states for a given orbital (1-state) state in *p–p* scattering. Protons obey Fermi statistics and as such they can stay in a state that has antisymmetric wave function. This means that a symmetric space wave function (*S*, *D*-states) can only be associated with an antisymmetric spin wave function and antisymmetric spin wave function (*P*, *F* states) can combine only with symmetric space wave functions (Triplet states). The two protons, therefore, cannot stay in a state that has symmetric wave function. This means that for incident proton energies below 10 MeV the nuclear scattering is s-wave scattering and the interaction is purely singlet-state interaction. There is only one phase shift, unlike two in *n-p* scattering (for singlet and triplet).

The Coulomb potential is a long range type of potential $\left(V \propto \frac{1}{r}\right)$ and so it makes itself felt no matter how distant apart are the colliding particles from each other. Every incident particle is scattered and effectively the total cross-section is infinite. As such for Coulomb potential one calculates differential scattering cross-section as a function of scattering angle.

Let us first examine the scattering of protons under Coulomb field alone. The Schrodinger equation in the relative co-ordinates of the two protons is

$$\nabla^2\psi + \left(k^2 - \frac{2kn}{r}\right)\psi = 0 \quad ...(2.16.1)$$

where, $$k = \frac{\mu v}{\hbar},\ n = \frac{e^2}{\hbar v} \quad ...(2.16.2)$$

μ is reduced mass and is equal to $\frac{M}{2}$ where M is mass of a nucleon (proton). v is relative velocity when the two protons are separated by large distance. The asymptotic solution of equation (2.16.1) is

$$\psi \underset{r\to\infty}{\to} \text{constant}\left[e^{-in\ln 2kr\sin^2\theta/2 + ikr\cos\theta} - \frac{n}{2kr\sin^2\theta/2} e^{ikr} e^{-in\ln 2kr\sin^2\theta/2} e^{2i\eta_o}\right] \quad ...(2.16.3)$$

where, $e^{2i\eta_o} = \frac{\Gamma(1+in)}{\Gamma(1-in)}$, and η_o is a real number.

From the equation (2.16.3) one finds that the Coulomb interaction modifies the phase of both incident and scattered waves. It is, therefore, not possible to obtain the asymptotic form of the scattered wave as a sum of incident plane wave and a scattered radial wave of certain angle dependent amplitude. However, the angle dependent term in the modified scattered wave gives the scattering amplitude $f(\theta)$ as,

$$f_c(\theta) = \frac{n}{2k\sin^2(\theta/2)} e^{-in \ln \sin^2(\theta/2)} \quad ...(2.16.4)$$

and the differential scattering cross-section in the centre of mass system is,

$$\sigma(\theta) = |f_c(\theta)|^2 = \frac{n^2}{4k^2 \sin^4(\theta/2)}$$

$$= \frac{e^2}{4\hbar^2 v^2} \cdot \frac{\hbar^2}{\mu^2 v^2} \operatorname{cosec}^4(\theta/2)$$

$$= \frac{e^4}{4\mu^2 v^4} \operatorname{cosec}^4(\theta/2) \quad ...(2.16.5)$$

This is Rutherford formula for the scattering of a proton ($Z = 1$) by a proton. Rutherford had initially obtained this formula using classical dynamics. The equivalence of classical and quantum mechanical results for the scattering is because of the nature of Coulomb forces.

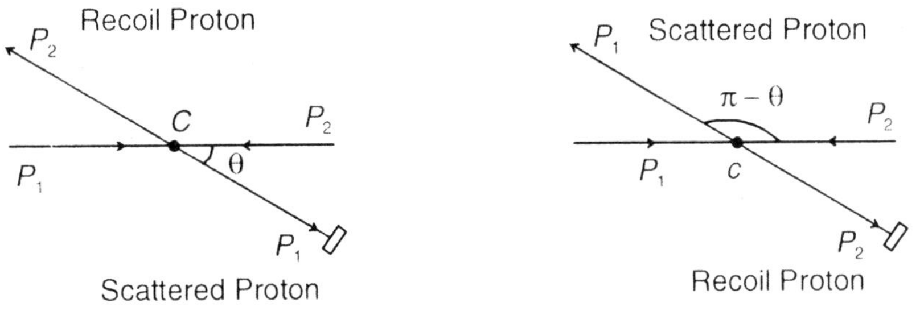

Fig. 2.14: Indistinguishability of scattered proton (p_1) and recoil proton (p_2)

In an actual experiment on proton scattering the protons scattered at an angle θ and those scattered at an angle $(\pi - \theta)$ can not be distinguished from the recoiled protons, particularly when the spin is neglected.

The detector in Fig. 2.14 cannot distinguish a proton scattered at an angle θ and a recoil proton, when the incident proton is scattered at an angle $(\pi - \theta)$. The scattering amplitude for the proton scattered at $(\pi - \theta)$ is given by

$$f_c(\pi - \theta) = \frac{n}{2k\cos^2\left(\frac{\theta}{2}\right)} e^{-in \ln \cos^2\left(\frac{\theta}{2}\right)} \quad ...(2.16.6)$$

The differential scattering cross-section will be

$$\sigma(\pi - \theta) = |f_c(\pi - \theta)|^2 \qquad ...(2.16.7)$$

or

$$\sigma(\pi - \theta) = \frac{e^4}{4\mu^2 v^4} \cos^4\left(\frac{\theta}{2}\right) \qquad ...(2.16.8)$$

According to classical theory the differential scattering cross-section for indistinguishable particles is the sum of $\sigma(\theta)$ and $\sigma(\pi - \theta)$ and therefore, we add (2.16.8) to (2.16.5) to get the differential scattering cross-section for Coulomb scattering of proton at an angle θ in CM system. Thus,

$$\sigma_c(\theta) = \frac{e^4}{4\mu^2 v^4}\left[\frac{1}{\sin^4\theta/2} + \frac{1}{\cos^4\theta/2}\right] \qquad ...(2.16.9)$$

But in quantum theory the indistinguishability of the protons puts another consideration. The wave functions for the protons should be suitably symmetrised. Instead of taking the sum of squares of the scattering amplitudes as done in equation (2.16.5) and (2.16.8). We have to take a linear combination of the scattering amplitudes $f(\theta)$ and $f(\pi - \theta)$. The wave functions are linearly combined such that the space part of the combination is symmetric (spin part antisymmetric) or antisymmetric (spin part symmetric). The total wave function, has to be antisymmetric, since protons obey *FD* statistics. Accordingly, the combination $f(\theta) - f(\pi - \theta)$ is antisymmetric in the triplet state ($^3S_1$) which has the statistical weight $2S + 1 = 3(S = 1)$. The combination $f(\theta) + f(\pi + \theta)$ is symmetric in singlet ($^1S_0$) state having statistical weight $2S + 1 = 1$ (S = 0). Thus the scattered waves at θ and $(\pi - \theta)$ are three times more frequent in triplet scattering than in singlet scattering. Hence the differential scattering cross-section is given by

$$\sigma(\theta) = \frac{1}{4}\{f(\theta) + f(\pi - \theta)\}^2 + \frac{3}{4}\{f(\theta) - f(\pi - \theta)\}^2 \qquad ...(2.16.9)$$

Using the expressions (2.16.4) and (2.16.6) respectively for $f(\theta)$ and $f(\pi - \theta)$, we get for the differential scattering cross-section

$$\sigma(\theta) = \frac{e^4}{4\mu^2 v^4}\left[\frac{1}{\sin^4(\theta/2)} + \frac{1}{\cos^4(\theta/2)} - \frac{\cos n \ln \tan^2(\theta/2)}{\sin^2(\theta/2)\cos^2(\theta/2)}\right] \qquad ...(2.16.10)$$

The equation (2.16.10) is the Mott scattering formula for protons under Coulomb forces. The third term arises from the quantum theoretical treatment of Coulomb scattering for indistinguishable charged particles (protons). For proton energies of interest (above 1 MeV), $n = \frac{e^2}{2\mu v^2}$ is small so that the cosine term in equation (2.16.10) is of the order of unity unless $\theta = 0$ or π. Thus away from $\theta = 0$ or π and for proton energies $E > 1$ MeV the Mott term is simplified and the equation (2.16.10) reduces to

$$\sigma(\theta) = \frac{e^4}{4\mu^2 v^4}\left[\frac{1}{\sin^4(\theta/2)} + \frac{1}{\cos^4(\theta/2)} - \frac{1}{\sin^2(\theta/2)\cos^2(\theta/2)}\right] \qquad ...(2.16.11)$$

When the distance between the protons is of the order of range of nuclear force, nuclear scattering becomes effective. In actual computation of *p-p* scattering, it is reasonable to assume that nuclear potential for *p–p* scattering is same as that for *n–p* scattering. Because of Pauli's exclusion principle, the triplet state is excluded for the identical particles and therefore, the lowest state, 1 = 0 is a singlet state. So at low energies, we have only 1 = 0 or *s*-wave scattering taking place, like in the case of *n–p* scattering. The differential scattering cross-section, when computed, in the CM system, under the effects of Coulomb and nuclear potentials, is given by

$$\sigma(\theta) = \frac{e^4}{4\mu^2 v^4}\left[\frac{1}{\sin^4(\theta/2)} + \frac{1}{\cos^4(\theta/2)} - \frac{1}{\sin^2(\theta/2)\cos^2(\theta/2)}\right]$$

$$+\frac{\sin^2\delta_o}{k^2} - \frac{e^2}{2\mu v^2}\cdot\frac{\sin\delta_o}{k}\left\{\frac{\cos(n.\ln\sin^2(\theta/2)+\delta_o)}{\sin^2(\theta/2)}\right.$$

$$\left.+\frac{1}{\cos^2(\theta/2)}\cdot\cos(n\ln\cos^2(\theta/2)+\delta_o)\right\} \quad \text{...(2.16.12)}$$

Since $n = \frac{e^2}{2\mu v^2}$ is small for energies of the order of 1 MeV and becomes still smaller at higher energies, the equation (2.16.12), under this condition becomes

$$\sigma(\theta) = \frac{e^4}{4\mu^2 v^4}\left[\frac{1}{\sin^4(\theta/2)} + \frac{1}{\cos^4(\theta/2)} - \frac{1}{\sin^2(\theta/2)\cos^2(\theta/2)}\right]$$

$$+\frac{\sin^2\delta_o}{k^2} - \frac{e^2}{2\mu v^2}\cdot\frac{\sin\delta_o\cos\delta_o}{k\sin^2(\theta/2)\cos^2(\theta/2)} \quad \text{...(2.16.13)}$$

In the equation (2.16.13), the first term is Mott scattering (pure Coulomb scattering), the second term is pure nuclear potential scattering (*s*-wave scattering) and the last term corresponds to the interference between the Coulomb and nuclear scattering, which depends linearly on the phase shift δ_o. From the experimentally known variation of $\sigma(\theta)$ with angle θ, it is possible to estimate δ_o and hence the scattering length. It is found that the phase shift is positive, which corresponds to negative scattering length. This negative sign of the scattering length shows that there cannot be any bound state for a *p-p* system. The positive sign for the phase shift also confirms the attractive nature of the nuclear potential. The interference is thus destructive for the *s*-wave and therefore, the magnitude of the cross-section is reduced.

The differential scattering cross-section $\sigma(\theta)$ for 2.4 MeV in CM system is shown in the Fig. 2.15.

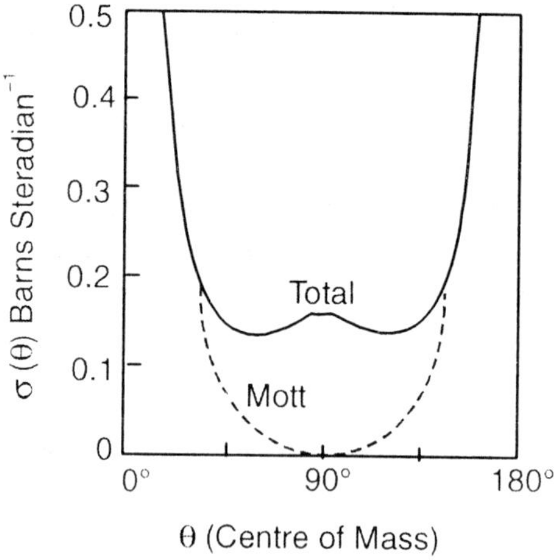

Fig. 2.15: The differential scattering cross-section s(q) in CM system as a function of scattering angle. The dotted line exhibits mott scattering

The dotted line exhibits Mott scattering. At $\theta \approx 90°$ (45° in Lab-system), the scattering is mainly nuclear. Except in the forward ($\theta \sim 0$) and backward directions ($\theta \sim 180°$), the scattering is almost isotropic (spherically symmetric). As stated above, the lowering of the value of $\sigma(\theta)$ is on account of the interference between nuclear and Coulomb functions.

If the nuclear potential between protons is assumed to be rectangular, the range and depth of the potential are found to be 2.58 *f* and 13.3 MeV respectively. These values are consistent with those obtained in the case of *n–p* scattering. The effective range theory for *p–p* scattering yields values for the effective range and scattering length.

$$r_o = (2.65) \times 10^{-15} \text{ m}$$

$$a_p = -7.69 \times 10^{-15} \text{ m}$$

The observed value of scattering length, when corrected for the effect of Coulomb force, becomes, $a_p = -17 \times 10^{-15}$ m. The *n–p* scattering length, for singlet state is -24×10^{-15} m.

The various parameters for *p-p* scattering and *n-p* scattering are given in Table 2.2.

TABLE 2.2

value $\times 10^{-15}$ m

	n-p	*p-p*
Triplet scattering length *a*	+5.38 ± 0.02	–
Singlet scattering length a_s	–23.69 ± 0.05	–17 ± 0.60
Triplet effective range r_{ot}	+1.70 ± 0.03	–
Singlet effective range r_{os}	+2.7 ± 0.5	2.65 ± 0.08

These studies, therefore, result into an evident observation that for the same angular momentum state (spin and orbital angular momentum) the (*n-p*) nuclear force is same as *p-p* nuclear force. This gives rise to the fact that the nuclear force is charge independent. If that is so, then it is reasonable to state that *n-n* interaction should be equal to *p-p* interaction except at very low energies (100 keV) where Coulomb effects are also present.

Neutron-Neutron Scattering

As stated earlier, no di-neutron bound system exists as such the nuclear force between two neutrons is studied indirectly. By bombarding D_2O and H_2O with neutron of 300 MeV, the values for differential scattering cross-sections are

$$\sigma_{nn}(\theta) = 3.7 \text{ mb per steradian}$$

$$\sigma_{pp}(\theta) = 3.6 \text{ mb per steradian}$$

$$\sigma_{nn} = 22 \text{ mb}$$

$$\sigma_{pp} = 23 \text{ mb}$$

Since the total cross-section for n-D_2O scattering system and (n-H_2O) scattering system be expressed as

$$\sigma_n, D_2O = \sigma_{nn} + \sigma_{np} + I$$

$$\sigma_n, H_2O = \sigma_{pp} + \sigma_{np} + I$$

where I is the interaction. Such studies show the equality of (*n-n*) and (*p-p*) interactions. One gets an idea of the equality of n-n and *p-p* interaction from a comparison of binding energies of lightest pair of mirror nuclei. However, in the case of *p-p* interactions, one also has Coulomb part which when omitted from the total interaction, will result in an interaction equal to the *n-n* interaction. The *B.E.* of the mirror nuclei H^3 and He^3 are respectively 8.482 MeV and 7.711 MeV. The difference of these B.E. is 0.77 MeV. The Coulomb energy of two discrete protons is 0.76 MeV$\left(\frac{3}{5}\frac{e^2}{R}\cdot z\,(z-1)\right)$. This, thus, supports the hypothesis that the nucleon-nucleon forces are charge independent *i.e.* the nuclear force between a neutron and a proton is same as between two protons or between two neutrons ($np = nn = pp$). The equality of nuclear force between two neutrons and between two protons shows that the nuclear force is charge symmetric ($nn = pp$).

The charge independence of nuclear forces, has led to entirely a new formalism. The two particles, proton and neutron, are regarded to be two charge states of a single particle, normally termed as nucleon. Heisenberg was the first to associate the idea of isotopic spin or iso-spin with this charge independence of nuclear forces.

This idea is analogous to ordinary spin. Just as a proton can have two spin states ($S_z = \pm \frac{1}{2}$) a nucleon has two states of iso-spin, ($T_3 = \pm \frac{1}{2}$) the $T_3 = \frac{1}{2}$ state representing a proton and $T_3 = -\frac{1}{2}$ the state representing a neutron. Thus a *p-p* system with $T_3 = +1$ is necessarily an iso-triplet state with $T = 1$ while an *n-p* system with $T_3 = 0$ can be an iso-singlet $T = 0$ or an iso-triplet $T = 1$.

Consider a 2-nucleon system with $T = 1$. It has three charge states namely $T_3 = 1$ (p-p system), $T_3 = 0$, (n-p system) and $T_3 = -1$ (n-n-system). If the nuclear forces are identical for the three systems, this means that the nuclear force does not depend on T_3 but only on T. A p-p system has iso-spin $T = 1$ and n-p system can have both iso-spin $T = 0$ and $T = 1$ states. (A detailed presentation of iso-spin formalism has been made in chapter one).

2.17 NON-CENTRAL FORCE (TENSOR FORCE)

It has been pointed out in earlier sections of this chapter that the theoretical description of the deuteron based on the central force characteristic of the two body nuclear interaction failed to produce its experimentally observed magnetic moment and a positive quadrupole moment. This discrepancy could be removed by adding a non-central (Tensor) component to the central force. Similar consideration was used to explain the difference in scattering of a neutron by para hydrogen (anti-parallel spin) and ortho-hydrogen (parallel spins). The nuclear force, therefore, should not only depend on the relative position vector $\vec{r}$ but also on the relative orientation of spins i.e. on the angle between the spin directions and the line joining the two particles. If one has to construct a non-central potential, it has to satisfy the following conditions:

1. It should be invariant under rotation of coordinates system used to describe the particles.
2. It should be invariant under reflection of coordinates system used to describe the particles (parity transformation). Hence the potential must be scalar. It has, however, to retain the basic requirement that its derivative should yield nuclear force and the force should be independent of velocity.

We now construct a tensor operator which satisfies these restrictions. We have the relative position vector $\vec{r}$ and the spins vectors $\vec{\sigma}_1$ and $\vec{\sigma}_2$ of the two particles. The spin operators σ_1 and σ_2 satisfy Pauli matrix algebra. Also $\vec{\sigma}_1$ and $\vec{\sigma}_2$ are invariant under reflection (space inversion but not invariant under rotation $\vec{\sigma}_1 \cdot \vec{\sigma}_2$ however, is invariant under rotation. The position vector $\vec{r}$ is not invariant under reflection ($\vec{r} \to -\vec{r}$) as such in the tensor potential r should occur an even number of times. This shows that $\vec{\sigma}_1$ and $\vec{\sigma}_2$ should appear in a bilinear combination. Moreover, any polynomial in $\vec{\sigma}$ can be reduced to an expression linear in $\vec{\sigma}$ by using spin identities. Thus the only linearly independent quantities which can be constructed from $\vec{r}$, $\vec{\sigma}_1$ and $\vec{\sigma}_2$ are

$$(\vec{\sigma}_1 \cdot \vec{\sigma}_2) \text{ and } (\vec{\sigma}_1 \cdot \vec{r})(\vec{\sigma}_2 \cdot \vec{r})$$

The tensor component of the potential, then have the form

$$V_T = V(r)\, S_{12}(r) \qquad \ldots(2.17.1)$$

S_{12} denotes the dependence on the angle between the vector $\vec{r}$ and the spin directions of the nucleons. The tensor operator S_{12} should be such that its average over all directions vanishes. When it is defined as,

$$S_{12} = \frac{3}{r^2}(\vec{\sigma}_1.\vec{r})(\vec{\sigma}_2.\vec{r}) - (\vec{\sigma}_1.\vec{\sigma}_2) \quad \text{...(2.17.2)}$$

$$= 3(\vec{\sigma}_1.\hat{r})(\vec{\sigma}_2.\hat{r}) - (\vec{\sigma}_1.\vec{\sigma}_2) \quad \text{...(2.17.3)}$$

where the unit vector is defined as

$$\hat{r} = \frac{\bar{r}}{|r|}$$

The second term in (2.17.2) or 2.17.3) is used so that the average of S_{12} over all directions of r becomes zero.

Now $$\frac{1}{4\pi}\int(\vec{\sigma}_1.\hat{r})(\vec{\sigma}_2.\hat{r})\, d\Omega = \frac{1}{3}(\vec{\sigma}_1.\vec{\sigma}_2) \quad \text{...(2.17.4)}$$

Thus the average of S_{12} over all directions r reduces to zero *i.e.*

$$\langle S_{12}\rangle = \frac{1}{3}(\vec{\sigma}_1.\vec{\sigma}_2).\, 3 - (\vec{\sigma}_1.\vec{\sigma}_2)$$
$$= 0$$

Wigner (1941) showed that the most general form of two body nuclear potential satisfying the above requirements can be expressed as

$$V(r) = V_R(r) + v_\sigma(r)\,\vec{\sigma}_1.\vec{\sigma}_2 + V_T(r)S_{12} \quad \text{...(2.17.5)}$$

The first two terms correspond to central potential and the third term gives non-central potential. It is clear that the second term depends only on the directions of spins of the two particles $\vec{\sigma}_1.\vec{\sigma}_2 = +1$ for triplet state (spins parallel) and $\vec{\sigma}_1.\vec{\sigma}_2 = -3$ for singlet state (spins antiparallel). Accordingly the central potential for the two states can be expressed as

For triplet state $\quad V_t(r) = V_R(r) + V_\sigma(r)$

and for singlet $\quad V_s(r) = V_R(r) - 3V_\sigma(r)$

From the theoretically obtained values of $V(r)$ for the two states assuming rectangular potential well (n–p scattering) it is possible to estimate $V_R(r)$ and $V_\sigma(r)$

$$V_t(r) = V_R(r) + V_\sigma(r) = 38 \text{ MeV}$$

$$V_s(r) = V_R(r) - 3V_\sigma(r) = 14 \text{ MeV}$$

This gives $V_R(r) = 32$ MeV and $V_\sigma(r) = 6$ MeV. Since different shapes of the potentials fit the experimentally known parameters like binding energy magnetic moment and quadrupole moment in deuteron as well as low energy n-p and p-p scattering, the estimation of the coefficients $V_R(r)$ and $V_\sigma(r)$ cannot be made uniquely.

It is our common experience that the shorter the wavelength of radiations (particles), the more detailed structure of the system could be studied. So far we have confined to low energies or longer wavelengths so that details of the nuclear force between two nucleons could not be observed. To do this we must use projectiles of shorter wavelengths or particles of high energies. Studies have been made with scattering of high energy protons, neutrons, etc. However, the high energy studies are not so simple as for low energies. Larger angular momenta are now possible so that many more states contribute and one has a large number of parameters making the interpretation of data more difficult and complicated. Nevertheless, useful information like the existence of a repulsive core and exchange nature of nuclear forces have come from such studies.

2.18 EXCHANGE FORCES

The angular distribution of high energy neutrons scattered by protons suggests that the nuclear forces contain a component with exchange characteristics. When a fast neutron hits a proton, very often the proton jumps forward with almost as much speed as the neutron had, while the neutron is stopped almost to a stand still. The simplest way to explain this is that the neutron snatches the positive charge from the proton and keeps on moving without transferring much of its momentum to proton. Such a characteristic of nuclear force was first suggested by Heisenberg (1932). An exchange interaction can be developed quantum mechanically when two particles can exist in a state in which some common property can be shared. Exchange forces are familiar in chemistry in the theory of homopolar bond, for example the sharing of an electron between two protons explains the binding of hydrogen molecular ion. Therefore, the property which is shared is the orbital electron and hence the problem could be solved exactly. Analogously, for the two nucleons, Heisenberg assumed, to begin with, that the property shared would be the charge. In the next section, on field theory of nuclear force, we shall discuss the mechanism of this exchange in terms of absorption and emission of virtual pi-mesons.

We have seen that the space and spin co-ordinates of the nucleons play an important role in characterising the nuclear forces. As such, besides charge, the space and spin co-ordinates also contribute in describing the exchange mechanism. Consequently, different types of exchange forces have been proposed. Since the nuclear forces have been found to be charge independent and charge symmetric, the exchange force acts between pairs of identical and non-identical nucleons. The central potentials does not involve any exchange property and hence known as ordinary force or Wigner force. In Heisenberg type of force only charge is exchanged. In Majorana type of exchange force, charge and spin are exchanged. We shall see that this is just equivalent to exchange of position of the nucleons and hence Majorana type of force is sometimes called as space-exchange force. When only spin directions take part in the exchange, the exchange force is called as Bartlett force. The exchange characteristic under these four potentials for a neutron-proton system ($^1S_0$) is shown in the Fig. 2.16.

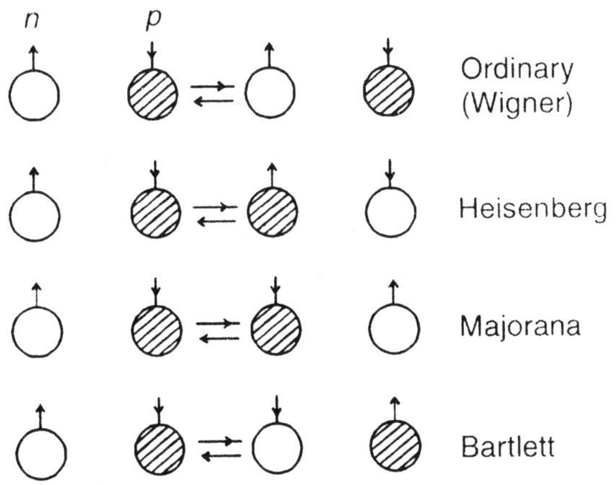

Fig. 2.16: The exchange characteristics of nuclear force between a neutron and a proton, under four different nuclear potentials

The potential energy for a pair of nucleons will be of the form

$$\langle V \rangle = \int \Psi^* VP\Psi d\tau$$

where Ψ is the wave function for the pair of nucleons.

If $\vec{r}_1, \vec{\sigma}_1$ and $\vec{r}_2, \vec{\sigma}_2$ are position vectors and spins of the particle 1 and 2 respectively then $\Psi = \Psi(\vec{r}_1\vec{\sigma}_1, \vec{r}_2\vec{\sigma}_2)$ and V is a radial function representing attractive potential (negative). P is an operator which describes the particular exchange force. Ψ can be expressed as a product of space wave function. $\psi(r_1, r_2)$ and spin function $\chi(\sigma_1, \sigma_2)$. i.e.

$$\Psi(r_1\sigma_1, r_2\sigma_2) = \psi(r_1 r_2)\ \chi(\sigma_1\sigma_2) = \psi_{12}\chi_{12}$$

we now examine the four exchange forces.

1. Ordinary or Wigner Force

$$V(r) = V(r)\ P_w$$

and $\quad P_w\Psi(\vec{r}_1, \vec{\sigma}_1\vec{r}_2\vec{\sigma}_2) = \Psi(\vec{r}_1\vec{\sigma}_1, \vec{r}_2, \vec{\sigma}_2)$

Here $P_w = 1$ (unit operator) and it is termed as Wigner operator. It is clear that in this type of force, no exchange is involved. $V(r)$ is always attractive.

2. Heisenberg Interaction or Charge Exchange Force

In this type of interaction, we find that the exchange of charge between the nucleons is same as the exchange of both the space and spin coordinates, as is clear from the Fig. 2.16. $V(r) = V_H(r)P_H$

$$P_H\Psi(r_1, \sigma_1, r_2, \sigma_2) = P_H\psi\ (r_1, r_2)\ \chi(\sigma_1, \sigma_2) = P_H\psi_{12}\chi_{12}$$

$$= \psi(r_2, r_1)\chi(\sigma_2, \sigma_1)$$
$$= \psi_{21}\,\chi_{21}$$

The space exchange or the parity is governed by the orbital angular momentum L and spin exchange by the spin angular momentum quantum number S. Space exchange is symmetric for even parity states (L-even) and anti symmetric for odd-parity states (L-odd) and spin triplet states (S = 1) are symmetric and spin singlet states (S =0) are antisymmetric. These characteristics when combined determine the Heisenberg exchange force.

$$P_H\psi_{12}\chi_{12} = (-1)^L\psi_{12}(-1)^{S+1}\chi_{12}$$
$$= (-1)^{L+S+1}\psi_{12}\chi_{12}$$

Thus

$V(r) = +V_H(r)$ for $L = 0$, even and $s = 1$. Triplet

or for L = odd and $s = 0$ singlet

and $V(r) = -V_H(r)$ for $L = 0$, even and $s = 0$

or for L = odd and $s = 1$

The potential is, thus, attractive for tiplet states ($s = 1$) with even parity ($L = 0$, even) and for singlet states (s = 0) with odd parity (L = odd). The wave function is symmetric for such states. On the other hand, the potential is repulsive for singlet states ($s = 0$) with even parity ($L = 0$, even) and for triplet states (s = 1) with odd parity (L = odd). The wave function for such states is antisymmetric.

(ii) Majorana Interaction

Majorana interaction is a space exchange force where exchange of space coordinates only takes place. The exchange potential is defined by

$$V(r) = V_M(r)\,P_M$$

The Majorana operator P_M is such that

$$P_M\Psi(r_1, \sigma_1, r_2\sigma_2) = P_M\psi(r_1, r_2)\,\chi(\sigma_1, \sigma_1) = P_M\psi_{12}\chi_{12}$$
$$= \Psi(r_2, r_1)\,\chi(\sigma_1\sigma_2)$$
$$= \psi_{21}\chi_{12}$$

we find that the Majorana operator is same as parity operator. Hence the Majorana operator can be expressed as

$$P_M\psi_{12}\chi_{12} = (-1)^L\psi_{12}\chi_{12}$$

and $V(\mathrm{r}) = +V_M(\mathrm{r})$ for $L = 0$, even

and $V(\mathrm{r}) = -V_M(\mathrm{r})$ for L = odd.

(iii) Bartlett Interaction: A Spin Excuange Force

As it is clear from the Fig. 2.19, in this interaction,

$$V(r) = V_B P_B$$

and the Bartlett operator is such that the spins of the nucleons are exchanged. That is

$$P_B\psi(r_1 r_2)\chi(\sigma_1\sigma_2) = P_B\psi_{12}\chi_{12}$$
$$= \psi(r_1 r_2)\chi(\sigma_2\sigma_1)$$
$$= \psi_{12}\chi_{21}$$

For triplet states, (s = 1) χ_{12} is symmetric so that $\chi_{12} = \chi_{21}$ and for singlet state (s = 0), χ_{12} is anti-symmetric, and $\chi_{12} = -\chi_{12}$. Therefore,

$$P_B\psi_{12}\chi_{12} = (-1)^{s+1}\psi_{12}\chi_{12}$$

and the potential $V(r)$ is attractive for triplet states and repulsive for singlet states. This shows that for a(n-p) system, nuclear force can not be totally of Bartlett type, since we have seen, that it is attractive for both singlet and triplet states. The characteristic of these operators are tabulated below:

TABLE 2.3

	Even Parity States		*Odd Parity States*	
	Triplet (S = 1)	*Singlet (S = 0)*	*Triplet (S = 1)*	*Singlet (S = 0)*
Exchange operator				
Wigner P_W	+1	+1	+1	+1
Heisenberg P_H	+1	−1	−1	+1
Majorana P_M	+1	+1	−1	−1
Bartlett P_B	+1	−1	+1	−1

Since Ψ is symmetric or antisymmetric when the exchange of coordinates (space, spin or both) is made, i.e. $p\Psi = \pm\Psi$, which means the eigen values of the operator is ± 1.

Exchange and Saturation of Nuclear Forces

In Chapter 1, we have come across two important properties of nuclei namely, giving a relation $R \propto A^{1/3}$ for nuclear radius R and mass number A and secondly, the binding energy of nuclei being proportional to the total number of nucleons inside a nuclei meaning thereby that binding energy per nucleon is constant for most of the nuclei. These two properties indicate the saturation properties of the nuclear force.

Now let us examine, how the four exchange forces are able to explain the saturation properties.

As mentioned earlier, the nuclear potential cannot be wholly of Bartlett type which accounts for spin exchange. It is attractive for triplet state and repulsive for singlet state and as such cannot explain the constant B.E. per nucleon. If Bartlett nuclear potential were to hold then $BE \propto A^2$, because in heavy nuclei there are number of nucleons of various orientations and the number of interacting pairs would be $\frac{1}{2}A(A-1)$ which leads to $BE \propto A^2$. Same is the case for He^4 where equal number of singlet and triplet pairs of nucleons exist and all nucleons are in s-states. Likewise the spin dependent Heisenberg interaction cannot account for saturation and is, yet adequate to describe deuteron. However, the Heisenberg force is attractive for nucleons of different charge states and repulsive between like nucleons. The Majorana forces are attractive in both triplet and singlet states for even parity and thus can explain large B.E. for alpha particle. Majorana force, therefore, can mostly account for the saturation property. We thus find that no single type of exchange interaction can account for some of the observed facts like saturation and constant B.E. per nucleon. It was, therefore, believed that a judicious mixture of the exchange forces will explain the saturation of nuclear densities and there was no need to assume that nucleons repel each other if they come too close. However, there is now very good evidence that the two nucleon interaction exhibits a strong repulsive core, at very close distances between them.

2.19 MESONS AND NUCLEAR FORCES (FIELD THEORY OF NUCLEAR FORCES)

So far we have described that the nuclear force between two nucleons is due to a potential which is mainly a function of distance between the nucleons. The force is propagated instantaneously and the description is non-relativistic. However, such a picture was found to be not so simple when attempts were made to explain many of the experimental phenomena, like magnetic moment and quadrupole moment of deuteron. It was, therefore, felt to examine the interaction between two nucleons empirically employing some more simpler concepts. The concept of field theory of nuclear forces, first put forward by H. Yukawa (1935) had been an attempt in this direction. In this description, a nucleon is a source of a force field in which the other nucleon finds itself. Yukawa argued that the Coulomb force between electric charges is of infinite range and electromagnetic interaction between charges takes place through the exchange of photons. In the similar fashion, it should be possible to describe the nuclear interaction in terms of a force field and the force may be conveyed through the exchange of a certain particle between the nucleons. However, there is a marked difference between the nuclear force and electromagnetic force, which lies in their range. The nuclear forces have finite range and as such any formalism for the nuclear interaction, when developed on the lines of electromagnetic forces must include a parameter accounting for the finite range.

The wave equation obtained from Maxwell's fields equations, is

$$\nabla^2 V - \frac{1}{c^2}\frac{\partial^2 V}{\partial t^2} = 0 \qquad \text{...(2.19.1)}$$

where V is a scalar potential. For static charges it reduces to the Laplace's equations in free space

$$\nabla^2 V = 0 \qquad \text{...(2.19.2)}$$

In the presence of a point source with charge q located at the origin, the equation takes the form of the Poisson's equation

$$\nabla^2 V = -\frac{1}{4\pi \in_o} 4\pi q \delta(r) \quad \text{...(2.19.3)}$$

where $\delta(r)$ is Dirac's delta function.

The solution of (2.19.3) is the familiar Coulomb potential

$$V(r) = -\frac{q}{4\pi\varepsilon_0 r} \quad \text{...(2.19.4)}$$

If there is a charge q at the point r, its potential energy will be $U = qV = q^2/4\pi\varepsilon_o r$.

Yukawa proposed a modification in the wave equation (2.19.1) in order to obtain a force of shorter range. The modified wave equation is expressed as

$$\nabla^2\phi - \frac{1}{c^2}\frac{\delta^2\phi}{\delta t^2} - \mu^2\phi = 0 \quad \text{...(2.19.5)}$$

where ϕ is the nuclear potential function. μ is a constant parameter having dimension of reciprocal of length. Using a wave like solution for eqn. (2.19.5) we get

$$-k^2 + \omega^2 / c^2 - \mu^2 = 0 \quad \text{...(2.19.6)}$$

Using de-Broglie relations for energy E and momentum P

$$E = \hbar\omega \quad \text{and} \quad p = \hbar k$$

where ω is angular frequency and k is wave-vector, the equation (2.19.6) reduces to

$$E^2 = c^2p^2 + M^2c^4 \quad \text{...(2.19.7)}$$

where $$M = \frac{\hbar\mu}{c} \quad \text{...(2.19.8)}$$

The equation (2.16.7) is the relativistic relationship between energy and momentum of a particle whose rest mass is M, as given by equation (2.19.8). Thus we find that the constant parameter μ defines the rest mass of the particle. The time independent equation is as obtained from (2.19.5) is

$$\nabla^2\phi - \mu^2\phi = 0 \quad \text{...(2.19.9)}$$

If the particle mass is zero, $\mu = 0$ and (2.19.9) reduces to the equation similar to the Laplace's equation, i.e. $\nabla^2\phi = 0$.

The equation corresponding to Poisson's equation is obtained by introducing a point source with strength g located at the origin. Such an equation is

$$\nabla^2\phi - \mu^2\phi = g\delta(r) \quad \text{...(2.19.10)}$$

The solution of the eqn. (2.19.10) is

$$\phi(r) = -\frac{ge^{-\mu r}}{4\pi r} \quad \text{...(2.19.11)}$$

where g is an arbitrary constant. From eqn. (2.9.11) we find that with proper choice of μ (or the mass M, of the particle), ϕ can be made to describe the short range of the nuclear force. The constant g behaves as a strength of a point source situated at the origin. If another source g is situated at r, then the potential energy or the interaction energy between such two sources is

$$U = g\phi(r)$$

$$= -\frac{g^2(r)}{4\pi}\frac{e^{-\mu r}}{r} \qquad ...(2.19.12)$$

The equation (2.19.11), shows that unlike Coulomb potential, the nuclear potential decreases more rapidly as we move away from the source, because of the exponential factor. This will, therefore, lead to the short range of the nuclear force. It is the well known Yukawa potential. It reduces to (2.19.4), if we put μ = 0 or $M = 0$ and $g = \frac{g}{\varepsilon_0}$. For a particle of finite mass, $M = \frac{\hbar\mu}{c}$, we find that the potential drops to around $\frac{1}{e}$ of its initial value at a distance $b = \frac{1}{\mu} = \frac{\hbar}{Mc}$. The quantity b may be taken as a measure of the range of the force. Since the range of the nuclear interaction 'b' is known from which it is possible to estimate the mass of the particle. For $b \approx 1.4$ Fm one gets.

$$\frac{1}{\mu} = 1.4 = \frac{\hbar}{Mc}$$

which gives $M \approx 280\ m_e$. The rest mass of the particle which is exchanged between nucleons is between the mass of the electron and mass of the proton, hence it is called as Meson. Therefore, like electric field around charges, there is meson field around the nucleons and during strong interaction these mesons are exchanged. Just as the charge 'e', associated with the electron, is the source of the electromagnetic field and photons emerge as field quanta, similarly, the source strength 'g' of the meson field is a property of the nucleon and is used to describe interaction between nucleons. We have seen that the interaction energy between two nucleons is $g^2 \frac{e^{-\mu r}}{4\pi r}$ (eqn. 2.19.12), the dimensionless parameter which may be regarded to characterise the strong interaction is the coupling constant $g^2 / \hbar c$ whose value is around 0.3. The corresponding coupling constant for electromagnetic field is $g^2 / \hbar c = \frac{1}{137}$. The strength of the meonic charge is thus much larger than that of the electric charge.

In cosmic ray experiments (1947) particles of mass around 280 m_e were observed. They were called as π-mesons or simply pions. There are three types of pions; electrically charge pions (π ±) and electrically neutral pion, (π°).

The pions are "carriers" of nuclear force field. The exchange processes involved between the nucleons are.

(i) $p \leftrightarrow n + \pi^+$

(ii) $n \leftrightarrow p + \pi^-$

(iii) $p \leftrightarrow p + \pi^\circ$

$n \leftrightarrow n + \pi^\circ$

Such exchange processes are not observable for free nucleons. (A free neutron, for example decays with a half life of about 12 minutes, $n \rightarrow p + e^- + \bar{\nu}$). A nucleon inside the nucleus cannot provide the necessary energy for the production of pions, hence the processes are forbidden classically. The law of conservation of energy is violated. However, quantum mechanically these exchange processes can be made possible provided that the energy $\Delta E = Mc^2$, required for the production of pions may be borrowed (from the nucleon) and return the same within a time Δt given by the uncertainty relation.

$$\Delta E\, \Delta t \approx \hbar$$

or $$\Delta t \approx \hbar / \Delta E = \frac{\hbar}{Mc^2}$$

Assuming that the pion travels, inside the nucleus, with a velocity of light (c), the maximum distance it can travel in time Δt is $c\Delta t$. If this distance is identified as the range of nuclear force, then

$$C\Delta t = b = c\frac{\hbar}{Mc^2} = \frac{\hbar}{Mc} = \frac{1}{\mu}$$

which is same as indicated by the potential ϕ (eqn. 2.19.11). The pions exist for such a short time 10^{-23} sec, that it cannot be observed. Hence such a pion is termed as virtual pion and the exchange process is termed as virtual process. In the virtual process, for example, a proton emits a π^+ gets transformed into a neutron and this pion (π^+) is absorbed by a neutron and transformed into a proton. This characterises strong interaction between a proton and a neutron. Similarly a π^- is emitted by a neutron, transformed into a proton and this π^- is absorbed by a proton and transformed into a neutron. The neutral pion (π°) is exchanged between like nucleons.

The Yukawa's meson field theory has been a step in the right direction towards the understanding of the nuclear forces. Yet the picture is not clear. The large value of the coupling constant has made the development of the theory more difficult and complex. The possibility of more than one pion being involved during interaction cannot be ruled out. Efforts are on to solve such mathematical difficulties.

2.20 GENERALISED PAULI'S EXCLUSION PRINCIPLE

The proton and neutron have been represented as two charge states of a single particle (each is often termed as a nucleon). The charge independence of nuclear forces has led to entirely a new concept of isobaric spin or iso-spin. A detailed discussion of iso-spin formalism has been given in Chapter 1. Like space and spin co-ordinates, the wave function of a two particle system also depends on the iso-spin co-ordinates of the particles (in iso-spin co-ordinates system). If r_1, r_2, σ_1, σ_2, and τ^1, τ^2 are space spin and iso-spin co-ordinates of two particles respectively, then the total wave function of the system will be

$$\Psi = \Psi_{space}\ \Psi_{spin}\ \Psi_{charge}$$

where

$$\Psi_{space} = \Psi(r_1 r_2),\ \Psi_{spin} = \chi(\sigma_1\sigma_2) \text{ and } \Psi_{charge} = \xi(\tau^1,\tau^2)$$

where χ is spin wave function with σ_1, σ_2 as the spin co-ordinates and ξ is iso-spin wave function having τ^1, τ^2 as iso-spin co-ordinates of the nucleon system.

$$\therefore \qquad \Psi_{Total} = \Psi(r_1, r_2)\ \chi(\sigma_1, \sigma_2)\ \xi(\tau^1, \tau^2)$$

Since the nucleons obey Fermi-Dirac statistics, the total wave-function must be anti-symmettric in the interchange of the parirs of co-ordinatess $r_1 r_2$, σ_1, σ_2, and τ^1, τ^2. However, if the state of the system is such that any of the two wave functions are symmetric then the third has to be anti symmetric so that the total wave function becomes anti symmetric. It has been shown (Chapter 1) that the mathematical development of iso-spin formalism is completely analogous to ordinary spin formalism. The iso-spin up state is regarded as the proton and iso-spin down state as the neutron. These two states may be represented by γ ans δ such that

$$\gamma = \binom{1}{0}, \text{ a proton}$$

$$\delta = \binom{0}{1}, \text{ a neutron}$$

The corresponding eigen values are +1 for the proton and −1 for the neutron.

It means that corresponding to two eigen values up and down there are two eigen functions γ and δ.

Analogous to Pauli's spin-operations $\vec{\sigma}_1$ and $\vec{\sigma}_2$, the iso-spin operators are $\vec{\tau}^{(1)}$ and $\vec{\tau}^{(2)}$.

$$\vec{\tau}^{(1)} \cdot \vec{\tau}^{(2)} = +1 \text{ for iso-spin triplet } (T = 1)$$

$$= -3 \text{ for iso-spin singlet } (T = 0)$$

In iso-spin space, the two nucleon system has the following symmetric (three) states and antisymmetric (ONE) state:

Symmetric Triplet

$(T = 1)$

$= \gamma_1\ \gamma_2$, two proton system

$$\xi(\tau^1,\tau^2) = \xi_{12} = \delta_1\delta_2$$

two neutron system

$$\frac{\gamma_1\delta_2 + \gamma_2\delta_1}{\sqrt{2}}$$

neutron proton system

with the third component having values

$T_3 = \pm 1$ and 0

Anti symmetric singlet

$(T = 0)$

$$\xi_{12} = \frac{\gamma_1\delta_2 - \gamma_2\delta_1}{\sqrt{2}}$$

neutron proton system

with $T_3 = 0$

We have seen that the interchange of space-co-ordinates multiplies the wave function $\Psi(r_1, r_2)$ by $(-1)^L$ the interchange of spin co-ordinates multiplies the spin wave function by $(-1)^{s+1}$. The interchange of iso-spin co-ordinates $\tau^{(1)}$ and $\tau^{(2)}$ of the nucleons will multiply the iso-spin wave function by $(-1)^{T+1}$ (analogous to ordinary spin formalism). Thus the total wave function, as determined by the interchange of the pairs of co-ordinates of the two particles is expressed as

$$\begin{aligned}\Psi_{21} &= \Psi_{21}\,\chi_{21}\,\xi_{21}\\ &= (-1)^L \Psi_{12}\,(-1)^{s+1}\chi_{12}\,(-1)^{T+1}\xi_{12}\\ &= (-1)^{L+s+T+2}\Psi_{12}\,\chi_{12}\xi_{12}\\ &= (-1)^{L+s+T+2}\Psi_{12}\end{aligned}$$

The nature of the wave function is thus decided by the L, S and T values of the state.

This represents the generalised Pauli's Exclusion Principle for the two particle system.

As an example let us consider the deuteron (n-p system) in the ground state, a mixture of a S-state ($L = 0$) and a D-state ($L = 2$). The space part of the wave functions is symmetric on the inter-change of space-coordinates. The spin wave function is symmetric, the two spins are parallel the total spin is 1 ($s = 1$). $\chi_{21} = (+)\chi_{12}$ and to have Ψ_{total} to be antisymmetric, the iso-spin wave function should be antisymmetric i.e. $\xi_{21} = (-1)\xi_{12}$, it should be an iso-spin singlet ($T = 0$) state. Thus the ground state of the deuteron (mixture of $L = 0$, $L = 2$) is an ordinary spin-triplet ($S = 1$) and iso-spin singlet ($T = 0$) state. If, however, the spins of neutron and proton in the deuteron are anti-parallel it is an ordinary spin singlet state (S =0) the spin wave function χ being antisymmetric, the iso-spin wave function ξ must be being symmetric ($T = 1$). Thus the singlet state of the deuteron is an iso-spin triplet.

EXERCISES

1. Show that ^3S, ^1P, ^3P and ^3D states of a neutron-proton system are the only one compatible with a nuclear spin equal to one.
2. Discuss how the scattering of neutrons from ortho and para hydrogen leads to the conclusion that nuclear forces are spin dependent. What is the significance of scattering length ?
3. A neutron is scattered by a deuteron in the laboratory calculate (*i*) the largest angle of scattering and (*ii*) maximum possible fractional energy loss per collision.
4. What is the importance of effective range approximation in neutron-proton scattering?
5. Mention necessary physical arguments supporting non-existence of an excited S-state in a deuteron.
6. Discuss the conditions under which the Heisending, Bertlett, and Majorana forces between nucleons are attractive or repulsive. What is the nature of force between neutron and proton in a deuteron in the singlet state?
7. What is the possible range of values for the depth of a one dimensional square well of width b, if that has only one bound state for a nucleon?

3

NUCLEAR TRANSFORMATIONS

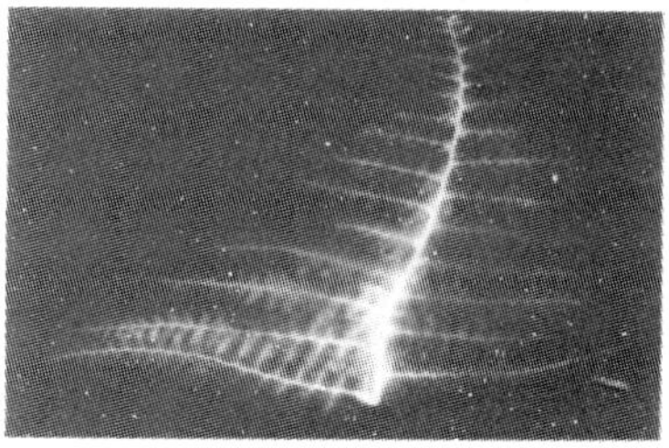

3.1 RADIOACTIVITY

While describing the basic properties of nucleus in Chapter one, we have seen that a certain ratio of protons and neutrons inside a nucleus gives it a stable configuration (Fig. 1.1). Some nuclei on the other hand, have such a ratio of protons and neutrons that they no longer stay stable. These nuclei are unstable and attain a stable structure only after giving out spontaneously certain particles. This phenomena of spontaneous emission of particles by certain nuclei of naturally occuring elements is termed as radioactivity. It was first discovered by a French Physicist Heny Becquerel in 1896. We shall see that this natural radioactivity is not different from the radioactivity being a property of many artificially produced unstable nuclei.

Like the discovery of X-rays by Rontgen in 1895, radioactivity was also discovered accidentally. H. Becquerel's father was a physicist interested in the studies of the phenomena of phospherescence and fluorescence. When X-rays were found to produce fluorescence in various materials, H. Becquerel thought about the possibility of producing X-rays by fluorescent materials when subjected to strong, intense, light. He placed a fluorescent uranium salt (potassium uranyl sulphate) on a photographic plate covered with black paper and exposed it to sun light. When the photographic plate was developed, Becquerel found that the plate was fogged or darkened. Becquerel, thought that the uranium salt has emitted radiation (rays) which could penetrate the black paper. He took these radiations as X-rays. He, then, tried to repeat the experiment, but failed to expose the arrangement to sun-light because of cloudy weather. He had to keep the arrangement for several days. He developed the photographic plate expecting it to be almost clear, but to his surprise, the plate was just as fogged as before even though the exciting sun light was absent. He repeated the experiment many times and ultimately came to the conclusion that the source of penetrating radiations was uranium in the fluorescent salt. He was also able to show that these radiations are capable of ionising gases and that some of these radiations are fast charged particles. He showed that the spontaneous emission of radiation or radioactivity was a property of the uranium atom itself. (It is now known that the portion of the radiations containing fast charged particles as studied by Becquerel, were the fast electrons emitted in β-disintegration of daughter product of nucleus U^{238}).

Madam Marie Curie and her husband Pierre Curie while working in Becquerel's laboratory, studied many other elements for radioactive emission and found that Thorium also emits radiations similar to those emitted by uranium. They also found that uranium ore known as Pitch-Blende was a stronger source of radiations than was expected and hence thought of the presence of other radioactive elements.

Starting with about a ton of Pitch-Blende, they first separated through chemical processes a radioactive element and named it as *Polonium* (after Madam Curie's native place, Poland) and later on another radioactive element and named it *Radium*. It would be exciting to know that out of nearly a ton of Pitch-Blende only a fraction of a gram (~0.1 gm) of radium was obtained, but it turned out to be several million times more radioactive than uranium. Both these elements, Polonium and Radium were found to emit spontaneously radiations, similar to those emitted by uranium.

The radiations from the radioactive elements became an exciting subject of investigations. Among the early experiments, Lord Rutherford and his co-workers were the first to show that these radiations contain components of different penetrating powers as measured by their absorption in matter. The less penetrating rays, which were completely stopped or absorbed by a thin sheet of card board or even a thin sheet (0.005 inch) of metal were called α-particles (alpha-particles). The more penetrating component which were absorbed by about 1 mm of lead were called β-particles (Beta-particles). The third component, was found to have much more penetrating power, it could not be stopped even by a sheet of lead (about 10 cm). These radiations were termed as γ-rays (Gamma rays). These radiations were later on identified as positively charged Helium nuclei (α-particles), negatively charged electrons (β-particles) and high energy photons (γ-rays) when allowed to pass through a magnetic field. The characteristic identifications of these radiations are shown in Figs. (3.1) and (3.2) respectively.

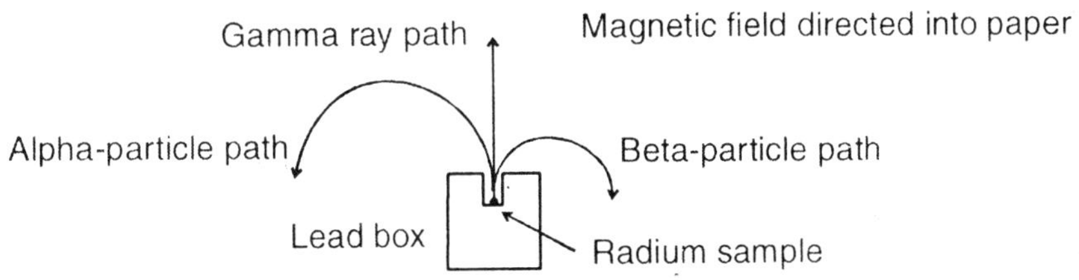

Fig. 3.1: Deflection of α-particles and β-particles when they travel through a region of magnetic field

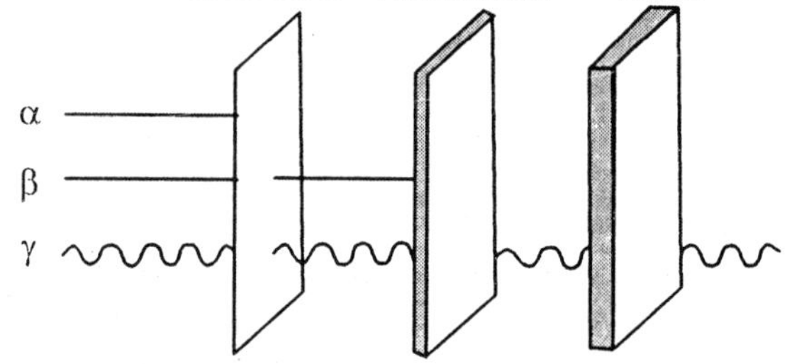

Fig. 3.2: Characteristic identifications of α-particles, β-particles and γ-rays, when they are allowed to pass throgh different materials

The deflections in magnetic field revealed their corpuscular behaviour, hence they are also called as alpha-particles, beta-particles and gamma-ray photons. The phenomena of natural radioactivity became a starting point for further investigations and understanding of the nucleus and nuclear processes (later on two more decay modes were added to the list of radioactive disintegrations, namely the positron emission and electron capture, more information about which is given later.

3.2 RADIOACTIVE TRANSFORMATIONS

3.2.1 Displacement Law

The hypothesis that there exists a genetic relation between some of the radioactive elements was first advanced by E. Rutherford and F. Soddy in 1903, on the basis of their work on the decay and production of chemically separable radioactive substances. A typical example is the study of γ-rays emitted from radium. A solution of radium evolves radon gas. When the radon gas was completely removed from radium, it was found that after a short time, the radon gas exhibited all the γ-activity while the solution none. This showed that by spontaneous transformation of one type of atoms, another type of atoms are produced.

Likewise α-particle ($^4_2$He) and β-particle ($_{-1}^{0}$e) emission also transform one type of atoms into other. By ejecting an alpha particle, the nucleus loses 2 positive charges, thereby droping down by 2 places in atomic number and it loses a mass of 4 units and so drops down four units in atomic weight. The β-particle emission increases the charge of the residual nucleus by one positive unit and negligible change in mass, since β-particle weighs only $\frac{1}{1836}$ part of proton. A gamma ray emission from a radioactive atom does not result into any change in the charge or the mass of the nucleus. Such displacements are symbolically represented as given below. α-particle emission from a radio-nuclide X is represented as

$$^A_ZX \xrightarrow[\alpha]{} {}^{A-4}_{Z-2}Y$$

Example:

$$^{238}_{92}\text{U} \longrightarrow {}^{234}_{90}\text{Th} + {}^4_2\text{He}$$

Similarly for β-emission

$$^{A}_{Z}X \longrightarrow {}^{A}_{Z+1}Y$$

$$^{234}_{90}Th \longrightarrow {}^{234}_{91}Pa + {}^{0}_{-1}e$$

Note: While representing the radioactive disintegration in the form of an equation, it is essential that the sum of superscripts and subscripts on the left side of the equation is equal to the corresponding sums on the right.

3.2.2 The Law of Radioactive Disintegrations

Since the radiations from radioactive materials were originally observed as scintillations on screens coated with fluorescent materials, attempts were made to formulate certain laws for the decay of such materials. It has been observed that all radioactive disintegrations occur in a random manner. The process is statistical and under such a behaviour it is posssible to presume that each radioactive atom has a definite probability of disintegration in a given time. This probability depends on the kind of an atom and is constant for all atoms of a given sample.

Let us have N_0 nuclei of a radioactive sample at time $t = 0$. Let there be a decay probability per unit time, λ, *i.e.* λ be the probability that one of these active nuclei will disintegrate in unit time. The probability that a nucleus will disintegrate in a time interval dt is λdt. If there are N nuclei present at a certain instant of time, then the number of nuclei which may disintegrate in time interval dt is $(\lambda dt)N$. If this number is denoted by dN then,

$$dN = -(\lambda dt)N \qquad \ldots(3.2.1)$$

The minus sign indicates that N decreases with time as a result of decay. It may be noted that N is very large and dN is small compared to N. From eqn. (3.2.1), we get the probable number of disintegrations per unit time as,

$$\frac{dN}{dt} = -\lambda N \qquad \ldots(3.2.2)$$

Equation (3.2.2) represents the rate at which nuclei disintegrate.

Though dN is an integer, but being very small compared to N, it may be considered that N will vary approximately in a continuous manner and then we may treat dN as a differential. On integrating eqn. (3.2.2), we get,

$$\frac{dN}{dt} = -\lambda N$$

or $$\frac{dN}{N} = -\int \lambda dt$$

∴ $$\ln N = -\lambda t + \text{constant}$$

Since $$N = N_0 \text{ at } t = 0$$

$\therefore \qquad \ln N_o = \text{constant}$

or $\qquad \ln N = -\lambda t + \ln N_o$

$$\ln N - \ln N_o = -\lambda t$$

$$\ln \frac{N}{N_O} = -\lambda t$$

$$N = N_o e^{-\lambda t} \qquad \ldots(3.2.3)$$

The equation (3.2.3), gives the number of unstable nuclei remaining after time t, when initially ($t = 0$) there were N_o radioactive nuclei. In obtaining equation (3.2.3), it is presumed that the sample is isolated. λ is a constant characteristic of each nuclide and is called as disintegration constant. It is expressed in (second)$^{-1}$. Equation (3.2.3) is represented as a graph in Fig. (3.3).

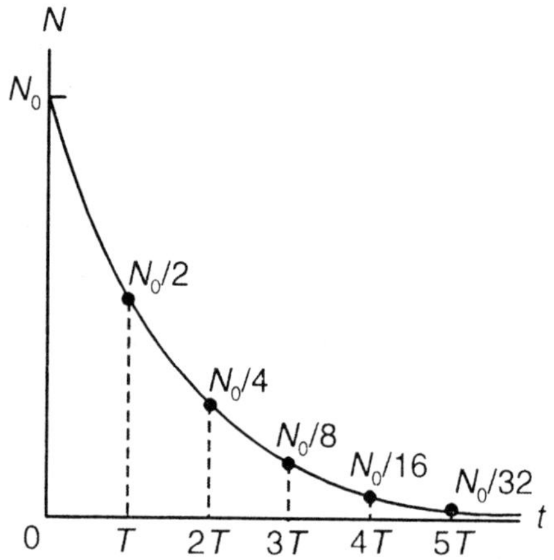

Fig. 3.3: Radioactive decay as a function of time elapsed

Half Life

In Fig. 3.3, we find that in time T, the number of radioactive nuclei has reduced to half the original number. Hence, for each radioactive substance, there is a fixed time interval T called the half life, during which half of the original nuclei have disintegrated. In other words, half life of a radioactive substance is the time in which half of the radioactive nuclei present intially are disintegrated. If there were N_0 nuclei (radioactive atoms) at time $t = 0$, than after time $t = T$, $N = \frac{N_0}{2}$ *i.e.* half of the atoms are left. To find the half life T, we put $t = T$ and $N = \frac{N_0}{2}$ in equation (3.2.3)

$$\frac{N_0}{2} = N_0 e^{-\lambda T} \text{ or } \frac{1}{2} = e^{-\lambda T}$$

or $$\ln\frac{1}{2} = -\lambda T$$

or $$\ln 2 = \lambda T$$

or $$\lambda = \frac{\ln 2}{T} = \frac{0.693}{T}$$

$$\therefore \quad T = \frac{0.693}{\lambda} \qquad ...(3.2.4)$$

Equation (3.2.4) relates the disintegration (decay) constant λ with the half-life T for a radioactive atom. Half life, as known, varies from 10^{-7} sec say for ^{212}Po to ~10^{16} secs (~10^9 years) for ^{238}U.

From equation (3.2.3), we can find (differentiating with time) the rate at which nuclei disintegrate.

$$\frac{dN}{dt} = -N_0\lambda e^{-\lambda t} = -\lambda N \qquad ...(3.2.5)$$

The equation (3.2.5) shows that the rate of disintegration $\frac{dN}{dt}$ is proportional to the number of radioactive atoms present. The rate of disintegration $\frac{dN}{dt}$ also decreases in the same manner (exponentially) and with the same half life as the number of nuclei present. The absolute value of disintegration rate $\left|\frac{dN}{dt}\right|$ is called as activity (A) of the substance.

Thus $$A = \left|\frac{dN}{dt}\right|$$

or $$A = N\lambda$$

If $N = N_0$ at $t = 0$, then at time $t = 0$, the activity is $A_0 = N_0$ and equation (3.25) then becomes.

$$A = A_0 e^{-\lambda t} \qquad ...(3.2.6)$$

This shows the exponential behaviour of the activity of a radioactive substance.

From equation (3.2.3), we have

$$\ln N = \ln N_0 - \lambda t$$

and converting it to ordianary logarithm to the base 10 gives

$$\log N = \text{Log}\, N_0 - 0.4343\lambda t \qquad ...(3.2.7)$$

A similar equation for the activity of a radioactive sample, (from eqn. 3.2.6) would be

$$\log A = \log N_0 - 0.4343\lambda t \qquad ...(3.2.8)$$

If we plot log N or log A against time, we would get a straight line, as shown in Fig. 3.4. The slope of such a line will be $-0.4343\,\lambda$.

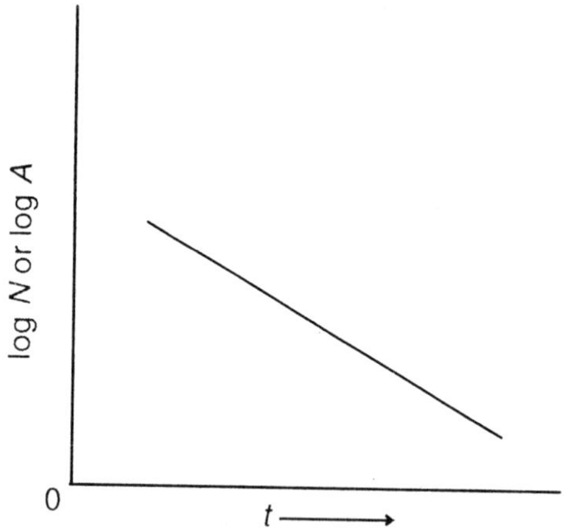

Fig. 3.4: Variation of $\log N$ or $\log A$ with time

We have the expression for activity of a radioactive sample as

$$A = N\lambda$$

Activity, thus, depends both on the total number of radioactive atoms, N, present and the decay constant λ. For this reason even if N is large, the activity A can be small, if the half life is long enough $\left(\lambda = \frac{0.693}{T_{1/2}}\right)$ and conversely, if for short lived sample, the activity A can be large, even if the sample is in small quantity (small value of N). The behaviour with time, in the two situations for same A and same N at $t = 0$, are shown in Fig. 3.5 below.

Since the process of disintegration is statistical, the same substance lives for different intervals of time before disintegration. As such it is useful to determine the average or mean life time of the active atom. It is equal to the sum of possible life times of all the atoms divided by the total number of atoms. If dN is the decrease in the number of active atoms during the time interval $dt(N \rightarrow N - dN, t \rightarrow t + dt)$, then it means that the members of the group (dN) have lived at least for time t. The sum of their ages (life times) at the time of disintegration will be tdN. Hence the mean or average life (τ) of an atom will be given by,

$$\tau = \frac{\int_{N_0}^{0} tdN}{\int_{N_0}^{0} dN} = \frac{\int_{0}^{N_0} tdN}{\int_{0}^{N_0} dN} = \frac{1}{N_0}\int_{0}^{N_0} tdN$$

But $$\ln\frac{N}{N_0} = -\lambda t \text{ or } t = -\frac{1}{\lambda}\cdot\ln = \frac{N}{N_0}$$

or $$\tau = -\frac{1}{N_0\lambda}\int_{0}^{N_0}\ln\frac{N}{N_0}\cdot dN = \frac{1}{\lambda}$$

$$\tau = \frac{1}{\lambda}. \qquad \ldots(3.2.9)$$

The average life of a radio-element is reciprocal of its decay constant.

$$\lambda = 0.693/T \qquad \text{from eqn. (3.2.4)}$$

$$\therefore \qquad \tau = \frac{T}{0.693} = 1.44T \qquad \ldots(3.2.10)$$

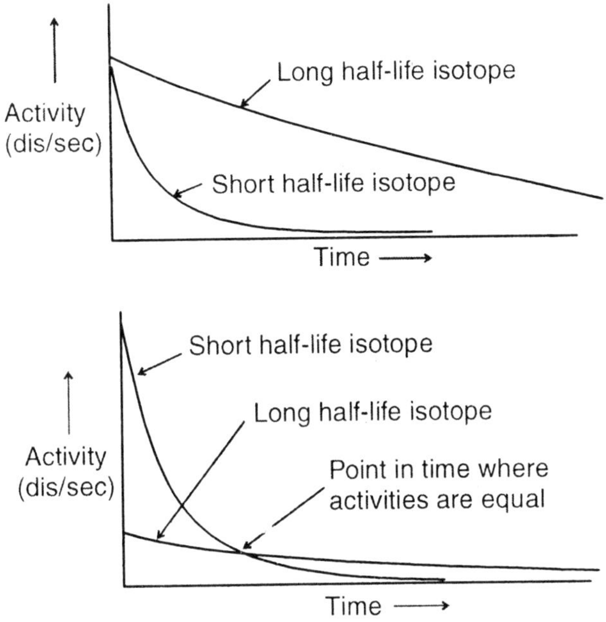

Fig. 3.5: Comparison of activities as function of time (*a*): when initial activities are equal (*b*): when number of active atoms at $t = 0$ are equal

Equation (3.2.10) gives the relation between half life (T) and mean life (τ) of a radioactive substance. From eqn. (3.2.6), we find that if the time t is equal to the average life time, the activity A falls to $\frac{1}{e}$ of its initial activity, as shown in the graph of Fig. (3.6).

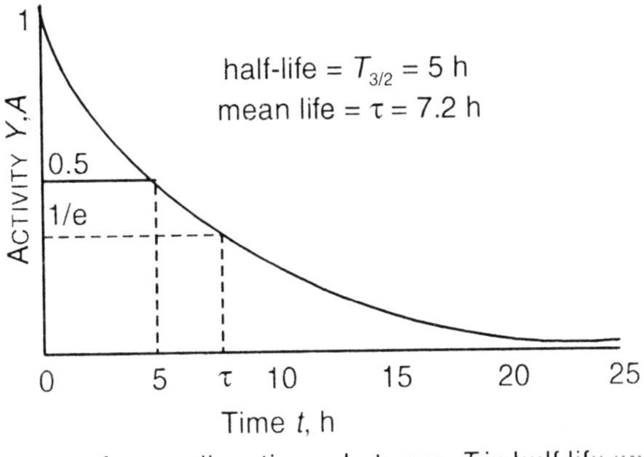

Fig. 3.6: Decay curve for a radioactive substance. T is half life and τ is mean life

Units of Radioactivity

Quantitative measurements on radioactive substances are, practically always, the determinations of the activity, which is expressed in terms of the number of disintegrations per unit time. The internationally accepted unit of radioactivity is the Curie, after the discoverer of Radium, Mme and P. Curie. One Curie is defined as the activity of a radioactive sample which has mean decay rate of 3.7 × 10^{10} disintegrations per sec (or dps). The Curie (Ci) is a large unit as such more commonly used units are milli-Curie (mCi) and micro-Curie (μCi). 1 mCi = 3.7 × 10^7 dps and 1 mCi = 3.7 × 10^4 dps, respectively.

In S.I. units the activity of a radioactive sample is measured in "Becquerel", after Antoine Becquerel who discovered radioactivity. One Becquerel of activity represents one disintegration per second in the radioactive sample. Usually bigger units mega Becquerel (1 mBq = 10^6 disintegrations/sec) and giga Becquerel (1 gBq = 10^9 disintegrations/sec) are used to measure radioactivity.

3.3 THE TRANSFORMATION LAWS FOR SUCCESSIVE DISINTEGRATIONS

In what we have discussed so far, it was assumed that a radioactive nucleus (usually called as parent nucleus) decays to a stable nucleus (daughter nucleus) OR the parent is isolated from the daughter. If, however, the daughter is also radioactive and the two are not isolated, then the daughter may or may not decay to a stable nucleus. In such a situation the parent and the daughter each will have their own decay constant. Thus in a successive transformation it becomes interesting to find the numbr of any one of the member at any instant.

Let us consider a simple situation. A radioactive sample A decays to B, which in turn decays to a stable C.

$$A \xrightarrow{\lambda_a} B \xrightarrow{\lambda_b} C \text{ (stable)}$$

where λ_a and λ_b are decay constants of A and B types of radioactive atoms respectively. Let N_0 be the number of atoms of type A at time $t = 0$. The rate at which it decays is given by

$$\frac{dN_a}{dt} = -\lambda_a N_a \qquad \text{...(3.3.1)}$$

where N_a is the number of active atoms of type A at a certain instant. The rate at which B is growing is $\lambda_a N_a$, but B also decays at the rate of $-\lambda_b N_b$. Therefore, the net change (increase) in the rate of number of active atoms of type B is given by

$$\frac{dN_b}{dt} = \lambda_a N_a - \lambda_b N_b \qquad \text{...(3.3.2)}$$

Since $N_a = N_0 e^{-\lambda_a t}$ ($N_b = 0$ at $t = 0$)

$$\frac{dN_b}{dt} = \lambda_a N_0 e^{-\lambda_a t} - \lambda_b N_b$$

or $$\frac{dN_b}{dt} + \lambda_b N_b = N_0 \lambda_a e^{-\lambda_a t}$$

Multiplying both sides by $e^{\lambda_b t}$

$$\frac{dN_b}{dt} e^{\lambda_b t} + \lambda_b N_b e^{\lambda_b t} = N_0 \lambda_a e^{-(\lambda_a - \lambda_b)t}$$

Integrating the above equation, we get

$$N_b e^{\lambda_b t} = \frac{N_0 \lambda_a}{\lambda_b - \lambda_a} e^{(\lambda_b - \lambda_a)t} + C \qquad \ldots(3.3.3)$$

where C is constant of integration.

Since $N_b = 0$ at $t = 0$, hence from (3.3.3)

$$C = \frac{N_0 \lambda_a}{\lambda_a - \lambda_b}$$

Using this value of C in (3.3.3), gives

$$N_b e^{\lambda_b t} = \frac{N_0 \lambda_a}{\lambda_b - \lambda_a} e^{(\lambda_b - \lambda_a)t} + \frac{N_0 \lambda_a}{\lambda_b - \lambda_a}$$

$$\therefore \quad N_b = \frac{N_0 \lambda_a}{\lambda_b - \lambda_a} \left[e^{-\lambda_a t} - e^{-\lambda_b t}\right] \qquad \ldots(3.3.4)$$

If there are more members of the chain of disintegrations, then in order to find the relative number of each member, we shall have to solve as many equations as there are radioactive samples. The procedure is same as adapted for the above simple case.

Here the sample B disintegrates into a stable substance C. The number of nuclei of C increases steadily until after a long time compared with the half lives of A and B, the number of nuclei of C becomes equal to the initial number of A *i.e.* N_0. The variation of N_a, N_b and N_c with time is shown in the Fig. 3.7. We find that N_b increases first, because it is produced faster than it disintegrates, then it reaches a miximum after a certain time after which it starts decreasing, because by then the activity of A has decreased to a large extent. Naturally occuring radioactive series can be examined mathematically through this technique.

Note: In the successive transformation, the various members are not removed at any instant.

From the eqution (3.3.4), one finds that $N_b = 0$ both at $t = 0$ and $t = \infty$. Hence the activity of B passes through a maximum value at some time between zero and infinity. If N_b attains a maximum at time t_{max}, then it can be obtained from the equation (3.3.4) by differentiating it with time and putting it equal to zero, *i.e.*

$$\left(\frac{dN_b}{dt}\right)_{t=t_m} = 0 \text{ gives}$$

or $$\frac{\lambda_a N_0}{\lambda_b - \lambda_a}\left[-\lambda_a e^{-\lambda_a t_{max}} + \lambda_b e^{-\lambda_b t_{max}}\right] = 0$$

or $$\lambda_a e^{-\lambda_a t_{max}} = \lambda_b e^{-\lambda_b t_{max}}$$

$$t_{max} = \frac{1}{\lambda_a - \lambda_b}\ln\left(\frac{\lambda_a}{\lambda_b}\right) \quad ...(3.3.5)$$

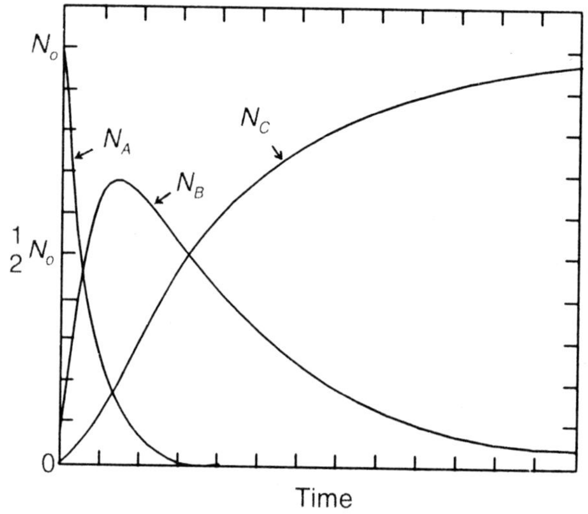

Fig. 3.7: Variation of the number of nuclei in a radioactive series $A \rightarrow B \rightarrow C$, with time, the third member C is stable nucleus

The equation (3.3.4) gives the number of daughter sample at any instant. However, for experimental purposes, it is the rate of disintegration or the activity which is required. By definition, the activities of A and B are respectively expressed as

$$\text{Activity of } A = \frac{dN_a}{dt} = N_a\lambda_a = \lambda_a N_0 e^{-\lambda_a t} \quad ...(3.3.6)$$

$$\text{Activity of } B = \frac{dN_b}{dt} = N_b\lambda_b = \frac{\lambda_b N_0 \lambda_a}{\lambda_b - \lambda_a}\left[e^{-\lambda_a t} - e^{-\lambda_b t}\right] \quad ...(3.3.7)$$

Note that here it is assumed that at $t = 0$, $N_b = 0$. If, however, N_b (0) is the number of B at $t = 0$, then the activity of B would be given by

$$N_b\lambda_b = \lambda_b \cdot N_b(0)\, e^{-\lambda_b t} + \frac{\lambda_b \lambda_a N_0}{\lambda_b - \lambda_a}\left[e^{-\lambda_a t} - e^{-\lambda_b t}\right] \quad ...(3.3.8)$$

From equations (3.3.6) and (3.3.7), the ratio of activities of B and A at any instant, is

$$\frac{\text{Activity of } B}{\text{Activity of } A} = \frac{A_b}{A_a} = \frac{\lambda_b N_0 \lambda_a e^{-\lambda_a t}\left[1 - e^{-(\lambda_b - \lambda_a)t}\right]}{(\lambda_b - \lambda_a)\cdot \lambda_a N_0 e^{-\lambda_a t}}$$

$$\frac{\text{Activity of } B}{\text{Activity of } A} = \frac{\lambda_b}{\lambda_b - \lambda_a}\left[1 - e^{-(\lambda_b - \lambda_a)t}\right] \qquad \ldots(3.3.9)$$

3.4 RADIOACTIVE EQUILIBRIM

The decay constants of different radioactive samples are usually different as such, the variation of N_b with time will be determined by the relative values of λ_a and λ_b. Two cases, therefore, arise (*i*) $\lambda_b > \lambda_a$ and (*ii*) $\lambda_b < \lambda_a$.

(*i*) When $\lambda_b > \lambda_a$ or $T_b < T_a$, which means the daughter sample is short lived compared to the parent. After a time *t*, large compared to the half life T_b of the daughter, $\lambda_b t >> 1$, and the ratio of activities in equation (3.3.9) tend to a constant value $\frac{\lambda_b}{\lambda_b - \lambda_a}$. When such a situation arises, the state is termed as *transient equilibrium*. Under such an equilibrium, the daughter and the parent ultimately decay with the half life of the parent.

From the equation (3.3.4) for time $t >> T_b$, $e^{-\lambda_b t}$ becomes negligible compared to $e^{-\lambda_a t}$ and then

$$N_b = \frac{N_0 \lambda_a e^{-\lambda_a t}}{\lambda_b - \lambda_a} \qquad \ldots(3.3.10)$$

But $N_0 e^{-\lambda_a t} = N_a$

$$N_b = \frac{N_a \lambda_a}{\lambda_b - \lambda_a}$$

or

$$\frac{N_b}{N_a} = \frac{\lambda_a}{\lambda_b - \lambda_a} \qquad \ldots(3.3.11)$$

The equation (3.3.10), shows that the number of active nuclei of daughter decreases exponentially with the half life of the parent. In the regime of transient equilibrium the ratio of N_b and N_a remains constant given by (3.3.11). The Fig. 3.8 shows the variation of activities of radio-thorium, R_dTh, ($\lambda_b = 1.2 \times 10^{-8}$ sec^{-1}) growing from radium ($^{228}_{88}$Ra) $\lambda_a = 3.35 \times 10^{-9}$ sec^{-1}. Thus in transient equilibrium, the ratio of activities of the parent and daughter remains constant and so is the ratio of their number of nuclei.

If, however $\lambda_b >> \lambda_a$, *i.e.* the parent nuclide A is long lived compared to its daughter nuclide B, then λ_a can be neglected compared to λ_b and $e^{-\lambda_a t} \approx 1$ and equation (3.3.4) then becomes,

$$N_b = N_0 \frac{\lambda_a}{\lambda_b}(1 - e^{-\lambda_b t}) \qquad \ldots(3.3.12)$$

After a time *t* long compared to T_b (half life of *B*), but short relative to T_a, (half life of *A*) *i.e.* $T_b < t < T_a$ $e^{-\lambda_b t} << 1$ and N_b approaches an equilibrium value

$$N_{b\,eq} = N_0 \frac{\lambda_a}{\lambda_b} \qquad \ldots(3.3.13)$$

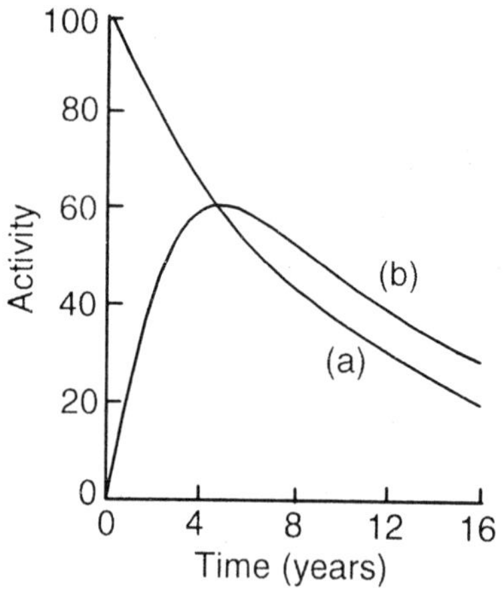

Fig. 3.8: (*a*) Decay of Ra^{228} (*b*) Growth of RdTh in equilibrium with Ra^{228}

Thus the ratio $\frac{N_{beq}}{N_0}$ is constant. This equilibrium is known as secular equilibrium. It is clear that under secular equilibrium, the activity ratio tends to unity. The above analysis of secular equilibrium can be extended to members of succesive transformations in a series, where the parent is long lived, such as the naturally occuring radioactive series. The number of successive products can be obtained from the following.

$$\lambda_a N_0 = \lambda_b N_b = \lambda_c N_c = \text{constant}$$

The activities of all the members are constant.

Thus secular equilibrium takes place whenever several short lived daughters are produced from succesive decays with a relatively long lived parent.

(*ii*) When $\lambda_b < \lambda_a$, that is the daughter is long lived compared to the parent $T_b > T_a$. The activity ratio given by equation (3.3.4) continuously varies with time. This is the situation of no equilibrium. The number of daughter can be calculated from the equation (3.3.4). In the regime of transient equilibrium, the daughter attains a maximum activity, where its activity is instantaneously same as the residual activity of the parent. However, if $\lambda_a >> \lambda_b$ the daughter decays independently with its own decay constant. For times $t >> T_a$, the number of daughter product is given by eqn. (3.3.4)

$$N_b = N_0 e^{-\lambda_b t}$$

The time t_{max} at which the daughter product attains maximum activity can be obtained from the equation (3.3.5).

3.5 RADIOACTIVE BRANCHING

When a radioactive sample does not decay through a single process, but decays through more than one process each having its own decay constant say $\lambda_1, \lambda_2, \lambda_3, \ldots$

This phenomena is termed as branching of the radioactivity. If λ is the decay constant for the radioactive substance, then

$$\lambda = \lambda_1 + \lambda_2 + \lambda_3 \quad ...(3.5.1)$$

where λ_1, λ_2, λ_3 are the decay constants of the respective processes.

Each process has a definite branching ratio, for example, the fraction of the decays or the branching ratio for the process 1 is $\frac{\lambda_1}{\lambda}$ and similarly for other processes. The branching ratio remains constant with time.

Since the half life $T_{½} = \frac{0.693}{\lambda}$, hence eqn. (3.5.1) gives

$$\frac{0.693}{T_{½}} = \frac{0.693}{(T_{½})_1} + \frac{0.693}{(T_{½})_2} + \ldots$$

or

$$\frac{1}{T_{½}} = \frac{1}{(T_{½})_1} + \frac{1}{(T_{½})_2} + \ldots$$

where $(T_{½})_1$ is half life of the substance if it were to decay through process 1 only.

Thus $(T_{½})_1$, $(T_{½})_2$ and so on are called as partial half-lives.

In naturally occuring radioactive series, such branching of radioactivity is observed. For instance in the uranium-series, the radio-nuclide ^{214}Bi transforms into ^{214}Po through the emission of β^- particle for 99.96% of the time of the decay while for 0.04% of the time, it transforms into ^{210}Tl through the emission of α-particle. if λ_α and λ_β are the decay constants, then the decay constant λ for ^{214}Bi-nucleus is $\lambda = \lambda_\alpha + \lambda_\beta$ and the corresponding partial half-lives are $\frac{1}{(T_\alpha)_{½}}$ and $\frac{1}{(T_\beta)_{½}}$ such that

$$\frac{1}{T_{½}} = \frac{1}{(T_{½})_\alpha} + \frac{1}{(T_{½})_\beta}$$

and the branching ratios are $\frac{\lambda_\alpha}{\lambda}$ and $\frac{\lambda_\beta}{\lambda}$ which are 0.04% and 99.96% respectively. $T_{½}$ for the ^{214}Bi radio nucleus is 19.7 minutes, from which the partial half lives $(T_{½})_\alpha$ and $(T_{½})_\beta$ can be obtanined.

3.6 DETERMINATION OF DECAY CONSTANT, λ, (HALF-LIFE $T_{1/2}$)

Since radioactive nuclides, both natural or artificially produced, mostly give out either one alpha-particle or one β-particle as such the rate of disintegration or the activity of the radio-nuclide can be used to determine λ. The average life time of most of the radio-nuclides ranges between 10^{-14} sec to 10^{16} years, as such the method of determination depends on the order of times involved. Thus various experimental methods have been devised depending on the half life of the radioactive substance and the type of particles emitted by it. We, however, describe here some of the important methods for the determination of decay constant.

1. For half-lives that are of the order of few seconds or days or about 10-15 years, it is possible to measure the decay constant making use of the usual decay equation (eqn. 3.2.2). The particles are generally counted by particle detectors. The rate of emission of particles can be determined at different intervals of time. If A_t is the rate of particles emitted after time t, then from equation (3.2.8), we can write,

$$\log A_t = \log A_0 - 0.4343\lambda t$$

where A_0 is the disintegration rate (activity) at some arbitrary zero time. Thus measuring A_t at different time intervals, a graph is plotted between $\log A_t$ and t which is a straight line (Fig. 3.5). The slope of such a line gives λ,

$$\lambda = \frac{\text{slope of the straight line}}{0.4343}$$

This method is known as direct method. The accuracy of the method is decided by the strength of the radioactive sample.

2. For very short half-lives, some special methods have been designed by different workers. Whenever a radio-nuclide decays emitting an alpha or a beta particle, the daughter nucleus recoils and if the daughter is also radioactive, it decays giving out radiations which are detected by radiation detectors. Moseley and Fajans devised a method known as rotating disc method. The schematic diagram of the method is shown in Fig. 3.9. Here the recoiled daughter nuclide (^{216}Po) are deposited continuously from the source on the surface of the rotating disc. Two radiation detectors are placed at suitable distance which receive the radiations from the decay of the recoiled radioactive nuclide. When the disc is rotated at uniform speed, the deposited radioactive nuclide passes by the detectors at regular intervals of time. The intensity of the radiations emitted from the deposited substance decreases as it passes from one detector to another and the difference in the intensity recorded by the detectors due to decay of the deposited radio-nuclide in the time interval corresponding to the passage of the point on the disc from one detector to the other. Half-life of ^{216}Po (ThA) has been measured by this method. Its value is $T_{1/2} = 0.158$ sec.

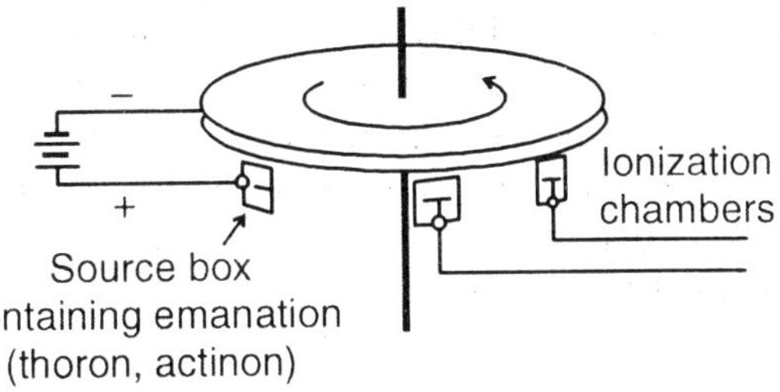

Fig. 3.9: Moseley and Fajan's rotating disc method for the determination of half lives

When the life-times are even shorter (~10^{-3} -10^{-4}s) then the recoil distance method devised by Jacobsen for ^{214}Po(Rac') is used. The daughter nucleus (recoil nucleus) from a parent nucleus moves down an evacuated tube (Fig. 3.10) at a certain valocity. During its motion it decays, the intensity thereby decreases. This decrease in the intensity of the radiation emitted is measured by the detector placed along the side of the evacuated tube from this intensity and the known recoil velocity of the daughter nucleus from the source, the decay constant is obtained. This method was used to measure the half life of recoil nucleus $^{214}_{84}$Po (Rac') from the parent source $^{214}_{84}$Bi (Rac). The recoil velocity was about 5 × 10^3 m/s. The half life of $^{214}_{84}$Po is 1.6 × 10^{-4} sec. For extremely short half life (10^{-7} -10^{-8} sec), delayed coincidence method has been found suitable. A schematic diagram is given in Fig. 3.11.

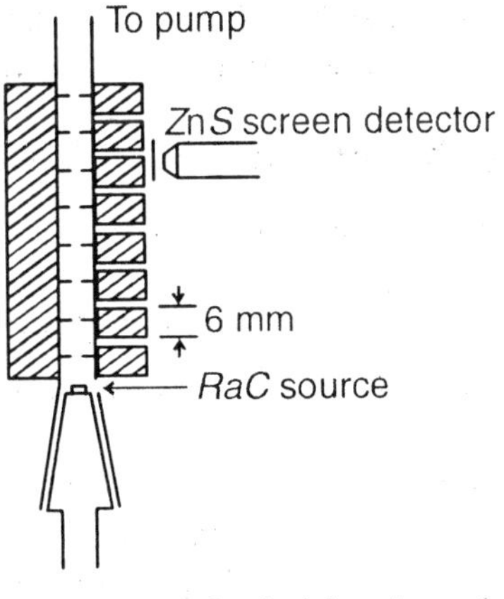

Fig. 3.10: Recoil distance method of Jacobsen for half life of ^{214}Po

From Fig. 3.11, we find that in this method there are two detectors, which detect two correlated signals occuring within a very short time interval between them. The parent or the source is ThC (^{212}Bi). It decays emitting a β^--particle which is detected by the β-detector. The daughter ThC' (^{212}Po) is an α-emitter. An α-particle from the daughter nucleus reaches the α-detector at a certain time after the β-signal. This interval of time between the two signals will be distributed according to decay law. The two pulses, so obtained in the two detectors are mixed according to delay coincidence technique and a single output pulse is recorded by a suitable counting device. The delay time for the β-emission can be varied and accordingly its dependence on the output signal for the β-α mixed pulse (coincidence output signal) can be determined and from this dependence λ can be determined. The half life for ThC' (^{212}Po) as found by the delayed coincidence method is 3×10^{-7}s.

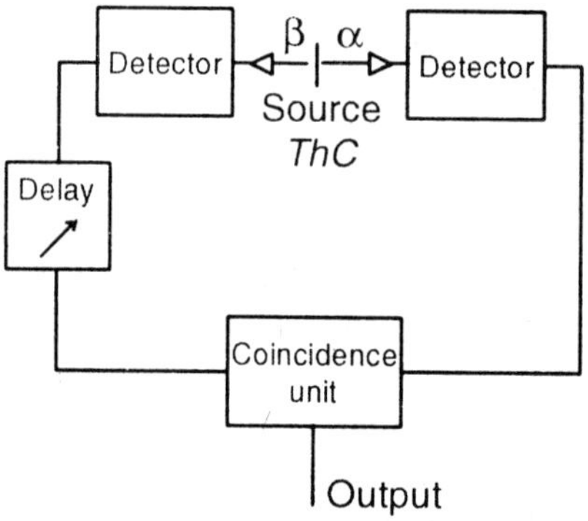

Fig. 3.11: Schematic diagram of delayed coincidence method for the half life ^{212}Po

When half life is long (~10^{15} years) and if there is secular equilibrium between the long lived parent and its daughter products then using eqn. (3.3.13), the half life of one can be calculated, if the half life of the other is known. For long half-lives, weighing method is also used where the number of active atoms in a sample can be calculated by the formula $N = x\frac{N_A}{A}$ where N_A is Avogadro number and A is atomic weight of the sample whose weight is x. Then the rate of disintegration $\frac{dN}{dt} = -N\lambda = -\frac{\lambda x N_A}{A}$. The rate of disintegration is determined by measuring the number of particles (α or β) emitted by the sample, from which λ can be calculated.

If the readioactive sample is not pure, but a mixture of number of radioactive elements, then the determination of decay constants of the components is not as straight forward as discussed above. The semi-log plot of the activity against time is not a straight line (Fig. 3.4) but a combination of such lines with different slopes (according to the decay constants of the constituents) merging into one another resulting into a curve, as shown in the Fig. 3.12. By analysing such curves, it is possible to determine the decay constants of the component radio-nuclides present in the sample. In the given figure, the curve is due to the mixture of two radio nuclides, such that the decay constant of one is greater than that of another ($\lambda_1 > \lambda_2$).

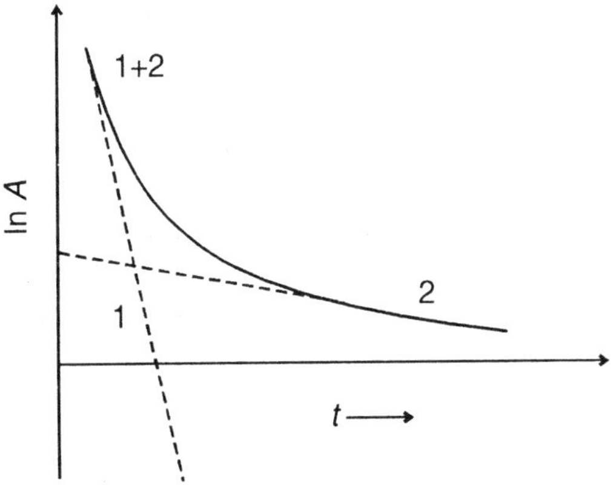

Fig. 3.12: Complex decay curve for a mixture of radio-elements of different activities (1 and 2)

If A_1 and A_2 are the activities of the radio-nuclides in the mixture, the resultant activity A at any instant of time t is

$$A = A_1 + A_2$$

$$= A_{10}e^{-\lambda_1 t} + A_{20}e^{-\lambda_2 t} \qquad ...(3.6.1)$$

A_{10} and A_{20} are the activities of nuclide 1 and nuclide 2 at $t = 0$. Since $\lambda_1 > \lambda_2$, $T_1 > T_2$ and as such in the initial stages of the decay, $t << T_2$, $\lambda_2 t$ is small, so that $e^{-\lambda_2 t} \approx 1$ and thus for small times,

$$A \simeq A_{10}e^{-\lambda_1 t} + A_{20}$$

or
$$\ln(A - A_{20}) \simeq -A_{10}\lambda_1 t \qquad ...(3.6.2)$$

When t is large enough, such that $t >> T_1$, the activity of first radio-nuclide almost disappears *i.e.* $A_{10}e^{-\lambda_1 t} \approx 0$ and then the activity of the sample is given by

$$A = A_{20}e^{-\lambda_2 t}$$

$$A / A_{20} \simeq \exp(^{-\lambda}{}_2 t)$$

or
$$\ln A - \ln A_{20} = -\lambda_2 t$$

Thus for large times ($t >> T_1$) when the activity of first reduces to almost zero, the semi-log plot is almost a straight line whose negative slope gives λ_2. When the straight line portion of the tail of the curve is extrapolated in the backward direction, it will cut the log$A$ axis at a point which will correspond to logA_{20} and thus A_{20} will be known. Then from the equation (3.6.1), a plot between log(A–A_{20}) and t, would be a straight line whose negative slope would give λ_1 and extrapolated value of the line for $t = 0$, would give A_{10}, the initial activity of the component 1.

Such an analysis can bc extended to the case of a mixture of more than two activities, provided the decay constants of the components are fairly different. If, however, the decay constants of the components are comparable, then such an analysis is not possible.

3.7 RADIOACTIVE DATING (AGE OF ROCKS AND MINERALS)

One of the earliest and most useful application of the radioactivity is to determine the ages of many specimens of geological and cosmological origin and hence to estimate the age of the earth. If we consider a radioactive mineral to be embedded in a rock at the time of its formation, then we would have the decay equation.

$$N_t = N_0 e^{-\lambda t} \qquad ...(3.7.1)$$

representing the number of radioactive atoms N_t left undecayed and t would then represent the time taken from the formation of the rock to say, the present times. N_0 corresponds to the number of radioactive atoms present initially at $t = 0$.

As the original mineral decays, it may adopt a course of producing a stable atom either in a single step or more then one step. Thus, a given sample as taken presently should contain atoms of a stable element and some atoms of un-decayed atoms of original sample. Assuming that the rock sample contains no other radioactive atoms and there is no loss of stable atoms, then the number of undecayed atoms plus the present number of stable atoms in the sample must be equal to the original number of radioactive atoms. If N_s is the number of stable atoms after time t then.

$$N_s + N_t = N_o \qquad ...(3.7.2)$$

or

$$N_s = N_o - N_t \qquad (N_o = N_t\, e^{\lambda t})$$

$$= N_t\, e^{\lambda t} - N_t$$

$$= N_t\, (e^{\lambda t} - 1)$$

$$\frac{N_s}{N_t} = (e^{\lambda t} - 1) \qquad ...(3.7.3)$$

$$\frac{N_s}{N_t} + 1 = e^{\lambda t}$$

$$t = \frac{1}{\lambda}\ln\left(\frac{N_s + N_t}{N_t}\right) \qquad ...(3.7.4)$$

Thus, knowing the decay constant of the original radioactive sample, and the ratio of stable atoms to the undecayed atoms after time t, the age (t) of the rock can be calculated.

At the time of solidification of the earth's crust, radioactive minerals were trapped in the rocks. Most of the radioactive nuclides are members of four radioactive series. If we consider the Uranium series, then the parent nuclide Uranium undergoes a succession of daughter products and ultimately ends up in stable atoms of lead. Assuming that no lead (A = 206) was present when uranium bearing rock was formed and whatever amount of lead is formed in the rock is all of radiogenic origin, then in the equation (3.7.4), N_s is the number of lead (A = 206) atoms and N_t is the number of undecayed uranium atoms present after time t.

$$t = \frac{1}{\lambda_U} \ln\left(\frac{N_{Pb} + N_U}{N_U}\right) \qquad \text{...(3.7.5)}$$

The number of lead and uranium atoms (N_{Pb} and N_U) or their respective masses can be determined by chemical or mass spectroscopic methods. λ_U being known, the age of the rock sample, can be calculated using equation (3.7.5). Similar calculations can be made for other natural radioactive series. Since in all the series, the end product is stable lead, hence this method of estimating the age of rocks is known as *lead method.*

Note: The decay process in the series takes place through more than one step (about 14 in the case of ^{238}U) till the stable end product is reached. The half-lives of the intermediate products are very short compared to the long half life of uranium and as such only the uranium and lead contents of a sample are considered in finding the ratio N_{Pb}/N_U.

The disintegration of the naturally occuring radioactive series is accompanied by the evolution of helium gas resulting from the emission of alpha particles. Presuming that no helium gas was present when the radioactive mineral say uranium bearing mineral was formed. During the successive disintegration of uranium, the emitted alpha particles turned into neutral helium gas atoms which are trapped in the rock containing uranium. For each radio nuclide of uranium-238 that reduces to stable lead atom (^{206}Pb), eight alpha particles are emitted $\left(^{238}U \rightarrow {}^{206}Pb + 8\,^4_2He + 6e^{-1} + \gamma\right)$. Hence the number of helium-gas atoms N_{He} formed at any instant of time t is eight times the number of uranium nuclides (N_0-N_U) decayed during this period, *i.e.*

$$N_{He} = 8(N_0 - N_U) = 8N_U\left(e^{\lambda_U t} - 1\right)$$

where N_0 is the number of uranium nuclides present initially (t = 0). The equation (3.7.3) then becomes

$$\frac{N_{He}}{N_U} = 8\left(e^{\lambda_U t} - 1\right) \qquad \text{...(3.7.6)}$$

where λ_U is decay constant of uranium. The time t is given by the expression

Thus by measuring the amount of the parent uranium present at time t and the amount of accumulated helium gas within the uranium bearing rock, the age 't' of the sample can be calculated. This method is termed as *Helium method.*

The natural uranium contains two radioactive isotopes ^{238}U and ^{235}U. The later decays with its own decay constant to a stable lead-atom (^{207}Pb) giving out seven alpha particles. Thus seven neutral helium atoms are produced for the decay of each ^{235}U atom. Thus the total number of helium atoms produced on the transformation of one ^{238}U and one ^{235}U atom each will be,

$$N_{He} = 8N_U\left(e^{\lambda_U t} - 1\right) + 7N_{U'}\left(e^{\lambda'_U t} - 1\right) \quad ...(3.7.7)$$

where N_U and $N_{U'}$ are the number of ^{238}U and ^{235}U atoms left after time t and λ_U and $\lambda_{U'}$ are the corresponding decay constants.

Since there is a chance of loss of some helium gas by diffusion as such there has to be some error in the estimation of the age of a rock by helium method.

Besides, Lead and Helium methods, there are two other methods known as Potassium method and Rubidium methods. These long lived radio nuclides can exist in the rocks at the time of their formation. Radioactive potassium nuclide ^{40}K has half-life of 1.3×10^9 years. It decays to ^{40}Ar. In calculating the age from the Argon-40 to Potassium-40 ratio, one has to take account of the fact that the decay of potassium has two decay-modes namely 11 percent of it decays to Argon-40 by orbital electron capture with a half life of about 1.2×10^{10} yrs. and the remaining 89% to Calcium-40 by beta emission (β^-) with a half life of around 1.5×10^9 yrs. The Argon to Potassium ratio (^{40}Ar/^{40}K) at time t is, therefore, given by

$$^{40}\text{Ar} / {}^{40}\text{K} = 0.11(e^{\lambda t} - 1)$$

where λ is the total decay constant of potassium-40.

Likewise, the rocks or minerals containing radioactive Rubidium (^{87}Rb) decays to stable Strontiim-87 emitting a β particle ($^{87}\text{Rb} \rightarrow {}^{87}\text{Sr} + \beta^-$) with a half life of about 4.7×10^{10} years. The age of a rock sample containing Rubidium-87 can be estimated by determining the ratio of active rubidium-87 atoms to stable strontium-87 atoms (N_{Sr}/N_{Rb}).

Thus the above methods have been utilised to determine ages of the order of 10^9 years, which is also the order of the age of the earth. The life times of Lunar rocks and stony meteorites have been estimated by the potassium-40 decay methods to the order of 4.7×10^9 years.

Carbon Dating

Radioactive carbon-14 (an isotope of carbon) has been found useful in estimating age of fossils, plants and archeological samples. Hence the name carbon-dating. Radioactive nuclides like uranium, thorium, etc. are mostly trapped in the rock minerals at the time of their formation, while radioactive carbon-14 is generally produced in the atmoshpere by cosmic rays. The high energy nuclei, mostly protons present in the cosmic rays collide with atoms in the atmosphere producing neutrons which may combine with nuclei of nitrogen-14 present in the atmosphere. Carbon-14 is formed ($^{14}\text{N} + {}^1\text{n} \rightarrow {}^{14}\text{C} + {}^1\text{H}$).

Once the ^{14}C is formed, it combines with oxygen molecule forming carbon-di-oxide. Green plants take up this carbon-di-oxide during the process of photo-synthesis. Animals eat plants and also give out carbon-di-oxide during respiration. The animal plant cycle and the production of ^{14}C in the atmosphere in the course of time, maintains the concentration of the ^{14}C. Due to the efficient mixing of ^{14}C, the ratio of radio carbon to ordinary carbon (^{12}C) is same every where. There exists an equilibrium between the two isotopes of carbon (^{14}C and ^{12}C).

When animals and plants die, the intake of ^{14}C and ^{12}C is stopped, but the ^{14}C present in them continues to decay with a half life of 5760 years producing ^{14}N. In other words after 5760 years, the sample would have only one half as much radio carbon-14 left relative to their total carbon content which was present at the time of the sample being a living matter. Thus by determining the ratio of radioactive carbon-14 to ordinary carbon in a sample, it is possible to estimate the age of the sample.

The methods of carbon dating has been extensively used to determine ages of ancient objects, remains of organic orgin, archeological specimens like wooden articles, textiles, mummies, leather goods and many other specimens from caves, and tombs. Quantitatively this method has been employed for measuring ages around 5000 to 50,000 years.

Another radioactive isotope tritium (^3H) having half life of 12.4 years has been used in determining ages of sample of glacial origins.

3.8 ALPHA-PARTICLES

3.8.1 Experimental Information About α-particles

The discovery of natural radioactivity and the identification of radiations emitted by radioactive elements as α, β and γ-rays made a starting point for the study and understaning of nucleus and its properties. Lord Rutherford was the first to identify these radiations through their deflections in electric and magnetic fields. Further, his co-workers, Geiger, Soddy, Marsden and others, measured the charge (q) on the α-particles, the charge (q) to mass (M) ratio (q/M) of these particles and subsequently the mass of alpha particles. In measuring the q/M of alpha particles, the principle of the method is same as adapted for e/m of cathode rays by J. J. Thomson.

The q/M value for the α-particles as obtained by Lord Rutherford and his co-worker Robinson is

$$\frac{q}{m} = 4.82 \times 10^7 \text{ Coulomb/kgm}$$

Rutherford and T. Royds showed experimentally that α-particles were doubly ionised helium atoms or helium nuclei and Rutherford and H. Geiger measured the charge on α-particles and found it to be twice the charge of the electron. The value of q/M when calculated for doubly ionised helium, it came out to be very close to the value obtained through experiments on α-particles. The numerical values of charge and mass of the α-particle are respectively,

$$q = 3.19 \times 10^{-19} \text{ Coulomb}$$

$$m = 6.62 \times 10^{-27} \text{ kgm.}$$

One of the important properties ot alpha particles is to produce ionisation as they travel through matter. The extent of ionisation depends on the energy of the α-particles emitted by a certain radioactive source. This property helped their detection in most of the nuclear detectors like ionisation chambers, counters, cloud chambers, scintillation counters, etc. (Chapter 8).

3.8.2 Range of Alpha-Particles

The α-particles when travel through matter lose energy through collision with the particles of matter (air or gas). Normally 34 eV of energy is needed to produce one ion-pair in air. Because of their large mass, the loss of energy per collision is relatively small. A large number of collisions, therefore, (around 10^6) take place along the path till energy falls below the ionisation energy of the medium concerned. After travelling a certain distance, the α-particles are absorbed. This distance upto which α-particles travel in a medium before being absorbed is known as Range of α-particles. The range of α-particles in air is given at a temprature of 15°C and 760 mm of mercury.

If the alpha particles emitted by a radioactive source are counted either in ionisation chamber or by scintillations produced on a zinc-sulphide screen and the counts per second are plotted as a function of distance from the source, one gets a curve shown in Fig. (3.13).

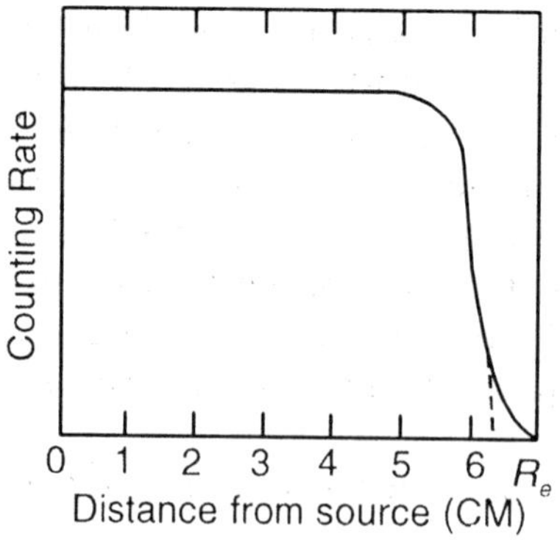

Fig. 3.13: Counting rate of α-particles as a function of distance from the source

The counting rate stays practicallv constant upto a certain distance from the source and then drops rapidly to zero. Since the curve does not fall to zero vertically, it is difficult to know the end of the path and hence the range. As such, one normally measures the extrapolated range R_e, which is obtained by drawing a tangent to the curve at its point of inflection. The distance at which the counting rate drops to half its original value is known as the range R of the α-particles.

Stopping Power

The energy lost by α-particle per unit distance as it travels through the media is known as stopping power.

$$S(E) = -\frac{dE}{dx}$$

For practical purposes, it is more convenient to use the relative stopping power (*R.S.P.*). For any material,

$$R.S.P = \frac{\text{Range of } \alpha \text{ - particle in air}}{\text{Range of } \alpha \text{ - particle in the material}}$$

In defining *R.S.P.*, air is being treated as a standard at 15°C and pressure of 1 atmosphere (760 mm of Hg) and the same source is used both for air and the material. Sometimes it becomes useful to express the stopping power of material by equivalent thickness expressed in mg/cm^2. Equivalent thickness in $\frac{\text{mg}}{\text{cm}^2}$ = Actual thickness (cm) or range × density (gm/cm^3) × 1000.

$$R_{eq}\left(\frac{\text{mg}}{\text{cm}^2}\right) = R(\text{cm})(\text{gm} / \text{cm}^3) \times 1000$$

The equivalent thickness thus gives the mass per unit area or the thickness of the material needed to absordb the α-particles. It is easier to determine the mass and density than thickness hence it is advantageous to express the range in this manner. The thickness of material which is equivalent in stopping power to 1 cm of air is given by

$$\text{Thickness in} \frac{\text{mg}}{\text{cm}^2} \text{ quivalent to 1 cm of air} = \frac{\text{Density } \times 1000}{\text{Relative Stopping power}}$$

The values of these quantities are listed in Table 3.1 for some of the most commonly used foil materials. These values are for α-particles from Rac' (^{214}Po) which have an extrapolated range 6.953 cm in air at 15°C and 760 mm of Hg.

TABEL 3.1

Substance	*Extrapolated range, cm*	*Relative stopping power*	*Density gm/cm^3*	*Equivalent thickness mg/cm^2*	*Thickness in mg/cm^2 equi-valent to 1 cm of air*
Mica	0.0036	1930	2.8	10.1	1.45
Aluminium	0.00406	1700	2.7	11.0	1.57
Copper	0.00183	1800	8.93	16.3	2.35
Gold	0.0014	4950	19.33	27.1	3.89

Bragg-Curve

Bragg and Kleeman were first to measure the range of α-particles in gases. The radioactive source of alpha particles was placed on a movable platform placed in a cylindrical tube. At the other end of the tube, an ionisation chamber, consisting of a plate and a grid, was placed. The ionisation produced by α-particles at different distances from the source was determined by the ionisation chamber. Th ionisation current was measured by the electrometer connected in series with the plate. A potential difference was maintained between the grid and the plate for this purpose. A schematic diagram of the experimental arrangement used by Bragg and Kleeman is shown in Fig. 3.14. Since α-particles collide with the atoms or molecules of the gas, a large number (~10^6 collisions) of collisions along its path take place till the energy of the α-particles falls below the ionisation energy of the gas. Specific ionisation *i.e.* ion-pairs or ionisation per unit path length, is determined at various distances from the source. When specific ionisation is plotted against distance from the source. One gets a curve known as *Bragg curve* shown in Fig. 3.15.

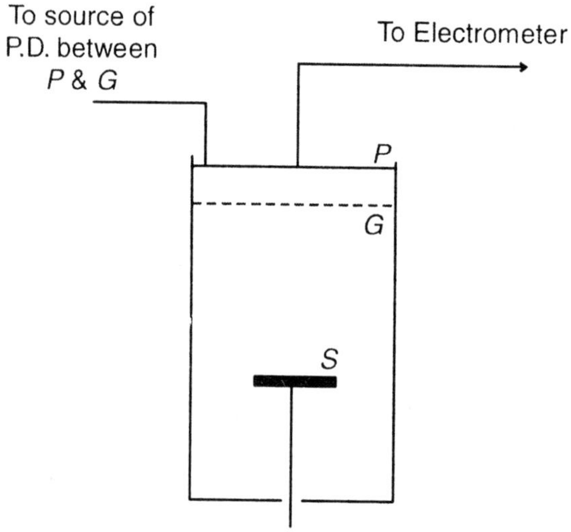

Fig. 3.14: Schematic arrangement of apparatus used by Bragg and Kleeman for Bragg curve (the range of α-particles)

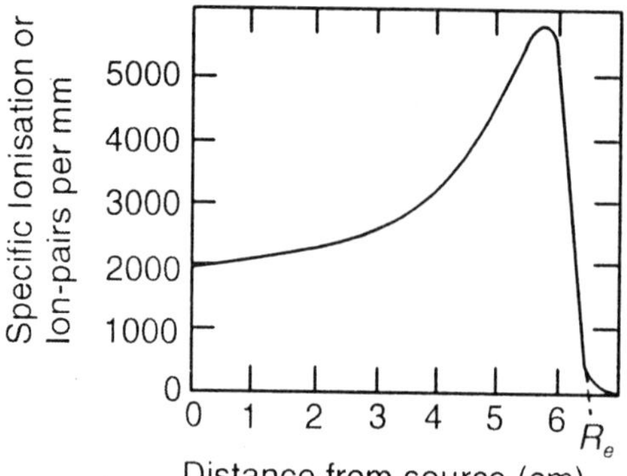

Fig. 3.15: Bragg-curve

The specific ionisation remains almost constant, upto a certain distance, it reaches a maximum and then gradually falls to zero. The range of α-particles is not precise, very near the end of the path. The curve exhibits a "tail" instead of vertically going to zero on the distance axis. Since the α-particles lose energy along their path due to collisions with the particles of the air or gas, their velocity thereby decreases and as such they are able to stay for more time towards the end of thier journey producing more ionisation and hence the curve shows maximum before going to zero. The tail could be explained on the basis of a phenomena known as straggling. The straggling is due to the random collisions of α-particles with the molecules of medium (air gas). Though all the α-particles start with the same initial velocity yet some may collide for more number of times than others. There is thus always a possibility that some α-particles may loss less energy than others and hence such particles will travel slightly farther. It is also possible that near the end of their path, the α-particles start getting neutralised, first some of them may absorb one electron, then He^+ may still continue to travel producing ionisation and finally it may absorb one more electron becoming a neutral Helium atom.

3.8.3 Disintegration Energy of α-particles

It is seen that α-particles emitted by a radioactive source have a well-defined range. This indicates that the α-particles also have a well defined energy (the range energy relationship is discussed in Section 3.8.4). The nucleus emitting α-particle is normally at rest (in laboratory system) having zero mementum. One, therefore, regards that in the process of α-emission, both the energy and momentum are conserved.

Symbolically the emission of an α-particle from a parent nucleus, say X is represented as,

$$^{A}_{Z}X \rightarrow {}^{A-4}_{Z-2}Y + {}^{4}_{2}He$$

where Y is the daughter nucleus and $^{4}_{2}He$ is Helium nucleus (α-particle).

$$^{238}_{92}U \rightarrow {}^{234}_{90}Th + {}^{4}_{2}He$$

Once the alpha particle is emitted, it carries certain kinetic energy and momentum and the daughter nucleus recoils with certain kinetic energy. The momentum it carries is equal and opposite to that carried by the α-particle. If M_α and m_f are the masses of α-particle and the daughter nucleus and v_α and v_f are their respective velocities then the momentum conservation condition is

$$m_f v_f = M_\alpha v_\alpha \qquad \text{...(3.8.1)}$$

and energy conservation condition is

$$Q = \frac{1}{2} m_f v_f^2 + \frac{1}{2} M_\alpha v_\alpha^2 \qquad \text{...(3.8.2)}$$

where Q is the energy released in the decay of the parent nucleus assumed at rest in the laboratory system. It is also known as the disintegration energy. Using (3.8.1) and (3.8.2), we get,

$$Q = \frac{1}{2} M_\alpha v_\alpha^2 \left(1 + \frac{M_\alpha}{m_f}\right)$$

Putting the kinetic energy of α-particle as K_α, then

$$K_\alpha = \frac{1}{2} M_\alpha v_\alpha^2$$

and $$Q = K_\alpha \left(1 + \frac{M_\alpha}{m_f}\right)$$

$$K_\alpha = \frac{m_f}{M_\alpha + m_f} Q \qquad \ldots(3.8.3)$$

The energy of the recoiled daughter nucleus will be

$$K_f = \frac{M_\alpha}{M_\alpha + m_f} Q$$

If A is mas number of the parent nucleus, and taking the masses, in units of a.m.u. to be of the order of mass numbers, we can write

$$K_\alpha = \frac{A-4}{A} Q \qquad \ldots(3.8.4)$$

$$M_\alpha + m_f \approx A \text{ and } m_f \sim A - M_\alpha \approx A - 4$$

The value of Q is obtained from the change in mass in the decay process *i.e.*

$$Q = (m_i - m_f - M_\alpha)\, c^2 \qquad \ldots(3.8.5)$$

$$Q = 931\,(m_i - m_f - M_\alpha) \text{ MeV} \qquad \ldots(3.8.6)$$

when masses are expressed in a.m.u.

Because, for the spontaneous emission of α-particles, it is necessary that Q must be positive or $m_i > m_f + m_\alpha$. In equ. (3.8.6) $m_i c^2$, $m_f c^2$ and $m_\alpha c^2$ are the rest mass energies of parent nucleus, daughter nucleus and the alpha particle respectively. It may be noted that for almost all the α-emitters in natural radio nuclides, the Q value for the emission of any other nuclear particle like a neutron, a proton or a deuteron comes out to be negative and hence α-emission is always preferred over emission of other nuclear particles, even though a nucleus contains only neutrons and protons. As an example, for the decay of ${}^{238}_{92}U \rightarrow {}^{234}_{90}Th + {}^{4}_{2}He$, the disintegration energy $Q = 5.40$ MeV as calculated from equation (3.8.6), while for neutron or proton emission $Q = -7.16$ MeV or $Q = -6.05$ MeV respectively. The emmission of α-particle is made possible due to its large binding energy.

The mass numbers of most of the α-emitters are around more than 200, the disintegration energy appears mostly as the kinetic energy of α-particle. (The daughter nucleus, being heavy compared to α-particles, carries very small amount of the kinetic energy). In the decay of ^{222}Rn, the disintegration energy Q = 5.587 MeV while the observed kinetic energy of α-particle K_α = 5.486 MeV.

3.8.4 Range-Energy Relationship for α-Particles

It is seen that as α-particles travel in a medium they lose energy, consequently their velocity decreases. Geiger (1910) while working in Rutherford's Laboratory, measured velocity of α-particles after they traversed through mica sheets of different thicknesses whose stopping power relative to air was known. Using Ra'C' as a source of α-particles, Geiger measured the velocity of alpha particles using a scintillation method of counting, and found that the velocity v of the α-particle at any point at a distance x from the source could be expressed as

$$v^3 = C(R - x) \qquad \ldots(3.8.7)$$

where C is a constant annd R is range in standard air. If $x = 0$, then the initial velocity of α-particles at the source will be

$$v_0^3 = CR \qquad \ldots(3.8.8)$$

This relationship between range R and initial velocity v_0 of α-particles is known as the Geiger formula or Geiger's Law. If different sources are used then the velocity of α-particles from different sources and their respective range are related according to Geiger formula. The constant C has same value in all cases, having value equal to 1.03×10^{27}.

$$v_0^3 = 1.03 \times 10^{27} R \qquad \ldots(3.8.9)$$

The energy E_0 of an alpha particle in MeV is

$$E_0 = \frac{1}{2} M v_0^2 = 0.28 \times 10^{-14} v_0^2$$

where v_0 is in meters/sec.

Hence it follows from eqn. (3.8.8), that

$$E_0^{3/2} = 3.09 \times 10^{-2} R \qquad \text{or} \qquad E_0 = 2.12 \times 10^2 R^{2/3} \qquad \ldots(3.8.10)$$

where E_0 is the initial energy of the alpha-particle in MeV and R is its range in meters of air.

Thus, it is possible to calculate the energy or the velocity of alpha particle emitted by a source, if its range is known. For example, the range of α-particle from Radium is 3.29×10^{-2} m of air, then the equation (3.8.10) gives the energy of such α-particles as 4.70 MeV, whereas the measured value is 4.79 MeV.

The Geiger Law or the Geiger formula is found to be applicable only for alpha particles having range between 3 to 7 cms of air. For lower ranges, the range R is approximately proportional to $v_0^{3/2}$ and at higher ranges, it is proportional to v_0^4. However, for α-particles emitted from different sources having varying kinetic energies, and mean Ranges R_0 in air, the value of $R_0 / E_0^{3/2}$ is almost constant, exhibiting the validity of Geiger formula.

3.8.5 Geiger-Nuttall-Law

We have seen that the range (in terms of standard air) of most of the naturally occuring radioactive elements vary between 2.8 cm (^{232}Th) and 8.6 cm (^{212}Po). Rutherford suggested (1907) that there should be a relationship between the range and life time of the alpha-emitting element. When decay constants λ or the half life T of the α-emitters were examined, it was found that the longest lived element has shorter range and a short lived one has long range e.g. for ^{232}Th, $R = 2.8$ cm and $T = 1.39 \times 10^{10}$ yrs and for ^{212}Po, $R = 8.6$ cm and $T = 3.0 \times 10^{-7}$ sec. However, large amount of data were necessary to establish an empirical relationship. After a span of about four years in 1911, Geiger and Nuttall, working in Rutherford's Laboratory experimented upon some 10-20 specimens from naturally occuring α-emitters and were able to put forward a relation between range R of α-particles and decay constant λ of the corresponding α-emitting element. When the logarithm of decay constant, log λ, against the logarithm of the alpha particle range in air, is plotted, an approximately straight line is obtained. Such lines are obtained for each of the natural radioactive series, all being parallel.

The empirical relation between λ and R is expressed as,

$$\log\lambda = C\log R + D \quad \ldots(3.8.11)$$

where C and D are constants.

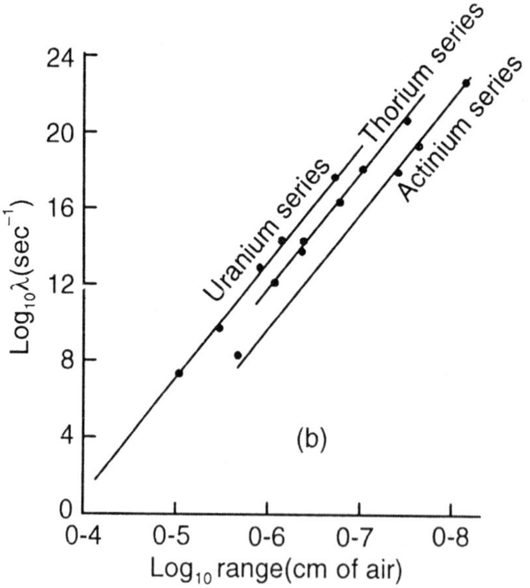

Fig. 3.16: Variation of log λ with log R for natural radioactive series

The constant 'C' has same value for the straight lines for all the α-emitters of the series of naturally occuring radioactive element. This confirms the parallel nature of the lines. The expression (3.8.11) is known as Geiger-Nuttall law.

Since the range of an alpha particle of a few MeV energy is related with its velocity by Geiger's rule (Eqn. 3.8.10),

$$R \propto v^3$$

or $$R \propto E^{3/2}$$

The Geiger-Nuttall law may be expressed as a relation between decay constant λ and energy E of the emitted alpha particle as

$$\log \lambda = C' \log E + D' \qquad \ldots(3.8.12)$$

where C' and D' are again constants.

It is now known that with the availability of a considerable amount of information relating decay constants (or half-lives) to alpha particle ranges (energies), the Geiger-Nuttall law is an approximate and is applicable only to even Z-even N nuclides. e.g. Uranium ($Z = 92$, $N = 146$) and Thorium ($Z = 90$, $N = 144$) series.

Since $\lambda = \dfrac{0.693}{T}$, the Fig. 3.17 shows the variation of log T with log E.

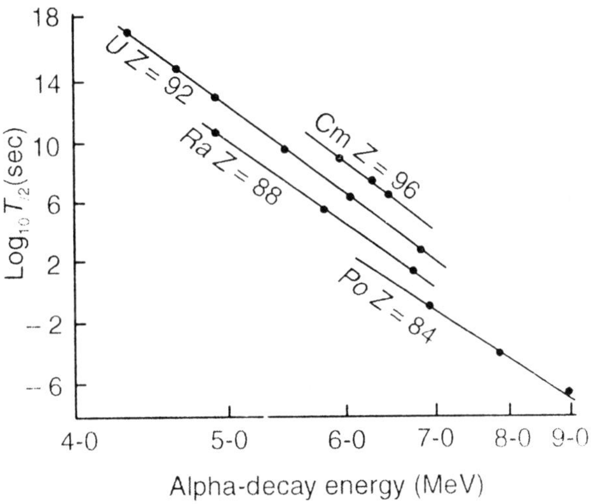

Fig. 3.17: Variation of log T with log E

3.8.6 FINE STRUCTURE OF α-PARTICLE SPECTRUM

So far we have regarded that all the α-particles emitted by a radioactive nucleus have same energy and same range (excluding straggling effect). But in reality, it has been observed that α-particles of different energies are emitted by an α-emitting source, e.g. alpha particles from ^{238}U have energies of 4.18 MeV and 4.13 MeV and ^{230}Th gives four groups having energies 4.682 MeV, 4.615 MeV and 4.471 MeV and 4.436 MeV. Such a fine structure could not have been observed with poor resolution experiments. However, E. Rutherford and A. B. Wood first observed fine structure in α-energies in 1916 and later on around 1930s. Rosenblum in France confirmed the emission of α-particles of different energies from the same radioactive substance. There can be two possibilities for such a fine structure in α-spectrum.

(*i*) The parent nucleus X staying in its ground state, when decays, the daughter nucleus may be formed in its ground state OR in one of the excited states as shown in Fig. 3.18. The excited daughter nucleus may lose its energy by emitting γ-rays. *The ground to ground transition emitting alpha particles, (group α_0) with energy E_0 constitute the main group (α_0).* α_1, α_2, α_3 are the other groups with energies E_{α_1}, E_{α_2}, E_{α_3}. The Fig 3.18 shows these transitions.

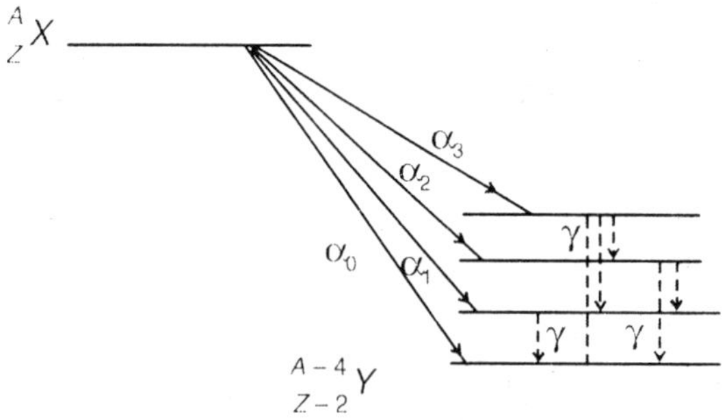

Fig. 3.18: Fine structure of alpha-particle energy spectrum. The nucleus X is in ground state

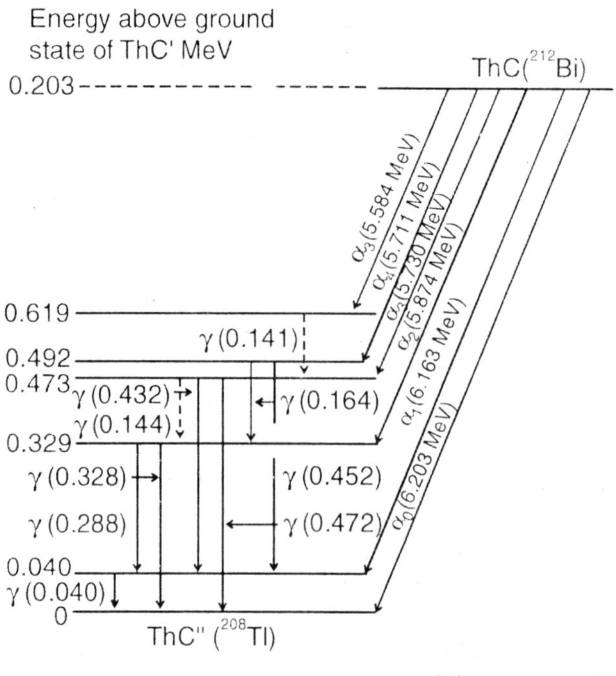

Fig. 3.19: Disintegration scheme for α-decay of ^{212}Bi (ThC) to ^{208}Tl (ThC)

Such a situation has been observed in the decay of ^{212}Bi (or ThC) through α-emission to ^{208}Tl (or ThC'). The disintegration scheme for ^{212}Bi (or ThC) is shown in Fig. 3.19, where six α-particle transitions as well as the subsequent γ-rays emitted are shown.

The energies of α-particles and their relative intensities (abundances) are given in the Table 3.2.

TABLE 3.2

Group		*α-Energy MeV*	*Relative Intensity (abundance)*
(main group)	α_0	6.203	27.2
	α_1	6.163	69.8
	α_2	5.874	1.70
	α_3	5.730	0.15
	α_4	5.711	1.10
	α_5	5.584	0.016

From the above table, we find that the transition from the ground state of ThC to the first excited state (or the second state) is most probable, because the α-particles arising out of these transitions constitute the largest proportion in the entire group.

(*ii*) In the other case, the parent nucleus X, itself is excited having various energy states and decays through transitions to the ground state of the daughter nucleus Y. These transitions are shown in Fig. 3.20. The main group α_0 has energy E_{α_0} and other groups α_1, α_2, α_3 have energies E_{α_1}, E_{α_2}, E_{α_3}. These energies are greater than the energy E_{α_0} of main group α_0. Therefore, these alpha particles having *large energies are known as long range alpha particles*. As an illustrative example of such a case is the decay of ^{212}Po (or ThC'), to ^{208}Pb (or ThD) the disintegration scheme of which is shown in Fig. 3.21 where the γ-ray transitions are also shown. The relative intensities (abundances) and energies of various groups are shown in the Table 3.3.

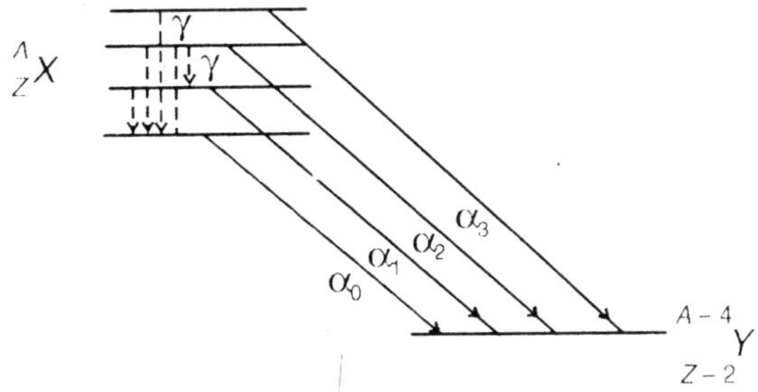

Fig. 3.20: Long range α-particles nucleus Y is in ground state

TABLE 3.3

Group	*Energy of particles MeV*	*Relative intensity (abundance)*
α_0	8.945	10^6
α_1	9.675	34
α_2	10.622	20
α_3	10.746	190

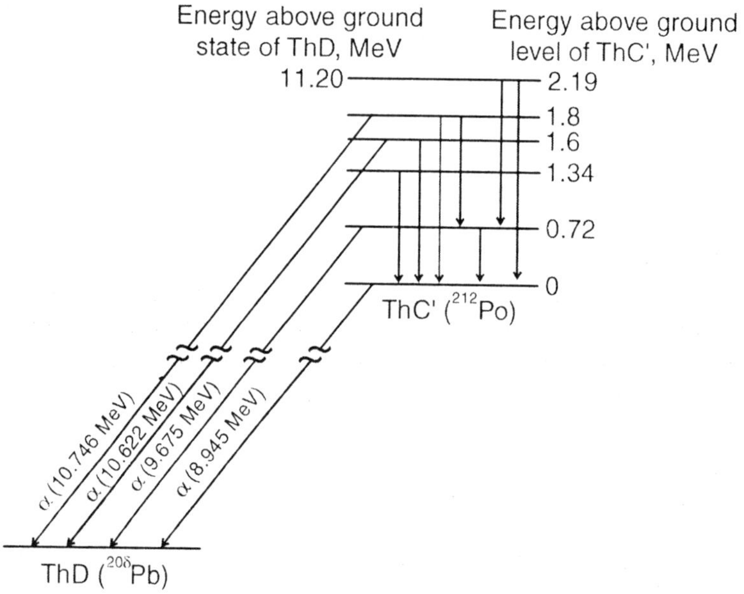

Fig. 3.21: Disintegration scheme for α-decay of ^{212}Po to ^{208}Pb

The fine structure in α-particle energy spectrum, thus depends on the nuclear energy levels of the parent and daughter nuclei involved in the α-particle disintegration. This is further confirmed by the strong correlation between the energies of the groups of alpha particles and the energies of the emitted gamma rays.

3.8.7 α-disintegration: Theoretical Explanation

We have seen that energetically it is possible for a nucleus of mass M_A to emit an alpha particle of mass m_α, whenever $M_A > M_{A-4} + m_\alpha$ where M_{A-4} is the mass of daughter nucleus. This is found to be so in naturally occuring radioactive elements for $Z \geq 82$ and that is why there does not exist permanently stable nuclei in nature above lead $\left({}^{208}_{82}\text{Pb}\right)$. Alpha emission also does not occur readily even when energetically it is possible, for example, the α-active nucleus ${}^{209}_{93}\text{Bi}$ waits for almost 10^{17} years before it emits an α-particle. Another characteristic feature of the alpha-emitting nuclei, is that shorter the life time of alpha emitter larger is the energy of emitted alpha particle. The energy of alpha particles from natural radioactive elements ranges between 4 MeV to 9 MeV while the half-lives associated with these emissions differ by a wide range. In the case of ^{238}U, the energy (E_α) of α-particle is 4 MeV and the half life (T) is 10^9 years, whereas for ^{212}Po, these values are E_α = 8 MeV and $T = 10^{-7}$ sec.

Normally one does not expect an alpha-particle to exist as an entity inside the nucleus. But we may assume that there is some possibility of two protons and two neutrons to combine (under the action of strong nuclear forces) to produce an alpha particle at a certain instant. Let us, then examine, how the potential energy of the α-particle varies as it escapes from the nucleus. Inside the nucleus, the alpha particle is under the influence of strong attractive short range nuclear forces which are zero outside the nucleus, as a result the alpha particle has some negative potential energy. Just outside the nucleus, the nuclear forces are no longer operative and the α-particle experiences a large repulsive (Coulombian) force between it and the daughter nucleus and thus gets a positive potential energy which decreases as the particle moves away from the nucleus till it becomes zero at infinity or even at very large distances from the centre of the nucleus. A graphical representation of the potential energy of the α-particle is shown in Fig. 3.22. The distance r is measured from the centre of the nucleus. In reality there is a discontinuity at $r = b$, the nuclear radius, where the potential energy abruptly jumps from $-V_a$ to $+ V_b$ and then it decreases as $\frac{1}{r}$ with r. The exact form of nuclear attractive forces is not known but it is customary to regard the nuclear potential, $-V_a$, to be constant and hence shown as a straight horizontal line (a square well potential). The vertical line through $r = b$ shows the abrupt change in the potential energy of the particle from $-V_a$ to $+V_b$. Thus the form of the potential is like a step potential of a certain height (like Volcano with a deep crater). The line AB represents the total energy, E_α of the alpha particle. From the figure we find that the total energy E_α is less than V_b and as such the alpha particle does not have sufficient energy to come out of the nucleus and appear at its outer surface. It is just like a ball bouncing inside a box to a height less than the wall of the box. In the same way, an alpha particle approaching a nucleus from large distance will experience a repulsive force according to Coulomb's law, and once it is able to enter the nucleus, it loses its identity and stays inside the nucleus under the effect of strong attractive nuclear forces. Nevertheless, suppose that the alpha particle has energy E_α inside the nucleus which is less than the potential barrier V_b, then according to classical physics it cannot come out of the nucleus. Similarly if it approaches the nucleus from outside with this energy ($E_\alpha < V_b$), it has to overcome the barrier, but again classically it cannot surmount the barrier. As a matter of fact, with $E_\alpha < V_b$, the alpha particle can reach only upto C at $r = r_c$ (Fig. 3.22) and gets reflected, and so will be the case for alpha particle inside the nucleus. It will be reflected back, whenever it strikes the wall of the potential at B (Fig. 3.22). Since most of the natural radioactive elements are heavy elements, Z > 82, which emit α-particles, and for such α-

emitters, V_b is around 25 to 30 MeV, while the energy of the emitted α-particles from these nuclei is around 4 MeV to 9 MeV. Besides, Rutherford had found that natural Uranium emits α-particles of 4 MeV energy whereas it scatters α-particles of 9 MeV energy. Such a paradoxial behaviour of α-particles from natural radioactive α-emitters presented a serious problem to be explained. Classical physics and even modern physics in the early days could not find suitable answer. Nevertheless, the Quantum mechanical tunnel effect could find a suitable explanation. It states that there is a finite, though small, probability that the particle can leak through the energy barrier even though $E_\alpha < V_b$. In other words, although the particle does not have enough energy to surmount the barrier, it can tunnel through it. Thus, alpha particles with energies 4 MeV to 9 MeV are able, so to speak, to tunnel through an energy barrier of greater magnitudes 25 MeV to 30 MeV. Such penetration of α-particle is known as Tunnel effect. G. Gamow (1928) and also Condon and Gurney, (independently) developed mathematical formalism for the confirmation of quantum mechanical ideas for α-decay. The results so obtained agree with the experimental observations on α-decay. We describe here Gamow's theory of α-decay. There are certain basic notions in the Gamow's theory of α-decay, which are mentioned below.

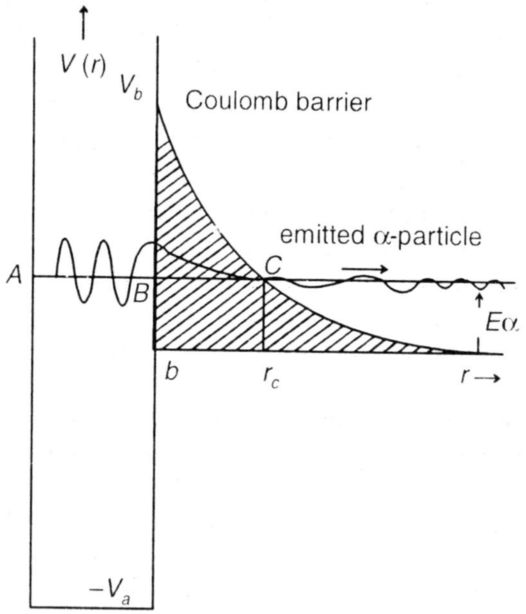

Fig. 3.22: Potential energy curve for α-particle. V_b is height of Coulomb potential, b denotes nuclear radius. The nuclear potential is denoted by $-V_a$

1. An alpha particle exists as a unit inside the nucleus.
2. It is contained in the nucleus by the surrounding potential barrier. It has certain potential energy and is in constant motion.

3. The problem is treated as one dimensional with no angular dependence ($l = 0$). It means that the angular momentum of the emitted α-particle and the daughter nucleus about their centre of mass is zero. (This will be so in most of the α-emitters which are even Z-even A nuclei).

Let us consider the Fig. 3.22. The alpha particle once formed, moves to and fro along the nuclear diameter, $2b$, (twice the width of nuclear potential) and in doing so, it strikes continuously the barrier at B. During such large number of collisions, there is a likelihood that once it leaks through the barrrier of thickness BC and appears at C, from where it travels with energy $(E_k - V_a)$ where E_k is its kinetic energy inside the nucleus. At large distances where V_a is zero, the energy E_k is same as the Energy E_α, the disintegration energy. For the sake of simplicity we shall take the energy of alpha particle as E_α even at finite distances beyond point C (same as inside the nucleus).

Let a radioactive nucleus $(Z + 2, A + 4)$ emits an alpha particle with decay constant λ. If P is the transmission coefficient per collision and ν is the frequency of collisions *i.e.* the number of collisions per sec the α-particle makes before it escapes the nucleus, then

$$\lambda = \nu P \qquad \text{...(3.8.13)}$$

If v is the velocity of alpha particle and $2b$ is the diameter of the nucleus, then the frequency of collisions is given by

$$\nu = \frac{v}{2b} = \frac{\sqrt{2E_\alpha / m_\alpha}}{2b} \qquad \text{...(3.8.14)}$$

Note that we have assumed $E_k = E_\alpha$ inside the nucleus. m_α is mass of the α-particle.

The radius of a nucleus is about 10^{-14} m and the velocity of α-particles inside the nucleus being of the order of 10^7 m/s. The frequency of collision, ν, will be $10^7/10^{-14}$ *i.e.* 10^{21} times per sec. The actual values of λ vary roughly from 10^7 sec^{-1} for ThC to 10^{-18} sec^{-1} for Thorium. With these values of λ, the probability of escape for the α-particles, thus, ranges from 10^{-14} to 10^{-39}. This means that even for the short lived α-emitter (ThC'), an alpha particle (energy about 9 MeV), has to make around 10^{14} attempts, before it is able to come out of the nucleus. It is this large number of escape attempts made by the α-particle, which make the radioactivity an observable effect. It is curious enough that even though the α-particle strikes the barrier for about 10^{21} times per sec, yet it may have to wait for almost 10^{10} years to escape from a nucleus of ^{238}U.

When an alpha particle approaches a nucleus from outside with energy less than the height of potential barrier ($E_\alpha < V_a$), the number of attempts at entry is very much less, and as such the effects are not so marked. The probability of entering the nucleus (P) is, however, the same for alpha particle as the probability of transmission for the same energy.

Transmission Coefficient

To calculate the transmission coefficient P, we write the Schrodinger wave equations and obtain their necessary solutions. For a given energy E_α of the alpha particle, the space is divided into three regions as marked in Fig. 3.22.

$$\left.\begin{array}{lll} \text{region I} & r < R, \; V = -V_a & \\ \text{region II} & b < r < r_c & \multirow{2}{*}{$\left.\right\} V = 2\,Ze^2/r$} \\ \text{region II} & r > r_c & \end{array}\right\} \quad \ldots(3.8.15)$$

The corresponding Schrodinger equations in these three regions are,

$$\left.\begin{array}{ll} \dfrac{d^2\psi_I}{dr^2} + \dfrac{2m_\alpha}{\hbar^2}(E_\alpha + V_a)\psi_I = 0 & \\ \dfrac{d^2\psi_{II}}{dr^2} + \dfrac{2m_\alpha}{\hbar^2}\left(\dfrac{2Ze^2}{r} - E_\alpha\right)\psi_{II} = 0 & r < r_c \\ \dfrac{d^2\psi_{III}}{dr^2} + \dfrac{2m_\alpha}{\hbar^2}\left(E_\alpha - \dfrac{2Ze^2}{r}\right)\psi_{III} = 0 & r > r_c \end{array}\right\} \quad \ldots(3.8.16)$$

Note: One has to use the reduced mass μ of the system; α-particle + residual nucleus, but the residual nucleus being heavier compared to alpha particle as such taking $\mu \approx m_\alpha$ will not lead to incorrect results

$$\left(\mu = \frac{m_\alpha M_A}{m_\alpha + M_A} \approx m_\alpha \text{ when } M_A >> M_\alpha\right)$$

Solutions of such equations have been described in books on Quantum Mechanics for problems on penetration through rectangular potential barriers. The solutions are subjected to the usual boundary conditions to be satisfied by the wave functions in the three regions. Presently our interest is in calculating the transmission coefficient, which by definition is given by the ratio between the intensity of the wave or the flux of particles that emerges from the barrrier and the intensity of the wave or the flux of the particles that arrives at the barrier. If particles whose energy E_α is less than the height V of a rectangular barrier of constant width, say L, then the transmission probability, P, for such a particle to tunnel through this barrier is given by

$$P = e^{-2L/\hbar}\{2m\,(V - E)\}^{1/2} \quad \ldots(3.8.17)$$

For the alpha particle, however, the height of the barrier is not constant, it varies as is clear from the Fig. 3.22, the potential varies from B to C as $\frac{1}{r}$, and at $r = b$, the potential is maximum and at $r = r_c$, the energy of the particle is same as the potential energy and beyond C, the energy is greater than the potential energy *i.e.* $E > V$ and as such when the particle appears at C, it will be able to leave the nucleus permanently. For such a varying potential between B and C the expression for transmission coefficient will become

$$P = e^{-2/h\int_b^{r_c}\{2m(V(r) - E_\alpha\}^{1/2}dr} \quad \ldots(3.8.18)$$

where $V(r) = \dfrac{2Ze^2}{Kr}$, $K = 4\pi\varepsilon_0$ and Z is the atomic number of daughter nucleus.

Using equations (3.8.14) and (3.8.18), the decay constant becomes

$$\lambda = \frac{1}{b}\sqrt{\frac{E_\alpha}{2m_\alpha}}\exp\left(-\frac{2}{\hbar}\int_b^{r_c}\{2m_\alpha(V(r)-E_\alpha)\}^{\frac{1}{2}}dr\right)$$

A small increase in the energy E_α will have the effect of decreasing the integrand and also the range of the integration (Fig. 3.22). Also the integral is an exponential function as such the decay constant is an extremely sensitive function of α-particle energies. The theory also predicts the same as shown below.

Note: From Fig. 3.22, one finds that r_c is the classical radius, the distance of closest approach.

To solve the integral in the equation (3.8.18) we introduce a new variable of integration defined by $x = r/r_c$. As r varies from b to r_c, the new variable varies from $x_c = \frac{b}{r_c}$ to 1. Also at $r = r_c$, $E = V(r_c)$, (the integrand becomes zero).

or $$E = \frac{2Ze^2}{K \cdot r_c}$$

or $$r_c = \frac{2Ze^2}{KE} \qquad ...(3.8.19)$$

Using the new variable and equation (3.8.19) for r_c, the integral in eqn. (3.8.18) becomes,

$$-\frac{4Ze^2}{K\hbar}\sqrt{\frac{2m_\alpha}{E}}\int_{x_c}^{+1}\left(\frac{1}{x}-1\right)^{1/2}dx$$

and the transmission coefficient is

$$P = e^{-4Ze^2/K\hbar}\sqrt{\frac{2m_\alpha}{E}}\int_{x}^{+1}\left(\frac{1}{x}-1\right)^{1/2}.dx \qquad ...(3.8.20)$$

Now

$$\int_{x_c}^{+1}\left(\frac{1}{x}-1\right)^{1/2}dx = \int_0^1\left(\frac{1}{x}-1\right)^{1/2}dx - \int_0^{x_c}\left(\frac{1}{x}-1\right)^{1/2}dx \qquad ...(3.8.21)$$

Since the potential barrier is relatively wide (compared to the range of nuclear forces) $r_c >> b$ or the quantity $x_c = b/r_c$ is a small quantity as such it is possible to approximate $\left(\frac{1}{x}-1\right)^{1/2} \approx \sqrt{\frac{1}{x}}$ in the second integral in eqn. (3.8.21). Hence

$$\int_{x_c}^{+1}\left(\frac{1}{x}-1\right)^{1/2}dx \approx \int_0^1\left(\frac{1}{x}-1\right)^{1/2}dx - \int_0^{x_c}\sqrt{\frac{1}{x}}dx \qquad ...(3.8.22)$$

The first integral on the right hand side of eqn. (3.8.22) can be evaluated by using a substitution.

$x = \sin^2\theta$ then $dx = 2\sin\theta\cos\theta\, d\theta$ and then evaluating the integral gives

$$\int_0^{+1}\left(\frac{1}{x}-1\right)^{1/2} dx = \frac{\pi}{2}$$

and

$$\int_{x_c}^{+1}\left(\frac{1}{x}-1\right)^{1/2} dx = \left(\frac{\pi}{2}-2\sqrt{x_c}\right) = \left(\frac{\pi}{2}-2\sqrt{\frac{b}{r_c}}\right) \quad ...(3.8.23)$$

The transmission coefficient P becomes

$$P = \exp\left(-\frac{4Ze^2}{K\hbar}\cdot\sqrt{\frac{2m_\alpha}{E_\alpha}}\left(\frac{\pi}{2}-2\sqrt{\frac{b}{r_c}}\right)\right) \quad ...(3.8.24)$$

$$\ln P = -\frac{4Ze^2}{K\hbar}\sqrt{\frac{2m_\alpha}{E_\alpha}}\left(\frac{\pi}{2}-2\sqrt{\frac{b}{r_c}}\right)$$

Since

$$\lambda = \nu P. = \frac{v}{2b}.P$$

$$\ln\lambda = \ln\frac{v}{2b} + \ln P$$

$$= \ln\frac{v}{2b} + \frac{8Ze^2}{K\hbar}\sqrt{\frac{2m_\alpha}{E_\alpha}\cdot\frac{b}{r_c}} - \frac{2Ze^2\pi}{K\hbar}\sqrt{\frac{2m_\alpha}{E_\alpha}} \quad ...(3.8.25)$$

Inserting the numerical values for various constants in the equation (3.8.25) we get,

$$\ln\lambda = \ln\frac{v}{2b} + 2.97\, Z^{\frac{1}{2}}\, b^{\frac{1}{2}} - 3.95\frac{Z}{\sqrt{E}} \quad (3.8.26)$$

Converting the expression for $\ln\lambda$ to the base 10, we obtain

$$\log\lambda = \log\frac{v}{2b} + 0.434\left(2.97\, Z^{\frac{1}{2}}\, b^{\frac{1}{2}} - 3.95\frac{Z}{\sqrt{E_\alpha}}\right)$$

$$= \log\frac{v}{2b} + 1.29\, Z^{\frac{1}{2}}\, b^{\frac{1}{2}} - 1.71\frac{Z}{\sqrt{E}} \quad ...(3.8.27)$$

Let us examine equation (3.8.27). For natural radioactive elements which emit α-particles, the energy of the α-particle lies between 4 MeV to 9 MeV, consequently the velocities lie in the narrow range of nearly 1×10^7 m/s to 1.4×10^7 m/s. The nuclear radius b, defined as $b = r_0 A^{1/3}$, where $r_0 = 1.2 \times 10^{-15}$ m, would be in the range of 1 and $1.4^{1/3}$ since A values lie between 208 and 238.

The Z-values lie between 84 and 92. These numerical values, therefore, indicate that the first two terms on the right hand side of equation (3.8.27) are almost constant and any small variation in them does not make a difference when the dependence of λ on these quantities is concerned. The last term, thus dominates the dependence of $\log\lambda$ on the energy or the velocity of the α-particle. One can thus express equation (3.8.27) as,

$$\log_{10}\lambda = C - \frac{D}{\sqrt{E_\alpha}} \qquad \ldots(3.8.28)$$

where, C and D are constants.

Comparison with Observations

The observed values of λ and E_α have been connected to give an empirical relation known as Geiger and Nuttall law. It is expressed as,

$$\log_{10}\lambda = A + B\log_{10}E_\alpha \qquad \ldots(3.8.29)$$

where A and B are constants. On the face of it equations (3.8.18) and (3.8.29) show no agreement. The theory gives a dependence of $\log\lambda$ on the reciprocal of velocity $\left(\frac{1}{\sqrt{E}}\right)$ of alpha particles, whereas the observation shows its dependence on logarithm of velocity ($\log\sqrt{E_\alpha}$). However, the disparity is not so marked. For the known α-emitters the difference between $\log_{10}E_\alpha$ and $\frac{1}{\sqrt{E_\alpha}}$ is small, for example, when E_α = 4 MeV, then $\log_{10}4 = 0.6021$ while $\frac{1}{\sqrt{4}} = 0.5$. Even for the range of α-energies, (4 to 9 MeV) the variation is small. That is $\log_{10}\left(\frac{3}{2}\right) = 0.18$ and $\left(\frac{1}{2} - \frac{1}{3}\right) = 0.166$. The Fig. (3.23) shows a graph between log $T_{1/2}$ and E_α or $\sqrt{E_\alpha}$. Since decay constant λ and half life $T_{1/2}$ are related as $\lambda = \frac{0.693}{T_{1/2}}$. The observed values for different α-emitters of each element are closer to straight lines. This is what one expected from Geiger-Nuttal law. Also for a large value of energy one would get a large value for λ according to equation (3.8.28) or low value for the half life, which is an observed fact. Short lived α-active nucleus emits α's of large energy. The theory though does not reproduce exactly the empirically obtained Geiger-Nuttall relation, yet the two expressions exhibit similar behaviuor. The Gamow's theory for the Quantum Mechanical explanation of the barrier penetration has achieved a success, though in a limited sense, particularly to explain almost large difference (about 30 orders of magnitude) in the half lives. A similar result would have been obtained employing more rigorous methods like WKB method for solving the penetration problem where the potential is a slowly verying function of distance. The result so obtained not the same as from Gamow's theory but are close to observed results. Besides giving the transmission coefficient, the theory can be applied to calculate the radius (b) of the nucleus from the known values of λ and E. This again is a check about the success of the simple theory. Since the radius snd the atomic number are related through $b = r_0 A^{1/3}$, the value of r_0 can be obtained which should be constant. Table 3.4 shows the values of r_0 obtained for different α-emitters. Any deviation from the constancy of r_0, as is the case for ^{208}Po and ^{210}Po, can be attributed to the approximations particularly the preformation of α-particle inside the nucleus, before emission.

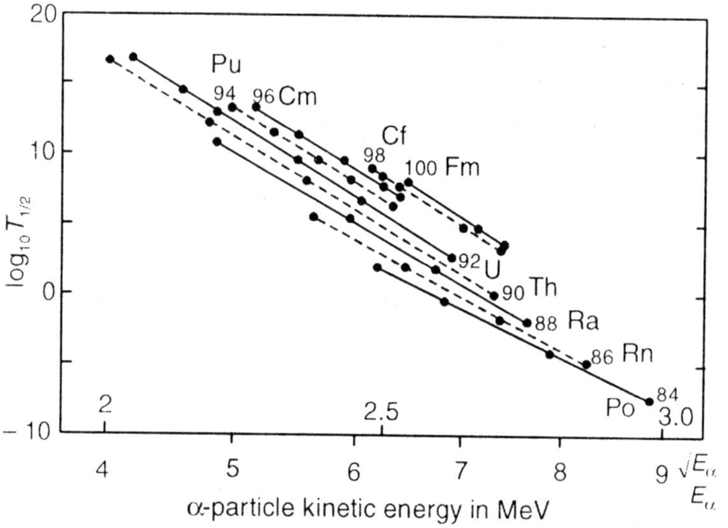

Fig. 3.23: Graph between $\log T_{1/2}$ and E_α or $\sqrt{E_\alpha}$

TABLE 3.4

Parent Nucleus	*E, MeV*	*$\log_{10}\lambda$*	*r_0 Fermi*
^{242}Cm	6.18	–7.282	1.55
^{240}Pu	5.20	–11.437	1.60
^{238}U	4.25	–17.312	1.59
^{236}Th	6.41	–3.426	1.58
^{210}Po	5.40	–8.765	1.43
^{208}Po	5.24	-9.867	1.43

The theory described so far is one dimensional where no change in angular momentum takes place. This is particularly true in the case of even-even nuclei. When transition from ground state of the parent nucleus takes place to the ground state of daughter nucleus ($l = 0$ to $l = 0$, no change in parity, *i.e.* both the states have even parity). If the emission of alpha particle takes place from a state with $l > 0$, or to a state $l > 0$, then the alpha particle carries an angular momentum. The Coulomb potential which, then is effective outside the range of nuclear force is modified and is expressed as,

$$V_{\text{eff}} = \frac{2Ze^2}{Kr} + \frac{\hbar^2}{2m_\alpha r^2} l(l+1) = V_c + V_l \qquad \ldots(3.8.30)$$

The first term in equation (3.8.30) is pure Coulomb potential and the second term, is due to the centrifugal effect and known as centrifugal potential. This reduces the value of transmission coefficient, because now the potential $V(r)$ in equation (3.8.18) is given by the effective potential V_{eff}. For a typical heavy nucleus, the effects of this extra potential (centrifugal potential) on the transmission coefficient is small as compared to the effects of energy E or even the radius b. V_l is much smaller than V_c as can be seen by the following:

$$\frac{V_l}{V_c} = \frac{\hbar^2}{2mr} \frac{l(l+1)Kr}{2Ze^2} \simeq 0.002\, l(l+1)$$

where, $Z = 88$ and $r \sim 10^{-14}$ m.

Since the effect of centrifugal barrier in heavy nuclei ($Z = 80$ to 92) is small, the decay constant or the life time does not depend strongly on l-values. Numerical values for the ratio of λ_l/λ_0 for certain l-values are given below for $Z = 88$ E_α = 4.88 MeV and $r = 10^{-14}$ m.

l	0	1	2	3	4	5	6
$\frac{\lambda_1}{\lambda_0}\left(\frac{T_{1/2,\,0}}{T_{1/2,\,1}}\right) \rightarrow$	1	0.699	0.37	0.137	0.037	0.0709	0.0011

The above calculations show that for small l-values (l = 1, 2, 3) the decay constants, calculated and observed, are of the same order of magnitudes.

The rartio of the observed half life to the calculated half life of even-even nuclei is called as hinderance factor.

$$\text{Hinderance factor (H.F.)} = \frac{T_{1/2,\,0}}{T_{1/2,\,\text{cal}}} = \frac{\lambda_{\text{cal}}}{\lambda_0}$$

The most important reason for hindered alpha disintegration is on account of the non-spherical nature of the nuclei.

The simple theoretical model of α-decay predicts nicely the experimental behaviour of decay constants with alpha particle energies, even when the decay constants vary 30 order of magnitudes for the small energy range (4 MeV to 9 MeV). However, the theory has certain weak points. It was assumed that the alpha particle actually existed inside the nucleus prior to its emission, while it is a known fact that only protons and neutrons exist inside the nucleus. Therefore, the decay constant λ, should also depend on the probability of finding an alpha particle inside the nucleus. If P_α is the probability for the existance of alpha particle inside the nucleus, then the decay constant should be,

$$\lambda = P_\alpha \nu P$$

It is not easy to make a simple estimate of P_α, the probability of finding an alpha particle inside the nucleus. It mainly depends on the wave function of a nuclear state, and as such it is expected to be different for different nuclei. In the Gamow's theory $P_\alpha = 1$. For heavy nuclei, which emit α-particles, it is reasonable to expect that P_α is of the same order of magnitude because there are only small differences in their masses. It is less than unity and as an estimate, for all heavy α-emitters, P_α is taken equal to 0.1.

We have assumed in the present discussion thet there is only a single energy of the emitted alpha particle from a nucleus. But it is a common experience that the same parent nucleus emits alpha particles of different energies. For example, ^{212}Bi emits 14 different decays each leaving the daughter nucleus in a state other than the ground state.

The explanation of α-decay became more realistic when large number of artificially produced α-active nuclei were produced.

3.9 BETA-PARTICLES

3.9.1 Experimental Information about β-Particles

Radioactive elements both natural and artificially produced emit beta-particles ($\beta^\pm$). The negatively charged beta-particles are electrons (β^-) and positively charged beta-particles are positrons (β^+). The positrons or the positive electrons are antiparticles to electrons, having the same mass as electrons but opposite charge. Positrons were predicted first by P.A.M. Dirac in 1927. Besides these two modes of beta emission, there is a third mode for beta-decay and is known as orbital electron-capture.

The charge (e) and charge to mass ratio (e/m) for electrons are determined experimentally and from these the mass of the electron is estimated. The magnitude of these quantities are,

$$\text{charge } (e) = 1.6 \times 10^{-19}\ \text{C},\ e/m = 1.742 \times 10^{10}\ \text{C/kg}$$

$$\text{mass } (m_e) = 9.1083 \times 10^{-31}\ \text{kg}.$$

A study of energy (momentum) distribution of β-particles from a parent beta-active source showed that the distribution is quite unlike to that of α-particle distribution. Instead of discrete groups of different energies like for alpha-particles, the β-particle energy distribution is continuous spectrum extending from a minimum, reaching a maximum and then it falls to zero at a certain energy known as end-point energy. The end-point energy, E_{max}, is the characteristic of the beta emitter. The shape of the curve is similar but end-point energy of different β-active radio-isotopes is different. Similar to α-particles, however, β–particles do have characteristic half-lives and all parts of the continuous spectrum decay with the same half life. The Fig. 3.24(a) corresponds to the β^- particles emitted from RaE ($^{211}_{83}$Bi) for which E_{max} = 1.17 MeV. The Fig. 3.24 (*b*) shows energy distribution of β^+ particles from the decay of ^{13}N with E_{max} = 1.24 Mev. The values of E_{max} for β-emitters vary, from 0.25 MeV to 2.15 MeV while for α-emitters, the energy varies between 4 to 8 MeV.

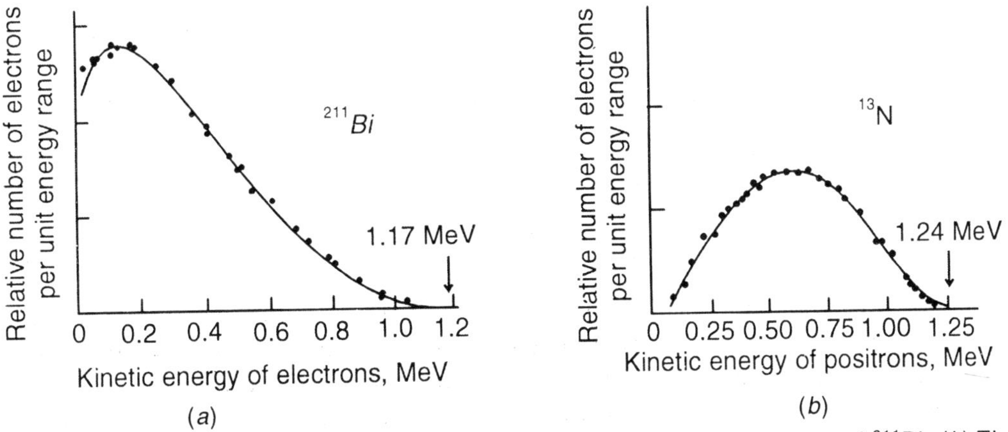

Fig. 3.24: (*a*) The energy spectrum of the electron emitted in the β^- decay of ^{211}Bi, (*b*) The energy spectrum of the positron emitted in the β^+ decay of ^{13}N

In some beta emitters the experimental momentum distribution shows sharp peaks superimposed over the continuous spectrum. Such a line spectrum shown in Fig. 3.25 is due to the electrons of difinite energies. These peaks are due to internal conversion of gamma rays emitted by the daughter nucleus formed in β-decay.

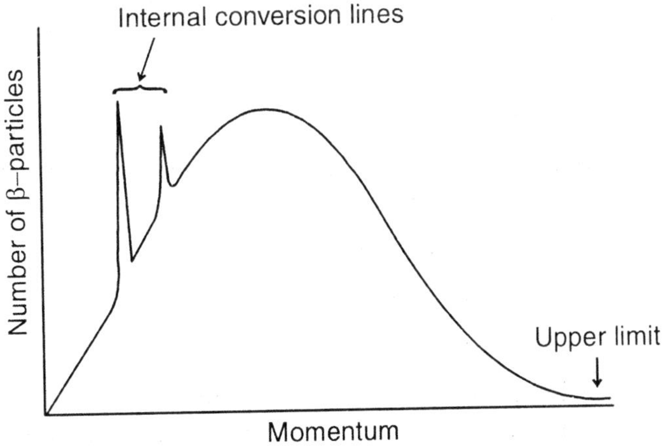

Fig. 3.25: Schematic representation of the momentum distribution of β-particles from a radioactive source. The peaks correspond to internal conversion lines

The three β-decay processes are all isobaric transitions. The mass number *A* of the parent nucleus remains the same after the decay. It is the atomic number,

Z, (charge) which increases or decreases by one. Since $A = Z + N$ where N is the neutron number, as such whenever β^- (electron) emission takes place, the Z-number increases by one, decreasing the neutron number by one *i.e.* in β^--decay Z-becomes $(Z + 1)$ and N becomes $(N - 1)$, thereby A remaining same. Likewise in β^+-process Z decreases by one and N increases bo one *i.e.* Z goes to $(Z - 1)$ and N to $(N + 1)$, again A remaining same. Symbolically these decay processes can be represented as follows:

If X and Y represent parent and daughter nuclei then for β^- emission,

$$^{A}_{Z}X_N \rightarrow \, ^{A}_{Z+1}Y_{N-1} + \beta^- \qquad \text{...(3.9.1)}$$

and for β^+ emission

$$^{A}_{Z}X_N \rightarrow \, ^{A}_{Z-1}Y_{N+1} + \beta^+ \qquad \text{...(3.9.2)}$$

These representations show that in actual β-decay process effectively it is a nucleon inside the nucleus which gets transformed. In electron emission a neutron transforms into a proton thereby increasing its number by one and in positron (β^+) emission, a proton transforms into a neutron, increasing the number of neutrons in the nucleus by one.

The third process, electron capture mostly taking place in the K-shell alongwith positron emission. Here the parent nucleus, instead of emitting a positron captures an electron from one of the innermost atomic orbits. Since the K-electrons are nearest to the nucleus, therefore, the probability of their being captured by a proton is relatively large and hence the name K-electron capture. The vacancy, thus caused is filled by the electrons from the higher atomic states resulting into the emission of X-ray photons. The wavelength of such photons will be the characteristics of the daughter atom. Symbolically K-capture is represented as

$$^{A}_{Z}X_N + \beta^- \rightarrow \, ^{A}_{Z-1}Y_{N+1} \qquad \text{...(3.9.3)}$$

Thus the three β-decay processes can be described as the following reaction equations

β^--decay	:	$n \rightarrow p + \beta^-$	...(3.9.4a)
β^+-decay	:	$p \rightarrow n + \beta^+$	...(3.9.4b)
Electron capture	:	$p + \beta^- \rightarrow n$	...(3.9.4c)

These transformations take place inside the nucleus. However, when a neutron is outside a nucleus (a free neutron) it undergoes a β^- decay with a half-life of about 12 minutes while no decay of a free proton has so far been observed. This is because of the fact that mass of a neutron is slightly greater than that of a proton. A neutron bound inside the nucleus does not decay spontaneously because the presence of other nucleons may make the process energetically impossible.

3.9.2 Q-Values (Energetics) in β-Decay Processes

It has been shown that electrons do not exist inside the nucleus. It was therefore, assumed that they are created in the process of decay, that is they immediately leave the nucleus after their creation. Here one finds a process analogous to what actually happens in atomic transitions (outside the nucleus). The photons do not exist inside the atoms but are produced only when an orbital electron in an atom jumps from a higher energy state to a lower energy state.

The energy involved in β-decay can be calculated from the mass of parent nucleus, daughter nucleus and the mass of the electron and using the Einstein's mass-energy relationship $E = mc^2$. Also we must include in the energy calculations the binding energy of the electrons in the atom. However, it is quite small compared to the energies involved in the β-decay processes and as such can be neglected.

(i) β^--Emission

Let a parent nucleus $X(A, Z)$ emits an electron leaving a daughter nucleus $Y(A, Z+1)$ *i.e.*

$${}^{A}_{Z}X \rightarrow {}_{Z+1}^{A}Y + \beta^- \quad \text{(eqn. 3.9.1)}$$

If $M_n(A, Z)$ is the nuclear mass of the parent nucleus, and $M_n(A, Z+1)$ is the nuclear mass of daughter nucleus, then the disintegration energy is given by,

$$Q_{\beta^-} = [M_n(A, Z) - \{M_n(A, Z+1) + m_e\}]c^2 \quad \text{...(3.9.5)}$$

where m_e is mass of an electron. c is the velocity of light. Q_{β^-} is the maximum kinetic energy available for the emitted electron (β^-). Generally, the mass and binding energy of a nucleus are defined in terms of those of neutral atoms. The atomic mass $M(A, Z)$ of atom is defined as,

$$M(A, Z) = M_n(A, Z) + Zm_e \quad \text{...(3.9.6)}$$

Here the binding energy of Z electrons in the atom has been neglected because of its negligible magnitude. In terms of the atomic masses, the energy expression for Q_{β^-} (eqn. 3.9.5) becomes,

$$Q_{\beta^-} = [M(A, Z) - Zm_e - \{M(A, Z+1) - (Z+1)\, m_e + m_e\}]C^2$$

$$= [M(A, Z) - M(A, Z+1)]c^2 \quad \text{...(3.9.7)}$$

If the masses are expressed in energy units, then (1 amu = 931 MeV)

$$Q_{\beta^-} = [M(A, Z) - M(A, Z+1)] \quad \text{...(3.9.8)}$$

It may be noted that the mass of the emitted electron does not appear in equations (3.9.7) or (3.9.8). The emitted electron, for energy calculations, may be used to compensate for the additional electron required to make the daughter atom neutral.

(ii) β^+ Emission

In this case we have,

$${}^{A}_{Z}X \rightarrow {}_{Z-1}^{A}Y + \beta^+ \quad (3.9.2)$$

The energy available for the positron is obtained as a difference of the mass of the parent nucleus and mass of the daughter nucleus plus the positron mass (being the same as electron mass)

$$Q_{\beta^+} = [M_n(A, Z) - \{M_n(A, Z-1) + m_e\}]c^2 \qquad ...(3.9.9)$$

In terms of the atomic masses (eqn. 3.9.6)

$$Q_{\beta^+} = [M(A, Z) - Zm_e - \{M(A, Z-1) - (Z-1)\,m_e + m_e\}]c^2$$

$$Q_{\beta^+} = [M(A, Z) - M(A, Z-1) - 2m_e]c^2 \qquad ...(3.9.10)$$

If the masses of the expressed in energy units, them

$$Q_{\beta^+} = [M(A, Z) - M(A, Z-1) - 2m_e] \qquad ...(3.9.11)$$

(*iii*) Orbital Electron Capture (EC)

The electron capture process is expressed as (eq. 3.9.3)

$${}^{A}_{Z}X + \beta^+ \rightarrow {}_{Z-1}^{A}Y$$

Since K-orbit is nearest to the nucleus, it is the electron in K-shell which has large probability of being captured by the nucleus. Then the decay process has to supply an energy equal to the binding energy of the K-electron enabling it to be captured by the nucleus. Hence,

$$Q_{EC} = [M_n(A, Z) + m_e - M_n(A, Z-1)]c^2 - B(K) \qquad ...(3.9.12)$$

where $B(K)$ is the binding energy of the K-electron in the parent atom (if electron from any other shell is captured then $B(K)$ will correspond to the binding enrgy of that electron). This much energy has to be supplied by the decay process and hence it is subtracted. In terms of the atomic masses

$$Q_{EC} = [M(A, Z) - Zm_e + m_e - M(A, Z-1) + (Z-1)\,m_e]c^2 - B(K)$$

$$= [M(A, Z) - M(A, Z-1)]c^2 - B(K) \qquad ...(3.9.12)$$

When masses are expressed in energy units

$$Q_{EC} = [M(A, Z) - M(A, Z-1)] - B(K) \qquad ...(3.9.13)$$

These expressions for the Q-values of the three β-decay processes indicate that:

(*a*) for β^- emission to take place, it is required that the mass of the parent atom should be greater than the mass of the daughter atom.

(*b*) For β^+ emission, the difference in the masses of parent and daughter atoms should be greater than twice the mass of an electron. Since an electron has rest mass energy of 0.511 MeV, as such the difference in the masses of parent and daughter atoms, expressed in energy units, should be greater than 1.022 MeV. In other words, the energy that is required to create two electrons (~1.022 MeV) should be supplied by the decay process, half of it for the positron emission and the other half for the atomic electron that must be ejected in going from a neutral atom of Z-electrons to one with $(Z-1)$ electrons.

(c) For K-electron capture, it is required that the difference in the atomic masses of parent and daughter atoms should be greater than the binding energy of the K-electron in the atom. In the lighter nuclides, small Z-value elements, the binding energy $B(K)$ is small and can be ignored, in that case the condition for electron capture just reduces to $M(A, Z)$ being greater than $M(A, Z-1)$ and both electron capture and positron emission can take place in the same nucleus. However, in heavy nuclides, large Z-values elements, electron capture occurs more often than positron emission.

Since we have neglected the effect due to the binding energy of the electrons in the atoms, the energy Q is shared (as kinetic energy) by the decay products. This Q-value gives the maximum kinetic energy of the β-particle (or the end point energy), provided the recoil energy of the daughter nucleus is neglected.

Example:

β^--emission

$$^{14}_{6}C \rightarrow {}^{14}_{7}N + \beta^- + \bar{\nu} \qquad ...(3.9.14)$$

Here

$$M_{Z+1}\left({}^{14}_{7}N\right) = 14.007515 \text{ amu and } M_Z\left({}^{14}_{6}C\right) = 14.007682 \text{ amu.}$$

Thus $M_{Z+1} > M_Z$ and hence β^- is possible

$$Q_{\beta^-} = 0.000167 \text{ amu} = 0.1556 \text{ MeV} = E_0$$

β^+-emission

$$^{11}_{6}C \rightarrow {}^{11}_{5}B + \beta^+ + \nu \qquad ...(3.9.15)$$

The mass difference of the masses of ${}^{11}_{5}B$ and ${}^{11}_{6}C$ in energy units comes around 1.985 which is greater than 1.022 MeV, hence positron emission can take place.

$$Q_{\beta^+} = 1.985 - 1.022 = 0.963 \text{ MeV.}$$

For electron capture (EC) let us consider the decay of ${}^{7}_{4}Be$ into ${}^{7}_{3}Li$ by capturing an K-electron

$$^{7}_{4}Be + \beta^- \rightarrow {}^{7}_{3}Li + \nu \qquad ...(3.9.16)$$

The atomic mass difference of the parent $\left({}^{7}_{4}Be\right)$ and daughter $\left({}^{7}_{3}Li\right)$ nuclei is 0.866 MeV which is less than $2m_eC^2$ or 1.022 MeV, hence positron emission is not possible. The decay, therefore, occurs only through electron capture (3.9.16).

Since electron capture is competitive with positron emission, there are nuclei which can undergo decay through both the processes, with different Q-values. For example, the decay of ${}^{80}_{35}Br$ to ${}^{80}_{34}Se$ can take place through both the modes

$$^{80}_{35}Br \rightarrow {}^{80}_{34}Se + \beta^+ + \nu \qquad ...(3.9.17)$$

and

$$^{80}_{35}Br + \beta^- \rightarrow ^{80}_{34}Se + \nu \quad ...(3.9.18)$$

Here the atomic mass difference between the parent and daughter nuclei in energy units is around 2.66 MeV which is greater than $2m_e$ = 1.022 MeV.

In the above examples of the three β-decay processes, the terms ν and $\bar{\nu}$, on the right hand side of each decay are known as neutrino and anti-neutrino, each being chargeless and (almost) zero mass. More about them is given in the next section.

3.9.3 Continuous β -Spectrum and Neutrino Hypothesis

The continuous energy (momentum) distribution of beta particles posed a puzzled yet an interesting problem. The maximum energy being a characteristic of the nucleus undergoing beta-decay but the maximum number of beta-particles emitted have energies of only 20% of the maximum energy. Also the maximum energy carried off by the beta particle is given by the energy equivalent of the mass difference between the parent nucleus (A, Z) and daughter nucleus, (A, $Z \pm 1$). Initially it was thought that the beta particles at the time of emission might have definite energy but during their passage they may collide with the atomic electrons and lose energy in a distributed manner. This hypothesis was found incorrect when an experiment was performed in 1927 by Ellis and Wooster for its verification. In the experiment a beta-active source, $^{210}_{83}Bi$ was placed inside a thick walled calorimeter designed to absorb all the radiations. The heat evolved after a given number of decays is measured. This heat energy divided by the number of decays gives the average energy per decay. In the case of $^{210}_{83}Bi$, it was found to be 0.35 MeV which is very near to the value 0.39 MeV, being average of the spectrum shown in Fig. 3.9(a) and not the maximum energy. This experiment, however, confirmed the continuous energy distribution of the beta-particle energies as actually observed. But this does disagree with the fact that in a beta-disintegration, only one beta particle is emitted whose energy is obtained from the difference of the masses of parent and daughter nuclei.

It was also presumed that there may be large number of excited states of daughter nucleus and transitions from the parent nucleus to these states of daughter nucleus may result in such distribution of energies. There should, then, result, into a corresponding continuum of γ-ray energies. But no such γ-ray energy distribution was observed in a β-decay. Besides, the low lying states of a nucleus are always discrete and well separated. These observed facts led to believe that either there is a violation of principle of conservation of energy or some other particle is also emitted along with a beta particle.

Likewise, if a beta particle only is emitted, the linear momentum and angular momentum are found to be not conserved. If the decay results into a beta particle and the daughter nucleus only, then the two must always move in exactly opposite directions so that the resultant momentum may be zero. This should be so because parent nucleus is at rest in centre of mass system. In actual decay process, the recoil nucleus and the beta particle never found to move in exactly opposite directions meaning thereby that the linear momentum is not coserved, if we take the decay process as a two body process (eqn. 3.9.1 or 3.9.2). Similarly, the angular momentum is also not conserved. Consider the beta decay where a neutron is transformed into a proton and an electron is emitted,

$$n \rightarrow p + \beta^{+}$$

It is well known that spins of neutron, proton and electron are ½ for each (spin angular momentum for each particle is ½ $\hbar$). If the angular momentum is to be conserved, then this decay cannot take place because here angular momentum is not conserved.

These physical laws cannot be violated. Such a discrepancy was removed by W. Pauli, who in 1930 proposed that β-decay process is not a two body phenomena. A new particle should also be emitted alongwith a beta particle. This particle should be chargeless and having no mass (almost zero rest mass). This particle was later named as NEUTRINO by E. Fermi. Also W. Pauli proposed that in a beta decay, if E_β is the energy of the beta particle and E_0 is the end-point energy then the energy taken away by the neutrino is $E_\nu = E_0 - E_\beta$. That is, in each β-decay, the disintegration energy is shared in a continuous manner by neutrino (ν), β-particle and the recoil nucleus. The neutrino is assigned a spin ½ or spin angular momentum of ½ $\hbar$. It is a Fermion. Thus the angular momentum is conserved. The spin of (β + ν) system would be 1 or 0.

The β-decay is now a three-body process. In the centre of mass frame of reference, the linear momenta of three particles; β-particle, neutrino and recoil nucleus must add to zero, as shown in Fig. 3.26. The κ-capture is a two body process in which a nucleus at rest captures an electron with effectively zero linear momentum. The decay products are then the daughter nucleus and a neutrino.

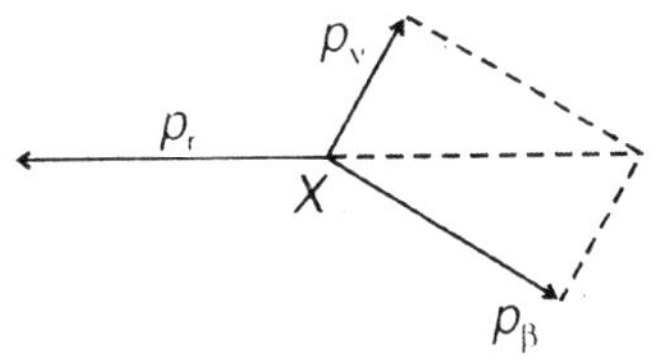

Fig. 3.26: Momentum conservation in β-decay. p_β, p_ν and p_r are the momentum of β-particle, neutrino and recoil nucleus respectively

Therefore, with the invention of neutrino the β-decay processes (3.9.1), (3.9.2) and (3.9.3) can be expressed as

β^- decay: $\quad {}^{A}_{Z}X_N \rightarrow {}^{A}_{Z+1}Y_{N-1} + \beta^- + \bar{\nu}$

β^+ decay: $\quad {}^{A}_{Z}X_N \rightarrow {}^{A}_{Z-1}Y_{N+1} + \beta^+ + \nu$

Electron Capture $\quad {}^{A}_{Z}X_N + \beta^- \rightarrow {}^{A}_{Z-1}Y_{N+1} + \nu \qquad$ ...(3.9.19)

And the corresponding examples would now look as follows:

$$^3_1H_2 \rightarrow {}^3_2He_1 + \beta^- + \bar{\nu}$$

$$^{11}_6C_5 \rightarrow {}^{11}_5B_6 + \beta^+ + \nu$$

$$^7_4Be_3 + \beta^- \rightarrow {}^7_3Li_4 + \nu$$

The reactions (3.9.4a), (3.9.4b) and (3.9.4c) would now be,

$$\left.\begin{array}{l} n \rightarrow p + \beta^- + \bar{\nu} \\ p \rightarrow n + \beta^+ + \nu \\ p + \beta^- \rightarrow n + \nu \end{array}\right\} \quad \text{...(3.9.20)}$$

It may be noted that in the above expressions (3.9.19) and (3.9.20), we have used $\bar{\nu}$ and ν to denote neutrinos in two cases. In the original treatment the two were taken to be same but later on it was found that they have to be different as shown in the next section.

3.9.4 Fermi Theory of β-decay

The Pauli's hypothesis of the emission of a chargeless and a massless particle, neutrino, alongwith beta particles in the nuclear decay could preserve the conservation of energy, linear and angular momenta. Under the classification of elementary particles, the beta particles and neutrinos fall under the category of Leptons (light particles). Also when antiparticles were discovered a new conservation law was enunciated. It is the law of conservation of leptons according to which the difference in the number of leptons and antileptons in the universe should remain constant. The electron has been regarded as lepton, the corresponding antilepton is the positron *i.e.* electron is particle and positron is antiparticle. Thus the conservation of leptonic number demand that alongwith the emission of an electron, an anti-neutrino ($\bar{\nu}$) must be emitted and a neutrino must be emitted alongwith the emission of a positron (eqns. 3.9.14 and 3.9.15). If the leptonic number for leptons is denoted by l then for antileption it will be $-l$. For nucleons $l = 0$, for electron and neutrino each, it is 1. Likewise for positron and anti-neutrino it is −1. In the β-decay, then, the conservation of leptons can be seen from the following β-decay processes:

(*i*) β^-–decay

$$\begin{array}{lcccc} & n \rightarrow & p + & \beta^- + & \bar{\nu} \\ l = 0, & & 0 & 1 & -1 \end{array}$$

(*ii*) β^+–decay

$$\begin{array}{lcccc} & p \rightarrow & n + & \beta^+ + & \nu \\ l = 0 & & 0 & -1 & 1 \end{array}$$

(*iii*) Electron capture

$$\begin{array}{lcccc} & p + & \beta^- \rightarrow & n + & \nu \\ l = 0 & & 1 = & 0 & 1 \end{array}$$

Thus l-values (leptonic number) are same on left hand side and right hand side of the above three decay processes hence law of lepton conservation holds. For the simple β-decay process, of our interest there will be no physical difference between ν and $\bar{\nu}$. We will be describing, more about these two particles, their properties and techniques for their detection and differentiation in the Section (3.9).

A formal theory of β-decay was first developed by E. Fermi (1934). His formalism was based on the fact that there was an analogy between the emission of beta particles and neutrinos during nucleonic transformation and photon emission when atomic (or nuclear) transitions take place between different energy levels. He assumed that a certain field of the light particles (β and ν) exists inside the nucleus, though the particles actually do not exist, just as electromagnetic field is present inthe atoms even though photons do not exist. Any interaction taking place between the electromagnatic field and atoms results into the transition between energy levels and photons are emitted. Fermi, therefore, assumed that the interaction between the leptonic field (combined field of β and ν) and the unstable nucleus leads to the transformation of one nucleon into another nucleon (change of charge state) inside the nucleus resulting in the emission of beta particles and neutrinos.

Fermi made use of the time dependent perturbation theory, according to which, the transition probability per unit time for the transition is given by,

$$\omega = \frac{2\pi}{\hbar}|H_\beta|^2_{fi}\,\rho_f(E) \qquad \text{...(3.9.21)}$$

where $|H_\beta|_{fi}$ is matrix element of the interaction (perturbation) H_β between the initial state defined by state function Ψ_i and the final state given by Ψ_f. $\rho_f(E)$ is the number of momentum states in the final system per unit energy range and is equal to $\rho_f(E) = \frac{dn_f}{dE_0}$ where E_0 is the total energy of transformation (same as E_{max}) and dn_f is the product of the number of states available to the electron and the neutrino. It is also called as density of states. The matrix element for the beta-interaction causing the transition is expressed as

$$|H_\beta|_{fi} = \int \Psi_f^* H_\beta \Psi_i d\tau \qquad \text{...(3.9.22)}$$

The final system constitutes daughter nucleus, electron and antineutrino, while the initial state contains only the parent nucleus.

Unlike the electromagnetic field, the leptonic field interacts weakly with the nucleus during the β-decay, as such it is termed as the weak interaction. This is so named because of the fact that the transition probabilities associated with these interactions are considerably smaller than those induced by the electromagnetic or strong (nuclear) interactions. In other words the mean life times of β-decay is much longer than the life time of γ-decay. The nature and form of the interaction H_β being not known, Fermi assumed a very simple form for it ($H_\beta = gQ$) where g is known as Fermi's coupling constant which determines the strength of the interaction for β-decay. Its value is of the order of 10^{-4} MeV fm^3. He further assumed that the matrix element depends on the wave functions of electron and neutrino ϕ_e, ϕ_ν respectively. The leptons are treated as free particle (since their wavelengths, at β-energies, are much larger than the nuclear dimensions). To begin with it was also presumed that the leptons (β's and ν's) do not carry any angular momentum, and there is no Coulomb interaction between the charges of the electron and the daughter nucleus. The interaction H_β will contain an operator responsible for the nucleonic transformation. The β-decay interaction has to be Lorentz-invariant. It should be hermitian, and that the electron-neutrino pair takes place as a single event right at the site of the nucleon, *i.e.* all the particles interact at a single point. The interaction matrix is then mathematically expressed as,

$$|H_\beta|_{fi} = g\int (u_f^* Q \cdot \phi_e^* \phi_\nu^* u_{in})\, d\tau + \text{complex conjugate} \qquad ...(3.9.23)$$

where u_f^*, ϕ_e^*, ϕ_ν^*, represent the wave functions for the daughter nucleus, electron and antineutrino respectively, *i.e.*

$$\Psi_f^* = u_f^* \phi_e^* \phi_\nu^*$$

u_{in} is the wave function for the parent nucleus *i.e.* $\Psi_{in} = u_{in}$ and g is the coupling constant. Q is the interaction operator responsible for the transformation of a neutron into a proton. The final state of the system (asterisked wave function) contain three particles while the initial system contains one particle only. However, to have a symmetry, it is proposed to make use of the fact that emission (creation) of a particle is equivalent to the absorption (annihilation) of an antiparticle and vice versa (a more rigorous statement of this fact is that a particle moving forward in time has the same state function as its antiparticle moving backward in time and vice versa). In view of this fact, the β-decay can be represented symbolically as

$$n + \nu = p + \beta^-$$

$$p + \bar{\nu} = n + \beta^+$$

The use of this fact simplifies mathematical treatment without affecting the physical ideas and results. In the interaction matrix, therefore, the wave function ϕ_ν^* for the antineutrino in the final state can be replaced by a wave function ϕ_ν for the neutrino in the initial state. Thus,

$$|H_\beta|_{fi} = g\int (u_f^* \phi_e^* Q \phi_\nu u_{in})\, d\tau + c.c. \qquad ...(3.9.24)$$

The equation (3.9.24) thus can be interpreted as a proton and an electron in the final state and a neutron and a neutrino in the initial state. The complex conjugate in (3.9.24) will describe the proton and an antineutrino in the initial state and a neutron and a positron in the final state. We are mainly interested in the electron emission and hence shall confine to the first term on the right hand side of equation (3.9.24). The interaction matrix for the electron emission is thus expressed as,

$$|H_\beta|_{fi} = g\int \left(u_f^* \phi_e^* Q \phi_\nu u_{in}\right) d\tau \qquad ...(3.9.25)$$

The nuclear wave functions are not known. The electron and neutrino have been regarded as free particle, the wave functions for electron and neutrino can be taken as plane waves normalised within a volume V.

$$\phi_\nu = \frac{1}{\sqrt{V}} e^{i(\vec{k}_\nu \cdot \vec{r})}$$

$$= \frac{1}{\sqrt{V}} e^{(i/\hbar)\vec{p}_\nu \cdot \vec{r}}$$

$$\phi_e = \frac{1}{\sqrt{V}} e^{i\vec{k}_e \cdot \vec{r}}$$

$$= \frac{1}{\sqrt{V}} e^{\left(\frac{i}{\hbar}\right)\vec{p}_e \cdot \vec{r}}$$

The c.c. of electron wave function and the wave function of neutrino are, therefore,

$$\phi_e^* = \frac{1}{\sqrt{V}} e^{-i(\vec{k}_e \cdot \vec{r})}$$

$$= \frac{1}{\sqrt{V}} e^{-\left(\frac{1}{\hbar}\right)(\vec{p}_e \cdot \vec{r})}$$

$$\phi_\nu = \frac{1}{\sqrt{V}} e^{-i(\vec{k}_\nu \cdot \vec{r})}$$

$$= \frac{1}{\sqrt{V}} e^{-\left(\frac{i}{\hbar}\right)\vec{p}_\nu \cdot \vec{r}}$$

Note: Neutrino momentum is opposite to the momentum of antineutrino, hence the negative sign.

$\vec{p}_e$ and $\vec{p}_\nu$ are momenta of electron and neutrino respectively. Neutrino being electrically neutral, does not interact with any of the particles. The electron, however, being a charged particle can interact electrostatically with the nucleus. But for small Z-values, the Coulomb interaction energy can be neglected in comparison with the kinetic energy of the electron. Therefore, the plane wave forms for the wave functions are justified. $\vec{r}$ is the position vector for the leptons.

Now, with the energies available for the electron and neutrino, the electron and neutrino momentum, $\vec{p}_e$ and $\vec{p}_\nu$ are of the order of mc and since the integration in the calculation of $|H_{\beta^-}|$ is to be carried out within the nuclear volume, the Compton wavelength for the electron $\frac{\hbar}{mc}$ is much larger than the nuclear dimension R, *i.e.*

$$\frac{R}{\hbar / mc} \cong 0.01$$

Note: If electron energy is of the order of 1 MeV, then $p_e \sim \frac{E}{c} \sim mc$ taking $E^2 = c^2p^2 + m_0^2c^4$ and neglacting $m_0^2c^4$, $E = cp$.

or $$p_e \sim \frac{10^{-13}}{10^8} = 10^{-21}$$

$$R \sim 10^{-15}$$

$$\therefore \quad \frac{R}{\hbar / mc} \sim \frac{R \cdot p_e}{\hbar} = \frac{10^{-21} \times 10^{-15}}{10^{-34}} \approx 10^{-2}$$

Thus, it is possible to expand the exponentials for p_e and p_ν and retain only the first term that is,

$$\phi_e^* = \frac{1}{\sqrt{V}}\left[1 - \left(\frac{i}{\hbar}\right)\vec{p}_e \cdot \vec{r} - \frac{1}{2\hbar^2}(\vec{p}_e \cdot \vec{r})^2 + \ldots\right]$$

$$\phi_\nu = \frac{1}{\sqrt{V}}\left[1 - \left(\frac{i}{\hbar}\right)\vec{p}_\nu \cdot \vec{r} - \frac{1}{2\hbar^2}(\vec{p}_\nu \cdot \vec{r})^2 + \ldots\right] \qquad \ldots(3.9.27)$$

As the integration is over a volume of radius R, and since $\vec{k} \cdot \vec{R} = \frac{\vec{p}_e \cdot \vec{R}}{\hbar} << 1$, the wave function is almost constant over the nuclear volume. One can then write $e^{i\vec{k}_e \cdot \vec{R}} \approx 1$. So that $\phi_e^* = \frac{1}{\sqrt{V}}$ and $\phi_\nu = \frac{1}{\sqrt{V}}$.

Under this approximation, retaining only first term in eqn. (3.9.27), we get,

$$\phi_e^* \sim \frac{1}{\sqrt{V}} \text{ and } \phi_\nu \sim \frac{1}{\sqrt{V}}$$

or $$\phi_e^* \phi_\nu \sim \frac{1}{V}$$

The matrix element then becomes

$$|H_{\beta^-}| = g\int u_f^* Q u_{in} \cdot \phi_e^* \phi_\nu d\tau = \frac{g}{V}\int u_f^* Q u_{in} d\tau$$

Putting $$M = \int u_f^* Q u_{in} d\tau \qquad \ldots(3.9.28)$$

$$|H_{\beta^-}|_{fi} = \frac{g}{V}|M|_{fi}$$

The transition probability eqn. (3.9.21) becomes,

$$\omega = \frac{2\pi}{\hbar}\frac{g^2}{V^2}|M|_{fi}^2 \rho_f(E) \qquad \ldots(3.9.30)$$

3.9.5 Calculation of Density of States or the Statistical Factor $\rho_f(E)$

As the electron and neutrino have independent existence in a volume V, the density of states ρ_f can be expressed as

$$\rho_f = \frac{dn}{dE_0} = \frac{dn_e dn_\nu}{dE_0}$$

where dn_e and dn_ν are respectively the number of states for electron and neutrino.

According to statistical mechanics, the number of momentum states of a particle with momentum p and $p + dp$ in a volume V is given by,

$$dn = \frac{4\pi Vp^2 dp}{(2\pi\hbar)^3} \quad ...(3.9.31)$$

Then the number of states for electron with momentum p_e and $p_e + dp_e$ will be

$$dn_e = \frac{4\pi Vp_e^2 dp_e}{(2\pi\hbar)^3} \quad ...(3.9.32)$$

and for neutrino it will be,

$$dn_\nu = \frac{4\pi Vp_\nu^2 dp_\nu}{(2\pi\hbar)^3} \quad ...(3.9.33)$$

Here volume V is same as that used for normalisation of wave functions.

The number of states available to both electron and neutrino are

$$dn = dn_e \cdot dn_\nu = \frac{4\pi Vp_e^2 dp_e}{(2\pi\hbar)^3} \cdot \frac{4\pi Vp_\nu^2 dp_\nu}{(2\pi\hbar)^3}$$

The number of states per unit energy in the final state is

$$\rho_f(E) = \frac{dn}{dE_0} = \frac{4\pi Vp_e^2 \cdot dp_e}{(2\pi\hbar)^3} \frac{4\pi Vp_\nu^2 dp_\nu}{(2\pi\hbar)^3 dE_0}$$

$$\rho_f(E) = \frac{16\pi^2 V^2 p_e^2 dp_e}{(2\pi\hbar)^6} p_\nu^2 \frac{dp_\nu}{dE_0} \quad ...(3.9.34)$$

Since the daughter nucleus is heavy (compared to leptons) it will receive negligible amount of kinetic energy by way of recoil and so to a good approximation the energy in β-decay is shared between electron and neutrino. The momentum, however, is conserved in three bodies namely the daughter nucleus, electron and neutrino. The neutrino momentum, therefore, is not determined when electron momentum is known. Hence the neutrino momentum is to be calculated from energy conservation and not the momentum conservation. The relativistic relationship between energy and momentum of electron and that of neutrino (mass being zero) are respectively given by,

$$\left.\begin{aligned} E_e^2 &= c^2 p_e^2 + m_0^2 c^4 \\ E_\nu &= cp_\nu = E_0 - E_e \end{aligned}\right\} \quad ...(3.9.35)$$

where E_e is energy of electron, p_e is its momentum m_0 is its rest mass and c is velocity of light E_ν, p_ν are respectively the energy and momentum of neutrino. The rest mass of neutrino is taken as zero. E_0 is the maximum energy of the emitted electron or the end point energy of the beta-spectrum.

Since the neutrino is not observable the momentum interval dp_ν for neutrino be converted in terms of dE_0 for a given (constant) electron energy E_e

$$\therefore \qquad dp_\nu = \frac{dE_0}{c} \qquad \text{...(3.9.36)}$$

Using equations (3.9.35) and (3.9.36) in equation (3.9.34) we get,

$$\rho_f = \frac{dn}{dE_0} = \frac{16\pi^2 V^2 (E_0 - E_e)^2 p_e^2 dp_e dE_0}{(2\pi\hbar)^6 c^2 dE_0 c}$$

or

$$\rho_f = \frac{16\pi^2 V^2}{(2\pi\hbar)^6} \frac{(E_0 - E_f)^2 p_e^2 dp_e}{c^3} \qquad \text{...(3.9.37)}$$

Using eqn. (3.9.37) in equation (3.9.30), we get for the probability of emission of an electron having momentum between p_e and $p_e + dp_e$ per unit time,

$$\omega(p_e)dp_e = \frac{g^2 |M|_{fi}^2 (E_0 - E_e)^2 p_e^2 dp_e}{2\pi^3 \hbar^7 c^3} \qquad \text{...(3.9.38)}$$

The equation (3.9.38) gives the momentum spectrum for electrons emitted in b-decay using the energy and momentum relationship for the electrons, it is possible to obtain the energy spectrum for the emitted electrons using

$$dp_e = \frac{E_e dE_e}{c\left(E_e^2 - m_o^2 c^4\right)^{1/2}}$$

The probability for the emission of electrons with energy E_e and $E_e + dE_e$ is given by

$$\omega(E_e)dE_e = \frac{g^2 |M|_{fi}^2}{2\pi^3 \hbar^7 c^6} E_e (E_0^2 - m_0^2 c^4)^{1/2} (E_0 - E_e)^2 dE_e \qquad \text{...(3.9.39)}$$

Fermi assumed that the operator Q is a scalar of the order one, the matrix element $|M|_{f_i}$ becomes of the order of unity as such $|M|_{fi}^2 \approx 1$ in equations (3.9.38) and (3.9.39) respectively. Hence, under Fermi's approximation, the number of electrons emittedd per second in the energy range E_e and $E_e + dE_e$ is

$$\omega(E_e)dE_e = C^2 E_e (E_0^2 - m_0^2 c^4)^{1/2} (E_0 - E_e)^2 dE_e, \qquad \text{...(3.9.40)}$$

where, $\qquad C^2 = g^2 / 2\pi^3 \hbar^7 c^6 \qquad$ ...(3.9.41)

These transitions for the electron emission under Fermi's approximations are known as Allowed transitions. The energy spectrum for allowed transitions is given by a plot of equation (3.9.40). The graph obtained by plotting $\frac{w}{C^2}$ as a function of E_e is shown in the Fig. (3.27). The distribution falls to zero for both high and low electron energies. This is contrary to what is actually observed, where more low energy electrons are found (A similar curve would be obtained for positrons also).

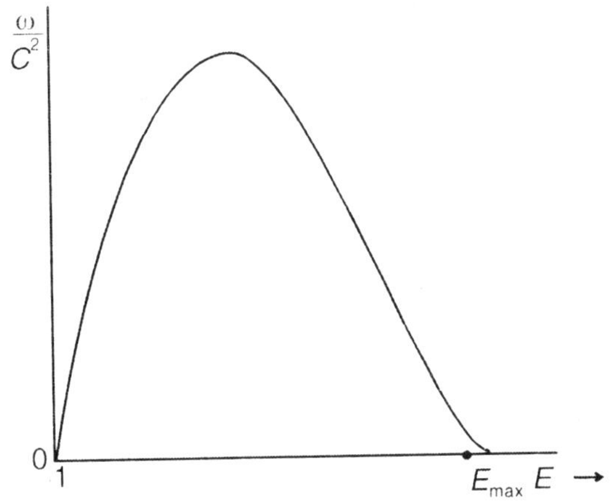

Fig. 3.27: Energy spectrum of β-particles for allowed transitions

3.9.6 Coulomb Correction

The disagreement of Fermi distribution for β-spectrum as obtained under approximation was partly due to the fact that no effect of Coulomb interaction between the electron and nuclear charge was considered. When Coulomb interaction is considered this discrepancy is removed. The attractive Coulomb potential between positive nuclear charge and the negative electron charge has to be overcome by the electrons, once they are created, to come out and this can happen only at the cost of electron's own energy resulting into decrease in their energy. Hence the number of low energy electrons is large. This, therefore, explains partly the maximum towards the low energy side of the β^--spectrum. Likewise, there will be repulsion for the positrons, they will get accelerated and as such more positrons of high energy will be emitted. This, thus, accounts for the maximum towards higher energy side of β^+-spectrum. The distribution for electron and positron emission, when Coulomb effect is considered shall look as shown in Fig. 3.28, where the curve under no Coulomb interaction is also shown.

In order to take account of the Coulomb interaction, which will be a function of the atomic number Z of the daughter nucleus and the momentum (energy) of the emitted electron. For this, therefore, the equations (3.9.38) is to be multiplied by a factor $F(z, p_e)$ and equation (3.9.39) by $F(z, E_e)$. The correction factor $F(z, p_e)$ or $F(z, E_e)$ is known as Fermi function or Fermi factor. It is essentially a barrier penetration factor and in the non-relativistic limit, when the velocity of beta-particles v is much less than velocity of the light, $v << c$, the Fermi function is related to the absolute square of the Coulomb wave function at the origin, and has the approximate form

$$F(z, E_e) = \frac{2\pi\eta}{1 - \exp(-2\pi\eta)} \qquad \ldots(3.9.42)$$

where $\eta = \frac{Ze^2}{4\pi\varepsilon_0 \hbar v}$ for positrons and $\eta = -\frac{Ze^2}{4\pi\varepsilon_0 \hbar v}$ for electrons. Extensive tabulated values of $F(z, E)$ are available which are mostly used. The momentum distribution of electrons with Coulomb correction is, given by,

$$\omega(p_e)dp_e = \frac{g^2|M|^2}{2\pi^3\hbar^7c^3} F(z, p_e)(E_0 - E_e)^2 \cdot p_e^2 dp_e \qquad \ldots(3.9.43)$$

and the corresponding energy distrioution for the electrons is given by,

$$\omega(E_e)dE_e = \frac{g^2|M|^2}{2\pi^3\hbar^7c^6} F(z, E_e)E_e\left(E_0^2 - m_0^2c^4\right)^{1/2}(E_0 - E_e)^2 dE_e \qquad \ldots(3.9.44)$$

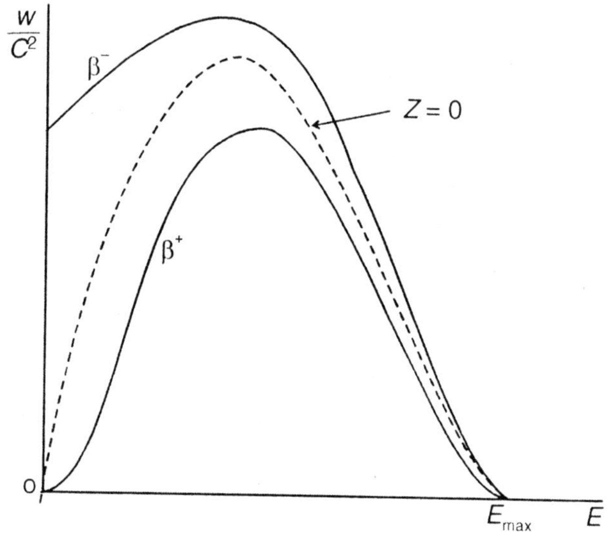

Fig. 3.28: Effect of Coulomb interaction on β-decay spectrum. The dotted line gives the allowed spectrum without Coulomb effect

The behaviour of the Fermi's function $F(z, E_e)$ has been found such that for low z-values ($z < 15$), it reduces to a constant value, which means that for low z-nuclei the β-spectrum is not much different than that one gets after neglecting the Coulomb effects. However, for large z, the Coulomb effects become more apparent. $F(z, E_e)$ now behaves such that the number of electrons does not become zero as energy tends to zero while the positron number becomes zero, as expected in principle (Fig. 3.12). The effect of Coulomb interactions is very nicely demonstrated by the decay of ^{64}Cu which emits both electrons and positrons of almost same end point energy.

$$^{64}_{29}\text{Cu} \rightarrow {}^{64}_{30}\text{Zn} + e^-, E_0 \sim 0.57 \text{ MeV}$$

and

$$^{64}_{29}\text{Cu} \rightarrow {}^{64}_{28}\text{Ni} + e^+, E_0 \sim 0.66 \text{ MeV}$$

The energy distribution of the electrons and positrons from ^{64}Cu is shown in Fig. 3.29. One observes that the maximum in the spectrum is shifted towards higher energy side for electrons whereas it is shifted towards higher energy side for positrons. This clearly shows the necessity of Coulomb correction and its different values for electrons and positrons emitted from the same Z-nucleus. The experimental data agree with Fermi's theory of β-decay.

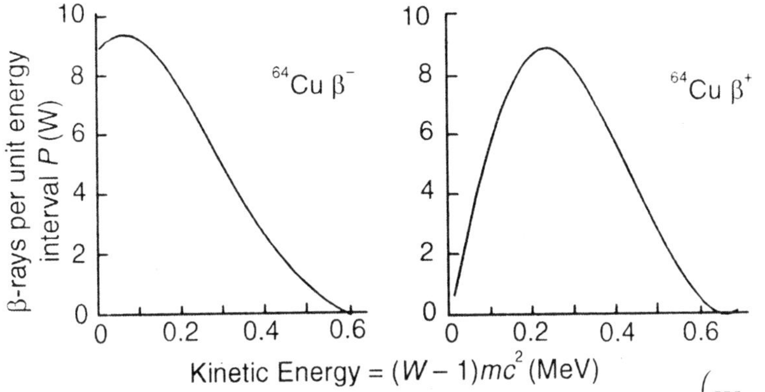

Fig. 3.29: Energy distribution of $\beta^{\pm}$ particles from ^{64}Cu $\left(W = E_e / m_c c^2\right)$

The corresponding momentum distributions from ^{64}Cu for electrons and positrons are shown in Fig. 3.30.

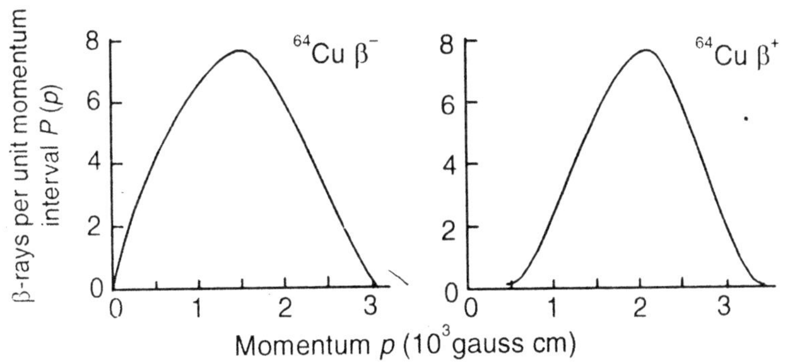

Fig. 3.30: Momentum distribution of $\beta^{\pm}$-particles from ^{64}Cu

3.9.7 Kurie Plot (Fermi-Kurie Plot)

The equation (3.9.44) can be expressed as,

$$\left[\frac{\omega(E_e)}{p_e^2 F(z, E_e)}\right]^{1/2} = K \cdot (E_0 - E_e) \qquad \ldots(3.9.45)$$

where, $$K = \left(\frac{g^2 |M|^2}{2\pi^3 \hbar^7 c^3}\right)^{\frac{1}{2}}$$

The number of electrons emitted per sec at different momentum values is known from experiments, the Fermi function is known from its tabulated values. Thus the left hand side of equation can be estimated for different E_e or p_e^2 values. When the values of the left hand side of eqn. (3.9.45) so obtained are plotted as a function of electron energy E_e, one gets a straight line of negative slope. The point at which the extra plotted line cuts the energy axis gives the end-point energy ($E_e = E_0$). Such a straight line graph known as Kurie plot (also known as Fermi-Kurie plot). The Fig. 3.31 shows a Kurie plot for β-active nucleus ^{32}P.

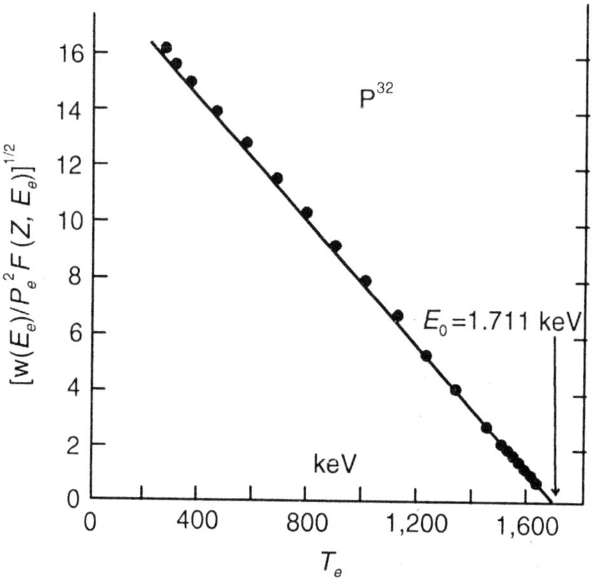

Fig. 3.31: Fermi-Kurie plot (a single transition) for ^{32}P

In some cases of β-decay, the Kurie plots are not just a single straight line but show a complex curve as shown in Fig. 3.32, where decay of ^{59}Fe nucleus takes place emitting two β-particles having end point energy as 271 KeV and 462 KeV respectively. An associated γ-rays of energy ($\beta_1 - \beta_2$) = 191 KeV is observed in the daughter nucleus. In Fig. 3.32, the straight line A for β_1 is fitted for high-energy data. When the data for straight line A are subtracted from the low energy, we get the line B for β_2.

The Kurie-plots not only give accurate determination of the end-point energy, they also allow one to know whether the spectrum is due to a transition to a single state of the daughter nucleus (Fig. 3.31) or to more than one final nuclear state of the daughter nucleus (Fig. 3.32). These transitions are allowed transitions. However, Fermi-Kurie plots show non-linearity below 200 KeV energy due to scattering of electrons in the source. The best one can go, therefore is upto 50 KeV.

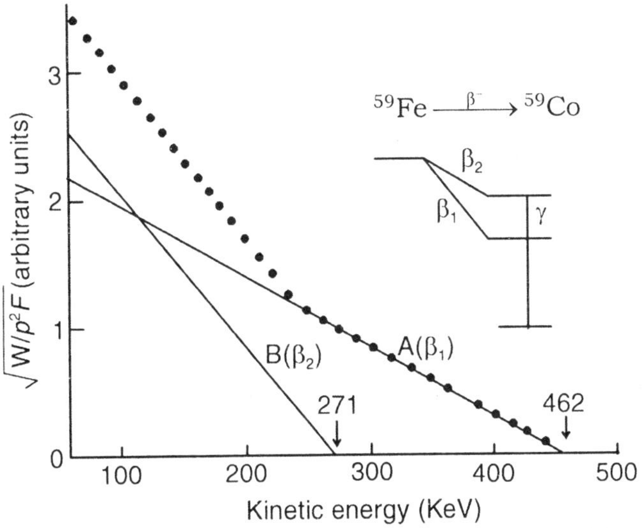

Fig. 3.32: Fermi-Kurie plotes of partial β-spectra of ^{59}Fe

3.9.8 Neutrino Mass

The shape of the curve for β-spectrum near the end point is determined by the energy conservation condition. If the rest mass of the neutrino (anti-neutrino) is zero, the momentum distribution of the electrons emitted in β-decay is given by the equation (3.9.33) and towards the end of the spectrum, the number of β-particles emitted per second approaches the upper limit of energy E_0 as $(E_0 - E_e)^2$. The tail of the distribution curve reaches the end point energy ($E_0 = E_e$) such that it has a horizontal tangent, as shown in the Fig. 3.33. On the other hand, if neutrino (anti-neutrino) has some finite rest mass ($m_\nu > 0$) the maximum energy carried by the electron will then be less than the end point energy equal to the amount of rest mass energy of the neutrino. In this case the density of states in the final system $\rho_f(E_e)$ is proportional to the neutrino momentum and the number of β -particles at the upper limit can be shown to be proportional to $(E_0 - E_e)^{1/2}$. The distribution curve, then reaches the end-point energy with a vertical tangent, as shown in Fig. 3.33.

The experimental studies of β^--decay of tritium (^3H) by Lubimov and others have put an upper limit for the mass of neutrino. These workers have used the Kurie plot or Fermi plot and found that the best fit of the experimental data with the theory is obtained with $m_\nu \leq 37$ eV/c^2 as its upper limit.

3.9.9 Allowed and Forbidden Transitions

We have seen that a β-decay is the result of interaction between beta-active nucleus and the field of leptons. The electron and neutrino emitted during the disintegration are regarded as free particles and neglecting Coulomb effects, the wave functions for them (ϕ_e, ϕ_ν) are defined by plane waves,

i.e.
$$\phi_e \sim e^{\frac{i}{\hbar}(\vec{p}_e \cdot \vec{r})}$$

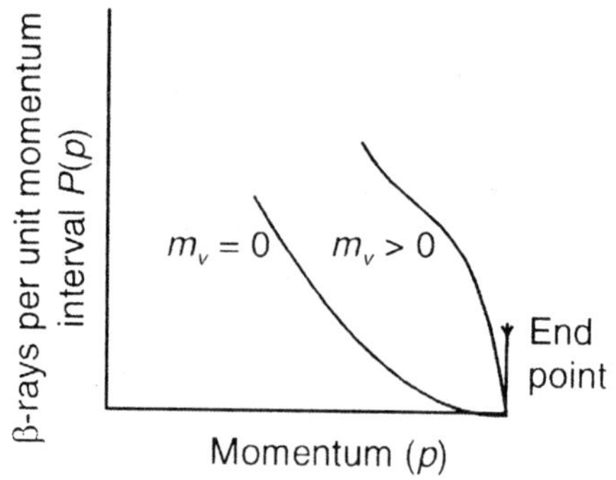

Fig. 3.33: Dependence of shape of β-spectrum near the upper limit (end point) on the neutrino mass m_ν (not to scale)

It is shown in Section (3.9 *d*) that exponentials in the plane waves can be expanded since for beta particles of around 1 MeV, $\frac{p \cdot R}{\hbar} = \frac{R}{\lambda\!\!\!^{-}}$, for medium size nucleus is of the order of 0.01, where R is nuclear radius and $\lambda = 2\pi\lambda\!\!\!^{-}$ is the wavelength of electron or neutrino. But (pr) is a measure of angular momentum carried by leptons, and according to quantum mechanics, the permissible values of angular momentum are $\sqrt{l(l+1)}\hbar$, where l is the orbital angular momentum quantum number of the emitted leptons. If only the first term in the expansion,

$$\phi = \exp\frac{i}{\hbar}\vec{p}\cdot\vec{r} = 1 + \frac{i}{\hbar}(\vec{p}\cdot\vec{r}) + \left(\frac{i}{\hbar}\frac{\vec{p}\cdot\vec{r}}{2}\right)^2 + \ldots$$

of the plane wave function is retained, neglecting all the higher order terms then this amounts to say that only $l = 0$ terms are considered for the leptones. Since the electron and the neutrino have, on an average approximately equal momenta, the inclusion of higher order terms in the expansion means considering terms with $l = 1$, $l = 2$ and so on. The rate of emission, therefore, is largest when the total orbital angular momentum, L, of electron-neutrino pair is zero. Beta-decay or beta-transitions with $L = 0$ are referred as Allowed transitions while those with $L > 0$ are called Forbidden transitions. It also follows from the order of magnitude of $\frac{pR}{\hbar}$ which is around 10^{-2}. Since then $\frac{p \times R}{\hbar} << 1$ and hence $L = 0$ is the most probable decay. However, if the decay is such that parity changes ($\Delta\pi = yes$) then the first term in the expansion of $e^{ikr} = e^{\frac{ipr}{\hbar}}$ contributes zero and higher order terms are then to be considered. First forbidden transition corresponds to $L = 1$, second forbidden with $L = 2$ and so on. Allowed transitions, are thus most intense and the intensity decreases in forbidden transitions by a factor $(R/\lambda\!\!\!^{-})^2$ compared to allowed transitions. Besides orbital angular momentum, both electron and neutrino have spin angular momentum of ½ each in units of $\hbar$. Their total spin S can be 0 or 1 depending on their spins being anti-parallel (↑↓or singlet state) or parallel (↑↑ or triplet state). The transitions are, therefore, also categorised. When spins are anti-parallel (S = 0) the transitions are Fermi transitions and those with parallel spins (S = 1) are known as Gammow-Teller transitions. Thus, there are Allowed and Forbidden transitions of both Fermi type and of Gammow-Teller type.

The interaction matrix $|M|_{fi}$ in the expression for transition probability (eqn. 3.9.28) is referred as Fermi interaction $|M_F|$ where the interaction operator Q which connects the initial nuclear state (parent nucleus state) and the final nuclear state (daughter nuclear state) is a unit operator and it is a scalar. $|M_F| \approx 1$ when $I_i = I_f$. Likewise the interaction is called Gammow-Taller interaction $|M_{GT}|$, when the interaction operator Q is spin operator $\vec{\sigma}$. The matrix elements for G.T operator cannot be evaluated unless the wave functions for the parent nucleus (initial state u_i) and that of daughter nucleus (final state u_f) are known. However, from the properties of the operator itself one finds that the initial and final spins of the nucleus are coupled vectorially by a unit vector. $\vec{\sigma}$ being an axial vector (Pseudo vector) cannot change the parity between initial and final states.

3.9.10 Selection Rules

It is known that each nuclear state is characterised by total angular momentum I and parity π. The parent and daughter nuclei after the β-transformation occupy certain definite states. Therefore, the conservation of angular momentum and parity will determine the selection rules for the type of transitions. In Fermi transitions (S = 0) the only angular momentum carried by the electron neutrino pair is their orbital angular momentum L. If I_i is the angular momentum of the parent (initial) nucleus and I_f that of the daughter nucleus (final) then

$$\vec{I}_f = \vec{I}_i + \vec{L} \quad \text{...(3.9.46)}$$

For Gammow-Teller transitions ($\vec{S}$ = 1), the electron-neutrino pair carries off spin angular momentum also, therefore, in this case

$$\vec{I}_f = \vec{I}_i + \vec{L} + \vec{S} = \vec{I}_i + \vec{L} + 1 \quad \text{...(3.9.47)}$$

Any change in the total angular momentum of the nucleus during β-decay is

$$\Delta\vec{I} = \vec{I}_f - \vec{I}_i$$

and this is determined by the changes ($\Delta\vec{L}$) and ($\Delta\vec{S}$) in the orbital angular momentum and spin angular momentum of the electron-neutrino pair, respectively. In the case of Allowed transitions, L = 0 as such ΔI must be decided by the spin angular momentum of electron-neutrino pair. For Fermi transitions ($\vec{S}$ = 0) therefore,

$$\Delta\vec{I} = \vec{I}_f - \vec{I}_i = 0$$

Here, $\vec{I}_i$ and $\vec{I}_f$ both can be zero *i.e.* $\vec{I}_i = \vec{I}_f = 0$ is possible.

For G-T allowed transitions ($L = 0$)

$$\vec{I}_f = \vec{I}_i + \vec{1}$$

$$|\Delta I| = |\vec{I}_f - \vec{I}_i| = \pm 1,\ 0$$

i.e. $I_f = I_i \pm 1$, and $I_f = I_i$

However, in G-T allowed transitions, there cannot be a transition between $I_f = I_i = 0$ because, such transitions cannot satisfy the condition of $\vec{S} = 1$ for the total spin angular momentum. However, $I_f = I_i = 0$ or $0^+ \to 0^+$ is a pure Fermi allowed transition. The conservation of parity demands that the parity of the initial wave function (of the parent nucleus) must be same as the product of the parities of the wave functions of the daughter nucleus, electron and neutrino that is

$$\pi_i = \pi_f(-1)^L$$

where π_i and π_f are the parities of the wave functions of parent nucleus and daughter nucleus respectively, and $(-1)^L$ is the parity for the electron-neutrino pair. For Fermi Allowed transitions $L = 0$, therefore, $\pi_i = \pi_f$ and hence there is no change in the parity. In the case of Gammow Teller (GT) transition also there is no change in the parity.

Thus for allowed transitions the selection rules are

(i) $\Delta I = 0$, No Fermi Type

(ii) $\Delta I = \pm 1, 0$, No Gammow-Teller Type (except $0 \to 0$)

Here the word 'No' stands for no change in parity.

Example of a Pure Fermi allowed transitions is

$$^{14}_{8}O \to ^{14}_{7}N^* + \beta^+ + \nu,\ (O^+ \to O^+,\ \text{No}) \qquad \ldots(3.9.48)$$

the daughter nucleus ^{14}N is in an excited state. Here ^{14}O and $^{14}N^*$, both have even parities and for ^{14}O, $I_i = O^+$ and for $^{14}N^*$, $I_f = O^+$, therefore $I_i = I_f = 0$ and no change in parity.

Likewise the decay of ^6He is an example of pure Gammow Teller type of allowed transition

$$^{6}_{2}He \to ^{6}_{3}Li + \beta^- + \bar{\nu},\ (O^+ \to 1^+,\ \text{No}) \qquad \ldots(3.9.49)$$

Here, there is no change in parity, as both the nuclei have even parities. But ^6Li has spin angular momentum 1 *i.e.* $I_f = 1^+$ and for ^6He, $I_i = 0^+$ as such $\Delta I = I_f - I_i = 1$.

Note: even parity is denoted by a superscript (+) over the angular momentum I, (I^+) and for the odd parity (–) is used (I^-).

It may be noted that a pure Gammow-Teller allowed transition is the decay of the 0^+ states of ^{14}O to the ground state 1^+ of ^{14}N. Here, $\Delta I = 1$, and no parity change.

In some allowed transitions both Fermi and Gammow-Teller selection rules are applicable. Such a situation of decays is possible in allowed decays where the initial and final states may have, both non-zero angular momentum *i.e.* $I_f = I_i \neq 0$. Some examples of such decays are:

(*i*) $\quad {}_0^1n \rightarrow {}_1^1H + \beta^- + \bar{\nu}$, ($\frac{1}{2}^+ \rightarrow \frac{1}{2}^+$, no)

(*ii*) $\quad {}_1^3H \rightarrow {}_2^3He + \beta^- + \bar{\nu}$, ($\frac{1}{2}^+ \rightarrow \frac{1}{2}^+$, no)

(*iii*) $\quad {}_{16}^{35}S \rightarrow {}_{17}^{35}Cl + \beta^- + \bar{\nu}, \left(\frac{3^+}{2} \rightarrow \frac{3^+}{2}, \text{no}\right)$

The applicability of the selection rules is confirmed by *ft*-values (discussed in the next Section).

3.9.11 Forbidden Transitions (or Forbidden Decays)

From the selection rules for allowed transitions, we find that spins of initial and final states can be different at most by unity and the parities must be same. But it has been found that transitions between different parities and $\Delta I > 1$, do take place. Such transitions are referred as Forbidden transitions. When transitions between nuclear states of different parities take place, say, between an s-state and a p-state a change in orbital angular momentum takes place and leptons cannot be emitted as s-waves. The matrix element $|M|_{fi}$ will then be zero and one has to consider the next higher terms in the expansion of plane-wave functions of the leptons (no Coulomb effect). As shown earlier the successive terms other than the first in the expansion corresponds to $l = 1$ $l = 2$ and so on indicating thereby, that the electron-neutrino pair carry the corresponding orbital angular momentum. The inclusion of term $\left\{\frac{i}{\hbar}(\vec{p}_e + \vec{p}_\nu)\cdot\vec{r}\right\}$ in the matrix element $|M_{fi}|$ renders it to be non-zero and the electron-neutrino pair now is emitted as p-waves ($l = 1$) which have odd-parity ($\pi = (-1)^l$). Such transitions are known as first forbidden transitions. When term for $l = 2$ is considered, leading to non-zero matrix element, the transitions are second forbidden transition and so on. Since the successive terms in the expansion of plane wave decrease by a factor $R/\lambda \approx 0.1$ for a nucleus of medium size and beta energies ~1 MeV the β^- emission probability decreases by a factor $(R/\lambda)^2$. Thus the first forbidden transition is weaker in intensity by a factor of about 1/100 compared to the allowed transition (s-wave emission) and the successive forbidden transitions become weaker.

Selection Rules for Forbidden Transitions

1. First forbidden transitions: ($l = 1$)

   (*a*) Fermi transitions: $\Delta I = \pm 1, 0$, Yes.
   (Singlet State, $\Delta S = 0$) (Except $0 \rightarrow 0$)

   (*b*) Gammow-Teller transitions: (Triplet State, $\Delta \vec{S} = 1$)
   $\Delta I = \pm 2, \pm 1, 0$, Yes
   (Except $\frac{1}{2} \rightarrow \frac{1}{2}$, $0 \rightarrow 1$)

In G-T selection rule, $0 \to 0$ transition may be allowed, since

$$\Delta\vec{I} = \vec{I}_f - \vec{I}_i = \vec{S} + \vec{L}$$

where $\vec{S} = 1$, $\vec{L} = 1$ are the spin and orbital angular momenta of electron-neutrino pair. Thus, the transition is possible for $I_f = I_i = 0$

Examples of first forbidden transition:

$$^{111}_{47}Ag \to {}^{111}_{48}Cd + \beta^- + \bar{\nu} \qquad \left(\frac{1}{2} \to \frac{1}{2}\right)$$

$$^{85}_{36}Kr \to {}^{85}_{37}Rb + \beta^- + \bar{\nu} \qquad \left(\frac{5}{2} \to \frac{3}{2}\right)$$

$$^{37}_{16}S \to {}^{37}_{17}Cl + \beta^- + \bar{\nu} \qquad \left(\frac{7}{2} \to \frac{3}{2}\right)$$

2. Second forbidden transitions:
   For these transitions $l = 2$,

   (*a*) Fermi Selection rules: $\Delta I = \pm 2, \ \pm 1,$ (Except $0 \to 1$) No

   (*b*) Gammow-Teller rules: $\Delta I = \pm 3, \ \pm 2,$ (Except $0 \to 2$) No

   Examples of second forbidden transitions:

$$^{10}_{4}Be \to {}^{10}_{5}B + \beta^- + \bar{\nu}, \qquad (0 \to 3)$$

$$^{135}Cs \to {}^{135}Ba + \beta^- + \bar{\nu}, \qquad \left(\frac{7}{2} \to \frac{3}{2}\right)$$

Note: In the above discussions of Fermi's simple theory of β-decay, the process of disintegration has been confined only to a single nucleon (neutron or a proton). For a β-active sample where number of nucleons take part, the theory becomes complicated which is beyond the scope of this book.

3.9.12 The Total Decay Rate for β-Disintegration

The total transition probability or the decay rate λ, for electron emission in a β-decay is obtained by integrating equation (3.9.43) over the electron momentum distribution from zero to a maximum momentum p_e. Thus,

$$\lambda = \int_0^{p_0} \omega(p_e) dp_e$$

$$= \int_0^{p_0} \frac{g^2}{2\pi^3} \frac{|M|_{fi}^2}{\hbar^7 c^3} F(z, E_e)(E_0 - E_e)^2 p_e^2 dp_e \qquad \ldots(3.9.50)$$

Assuming that $|M|_{fi}^2$ is independent of energy

$$\lambda = \frac{1}{\tau} = \frac{g^2 |M|_{fi}^2}{2\pi^3 \hbar^7 c^3} \int_0^{p_0} F(z, E_e)(E_0 - E_e)^2 p_e^2 dp_e \qquad \ldots(3.9.51)$$

where τ is the mean life for the decay (electron emission). If $T_{1/2}$ is half life for β^- decay, then

$$\lambda = \frac{\ln 2}{T_{1/2}} = \frac{1}{\tau}$$

$$= \frac{g^2|M|_{fi}^2 m_e^5 c^4}{2\pi^3\hbar^7}\int_0^{p_o/m_e c} F(z,E_e)\left\{\frac{E_0 - E_e}{m_e c^2}\right\}^2 \frac{p_e^2}{m_e c^2}\frac{dp_e}{m_e c}$$

or
$$\lambda = \frac{g^2|M|_{fi}^2 m_e^5 c^4}{2\pi^3\hbar^7} f(z,E_0) \qquad \text{...(3.9.52)}$$

where $f(z, E_0)$ is the dimensionless function given by

$$f(z,E_o) = \frac{1}{m_e^4 c^7}\int_o^{p_o/m_e c} F(z,E_o)(E_o - E_e)^2 p_e^2 dp_e \qquad \text{...(3.9.53)}$$

The function $f(z, E_o)$ is known as the Fermi integral. Except in the case $Z=1$ for the daughter nucleus, the integral must be evaluated numerically. Since half-life $T_{1/2} = \ln 2/\lambda$, we can obtain β-decay half-life from the transition probability. Using equation (3.9.52), we get

$$T_{1/2} = \frac{\ln 2}{\lambda} = \frac{2\pi^3\hbar^7}{g^2|M|_{fi}^2 m_e^5 c^4}\frac{\ln 2}{f(z,E_0)} \qquad \text{...(3.9.54)}$$

Instead of half-lives, nuclear β-decay rates are often quoted in terms of *ft*-values, the product of Fermi integral $f(z, E_0)$ (abbreviated to f) and $T_{1/2}$ being used as t.

i.e.
$$ft = f(z,E_o)T_{1/2} = \frac{2\pi^3\hbar^7 \cdot \ln 2}{g^2|M|_{fi}^2 m_e^5 c^4}$$

or
$$ft = \frac{\tau_0 \ln 2}{|M|_{fi}^2} \qquad \text{...(3.9.55)}$$

where $\tau_0 = 2\pi^3\hbar^7/g^2 m_e^5 c^4$ is a constant known as universal time constant.

Though it is the half life which is measured experimentally, the *ft* value for a β-decay is a more meaningful and significant physical quantity because it is related to the square of the nuclear matrix element $|M|_{fi}^2$. The quantity *ft* is also termed as comparative half life.

The measured *ft* values are found to vary over many orders of magnitude (10^3 sec to 10^{18} sec) particularly when both the allowed and forbidden transitions are included. It is often more convenient to use $\log_{10} ft$ values. Since *ft* is inversely proportional to the square of the matrix element $(|M|_{fi}^2)$, as such for all allowed transitions $|M|_{fi}$ being almost same and successively smaller for forbidden transitions of increasing order. If the values of $\log_{10} ft$ values are calculated for different β-emitters from their known values of end point energy E_0 and half lives $T_{1/2}$, the values so found are expected to fall into groups according to allowed, first, second, etc. forbidden transitions. Such groups for odd-mass nuclei are shown in the Fig. 3.34. The different groups of values of $\log_{10} ft$ help in categorising the β-emitters. Since the *ft*-value would give its order of forbiddenness it must also correspond to ΔI values (selection rules) which govern the β-decay. The groupings of the $\log_{10} ft$ values are classified as given below.

1. The smallest $\log_{10} ft$ values are found in the case of Mirror nuclei. In these nuclei the ground state wave functions of the parent (initial) and daughter nuclei (final) are alike. Such decays have high probability to take place. They are known as allowed and favoured or super allowed transitions. The log *ft* values lie between 3 to 4. Examples of such transitions are:

Decay	Half life	log *ft*
${}^1_0n \rightarrow {}^1p + \beta^- + \bar{\nu}$,	750 sec	3
${}^3H \rightarrow {}^3He + \beta^- + \bar{\nu}$,	319 × 10⁸ sec	3
${}^{11}C \rightarrow {}^{11}B + \beta^+ + \nu$,	1230 sec	3.6
${}^{23}Mg \rightarrow {}^{23}Na + \beta^+ + \nu$,	11.6 sec	3.5

From these examples one finds that though the half life of decays have a wide range, log *ft* values are almost constant.

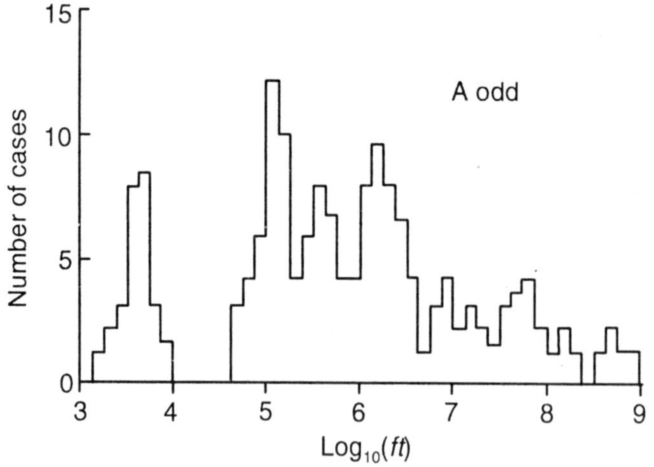

Fig. 3.34: Comparative half-life for β-transitions between ground states of odd mass nuclei as represented by log*ft* values

2. The next range of log *ft* value can be grouped between 4 to 6. Here the wave functions u_i and u_f may not be strictly identical, such that $|M|_{fi}$ is less than unity. These transitions are allowed but are not favoured like super allowed transitions. Examples of this group are:

Decay	T½	log ft
$^{24}Na \rightarrow {}^{24}Mg + \beta^+ + \bar{\nu}$,	5.4×10^4 sec	6
$^{35}S \rightarrow {}^{35}Cl + \beta^- + \bar{\nu}$,	7.5×10^6 sec	5

3. The log ft values between 6 to 10 correspond to first forbidden transitions. The second forbidden transitions fall under the category of log ft values between 10 and 14. The higher order transitions will have higher values for log ft. The large values of log ft values for forbidden transitions in β-decays is for the fact that in forbidden transitions, $l > 0$, the emission of beta particles have to face an angular momentum barrier. This results in the small size of the nuclear matrix element $|M|_{fi}$ amd hence an increase in the ft-values (eqn. 3.9.55).

Though the various β-decays could be classified in various groups depending on the log ft values, but as can be seen from the Fig. 3.19 this grouping cannot be sharp and well-separated. In the case of pure Fermi and pure Gammow Teller transitions, the value of $|M|_{fi}$ is maximum. But when ft values are considered such a behaviour of $|M|_{fi}$ occurs in certain cases categorised as super allowed transitions. In general therefore, this is not so, as is observed by the spread of ft values for allowed transitions. Whenever, allowed transitions are mixed, where both Fermi and G-T transitions are favoured, the matrix element $|M|_{fi}$ can be expressed as

$$|M|^2_{fi} = |C_F|^2 \, |M_F|^2_{fi} + |C_{GT}|^2 \, |M_{GT}|^2_{fi} \qquad \text{...(3.9.56)}$$

where C_F and C_{GT} are coupling constants expressed in units of g, and are such that

$$|C_F|^2 + |C_{GT}|^2 = 1 \qquad \text{...(3.9.57)}$$

and $|M_F|_{fi}$ and $|M_{GT}|_{fi}$ are the matrix elements for Fermi and Gammow-Teller interactions. The equation (3.9.55) for ft value then can be expressed as,

$$ft = \frac{B}{(1-x)\,|M_F|^2 + x\,|M_{GT}|^2} \qquad \text{...(3.9.58)}$$

where $x = C_{GT}^2$ and the constant B is given by

$$B = \ln 2 \cdot \tau_0 = \frac{2\pi^3 \hbar^7 \cdot \ln 2}{g^2 c^4 m_e^5} \qquad \text{...(3.9.59)}$$

The experimental results on the decay of nuclei like ^1n, ^3H, ^6He, and ^{14}O for which M_F and M_{GT} are known, have been used to give the constants in the equation (3.9.58), since ft values for these nuclei are also known. It is found that the value of B is 2787 ± 70 sec and $x = 0.56$. From the value of B the value of coupling constant for Fermi interaction is calculated which is $g_F = 1.435 \times 10^{-49}$ erg cm^3 or $g_F = 1.435 \times 10^{-62}$ J-m^3.

3.9.13 Sargent Curves

It was pointed out by Sargent that there exists a relation between the decay constant λ (or life time) of a β-disintegeration and its maximum energy E_0. Such relationship exists in the case of α-emitters between the decay constant and energy. If $\log \lambda$ for various β-emitters (naturally occuring radio elements) is plotted against $\log E_0$, the points fall on straight lines as shown in Fig. 3.35. These lines are known as Sargent curves or Sargent diagrams. These lines could be represented by an expression

$$\lambda = KE_0^5 \qquad \ldots(3.9.60)$$

K is a constant. It differs by a factor of 100 for the two curves for the same maximum energy E_0. It means that the decay constants or the half lives for the radio-elements also differ by a factor of about 100. This difference in the behaviour of decay constants was explained by Gammow in terms of transitions. Accordingly the upper curve (short lived or large λ) corresponds to the allowed transitions, while the longlived (or small λ) radioactive elements correspond to the forbidden transitions, and lie on the lower graph. However, on the availability of large number of artificially produced radio-elements (β-emitters) such a sharp distinction was not possible. The points scatter around such almost parallel lines.

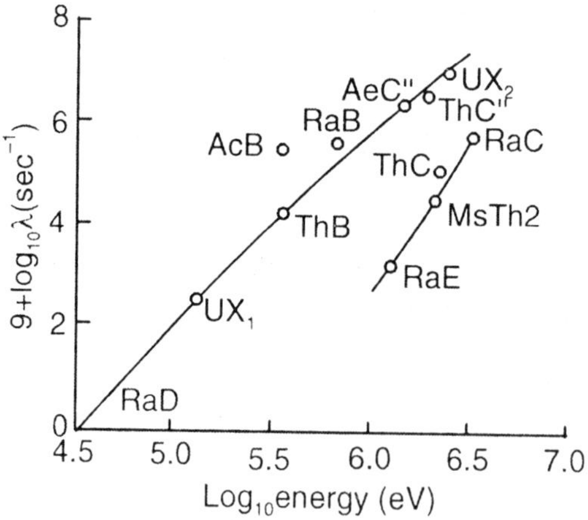

Fig. 3.35: Sargent curves for some heavy radioactive nuclei

Using Fermi's theory of β-decay, it is possible to establish theoretically the Sargent's fifth power law ($\lambda \propto E_0^5$). For allowed transitions, the total transition probability or the decay constant, neglecting Coulomb effects, is given by

$$\lambda = \text{Constant} \times f(E_0) \qquad \ldots(3.9.61)$$

where the Fermi's integral is

$$f(E_0) = \frac{1}{m_e^5 c^7} \int_0^{p_0} (E_0 - E_e)^2 p_e^2 dp_e \qquad \ldots(3.9.62)$$

Since for all radio nuclides $E_0 >> m_e c^2$, $p_0 >> m_e c$, then the Fermi's integral can be evaluated. One will then find that

$$\lambda \propto E_0^5 \qquad \text{...(3.9.63)}$$

Sargent's observations, thus agree with theory particularly in the case of light nuclei (low Z-nuclei). For heavy nuclei Coulomb effects introduce certain distortions, as such theory becomes complicated.

3.9.14 Electron-Capture Process

It has been pointed out that there is a competition between positron emission and capture of orbital electrons (K-electron or L-electron, etc.) by the nucleus. Besides, it was also indicated that, during the transformation of a proton into a neutron as a result of weak interaction, inside a β-active nucleus (high Z-values) if the difference between the atomic mass between the parent nucleus and daughter nucleus is less than $2m_e c^2$, capture of an orbital electron is favoured. The rest mass energy of the captured electron, is then added to the energy released. As the energy of the electron (K-electron or L-electron) is fixed, the energy released must be carried, by the neutrino, and hence all the neutrinos come out with the same energy. Once the orbital electron is captured, a vacancy is caused in the corresponding orbit which is filled by the jumping of an electron to it from the next higher orbit and characteristic X-ray photon is emitted. For example, if K-electron is captured and the vacancy is filled by an electron from the L-orbit, then characteristic K-x-rays from the daughter nucleus is emitted. The observation of these X-ray photons establishes the capture of K-electron.

Theory of Electron Capture

The transition probability per unit time for the electron capture process would be just equal to the decay constant for the process since both the captured electron and the emitted neutrino have definite energies. One can use the same expression for the probability as used for $\beta^\pm$ decay. The interaction of the field (weak interaction) being same here and so $|M|^2_{fi}$ is same as for the electron emission. Thus for allowed transitions, the decay constant for K-electron capture will be,

$$\lambda_K = \frac{2\pi}{\hbar} g^2 |M|^2_{fi} |\phi_\nu \phi_e|^2 \rho_\nu(E) \qquad \text{...(3.9.64)}$$

where $\rho_\nu(E)$ is the density of neutrino states. As before the neutrino wave function ϕ_ν is taken as a plane wave

$$\phi_\nu = \frac{e^{i\vec{k}\cdot\vec{r}}}{\sqrt{V}} = \frac{1}{\sqrt{V}} e^{\frac{i}{\hbar}(\vec{p}\cdot\vec{r})} = \frac{1}{\sqrt{V}} \left\{1 + \frac{i}{\hbar} \vec{p} \cdot \vec{r} + \ldots\right\}$$

which for allowed transitions at $r = 0$, gives

$$\phi_\nu(0) = \frac{1}{\sqrt{V}} \qquad \text{...(3.9.65)}$$

For the electron wave function, we can use the K-orbit wave function for hydrogen like atoms, since the wave function ϕ_e must characterise the bound state of the initial system.

$$\phi_e = \phi_k = \frac{1}{\sqrt{\pi}}\left(\frac{Z}{a_0}\right)^{3/2} \exp\left(-\frac{Zr}{a_0}\right) \quad \text{...(3.9.66)}$$

where $a_0 = \dfrac{4\pi\varepsilon_0\hbar^2}{m_e e^2}$ is the Bohr radius and Z is the atomic number of parent nucleus.

As the wave functions will have appreciable values only for short distances, we may evaluate ϕ_e for (approximately) $r = 0$

i.e.
$$\phi_e(0) = \phi_k(0) = \frac{1}{\sqrt{\pi}}\left(\frac{Z}{a_0}\right)^{3/2}$$

or
$$\phi_e(0) = \frac{1}{\sqrt{\pi}}\left\{\frac{Z \cdot m_e e^2}{4\pi\varepsilon_0\hbar^2}\right\}^{3/2} \quad \text{...(3.9.67)}$$

The statistical factor or the density of states ρ_ν for the neutrino is to be given by the number of neutrino states in the momentum range p_ν and $p_\nu + dp_\nu$. Therefore,

$$\rho_\nu = \frac{dn_\nu}{dE_\nu} = \frac{4\pi V \cdot p_\nu^2 dp_\nu}{(2\pi\hbar)^3 dE_\nu} \quad \text{...(3.9.68)}$$

The energy E_ν of the neutrino and its momentum p_ν are related as $E_\nu = cp_\nu$, since neutrino mass is zero. c is velocity of light

$\therefore$
$$dE_\nu = c\,dp_\nu \quad \text{...(3.9.69)}$$

Also the neutrino energy in terms of the binding energy B_k of the K-electron and the maximum available energy E_0 is given by (E_0 is difference of nuclear masses of parent and daughter nucleus)

$$E_\nu = E_0 + m_e c^2 - B_k \quad \text{...(3.9.70)}$$

Using eqn. (3.9.69) in eqn. (3.9.68)

$$\rho_\nu(E) = \frac{4\pi V E_\nu^2}{(2\pi\hbar)^3 c^2}\,\frac{1}{c} \quad \text{...(3.9.71)}$$

Using equations (3.9.65), (3.9.67) and (3.9.71) in equation (3.9.64) we get for the decay constant λ_k for the k-electron capture as,

$$\lambda_k = \frac{2\pi}{\hbar} g^2 |M|_{fi}^2 \cdot \frac{1}{V} \cdot \frac{1}{\pi}\left\{\frac{Zm_e e^2}{4\pi\varepsilon_o\hbar^2}\right\}^3 \cdot \frac{4\pi V E_\nu^2}{(2\pi\hbar)^3 \cdot c^3}$$

$$\lambda_k = \frac{2\pi}{\hbar} g^2 |M|_{fi}^2 \cdot \frac{4E_\nu^2}{(2\pi\hbar)^3 c^3}\left\{\frac{Zm_e e^2}{4\pi\varepsilon_o\hbar^2}\right\}^3 \quad \text{...(3.9.72)}$$

Using equation (3.9.70) for neutrino energy E_ν, we get,

$$\lambda_k = \frac{g^2 \cdot m_e^3 e^6 Z^3 |M|_{fi}^2}{64 \cdot \pi^2 \varepsilon_0^3 \hbar^{10} c^3} (E_0 + m_e c^2 - B_k)^2 \quad \ldots(3.9.73)$$

Note: The equation (3.9.73) for λ_k can be multiplied by a factor 2, to account for the presence of two electrons in *K*-orbit.

Putting $W_0 = \frac{E_0}{m_e c^2}$ and $W_B = \frac{B_k}{m_e c^2}$, we get,

$$\lambda_k = \frac{g^2 \cdot m_e^5 c^4}{\pi^2 \hbar^7} \left\{ \frac{Ze^2}{4\pi\varepsilon_0 \hbar c} \right\}^3 \cdot |M|_{fi}^2 (W_0 + 1 - W_B) \quad \ldots(3.9.74)$$

Putting

$$f_k = 2\pi (\alpha z)^3 (W_0 + 1 - W_B)^2 \quad \ldots(3.975)$$

where $\alpha = \frac{e^2}{4\pi\varepsilon_0 \hbar c} = \frac{1}{137}$ is Sommerfeld's fine structure constant. We get for λ_k

$$\lambda_k = \frac{g^2 m_e^2 c^4}{2\pi^3 \hbar^7} |M|_{fi}^2 \cdot f_k \quad \ldots(3.9.76)$$

Now using $\tau_0 = \frac{2\pi^3 \hbar^7}{g^2 m_e^5 c^4}$ being a universal time constant, we get,

$$\lambda_k = \frac{|M|_{fi}^2 f_k}{\tau_0} \quad \ldots(3.9.77)$$

If t_k is half-life for *K*-capture, then

$$\lambda_k = \frac{\ln 2}{t_k} = \frac{|M|_{fi}^2 f_k}{\tau_0}$$

or

$$f_k t_k = \frac{\tau_0 \ln 2}{|M|_{fi}^2} \quad \ldots(3.9.78)$$

The expression (3.9.69) for λ_k should be corrected for the screening of the nuclear charge by the orbital electrons. The effect of these electrons is to reduce the value of *Z*. For *K*-electrons, the reduction in *Z* is by about 0.35 and for *L*-electrons it is by about 4.15.

As stated earlier that in certain β-active nuclei there is competition between positron emission and *K*-capture. It is therefore possible to calculate the branchimg ratio $\frac{\lambda_k}{\lambda_{\beta^+}}$, which will be independent of $|M|_{fi}^2$, since it is same for the two competing processes. From equation (3.9.74) and equation (3.9.52) the branching ratio is

$$\frac{\lambda_k}{\lambda_{\beta^+}} = \frac{2\pi^2 \hbar^7}{g^2 m_e^5 c^4 |M|_{fi}^2} \cdot \frac{g^2 m_e^5 c^4 |M|_{fi}^2}{\pi^2 \hbar^7 f(Z, E_0)} \times (W_0 + 1 - W_B)^2 \quad \ldots(3.9.79)$$

or $$\frac{\lambda_k}{\lambda_{\beta^+}} = \frac{2\pi(\alpha Z)^3(W_0 + 1 - W_B)^2}{f(Z,\ E_0)} \qquad \text{...(3.9.80)}$$

From the above ratio we find that λ_k increases reapidly with increasing Z. (It increases as Z^3), whereas for high Z, the positron has to penetrate a Coulomb barrier and so decreases with increasing Z. The Coulomb effect term is included in the Fermi integral $f(z, E_0)$ (eqn. 3.9.53). Electron capture therefore becomes more probable in high Z-elements. For light elements, (low Z-values), positron emission predominates. In light nuclei positron emission is favourable for high end-point energies. The experimental observation of the branching ratio in a number of β-active nuclei has shown good agreement with the theoretical findings.

The binding energy of the K-electron is

$$B_k = \frac{1}{2}\frac{m_e e^4 Z^2}{(4\pi\varepsilon_0 \hbar)^2} \quad \text{and} \quad W_B = \frac{B_k}{m_e e^2} = \frac{1}{2}\frac{m_e e^4 Z^2}{m_e e^2 (4\pi\varepsilon_0 \hbar)^2}$$

$$W_B = ½\frac{e^4 Z^2}{(4\pi\varepsilon_0 \hbar c)^2} = \frac{1}{2}(\alpha z)^2 \qquad \text{...(3.9.81)}$$

where $$\alpha = \frac{e^2}{4\pi\varepsilon_0 \hbar c} = \frac{1}{137}$$

Using eqn. (3.9.81) in eqn. (3.9.75) we get,

$$f_k = 2\pi\ (\alpha Z)^3\ \{W_0 + 1 - \frac{1}{2}(\alpha Z)^3\} \qquad \text{...(3.9.82)}$$

We know that the numerical value of the product f.t. determines the forbiddenness of a transition as such in a given K-capture process, it is possible to know whether K-capture transition is an allowed transition or a forbidden transition. As an example, consider the decay of ^{7}Be by K-capture

$$^7_4\text{Be} + e_k^- \rightarrow {}^7_3\text{Li} + \nu \qquad \text{...(3.9.83)}$$

The maximum energy available for the transition E_0 being the difference in the nuclear masses of ^{7}Be and ^{7}Li or E_0 = 0.35 MeV, thus

$$W_0 = \frac{0.35}{0.51} = 0.68, \qquad m_e c^2 = 0.51 \text{ MeV}$$

The half life for the process, t_k = 53.4 days. The binding energy of K-electron in beryllium (Z = 4) is around 212 eV which is negligible compared to E_0. We, therefore, get,

$$f_k = 4\pi(\alpha Z)^3\ (W_0 + 1) \qquad \text{...(3.9.84)}$$

Taking into account the two electrons in K-shell

$$f_k = 4\pi(\alpha Z)^3 (W_0 + 1)^2$$

$$= 4\pi\left(\frac{4}{137}\right)^3 (1.68)^2 = 8.6 \times 10^{-4}$$

and $t_k = 53.4$ days $= 4.6 \times 10^6$ sec

$$f_k t_k = 8.6 \times 10^{-4} \times 4 \times 10^6 = 3956$$

$$\log_{10} f_k t_k = 3.596 \approx 3.6$$

Thus the decay of Be through K-electron capture is a super allowed transition. This also establishes the fact that in K-capture and in β-decay, the interaction is same (same $|M|^2_{fi}$).

Since neutrinos cannot be detected, the measurement of daughter nucleus recoiling in a direction opposite to the direction of emitted neutrino, gives an idea of energy change involved in the process. This is actually done by applying retarding potential to stop the recoiled nucleus.

Instead of K-capture, if an electron from the L-orbit is captured by a radioactive nucleus (β-active nucleus), then same theory can be applied to calculate the decay constant. As the L-electron is farther from the nucleus compared to the K-electron, the probability for the capture of L-electron is smaller than for K-electron. For that matter, the probability for the electrons from successive higher orbitals to be captured will decrease.

3.9.15 Forms of β-Interactions

The interaction giving rise to β-decay takes place between a nucleon and the combined field of electron-antineutrino (positron-neutrino). All the particles involved in the process are Fermions (Spin $\frac{1}{2}\hbar$). So far we have discussed the process non-relativistically and obtained transition rates for pure Fermi allowed and pure Gammow-Teller transitions. However, there is no justification to treat neutrino under non-relativistic limits. It becomes, therefore, necessary to deal the interaction between Fermions relativistically and apply non-relativistic limits wherever possible. Nevertheless, Dirac has developed an elegant relativistic theory of Fermions and the interaction between them has been defined as relativitistic interaction operators. These operators have been obtained as a set of possible linearly independent products of the well known Dirac γ-matrices. It is not possible to give here the details of the γ-matrices and their transformation properties. We, however, give here the salient features required for understanding the forms of β-interactions. The Dirac spinors, defined to represent the Fermion wave functions, alongwith the Dirac matrices form five forms of interaction, namely, (1) Scalar (S), (2) Polar vector (V), (3) Antisymmetric tensor (T), (4) Axial vector (A) and (5) Pseudo scalar (P). All these five forms of interactions have definite transformation properties, that is they are invariant under proper Lorentz transformations. Each of these interactions has a definite form of the Hamiltonian operator H defined in terms of the γ-matrices and which conserves parity and angular momentum. One can, however, express the interacting Hamiltonian H_β as a linear combination of the five interactions: $H_\beta = \sum_i C_i H_i$, where C_i's are the coefficients which include the coupling constants and H_i's are the Hamiltonian operator for the corresponding form of the interaction, i stands for the five forms, namely,

$$i = S, V, T, A, P$$

The interaction Hamiltonian H_β contains ten complex constants which can be reduced to five real coefficients using the parity and time reversal operations. Under the assumption of plane wave functions for leptons and slow nucleon limit for the nucleons and neglecting Coulomb effects, it has been found possible to arrive at selection rules with each form of interaction in the case of allowed transitions. The Scalar (S) and Vector (V) forms give Fermi selection rules and the tensor (T) and Axial vector (A) forms give Gammow-Teller selection rules. The Pseudo scalar (P) cannot give allowed transitions, since it gives rise to a change in parity even when $\Delta I = 0$. Thus the selection rules under the remaining four forms of interactions are

$$\left.\begin{matrix} S \\ V \end{matrix}\right\} \Delta I = 0, \text{ No (Fermi)}$$

$$\left.\begin{matrix} T \\ A \end{matrix}\right\} \Delta I = 0, \pm 1, \text{ No (Gammow-Teller)}$$

$$(\text{except } 0 \to 0)$$

One can thus have the four combinations for the allowed transitions in a β-decay process. These are *ST*, *SA*, *VT* and *VA*. The experimental data on β-decay show a good agreement with the combination of Vector (V) and Axial vector (A) form of interactions, which is commonly known as *V*-A theory.

An operator made of a mixture of scalars and pseudo scalars or a mixture of vectors and Axial vectors does not have a definite parity and as a result parity is not conserved under its action. Since parity is not conserved, a suitable mixture of these operators is to be considered.

Operators that are scalars, pseudo scalars and tensors under parity operation produce leptons (anti-leptons) of both helicities and as such have to be discarded. Only vector V and Axial vector operator A can accommodate the observed result in that leptons are of one helicity and antileptons the other value. Further, since V and A are of different parity, a linear combination of V and A is required as the operator for the β-decay. This leads to the (V-A) operator for the β-decay. The minus sign takes care of the angular distribution of electrons

$$[W(\theta) = (1 + \sigma \times pc/E) = 1 + a\frac{v}{c}\cos\theta].$$

θ is the angle with respect to I through which electron of energy E is emitted.

3.9.16 Neutrino and Antineutrino

While describing the β-disintegration process, we had discussed Pauli's hypothesis favouring the emission of an electrically neutral massless particle, which avoided the violation of the conservation of physical laws. This particle was named as neutrino. The β-decay process was described by the following reactions (eqn. 3.9.4 *a*, *b*, *c*).

(*a*) $n \rightarrow p + \beta^- + \bar{\nu}$

(*b*) $p \rightarrow n + \beta^+ + \nu$

(*c*) $p + \beta^- \rightarrow n + \nu$

Here ν represents a neutrino and $\bar{\nu}$ represents an anti-neutrino, which is an anti-particle for neutrino. The emission of antineutrino alongwith the emission of an electron is in accordance with the law of conservation of leptons. The various properties assigned to the neutrino are:

(*i*) zero electric charge
(*ii*) zero or almost zero rest mass (less than $10^{-4}\, m_e$)
(*iii*) Half integral spin angular momentum (Fermion)
(*iv*) Extremely small interaction with matter and hence zero magnetic moment,
(*v*) Infinite mean life (stable)
(*vi*) A definite helicity which implies that the intrinsic spin angular momentum of a particle is parallel or antiparallel to its direction of motion.

In the above reactions (*b*) and (*c*) are equivalent and as a matter of fact they compete for the decay of the same nuclear state (discussed in Section 3.9). This is so because an emission of an antiparticle is equivalent to the absorption of a particle.

It may be noted that neutrino differs from photons in respect of its properties (*iii*), (*iv*) and (*vi*). Photon is its own antiparticle and has electromagnetic interaction with matter while neutrino has no such interaction with matter and has an antiparticle (antineutrino, $\bar{\nu}$).

Though the theory put forward by Majorana for chargeless elementary particles favours neutrino and antineutrino to be same. But there are evidences which establish the fact that neutrino and antineutrino are different particles.

The phenomenon of β^+-disintegration itself, is a most convincing evidence for the existence of neutrino and anti-neutrino. The availability of β-active nuclei has made it possible to look for these particles directly. In the cloud-chamber photographs for the decay of ^{6}He ($^6\text{He} \rightarrow {}^6\text{Li} + \beta^- + \bar{\nu}$), it was found that the missing linear momentum required for the conservation between the recoil nucleus and the β^- particle must be carried by the anti-neuterino ($\bar{\nu}$). Such an association of finite momentum supports the existence of neutrinos.

3.9.17 Double-β-decay

We have seen that Dirac's postulation for the existence of antiparticles and the conservation of leptons demonstrated that in a β-decay process, the particle which is emitted alongwith electron, should be anti-neutrino. The antineutrino as shown below, is different from a neutrino. The double β–decay process helps in establishing the difference between a neutrino and an antineutrino. A double β–decay is a process where two electrons or two positrons are emitted.

$$A(Z, N) \rightarrow A(Z + 2, N - 2) + 2\beta^- + 2\bar{\nu} \quad \ldots(3.9.85)$$

$$A(Z, N) \rightarrow A(Z-2, N+2) + 2\beta^+ + 2\bar{\nu} \quad \text{... (3.9.86)}$$

This means that two neutrons are transformed into two protons emitting two electrons and two anti neutrinos that is, a reaction equation can be rxpressed as,

$$2n \rightarrow 2p + 2\beta^- + 2\bar{\nu}$$

and $$2p \rightarrow 2n + 2\beta^+ + 2\nu \quad \text{...(3.9.87)}$$

The above equations hold, if we regard neutrino being different from anti neutrino ($\nu \neq \bar{\nu}$). If, however, neutrino and antineutrino are same ($\nu = \bar{\nu}$), *i.e.* the two are Majorana particles and not Dirac particles. In this case, then a neutrino from first β-decay in a double decay process (first reaction in eqn. 3.9.87) is absorbed by second neutron, and in the double β-decay process no neutrino will be emitted and the reaction will then be,

$$2n \rightarrow 2p + 2\beta^- \quad \text{...(3.9.88)}$$

It may be mentioned here, that such neutrinoless double β-decay process is strictly forbidden if the neutrinos are Dirac particles.

The two possible double β-decay have however, markedly different probabilities of occurance. In a neutrinoless double β-decay process, the neutrinos are virtual particles (not emission of a real neutrino). The process can proceed at a much faster rate or having large probability for the process to occur. An important fact in support of this faster rate is that the phase space available for the final states for the neutrinoless double β-decay process is much larger than when the neutrinos are actually emitted. Another way to distinguish the two types of double β-decay, is the spectrum of the emitted electrons. If no neutrinos ($\nu = \bar{\nu}$) are emitted, the total energy carried by the two electrons is equal to the Q-value of the decay (neglecting the small amount of energy associated with recoil nucleus). The spectrum shows a single line. On the other hand if two neutrinos ($\nu = \bar{\nu}$) are also emitted, the sum of the energies of the two electrons will have a continuous distribution given by the energy and momentum conservations of two electrons, two neutrinos and recoil nucleus. In one of the experiments of double β-decay of even-even nucleus ^{82}Se to ^{82}Kr by Elliot, Hahn and Moe in 1987, gives a limit of half life of the decay to be around 10^{20} years and an energy distribution spectrum of the two electrons as expected for $\nu \neq \bar{\nu}$. The theoretical values for the half-life $\nu \neq \bar{\nu}$ was expected to be around 10^{21} years, while for the case if ν and $\bar{\nu}$ were to be same, was around 10^{16} years.

$${}^{82}_{34}\text{Se} \rightarrow {}^{82}_{36}\text{Kr} + 2\beta^- + 2\bar{\nu} \quad \text{...(3.9.89)}$$

A similar experiment on double β-decay was in the case of deay of ^{130}Te to ^{130}Xe.

$${}^{130}_{52}\text{Te} \rightarrow {}^{130}_{54}\text{Xe} + 2\beta^- + 2\bar{\nu} \quad \text{...(3.9.90)}$$

The calulated value for half life is around 10^{21} years while if $\nu = \bar{\nu}$, then the decay scheme would be

$${}^{130}_{52}\text{Te} \rightarrow {}^{130}_{54}\text{Xe} + 2\beta^- \quad \text{...(3.9.91)}$$

and the half life is expected to be around 10^{16} years while the experimental value puts the limit to around 10^{21} years. Thus the double β-decay processes in even-even nuclei support the fact that neutrinos are Dirac particles; (neutrino and antineutrino are different particles).

3.9.18 Experimental Detection of Neutrino

Since neutrino, being electrically neutral, having no magnetic moment, cannot interact directly with matter (except weak interaction) it is not possible to identify the particle through experimental observation. Bethe and Becker pointed out that the so called inverse β-decay is the only action of free neutrinos which can be predicted with certainty. According to this a neutrino is captured by a nucleon accompanied by the emission of a beta particle. The two reactions of inverse beta-decay are

$$p + \bar{\nu} \rightarrow n + \beta^+ \quad \text{...(3.9.92)}$$

$$n + \nu \rightarrow p + \beta^- \quad \text{...(3.9.93)}$$

The interaction is very weak. The probability for the inverse β-decay is very low, since the cross-section for the interaction being of the order of 10^{-44} cm^2. It is because of this low probability of interaction, the neutrinos are able to travel through large amounts of matter. However, nuclear reactors provide an intense flux of antineutrinos. If these are used to bombard a suitable target, it is possible to have a detectable result.

During nuclear fission in reactors large number of β^--emitters are produced. These β^--emitters emit electrons and antineutrinos (eqn. 3.9.14). If the antineutrino ($\bar{\nu}$) from a reactor enter a liquid scintillator (a hydrogeneous material), then the proton in the material may capture this antineutrino and a process of the type $p + \bar{\nu} = n + \beta^+$ may take place whenever the antineutrino energy is above a threshold. The positron (β^+) so produced may collide with an electron in the material, the two annihilate and produce two γ-ray photons of energy 0.51 MeV each, which can be detected by scintillation counters (Chapter 8) as a prompt signal. The neutron on the other hand travels in the material losing its energy through elastic collisions with the particles of the scintillating liquid. This slowed down neutron may then be absorbed by a cadmium nucleus present in the liquid in the form of a certain compound of cadmium. On being captured, excitation energy is released in the form of γ-rays. There will be a delay of a few micro-seconds in the production of γ-rays through these two processes. Based on this principle, F. Reins and C.L. Cowan in 1959 designed an experiment in U.S.A. A schematic diagram of the experimental set up is shown in Fig. 3.36. A tank containing about 1.4×10^3 litres of water (hydrogeneous material) used as liquid scintillator was used. In it certain amount of $CdCl_2$ was added. Antineutrinos from a powerful reactor entered the tank. The moment a proton in the liquid scintillator captured an antineutrino, a positron and a neutron are produced simultaneously. A pair of γ-ray photons (0.51MeV, each) are produced, as soon as the positrons encountered an electron in the liquid, the

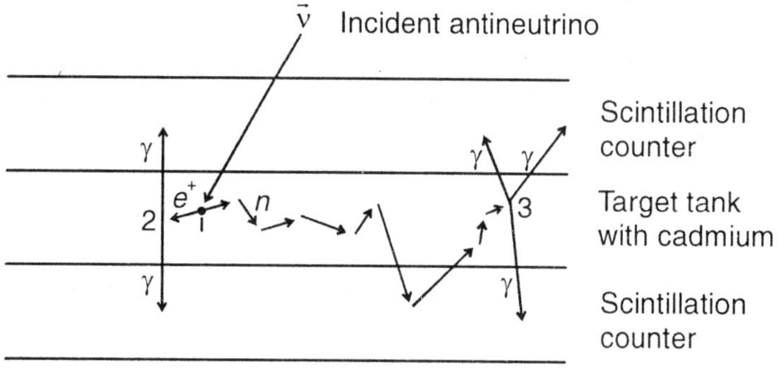

Fig. 3.36: Schematic diagram of the neutrino detection experiment. Neutrino reacts with proton at 1. Ejected β^+ annihilates at 2. The neutron is captured at 3 emitting high energy gamma-rays

two being annihilated. A prompt signal is detected by the scintillation counters placed around the tank. The γ-rays produced by the capture of slowed down neutron by a cadmium nucleus are detected by these counters after a delay time of about of 30 μsec. Thus the neutron and positron signals are detected in delayed coincidence circuits (Chap. 8). In order to establish the fact that a reaction due to an antineutrino has actually taken place, the reactor was put on and off and the difference in counting rates was measured. The anti-neutrino flux provided by the reactor was about 10^{20} m^{-2}sec^{-1} and the number of events registered was about three to four per hour. The cross-section for the inverse β-decay was found to be around 10^{-44} cm^2 or 10^{-20} barns. Even though the cross-section for the process was very small, its measurement confirmed the existence of antineutrino. This is a proof in favour of the two particles being different. The experimental demonstration of the over throw of parity conservation in β-decay assigned a new property named Helicity (Section 3.9.16) to the neutrinos. It is a two valued quantity (±1), which implies that the intrinsic spin momentum of a particle is parallel or antiparallel to the direction of motion. Neutrino has helicity (–1) or left handed and anti-neutrino (+1) or right handed. The Fig. 3.37 shows the helicities for neutrino and antineutrino.

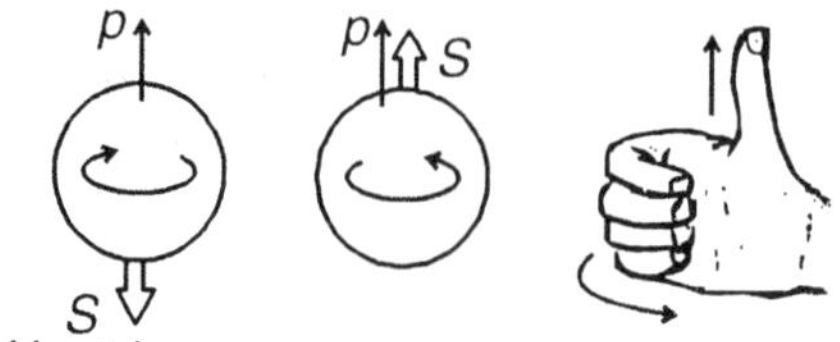

Neutrino, ν Antineutrino, ν

Fig. 3.37: Representation of neutrino and anti-neutrino helicities

3.9.19 Non-Conservation of Parity in β-decay

It is known that parity is conserved in strong (nuclear) interactions and electromagnetic interactions. While describing the forms of interaction operators in β-decay, we have seen that only those forms are retained where there is no change in parity, showing thereby that in β-decay (weak interactions) parity is conserved. It was since then believed that parity is not only conserved in β-decay, it operates in the decay of some of the elementary particles as well. However, around 1955 two mesons θ and τ discovered in cosmic rays were found to decay in two different modes, even when both have same mass, spin and half life. It, therefore, posed a problem. The θ-meson decays into two pions and τ-meson into three pions. It was therefore, thought that only one of these decays can conserve parity and in the other case it may be violated. In these modes of decays neutrinos are not emitted as such it was believed that the violation of parity conservation should be in the nature of the interaction itself.

In 1956 T. D. Lee and C. N. Yang in U.S.A. got motivated by this θ-τ puzzle and proposed that in weak interactions parity may not be conserved. Since β-disintegration is a result of weak interaction and hence in β-decay also parity may not be conserved. In order to varify this idea, Lee and Yang argued that if parity is not conserved in weak interactions like in β-decay, then the angular disribution of the electrons emitted from oriented (polarised) nuclei should be asymmetrical with respect to the angle between the polarisation of the parent nucleus and the momentum of the electron. In order to have an understanding of this fact (non-conservation of parity), we must have an idea about parity transformation. We know that parity transformation is an operation under which the spatial co-ordinates are inverted namely

$$(x, y, z) \xrightarrow[p]{} (-x)(-y)(-z) \qquad \ldots(3.9.94)$$

$$(r, \theta, \phi) \xrightarrow[p]{} (r)(\pi - \theta)(\pi + \phi) \qquad \ldots(3.9.95)$$

The parity operation is thus like taking a mirror image of co-ordinates. Under a parity operation a usual scalar (S) is unchanged but a usual vector (V), normally called as a Polar vector changes sign. The spatial location or position vector $\vec{r}$ and linear momentum $\vec{p}$ are examples of polar vectors. It is possible to construct vectors that do not change sign under a parity operation. For example, angular momentum $\vec{L} = \vec{r} \times \vec{p}$ does not change sign under parity operation, because both $\vec{r}$ and $\vec{P}$ change sign under parity operation. Spin angular momentum, Torque are some of the examples of vectors that do not change sign under mirror reflection of co-ordinates. These vectors are called Axial vectors (A). The scalar product of an axial vector and a polar vector (V) is a scalar but it changes sign under space inversion. Such scalars are known as Pseudo scalars (p). A scalar product of three polar vectors $\vec{a}$, $\vec{b}$, and $\vec{c}$ of the type $\{\vec{a}\,.(\vec{b} \times \vec{c})\}$ is a Pseudo scalar. Likewise the scalar product $(\vec{I}, \vec{p})$ of intrinsic angular momentum $\vec{I}$ and linear momentum $\vec{P}$ is a Pseudo scalar.

Whenever, an operator is made as a mixture of scalar and Pseudo scalar or as a mixture of vector and axial vector, then such an operator does not have a definite parity. For such an interaction operator parity will not be conserved.

In a normal β^--decay experiment one measures the linear momentum of the electron p_e and the linear momentum of neutrino $\vec{p}_\nu$ is determined from the recoil of the daughter nucleus, according to the law of conservation of momentum. From the knowledge of $\vec{p}_e$ and $\vec{p}_\nu$, one can get information about a term in the production operator which depends on the scalar product $\vec{p}_e \cdot \vec{p}_\nu$. If all the β-active nuclei are polarised such that the intrinsic angular momentum $\vec{I}$ of all the nuclei point in the same direction, then there will be an additional vector entering in the measurements. As discussed above the intrinsic angular momentum $\vec{I}$ is an axial vector. Because $\vec{I} = \vec{L} + \vec{S} = \vec{r} \times \vec{p} + \vec{S}$ and $\vec{r} \times \vec{p}$ is an axial vector and spin $\vec{S}$ is an axial vector, hence $\vec{I}$ is an axial vector. Therefore, the scalar product $\vec{I} \cdot \vec{p}_e$ is a pseudo scalar. If such a term is present in the β-interaction operator, then the interaction will not conserve parity. In an experiment, one measures the angle θ between $\vec{I}$ and $\vec{p}_e$. Under parity operation $\vec{p}_e$ changes to $-\vec{p}_e$ and $\vec{I}$ remains unchanged, but θ changes to π–θ. Form this it is clear that if the parity is conserved, the electron emission will be symmetric about a plane perpendicular to the direction of polarisation *i.e.* to the direction of polarisation of *I* (parent nuclei). In an actual experiment (described below) the β^- emission was found to be asymmetric.

Based on this proposition of Lee and Yang, the first experiment was performed in 1957 by C.S. Wu and her collaborators at the National Bureau of Standards, Washinton (U.S.A.). A sample of ^{60}Co was taken as a source of β^- emitter: $\left({}^{60}_{27}\text{Co} \rightarrow {}^{60}_{28}\text{Ni} + \beta^- + \bar{\nu}\right)$. The decay scheme is shown in Fig. 3.38 (*b*).

The ground state of ^{60}Co nucleus has a spin-parity 5^+. It decays predominately (99% of the time) to the spin-parity 4^+ state of ${}^{60}_{28}$Ni at an excitation energy 2.5 MeV and β^- of maximum energy 0.31 MeV are emitted. The decay is purely of the Gammow-Teller type allowed transition, since $\Delta I = 1$. In the experimental set up, the Cobalt source (Fig. 3.38 *a*) in the form of a thin film deposited on a crystal of cerium magnesium nitrate, which is paramagnetic. It is cooled by adiabatic demagnetisation to a tempreature 0.01 K. A magnetic field is applied through the solenoid to polarise the electrons of the paramagnetic crystal. The field so produced is strong enough to orient the cobalt nuclei. The cobalt nuclei having non-zero spin get polarised along the direction of the magnetic field. The direction of the polarisation could be reversed by reversing this external magnetic field. The degree of polarisation of the ^{60}Co nuclei is determined by measuring the anisotropy of the γ-rays that are emitted by the NaI scintillation counters *A* and *B* shown in Fig. 3.38 (*a*). The counter *A* is perpendicular and *B* is parallel to the direction of polarisation and the results shown in Fig. 3.38 (*c*) indicates the γ-ray anisotropy. The intensity is much greater initially in the horizontal direction (counter *B*) than in vertical direction (counter *A*). Likewise intensity of β^--emission is shown in Fig. 3.38 (*d*). It may be noted that the results clearly shows asymmetrical emission of β^- particles with respect to the direction of the field. The anisotropy of γ-emission and asymmetry of β^--emission both disappear (in about 8 min) as the crystal warms up.

The angular dependence of β^--particles emitted in the experiment of C.S. Wu and others can be expressed as

$$(a + b\cos\theta)$$

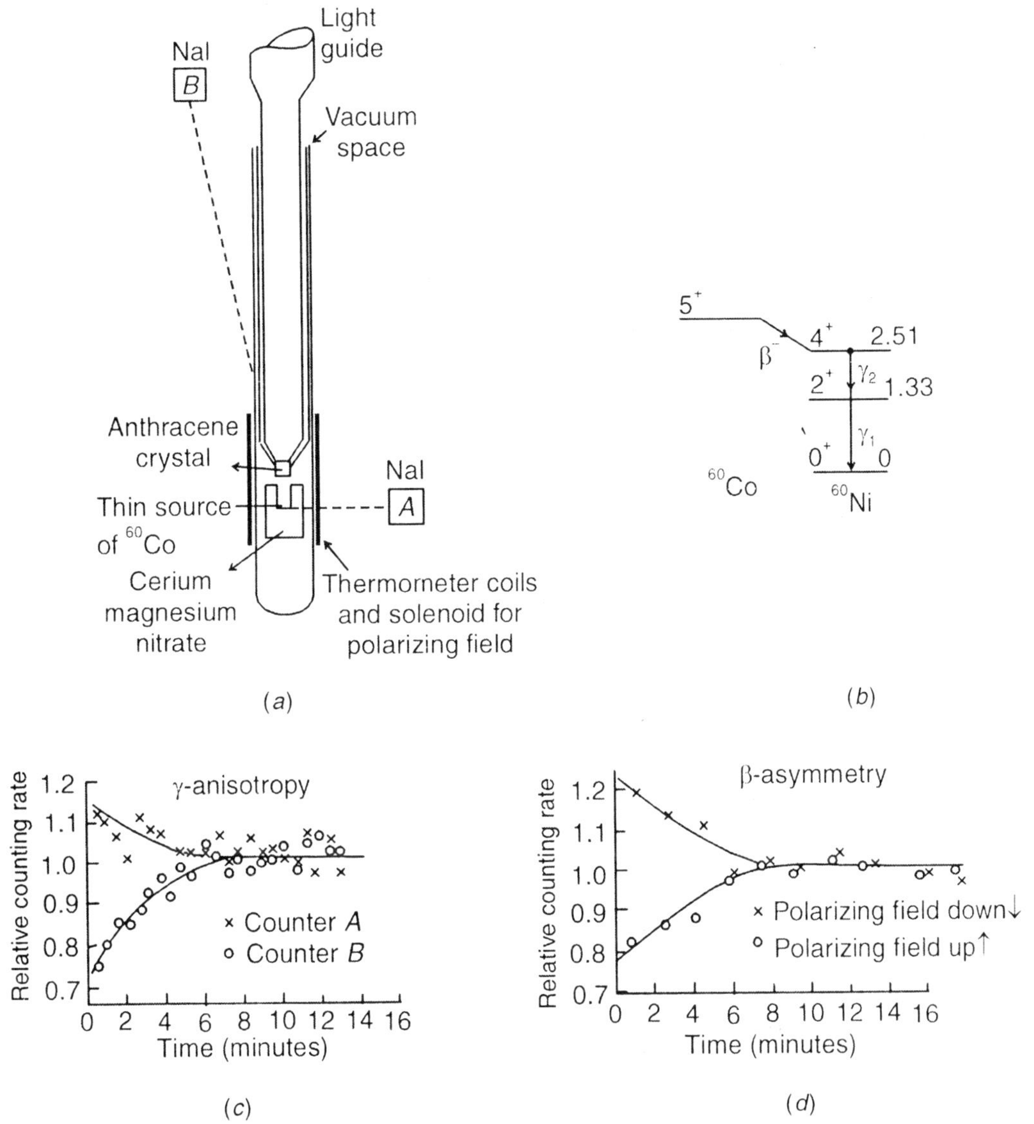

Fig. 3.38: Experiment for the verification of non-conservation of parity in β-decay.
(*a*) Arrangement of the apparatus
(*b*) Decay scheme of ^{60}Co (G-T-Transition)
(*c*) Anisotropy in the gama rays as obtained from the counters *A* and *B* at different times as the source warms up.
(*d*) β-particle asymmetry for two directions of polarising field

where θ is the angle between the direction of emission of β^--particles (or $\vec{p}_e$) and the direction of polarisation of ^{60}Co nuclei (or $\vec{I}$), and *a* and *b* are constants. If the parity is to be conserved, this intensity should be unchanged under parity operation. But we can see that the parity changes θ to π−θ and then the intensity will change. This asymmetry may also be looked upon as if the electrons have preference of direction for their emission. More electrons are emitted in a direction opposite to the direction of polarisation (direction of orientation of nuclear spin) with their spins opposite to their direction of motion. Also parity conservation demands that a physical system and its mirror image should be identical. This is, however, not so in the case of β^- decay, as is clear from Fig. 3.39 (*a*). The experiment therefore, confirmed the non-conservation of parity in β^- decay (weak interactions). If positron emission is considered, then also similar asymmetry would be observed supporting the non-conservation of parity. The spin direction of positrons would be parallel to their direction of motion.

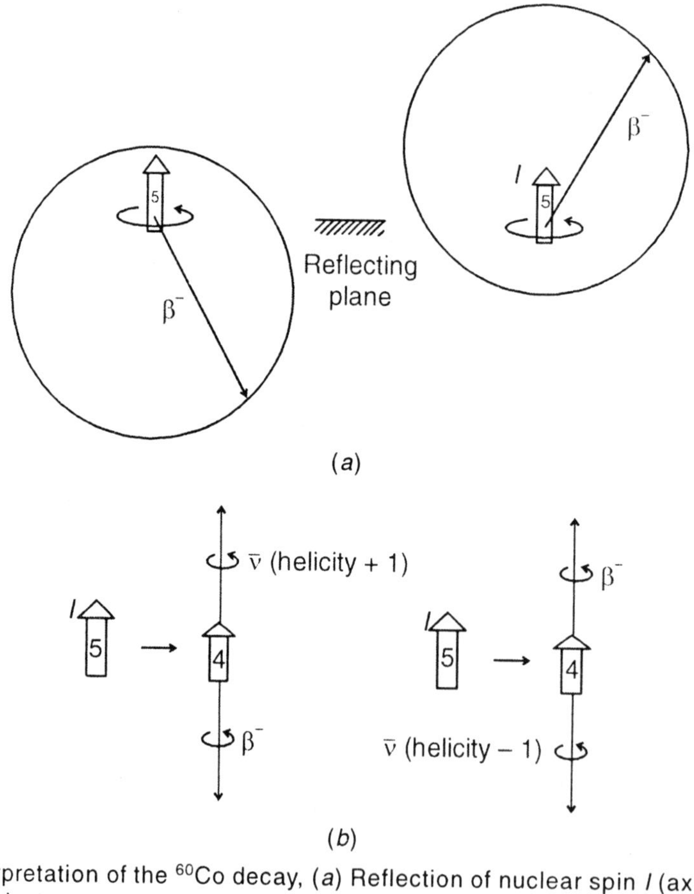

Fig. 3.39: Interpretation of the ^{60}Co decay, (*a*) Reflection of nuclear spin *I* (axial-vector) and the electron momentum (polar vector) in a horizontal plane (*b*) Explanation of the asymmetry in terms of a two component theory

The two component theory of neutrino (described in the next section) also favours the non-conservation of the parity. From Fig. 3.39 (b), it is clear that under parity operation (reflection in a mirror). The spin direction of leptons would remain unchanged, being axial vectors, while the direction of their motion (the direction of moment $\vec{P}_e$ and $\vec{P}_\nu$) is changed because the two are polar vectors. In the mirror image the direction of motion of the antineutrino has become antiparallel to its spin as such its helicity has changed to −1. In other words on mirror reflection or under parity, an anti-neutrino of helicity (+1) has changed to a neutrino of helicity (−1). But parity conservation demands that the two systems should be identical whereas the mirror image in the present case would not be physically possible. Hence the asymmetry in β^- emission indicates the violation of parity conservation.

3.9.20 Helicity of Neutrino

It was pointed out in Section (3.9.18) that neutrino and antineutrino are different particles and the property which makes this distinction is the helicity. The helicity, h of a particle is defined as the projection of $\vec{\sigma}$, along its direction of motion (given by its momentum p)

$$h = \frac{\vec{\sigma} \cdot \vec{p}}{|\vec{\sigma}||\vec{p}|} \quad \text{...(3.9.96)}$$

For a neutrino, therefore, $\vec{\sigma}$ is its spin and $\vec{p}$ is its momentum. For a massless particle, neutrino, the eigen values of h can only be ± 1. We are familiar with a massless particle photon which has only two possible orientation in terms of two linearly independent transverse polarisation. But unlike neutrino, photon is its own antiparticle. As such the particles with positive helicity (+1) are referred as "right handed" and with negative helicity (−1) are termed as "left handed" particles. This idea of right-left handedness was first proposed on theoretical grounds by H. Weyl in 1929. He showed that for a massless and a chargeless particle, the Dirac relativistic equations have two possible solutions. Accordingly particles and their antiparticles will have opposite spins when two travel in the same direction.

This would, therefore, mean that the particles will be either right handed or left handed. Such a characteristic property is associated with name 'Helicity'. A right handed particle has a positive helicity or $h = +1$ and a left handed particle is given negative helicity or $h = -1$. When the over-throw of parity was confirmed experimentally, Abdus Salam in U.K., Landau in Russia and Lee and Yang in U.S.A. showed that helicity is a fundamental property of neutrinos. The nuetrino has negative helicity or $h = -1$, meaning thereby that its spin is antiparallel to its ditrection of motion and anrineutrino has positive helicity of $h = +1$, its spin is parallel to its direction of motion. The screw-sense of the two particles is shown in Fig. 3.37.

The helicity of neutrino was first determined experimentally by *M.* Goldhaber and his associates in 1958 where they employed the K-electron capture in the 0^+ ground state of ^{152}Eu decaying to the 1^- excited state of ^{152}Sm at 963 KeV.

$$\underset{(0^+)}{^{152}\text{Eu}} + e_k^- \rightarrow \underset{(1^-)}{^{152}\text{Sm}^*} + \nu \quad \text{...(3.9.97)}$$

$$\begin{matrix} ^{152}Sm^* & \to & ^{152}Sm + \gamma \\ (1^-) & & (0^+) \end{matrix} \qquad ...(3.9.98)$$

The conservation of angular momentum requires that in the first stage of decay of ^{152}Eu capturing an *K*-electron, the spin of excited $^{152}Sm^*$ must be opposite to the spin of neutrino and the two must move in opposite directions, to conserve linear momentum. The recoil nucleus $^{152}Sm^*$ and the neutrino must have same helicity or same polarisation. When the excited $^{152}Sm^*$ decays to its ground state with zero spin, the emitted photon takes away the angular momentum, which should be same as that of $^{152}Sm^*$. Thus the sign of the circular polarisation of the photon must be same as the sign of longitudinal polarisation of the emitting nucleus $^{152}Sm^*$ and, therefore, the sign of neutrino helicity be same as the sign of circular polarisation of the γ-rays. By measuring the polarisation of γ-rays, it was found that the helicity of neutrino is negative *i.e.* $h = -1$ for neutrino. From various experiments on β-decay it was concluded that helicity of anti-neutrino is positive *i.e.* for antineutrino $h = +1$.[ft]

For particles of non-zero mass like electrons, the helicity is $-v/c$ and for antiparticles, like positron the helicity is $+v/c$, where v is their velocity.

3.10 ELECTROMAGNETIC TRANSITIONS IN NUCLEI

A nucleus may be formed in one of its excited states after the decay of its parent or as a result of a nuclear reaction. The excited nucleus makes transition to its lower energy states or ground state in one of the following ways:

(*i*) Emission of electromagnetic radiations as photons of definite energy-known as γ-rays.

(*ii*) Transforming the excitation energy to one of the atomic electrons so that it gets ejected from the atom. The process is known as internal conversion of γ-rays.

(*iii*) Creation of an internal electron-positron pair.

If emission of β-particles from the excited state is allowed by conservation laws, it may compete with the above processes. At excitation energies above about 8 MeV (the binding energy of a nucleon in the nucleus), the emission of a nucleon dominates all the other processes.

3.10.1 Electromagnetic Transitions

The quantum of electromagnetic radiation—photon—has a spin angular momentum quantum (S) of one (S = 1). Along a fixed direction (taken as *Z*-direction) the component of the spin angular momentum is $S\hbar = \hbar$. When a

---

[ft]From these experimental observations on helicities, it emerges that all leptons emitted in β-decay have negative helicity ($h < 0$) and antileptons have positive helicity ($h > 0$). The operators that are scalars (*s*), tensors (*T*) and pseudoscalars (*p*) under parity operations produce leptons (as well as antileptons) of both helicities.

nucleus emits a photon, its total angular momentum along the Z-direction must change by at least one unit. In fact the emitted photon has to take away an angular momentum equal to the difference between the angular momenta of the nucleus in the two states between which the transition takes place. The angular momentum (along Z-direction) carried away by a photon defines the multipolarity of the transition. If the angular momentum (along Z-direction) carried away by the photon is $L = 1$ the transition is dipole and radiation is known as dipole radiation. For $L = 2$ the transition is quadrupole and for $L = 3$ it is octopole. For angular momentum L of radiation field the mutipolarity of the γ-transition is 2^L. The γ-transition may (or may not) cause a change in the parity of the nuclear state (denoted by π). If in a γ-transition with $L = 1$ the parity π_i; of the initial state and parity π_f of the final state are different *i.e.* $\Delta\pi$ = yes or $\frac{\pi_i}{\pi_f} = (-1)^1$, the emitted radiation is known as electric dipole radiation. On the other hand if $\Delta\pi$ = No or $\frac{\pi_i}{\pi_f} = (-1)^0$ the nucleus emits a magnetic dipole radiation. The electric or magnetic nature of a transition in which an angular momentum L is carried away by the radiation field, is defined as below:

L =	1	2	3	4	$\frac{\pi_i}{\pi_f}$
Electric Transition $\Delta\pi$ =	Yes	No	Yes	No	$(-1)^L$
Magnetic Transition $\Delta\pi$ =	No	yes	No	Yes	$(-1)^{L+1}$

If the total angular momentum quantum number or spin of a nucleus is I_i in the initial state and I_f in the final state, the radiation field can carry away any angular momentum L between $|I_i + I_f|$ and $|I_i - I_f|$ so that

$$|I_i - I_f| \leq L \leq |I_i + I_f|$$

If for example $I_i = 1$ and $I_f = 2$ the possible values of L = 1, 2 and 3. As the change in parity is fixed (yes or no), the γ-transition is a mixture of alternate electric and magnetic multipoles. When there is a change in parity ($\Delta\pi$ = Yes) the γ-transition is a mixture of electric dipole ($E1$), magnetic quadrupole ($M2$) and electric octa pole ($E3$) transitions. If there is no change in parity ($\Delta\pi$ = No) the transition is a mixture of $M1$, $E2$ and $M3$.

The probability of a γ-transition varies as $(R/\lambda\!\!\!^-)^{2L}$ where $\lambda = 2\pi\lambda\!\!\!^-$ is the wavelength of the γ-radiations and R the radius of the nucleus emitting it. We have,

$$\frac{R}{\lambda\!\!\!^-} = \frac{R\hbar\omega}{\hbar c} = \frac{RE_\gamma}{\hbar c} \quad \text{...(3.10.1)}$$

where E_γ is the energy of the γ-ray photon. For $E_\gamma << \frac{\hbar c}{R}$ (10 Mev) $R/\lambda\!\!\!^- << 1$ and the transition probability for higher values of L is very small. One, therefore, observes only the lowest multipole transitions.

While the electric multipole transitions can be visualised as due to vibrating charges, the magnetic transitions are visualised as due to fluctuating currents inside a nucleus. The probability of a magnetic transition is less than that of an electric transition of the same multipolarity by a factor of about v^2/c^2 (≈ 0.05), where v is the velocity of the nucleons inside the nucleus. Thus when the lowest multipole transition is electric, the higher multipole transitions-magnetic or electric are seldom observed. If the lowest multipole transition is magnetic, the next higher electric multipole transition may sometimes be mixed with it. One may thus observe a mixed M_1 and E_2 transition, but never a M_2 transition mixed with E_1.

The minimum angular momentum carried away by a photon field is $L = 1$. It is, therefore, not possible to have a γ-transition when the initial and final state spins of the excited nucleus are both zero ($I_i = I_f = 0$).

3.10.2 Radiation Field Multipolarity

The multipole expansion of the radiation field should ideally be done employing quantum electro-dynamics. However, a semi-classical theory gives a good insight about the electromagnetic transitions.

The Maxwells equations in vacuum are written as

(a) $$\vec{\nabla} \times \vec{H} = \varepsilon_o \frac{\partial \vec{E}}{\partial t} \quad \text{...(3.10.2)}$$

(b) $$\vec{\nabla} \times \vec{E} = -\mu_o \frac{\partial \vec{H}}{\partial t} \quad \text{...(3.10.3)}$$

(c) $$\vec{\nabla} \cdot \vec{H} = 0 \quad \text{...(3.10.4)}$$

(d) $$\vec{\nabla} \cdot \vec{E} = 0 \quad \text{...(3.10.5)}$$

The symbols have their usual meaning. The solution of these equations describes the electromagnetic waves with superposition of the incoming and outgoing waves of different frequencies. The electric and magnetic fields satisfying the above equations can be written as

$$\vec{E} = \int_0^{\infty} \left[\vec{E}_r(r, \omega)_e^{i\omega t} + \vec{E}_r^*(r, \omega)_e^{-i\omega t}\right] d\omega \quad \text{...(3.10.6a)}$$

$$\vec{H} = \int_0^{\infty} \left[\vec{H}_r(r, \omega)_e^{i\omega t} + \vec{H}_r^*(r, \omega)_e^{-i\omega t}\right] d\omega \quad \text{...(3.10.6b)}$$

where $\vec{E}_r$ and $\vec{H}_r$ represent the electric and the magnetic fields respectively of the radiations of different frequencies.

Substituting the values of $\vec{E}$ and $\vec{H}$ in equation (3.10.2) and (3.10.3) gives

$$\vec{\nabla} \times \vec{H} = -iw\varepsilon_o \vec{E} \quad \text{...(3.10.7a)}$$

$$\vec{\nabla} \times \vec{E} = iw\mu_0 \vec{H} \quad \text{...(3.10.7b)}$$

Taking curl of the above equations gives

$$\vec{\nabla} \times \vec{\nabla} \times \vec{H} = w^2 \varepsilon_o \mu_o \vec{H} \quad \text{...(3.10.8a)}$$

$$\vec{\nabla} \times \vec{\nabla} \times \vec{E} = w^2 \varepsilon_o \mu_o \vec{E} \quad \text{...(3.10.8b)}$$

Putting, $\omega^2 \varepsilon_o \mu_o = k^2$ we get

$$\left[\vec{\nabla} \times \vec{\nabla} \times \; - k^2\right]\vec{H} = 0 \quad \text{...(3.10.9a)}$$

$$\left[\vec{\nabla} \times \vec{\nabla} \times \; - k^2\right]\vec{E} = 0 \quad \text{...(3.10.9b)}$$

As the photon carries away momentum, the fields $\vec{H}$ and $\vec{E}$ should be angular momentum eigen functions. One must find some angular momentum operator which commutes with the operator $\vec{\nabla} \times \vec{\nabla}$. The eigen function of such an operator can represent the magnetic and electric fields of equations (3.10.9).

If $\vec{L}$ is the orbital angular momentum operator defined as

$$\vec{L} = -i\hbar \vec{r} \times \vec{\nabla} \quad \text{...(3.10.10)}$$

then the z component of the operator is

$$L_z = -i\hbar\left(x\frac{\partial}{\partial y} - y\frac{\partial}{\partial x}\right) \quad \text{...(3.10.11)}$$

and $$L^2 = -\hbar^2(\vec{r} \times \vec{\nabla})^2 \quad \text{...(3.10.12)}$$

It is observed, see Appendix C, that the operator L_Z does not commute with $\vec{\nabla} \times \vec{\nabla}$. However, if an operator S_Z defined as

$$S_z = -\,i\,\hbar \vec{k}\, x \quad \text{...(3.10.13)}$$

is added to L_Z then the operator $L_Z + S_Z$ commutes with $\vec{\nabla} \times \vec{\nabla}$. Putting

$$J_Z = L_Z + S_Z \quad \text{...(3.10.14)}$$

the eigen function of the operator J_z should be able to represent the fields $\vec{E}$ and $\vec{H}$ of equation (3.10.8). The operator J^2 defined as

$$J^2 = J_x^2 + J_y^2 + J_z^2 \quad \text{...(3.10.14)}$$

is also found to commute with $\vec{\nabla} \times \vec{\nabla}$. If $\vec{B}$ is an eigen function defined as

$$\vec{B} = LR_L(r)\,Y_{LM}(\theta, \phi) \quad \text{...(3.10.15)}$$

then it can be seen that

$$J_z\vec{B} = M\hbar\vec{B} \quad \text{...(3.10.16)}$$

where, $$L_z = M\hbar$$

and $$J^2\vec{B} = \hbar^2 L(L + 1)\,\vec{B} \quad \text{...(3.10.17)}$$

Thus $\vec{B}$ is an eigen function of the operators J_z and J^2 and can represent the fields $\vec{E}$ and $\vec{H}$.

Equation (3.10.8) can be written as,

$$(\nabla \times \nabla \times -k^2)\vec{B} = 0$$

or $$\vec{\nabla}\left(\vec{\nabla} \cdot \vec{B}\right) - \nabla^2\vec{B} - k^2\vec{B} = 0$$

As in free space $\vec{\nabla} \cdot \vec{H}$ and $\vec{\nabla} \cdot \vec{E}$ are both zero and $\vec{B}$ can represent either $\vec{H}$ or $\vec{E}$, the above equation can be written as,

$$(\nabla^2 + k^2)\vec{B} = 0 \qquad ...(3.10.18)$$

This is similar to the Schrodinger equation. The eigen function in polar coordinates can be written as $R(r)Y_{LM}(\theta, \phi)$ where the equation satisfied by the radial part is

$$\left(\frac{d^2}{dr^2} - \frac{L(L+1)}{r^2} + k^2\right)R_L(r) = 0$$

The solution of the above equation are the spherical Bessel functions $j_L(kr)$. The eigen function satisfying the equations (3.10.8) can thus be written as,

$$\vec{B} = C\vec{L}j_L(kr)Y_{LM}(\theta, \phi) \qquad ...(3.10.19)$$

where C is a constant.

If the vector $\vec{B}$ represents the magnetic field vector satisfying equation (3.10.8a) the corresponding electric field vector $\vec{E}$ is obtained from equation (3.10.7a). If $\vec{B}$ represents the electric field vector of equation (3.10.8b) the corresponding magnetic field vector is obtained using equation (3.10.7b).

If $\vec{B}$ is chosen to represent the magnetic field vector $\vec{H}^e_{LM}$ then,

$$\vec{H}^e_{LM} = C^e\vec{L}j_L(kr)Y_{LM}(\theta, \phi)$$

It is seen that the radial component of $\vec{H}^e_{LM}$ is zero as $L_r = 0$ $\left(\vec{L} = -i\hbar\vec{r} \times \vec{\nabla}\right)$. The fields in which the radial component of the magnetic field is zero are known as electric multipole fields and are denoted by a superscript e. The corresponding electric field E^e_{LM} is obtained from equation (3.10.7a) as,

$$E^e_{LM} = -\frac{1}{i\omega\varepsilon_o}\nabla \times \vec{H}^e_{LM}$$

$$= -\frac{c^e}{i\omega\varepsilon_o}\nabla \times Lj_L(kr)Y_{LM}(\theta, \phi) \qquad ...(3.10.21)$$

If the vecter $\vec{B}$ represents the electric field vector E^m_{LM} of equation (3.10.8b) then E^m_{LM} is given as

$$E^m_{LM} = C^m Lj_r(kr)Y_{LM}(\theta, \phi) \qquad ...(3.10.22)$$

It is seen that the radial component of the electric field is zero. Such fields are called magnetic multipole fields and are represented with the superscript m.

The corresponding magnetic field vector for a magnetic multipole field is given as,

$$H_{LM}^{m} = \frac{C^m}{i\omega\mu_o}\vec{\nabla} \times \vec{L}j_L(kr)Y_{LM}(\theta, \phi) \qquad \ldots(3.10.23)$$

The constants C^e and C^m are evaluated by normalising the radiation fields such that after the emission of a photon in which a radiation field of multipolarity L and frequency ω appears, the field energy is $\hbar\omega$. In an electromagnetic field half the average energy is in electric field and half in magnetic field, we have,

$$\varepsilon_o|\vec{E}|^2 = \mu_o|\vec{H}|^2 = \frac{\hbar ck}{2} \qquad \ldots(3.10.24)$$

A single frequency of the radiation wave is obtained if the wave is enclosed in a large sphere of radius $R_0(R_0 \to \infty)$ so that $kR_0 >> 1$. At $r = R_0$ the radiation field is zero so that

$$j_L(kR_0) = 0$$

Putting the value of $\vec{H}$ in equation (3.10.24), we get,

$$\frac{\hbar ck}{2\mu_o} = C_e^2 \int_0^{R_0} |j_L(kr)|^2 r^2 dr \int (\vec{L}Y_{LM})^* \vec{L}Y_{LM} d\Omega$$

Since L is Hermitian operator we have

$$\int (\vec{L}Y_{LM})^* \vec{L}Y_{LM} d\Omega = \int Y_{LM}^* \vec{L}\vec{L}Y_{LM} d\Omega$$

$$= \int Y_{LM}^* L^2 Y_{LM} d\Omega$$

$$= L(L+1)\hbar^2 \qquad \ldots(3.10.25)$$

for large values of r we have

$$j_L(kr) = \frac{1}{r}\cos\left(kr - \frac{\pi}{2}(L+1)\right)$$

so that

$$\int_0^{R_0} |j_L(kr)|^2 r^2 dr \approx \frac{1}{k^2}\int_0^{R_0} |\cos(kr - \frac{\pi}{2}(L+1)|^2 dr$$

$$\approx \frac{R_0}{2k^2} \qquad \ldots(3.10.26)$$

Substituting from equations (3.3.25) and (3.10.26) we have,

$$\frac{\hbar ck}{2\mu_0} = C_e^2 L(L+1)\hbar^2 \frac{R_0}{2k^2}$$

so that

$$(C^e)^2 = \frac{ck^3}{\mu_0 \hbar R_0 L(L+1)} \qquad \ldots(3.10.27)$$

Similarly putting the value of $\vec{E}$ in equation (3.10.24) gives for the magnetic multipole fields

$$(C^m)^2 = \frac{ck^3}{\varepsilon_0 \hbar R_0 L(L+1)} \qquad ...(3.10.28)$$

Substituting the value of C^e and C^m the multipole expansion of the radiation field is obtained as,

$$E^e_{LM} = i\left[\frac{ck}{\varepsilon_0 \hbar R_0 L(L+1)}\right]^{1/2} \vec{\nabla} \times \vec{L} j_L(kr) Y_{LM}(\theta, \phi) \qquad ...(3.10.29)$$

$$H^e_{LM} = k\left[\frac{ck}{\mu_0 \hbar R_0 L(L+1)}\right]^{1/2} \vec{L} j_L(kr) Y_{LM}(\theta, \phi) \qquad ...(3.10.30)$$

$$E^m_{LM} = k\left[\frac{ck}{\varepsilon_0 \hbar R_0 L(L+1)}\right]^{1/2} \vec{L} j_L(kr) Y_{LM}(\theta, \phi) \qquad ...(3.10.31)$$

$$H^m_{LM} = -i\left[\frac{ck}{\mu_0 \hbar R_0 L(L+1)}\right]^{1/2} \vec{\nabla} \times \vec{L} j_L(kr) Y_{LM}(\theta, \phi) \qquad ...(3.10.32)$$

It can be shown that operator

$$\vec{\nabla} \times \vec{L} = \frac{\hbar}{i}\left\{\vec{r}\nabla^2 - \vec{\nabla}\left(1 + r\frac{\partial}{\partial r}\right)\right\}$$

The above expression can be evaluated using the properties of the Bessel function. We have,

$$\nabla^2 j_L(kr)\, Y_{LM}(\theta, \phi) = -\, k^2 j_L(kr)\, Y_{LM}(\theta, \phi)$$

and

$$r\frac{\partial}{\partial r} j_L(kr) = kr j_{L+1}(kr) - (L+1) j_L(kr)$$

$$= L j_L(kr) - kr\, j_{L+1}(kr)$$

Substituting these relations we get

$$\vec{\nabla} \times \vec{L} = j_L(kr) Y_{LM}(\theta, \phi) = \frac{\hbar}{i}\left(-k^2 r - (L+1)\vec{\nabla}\right) j_L(kr) Y_{LM}(\theta, \phi)$$

$$+ \frac{\hbar}{i}\vec{\nabla}(kr) j_{L+1}(kr) Y_{LM}(\theta, \phi)$$

The wave function $R(r)$ is insignificant outside the nuclear radius so that r is so small that $kr \rightarrow 0$ the above relation then yields

$$\vec{\nabla} \times \vec{L} = -\frac{\hbar}{i}(L+1)\vec{\nabla} j_L(kr) Y_{LM}(\theta, \phi) \qquad ...(3.10.33)$$

The parity of the function Y_{LM} is $(-1)^L$. The parity of the operator L is +1 while that of the operator $\vec{\nabla}\times$ is (–1). Thus the parity of the function E^e_{LM} and E^m_{LM} is $(-1)^L$ while the parity of the function H^m_{LM} and E^e_{LM} is $(-1)^{L+1}$. Further, as $Y_{o,o} = 0$ and $\vec{\nabla}\, Y_{o,o} = 0$, there is no radiation field with $L = 0$. This is in conformity with the transverse nature of the electromagnetic waves. The angular momentum carried away by the electromagnetic field is $\hbar\sqrt{L(L+1)}$ with $L \geq 1$. Any electromagnetic field satisfying Maxwells equations can be built up by the mixture of various multipole fields. Thus in general we have,

$$\vec{E} = \sum_{M=-L}^{+L} \sum_{L=1}^{\infty} \left(a^e_{LM}\, \vec{E}^e_{LM} + a^m_{LM}\, \vec{E}^m_{LM}\right) \quad ...(3.10.34)$$

$$\vec{H} = \sum_{M=-L}^{+L} \sum_{L=1}^{\infty} (b^e_{LM}\, \vec{H}^e_{LM} + b^m_{LM}\vec{H}^m_{LM}) \quad ...(3.10.35)$$

where a and b are the relative amplitudes of the corresponding fields. During the emission of electromagnetic radiations the parity of the system is conserved. The parity of the initial and the final state of the nucleus emitting radiation is definite, the partiy of the electromagnetic field is also definite. The partities of the radiation fields mixing together must, therefore, be same. It is thus possible that the fields $\vec{E}^e_{LM}$ and $\vec{E}^m_{L+1,\,M}$, may be mixed as they both have the same parity. Similarly fields $\vec{H}^e_{LM}$ may be mixed with $\vec{H}^m_{L+1,\,M}$.

The electric and the magnetic fields can be described in terms of a vector potential $\vec{A}$ and a scalar potential ϕ such that,

$$\vec{H} = \vec{\nabla} \times \vec{A} \quad ...(3.10.36)$$

Using equation (3.10.7b) we have

$$\vec{A} = \frac{1}{iw\mu_0}\vec{E} - \vec{\nabla}\phi \quad ...(3.10.37)$$

One can choose $\vec{A}$ such that $\vec{\nabla}\cdot\vec{A} = 0$. The scalar potential ϕ may be taken as zero at $r \to \infty$. Also as $\nabla^2\phi = 0$ the scalar potential vanishes everywhere and the vector potential is then given as,

$$\vec{A} = \frac{1}{iw\mu_0}\vec{E} \quad ...(3.10.38)$$

3.10.3 γ -ray Transition Probability

Nucleus is a system composed of neutrons and protons. Any excited state of nucleus is a state of the whole nucleus. A nuclear excited state due to the excitation of a single nucleon is not justified. However, to simplify the semi-classical calculations of γ-ray transition probability it is assumed that the excited state is due to a proton. When the proton interacts with the electromagnetic field, a photon is emitted. In an electromagnetic field the Hamiltonian $\mathcal{H}$ of a proton is given as

$$\mathcal{H} = \frac{1}{2m_p}(\vec{p} - e\mu_0\vec{A})^2 + u(r) + \frac{\mu_N}{h}\, g_{sp}(\vec{S}_p \cdot \vec{H}) \quad ...(3.10.39)$$

where μ_N is nuclear magneton, $u(r)$ the spherically symmetric nuclear potential, S_p the spin of the proton and $\vec{H}$ the magnetic component of the electromagnetic field. The last term is the energy of interaction of the intrinsic magnetic moment of the proton with the magnetic field of the radiations. The first term in the above expression can be expanded as follows:

$$\frac{1}{2m_p}(\vec{p} - e\mu_0\vec{A})^2 = \frac{1}{2m_p}\left[p^2 + e^2\mu_0^2A^2 - e\mu_0(\vec{p}\vec{A} + \vec{A}\vec{p}\right]$$

The term A^2 is the second order term and represents emission of two photons. This term can be safely neglected. The Hamiltonian is then given as,

$$\mathcal{H} = \left[\frac{p^2}{2m_p} + u(r)\right] + \left[\frac{e\mu_0}{2m_p}(\vec{p}\cdot\vec{A} + \vec{A}\cdot\vec{p}\right] + \frac{\mu_N}{\hbar}g_{sp}(\vec{S}_p \cdot \vec{H})$$

$$= \mathcal{H}_0 + \mathcal{H}_1 \qquad \text{...(3.10.40)}$$

$\mathcal{H}_0$ represents the Hamiltonian of the proton in the absence of the electromagnetic field. $\mathcal{H}_1$ is the perturbation caused by the presence of the radiation field. The energy levels of the proton are given by the intrinsic Hamiltonian $\mathcal{H}_0$ as the eigen values of the equation

$$\mathcal{H}_0\psi = E_n\psi_n \qquad \text{...(3.10.41)}$$

where n is an integral number.

The transition from one energy state of the proton to another is due to the perturbation $\mathcal{H}_1$ which can be written as

$$\mathcal{H}_1 = \mathcal{H}_1 e^{i\omega t} + \mathcal{H}_1^* e^{-i\omega t}$$

$$= \frac{e\mu_0}{2m_p}(\vec{p}\cdot\vec{A} + \vec{A}\cdot\vec{p}) + \frac{\mu_N}{\hbar}g_{sp}(\vec{s}_p \cdot \vec{H}) \qquad \text{...(3.10.42)}$$

The transition probability from the state ψ_i to state ψ_f is given as

$$P = \frac{2\pi}{\hbar}\left|\int \psi_f^*\mathcal{H}_1\psi_i dv\right|^2 \frac{dn}{dE} \qquad \text{...(3.10.43)}$$

where $\frac{dn}{dE}$ is the density of final states.

The transition from a higher energy state to a lower energy state with the emission of radiation takes place due to the operator $\mathcal{H}_1 e^{i\omega t}$. The operator $\mathcal{H}_1^* e^{-i\omega t}$ gives rise to an absorption of radiation with transition fron ψ_f to ψ_i. To calculate the energy density of the final states, consider the surface of a large sphere of radius R_0 so that $kR_0 >> 1$. The assymptotic form of the Bessel function is

$$j_L(kr) \approx \frac{1}{kr}\cos\left(kr - \frac{\pi}{2}(L+1)\right)$$

as

$$= \frac{1}{kr}\sin\left(kr - \frac{L\pi}{2}\right)$$

as $\quad r \to R_0, \quad \sin\left(kr_0 - \frac{L\pi}{2}\right) \to 0$

so that $\quad kR_0 - \frac{L\pi}{2} = n\pi$

where n is an integer. As $k = \frac{E_r}{\hbar c}$

one gets $\quad E_\gamma = \frac{\pi\hbar c}{R_0}\left(n + \frac{L}{2}\right)$

on differentiation

$$\frac{dn}{dE} = \frac{R_0}{\pi\hbar c} \qquad \text{...(3.10.44)}$$

The proton wave functions ψ_i and ψ_f are zero outside the nuclear radius r_0, where $kr_0 << 1$. The interaction of proton with the radiation field is significant only for $r \leq r_0$.

For $kr << 1$

$$j_L(kr) = \frac{2^L L!}{(2L+1)!}(kr)^L$$

$$= \frac{(kr)^L}{(2L+1)!!} \qquad \text{...(3.10.45)}$$

where $\quad (2L + 1)!! = 1, 3, 5, \ldots(2L + 1)$.

Substituting the value of E^e_{LM}, from equation (3.10.29) in equation (3.10.38) one gets the value of A as

$$\vec{A}^e_{LM} = \frac{1}{\mu_0 ck}\left[\frac{c_k}{\varepsilon_0 R_0 \hbar L(L+1)}\right]^{\frac{1}{2}} \vec{\nabla} \times \vec{L} j_L(kr) Y_{LM}(\theta, \phi)$$

Putting the value of $\vec{\nabla} \cdot \vec{L}$ and $j_L(kr)$ from equation (3.10.33) and (3.10.45) respectively gives

$$A^e_{LM} = -\frac{1}{\mu_0 ck}\left[\frac{c_k}{\varepsilon_0 R_0 \hbar L(L+1)}\right]^{1/2} \frac{1}{(2L+1)!!}\frac{\hbar}{i}\vec{\nabla}F(kr, \theta, \phi) \qquad \text{...(3.10.46)}$$

where,

$$F(kr, \theta, \phi) = (kr)^L Y_{LM}(\theta, \phi) \qquad \text{...(3.10.47)}$$

For calculating the matrix element of equation (3.10.43), we see that

$$\int \psi_f^*(\vec{p}\cdot\vec{A}+\vec{A}\cdot\vec{p})\psi_i dV = \int \psi_f^*(\vec{p}\cdot\vec{\nabla}F^* + \vec{\nabla}F^*\cdot\vec{p})\psi_i dV$$

$$= \frac{i}{\hbar}\int \psi^*(\vec{p}\cdot\vec{p}F^* + \vec{p}F^*\ \vec{p})dV$$

as $\quad \vec{\nabla} = \frac{i}{\hbar}\vec{p}$

and absorbing $\frac{i}{\hbar}$ in the constant part, we get

$$\int \psi^*(\vec{p}\cdot\vec{A}+\vec{A}\cdot\vec{p})\psi_i dV = \int \vec{p}\psi^*\ \vec{p}F^*\ \psi_i dV + \int \psi_f^*(\vec{p}F^*)p\psi_i dV \qquad \text{...(3.10.48)}$$

The following relations are generally valid

$$(\vec{p}F^*)\psi = \vec{p}(F^*\psi) - F^*\vec{p}\psi$$

and $\quad (\vec{p}F^*)\vec{p}\psi = \vec{p}(F^*\vec{p}\psi) - F^*p^2\psi$

Using these relations the right hand side of equation (3.10.48) gives

R.H.S. $= \int \vec{p}\psi_f^*[\vec{p}F^*\psi_i - F^*p\psi_i]\,dV + \int \psi^*\ \vec{p}(F^*\vec{p}\psi)d\,V - \int \psi^* F^* p^2\psi_i\,dV$

$$= \int p^2\psi_f^*F^*\psi_i dV - \int \vec{p}\psi_f^*F^*(\vec{p}\psi_i)dV + \int \vec{p}\psi_f^*F^*\psi_i dV - \int \psi_f^* p^2\psi_i F^* dV$$

$$= \int p^2\psi_f^*F^*\psi_i dV - \int \psi_f^*(p^2\psi_i)F^* dV$$

Adding and subtracting the term $2m_p\int \psi_f^* u_r F^*\psi_i dV$ we get

R.H.S. $= \int 2m_p\left(\frac{p^2}{2m_p}+u(r)\right)F^*\psi_f^*\psi_i dV - \int \psi^*F^*2m_p\left(\frac{p^2}{2m_p}+u(r)\right)\psi_i\,dV$

$$= 2m_p(E_i - E_f)\int \psi_f^*F^*\psi_i dV$$

$$= 2m_p\hbar\omega\int \psi_f^*F^*\psi_i dV \qquad \text{...(3.10.49)}$$

As $p^* = -p$, $\vec{A}^* = -A$ and $\mu_N = \frac{e\hbar}{2m_p}$ one gets,

$$-\frac{e\mu_0}{2m_p}\int \psi_f^*(\vec{p}\cdot\vec{A}+\vec{A}\cdot\vec{P})\psi_i dV = e\hbar\left[\frac{c_k}{\varepsilon_0\hbar R_0 L(L+1)}\right]^{1/2}\frac{L+1}{(2L+1)!!}\int \psi_f^*F^*\psi_i dV \qquad \text{...(3.10.50)}$$

and $\int \psi^*\frac{\mu_N g_{sp}}{\hbar}(\vec{s}_p\cdot\vec{H})dV = \frac{\mu_0 e}{2m_p}g_{sp}k\left[\frac{c_k}{\mu_0\hbar R_0 L(L+1)}\right]^{1/2}\cdot\frac{1}{(2L+1)!!}$

$$\cdot\times\hbar\int \psi_f^*\left(\frac{\vec{S}_p\cdot\vec{L}F}{\hbar^2}\right)^*\psi_i dV\ldots \qquad \text{...(3.10.51)}$$

Substituting in equation (3.10.43) the γ-ray transition probabiltiy is obtained as,

$$P = \frac{2\pi}{\hbar}\frac{R_0}{\pi\hbar c}\frac{e^2\hbar^2 ck}{\varepsilon_0\hbar R_0 L(L+1)}\frac{1}{\{(2L+1)!!\}^2}$$

$$\times (L+1)\int \psi_f^* F^* \psi_i dV + g_{sp}\frac{\hbar\omega}{2m_p c^2}\int \psi_f^* \frac{\vec{s}_p \cdot \vec{L}F^*}{\hbar^2}\psi_i dV \qquad \ldots(3.10.52)$$

The term $\frac{\hbar\omega}{2m_p c^2}$ is very small and can be neglected, so that

$$P_{LM}^l = \frac{2ke^2\hbar^2}{\hbar^3\varepsilon_0}\frac{1}{[(2L+1)!!]^2}\frac{L+1}{L}\left|\int \psi_f^* F^* \psi_i dV\right|^2$$

$$= \frac{2\omega e^2}{\hbar c\,\varepsilon_0}\frac{L+1}{L}\frac{1}{[(2L+1)!!]^2}\left|\int \psi_f^* F^* \psi_i dV\right|^2$$

$$= \frac{8\pi d\omega(L+1)\alpha}{L}\frac{1}{[(2L+1)!!]^2}\left|\int \psi_f^*(kr)^L Y_{LM}^*(\theta,\phi)\psi_i dV\right|^2 \qquad \ldots(3.10.53)$$

where $\qquad \alpha = \frac{e^2}{4\pi\varepsilon_0\hbar c}$

The term $r^L Y_{LM}(\theta, \phi)$ is the multipole operator. The term represents the dynamic multipole moment giving rise to the transition.

For a spherically symmetric nuclear potential the nuclear wave functions ψ_i and ψ_f can be written as product of radial wave function $\frac{u(r)}{r}$ spin wave function χ_s and angular wave function $Y_{LM}(\theta, \phi)$. One can write

$$\psi = \frac{u_l(r)}{r}\phi_{jlm_j}$$

$$= \frac{u_l(r)}{r}\sum_{m_s=-\frac{1}{2}}^{m_s=+\frac{1}{2}} C\left(\frac{1}{2}, l, j; m_s m_j\right)\chi_{\frac{1}{2}m_s} Y_{l,m_l} \qquad \ldots(3.10.54)$$

The proton state in the nucleus is specified by its total angular momentum j, orbital angular momentum l and spin angular momentum $S = ½$, with the corresponding projection quantum numbers m_j m_l and m_S. The state can thus be specified $\psi(j, l, m_j)$. Specifying the initial state $\psi_i = \psi(j', l', m'_j)$ and final state $\psi_f = \psi(j l m_j)$. As there is no preference of direction of emission of the photon the initial state and the final states are $(2j'+1)$ and $(2j+1)$ fold degenerate respectively. The transition probability is summed over all the $(2j'+1)$ and $(2j+1)$ states giving

$$P^e(j'l' \to jl) - 8\pi\alpha\omega\frac{L+1}{L}\frac{1}{[(2L+1)!!]^2}\left|\int u_l u_{l'}(kr)^L dr\right|^2 \times S_{j'l'Ljl} \qquad \ldots(3.10.55)$$

where, $$S_{j'l'Ljl} = \frac{1}{2j+1}\left|\sum_{m_j m_{j'}} \int \phi^*_{jlm_j} Y^*_{LM} \phi_{j'l'm'_j}\, d\Omega ds\right|^2 \quad \text{...(3.10.56)}$$

In this equation the integration is over the angles ($d\Omega$) and the spins (ds). The integral can be written as

$$\int \phi^*_{jlm_j} Y^*_{j'l'm'Lj}\, d\Omega ds = \sum_{m_s m_{s'}} C\left(\frac{1}{2} l\, j, m_s m_j\right) C\left(\frac{1}{2} l'\, j'\, m_{s'} m_{j'}\right) \times \int \chi_{\frac{1}{2}m_s} \chi_{\frac{1}{2}m_{s'}}\, ds \int Y^*_{lm_l} Y^*_{LM} Y_{l'm'_l}\, d\Omega$$

... (3.10.75)

The above integral gives a non-vanishing value.

If $\quad m_s = m_{s'}\quad$ and $m_{l'} - m_l = M$

and $\quad (l - L) < l' < (l + L),\ L \neq 0$ ...(3.10.58)

The integral over spherical wave function in eqn. (3.10.66) is zero unless $l + L + l'$ is even, which means that for electric radiation with $L = 0$ (dipole) $l + l'$ must be odd. As the parity of a state is given by $(-1)^l$, then if $l + l'$ is odd the parities of the two states must be different. For $L = 2$ (electric quadrupole radiation) $l + l'$ is even both l and l' must either be odd or even. It means that the parity of both the initial and final state is similar (either odd or even).

In the absence of the knowledge of nuclear potential the wave function $u_l(r)$ depends upon the model employed. In a very simple model one assumes that in both initial and final states the proton is uniformly distributed over the nuclear volume which is spherical with radius R. In this picture

$$u_l = u_{l'} = r\sqrt{4\pi\rho_p} \text{ for } 0 \leq r \leq R$$
$$= 0 \text{ for } r > R$$

The proton density in the nucleus ρ_p for a single proton is $\frac{3}{4\pi R^3}$. The integral over radial wave function in eqn. (3.10.55) then yields

$$4\pi \int_o^R \rho (kr)^L r^2 dr = (kr)^L \frac{3}{L+3} \quad \text{...(3.10.60)}$$

If one puts the integral $S_{j'l'L'jL}$ equal to unity the transition probability is known as Weisskopf's estimates. Putting the value of $k = \frac{w}{c}$, R in units of Fermi (10^{-15} m) and γ-ray transition energy in MeV, the transition probability for electric multipole transitions is given as

$$P_L^e = \eta^2 \frac{4.4(L+1)}{L[(2L+1)!!]^2}\left(\frac{3}{L+3}\right)^2 \left(\frac{\hbar\omega(\text{MeV})}{197}\right)^{2L+1} R^{2L} \times 10^{21}\,\text{sec}^{-1} \quad \text{...(3.10.61)}$$

In a similar way if one starts with the magnetic multipole field, the transition probability for a magnetic multipole radiation is given as

$$P_L^m = \eta^2 \frac{1.9(L+1)}{L[(2L+1)!!]^2}\left(\frac{3}{L+3}\right)^2 \left(\frac{\hbar\omega(\text{MeV})}{197}\right)^{2L+1} R^{2L-2} \times 10^{21}\,\text{sec}^{-1} \quad \text{...(3.10.62)}$$

The factor η^2 gives the effective charge of the proton in the nucleus and is given as

$$\eta^2 = \left[\varepsilon\left(\frac{A-1}{A}\right)^2 + (-1)^L\left(\frac{Z-\varepsilon}{A^2}\right)\right] \quad ...(3.10.63)$$

where $\varepsilon = 0$ for neutron and $\varepsilon = 1$ for a proton, A the mass number and Z the atomic number of the nucleus emitting γ-rays.

The transition probability has been calcualted for a proton making a translation. A neutron can also lead to the transition except that its effective charge is given by equation (3.10.63). In fact the transition takes place between two nuclear states and not individual proton or neutron states. This implies that a more realistic wavefucntion $u(r)$ should be employed. The Weisskopf's estimate of transition probability describes the γ-ray life times only within a factor of about 10. Realistic nuclear wave functions are available only for the very light nuclei.

The γ-ray transition probability varies as $(E_r/197)^{2L+1}$. As $E_r < 1$ MeV the transition probability reduces strongly as the multipolarity L increases and the transition energy decreases. This inhibits strongly the mixing of higher multipoles. This effect of the effective charge correction is that in heavy nuclei transitions with $L > 1$ take place preferably due to protons. The reduced width of the transition is obtained by dividing the transition probability by the right hand side of equation (3.10.61). According to equation (3.10.55) the reduced width gives the value of $S_{j'l'Ljl}$ and correction to the radial matrix element due to proper wave functions $u(r)$. The reduced width can be equated to a matrix element $|m|^2$ which takes into account both the above factors. In the Weisskopf's estimates this matrix element is taken to be unity. A comparison of the experimental values of the γ-ray life times with Weisskopf's estimates gives the value of the matrix element $|m|^2$.

Due to the strong dependence on L, the γ-ray transition takes place only by the lowest allowed multipole. The electric multipole transitions are depicted as $E1$ (electric dipole), $E2$ (electric quadrupole), $E3$ (electric octopole) and so on. The magnetic transitions in a similar way are represented as $M1$, $M2$, $M3$, $M4$ and so on. On considration of parity the multipoles that can mixed are $E1$, $M2$, $E3$ and $M4$ transitions or $M1$, $E2$, $M3$, $E4$ transitions. One never finds $E1$ mixed with $M2$. Normally the transition probability for $E2$ transition is 10^{-4}–10^{-5} of that of $M1$ transition. However, a number of cases have been found where $E2$ is mixed with $M1$. The $E2$ transition takes place due to the quadrupole moment operator. In a deformed nucleus the quadrupole moment is large which makes the matrix element $|m|^2$ very large. This permits a comparable transition probability for $M1$ and $E2$ transitions.

Selection Rules for γ-Transitions

The main selection rules for γ-transitions are given in equation (3.10.58) which are,

$$m_s = m_{s'}$$

$$|m_{l'} - m_l| = M$$
$$|l' - l| \leq L \leq |l' + l|$$

M is the projection of L along the z direction.

Apart from the above selection rules there are selection rules involving the isotopic spins of the states involved. According to these selection rules, γ-transition takes place if

$$\Delta T_3 = 0 \text{ and } \Delta T = 0 \pm 1$$

ΔT_3 represents a change in the charge of the nucleaus, which obviously does not happen. The isotopic spin selection rules are applicable to the very light nuclei where states of different isotopic spins are observed. Even in light nuclei, the excited states do not differ in isotopic spin T by more than one. This selection rule is therefore of little practical value.

In nuclei with equal number of neutrons and protons the isotopic spin selection rule is $\Delta T = \pm 1$. However, isotopic spin is not a good quantum number and one can have transition with $\Delta T = 0$. The reduced width for such transition is much less than for $\Delta T = \pm 1$ transitions.

As discussed in Chapter 1 a deformed nucleus has an angular momentum $\vec{R}$ associated with its rotational motion. If the angular momentum of the nucleus due to unpaired ncleons is $\vec{J}$ the total angular momentum of the nucleus is $\vec{I}$ where,

$$\vec{I} = \vec{J} + \vec{R}$$

K is the quantum number which gives the projection of I along the symmetry axis. For γ-ray transitions in deformed nuclei the selection rule is

$$\Delta K < L$$

In ^{180}Hf nucleus there is an excited state with $J = 9^-$ and $K = 9$ with negative parity. There is another state which is 8^+ (parity +1) which has $K = 0$. The E_1 transition between the two states has not been observed as it is forbidden according to the K selection rule ($\Delta K = 0$). Experimentally the reduced width of the transition is 10^{-15}.

3.10.4 Comparision with Experiments

There is a considerable amount of data on the life time of γ-ray transitions of different multipoles. This can be compared with the Weisskopf's estimates to give the reduced width or the matrix element $|m|^2$ of the transition.

In light nuclei the calculated value of matrix element is about 0.05 whereas the measured value is 0.044 which is in good agreement. For M_1 transitions also the agreement is fairly good with $|m|^2 \simeq 0.15$. For E_2 transitions experimental values give $|m|^2 \simeq 5$. For the only known $E3$ transition from the 6.14 MeV level of ^{16}O the value of $|m|^2 \simeq 30$.

In heavy nuclei matrix element $|m|^2 \simeq 10^{-5}$ for $E1$ transitions. In deformed nuclei the value of the matrix element $|m|^2$ is about 10^2. In other nuclei the value of $|m|^2$ for $E2$ is 10^{-2}. The matrix element for $E3$, $E4$, $M1$ and $M2$ transitions in heavy nuclei is of the order of 10^{-2}. For $M3$ transition $|m|^2 \sim 1$ and for $M4$ transition $|m|^2 \sim 5$.

The measurement of life time of nuclear states gives information about the multipolarity of the transition. In the deformed nuclei it gives approximately the quadrupole moment.

3.10.5 Isomerism

When life time of an excited state of a nucleus is 10^{-6} seconds or more it is usually termed as an isomeric state. From the Weisskopf estimates one can see that most of the $E2$, $M3$, $E4$ and $M4$ transitions have life times greater than 10^{-6} s and thus from isomeric states.

As discussed in the shell model of the nucleus (Chapter 4) there are number of nuclei which have isomeric excited states just before the closure of a shell (magic number). The reason for these islands of isomerism has been discussed in detail in the shell model.

3.11 INTERNAL CONVERSION OF γ-RAYS

As discussed in the beginning, an excited nucleus looses its excitation energy either by emitting γ-ray photon or alternately imparting the energy to one of the atomic shell electrons. The process is known as the internal conversion of γ-rays. The internal conversion coefficient (ICC) α is defined as the ratio of the probability λ_e of emission of an electron to that of emission of a photon *i.e.* λ_γ. If the number of internal conversion electrons and the γ-ray photons emitted per second from a radioactive source are Ne and N_γ respectively then

$$\alpha = \frac{\lambda_e}{\lambda_\gamma} = \frac{N_e}{N_\gamma} \qquad \ldots(3.11.1)$$

The total probability λ of the decay of the excited state is the sum of probabilities for decay by emission of an electron and a photon per second so that

$$\lambda = \lambda_e + \lambda_\gamma$$

$$= \lambda_\gamma\left(1 + \frac{\lambda_e}{\lambda_\gamma}\right)$$

$$= \lambda_\gamma(1 + \alpha) \qquad \ldots(3.11.2)$$

The life time T of the excited state and the partial life times for emission of a conversion electron and that for emission of a photon are related as

$$\frac{1}{T} = \frac{1}{T_e} + \frac{1}{T_\gamma} \qquad \ldots(3.11.3)$$

The partial life time for γ-ray emission is related to life time of the excited state as

$$T_r = (1 + \alpha)T \qquad \ldots(3.11.4)$$

Internal conversion of γ-rays can take place in any of the atomic shells or subshells, so that

$$\alpha = \alpha_k + \alpha_{LI} + \alpha_{LII} + \alpha_{LIII} + \alpha_M$$

where α_k, α_{LI}, α_{LII}, α_{LIII} and α_M are the internal conversion coefficients in the respective shells/subshells. For the L shell

$$\alpha_L = \alpha_{LI} + \alpha_{aII} + \alpha_{LIII}$$

Experimentally one can measure α_k or the ratios $\frac{\alpha_k}{\alpha_L}$ and $\frac{\alpha_{L\,III}}{\alpha_{L\,I}}$.

If the energy of the γ-ray transition is E_r the energy of the internal conversion electron E_e is given as

$$E_e = E_\gamma - B.E.$$

where B.E. is the binding energy of the electron in the shell from which it is ejected. If the γ-ray energy is less than the binding energy of the electron in a particular shell, the internal conversion takes place in the next higher shells only. After the ejection of the electron the atom is left in an ionized state. It returns to its ground state with the emission of X-rays or Auger electrons.

3.11.1 Probability of Internal Conversion of γ-rays

The internal conversion coefficient is found to depend upon:

(*a*) Energy of the γ-ray transition E_r.
(*b*) Multipolarity L of the γ-ray transition.
(*c*) Atomic number of the atom.
(*d*) The shell or subshell of the atom in which internal conversion takes place.

The K-shell electron of an atom has the largest probability of all the other electrons, to be close to the nucleus. If γ-ray energy is large enough, the internal conversion takes place predominately in the K-shell of the atom. If the γ-ray transition has a mixed multipolarity and α_L, the ICC in a particular shell for transition of multipolarity L, then the total internal conversion coefficient in the shell is

$$\alpha = \Sigma a_L \alpha_L$$

where a_L is the intensity of the γ-transition of multipolarity L.

The calcualtion of the internal conversion coefficient can be done under the following simplifying assumptions:

(*a*) Internal conversion takes place in the K-shell of the atom.
(*b*) The interaction between the excited nucleus and the K-shell electron is purely electrostatic.
(*c*) The wave function of the K-shell electron is hydrogen like, *i.e.*, the effect of the electrons in the outer shells of the atom is neglected.
(*d*) After ejection, the electrons does not experience any electromagnetic interaction and it behaves as a plane wave.
(*e*) The kinetic energy of the electron is much larger than its binding energy.

With the above simplifications the probability for the internal conversion can be calculated which when divided by the probability for γ-ray emission gives the internal conversion coefficient of the γ-ray.

The wave functions ψ_i and ψ_f of the nucleus in the initial and final states respectively depend only upon the nuclear coordinates. ϕ_i is the wave function of the electron in K-shell while ϕ_f is the wave function of the free electron. For the K-shell electron of a hydrogen like atom the wave function can be written as,

$$\phi_i = (\pi a^3)^{-1/2} e^{-R/a} \quad \text{...(3.11.5)}$$

where, $a = \frac{a_0}{Z}$ and a_0 is Bohr radius and Z the atomic number of the atom.

The wave function ϕ_f of the emitted electron is a plane wave given by

$$\phi_f = V^{-1/2} e^{i\vec{k}_e \cdot \vec{R}} \quad \text{...(3.11.6)}$$

where $\vec{k}_e$ is the wave vector and R the position vector of the electron enclosed in a volume V.

The transition probability for the nucleus to go from state ψ_i to ψ_f and one of the two K-shell electrons to go from state ϕ_i to ϕ_f is given as,

$$P = 2 \cdot \frac{2\pi}{\hbar} |H_{if}|^2 \frac{dn}{dE} d\Omega \quad \text{...(3.11.7)}$$

where, $\frac{dn}{dE} d\Omega$ is the number of states of the ejected electrons per unit energy range for electron directions within the solid angle element $d\Omega$. The number of states in volume V and with momentum between p and $p + dp$ moving in solid angle $d\Omega$ is

$$dn = \frac{p^2 dp V}{(2\pi\hbar)^3} d\Omega$$

$$E = p^2/2m \therefore dE = \frac{pdp}{m}$$

$$\frac{dn}{dE} = \frac{mpV}{(2\pi\hbar)^3} = \frac{m\hbar k_e V}{(2\pi\hbar)^3} \quad \text{...(3.11.8)}$$

The matrix element can be written as

$$H_{if} = \sum_{i=1}^{z} \int \phi_f^* \psi_f^* \frac{e^2}{4\pi\varepsilon_0 |\vec{R} - \vec{r}_i|} \phi_i \psi_i dV \quad \text{...(3.11.9)}$$

The electrostatic interaction is between a proton at position r_i and the electron at position R. The summation is over all the protons in the nucleus. Putting values of ϕ_i and ϕ_f we have,

$$H_{if} = \frac{(\pi a^3 V)^{-1/2}}{4\pi\varepsilon_0} \sum_{i=1}^{z} \int d^3R \int d^3r e^{-i\vec{k}_e \cdot \vec{R}} \psi_f^* \times \left| \frac{e^2}{\vec{R} - \vec{r}_i} \right| e^{-R/a} \psi_i \quad \text{...(3.11.10)}$$

The integration is over the volume of the nucleus (d^3r) and the volume of the electron (d^3R). The major contribution to the integral is when $\vec{R} > \vec{r}_i$ so that

$$\frac{1}{|\vec{R} - \vec{r}_i|} = \sum_{L=0}^{\infty} \sum_{m=-L}^{+L} \frac{4\pi}{2L+1} \frac{r_i^L}{R^{L+1}} Y_{LM}(\Theta, \Phi) Y_{LM}(\theta_i\ \phi_i) \quad \text{...(3.11.11)}$$

where Θ, Φ refer to polar angles of vector $\vec{R}$ and $\theta_i\, \phi_i$ to the polar angles of vector $\vec{r}_i$. Substituting in equation (3.11.10) and separating the nuclear and electronic parts we have,

$$H_{if} = \sum_{L=0}^{M} \sum_{M=-L}^{+L} \frac{4\pi e}{(\pi a^3 V)^{1/2} 4\pi\varepsilon_0 (2L+1)} H_{LM} J_{LM} \qquad \ldots(3.11.12)$$

where H_{LM} is the electric multipole matrix element for transition from the nuclear state ψ_i to state ψ_f. J_{LM} is the matrix element for the change of state of the electron. We have,

$$H_{LM} = \sum_{z=1}^{z} \sum_{L=1}^{\infty} \sum_{M=-L}^{+L} e\int \psi_f^* r_i^L Y_{LM}(\theta_i\, \phi_i)\psi_i d^3r \qquad \ldots(3.11.13)$$

and
$$J_{LM} = \sum_{L=1}^{\infty} \sum_{M=-L}^{+L} \int e^{i\vec{k}_i \cdot \vec{R}} Y_{LM}(\Theta\, \Phi) e^{-\frac{R}{a}} R^{-(L+1)} d^3R \qquad \ldots(3.11.14)$$

As the energy imparted to the electron is much larger than its binding energy in the K-shell the wavelength (k_e^{-1}) of the outgoing electron is much smaller than the radius a of the K-shell so that $ka >> 1$. In this approximation $e^{-R/a} \approx 1$ and the integral J_{LM} can be evaluated using addition theorem of spherical harmonics so that,

$$J_{LM} = 4\pi(i)^{-L} \frac{k_e^{L-2}}{(2L-1)!!} Y_{LM}(\Theta\Phi) \qquad \ldots(3.11.15)$$

where Θ, Φ are the polar angles made by vector $\vec{k}_e$ with respect to an arbitrary z-direction. The matrix element for emission of an internally converted K-shell electron is

$$H_{if} = \frac{(4\pi)^2 e(i)^{-L} k_e^{L-2}}{4\pi\varepsilon_0 (\pi a^3 V)^{1/2} (2L+1)(2L-1)!!} Y_{LM}(\Theta\Phi) H_{LM}$$

$$H_{if} = \frac{(4\pi)^2 e(i)^{L}}{4\pi\varepsilon_0 (\pi a^3 V)^{1/2} \{(2L+1)!!\}} Y_{LM}(\Theta\Phi) H_{LM}(\theta_i \phi_i)$$

$$|H_{if}|^2 = \frac{256 e^2 \pi^4}{(4\pi\varepsilon_0)^2 \pi a^3 V \{(2L+1)!!\}^2} |H_{LM}|^2 \times \int Y_{LM}^* Y_{LM} d\Omega \qquad \ldots(3.11.16)$$

The probability for internal conversion of the γ-ray is thus given as

$$P_e = 2 \cdot \frac{2\pi}{2} \frac{dn}{dE} \int |H_{if}|^2 d\Omega$$

$$= \frac{4\pi}{\hbar} \frac{Vm\hbar k_e}{(2\pi\hbar)^3} \cdot \frac{256\pi^4 e^2}{\pi a^3 V} \frac{k_c^{2L-4}}{\{(2L+1)!!\}^2} |H_{LM}|^2$$

$$P_e = \frac{128\pi m^2 z^2 k_e^{2L-3}}{\hbar^3 a_0^3 \{(2L+1)!!\}^2} |H_{LM}|^2 \qquad \text{...(3.11.17)}$$

where, $a = a_0/Z$

The probability of emission of a γ-ray photon is given by equation (3.10.53) as

$$P_\gamma = 8\pi\alpha\omega \frac{L+1}{L} \frac{1}{\{(2L+1)!!\}^2} \left| \int \psi_f (k_\gamma r)^L Y^*_{LM}(\theta\phi)\psi_i d^3r \right|^2$$

dividing $P_e(k)$ by P_γ the internal conversion coefficient is obtained by putting the values of

$$a_0 = \frac{4\pi\varepsilon_0\hbar^2}{me^2}$$

and $$\alpha = \frac{c^2}{4\pi\varepsilon_0\hbar c}$$

we get for multipolarity L

$$\alpha_k^L = \frac{P_L(k)}{P_L(h\nu_0)} = \frac{16Z^3}{a_0^4} \cdot \frac{L}{L+1} \frac{k_e^{2L-3}}{k^{2L+1}}$$

The wave vector k for photon is $k = \frac{\omega}{c}$ and the wave vector k_e for electron is given by the relation

$$\frac{\hbar^2 k_e^2}{2m} = \hbar\omega - B.E. \approx \hbar\omega$$

so that

$$k_e = \left(\frac{2m\omega}{\hbar}\right)^{1/2}$$

Substituting values of k and k_e we get

$$\alpha_k^L = \frac{16Z^3}{a_0^4} \frac{L}{L+1} \left(\frac{2m\omega}{\hbar}\right)^{\frac{2L-3}{2}} \left(\frac{c}{\omega}\right)^{2L+1}$$

$$= 16Z^3 \left(\frac{me^2}{\hbar^2 4\pi\varepsilon_0}\right)^4 \frac{L}{L+1} \left(\frac{2mc^2}{\hbar\omega}\right)^{L+5/2} \cdot \frac{\hbar^4}{2^4 m^4 c^4}$$

$$= Z^3 \left(\frac{e^2}{4\pi\varepsilon_0\hbar c}\right)^4 \frac{L}{L+1} \left(\frac{2mc^2}{\hbar\omega}\right)^{L+5/2}$$

$$= Z^3 \alpha^4 \frac{L}{L+1} \left(\frac{2mc^2}{E_\gamma}\right)^{L+5/2}$$

This expression for internal conversion coefficient for electric multipole transitions in the K-shell of an atom is only approximate as the electron wave function should be treated relativistically. It is seen that the internal conversion coefficient increases strongly with the atomic number Z and the multipolarity; L, of the gamma ray transition but decreases with increasing transition energy $\hbar\omega$, the internal conversion is an important phenomenon for nuclei of high atomic number and low energy transitions of high multipole order L. The calculation of the interanal conversion coefficient does not take into account the effect of outer shell electrons on the wave function of the K-shell electron. The screening effect of the outer shell electrons is significant in high atomic number atoms.

To obtain the internal conversion coefficient in the magnetic multipole transitions one has to take the electromagnetic interaction between the nucleus and the atomic electrons.

Rose has calculated the internal conversion coefficient for different energy gamma rays, in different shells of almost all the elements for different electric and magnetic multipole radiations. He has used relativistic Dirac equations to describe the electron state and suitable theory to take the screening effects and the finite size of the nucleus. The values of the various internal conversions coefficients have been tabulated.

Figure 3.40 shows the variation of the K-shell internal conversion coefficients with transition energy and multipole order in atoms with $Z = 40$. It may be observed that the internal conversion coefficient for magnetic transition is about ten times the internal conversion coefficient of an electric transition of the same multipolarity. It may also be noted that the internal conversion coefficient is large for transitions for which gamma ray emission probability is low (small energy $\hbar\omega$ and large multipolarity L).

Experimental determination of the internal conversion coefficient is made by measuring the number of electrons emitted from different shells of an atom using a β-particle spectrometer. The number of gamma ray photons emitted from the same source is many times measured by the β-particle spectrometer simultaneously by counting the number of photoelectron ejected from a very thin gold foil. The measured values of internal conversion coefficient is of great help in determining the multipolarity of the gamma ray transition.

3.12 ANGULAR DISTRIBUTION OF γ-RAYS

Consider a γ-transition from a nuclear excited state characterised by spin $\vec{I}_i$ to another state with spin $\vec{I}_f$ emitting a photon with angular momentum L. As mentioned earlier

$$\vec{I}_i - \vec{I}_f = \vec{L}$$

or $$|I_i - I_f| \le L \le |I_i + I_f|$$

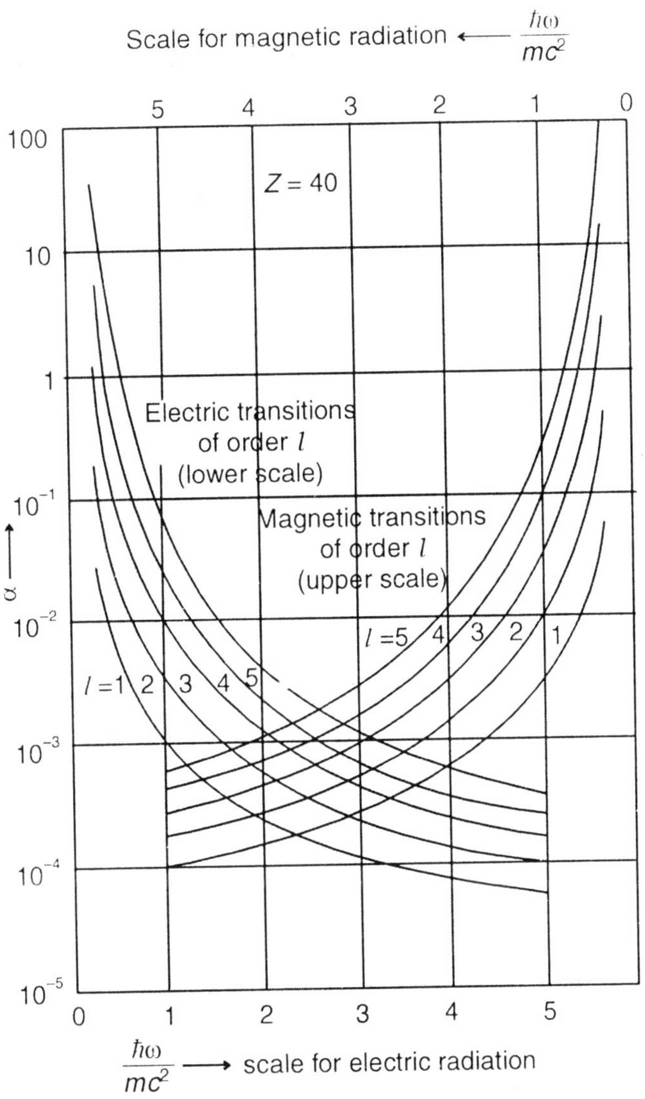

Fig. 3.40: Internal conversion coefficients for an element with Z = 40. The lower scale applies to electric multipole radiations the upper one to magnetic multipole radiation. *l* indicates the multipolarity 2^l. (M. E. Rose *et al.*, (1951). *Phy. Rev.*, **83**: 79).

The component of angular momentum $\vec{L}$ along some fixed z-direction is $L_z = Mh$. Whille $L^2 = L(L + 1)\hbar^2$. Each nuclear state with spin I is composed of $(2I + 1)$ magnetic substates which depending upon the environment of the nucleus may or may not be degenerate. The magnetic sublevels have magnetic quantum number m_i in the initial state and m_f in the state of the nucleus. The selection rules for γ-transition require $m_i - m_f = M$.

The photons emitted in a transition from a specific substate m_i to another specific substate m_f possesses a characteristic angular distribution $F_L^M(\theta)$ which is independent of I_i and I_f. Here θ is the angle between the z-axis and the direction of emission of the photon. For a dipole radiation with $L = 1$ and $M = 0$ or ± 1 the angular distribution functions are given as,

$$F_1^0(\theta) = 3\sin^2\theta = 2 - 2P_2(\cos\theta)$$

$$F_1^{\pm 1}(\theta) = \frac{3}{2}(1 + \cos^2\theta) = 2 + 2P_2(\cos\theta)$$

where P_2 are the Legendre polynomial of second order.

The energy separation between the various magnetic sublevels of a nuclear state is very small and can not be observed by any γ-ray spectrometer. For calculating the angular distribution of γ-rays emitted from state I_i to state I_f the angular distribution for each transition $m_i - m_f$ and its relative intensity has to be taken into account. If the relative population of each of the m_i states is $P(m_i)$ and the transition probability from state m_i to state m_f is $G(m_i m_f)$, the angular distribution of the γ-rays $F_L(\theta)$ for state I_i to state I_f is

$$F_L(\theta) = \sum_{m_i m_f} P(m_i)G(m_i m_f)F_L^M(\theta) \qquad \text{...(3.12.1)}$$

The probability for the transition $m_i \rightarrow m_f$ is a product of a nuclear factor and a geometrical factor. The nuclear factor is independent of m_i and m_f and is a constant. The geometrical factor gives the transition probability $G(m_i m_f)$. The state $|I_i, m_f>$ can be expanded in terms of the state $|I_f, m_f>|L_1 M>$ using the Clebsch-Gordon coefficients as

$$|I_i m_i> = \sum_{m_f, M} |I_f, m_f> |L, M> <I_f m_f LM|I_i m_i>$$

The last term is the Clebsch-Gorden coefficient. For a given value of I_i, I_f, m_i, m_f and L, transition probability is given as

$$G(m_i, m_f) = <I_f m_f LM| I_i m_i> \qquad \text{...(3.12.3)}$$

The relative population of $P(m_i)$ of the substates depends upon the energies of the substates and on the manner in which state I_i was created. If states m_i are all equally populated, the angular distribution $F_2(\theta)$ is isotropic. By cooling the nucleus to a temperature of less than 1 K in a magnetic field only one value of m_i is possible. In such a case the distribution of the emitted γ-rays is anisotropic and is given as

$$F_L(\theta) = G(m_i m_f)\, F_L^M(\theta) \qquad \text{...(3.12.4)}$$

3.13 ANGULAR CORRELATION IN A γ-γ-CASCADE

Consider a nucleus in an excited state I_i making transition to state I with the emission of γ-ray (γ_1) of multipolarity L_i and the state I decaying in cascade to state I_f with the emission of γ-ray (γ_2) of multipolarity L_2 (Fig. 3.41). Let the direction of emission of γ_1 be along the z-axis the angular distribution $F_L(\theta)$ of the second γ-ray with respect to z-axis represents the angular correlation $\omega(\theta)$ between the two γ-rays so that

$$\omega(\theta) = F_L(\theta)$$

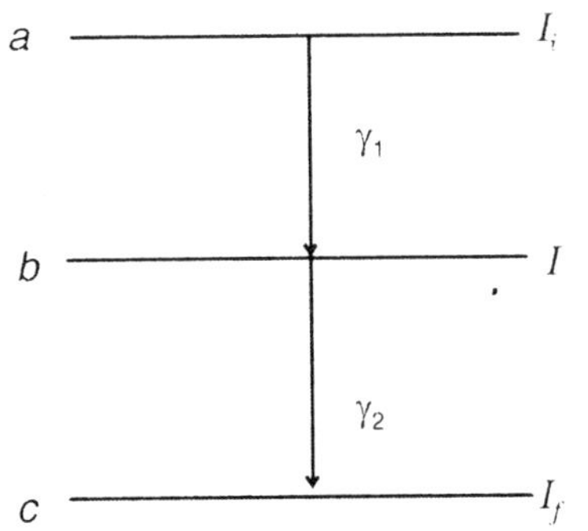

Fig. 3.41: Cascade γ-radiations

It is assumed that when the excited nucleus in state I_i is formed, all the substates m_i are equally populated. The probability of forming a substate m of the state I after transition from state I_i is given as,

$$P(m) = \sum_{m_i} G(m_i m) F_{L_1}^M(\theta = 0) \qquad \text{...(3.13.1)}$$

As the γ-ray γ_1 is emitted in z-direction, the effective components of L_1 is only $L_{lz} = M_1 = \pm 1$. The angular correlation function $\omega(\theta)$ is then given as

$$\omega(\theta) \propto \sum_{m_f m m_i} \langle \mathrm{Im}\, L_1 \pm 1\rangle^2 F_{L1}^{\pm 1}(\theta)\langle I_f m_f L_z m_z|\mathrm{Im}\rangle^2 F_{Lz}^M(\theta) \qquad \text{...(3.13.2)}$$

The value $L_{1z} = \pm 1$ represents the left and the right circularly polarisation of the first γ-ray. There is no interference term for the various transitions $m_i \to m \to m_f$. The above equation gives $\omega(\theta)$ only for pure multipole transitions where L_1 and L_2 have only one value.

A rigorous derivation of the angular correlation of two cascade γ-rays is given as,

$$\omega(\theta) = \sum_k 1 + A_{kk} P_k(\cos\theta) \qquad \text{...(3.13.3)}$$

where K_{max} = min value of $2I$, $2L_1$ and $2L_2$.

The coefficient A_{kk} can be broken up in two factors each depending upon one γ-ray transition only so that

$$A_{kk} = F_k(L_1 L\, I_i I)\, F_k(L_2 L\, I_f I) \qquad \text{...(3.13.4)}$$

where

$$F_k(L_1 L\, I_i I) = (-1)^{I_i + I - 1}(2L_1 + 1)(2I + 1)(2K + 1)^{1/2} \begin{pmatrix} L_1 & L_1 & k \\ 1 & -1 & 0 \end{pmatrix} \cdot \begin{Bmatrix} L_1 & L_1 & k \\ I & I & I_i \end{Bmatrix} \qquad \text{...(3.13.5)}$$

where the last two factors represent the Clebsch-Gorden and Racah coefficients. The values of F_k for different values of L I_i and I are tabulated from which the angular correlation function $\omega(\theta)$ can be calculated.

In one or both the γ-ray transitions are a mixed multipole of L_1 and L'_1 and L_2 and L'_2 then the angular correlation function becomes

$$\omega(\theta) = 1 + A_{kk}P_k (\cos\theta)$$

where $\quad A_{kk} = A_k(L_1L'_1I_iI)\ A_k(L_2L'_2I_fI)$

and

$$A_k(L_1L'_1\ I_2I) = \frac{F_k(L_1L_1I_iI) + 2\delta_1F_k(L_1L'_1\ I_1I) + \delta_1^2F_k(L'_1\ L'_1\ I_iI)}{1 + \delta_1^2} \qquad ...(3.13.6)$$

δ_1 is the mixing ratio of the L'_1 and L_1 multipoles in the first γ-ray. Similar expression holds for the second factor.

The angular correlation function gives information regarding the spins I_i and I as well as the mixture of multipole radiations. The angular correlation between two γ-rays is measured by detecting one γ-ray in one detector placed at different angles θ with respect to the first detector. To ensure that the two γ-rays are emitted from the same nucleus, a coincidence between the pulses from the two detectors is done with a short resolving time. Figure 3.42 shows a schematic diagram of the apparatus.

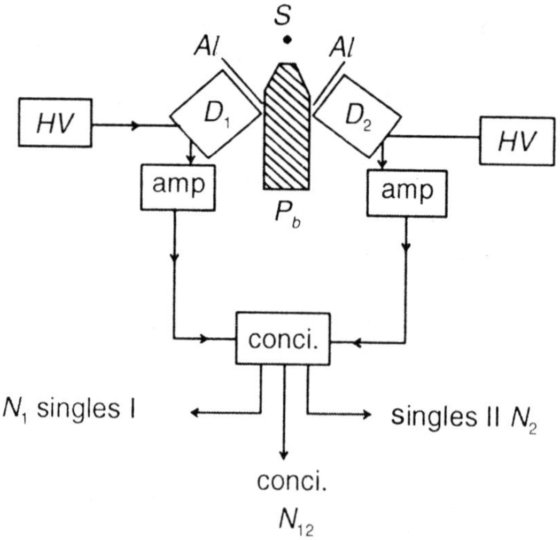

Fig. 3.42: Simple coincidence arrangement

The first measurement of the angular correlation of the cascade γ-rays of ^{60}Ni observed in the decay of ^{60}Co was studied by Brady and Deutsch in 1948.

The decay scheme of ^{60}Co is shown in Fig. 3.43. The radioisotope ^{60}Co decays to the 4 MeV excited state of ^{60}Ni. The state decays with the emission of two γ-

rays in cascade to the ground state of ^{60}Ni. The spins of the ground, first and second excited states are 0, 2 and 4 respectively. Each of the γ-ray transitions is pure multipole with angular momentum $L = 2$. The angular correlation function expected for the spin sequence is

$$\omega(\theta) = 1 + 1/8\, P_2(\cos\theta) + 1/24 P_4(\cos\theta) \qquad ...(3.13.7)$$

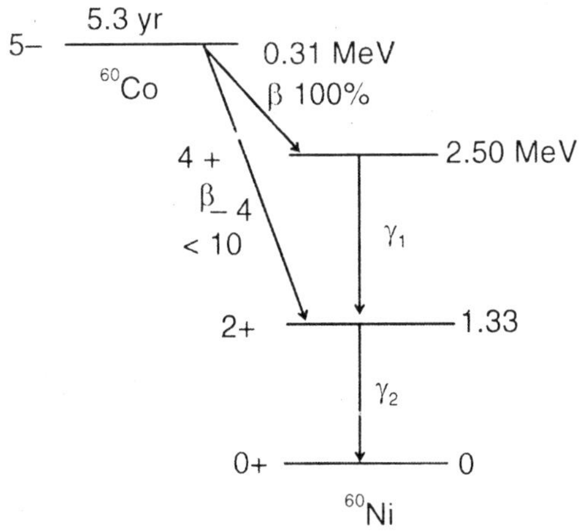

Fig. 3.43: Decay scheme of ^{60}Co

Figure 3.44 shows the measured angular correlation function, which is in conformity to that predicted theoretically.

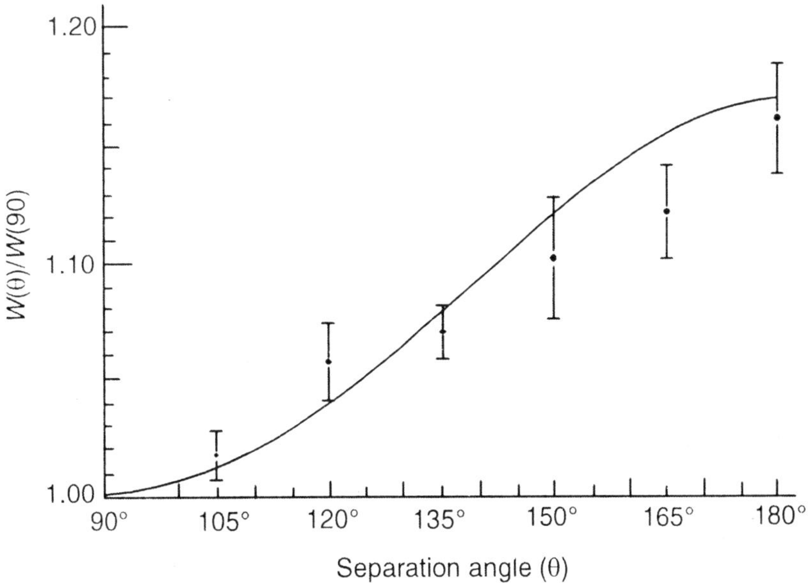

Fig. 3.44: Angular correlation function $W(\theta)$ for ^{60}Co γ-ray cascade

3.14 RESONANCE SCATTERING AND ABSORPTION OF γ-RAYS

According to the electromagnetic theory an oscillating electron emits electromagnetic radiations. The electron experiences a radiation resistance due to which the oscillations are damped. For such an oscillator the radiations have a distribution in its frequency ω. The intensity $I(\omega)$ of radiations of frequency ω is given as

$$I(\omega) = \frac{A^2\Gamma}{(\omega - \omega_0)^2 + (\Gamma / 2\hbar)^2} \qquad \ldots(3.14.1)$$

where A is a constant, ω_0 the frequency of the undamped oscillator and Γ the width in the frequency due to damping. It is seen that

$$I(\omega) = 1/2I(\omega_0) \text{ for } \omega = \omega_0 \pm \Gamma/2$$

Thus the frequency width of the radiations is Γ.

Based on the above, Rayleigh in 1890 predicted resonance absorption of atomic radiations. R. W. Wood observed the resonance absorption of sodium light by sodium vapours in 1904. Using Dirac theory of electron, Weisskopf in 1930, explained the emission and resonance absorption of atomic radiations.

Quantum electrodynamics predicted similar results in a system where energy levels are quantised. The efforts of Kuhn in 1929 to observe resonance absorption of nuclear γ-rays could not yield any results.

Consider a nucleus of an atom in state a making transition to state b. Let energy differences between the levels be E_0, and the energy of the photon emitted be E_γ Fig. (3.45).

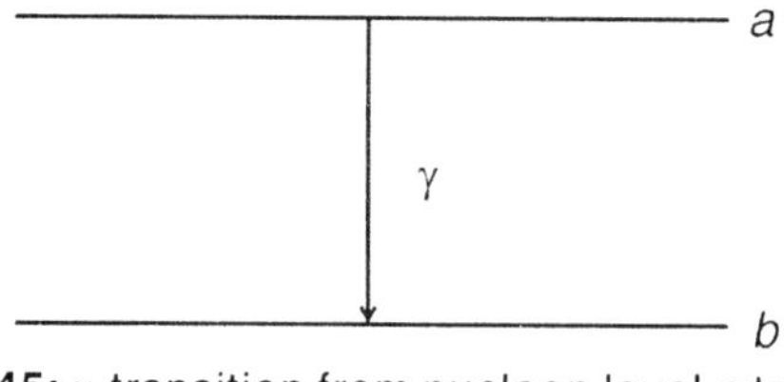

Fig. 3.45: γ-transition from nucleon level a to level b

As the photon is emitted with momentum E_γ/c the atom emitting it, recoils with the same momentum E_γ/c. The kinetic energy of recoiling atom is $\frac{p^2}{2M} = \frac{E_\gamma^2}{2Mc^2}$, where M is the mass of the atom and c velocity of light. As the recoil energy R is very small one can write

$$R = \frac{E_\gamma}{2Mc^2} \approx \frac{E_0^2}{2Mc^2}$$

This recoil energy has to come out of the excitation energy E_0 of the atom. The energy of the emitted photon is thus

$$E_\gamma = E_0 - R$$

$$E_\gamma = E_0 - \frac{E_0^2}{2Mc^2} \qquad \text{...(3.14.2)}$$

When the photon is absorbed by the nucleus of an atom it has to provide the recoil energy equal to $E_0^2 / 2Mc^2$ to the nucleus. Thus for resonance absorption of the photon to take place, the incident photon shall have an energy $E_0 + E_0^2 / 2Mc^2$. The photon emitted from the nucleus thus falls short of the resonance energy by an amount $E_0^2 / 2Mc^2$. The recoil energy of the atom of mass 100 amu can be calculated for different transition energies:

E_0	1-2 eV (optical)	1 keV	100 keV	1 MeV
R	10^{-11} eV	10^{-5} eV	10^{-1} eV	10 eV

The short fall of energy from resonance inceases as the square of the transition energy.

According to Heisenberg's uncertainty principle, the uncertainty in energy ΔE and the life time ΔT of a state are related as

$$\Delta E \Delta T = \hbar$$

The uncertainty in life time is the mean life of the state τ and the uncertainty in energy is the width Γ of the level so that $\tau\Gamma = \hbar$. Due to the width in the energy of the level, the energy of the emitted photon has a spread Γ. According to quantum electrodynamics, the emitted photons have an energy spectrum exactly as predicted classically equation (3.14.1), having a peak at $E_0 = \hbar\omega_0$ and a width Γ. For a mean life of 10^{-8} seconds the width of the γ-ray energy distribution is $\Gamma = \hbar/\tau = 6.5 \times 10^{-8}$ eV.

It is thus observed that a photon emitted in a transition from state a to state b falls short of resonance energy by an amount $2R$. Further the emitted photons as well as the energy required for resonance each have a width Γ. In a normal case the situation is as shown in Fig. 3.46. The resonance absorption of the photon is not possible.

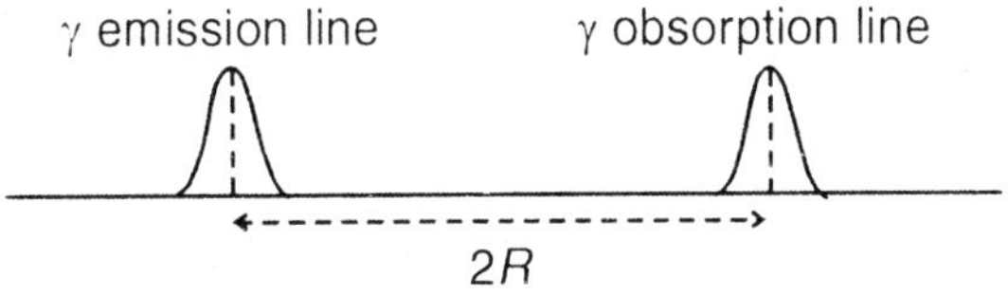

Fig. 3.46: γ-ray emission and absorption lines

If, however the energy shift, $2R$, is comparable to the level width Γ, there would be overlap of the resonance energy and the photon energy distributions. Some resonance absorption would be possible in such a case. In the case of the yellow line of sodium, Γ is about 6×10^{-8} eV while $2R$ is about 5×10^{-10} eV. In this case $\Gamma >> 2R$ and resonance absorption is possible. In the case of a nucleus of mass 100 amu emitting γ-ray of energy 1 MeV with a life time of 10^{-12} sec the values of Γ and $2R$ are 6.5×10^{-4} eV and 10 eV. The value of $2R >> \Gamma$ and resonance can not be observed.

Apart from the natural line widths of the γ-ray there is Doppler broadening of the γ-ray line. If the atom is moving in the direction of emission of the photon, the photon energy is increased. If the velocity of the atom is in opposite direction, the emitted photon has lower energy.

If an atom moving with momentum $\vec{P}_i$ emits a photon of momentum $\vec{P}$, the momentum of the atom after emitting the photon would be $\vec{P}_i - \vec{P}$. The energy associated with this change in momentum is given as

$$\Delta E = \frac{(\vec{p}_i - \vec{p})^2}{2M} - \frac{p^2}{2M} = \frac{p^2}{2M} - \frac{2\vec{p}\cdot\vec{p}_i}{2M}$$

The first term is the recoil energy R of the atom and the second term is the energy change due to Doppler effect. The second term can be written as,

$$\frac{2\vec{p}\cdot\vec{p}_i}{2M} = 2\cdot\frac{pp_i\cos\phi}{2M}$$

$$= 2(\varepsilon R)^{1/2}\cos\phi = D\cos\phi \qquad \text{...(3.14.3)}$$

where $\varepsilon = p_i^2 / 2M$ is the kinetic energy of the atom and ϕ is the angle between the momentum of the atom $\vec{p}_i$ and the photon momentum $\vec{p}$. The energy of the emitted photon is thus

$$E_r = E_0 - \Delta E = E_0 - R - D\cos\phi \qquad \text{...(3.14.4)}$$

For random velocity of the emitting atom the angle ϕ varies between 0 and π.

The kinetic energy of the atom may arise due to (*a*) thermal motion (*b*) previous emission of β-particle or a γ-ray (*c*) previous capture of a bombarding particle e.g. a proton or (*d*) motion given to the atom mechanically. In cases (*b*) and (*c*) one can choose the angle ϕ in such a way that $D\cos\phi = 2R$ so that the absorption and emission lines overlap in energy and there is resonance absorption. In the case (*d*) the atom emitting the photon is placed at the rim of a circular disc, which is rotated at a very high speed. The tangential velocity of the atom provides the necessary Doppler shift to emitted γ-ray such that $D\cos\phi \approx R$, giving rise to resonance absorption. In the case of thermal motion angle ϕ changes from 0 to 2π so that $D\cos\phi$ have both negative and positive values. This gives rise to a thermal energy distribution of the emitted γ-ray. The effect is appreciable in a gaseous source. The atoms in a gas have Maxwellian velocity distribution and one can write the average Doppler energy shift D as

$$\overline{D} = 2(\overline{\varepsilon}R)^{1/2} \qquad \text{...(3.14.5)}$$

where $\overline{\varepsilon}$ is the average kinetic energy of the emitting atom. At room temperature 300K the average kinetic energy $KT = 0.025$ eV so that average Doppler shift D can be claculated.

E_r	1 keV	100 keV	1 MeV
R	10^{-5} eV	10^{-1} eV	10^{+} eV
$\overline{D}$	10^{-3} eV	10^{-1} eV	1 eV

It is thus seen that for photon energies upto 100 KeV the thermal Doppler broadening is about equal to the recoil energy. One can observe some resonance absorption as depicted in Fig. 3.47. If source of photons of energy $E_\nu < 100$ KeV is cold and absorber is hot or vice versa there is only a small amount of overlap of the absorption and absorption lines (Fig. 3.47). The overlap of the two absorption and emission lines increases if both the source and absorbers are at high temperature. Generally in an experiment the source is heated and absorber is at room temperature. As the temperature increases the Doppler broadening increases and the resonance absorption increases accordingly.

The thermal width is superimposed over the natural line width Γ of the γ-lines. By studying the resonance absorption of γ-rays as a function of temperature of the source, keeping absorber temperature constant, one can estimate the natural line width Γ from which the life time of the γ-transition can be estimated. The method has been extensively used in measuring γ-ray life times.

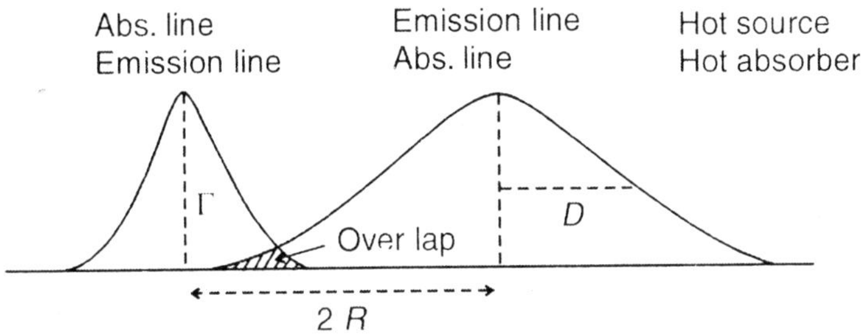

Fig. 3.47: One of the emission or absorption γ-ray line may be Doppler broadened by heating to give an overlap

Cross Sections

Scattering and absorption are complimentary processes. Whenever there is absorption there is scattering as well. Photon of wavelength $2\pi\lambda$ and energy E is absorbed and scattered by an absorber with cross sections σ_a and σ_s respectively where,

$$\sigma_a(E) = \sigma_0 \frac{\Gamma\Gamma_\gamma}{4(E - E_0)^2 + \Gamma^2} \quad \ldots(3.14.6)$$

$$\sigma_s(E) = \sigma_0 \frac{\Gamma_\gamma^2}{4(E - E_0)^2 + \Gamma^2} \quad \ldots(3.14.7)$$

where Γ is the total width of the excited state and Γ_γ the partial γ-ray width of the level. If α is the internal conversion coefficient of the γ-ray

$$\Gamma_\gamma = \frac{\Gamma}{1+\alpha} \text{ and } \sigma_s = \frac{\sigma_a}{1+\alpha}$$

$$\sigma_0 = \frac{2I_a + 1}{2I_b + 1} \cdot 2\pi\lambda^2 \quad \ldots(3.14.8)$$

I_b and I_a are the spins of ground state and excited state respectively. The energy distribution of the γ-rays emitted from the source or scattered from an absorber has a Lorenzian shape *i.e.*

$$I(E) = \frac{\Gamma}{2\pi} \frac{A^2}{(E - E_0)^2 + (\Gamma / 2)^2} \qquad ...(3.14.9)$$

A is the normalising factor.

The intensity $I(E)$ is normalised such that

$$\int_0^\infty I(E)dE = 1$$

The expressions (3.14.6; 3.14.7) for absorption and scattering cross section hold for a definite energy E of the photon. If the intensity distribution of the photon is given as in equation (3.14.1), the observed cross section for absorption and scattering would be given as

$$\sigma_{\text{obs}} = \frac{\int_0^\infty I(E)\sigma(E)dE}{\int_0^\infty I(E)dE} \qquad ...(3.14.10)$$

In the case of Doppler broadening the photon energy distribution $I(E)$ is no longer Lorenzian but Maxwellian and the observed cross section has to be calculated accordingly. The maximum absorption cross section is observed when photon energy is E_0 in that case.

$$\sigma_a = \frac{\sigma_0 \Gamma_r}{\Gamma} = \frac{\sigma_0}{1 + \alpha} \qquad ...(3.14.11)$$

If the incoming photons have a Doppler broadening having a width D and centred at the resonance energy E_0 the observed cross section is

$$\sigma_a = \sigma_0 \frac{\Gamma_r}{\Gamma} \frac{\Gamma}{D} = \sigma_0 \frac{\Gamma_r}{D} \qquad ...(3.14.12)$$

In optical transitions at room temperature $\frac{\Gamma_r}{D} \approx 10^{-1}$ and resonance absorption can be observed. In the case of γ-rays usually $\frac{\Gamma_r}{D} \approx 10^{-4} - 10^{-7}$, so even when the γ-ray line is centered at E_0 the effective cross section becomes very low.

3.15 MÖSSBAUER EFFECT

While studying the resonance absorption of 129 keV γ-rays of ^{191}Ir, Mössbauer in 1958 discovered that if the source and the absorber are both cooled, the resonance absorption increases. Normally the resonance absorption should go down as the overlap due to Doppler broadening should reduce. The effect was explained by Mössbauer by the fact that when the source of γ-rays and the absorbing atom are bound in a crystal lattice there is no recoil of either of the atoms. The number of atoms which do not show recoil, increases as the temperature of the source or the absorber is lowered. The resonance absorption which may be very small at room temperature increases sharply as the source and absorber are cooled to about liquid nitrogen temperature of 77K. At low temperature the whole crystal takes up the recoil momentum and there is no loss of energy of the emitted photon.

It is known that the atoms in a crystal lattice can vibrate about their mean position with some frequency ω. The frequency ν of the emitted γ-ray should be modulated to give the possible frequencies of the γ-ray as ν, $\nu \pm \omega$, $\nu \pm 2\omega$ .... In 1907 Einstein explained the specific heat of solids on the assumption that in a crystal the atoms vibrate with only one frequency. However, Debye was able to explain the T^3 term in the specific heat by assuming all frequencies upto a maximum frequency ω_D and obeying the distribution function.

$$c(\omega) = \text{constant} \cdot \omega^2$$

The frequency distribution has a sharp cut off at frequency ω_D such that

$$\hbar\omega_D = k\theta_D \qquad \text{...(3.15.1)}$$

where θ_D is the Debye temperature of the crystal and k the Boltzmann constant.

The wavelength of the vibrations corresponding to the Debye frequency ω_D is $\lambda = 2d$ where d is the lattice spacing. This is the shortest wavelength. There are larger wavelengths *i.e.* lower ω also present in the crystal. Recoil of the radioactive atom after emission of a photon can excite these vibrations, but its probability is small because exciting such vibrations involves moving of a large number of atoms in the lattice. According to Debye theory the number of photons emitted without any energy loss by recoil is given by the recoil free fraction f as

$$f = e^{-3R/2k\theta_D} \qquad \text{...(3.15.2)}$$

R is the recoil energy of the free atom. The above equation is true at 0 K. At a finite temperature some frequencies can be excited and the photon can loose energy. At a finite temperature T the fraction of photons emitted without loss of energy is given as

$$f = e^{-2W}$$

where

$$W = \frac{3R}{k\Theta_D}\left(\frac{1}{4} + \frac{T}{\Theta_D}\right)^2 \int_0^{\Theta_{D/T}} \frac{x}{e^x - 1}\,dx \qquad \text{...(3.15.3)}$$

The factor f is known as Lamb-Mössbauer factor and is similar to the Debye Waller factor which determines the intensity of scattered X-rays emitted at any temperature without change of wavelength.

A large value of the f-factor would enhance the recoil free emission and absorption of γ-rays. This can be realised when recoil energy R is small, Debye temperature Θ_D is large and the temperature T of the source and the absorber is low.

The most commonly used source for the study of Mössbauer effect is the 14.4 keV excited state of ^{57}Fe formed in the decay of 270 day activity of ^{57}Co. The decay scheme of ^{57}Co is shown in Fig. 3.48. The life time of the 14.4 KeV state is 1.4×10^{-7} seconds corresponding to a line width of 4.6×10^{-9} eV. The energy of recoil of the atom on emitting 14.4 KeV photon is 0.002 eV.

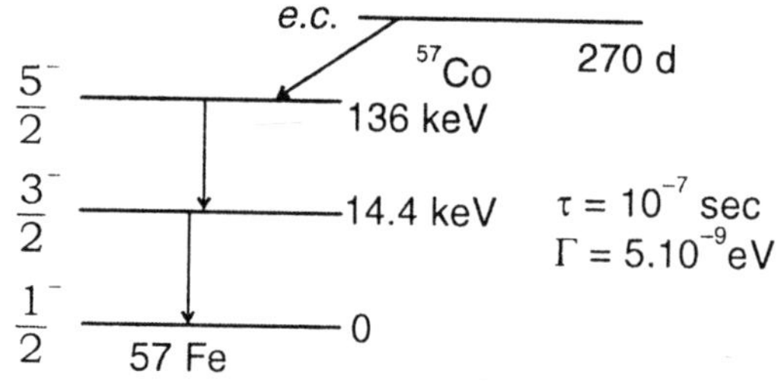

Fig. 3.48: Decay scheme of ^{57}Co to ^{57}Fe

Usually a Mössbauer source is prepared by incorporating radioactive ^{57}Co in the lattice of rhodium metal having f-factor of 0.7 at room temperature. The line width from the source is close to the natural line width of the 14.4 keV γ-line. The absorption of the 14.4 keV γ-ray emitted from the Mössbauer source can be studied in any absorber which has iron as one of the constituents. When an iron foil is used as an absorber, one observes the Zeeman splitting of the ground state and the 14.4 keV excited state as there is an internal magnetic field of about 330 kilogauss in iron metal. The Mössbauer effect is studied by changing the energy of the emitted γ-ray. This is achieved by giving a velocity to the source with respect to the absorber. If the source is moved with velocity v in the direction of emission of the γ-rays, the Doppler energy shift of the photon is

$$\Delta E = E_\gamma \frac{v}{c}$$

Using an electromechanical transducer, the source can be moved either with constant velocity or with constant acceleration. In the first case the change in photon energy is fixed. In the second case the photon energy is varied continuously from E_0 to $E_0 + \Delta E_{max}$ where $\Delta E = E_\gamma v_{max}/c$.

The γ-rays etted from the moving source are allowed to pass through a thin absorber containing 3 to 10 mg of iron per cm^2 area. The γ-rays transmitted through the absorber foil are detected in a suitable detector (usually a Xenon filled proportional counter) and counted in a multiscalar. A multichannel pulse height analyser can usually be converted into a multiscalar. Each channel of the multiscalar counts the number of photons with some Doppler shifted energy passing the absorber. The experimental arrangement is shown in Fig. 3.49. As the time for which the source velocity is between v and $v + dv$ is small, the number of counts collected in a single sweep of velocity, in the corresponding channel is quite small. The source velocity is given a large number of sweeps to collect enough number of counts. The absorption spectrum gives information regarding the energies of the levels in the absorbing nuclei. The Mössbauer absorption spectrum of 14.4 keV γ-ray line emitted from a ^{57}Co source in rhodium matrix in an iron absorber is shown in Fig. 3.50. The six absorption lines are clearly visible. From the separation of the lines, it is possible to obtain the amount of Zeeman splitting of the ground state (1/2 +) and the 14.4 keV excited state(I = 3/2+) ^{57}Fe nuclei. From the measured value of the magnetic moment of the ground state, the magnitude of the internal magnetic field acting on an iron nucleus in an iron foil is calculated as 330 KG.

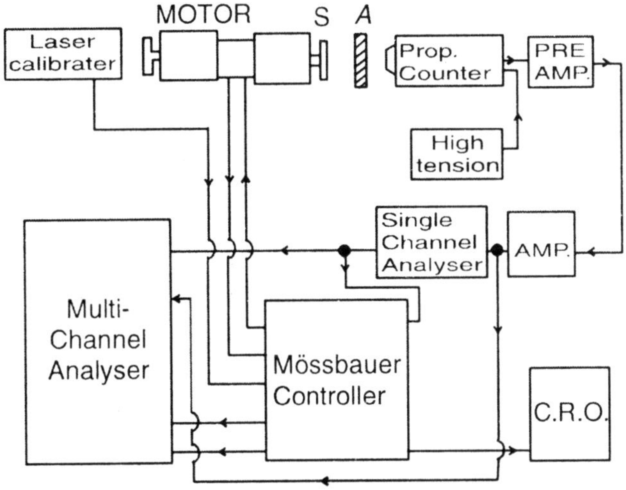

Fig. 3.49: Schematic diagram of the apparatus to study Mössbauer effect in a constant acceleration mode of the source

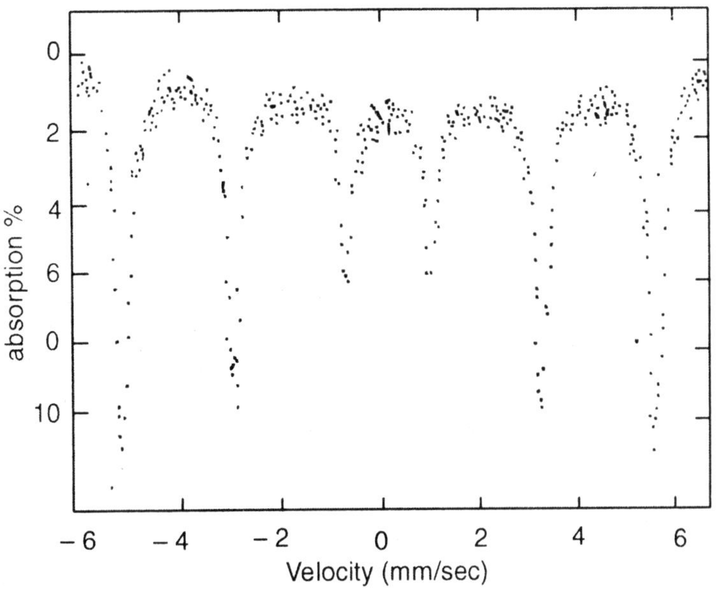

Fig. 3.50: Zeeman splitting of the 14.4 keV excited and ground states of ^{57}Fe gives rise to six line Mössbauer pattern

Using the Mössbauer effect one can measure the splitting of the ground state and the excited state of the nucleus ^{57}Fe in different environments. Usually in a chemical compound there is an electric field gradient acting at the iron nucleus giving rise to the quadrupole splitting of the nuclear states. An example of quadrupole splitting is the Mössbauer effect in the salt sodium ntiroprusside shown in Fig. 3.51.

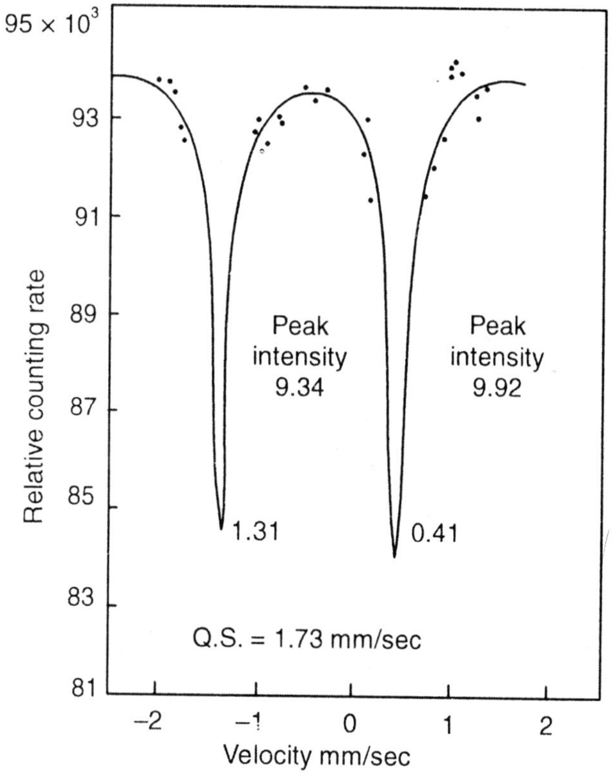

Absorber: Sodium Nitroprusside

Source: Co^{57}/Pd at room temperature

Fig. 3.51: Mössbauer absorption spectrum of sodium nitro prusside. Quadrupole splitting (Q.S.) of the ^{57}Fe states give rise to two line pattern. The molecular fields in sodium nitro prusside ($Na_2Fe(CN)_5NO)2H_2O$ give rise to strong electric quadrupole field

Mössbauer effect is extensively used in the study of hyperfine fields acting on ^{57}Fe and other Mössbauer nuclei (which can be used as source in the study of the effect) in different compounds and environments.

3.16 MEASUREMENT OF LIFE TIME OF NUCLEAR EXCITED STATES

The life time of an excited state is related to the transition probability from the state to lower energy states. Life time measurement help in determining the multipolarity of the transition and its matrix element.

The life times of nuclear excited states vary from 10^{-6} seconds to 10^{-16} seconds. Electronic method can be used in measuring life times from 10^{-7} sec to 10^{-10} sec. For life time in the range of 10^{-11} sec to 10^{-14} seconds it is possible to employ, in some cases, the resonance scattering of the emitted γ-rays. If in a nuclear reaction with a projectile, accelerated in an accelerator, the excited nucleus recoils with large velocity, the energy of the emitted γ-ray changes due to Doppler effect. Measurement of this Doppler shift in γ-ray energy with the flight time of the excited nucleus gives the lifetime of the γ-ray.

The above methods of life time measurement are discussed here briefly.

3.16.1 Electronic Method

When two γ-rays are emitted in cascade as shown in Fig. 3.41, it is possible to measure the life time of the intermediate state. The two γ-rays γ_1 and γ_2 are detected in detectors 1 and 2 respectively. The output of the two detectors is fed to a fast coincidence circuit. The number of output pulses as a function of the delay in the path of detector 2 gives in principle the life time of the intermediate state.

With the advances in electronics, the technique of measurement of life times of nuclear states has also undergone big changes. At present the most preferred method is what is known as slow-fast coincidence technique. A typical slow fast coincidence set up is shown in Fig. 3.52. The scintillator used in most experiments is NaI (Ti), as it permits measurement of the energies of the γ-rays between which time delay is being measured. If measurement of γ-ray energy is not critical, one can use a BF_2 crystal or a plastic scintillator, as detector of γ-rays. In experiments where energy of one of the radiations has to be precisely known, a combination of a semiconductor detector and a NaI (TI) scintillator can be used. The lowest life time that can be measured depends upon the detector used and the energies of the γ-rays.

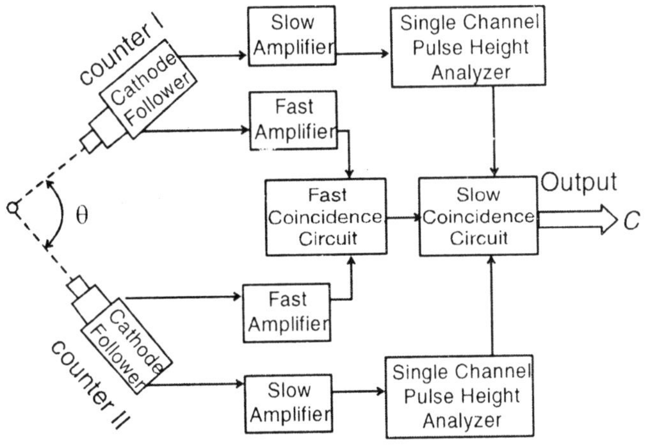

Fig. 3.52: The slow-fast coincidence arrangement. For life time measurements the fast coincidence circuit is replaced by a time to pulse heights converter

The time measuring pulse from the detector should have a very samll rise time and as little a time jitter as possible. In a scintillation detector, a photo multiplier tube with as small a time jitter as possible must be employed. The photomultiplier tube RCA 8575 has a time jitter in the transit time of the electrons of only 0.3 ns. The timing pulse from the photomultiplier tube is usually taken from its anode with a load rasistance of 50 ohms. With such a load resistance, the time for collection of charge at the anode is extremely small. Taking the anode capacitance of 50×10^{-12} farads, the time constant CR is only 2.5 n sec. The rise time of the pulse at the anode is only about 2.5 ns. This pulse can be shaped in a zero cross over discriminater to give a standard pulse with minimum of time jitter. The zero cross over discriminater has a time jitter which is much less than a leading edge discriminater (an ordinary biased univibrator). The output pulse of the discriminater is fed to a time to pulse height converter through a variable delay. The variable delay is a number of 50 ohm cable whose length can be changed for changing the delay. The time to pulse height converter (TPHC) is an electronic circuit which gives an pulse whose height is proportional to the delay between the two pulses at the input of the TPHC. There is a start input which is triggered by the pulse from the detector of γ_1. A condenser is charged by a fixed current going into it. The pulse from detector of γ_2 fed to the 'stop' input stops the charging of the condenser. The voltage developed across the condenser is proportional to the time delay between γ_1 and γ_2 pulses respectively. If the intermediate state has a life time comparable to the resolving time of the circuit, the pulse height spectrum of TPHC shows the exponential decay of the intermediate state.

To ensure that one is looking at coincidences between the pulses due to γ_1 and γ_2 in the two detectors only, an energy selection is made in the slow channels. The peak due to γ_1 is selected in the single channel analyser in one channel. In the other channel the peak due of γ_2 is selected. The pulses from two single channel analysers are fed to a slow coincidence circuit whose output gates the THPC pulses going to the multichannel pulse height analyser. The prompt coincidence curve due to γ-rays from a ^{60}Co source and the delayed coincidence curve due to γ–rays from a ^{181}Hf source are shown in Fig. 3.53. The delay coincidence spectrum gives a half life of 10.8 ns for the 133 KeV state in ^{181}Ta formed by the β^-decay in ^{181}Hf.

If the half life of the intermediate state is less than the resolving time, it is not possible to differentiate the slope of the prompt and delayed coincidence spectrum. In such a case the centroid of the delayed coincidence spectrum is shifted from the centroid of the prompt spectrum. The centroid shift can be calculated from the data and a half life of γ_2 as small as one tenth the resolving time can be measured.

3.16.2 Resonance Scattering using Thermal Doppler Broadening

We have seen in Section 3.7 that a photon emitted from a nucleus losses some of its energy to the recoiling nucleus. Similarly if the photon is to be absorbed by a nucleus, it must provide recoil kinetic energy to the absorbing nucleus. Thus for resonance absorption or scattering of γ-rays there is a short fall of energy E_0^2 / Mc^2. Usually this energy is much larger than the level width Γ and one does not observe resonance scattering of γ-rays. It is also shown that if the source of γ-rays is heated there is a Doppler broadening of the γ-ray line given as (eqn. 3.14.5)

$$\overline{D} = 2(\varepsilon R)^{1/2}$$

If this broadening is equal to $2R$ there is some overlap (Fig. 3.47a) of the absorption and emission lines and resonance scattering can take place.

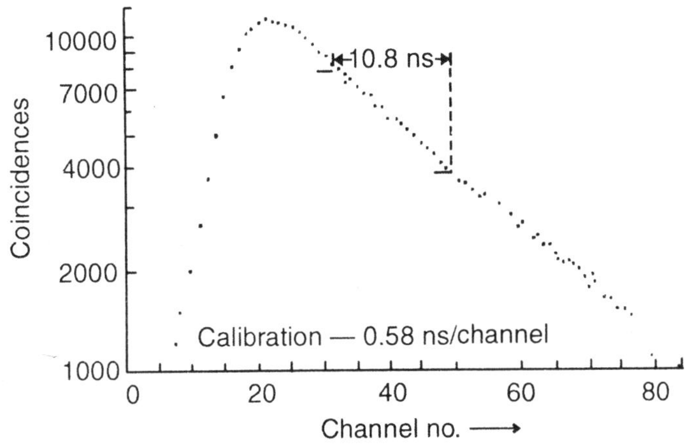

Fig. 3.53: Time delay spectrum of ^{181}Hf-γ-rays giving a half life of 10.8 ns for the 133 KeV state of ^{181}Ta

If v is the relative velocity between the emitter of γ-ray (at temperature T_1) and the absorbing atom of the same nuclide as the emitter at temperature T_2, then assuming a Maxwellian distribution of velocities of both the atoms, the distribution of the relative velocity is given as,

$$P(v)dv = \left\{\frac{M}{2\pi k(T_1 + T_2)}\right\}^{1/2} \exp\cdot\left\{-\frac{Mv^2}{2k(T_1 + T_2)}\right\}$$

The resonance scattering cross section is

$$\sigma_{\text{scatt.}} = \frac{2I_i + 1}{2I_f + 1}\frac{\lambda^2}{8\pi}\Gamma_r^2 \int_{-\infty}^{\infty} \frac{P(v)dv}{\frac{E^2}{c^2}\left(v + u - \frac{E_0}{Mc}\right)^2 + \frac{\Gamma^2}{4}}$$

where u is the fixed relative velocity between the emitting and the absorbing atoms. The integral has a maxima at a velocity $v = E_0/Mc - u$. The function $P(v)$ is a slowly varying function over the resonance region. One can replace $P(v)$ by $P(E_0/Mc - u)$ giving

$$\sigma_{\text{scatt.}} = \frac{2I_i+1}{2I_f+1}\frac{c^2\hbar^2}{4E_0^3}\frac{\Gamma_r^2}{\Gamma}\left(\frac{Mc^2}{2\pi k(T_1+T_2)}\right)^{\frac{1}{2}}\exp\left(-\frac{M\left(\frac{E_0}{Mc}-u\right)}{2k(T_1+T_2)}\right)$$

In the above expression Γ_r/Γ is known from the measurement of internal conversion coefficient of the γ-ray as

$$\frac{\Gamma_r}{\Gamma} = \frac{1}{1+\alpha}$$

The measurement of the scattering cross section when the source is at temperature T_1 and absorber at temperature T_2 gives the level width Γ of the γ-ray.

The experimental arrangement for studying the resonance scattering of γ-rays is shown in Fig. 3.54. A NaI (Tl) scintillation detector is shielded from the source of γ-rays by a conical lead shield. A suitable scatterer of known thickness surrounds the detector such that the γ-rays from the source illuminate the whole scatterer. The scatterer is at room temperature while the source is placed inside the heater coil whose temperature can be increased upto 1200°C. The pulse height spectrum of the γ-rays in the scintillation crystal is recorded in a multichannel analyser. When both the source and the scatterer are at room temperature, the pulse height spectrum from the scintillation counter represents the scattered γ-rays and the general background. The intensity of this pulse height spectrum is not affected by the temperature of the γ-ray source. When the source is heated to sufficiently high temperature the resonance scattering of γ-rays also takes place and full energy peak of the γ-ray are observed superimposed over the background spectrum. The difference of the spectra recorded for the same time when the source is at room temperature and at high temperature gives the contribution due to resonance scattering. From the measured value of the source strengths and the solid angle subtended by the scatterer at the source and the detector, the cross section for resonance scattering can be calculated giving the γ-ray line width Γ.

In the above experiment the number of resonantly scattered photons is much smaller than the number of incoherently scattered photons and for resonable accuracy observations have to be taken over a large period of time.

3.16.3 Doppler Shift Method

The high energy heavy ion beams available from accelerators can be used to induce nuclear reactions in which the excited nucleus formed moves in the direction of the beam with large velocity. The velocity of the excited nucleus can be calculated from the energy of the heavy projectile, the Q value of the reaction and masses of the particles emitted in the reaction. The target used for the reaction is so thin that the recoiling nucleus comes out of the target without much loss of energy and moves in vacuum. If the excited nucleus decays in flight, the energy of the emitted γ-ray is modified by Doppler effect.

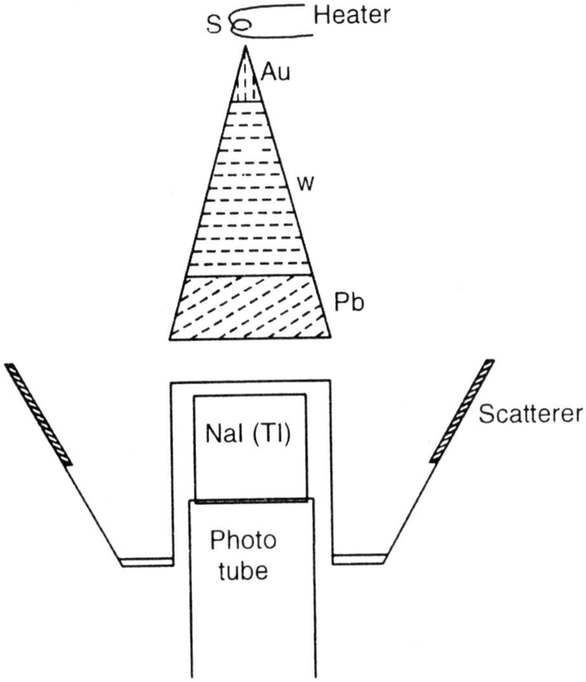

Fig. 3.54: Schematic diagram of the apparatus used for the study of resonance flourescence of γ-rays. The source S is heated with a heater. The cone made of gold, tungsten and lead absorb the γ-rays reaching the scintillation detector directly. γ-rays scattered resonantly from the scatterer are detected in the scintillation spectrometer

If the γ-rays are observed at an angle θ with the beam the change in its energy due to Doppler effect is

$$\Delta E = v/c \cdot E_r \cos\theta$$

where v is the velocity of the recoiled nucleus.

This change in energy is large enough to be measured with a NaI(TI) scintillater or a HPGe detector. An absorber is placed at a distance d from the target. The ion takes about 10^{-14} sec to come to rest in the absorber. The Doppler shift of the γ-rays will be observed if the excited nucleus decays in a time span of d/v seconds where d is the distance traversed before decay. If the recoil velocity is 10^9 cm/sec and life time of the excited state is 10^{-12} sec the distance the ion traverses in vacuum during the life time is $10^{-12} \times 10^9 = 10^{-3}$ cm or 10 microns. It is seen that the absorber has to be placed at different distances varying from 1 μ to 10 μ from the target. The placing of the absorber must be precise. The distance between the target and the absorber is measured by using the capacitance between the two in the tank circuit of an oscillator. The frequency of the oscillator depends upon the distance of the absorber (plunger) from the target.

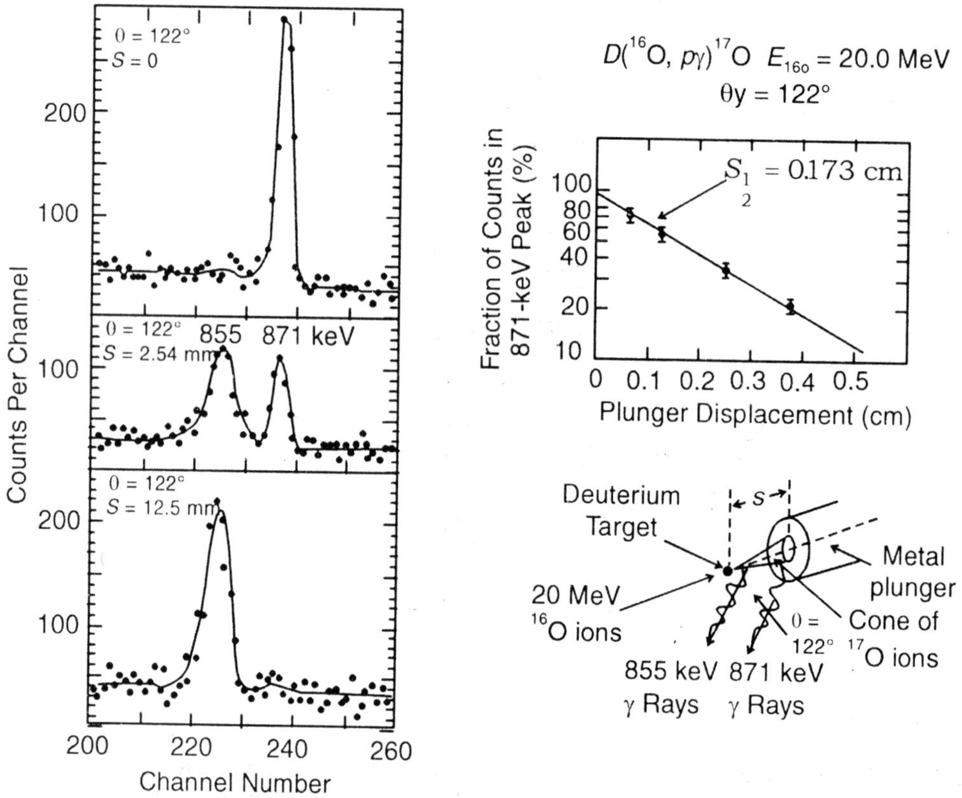

Fig. 3.55: Measurement of the life time of the 871 KeV level of ^{17}O by recoil distance method. The γ-ray spectra show the relative change in the intensity of the 871 and 855 (Doppler shifted) peaks as the distance S between the target and the absorber is increased from zero. In the top spectrum the γ-rays are all emitted after the ^{17}O nucleus formed is stopped in the absorber. In the bottom figure when S = 12.5 mm all the γ-rays are emitted durihg flight. The fractional intensity of the stopped peak is plotted as a function of plunger (absorber) distance at the top right. The bottom right figure shows the arrangement used

The Doppler shiffed γ-ray spectrum as a function of the distance of absorber from the target is shown in Fig. 3.55. The method can be used for measuring life time in the range of 10^{-12} to 10^{-14} seconds.

EXERCISES

1. Calculate the mass of 1 curie of ^{14}C. Assume the half life of ^{14}C is 6000 years. (0.075 gm)

2. Calculate the kinetic energy of α-particles emitted by $^{235}_{92}U$ nucleus. (5.30 MeV)
3. A substance (non-active) is bombarded with certain accelerated charged particles, producing a radioactive substance at the rate of g nuclei per sec. Show that the maximum number of active nuclei that can be produced is g/λ, where λ is the decay constant of the active nuclei.
4. Why is α-decay observed more often than nucleon decay?
5. Differentiate between the Fermi and Gammow-Teller transition and explain how the two correspond to allowed transitions.
6. Explain why low energy β^- particles are more in number than β^+ particles in a β decay process.
7. How do experimental results on α-decay compare with those obtained from Gammow's theory.
8. Enumerate the arguments which support the violation of parity conservation in a β–decay.
9. Calculate the recoil energy of the proton producted in the decay of a free neutron when the β^-–particle is emitted with the maximum kinetic energy. (780 eV)
10. Obtain expressions for recoil energy of the nucleus in γ-decay and in electron capture type of β-decay.
11. The end point energy of β^-–particle emitted by the decay of ^6He is 3.50 MeV. Determine the recoil energy of the produced nucleus. If a β^--particle is emitted with a kinetic energy 1.5 MeV at an angle of 90° to the direction of motion of the recoil nucleus. Calculate the momenta of the β^--particle, neutrino and the recoil nucleus. **Ans.** (1.42 KeV, p_e = 1.94 MeV/C, p_ν = 2 MeV/C, p_r = 0.48 MeV/C).
12. List all the possible multipole transitions from an excited state $\frac{3}{2}^+$ to the ground state $\frac{1}{2}^+$.
13. Explain why it is not possible to have a γ-transition from an excited state 0+ to ground state 0+.
14. How is the multipolarity of a γ-transition in a nucleus related to its multipole moment?
15. What is reduced width of a γ-transition.
16. Why is Weisskopf's estimate of γ-transition probability differ from the measured values.
17. What is internal conversion coefficient of a γ-transition.
18. With increasing multipolarity the transition probability for γ-transition decreases but the internal conversion coefficient increases. Why?
19. What factors must be taken into consideration to calculate the internal conversion coefficient of a γ-ray.
20. Resonance scattering of sodium light from sodium atoms can be easily demonstrated but not so for the resonance scattering of γ-rays. Why?
21. What is the origin of the natural linewidth of a γ-ray. How is it related with its life time.
22. Why there is an angular correlation between the two γ-rays emitted in cascade.
23. What information is obtained from the measurement of angular correlation of cascade γ-rays.
24. Discuss the measurement of the life time of a γ-ray transition which follows another-transition.
25. How Doppler effect is used to measure the life time of transition.
26. An excited state of nucleus with orbital 5/2+ decays to an excited 3/2+ emitting a 140 KeV photon. It also decays to the ground state with emission of 350 KeV γ-ray. What would be the approximate relative intensity of the two γ-transitions.

4

NUCLEAR MODELS

4.1 INTRODUCTION

A nucleus consists of a number of nucleons, which interact with each other due to the strong forces. The nature of the strong interaction is not known exactly. This makes the study of nuclear structure a complex problem. One can not write the Schrodinger equation for a nucleus and solve it. One has to rely on some simple model of the nucleus to understand its properties. The tremendous amount of accumulated experimental data has been used to find the similarities and patterns to develop a variety of models of the nucleus. Niels Bohr who gave the first model of the atom proposed the liquid drop model of the nucleus. This model has been discussed in the first chapter. We shall discuss the shell model and unified collective model of the nucleus. The shell model could explain many properties of a nucleus, such as spin, magnetic moment and excited states. This model fails to explain large quadrupole moments, rotational and vibrational spectra of the nucleus. These properties can be understood as a result of collective motion of many nucleons in the nucleus. To explain the collective properties of a nucleus A. Bohr and Mottelson developed the unified collective model, which has its origin in the liquid drop model.

4.2 SHELL MODEL OF THE NUCLEUS

4.2.1 Fermi Gas Model

In this model the nucleons in a nucleus are assumed to move in a central potential of the square well type with a depth of about 50 MeV. It is assumed that there is no interaction between the individual nucleons. The nucleus is thus assumed to consist of Fermi gas of neutrons and protons obeying independently the Pauli's exclusion principle. The highest occupied state for each type of particle is called the Fermi level.

For a nucleon with a momentum between p and $p + dp$ and confined in a volume V, the volume in the phase space is given by $4\pi p^2 dpV$. The number of nucleon states dn available to each nucleon is given by

$$dn = 4\pi p^2 dpV/h^3 \quad \text{... (4.2.1)}$$

Assuming the Coulomb interaction between the protons to be negligible and integrating the above equation up to the Fermi level momentum ($P_F = (2ME_F)^{1/2}$ where M is the mass of the nucleon) and equating n to the total number of nucleon A, we get,

$$P_F^3 = \frac{3Ah^3}{4\pi V}$$

Each state can accomodate two neutrons and two protons (spin up and spin down) and the nuclear volume V can be written as $V = 4\pi r_o^3 A/3$. Hence,

$$P_F^3 = \frac{9\pi\hbar^3}{8r_0^3}$$

and
$$E_F = \frac{\hbar(9\pi)^{2/3}}{8Mr_0^2} \quad \text{... (4.2.2)}$$

And the Fermi energy E_F comes out to be 33 MeV for a spherical well of radius 1.2 Fm. Adding binding energy of the nucleon (= 8 MeV) gives the depth of the potential well as 41 MeV. Due to the Coulomb repulsive potential the well depth for protons will be less than that for the neutrons. The asymmetry energy observed in the liquid drop model is explained on this model. This model can be used to calculate the level density of a nucleus but not the exact level energies.

4.2.2 Single Particle Shell Model

The following experimental observations are the basis on which the single particle shell model was proposed.

(*a*) Nuclei having certain values (2, 8, 20, 28, 50, 82, 126) of neutron or proton number show unusually large binding energies. These numbers are called magic numbers.

(*b*) There are a large number of isotopes of elements in which the proton number is magic.

(*c*) The binding energy of the nucleon just above a magic number is particularly low.

(*d*) The nuclei with even number of protons and neutrons exhibit zero spin and positive parity in their ground state.

(*e*) The separation energy of a neutron or a proton in a magic nucleus (having neutron or proton number magic) shows a sudden increase.

(*f*) Nuclei having neutron or proton number just below a magic number show excited states, which have a large life times. These nuclei are known as isomers. There are a number of isomers just below magic numbers. These are sometimes known as islands of isomerism.

(g) Magic nuclei show almost zero electric quadrupole moment, implying spherical shape, while nuclei away from magic number show large deformation.

The experimental observations suggest some sort of a shell structure as observed in atoms. In the case of atoms the shell closure produces very stable noble gas atoms. In a similar fashion, when the neutron or the proton number in a nucleus is equal to a magic number, it is particularly more stable than its neighbours. It was proposed that in a nucleus the nucleons form pairs with opposite spins and orbital angular momenta. The last odd unpaired nucleon determines the properties of the nucleus, while the remaining even nucleons only provide a central potential for the last nucleon. The potential has to be such as to provide shell closures at the magic numbers.

The simplest potentials in which the last nucleon can be assumed to be moving are the finite square well potential, the harmonic oscillator potential, and the Wood-Saxon potential. In the first two cases the Schrodinger equation can be solved exactly while numerical methods have to used to obtain the results in the third. The central potentials mentioned above can lead to only spherically symmetric nuclear states. It cannot explain the electric quadrupole moment of nuclei, which arises due to the non-spherical shape.

Because of its simplicity, the harmonic oscillator potential is generally used for shell model calculations. The Schrodinger equation for a nucleon of mass M moving in a spherically symmetric potential $V(r) = (1/2)Mr^2\omega^2$, and having energy eigen value E can be written as,

$$\left[-\frac{\hbar^2}{2M}\nabla^2 + V(r)\right]\psi(r) = E\psi(r) \qquad \ldots(4.2.3)$$

or

$$\left[-\frac{\hbar^2}{2M}\nabla^2 - \left(E + \frac{1}{2}Mr^2\omega^2\right)\right]\psi(r) = 0 \qquad \ldots(4.2.4)$$

The wave function can be decomposed into the radial and spherical parts as

$$\psi(\vec{r}) = R_{lm}(r)\, Y_{lm}(\theta, \phi)$$

$Y_{lm}(\theta, \phi)$ are the well-known spherical harmonics. The equation for the radial part $R(r)$ is solved in the usual manner, giving the energy eigen values as:

$$E = \hbar\omega(N + 3/2) \qquad \ldots(4.2.5)$$

where, $N = 2(n-1) + l$

and $n = 1, 2, 3, \ldots$

and $l = 0, 1, 2, \ldots n$

In equation (4.2.5) N represents the total number of oscillator quanta excited. $N = 0, 1, 2.$ *etc.* correspond to the ground, first and second excited states with energy eigen values starting from $E = (3/2)\hbar\omega$. Each state has its energy separated by an energy $\hbar\omega$ from the previous one. The degeneracy of the state N is $(1/2)(N+1)(N+2)$ and according to the Pauli principle the state can accomodate $(N+1)(N+2)$ neutrons or the same number of protons. Each value of N represents a shell.

For a closed shell nucleus where all the shells up to a given N_0 (maximum number of excited quanta) are filled, the total number of nucleons of a particular type will be.

$$\sum_{N=1}^{N_0}(N+1)(N+2) = (1/3)(N_0+1)(N_0+2)(N_0+3) \quad \ldots(4.2.6)$$

The level order and the spectroscopic notation of the states are given in Table 4.1. Larger values of l within a shell have lower energies.

TABLE 4.1: ORDER OF LEVELS IN AN OSCILLATOR POTENTIAL AND NUMBER OF PARTICLES OF EACH KIND (PROTONS OR NEUTRONS) IN A SHELL.

Energies E_n	*n*	*l*	*nl*	l_j	*Number of particles in a shell* $(N+1)(N+2)$	*Number of particles upto and including the shell* $1/3(N_o+1)(N_o+2)(N_o+3)$
$\frac{3}{2}\hbar\omega$	1	0	1*s*	$s_{1/2}$	2	2
$\frac{6}{2}\hbar\omega$	1	1	1*p*	$p_{3/2}, p_{1/2}$	6	8
$\frac{7}{2}\hbar\omega$	1, 2	2, 0	1*d*, 2*s*	$d_{5/2}, d_{3/2}, s_{1/2}$	12	20
$\frac{9}{2}\hbar\omega$	1, 2	3, 1	1*f*, 2*p*	$f_{7/2}, f_{5/2}, p_{3/2}$	20	40
$\frac{11}{2}\hbar\omega$	1, 2, 3	4, 2, 0	1*g*, 2*d*, 3*s*	$g_{9/2}, g_{7/2}, d_{6/2}$	30	70
				$d_{3/2}, s_{1/2}$		
$\frac{13}{2}\hbar\omega$	1, 2, 3	5, 3, 1	1*h*, 2*f*, 3*p*	$h_{11/2}, h_{9/2}, f_{7/2}$	42	112
				$f_{5/2}, p_{3/2}, p_{1/2}$		
$\frac{15}{2}\hbar\omega$	1, 2, 3, 4	6, 4, 2, 0	1*i*, 2*g*, 3*d*, 4*s*	$i_{13/2}, i_{11/2}, g_{9/2}$	56	168
				$g_{7/2}, d_{5/2}, d_{3/2}, s_{1/2}$		

The fourth and the fifth columns give the group of states corresponding to different values of N_o. In the harmonic oscillator potential these states are degenerate. As the nucleon has spin angular momentum (½) $\hbar$, the fifth column lists the total angular momentum j also where $j = l + 1/2$ or $j = l - 1/2$, with the exception of $l = 0$. In the last two columns the number of particles that can be accomodated in each shell and up-to the shell respectively are indicated. As is evident from the last column, the shell closure takes place at nucleon numbers 2, 8, 20, 70, 112, etc. Only the first three numbers match with the magic numbers. When one uses an infinitely deep square well potential in equation (4.2.2) the degeneracy of the levels is lifted and the shell closure occurs at nucleon numbers 2, 8, 18, 20, 34, 40, etc., which shows agreement with only the first two magic numbers.

To explain the observed magic numbers Mayer and Hansel, Jensen and Suess proposed that there is an interaction between the spin and the orbital angular motion of a nucleon and strength of this interaction increases with the increasing orbital angular momentum. Thus a term representing the spin-orbit interaction added to harmonic oscillator or the square well potential so that the potential becomes $-V(r)\,(1 + a\vec{L} \cdot \vec{S})$ in such a way that the states with $j = l + 1/2$ and $j = l - 1/2$ terms are split. The higher j value state (for a given value of l) lies lower in energy. The decrease in energy of the $l + 1/2$ state is such that at $l = 3$ it is almost equal to the difference between the two shells. The arrangement of the shells, with the spin-orbit interaction introduced, is shown in Table 4.2. It may be observed that the l-s interaction reduces the energy of the $1f\,(l = 3)$ state to such an extent that it forms a separate shell. As l increases further the $l + s$ state gets depressed so much that it falls at the top of the earlier shell. The orbitals in Table 4.2 are listed in the order of increasing energy and each sub shell can accomodate upto $(2j + 1)$ protons or neutrons. The proton and the neutron levels are independent of each other. The last column reproduces all the experimentally observed magic numbers, which is a striking success for the simple model. The level scheme for a nucleon in a harmonic oscillator and square well potential (infinite, finite and with rounded edges) with and whitout inclusion of spin-orbit coupling is shown in Fig. 4.1.

TABLE 4.2: THE ARRANGEMENT OF SHELLS AFTER INCLUDING SPIN-ORBIT INTERACTION IN THE OSCILLATORY POTENTIAL.

Shell Number	*State in a shell (lj)*	*Number of Particles in a shell*	*Total particles upto the shell closure*
1	$1s_{1/2}$	2	2
2	$1p_{3/2}$, $1p_{1/2}$	6	8
3	$1d_{5/2}$, $2s_{1/2}$, $1d_{3/2}$	12	20
3*a*	$1f_{7/2}$	8	28
4	$2p_{3/2}$,$1f_{5/2}$$2p_{1/2}$$1g_{9/2}$	22	50
5	$1g_{7/2}$, $2d_{5/2}$, $2d_{3/2}$, $3s_{1/2}$,$1h_{11/2}$	32	82
6	$1h_{9/2}$, $2f_{7/2}$ $2f_{5/2}$, $3p_{3/2}$, $3p_{1/2}$, $1i_{13/2}$	44	126
7	$2g_{9/2}$, $3d_{5/2}$, $1i_{11/2}$, $2g_{7/2}$, $4s_{1/2}$, $3d_{3/2}$, $1j_{15/2}$	58	184

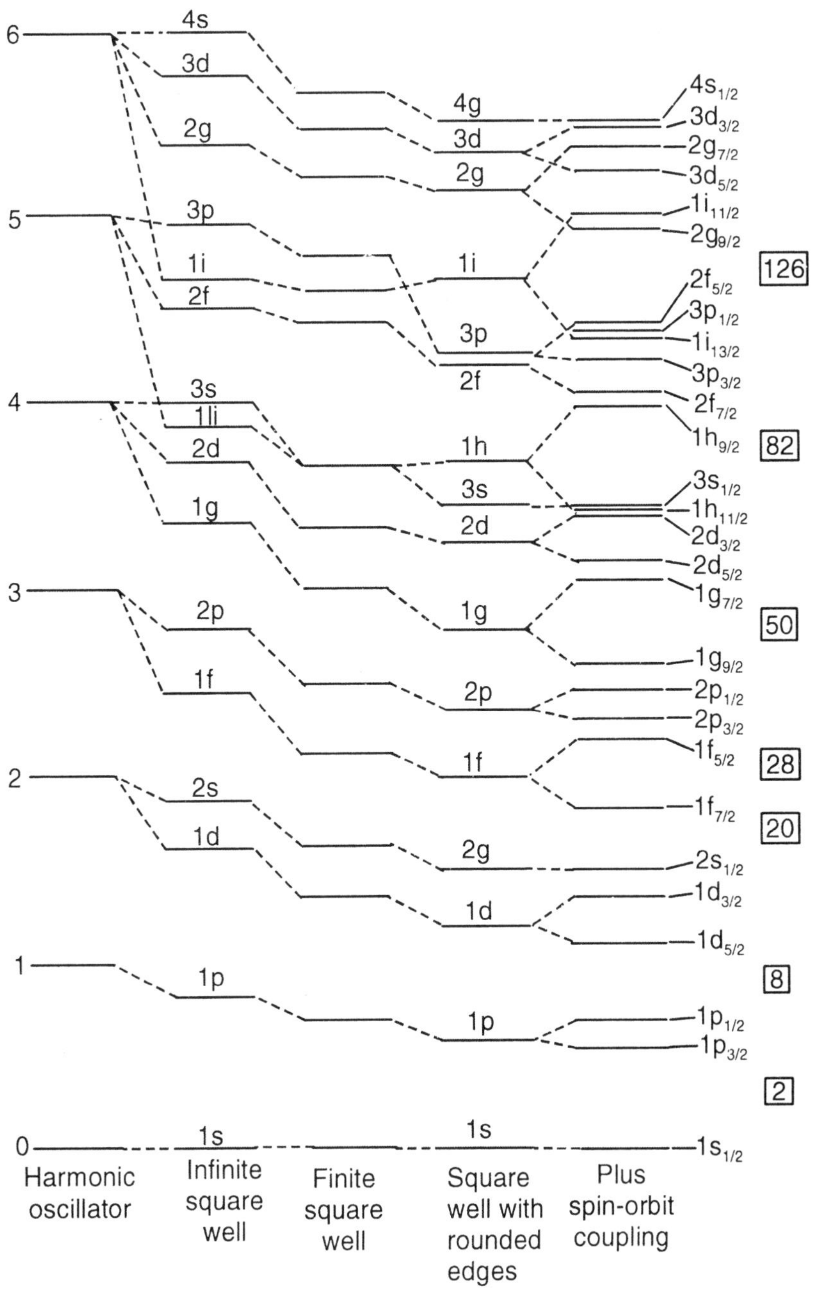

Fig. 4.1: Single particle levels in different potentials

4.2.3 Explanation of Nuclear Data by Shell Model

It will be interesting to see how the above scheme of shell structure which is produced with the introduction of spin orbit interaction to the central potential helps to obtain the large amount of nuclear data. As a first important evidence we can see that it explains correctly the extra stability of nuclei with magic value of N and Z as a result of a large energy gap between levels belonging to different major shells. Also this model predicts a large number of phenomena in odd mass nuclei in their ground states. It is known that the ground states of all even-even nuclei have spin zero e.g. calcium isotopes, with mass numbers equal to 40, 42, 44, 46, 48, 50 and oxygen isotopes with A = 16, 18, 20, 22. It is evident that the addition of each pair of neutrons contributes only zero spin. But if we add a proton and a neutron to an even-even nucleus e.g. ^{18}F or ^{20}F the spin is no longer zero. This leads us to conclude that in a nucleus, pairs of protons or pairs of neutrons tend to align themselves to give zero spin angular momentum. This is the so called pairing property of nucleon-nucleon interaction and has many interesting results for the nuclear structures. This model has been very successful in predicting the ground state spins of odd A nuclei. As the even protons and even neutrons pair off their spins to zero, the odd particle determines the angular momentum of odd A nucleus. Thus all even Z-even N nuclei in their ground state have a spin I = 0. Experimentally also this has been observed to be so. For the odd A nucleus (i.e.) odd Z-even N or even Z-odd N, the ground state angular momentum is determined by the level occupied by the last odd nucleon. However for the odd Z-odd N nuclei the shell model does not directly specify how their odd particle spins will couple to each other.

The parities of the nuclear states can also be predicted by the shell model. The parity defines the space inversion symmetry of the wave function of the nuclear state. The parity of a nucleus as a whole is the product of the parities of the neutrons and protons and is therefore determined by the occupation of the single particle states predicted by shell model and is given by $(-1)^l$, l being the orbital angular momentum quantum number of the last odd particle state. If l is even the parity is considered as even (+) and if l is odd the parity is odd (–).

The ground state spins I and parities (π) of typical nuclei are considered in the following:

The level configurations for the simple odd proton nuclei ${}^7_3Li_4$ and ${}^{15}_7N_8$ in their ground states can be written as:

$$ {}^7_3Li_4 : (IS_{1/2})^2, (1P_{3/2})^1 \text{ and} $$

$$ {}^{15}_7N_8 : (1S_{1/2})^2, (1P_{3/2})^4, (1P_{1/2})^1 $$

After putting pairs of particles in the earlier shells, they give rise to the spins of last odd protons as 3/2 and 1/2 respectively. The parities of both the states are $(-1)^l$ and is odd (–) since l = 1 for the last odd protons in both the cases. Hence the states can be specified as I^π equal to $3/2^-$ and $1/2^-$ respectively. The neutron number being even they contribute only zero spin. The experimentally observed values of the spins of the above nuclei are the same as those predicted by the shell model. The odd neutron nuclei, $^{33}S_{17}$ and $^{29}Si_{15}$, have their neutron level configurations.

$$^{33}_{16}S_{17} : (1S_{1/2})^2 \mid (1P_{3/2})^4 (1P_{1/2})^2 \mid (1d_{5/2})^6 (2S_{1/2})^2 (1d_{3/2})^1$$

$$^{29}_{14}Si_{15} : (1S_{1/2})^2 \mid 1P_{3/2})^4 (1P_{1/2})^2 \mid (1d_{5/2})^6 (2S_{1/2})^1$$

and have spins 3/2 and 1/2 respectively.

We can see for the mirror nuclei, $^{13}C_7$ and $^{13}N_6$, the level configuration for the 7 odd protons in $^{13}N_6$ or 7 odd neutrons in $^{13}C_7$ is the same and is given by

$$(1S_{1/2})^2, (1P_{3/2})^4, (1P_{1/2})^1$$

and both have spin $\frac{1^-}{2}$ as predicted by the shell model and confirmed by the experiment. As another example of the mirror nuclei, consider $^{17}_{8}O_9$ and $^{17}_{9}F_8$ which have the configuration for the odd neutron or odd proton as

$$(1S_{1/2})^2 \mid (1P_{3/2})^4 (1P_{1/2})^2 \mid 1d_{5/2})^1$$

Thus the model predicts $5/2^+$ for both these nuclei and was confirmed by the experiment.

4.2.4 Nordheim's Rules for Odd *Z*-odd *N* Nuclei

For finding the spins of odd *Z*-odd *N* nuclei, Nordheim proposed two empirical coupling rules. They are used to predict the ground state spins, J, of odd Z-odd N nuclei. They are stated as

(1) $I = |J_1 - J_2|$ for $J_1 = l_1 \pm ½$ and $J_2 = l_2 \mp ½$

(2) $|J_1 - J_2| \leq I \leq J_1 + J_2$ for $J_1 = l_1 \pm ½$ and $J_2 = l_2 \pm ½$

l_1, l_2 and J_1, J_2, are orbital and total angular momenta respectively of odd proton and odd neutron. I is the total angular momentum of the odd *Z*-odd *N* nucleus.

Modified Nordheim's Rules

Nordheim's rules were modified by Brennen and Bernstein after comparing with experimental results. The modified rules are

(1) $I = |J_1 - J_2|$ for $J_1 = l_1 \pm ½$ and $J_2 = l_2 \mp ½$

(2) $I = |J_1 \pm J_2|$ for $J_1 = l_1 \pm ½$ and $J_2 = l_2 \pm ½$

(3) $I = J_1 + J_2 - 1$

Using the above rules the angular momentum of the deuteron which contains one proton and one neutron can be calculated to be either $1\hbar$ or 0. Experimentally it was measured to be $1\hbar$.

However, a very simple model like this has a few exceptions. The nuclei $^{47}_{22}Ti_{25}$ and $^{55}_{25}Mn_{30}$ have their level configuration

$$(1s_{1/2})^2 \mid (1p_{3/2})^4 (1p_{1/2})^2 \mid (1d_{5/2})^6 (2s_{1/2})^2 (1d_{3/2})^4 (1f_{7/2})^5$$

The 25th proton or neutron in these nuclei predicts 7/2 instead of the experimentally observed 5/2. Consider the nuclei $^{75}_{33}As_{42}$ and $^{61}_{28}Ni_{33}$, their odd particle configurations are

$$(1s_{1/2})^2 (1p_{3/2})^4 (1p_{1/2})^2 (1d_{5/2})^6 (2s_{1/2})^2 (1d_{3/2})^4 (1f_{7/2})^8 (2p_{3/2})^4 (1f_{5/2})^1$$

These two nuclei with 33 protons and 33 neutrons respectively are expected to have spin $\frac{5}{2}^-$ whereas experimentally they were found to have $\frac{3}{2}^-$ spin. Also exceptions were found for neutron numbers 57, 59, and 61. These exceptions were explained by putting pairs of particles into high spin shell before the low spin shell completes. This is due to the so called 'pairing Energy' which is a negative potential energy connected with double occupancy of a level, and it is larger, the higher the l value of the level. Thus in $^{75}_{33}As$ and $^{61}Ni_{33}$ the pairing energy favours the occupancy of the $f_{5/2}$ level (l = 3) by the proton or neutron pairs leaving $p_{3/2}$ level for the unpaired proton or neutron respectively. Hence the spin for these two nuclei in their ground states is $\frac{3}{2}^-$.

The pairing energy also explains why high spins such as 11/2 and 13/2 are not found in the ground states, It is because the high spin levels are filled before the low spin levels start to fill as a result of pairing energy.

4.2.5 Islands of Isomerism

Islands of isomerism can be explained by the shell model. Nuclear isomeric states are those excited levels whose life times are large. This happens for the nuclei in the end of the shells V and VI where the successive subshells have larger difference in spin. An odd nucleon in an unfilled shell having a low spin makes a transition to a subsequent state of high spin in the same shell which leads to low probability of transition and the consequent long life. Thus in the shells V and VI, where spin changes from $s_{1/2}$ to $h_{11/2}$, $p_{1/2}$ to $i_{13/2}$ take place in the transition, we can find such isomeric transitions.

4.2.6 Magnetic Moments of Nuclei

Magnetic moments (μ) of nuclei can be predicted from the single particle shell model. For the even Z-even N nuclei, as the angular momentum of the nucleus is zero, the magnetic dipole moment is also zero. For an odd number of nucleons the μ is expected to come from the magnetic moments of the unpaired nucleons. The Schmidt model to calculate the magnetic moment of the odd A nuclie discussed in chapter 1 is in conformity with the shell model. The deviation from the Schmidt model calculation of the magnetic moment could be explained if more than one nucleon is responsible for the same.

4.2.7 Configuration Mixing

The single particle potentials are assumed to be not disturbed by the interaction between the nucleons. However it is possible that a strong inter particle interaction can modify the shell model wave function and the modified wave functions can be produced by the superposition of two or more configurations and this is called configuration mixing. For example, the ground state of $^{18}_{7}\mathrm{F}$ is found to show an admixture of $^{1}d_{5/2}$, $^{2}s_{1/2}$ and $^{1}d_{3/2}$ states.

4.2.8 Electric Quadrupole Moments

Electric Quadrupole moment is the measure of the deviation of the nuclear charge distribution from the spherical symmetry. The electric quadrupole moment of a nuclues is defined as

$$Q(\Psi) = \frac{1}{e}\sum_i \int \Psi^* q_i (3z_i^2 - r_i^2)\Psi dr \qquad ...(4.2.7)$$

The quadrupole moment (Q), depends on the state Ψ of the nucleus and for a nuclear state with spin I, can be written as

$$Q = -\frac{2I-1}{2I+2} < r^2 > \qquad ...(4.2.8)$$

where, $<r^2>$ is the mean square radius of the nucleus. Figure 4.2 shows the measured quadrupole moments Q for odd proton nuclei vs the proton number Z and for odd neutron nuclei vs the neutron number N. The values are generally much larger than the values predicted by the shell model. The arrows indicate closed shell nuclei as predicted by the shell model. Shell model also predicts a change of sign for Q at the closed shells. If the quadrupole moments are due to the odd protons, then the quadrupole moments of odd N nuclei should be significantly smaller than those of odd proton nuclei, which in fact are not very different whereas the quadrupole moments of odd Z nuclei with only a few nucleons outside the closed shell are very different and are of the order of $-R^2$. This can be understood only if many protons are in non spherical symmetric states. This makes the whole nucleus to appear as non-spherical nucleus. The liquid drop model of the nucleus cannot be applied to explain the large quadrupole moments and the non spherical shape as a result of the cooperative effects of many nucleons, because the most stable state of the liquid drop is the spherical shape. Rainwater in 1950 showed that using the single particle shell model in combination with the liquid drop model can predict that nuclei are in general deformed and that the nucleons move in a deformed non-spherical potential. This idea was later developed into the so called collective model by Aage Bohr and Ben Mottelson.

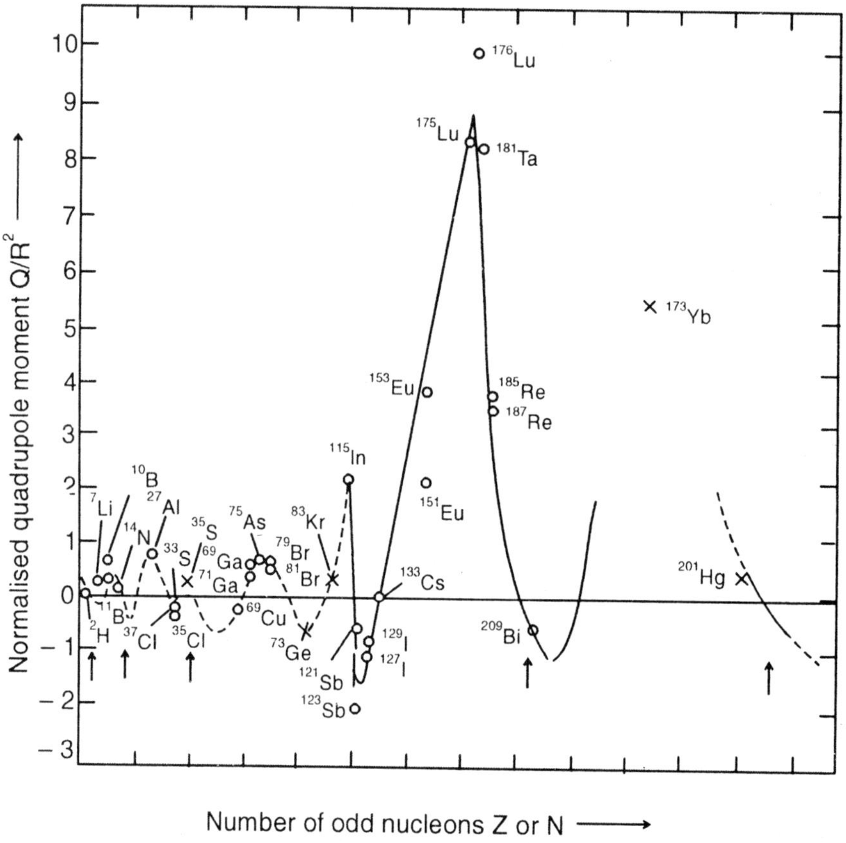

Fig. 4.2: Experimenrtal values of quadrupole moments for odd proton nuclei plotted versus *Z* and *N*. Circles are for proton numbers and crosses are for neutron numbers.

4.3 COLLECTIVE MODEL OF THE NUCLEUS

4.3.1 Rotational Motion of the Nucleus

The extreme single particle shell model was improved by considering several particles outside the magic core and the consequent configuration mixing produced the independent particle model. The independent particle model has improved the agreement between the predicted Schmidt values of the magnetic moments and the observed magnetic moments. Yet in many cases the agreement could not be obtained. As was mentioned, this independent particle model was not able to explain the enhanced quadrupole moments e.g. of rare earth nuclei, for which a stronger cooperative effect is required. It was realised that this effect must involve the nucleons not only of unfilled shells but also of the closed shells that have been disregarded, except for their role in providing the central potential. This idea combined with the liquid drop model of the nucleus essentially is the theory of nuclear deformations. A deformed potential well is assumed for a deformed nucleus and the shell model calculations are done based on this potential. This picture leads to a nuclear model resembling that of a molecule, in which the nuclear core possesses vibrational and rotational degrees of freedom. For the rotational motion there seemed no reason to expect the classical rigid body value, however the large number of nucleons participating in the deformation suggested that the rotational frequency would be small compared with those associated with the motion of the individual particles. It was realized that a non spherical equilibrium shape would arise as a direct consequence of single particle motion in anisotropic orbits when the deformability of the nucleus as a whole is assumed as in the liquid drop model.

In this picture it is easy to understand the pattern of the low energy excitation spectra in terms of competition between the pairing effect and the tendency toward deformation implied by the anisotropy of the single particle orbits. The outcome of this competition depends on the number of particles in unfilled shells. For few particles, the deformation in the absence of interaction is relatively small and can easily be dominated by the tendency to form spherical pairs. But with increasing number of particles the spherical equilibrium shape becomes less stable and eventually a transition takes place to a deformed equilibrium shape. This effect can be seen in the potential energy surfaces shown in Fig. 4.3. In such a coupled system of motion of a single particle in a deformable core, the rotational motion emerges as a low frequency component of the vibrational degrees of freedom for a sufficiently strong coupling. The rotational motion resembles a wave travelling across the nuclear surface and the moment of inertia is much smaller than for rigid rotation.

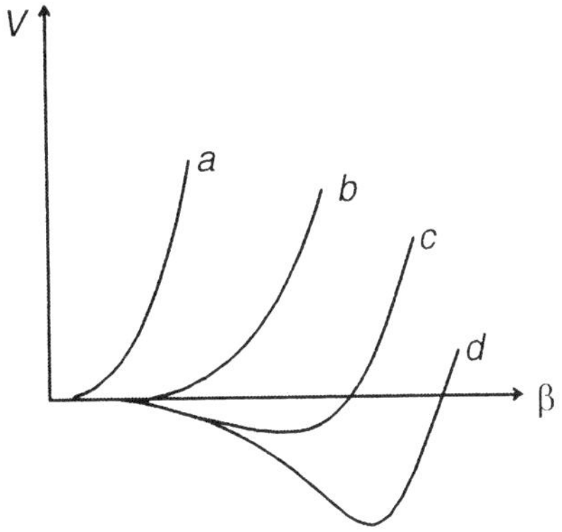

Fig. 4.3: Nuclear potential energy surfaces for an even-even nucleus vs deformation parameter β

Figure 4.4 shows the coupling scheme for a particle in a slowly rotating spheroidal nucleus. The intrinsic quantum number Ω represents the projection of the particle angular momentum (J) along the nuclear symmetry axis Z' and K the projection of I along Z' (and $K = \Omega$) while R is the collective angular momentum of the nuclear core and is directed perpendicular to the symmetry axis, since the component along Z', which is a constant of motion, vanishes in the nuclear ground state. The total angular momentum is denoted by I and M denotes the projection of I along Z axis.

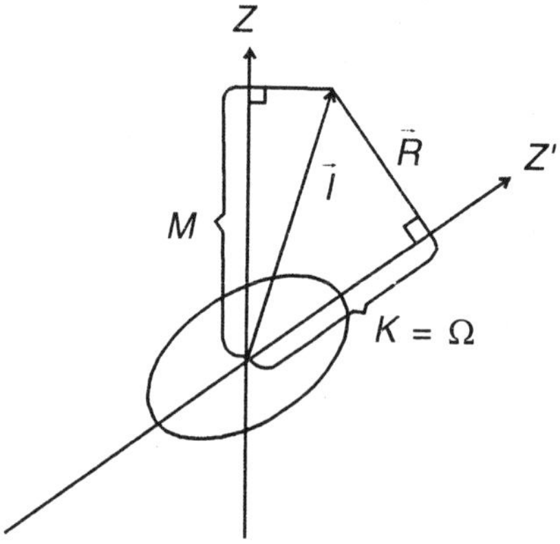

Fig. 4.4: The coupling scheme for a particle in a rotating spheroidal nucleus. K is the projection of nuclear spin I on nuclear symmetry axis Z'. R is the collective angular momentum of the core of the nucleus perpendicular to Z' axis.

Figure 4.5 shows the velocity fields for rotational motion. For the rotation generated by the irrotational flow, the velocity is proportional to the nuclear deformation (amplitude of the travelling wave). Thus for a spheroidal shape the moment of inertia $\mathcal{J} = \mathcal{J}_{\text{rigid}} (\Delta R/R)^2$ where $\mathcal{J}_{\text{rigid}}$ is the moment of inertia for the rigid rotation, while R is the mean radius and ΔR is the difference between major and minor semiaxes.

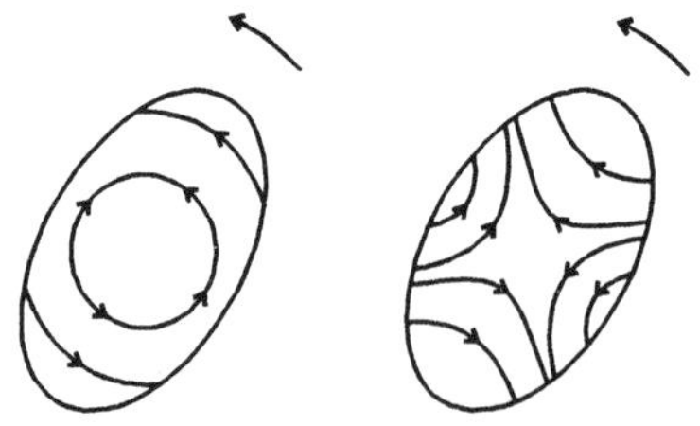

(*a*) Rigid rotation (*b*) Irrotational Flow

Fig. 4.5: The velocity fields for rotational motion (*a*) Shows rigid rotation (*b*) Shows the irrotational flow

The collective rotational energy can be written as

$$E_R = (1/2)\mathcal{J}\omega^2 \qquad ...(4.3.1)$$

where, $\mathcal{J}$ is moment of inertia and ω is the angular velocity about the axis of rotation which is perpendicular to the symmetry axis. If the rotational angular momentum $\mathcal{I}\omega = |R|$ is used in the above equation we find (for $\Omega = K\hbar$) the component of intrinsic angular momentum along the symmetry axis) for $K > 1/2$.

$$E_R = (\hbar^2/2\mathcal{I})\,[I(I+1) + j(j+1) - 2k] \qquad \text{...(4.3.2)}$$

Using the eigen values for I^2 and j^2 for the rotational state the rotational energy can be written as

$$E_{rot} = \frac{\hbar^2}{2\mathcal{I}}[I(I+1) - I_g(I_g+1)] \qquad \text{...(4.3.3)}$$

where I and I_g are the nuclear spins of the rotational state and the ground state.

Intrinsic quantum number K determines the rotational band of levels. For even Z-even N nuclei in their ground state K value is zero and has even parity Because of the symmetry considerations only even I values occur and the spin parity values for the rotational band are written as

$$K = 0 \text{ and } I = 0^+, 2^+, 4^+, 6^+$$

where the total angular momentum arises solely from the collective rotation and energies proportional to $I(I+1)$. For non zero values of K, I takes all the values K, $K+1$, $K+2$...

Very well defined rotational bands are found in the even Z-even N nuclei with N and Z values lying within the ranges shown in the Fig. 4.6. The figure shows the region of nuclei in which the rotational band structure has been identified. The vertical and horizontal lines indicate neutron and proton numbers that form the closed shells. The strongly deformed nuclei are seen to occur in the regions where there are many particles in the unfilled shells that can contribute to the deformation.

Figure 4.7 shows the rotational band structure of $^{238}_{92}U$ for $K = 0$. The levels of the rotational band may be excited by different processes and notable among them is the Coulomb excitation. The decay is mainly through the emission of electric quadrupole radiation and the ratios of the successive excited states to the first excited state energies are

$$\frac{E_4^+}{E_2^+} = 10/3, \quad \frac{E_6^+}{E_2^+} = 7, \quad \frac{E_8^+}{E_2^+} = 12\ldots$$

which is independent of the moment of inertia and the rotational frequency.

The nuclei with large nuclear deformations are characterised by small values of excitation energies E_2 and are found in the light elements with $A \simeq 8$ and 24 and in the heavier elements with $150 < A < 190$ and $A > 222$. These regions include the nuclei for which the number of nucleons in the unfilled shells are especially large compared to those in the filled shells. In the mass region $40 < A < 150$ the conditions for the occurrence of large deformations are less favourable due to spin-orbit coupling.

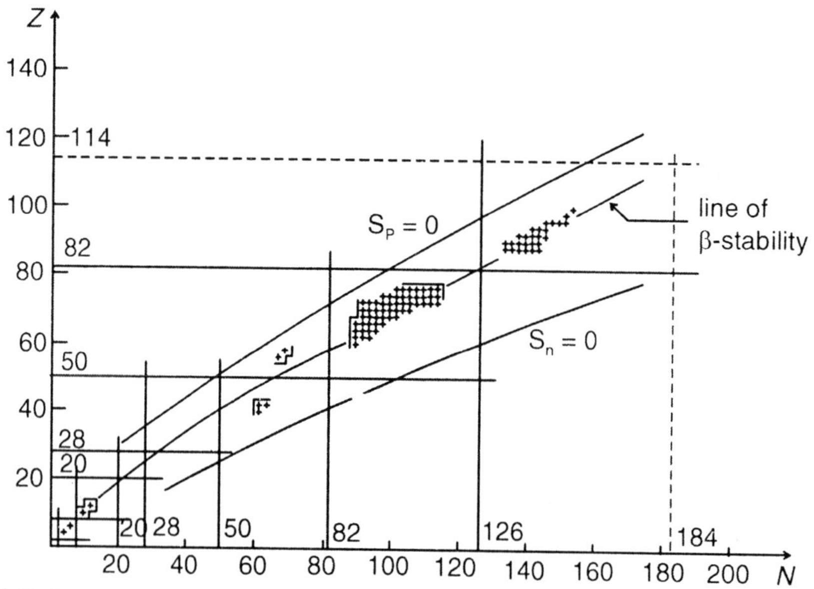

Fig. 4.6: N-Z diagram showing Regions of deformed nuclei. The Plus signs represent even-even nuclei. The curves labeled $S_n = 0$ and $S_p = 0$ are stability limits for neutron and proton emission. (Nuclear structure vol-2., Aege Bohr and B.M. Mottelson)

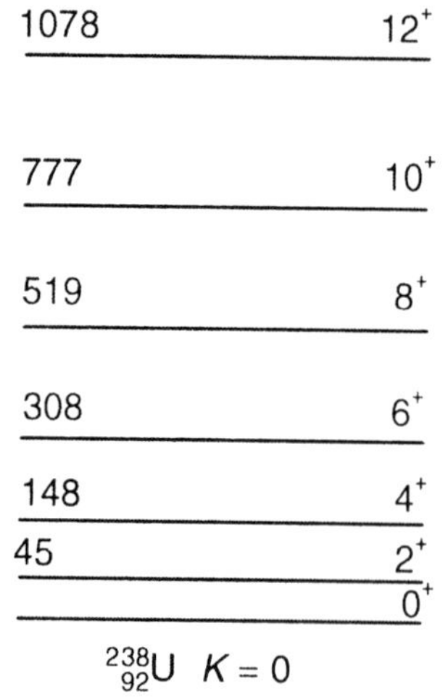

Fig. 4.7: Rotational bands for $K = 0$ in $^{238}_{92}U$

If the nucleus is assumed to have a spheroidal shape, the deformation parameter β can be written as

$$\beta = 1.06(\Delta R/R_0)$$

where R_0 is mean nuclear radius and ΔR is the difference between the semi-major and semi-minor axis of the spheroid. The values of β can be obtained from the E_2 transition probabilities which determine the quadrupole moment of the nuclear shape.

For a uniformly charged nucleus of spheroidal shape the quadrupole moment of the intrinsic motion, Q_0, can be expressed in terms of the deformation parameter β as

$$Q_0 = \frac{3}{(5\pi)^{\frac{1}{2}}} ZR_0^2\beta(1 + 0.16\beta + \ldots) \qquad \ldots(4.3.4)$$

where Z is the nuclear charge. For strongly deformed nuclei the quadrupole moments are an order of magnitude larger than those associated with a single proton. In some cases the transition probabilities are found to be larger by more than a factor of 100.

For odd A nuclei a measurement of $M1$ transition probability in the decay of first excited state can be used to evaluate their magnetic dipole moments. The magnitude of these transition probabilities between rotation states can be related to the gyromagnetic ratios g_k and g_R for the intrinsic and collective motion respectively.

The static magnetic moment μ of a state in the rotational band can be expressed in terms of g_k and g_R. Hence for $K = 1/2$, it can be written as

$$\mu = \frac{K^2}{I+1}(g_K - g_R) + Ig_R \qquad \ldots(4.3.5)$$

g_R gives information on the rotational motion and for a flow of uniformly charged nuclear matter it can be written as

$$g_R = Z/A$$

The moment of inertia of a rigid rotation can be written as

$$\mathcal{J}_{\text{rigid}} = \frac{2}{5} AMR_0^2(1 + 0.31\beta) \qquad \ldots(4.3.6)$$

where AM is the mass of spheroid.

The moment of inertia of spheroidal nuclei are very much smaller than $\mathcal{J}_{\text{rigid}}$ and are found to increase with increasing β. This rotational motion can be understood by a simple model of a wave travelling on the surface of a liquid drop. Assuming irrotational flow, this model yields the moment of inertia for nucleus of spheroid shape as

$$\mathcal{J}_{\text{rigid}} = \frac{2}{5} AMR_0^2\beta^2 \qquad \ldots(4.3.7)$$

The observed moments are appreciably larger than $\mathcal{J}_{\text{rot}}$.

For independent particle motion, the effective moment of inertia would be approximately that corresponding to the rigid rotation, but the residual interactions between the nucleons reduce the moment, which then exhibits a dependence on β. The value of β decreases as one reaches the closed shell configuration and eventually the nuclear deformation collapses and the equilibrium shape becomes spherical. The nucleus then no longer possesses a rotational spectrum and the collective excitations correspond to vibrations about the spherical equilibrium.

4.3.2 Vibration of Spherical Nuclei

The closed shell nuclei with a few nucleons outside the closed shell have spherical shape. A simple harmonic vibration of the surface about the equilibrium shape can be the simplest collective motion of such spherical nuclei. The corresponding energies give rise to levels in the excitation spectrum of the nucleus which have a greater spacing than the levels of a rotational band. The surface oscillations of a spherical nucleus may be characterised according to their multipole order. The excitation quanta are called phonons and have total angular momentum λ and parity $(-1)^{\lambda}$. They can be further characterised by their component of angular momentum μ along a space fixed axis. μ can take values from $-\lambda$ to $+\lambda$ and $\mu = 0$ represents an axially symmetric nuclear shape. The vibrational motion is associated with an oscillating electric multiple moment and the vibrational amplitude $\alpha_{\lambda\mu}$ will define the nuclear surface for an ideal nucleus of constant density and sharp surface as

$$R(\theta, \phi) = R_o\left[1 + \sum_{\lambda\mu} \alpha_{\lambda\mu} Y_{\lambda\mu}(\theta, \phi)\right] \quad \ldots(4.3.8)$$

$Y_{\lambda\mu}(\theta, \phi)$ are the spherical harmonics which describe successive modes of surface disturbances. The kinetic energy of oscillation can be expressed as

$$H = \sum_{\lambda\mu}\left(\frac{1}{2} B_{\lambda} |\alpha_{\lambda\mu}|^2 + \frac{1}{2} C_{\lambda} |\alpha_{\lambda\mu}|^2\right) \quad \ldots(4.3.9)$$

with energy quanta

$$\hbar\omega_{\lambda} = \hbar\left(\frac{C_{\lambda}}{B_{\lambda}}\right)^{1/2}$$

The parameters B_{λ} and C_{λ} represent mass transport associated with the vibration and effective surface tension of the liquid drop respectively. The B_{λ} is given by

$$(B_{\lambda})_{irrot} = \frac{1}{\lambda}\frac{3}{4\pi} AMR^2 \quad \ldots(4.3.10)$$

which correspond to surface oscillations of an irrotational and incompressible liquid drop. C_{λ} may be obtained from the surface energy appearing in the semi empirical mass formula.

The lowest frequencies of collective vibration are quadrupole type ($\lambda = 2$) since surface deformation $\lambda = 1$ simply represents a centre of mass displacement. The quadrupole vibrational spectrum for an even-even nucleus is illustrated schematically in Fig. 4.8.

The quadrupole vibrational quanta each have an energy $h\omega_2$ and carry two units of angular momentum, which is shown on the right. The equality of energy spacing and the degeneracy of the different spin values are a consequence of the harmonic oscillator approximation and will be removed by higher order terms in nuclear energy. The experimental observation of 2^+ single phonon vibrational state and the 4^+ two phonon state at double the energy of single phonon state in the even-even nuclei is an outstanding feature of these vibrational nuclei.

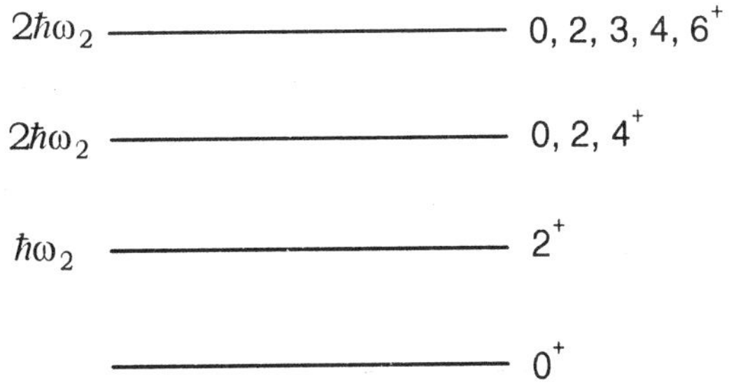

Fig. 4.8: Quadrupole vibrational spectrum for spherical even-even Nuclei

The vibrational excitations are characterised by the enhanced electric transition probabilities. The parameters B_λ and C_λ characterise the vibrational excitation. The static moments of the vibrational excitations vanish to lowest order and hence are very small. The smallness of static quadrupole moments is a chacteristic features of excited states of even-even nuclei in the vibrational spectrum in contrast to the rotational excitations. $M1$ transition are forbidden in the decay of vibrational states even when $\Delta I = 0$ or 1 e.g. transitions between second and first 2^+ excited states.

The magnetic moment associated with the collective vibrational motion is proportional to the angular momentum. The static moment of a vibrational excitation is given by

$$\mu = g_R \cdot I \qquad \text{...(4.3.11)}$$

where g_R the collective g-factor is similar to that associated with rotational motion of the deformed nuclei.

4.3.3 Classification of Vibrations of Spheroidal Nucleus

For stable deformed shapes of the nuclei, the lowest collective excitations correspond to rotation without altering the shape and also the vibration about the equilibrium shape. The symmetry of the vibrations may be characterised by a quantum number λ which represents the number of nodal surfaces and corresponds to the multipole order. The main types of vibrations for a quadrupole deformation, may be described in terms of phonons. The projection ν of the photon angular momentum along the symmetry axis may be specified and the lowest order shape vibrations have $\lambda = 2$ and are of quadrupolar type. A deformation of the order $\lambda = 2$ and $\nu = 1$ is equivalent to a rotation and the only occurring quadrupole vibration will have $\nu = 0$ (β vibration) and $\nu = \pm 2$ (γ vibration). The quadrupole vibrational excitation pattern expected for an even-even nucleus is given in the Fig. 4.9. The quadrupole vibrational energies for the very strongly deformed nuclei will be of the order of a MeV in the heavy nuclei. λ = 3 term corresponds to octupole deformation.

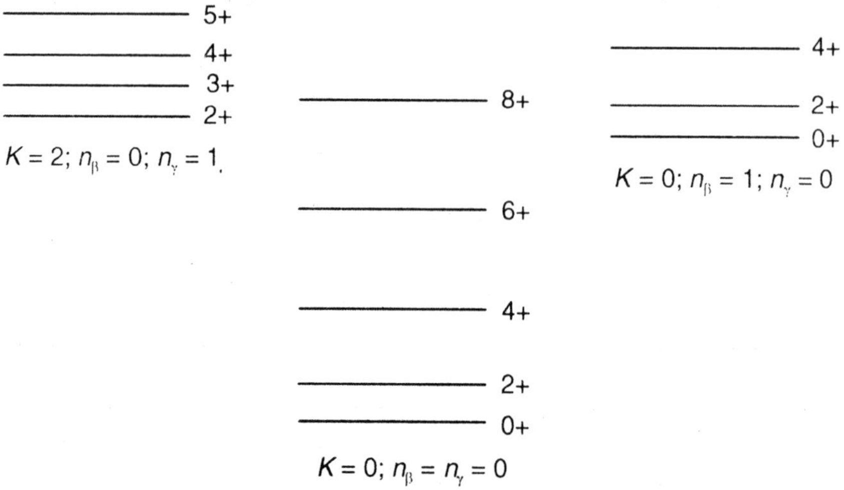

Fig. 4.9: Quadrupole vibrations of an even-even nucleus with spheroidal equillibrium shape. $\nu = 0$ (β vibration) and $\nu = 2$(γ vibration)

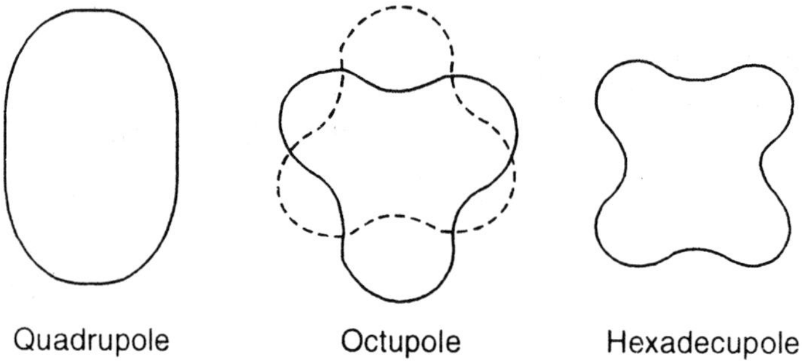

Fig. 4.10: Some typical shapes of the vibrations for λ = 2, 3, 4

Photon numbers n_β and n_γ together with the K-value define the spin of the vibrational state. Rotational bands are built upon each such state. Fig. 4.10 shows some typical shapes of these vibrations for λ = 2, 3, 4.

EXERCISES

1. Show that the average value of $<l^2>$ within one shell is equal to $N(N + 3)/2$.
2. Calculate the magnetic moment for the closed shell ± 1 nucleus $^{15}N(p_{1/2})^{-1}$

   Ans: [–0.28]
3. Calculate the angular momentum and parities of ^{16}O ^{17}O.
4. Write the single particle configuration for the ground state spin and parity given in the brackets of the following nuclei

   ^{3}He, ($½^+$), ^{15}N ($½^-$), ^{103}Rh($½^-$), $^{210}Bi(1^-)$

5. Is the average potential for neutrons and protons the same or different? Why?
6. Obtain the series of magic numbers without the spin-orbit interaction.
7. Calculate the rotational spectrum of ^{178}Hf if the moment of inertia of the band is 15 KeV $\left(=\frac{\hbar}{2\mathcal{I}}\right)$
8. Calculate the rotational frequency for rigid rotation of spherical ^{20}Ne nucleus at angular momentum $I = 8$. **Ans:** $\omega = 5.9 \times 10^{21}\ s^{-1}$

5

NUCLEAR INTERACTIONS
(Nuclear Reactions)

5.1 NUCLEAR REACTIONS

When a nuclear particle comes in close contact with another nuclear particle, there is usually an exchange of energy and momentum. As a result of such nuclear interaction one, two or more nuclear particles are produced. A nuclear reaction can be expressed as

$$A + a = B + b \qquad ...(5.1.1)$$

Here a nuclear particle 'a' moving with high speed, called the projectile hits the target nuclear particle A and produces nuclear particles B and b. Generally A is a target nucleus bombarded by particle 'a' accelerated in an accelerator to produce another nucleus B and an outgoing particle b. In a compact notation the reaction is written as A(a, b)B. Particle 'a' could be neutron, proton, α-particle or even heavy nuclei. Similary particle b could be any particle. In some reactions the outgoing particle b could be more than one. It could even be a photon. If the outgoing particle b is same as incoming particle,. without a change of energy, the reaction represents elastic scattering.

$$A + a \rightarrow A + a$$

If the bombarding particle excites the nucleus A and moves with a reduced energy, it is a case of inelastic scattering.

$$A + a \rightarrow A^* + a$$

where A* represents the nucleus A in an excited state. In general the target nucleus could be in one of its quantum states $A_{\alpha'}$ and the projectile in state $a_{\alpha''}$. The reaction products could also be in different quantum states, β, γ, and so on, so that the reactions could be written as,

$$A_{\alpha'} + a_{\alpha'} \rightarrow A_{\alpha'} + a_{\alpha''}$$

$$\rightarrow A_{\beta'} + a_{\beta''}$$

$$\rightarrow B_{\gamma'} + b_{\gamma''} \qquad \text{etc.} \qquad ...(5.1.2)$$

Here α' and α" are quantum states of the target nucleus and the incident particle respectively while β' and β" and γ' and γ" could be the quantum states of the reaction products. The conservation of energy, angular momentum and parity restrict the possible choices of the states β', β", γ' and γ". Any possible pair of final nucleus and the emitted particle each in some definite quantum state is called a reaction channel. The channel $A_{\alpha'} + a_{\alpha''}$ is called the entrance channel. For the same entrance channel, the reaction channel would be different for different quantum state of the product nucleus.

Usually the projectile moving with high speed, interacts with the target nucleus at rest. After interaction the projectile may change its direction (elastic scattering) and the target nucleus may recoil in some other direction. In the center of mass system both the target nucleus and the projectile approach each other with equal and opposite momentum and after interaction change the direction of their motion, again moving with equal and opposite momentum. If as a result of interaction, nucleus B and outgoing particle b are created, they move with equal and opposite momentum in the center of mass system. In all nuclear reactions, the laws of conservation of energy, momentum, angular momentum, charge and parity are necessarily followed. Except at every high energies—in the domain of particle physics—the number of neutrons and protons individually is also conserved.

5.2 ENERGY CONSIDERATIONS IN NUCLEAR REACTIONS

If the projectile a and the traget nucleus A are far apart, there is no interaction between the two and the total energy in the entrance channel of the reactions is the sum of the kinetic energy of projectile and internal energies of the projectile and the target nucleus. If a projectile hits a nucleus at rest with a kinetic energy T_Q, the energy available for the reaction in the center of mass system is

$$E_a = T_Q \frac{M_A}{M_a + M_A} \quad ...(5.2.1)$$

where M_a and M_A are the masses of the projectile and the target nucleus. E_a is referred as the channel energy of the entrance channel. The total energy in the entrance channel can be written as

$$E = E_a + \varepsilon_a + \varepsilon_A \quad ...(5.2.2)$$

where ε_a and ε_A are the internal energies of the two reacting particles.

If the nuclear reaction results in nucleus B and an outgoing particle b then the total energy in the reaction channel is

$$E = E_b + \varepsilon_b + \varepsilon_B \quad ...(5.2.3)$$

where E_b is the sum of the kinetic energies of the outgoing particle b and the residual nucleus. Since the number of neutrons and protons remains constant during the reaction, it is not necessary to include the contribution of the rest mass of these particles. The internal energies ε_a, ε_A, ε_b and ε_B are equal to the negative of the binding energy of the respective particle. As total energy is conserved, the entrance channel energy E_a and the reaction channel energy E_b are related as

$$E_b = E_a + Q_{ab}$$

where, $$Q_{ab} = \varepsilon_a + \varepsilon_A - \varepsilon_b - \varepsilon_B \qquad ...(5.2.4)$$

Q_{ab} is the "Q-value" of the reaction $A(a, b)B$. The Q value is the difference of the energies in the outgoing and the entrance channel and is characteristic of the channels a and b. Q_{ab} is independent of the channel energies. If in a nuclear reaction Q_{ab} is positive, it is known as exothermic reaction. If Q_{ab} is negative, it is an endothermic reaction. For an endothermic reaction, energy difference is provided by the kinetic energy of the projectile, while an exothermic reaction can take place even if the energy of the projectile is almost zero. If the residual nucleus B is formed in its excited state, the excitation energy is derived from Q_{ab}.

5.3 CROSS SECTION FOR NUCLEAR REACTION (STATISTICAL CONSIDERATIONS)

In the study of nuclear reaction a beam of particles 'a' with kinetic energy T_a bombards a target consisting of atoms of A. The individual nuclei in the target act as independent reacting centers. The cross section for a particular type of nuclear reaction is defined as

$$\sigma = \frac{\text{Number of events of given type per unit time per target nucleus}}{\text{Number of incident particles per unit area per unit time}}$$

The perturbation theory can be used to calculate the probability for a particular type of reaction to take place. If ψ_i and ψ_f are the wave functions of the entrance and the outgoing channels, the probability of transition from state i to state f is given as

$$W = \frac{2\pi}{\hbar}|H_{if}|^2 \frac{dn}{dE} \qquad ...(5.3.1)$$

where $|H_{if}|$ is the matrix element for the transition and dn/dE is the number of states possible for the reaction channel per unit energy. In a finite volume V the energy density of the states is given as

$$\frac{dn}{dE} = \frac{V4\pi P_b^2}{(2\pi\hbar)^3}\frac{dp_b}{dE} \qquad ...(5.3.2)$$

For a non relativistic case the momentum P_b of the particle b is

$$\frac{P_b^{\,2}}{2m_b} = E \qquad \therefore \qquad \frac{2P_b dP_b}{2m_b} = dE$$

so that $$\frac{dP_b}{dE} = \frac{1}{v_b} \qquad ...(5.3.3)$$

In the final state, due to its spin, the degeneracy of the state of particle b is $(2I_b + 1)$ and of the nucleus B is $(2I_B + 1)$, so that

$$\frac{dn}{dE} = \frac{4\pi P_b^2 V}{(2\pi\hbar)^3 v_b}(2I_b + 1)(2I_B + 1) \qquad ...(5.3.4)$$

The probability of transition from state i to state f per unit time is thus,

$$W = \frac{1}{\pi\hbar^4}\frac{P_b^2}{v_b}\cdot V.|H_{if}|^2(2I_B+1)(2I_b+1) \quad ...(5.3.5)$$

If the projectiles 'a' have density of n_a particles per unit volume and move with velocity v_a, then the number of particles hitting the target per unit area per unit time is $n_a v_a$. The probability of reaction taking place per unit time is

$$W = n_a v_a \sigma_r$$

Taking the density n_a to be equal to $1/V$ *i.e.* one particle in volume V we have

$$\sigma_r = \frac{1}{\pi\hbar^4}\frac{P_b^2}{v_a v_b}(2I_B+1)(2I_b+1)|VH_{if}|^2 \quad ...(5.3.6)$$

If U is the interaction energy between the initial state $A + a$ and the final state $B + b$ responsible for the transition then

$$H_{if} = \frac{1}{V}\int \Psi_f^* U \Psi_i \cdot d\tau \quad ...(5.3.7)$$

Here integration is extended over the nuclear volume and Ψ_i and Ψ_f are functions of only internal coordinates.

If the projectile is a positively charged particle as is usually the case, it experiences a Coulomb repulsion in the vicinity of the nucleus. This reduces the wave function by a factor e^{-G_a} where

$$G_a = \frac{1}{\hbar}\int_{R_1}^{R_2}[2m_a(V_a - E_a)]^{1/2}\, dr$$

$$\simeq \frac{\pi Z_A Z_a e^2}{\hbar v_a} \quad ...(5.3.8)$$

where eZ_A and eZ_a are the charges on particles A and a respectively.

This would give a factor e^{-2G_a} in $|H_{if}|^2$. Similarly if the outgoing particle is also positively charged, it has to overcome a Coulomb barrier causing a reduction in Ψ_f^* by a factor e^{-G_b}. Thus when the projectile and the outgoing particles are both positively charged we have

$$H_{if} = \frac{1}{V}\int \Psi_f^* U \Psi_i e^{-G_a} e^{-G_b} d\tau \quad ...(5.3.9)$$

This can be written approximately as

$$H_{if} = \frac{<U> \times \text{(Volume of nucleus)}}{V} e^{-G_a}e^{-G_b} \quad ...(5.3.10)$$

where $<U>$ is the average of the interaction over the nuclear volume.

Putting the value of H_{if} in equation (5.3.6) gives

$$\sigma_r = \frac{1}{\pi\hbar^4}\frac{P_b^2}{v_a v_b}[|VH_{if}|^2(2I_B+1)(2I_b+1)] \quad ...(5.3.11)$$

Equation (5.3.11) permits us to predict the general behaviour of the variation of reaction cross section with projectile energy (excitation function).

(1) Elastic Scattering; both particles uncharged: This is the case of neutron scattering. Neutron being neutral does not experience any Coulomb repulsion so $G_a = G_b = 0$

In elastic scattering $v_a = v_b$ hence

$$\frac{P_b^2}{v_a v_b} = \frac{P_b^2}{v_b^2} = M_n^2$$

where M_n is mass of neutron. At low projectile enrgies H_{if} is approximately constant and the cross section is independent of energy as shown in Fig. 5.1.

At higher bombarding (>10 MeV) energies angular momentum states with $L > 0$ are possible and the cross section may change with energy.

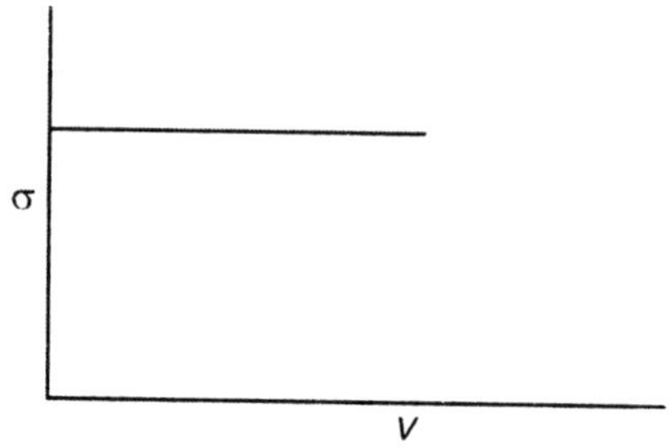

Fig. 5.1: Elastic scattering cross section vs projectile velocity for neutrons

(2) Exothermic reactions in which the projectile is neutral as represented by (n, α), (n, p), (n, γ) and neutron induced fission (n, f) reactions. In such reactions Q is usually positive and about a few MeV. The incoming neutron energy is generally low (< 1 eV) so that v_a is almost constant and for the outgoing particle b, v_b is constant, G_b is also independent of projectile energy. So that the cross section varies as $1/v_a$. This is the famous '$1/v$ law' for neutron induced exothermic reactions (Fig. 5.2).

(3) Exothermic reactions with charged projectiles: In reactions like (p, γ), (p, n), (α, n) and (α, γ) the outgoing particle is uncharged and hence $G_b = 0$. As v_b is almost constant, so that

$$\sigma \propto \frac{1}{v_a} e^{-G_a}$$

The variation of cross section with projectile velocity is shown in Fig. 5.3.

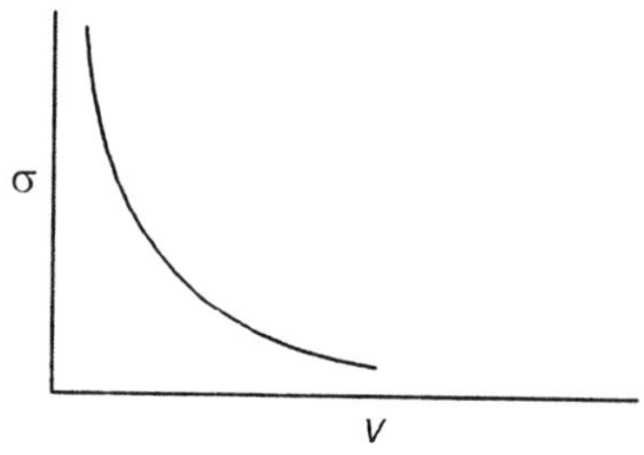

Fig. 5.2: Exothermic reaction cross-section vs velocity of the incident neutron

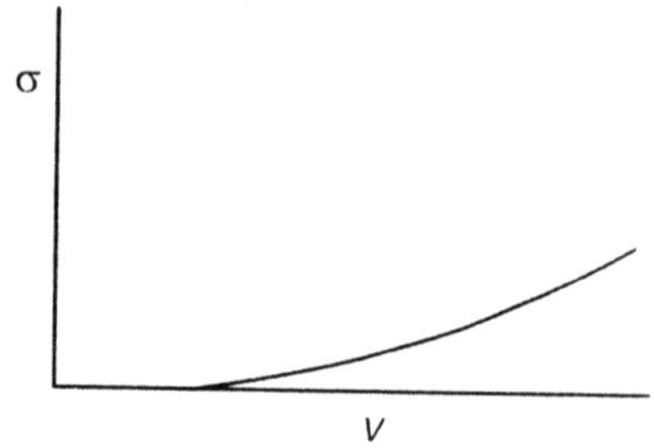

Fig. 5.3: Reaction cross-section vs projectile velocity for charged particles

(4) Inelastic scattering (n, n') reaction: In such a case $G_a = G_b = 0$. As the residual nucleus is left in an excited state the reaction has a negative Q value (Endothermic). The incident projectile energy has to be above the threshold energy of the reaction. With variation of energy near threshold, v_a does not change much and $P_b^2 / v_b \propto$ (excess energy)$^{1/2}$. Near the threshold v_b changes rapidly, hence the cross section increases sharply above the threshold (Fig. 5.4).

(5) Endothermic reaction with outgoing charged particle as in (n, α) and (n, p) reactions. Here the situation is similar to the (n, n') reactions except that the factor e^{-G_b} becomes significant. The low energy outgoing particle is not able to come out due to the Coulomb barrier. The cross section varies as

$$\sigma \propto \sqrt{\text{excess energy}}\,.\,e^{-G_b}$$

The variation of cross section with projectile energy is shown in Fig. 5.5. In the above discussion we have not taken into consideration the effect of the orbital angular momentum of the incoming particle 'a' which may affect the value of the matrix element significantly.

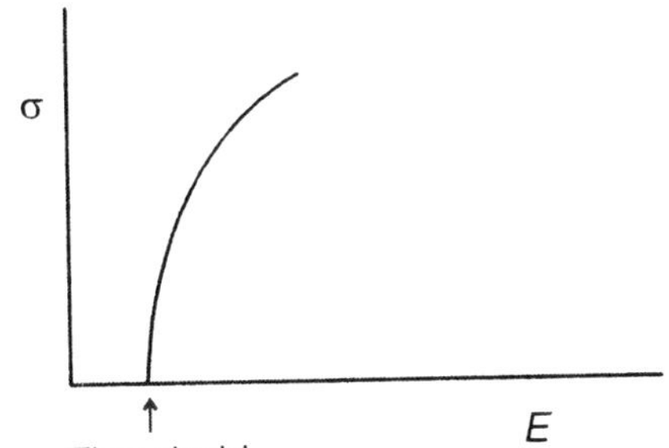

Fig. 5.4: Behaviour of cross-section near threshold for inelastic neutron scattering

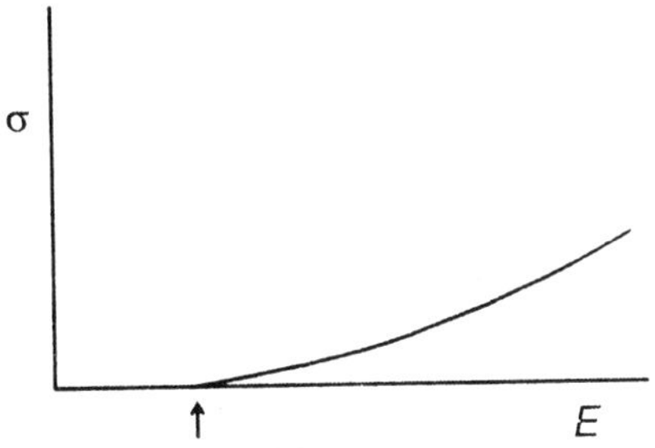

Fig. 5.5: Behaviour of cross-section near threshold for endothermic neutron reaction with charged particle emission

5.4 INVERSE REACTION CROSS SECTION

Equation 5.3.6 permits us to write an important relation between the reaction 5.1 going from left to right and from right to left. Let σ_{AB} represent cross section for reaction, A + a ⇔ B + b going from left to right and σ_{BA} for reaction going from right to left. The matrix element H_{if} is hermitian so that $H_{if} = H_{fi}^*$, so that

$$\langle |H_{if}|^2 \rangle = \langle |H_{fi}|^2 \rangle$$

equation (5.3.6) for forward and inverse reaction can be written as,

$$\frac{\sigma_{AB}}{\sigma_{BA}} = \frac{P_b^2}{P_a^2} \cdot \frac{(2I_B + 1)(2I_b + 1)}{(2I_A + 1)(2I_a + 1)} \qquad \ldots(5.4.1)$$

Both the reactions are considered in the center of mass system. Equation (5.4.1) has been employed to determine the spin of the π meson from the reaction.

$$p + p \rightarrow \Leftrightarrow \leftarrow \pi^+ + d - 137 \text{ MeV}$$

The statistical weight $(2I_A + 1)\,(2I_a + 1)$ is equal to 4 for the left hand side. However Pauli principle excluded indentical states of the two protons so the statistical weight is only 2. On the right hand side the statistical weight factor is $3(2I_\pi + 1)$. So in the centre of mass system.

$$\frac{\sigma_{pp\rightarrow\pi d}}{\sigma_{\pi d\rightarrow pp}} = \frac{3(2I_\pi^2 + 1)}{2}\frac{P_\pi^2}{p_p^2} \qquad \ldots(5.4.2)$$

From the measured values of the cross section of the two reactions at known energies the spin of the π meson has found to be zero ($I_\pi = 0$).

5.5 NUCLEAR REACTION: GEOMETRICAL CONSIDERATIONS

Let us consider a plane wave of the projectile particles 'a' incident upon the nucleus A. The particle 'a' after reaction may be re-emitted in the same channel A + a or it can lead to another channel B + b. We usually separate the elastic scattering events from the nuclear reaction events. The total cresss section σ_t is

$$\sigma_t = \sigma_{sc} + \sigma_r$$

σ_{sc} is elastic scattering cross section and σ_t the reaction cross section. Depending upon the distance at which the particle 'a' passes from the center of the nucleus A, it will have different orbital angular momentum l. If $\vec{p}$ is the momentum of the projectile and $\vec{r}$ the distance of its trajectory from the center of nucleus A, the orbital angular momentum of the particle is $\vec{L} = \vec{r} \times \vec{p}$. The quantum number l gives the projection of vector $\vec{L}$ along some fixed direction z as $l\hbar$. It is seen that projectile passing at different distances from the nucleus will have different orbital angular momenta. However, as angular momentum along the z-axis is quantised, l can have values 0, 1, 2, 3, ... If λ is the de Broglie wavelength of the projectile then

$$|\vec{r} \times \vec{p}| = \frac{r\hbar}{\lambda} = l\hbar$$

$$\therefore \qquad l = \frac{r \cdot 2\pi}{\lambda} = \frac{r}{\bar{\lambda}} \qquad \left(\bar{\lambda} = \frac{\lambda}{2\pi}\right)$$

If $r < \bar{\lambda}$ then we have only orbital angular momentum quantum number $l = 0$. As the distance of the trajectory r increases the value of l also increases. The area of the zone between the lth zone and the $(l + 1)$th zone is $(2l + 1)\pi\bar{\lambda}^2$.

It is easy to see that the cross section for scattering and absorption of the projectile should depend upon the distance at which it passes the scattering center—the nucleus A. Thus the scattering and reaction cross section can be expresed as the sum of cross sections for different values of l, so that

$$\sigma = \sum_{l=0}^{\infty} \sigma_{sc,l} + \sum_{l=0}^{\infty} \sigma_{r,l} \qquad \ldots(5.5.1)$$

The incident beam of particles is represented as a plane wave. If $\vec{r}$ is the distance between the particle 'a' and the target nucleus A, then the plane wave can be written as

$$\psi = e^{i\vec{k}\cdot\vec{r}} = e^{ikz}$$

where z-axis is parallel to the momentum of the particle 'a'.

The momentum vector $\vec{k}$ is related to the particle velocity $\vec{v}$ and the de-Broglie wavelength $\lambda(\lambda = 2\pi \lambda\!\!\!^{-})$ as

$$|\vec{k}| = \left|\frac{M\vec{v}}{\hbar}\right| = \frac{1}{\lambda\!\!\!^{-}}$$

where, $$M = \frac{M_a M_A}{M_a + M_A}$$

the channel mass M is the reduced mass of 'a' and A. The incoming plane wave e^{ikz} can be expressed in terms of the spherical harmonics as

$$e^{ikz} = \frac{\pi\sqrt{2}}{(kr)^{1/2}} \sum_{l=0}^{\infty} i^l (2l+1)^{1/2} J_{l+1/2}(kr) Y_{l,0} \quad \text{...(5.5.2)}$$

where, $$Y_{l,0} = \frac{(2l+1)^{1/2}}{(4\pi)^{1/2}} P_l(\cos\theta)$$

and $\int Y_{l,0} Y_{l',0} d\Omega = \delta_{ll'}$

The Bessel function $j_{1+1/2}(kr)$ have their asymptotic expression as

$$J_{l+1/2}(kr) = \frac{\sqrt{2}}{\sqrt{\pi}} \frac{(kr)^{l+1/2}}{(2l+1)!!} \quad \text{for } kr << 1$$

$$J_{l+1/2}(kr) = \left(\frac{2}{\pi kr}\right)^{1/2} \sin\left(kr - \frac{l\pi}{2}\right) \quad \text{for } kr >> 1$$

$$= \left(\frac{2}{\pi kr}\right)^{1/2} \frac{\left(e^{i\left(kr-\frac{l\pi}{2}\right)} - e^{-i\left(kr-\frac{l\pi}{2}\right)}\right)}{2i}$$

We are interested in the wave function at large distances $kr >> 1$. Substituting value of $j_{1+1/2}(kr)$ for $kr >> 1$, we get

$$e^{ikz} = \frac{\pi^{1/2}}{kr} \sum_{l=0}^{\infty} i^{l+1} (2l+1)^{1/2} Y_{l,0} \left\{ e^{-i\left(kr-\frac{l\pi}{2}\right)} - e^{i\left(kr-\frac{l\pi}{2}\right)} \right\} \quad \text{...(5.5.3)}$$

The above expression represents a spherical outgoing and a spherical incoming wave. The expression holds for projectile and targets both having zero spin.

When the reaction takes place, only the outgoing wave e^{ikr} is modified by a factor η_l. The wave function in the incident channel for $kr >> 1$ is given as

$$\psi(\vec{r}) = \frac{\pi^{1/2}}{kr}\sum_{l=0}^{\infty}(2l+1)^{1/2}\cdot i^{l+1}\left\{\exp\left[-i\left(kr-\frac{l\pi}{2}\right)\right]-\eta_l\exp\left[i\left(kr-\frac{l\pi}{2}\right)\right]\right\}Y_{l,0} \quad \text{...(5.5.4)}$$

Here η_l is a complex number and modifies the outgoing wave of angular momentum l. The scattered wave is the difference between the actual wave interacting with the target and the incident wave (5.5.4) so that

$$\psi_{sc} = \psi(\vec{r}) - e^{ikz}$$

$$= \frac{\pi^{1/2}}{kr}\sum_{l=0}^{\infty}(2l+1)^{1/2}\cdot i^{l+1}(1-\eta_l)\exp\left[i\left(kr-\frac{l\pi}{2}\right)\right]\cdot Y_{l,0} \quad \text{...(5.5.5)}$$

To obtain the scattering cross section we divide the number of scattered particles N_{sc} by the number of incident particles N per unit area per second. To find the number of scattered particles N_{sc} we enclose the scattering centre by a large sphere of radius R_0 and equate N_{sc} to the flux of ψ_{sc} through this sphere so that

$$N_{sc} = \frac{\hbar}{2iM}\int\left(\frac{\partial\psi_{sc}}{\partial r}\psi^*_{sc} - \frac{\partial\psi^*_{sc}}{\partial r}\psi_{sc}\right)_{r=R_0} R_0^2\sin\theta d\theta d\phi \quad \text{...(5.5.6)}$$

As $Y_{l,0}$ are orthogonal so we have,

$$\int Y^*_{l,0}Y_{l,0}\sin\theta d\theta d\phi = 1$$

The equation (5.5.5) gives

$$N_{sc} = \frac{v\pi}{k^2}\sum_{l=0}^{l=\infty}(2l+1)\,|1-\eta_l|^2 \quad \text{...(5.5.7)}$$

where v is the velocity of the incident particles. The value of flux in the plane wave e^{ikz} is v so that the scattering cross section is given as

$$\sigma_{sc,l} = \pi\lambda\!\!\!^{-2}\sum_{l=0}^{\infty}(2l+1)\,|1-\eta_l|^2 \quad \text{...(5.5.8)}$$

The reaction cross section is determined by the number N_r taken out of the beam per second. This is the number of particles which enter a large sphere of radius R_0 but do not come out.

$$N = -\frac{\hbar}{2iM}\int\left(\frac{\partial\psi}{\partial r}\psi^* - \frac{\partial\psi^*}{\partial r}\psi\right)R_0^2\sin\theta d\theta d\phi$$

This gives the reaction cross section as

$$\sigma_{r,l} = \pi\lambda\!\!\!^{-2}\sum_{l=0}^{\infty}(2l+1)(1-|\eta_l|^2) \quad \text{...(5.5.9)}$$

It may be observed that the scattering cross section is largest when $\eta_l = -1$. This implies that $\sigma_{r,\,l} = 0$. On the other hand when $\eta_l = 0$ then reaction cross section is maximum but equal to scattering cross section $\sigma_{sc} = (2l + 1)\pi \lambda\!\!\!^{-2}$. It is thus seen that we can not have reaction cross section without scattering, while we can have scattering without reaction. Further the maximum value of scattering cross section is $4(2l + 1)\pi \lambda\!\!\!^{-2}$, which is four times the maximum value of reaction cross section.

It is further observed that the scattering cross section depends upon $|1 - \eta_l|^2$ which means that the incoming and the outgoing wave amplitudes are added, the incoming and the outgoing waves are coherent and can interfere with each other. In reaction cross section we have term $(1 - |\eta_l^2|)$ which implies incoherence.

The angular distribution of the scattered beam is also determined by η_l. The cross section for scattering at an angle θ with respect to the incident beam in a solid angle $d\Omega$ is $\sigma_{sc}(\theta)d\Omega$. As before we determine the number of scattered particles $N_{sc}(\theta)d\Omega$ by equating it to the flux through a surface element of the big sphere of radius R_0.

$$N_{sc}(\theta)d\Omega = \frac{\hbar}{2iM}\left(\frac{\partial \psi_{sc}}{\partial r}\psi_{sc}^* - \frac{\partial \psi_{sc}^*}{\partial r}\psi_{sc}\right)R_0^2 d\Omega$$

Using Ψ_{sc}, from equation 5.5.5 and dividing by flux $N = v$, we get

$$\sigma_{sc}(\theta)d\Omega = \frac{\pi}{k^2}\left|\sum_{l=0}^{\infty}(2l+1)^{1/2}(1-\eta_l)Y_{l,0}(\theta)\right|^2 d\Omega \qquad \text{...(5.5.10)}$$

Here the contribution from different angular momenta l interfere and it is no longer possible to write $\sigma_{sc}(\theta)$ as sum of contributions for different values of l.

A target nucleus of radius $R >> \lambda\!\!\!^{-}$ may absorb all the particles that may impinge on it. The nucleus can be regared as a "black" nucleus. All particles with $l \leq R/\lambda\!\!\!^{-}$ strike the nucleus and for a black nucleus,

$$\eta_l = 0 \quad \text{for} \quad R > l\lambda\!\!\!^{-}$$

$$\eta_l = 1 \quad \text{for} \quad R < l\lambda\!\!\!^{-}$$

For such a situation equation (5.5.8) and (5.5.9) give

$$\sigma_r = \sigma_{sc} = \sum_{l=0}^{R/\lambda\!\!\!^{-}}(2l+1)\pi\lambda\!\!\!^{-2}$$

$$= \pi(R + \lambda\!\!\!^{-})^2 \simeq \pi R^2 \qquad \text{...(5.5.11)}$$

and $$\sigma_t = \sigma_r + \sigma_{sc} = 2\pi R^2 \qquad \text{...(5.5.12)}$$

This result has been confirmed experimentally.

A projectile wave scattered by a "black" nucleus may be compared with the diffraction of light by a circular disc. Fig. 5.6 shows the angular distribution of elastically scattered neutrons of energy 7 MeV. The data has been fitted for diffraction by a sphere of radius $R = 1.4 \times A^{1/3}$F having a fuzzy edge.

5.6 NUCLEAR REACTION MECHANISM

We shall now discuss the mechanism by which a nuclear reaction takes place. A nucleus is like a charged sphere with a Coulomb potential and a nuclear potential surrounding it. While the Coulomb potential varies as $1/r$, the nuclear potential is confined upto a distance r_0 from the nucleus. If the incident particle is charged it first sees the Coulomb potential. If the particle is able to overcome the Coulomb repulsion (due to its high kinetic energy), it may come close enough to the nucleus to experience the nuclear potential. A neutron being uncharged is not affected by the Coulomb potential.

If the incident particle enters the target nucleus and its wavelength is such that it can form standing waves, the particle can stay inside the nucleus for a long time. If the particle does not form standing waves, it may interact with the nuclear potential and get scattered. The time span for such an event is about 10^{-22} seconds. Sometimes the incident particle may interact with only one of the nucleons in the nucleus and may exchange energy and momentum. Such events give rise to what are known as direct reactions. The outgoing particle is ejected with high energy preferentially in the foreward direction. The time span of direct reactions is also about 10^{-20} to 10^{-22} seconds.

If the incident particle remains inside the target nucleus for long time, it looses its kinetic energy to other nucleons. The particle becomes an integral part of the new nucleus, which is called a compound nucleus. A nuclear reaction can thus be represented as

$$\mathrm{A} + \mathrm{a} \rightarrow \mathrm{C}^* \rightarrow \mathrm{B} + \mathrm{b}$$

In the case of potential scattering and direct reactions C^* is only a compound system which disintegrates in a time span of 10^{-20} seconds. In the case of compound nucleus formation, the time span for the reaction to take place is about 10^{-16} sec. The compound nucleus formed in a highly excited state has no memory as to how it was formed and decays to different reaction channels B + b irrespective of the incident channels. The incident particle may make 10^6 to 10^{10} collisions inside the nucleus and may loose its separate identity. As an example $^{64}Zn^*$ can be formed by the reaction ($^{60}Ni + \alpha$) and also by the reaction (^{63}Cu + p). If the excitation energy of ^{64}Zn nucleus so formed is same, the end product in the two reactions also remains the same.

When a nucleus C^* is formed in an excited state it has a certain probability to decay in any one of the reaction channels (like B + b) permitted by conservation laws. If Γ_i is the probability of transition per unit time to decay into the ith channel, the total probability Γ of the decay of the nucleus C^* is given as,

$$\Gamma = \sum_i \Gamma_i \qquad \text{...(5.6.1)}$$

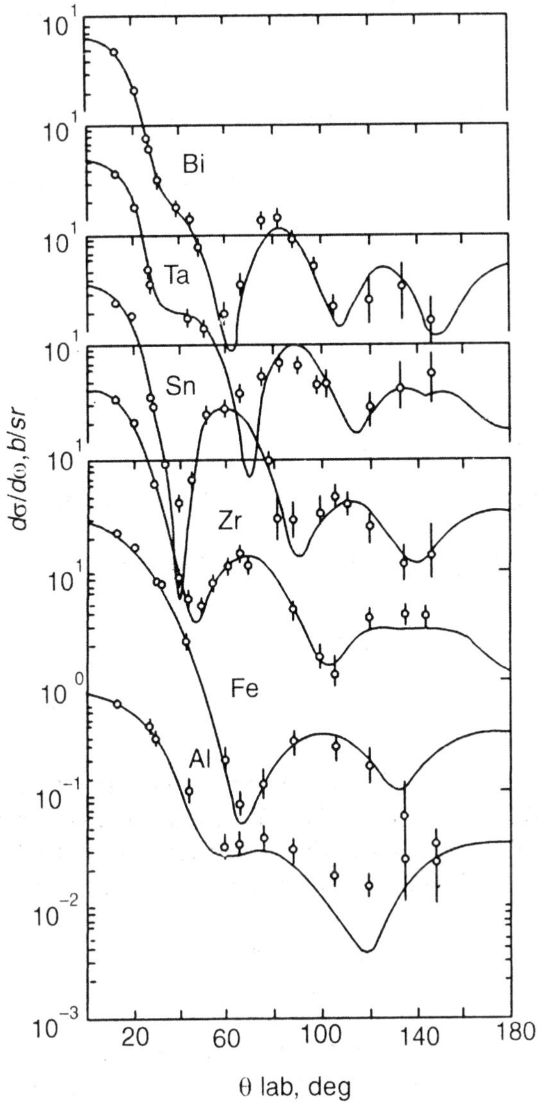

Fig. 5.6: Differential elastic scattering cross-section for neutrons as a function of angle

If σ_r is the cross section for forming the compound nucleus C^* the cross section for the nuclear reactions to lead to the ith channel is

$$\sigma_i = \sigma_r \cdot \frac{\Gamma_i}{\Gamma} \quad \ldots(5.6.2)$$

We know that the transition probability is related to the liferime τ and the energy width ΔE of the state as

$$\Gamma = \hbar/\tau = \Delta E$$

Thus energy state which decays with life time τ has a width of its energy state as ΔE.

In a direct reaction A(a, b)B we have derived the cross section for the reaction as (eqn. 5.3.6)

$$\sigma_r = \frac{1}{\pi\hbar^4}\frac{p_b^2}{v_a v_b}(2I_B+1)(2I_b+1)\,|VH_{if}|^2 \qquad \ldots(5.6.3)$$

When a reaction goes through an intermediate state C the matrix element is replaced by

$$H_{if} = \left|\frac{H_{CA}H_{BC}}{E_i - E_C^*}\right|^2$$

so that, neglecting the spin part, we have

$$\sigma_r = \frac{1}{\pi\hbar^4}\frac{P_b^2}{v_a v_b}\left|\frac{H_{CA}H_{BC}}{E_i - E_C^*}\right| \qquad \ldots(5.6.4)$$

where E_i is the energy in the entrance channel and E_C^* the energy eigen value in the compound nucleus.

We shall for simplicity consider the neutron capture (n, γ) reaction. In the above relation.

$$E_i = E_a + \varepsilon_a + \varepsilon_A$$

$$E_C{}^* = \varepsilon_c + E^*$$

$$E_f = \varepsilon_b + \varepsilon_B + \varepsilon_b$$

Here, as discussed in 5.1, E_a is the kinetic energy of the projectile 'a' in the center of mass system, ε_A, ε_C, ε_b and ε_B are the internal energies of the respective particles in their ground states which are negative of the binding energy. E^* is the excitation energy of the compound nucleus. From the above energy equations one gets

$$E_i - E_C^* = (\varepsilon_a + \varepsilon_A - \varepsilon_c) + (E_a - E^*)$$

$$= \Delta\varepsilon + E_a - E^* \qquad \ldots(5.6.5)$$

The term $(\varepsilon_a + \varepsilon_A - \varepsilon_c)$ is the difference between the binding energies of the incident channel and the compound nucleus and can be denoted as $\Delta\varepsilon$. It is seen that when $\Delta\varepsilon + E_a = E^*$, the denominator in equation 5.6.4 becomes zero and the cross section becomes very large. This is the case of resonance. Once the compound nucleus is formed it can decay back to a neutron and nucleus A (elastic scattering) or it can deexcite to its ground state, emitting γ-rays. If excitation energy E^* is large enough, other particles *e.g.* proton, deuteron or α-particle can also be emitted. We shall consider here the case of γ-ray emission to avoid the effect of Coulomb potential.

The probability ω_{cn} of emission of neutron by compound nucleus C* is given as

$$\omega_{cn} = \frac{2\pi}{\hbar}|H_{cn}|^2\,\rho(E_n) \qquad \ldots(5.6.6)$$

$|H_{cn}|$ is the matrix element for neutron emission and $\rho(E_n)$ the density of states as a function of neutron energy. The matrix element $|H_{cn}|$ is given as

$$H_{cn} = \int \psi_c^* H' \psi_n d\tau$$

where H' is the interaction Hamiltonian. Let us consider the case of slow neutrons for which $\psi_n = e^{ikr} = 1$. As such H_{cn} becomes independent of neutron energy; the only dependence comes from the term $\rho(E_n)$, the density of states.

The probability for emission of gamma rays *i.e.* (n, γ) reaction is given as

$$\omega_{cr} = \frac{2\pi}{\hbar}|H_{cr}|^2 \rho(E_r) \qquad \text{...(5.6.7)}$$

where $\rho(E_r)$ is the energy density of states of the photons. When both the modes of decay of the compound nucleus, emission of gamma rays and elastic scattering are possible, the probability per second for the reaction $A(n, n)$ or $A(n, \gamma)$ is given as

$$\omega_{if} = \frac{2\pi}{\hbar}\frac{|H_{nc}|^2 \{|H_{cn}|^2 \rho(E_n)+|H_{cr}|^2 \rho(E_r)\}}{(E - E^*)^2} \qquad \text{...(5.6.8)}$$

As the excited state C^* decays with time, its wave function ψ_{C^*} should have a time dependent factor so that

$$\psi_{C^*} = \psi_C(r) e^{-\frac{iE^* t}{\hbar}} \cdot e^{-\frac{\Gamma t}{2i\hbar}}$$

$$= \psi_c(r) e^{-i/\hbar\left(E^* - \frac{i\Gamma}{2}\right)t} \qquad \text{...(5.6.9)}$$

This equation implies that the excitation energy E^* is replaced $E^* - i\Gamma/2$. Substituting the value of E^* in equation (5.6.8) gives

$$\omega_{if} = \frac{2\pi}{\hbar}\frac{|H_{nc}|^2 \{|H_{cn}|^2 \rho(E_n)+|H_{cr}|^2 \rho(E_r)\}}{(E_i - E^*)^2 + \frac{\Gamma^2}{4}} \qquad \text{...(5.6.10)}$$

Putting, $\omega_{cn} = \frac{\Gamma_n}{\hbar}$ and $\omega_{cr} = \frac{\Gamma_r}{\hbar}$ we have

$$\Gamma = \Gamma_n + \Gamma_r$$

and

$$\omega_{if} = \frac{1}{\hbar}\frac{|H_{nc}|^2 \Gamma}{(E - E^*)^2 + \Gamma^2/4} \qquad \text{...(5.6.11)}$$

The energy density $\rho(E_n)$ is given as

$$\rho(E_n) = \frac{4\pi V p^2}{h^3} \cdot \frac{dp}{dE}$$

$$\rho(E_n) = \frac{4\pi V}{h^3}\frac{h^2}{\lambda^2}\cdot\frac{1}{\upsilon_n}$$

$$\frac{1}{\upsilon_n} = \frac{h\lambda^2}{4\pi V}\rho(E_n) \quad \text{...(5.6.12)}$$

In elastic scattering the velocity of outgoing neutron is same as that of the incident neutron so that

$$\upsilon_n = \upsilon_{inc}$$

We have

$$\omega_{if} = \eta_n \upsilon_{inc}\sigma_{if}$$

where n_n is the density of neutron in the incident beam thus taken equal to $1/V$. Putting the value of υ_{inc} from equation (5.6.12) gives

$$\sigma_{if} = \frac{\omega_{if}\cdot V}{\upsilon_{inc}}$$

$$= V\cdot\omega_{if}\cdot\frac{h}{4\pi V}\lambda^2\rho(E_n)$$

$$= \frac{\omega_{if}\cdot\hbar\lambda^2\rho(E_n)}{2}$$

Now substituting the value of $\rho(E_n)$ from the equation (5.6.6) gives

$$\sigma_{if} = \omega_{if}\cdot\frac{\hbar}{2}\lambda^2\cdot\frac{\omega_{cn}}{|H_{cn}|^2}\cdot\frac{\hbar}{2\pi}$$

$$= \omega_{if}\cdot\hbar^2\pi\bar{\lambda}^2\cdot\omega_{cn}\frac{1}{|H_{cn}|^2}$$

where $\bar{\lambda} = \lambda/2\pi$

$$= \frac{|H_{nc}|^2\cdot\Gamma\cdot\hbar^2\pi\bar{\lambda}^2}{\hbar(E-E^*)^2+\Gamma^2/4}\cdot\frac{\Gamma_n}{\hbar}\cdot\frac{1}{|H_{cn}|^2}$$

$$= \frac{\pi\bar{\lambda}^2\cdot\Gamma\Gamma_n}{(E-E^*)^2+\frac{\Gamma^2}{4}}\cdot\frac{|H_{nc}|^2}{|H_{cn}|^2} \quad \text{...(5.6.13)}$$

As interaction Hamiltonian for the formation of the compound nucleus and its decay in the same channel is same, we have $|H_{nc}|^2 = |H_{cn}|^2$ so that

$$\sigma_{if} = \pi\bar{\lambda}^2\frac{\Gamma\cdot\Gamma_n}{(E-E^*)^2+\Gamma^2/4} \quad \text{...(5.6.14)}$$

The cross section for elastic scattering is

$$\sigma_n = \sigma_{if}\frac{\Gamma_n}{\Gamma} \quad \text{...(5.6.15)}$$

$$\sigma_n = \pi \lambda^2 \frac{\Gamma_n^2}{(E - E^*)^2 + \Gamma^2/4}$$

and

$$\sigma_r = \sigma_{if} \frac{\Gamma_r}{\Gamma} = \pi \lambda^2 \frac{\Gamma_r \Gamma_n}{(E - E^*)^2 + \dfrac{\Gamma^2}{4}} \qquad \ldots(5.6.16)$$

The above equations represent the Breit-Wigner or Kapoor-Peirls relations for resonance reactions. When the incident neutron energy in the centre of mass is such that the difference in the binding energies plus the kinetic energy of the neutron is equal to the excitation energy of the compound nucleus C, the reaction and the scattering cross sections increase suddenly. The variation of the cross section with neutron energy shows a sharp peak with a width Γ.

Figure 5.7 shows the total cross section for neutron induced reaction (both elastic scattering and γ-ray emission) in cadmium metal with natural isotopic abundance. Cadmium metal consists of two isotopes—^{113}Cd and ^{114}Cd. The isotope ^{114}Cd has no resonance absorption and the cross section follows the $1/v$ law as discussed in Section 5.1. The isotope ^{113}Cd follows the $1/v$ law and also shows a resonance absorption. The energy of neutron at which resonance absorption takes place is 0.176 eV. The peak resonance absorption cross section is 7200 barns (1 barn = 10^{-24} cm^2) and the width of the resonance peak is Γ = 0.115 eV.

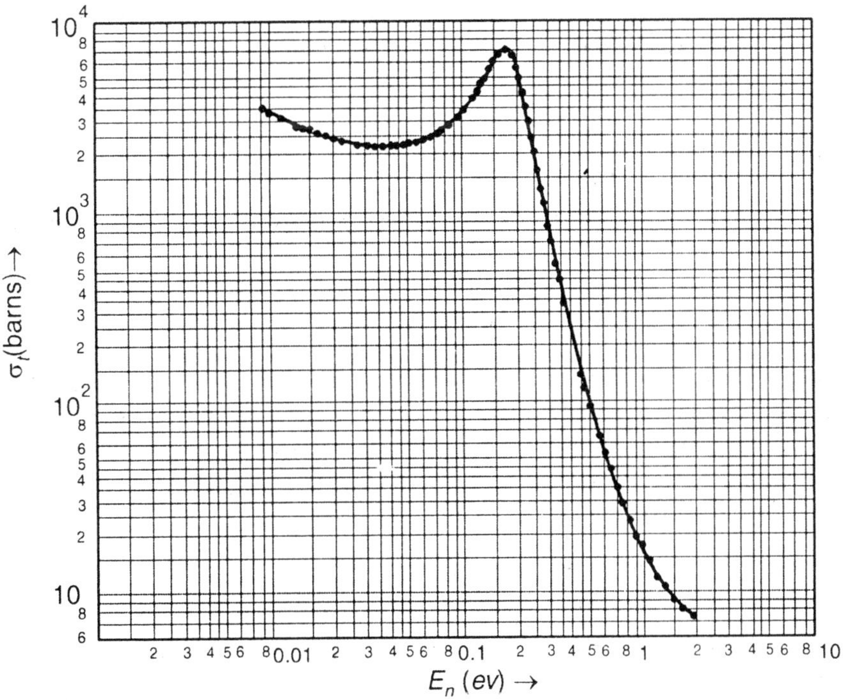

Fig. 5.7: Total cross-section for neutron induced reaction in cadmium metal

5.7 NEUTRON SCATTERING CROSS-SECTION

Nuclear reactions cross section can be calculated if we have complete knowledge about the structure of the nucleus. The cross section given in eqn. 5.5.11 and 5.5.12 show that the reaction and scattering cross sections are determined by the factor η_l, which depends upon the the internal properties of the nucleus. As discussed in Chapter 1 the structure of the nucleus is such that it has a fuzzy surface. However for simplicity it is assumed that a nucleus A is like a rigid ball of radius R_A. It is further assumed that the projectile 'a' interacts with a nucleus only when the distance between the two is less than $R = R_A + R_a$, where R_a is the radius of the projectile. For distance $r > R$ the projectile does not interact with the nucleus. R is known as the channel radius. For neutrons of kinetic energy upto a few MeV, the angular momentum $l = 0$. As such we will restrict our discussion to neutron with $l = 0$. We shall neglect the spin of the projectile and the target nucleus.

For the projectile a at a distance $r > R$ from the target nucleus, the wave function $\psi(r)$ corresponds to the ralative motion of the two particles with no interaction between them, so that fot $r > R$,

$$\nabla^2\psi + k^2\psi = 0 \qquad ...(5.7.1)$$

k is the entrance channel wave number. For large values of r the wave function ψ consists of the superimposition of plane incident wave e^{ikz} as discussed earlier. After interaction with the nucleus at a distance R the modified wave function is as given in eqn. (5.5.4). We are now interested in the behaviour of the wave function at a distance $r = R$ the channel radius. The wave function of the incident particle for $r > R$ can be written as

$$\psi(\vec{r}) = \psi(r, \theta) = \sum_{l=0}^{\infty} \frac{u_l(r)}{r} \cdot Y_{l,0} \qquad ...(5.7.2)$$

We shall restrict the discussion only to $l = 0$ neutrons. For $r \geq R$ the wave function can be written as

$$u_0 = \frac{i\sqrt{\pi}}{k}(e^{-ikr} - \eta_0 e^{ikr}) \qquad ...(5.7.3)$$

The factor η_0 determines the scattering (eqn. 5.5.8) and the reaction (eqn. 5.5.10) cross sections. However the value of η_0 has to be determined from the boundary conditions at the surface of the nucleus.

The behaviour of the wave function u_0 (corresponding to $l = 0$) just outside the nuclear surface ($r = R$) can be described by the quantity.

$$f_0 = R\left.\frac{du_0/dr}{u_0}\right|_{r=R} \qquad ...(5.7.4)$$

If the wave function for the system for $r < R$ can be written as

$$\psi_i = u_i/r \qquad ...(5.7.5)$$

then at the surface $r = R$ the wave function and its derivative should be equal to the wave function and its derivative for $r > R$, so that the value for f_0 at $r = R$ must be the same for u_I and u_0 both.

Now $$u_0 = \frac{\sqrt{\pi}}{ik}(\eta_0 e^{ikr} - e^{-ikr})$$

$$\left.\frac{du_0}{dr}\right|_{r=R} = \sqrt{\pi}(\eta_0 e^{ikr} + e^{-ikr})$$

$$\therefore \quad f_0 = ikr\frac{\eta_0 e^{ikr} + e^{-ikr}}{\eta_0 e^{ikr} - e^{-ikr}} \quad \ldots(5.7.6)$$

or $\eta_0(f_0 - ikR)e^{ikr} = (f + ikR)\, e^{-ikr}$

$$\therefore \quad \eta_0 = \frac{f_0 + ikR}{f_0 - ikR}e^{-2ikR} \quad \ldots(5.7.7)$$

If f_0 is a real number then from equation 5.7.7 we have $|\eta_0|^2 = 1$ hence the reaction cross section vanishes. This is the case of pure scattering without reaction. To obtain the value of the scattering and reaction cross section, we assume that f_0 is a complex number and have

$$f_0 = (a + ib)kR \quad \ldots(5.7.8)$$

so that

$$\eta_o = \frac{a + i(b+1)}{a - i(b-1)}e^{-2ikR} \quad \ldots(5.7.9)$$

and we get

$$|\eta_0|^2 = \eta_0\eta_0^* = \frac{\{a + i(b+1)\}\,\{a - i(b+1)\}}{\{a - i(b-1)\}\,\{a + i(b+1)\}}e^{-2ikR} \cdot e^{2ikR}$$

$$= \frac{a^2 + (b+1)^2}{a^2 + (b-1)^2} \quad \ldots(5.7.10)$$

so that

$$1 - |\eta_o|^2 = 1 - \frac{a^2 + (b+1)^2}{a^2 + (b-1)^2}$$

$$= \frac{-4b}{a^2 + (b-1)^2}$$

This gives

$$\sigma_r = 4\pi\lambda\!\!\!^{-2}\frac{-b}{a^2 + (b-1)^2} \quad \ldots(5.7.11)$$

Also we have

$$|1 - \eta_0| = 1 - \frac{f_0 + ikR}{f_0 - ikR}e^{-2ikR}$$

$$|1-\eta_0| = \frac{f_0 - ikR - (f_0 + ikR)e^{-2ikR}}{f_0 - ikR}$$

$$= e^{-ikR}\frac{f_0(e^{ikR} - e^{-ikR}) - ikR(e^{ikR} + e^{-ikR})}{f_0 - ikR}$$

$$= e^{-ikR}\frac{2if_0 \sin kR - ikR \cdot 2\cos kR}{f_0 - ikR}$$

$$= 2ie^{-ikR}\frac{f_0 \sin kR - kR\cos kR}{f_0 - ikR} \qquad ...(5.7.12)$$

$$|1-\eta_0|^2 = 4\frac{(f_0 \sin kR - kR\cos kR)(f_0^* \sin kR - kR\cos kR)}{(f_0 - ikR)(f_0^* + ikR)}$$

$$= \frac{4f_0 f_0^* \sin^2 kR - f_0 kR\sin kR\cos kR - f_0^* kR\sin kR\cos kR + k^2R^2\cos^2 kR}{f_0 f_0^* - ikR(f_0^* - f_0) + k^2R^2} \qquad ...(5.7.13)$$

Putting $f_0 = kR(a + ib)$

$$|1-\eta_0|^2 = \frac{4k^2R^2\{(a^2+b^2)\sin^2 kR - 2a\sin kR\cos kR + \cos^2 kR\}}{k^2R^2\{a^2 + b^2 - 2b + 1\}} \qquad ...(5.7.14)$$

The denominator in equation (5.7.14) is D when

$$D = a^2 + b^2 - 2b + 1$$
$$= a^2 + (b-1)^2 \qquad ...(5.7.15)$$

The numerator N in equation (5.7.14) is

$$N = (a^2 + b^2 - 2b + 1)\sin^2 kR + (2b-1)\sin^2 kR - \sin^2 kR + \sin^2 kR$$
$$- 2a\sin kR\cos kR + \cos^2 kR$$
$$= (a^2 + (b-1)^2)\sin^2 kR + 2(b-1)\sin^2 kR - 2a\sin(kR)\cos(kR) + 1$$
$$= \{a^2 + (b-1)^2\}\sin^2 kR + 2\sin kR((b-1)\sin(kR) - a\cos kR) + 1$$
$$...(5.7.16)$$

Combining (5.7.15) and (5.7.16) gives

$$|1-\eta_0|^2 = 4\left[\sin^2 kR + 2\sin kR\frac{(b-1)\sin kR - a\cos kR}{a^2 + (b-1)^2} + \frac{1}{a^2 + (b-1)^2}\right] \qquad ...(5.7.17)$$

The scattering cross section is then given as (for $l = 0$)

$$\sigma_{osc} = 4\pi\lambda^2\left[\sin^2 kR + 2\sin kR\frac{(b-1)\sin kR - a\cos kR}{a^2 + (b-1)^2} + \frac{1}{a^2 + (b-1)^2}\right] \qquad ...(5.7.18)$$

In the above expression for elastic scattering cross section the first term represents the potential scattering. The second term is the interference term between potential scattering and resonance scattering. The third term represents the resonance scattering. Figure 5.8 shows the effect of the interference term on neutron scattering. It may be observed from equation 5.7.11 that b can not be positive else the reaction cross section becomes negative so that $b \leq 0$. If $b = 0$ there is no reaction, we shall have a case of only scattering. If wave function $u_R = 0$ then

$$f_0 = \left(\frac{R}{u}\frac{du}{dr}\right)_{r=k} = \infty$$

which means that the term $a^2 + (b-1)^2$ is ∞ and we have

$$\sigma_{sc} = 4\pi ƛ^2 \sin^2 kR \qquad ...(5.7.19)$$

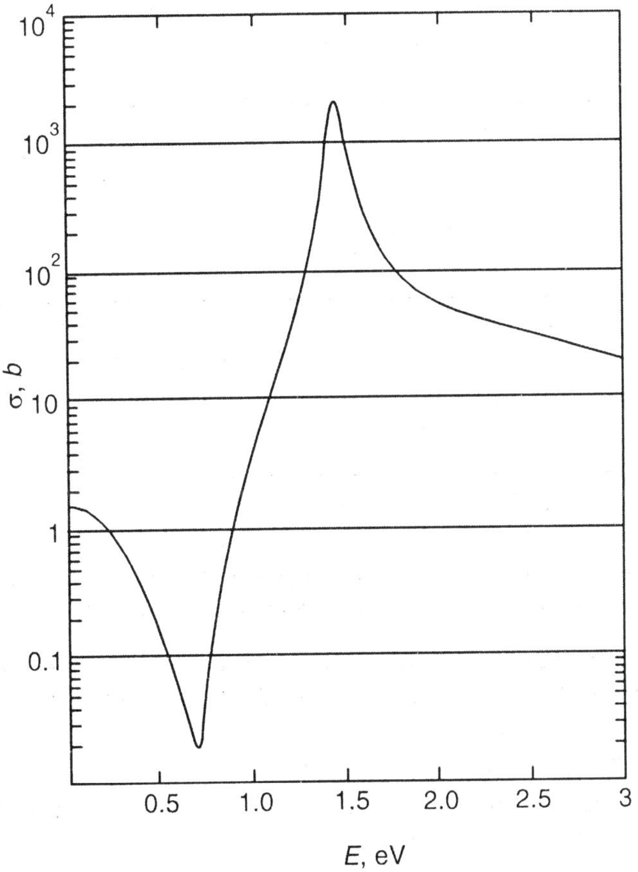

Fig. 5.8: Effect of interference between the negative resonance amplitude and positive potential amplitude on the neutron scattering cross-section as a function of incident energy

As the wave function at the channel radius $r = R$ becomes zero, equation (5.7.19) represents elastic scattering by a hard sphere which the incoming wave can not penctratc.

If $kR \to 0$ both the first term and the second term become negligible and we have a case of pure resonance scattering. At resonance $f = 0$ ($a = 0$, $b = 0$) which requires $\frac{du}{dr} = 0$. The wave function at the channel radius meets the surface horizontally and the amplitude of the wave function inside and outside the nucleus is same (Fig. 5.9). This means that the incoming particle wave enters the nucleus completely forming the compound nucleus. If $|a|$ and $|b|$ have a large value then resonance scattering is small and the last two terms in equation 5.7.17 are small so that

$$\begin{aligned}\sigma_{sc} &= 4\pi \lambda\!\!\!^{-2} \sin^2 kR \\ &= 4\pi \lambda\!\!\!^{-2} (kR)^2 \\ &= 4\pi R^2\end{aligned} \qquad ...(5.7.20)$$

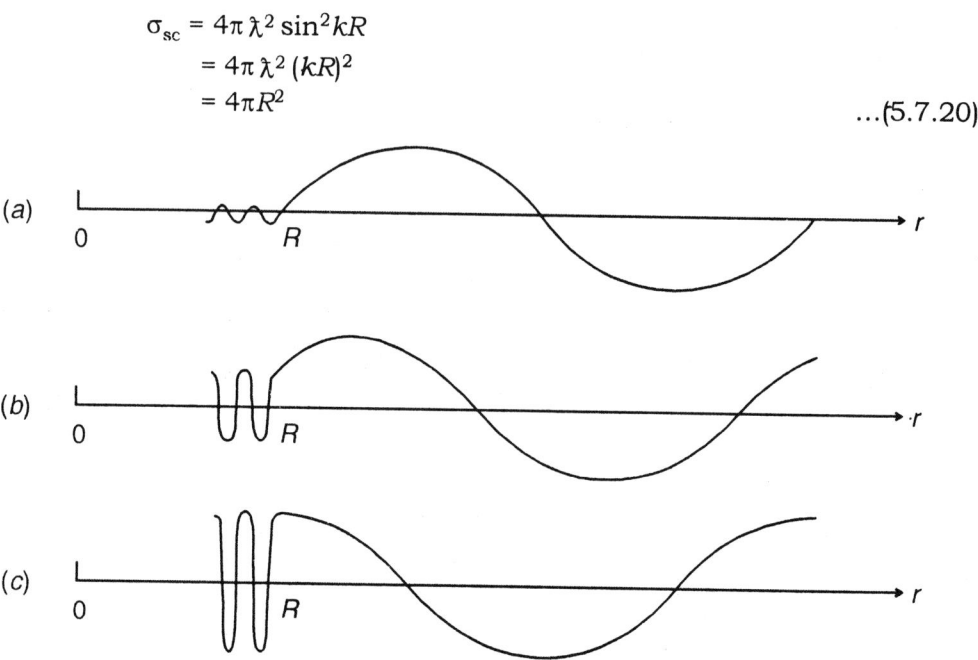

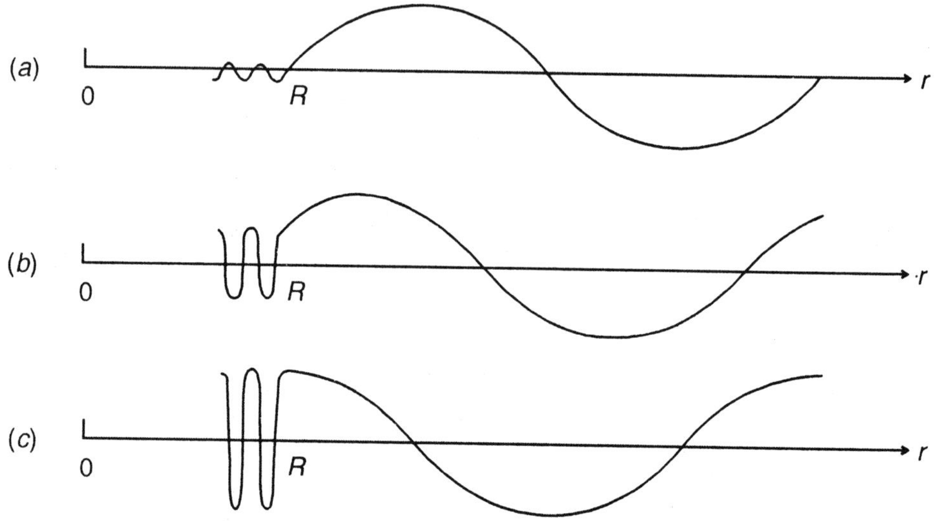

Fig. 5.9: Neutron wave function as a function of distance r from the centre of nucleus (a) neutron energy between resonances (b) near resonances and (c) at the resonance

When the incident particle penetrates the nucleus *i.e.* $r < R$, it experiences an attractive potential $-V_0$. Inside the nucleus the wave vector is

$$K = \frac{\sqrt{2M(E + V_0)}}{\hbar} \qquad ...(5.7.21)$$

The wave function inside the nucleus can be written as superposition of an incoming and an outgoing wave as

$$\psi_i(r) = A\frac{e^{iKR}}{r} + B\frac{e^{-iKR}}{r} \qquad ...(5.7.22)$$

the logarithmic derivative for the inside wave function ψ_i at $r = R$ is equal to the logarithmic derivative for the outside wave function, so that

$$f_0 = iKR\frac{Ae^{iKR} - Be^{-iKR}}{Ae^{iKR} + Be^{-iKR}} \qquad ...(5.7.23)$$

As the value of K is determined by V_a and as $V_0 >> E$, the expression (5.7.23) does not vary with the incident energy. This implies that the factors a and b of expression (5.7.8) also do not vary with incident energy and are proportional to f_0/kR. The reaction cross section σ_r of equation (5.7.11) is thus

$$\sigma_r \propto \frac{\lambda\!\!\!^{-2}}{\lambda\!\!\!^{-}} \propto \lambda\!\!\!^{-} \propto \frac{1}{v}$$

This is the l/v law for reaction cross section at low neutron energies as we have derived earlier also.

If the incoming neutron is absorbed, there is no outgoing wave so that in equation (5.7.22) $A = 0$. In that case

$$f_0 = iKR = kR\,(a + ib) \quad \text{...(5.7.24)}$$

so that $a = 0$ $b = -\dfrac{K}{k}$. Substituting the above values in eqn. (5.7.18) gives

$$\sigma_{sc} = 4\pi\lambda\!\!\!^{-2}\left\{\sin^2 kR + \frac{2\sin^2 kR}{b-1} + \frac{1}{(b-1)^2}\right\}$$

$$= 4\pi\lambda\!\!\!^{-2}\left\{\frac{b+1}{b-1}\sin^2 kR + \frac{1}{(b-1)^2}\right\}$$

so that

$$\sigma_{sc} = 4\pi\lambda\!\!\!^{-2}\left\{\frac{K-k}{K+k}\sin^2 kR + \frac{k^2}{(K+k)^2}\right\} \quad \text{...(5.7.25)}$$

and

$$\sigma_r = 4\pi\lambda\!\!\!^{-2}\frac{Kk}{(K+k)^2} \quad \text{...(5.7.26)}$$

The above relations have been derived assuming no outgoing wave ($A = 0$). However, a nucleus is seldom black. In the case of elastic scattering where there is no reaction $|A| = |B|$ (equation 5.7.14). There can be a phase difference between the incoming and the scattered wave, so we can put

$$A = -Be^{2i\xi} \quad \text{...(5.7.27)}$$

and
$$\psi_i(r) = \frac{-B}{r}e^{i\xi}(e^{iKr+i\xi} - e^{-iKr-i\xi})$$

$$= -2iBe^{i\xi}\frac{\sin(Kr+\xi)}{r} \quad \text{...(5.7.28)}$$

This gives the logarithmic derivative f_0 as

$$f_0 = KR\cot(KR+\xi) \quad \text{...(5.7.29)}$$

$$= kR\,(a + ib) \text{ at the bounding } r = R$$

so that

$$a = \frac{K}{k}\cot(KR + \xi) \text{ and } b = 0 \qquad ...(5.7.30)$$

This gives $\sigma_r = 0$ ...(5.7.31)

and

$$\sigma_{sc} = 4\pi\lambda^2\left[\sin^2 kR - 2\sin kR\left(\frac{\sin kR - \frac{K}{k}\cot(KR+\xi)\cos kR}{K^2/k^2(\cot^2 KR + \xi) + 1}\right) + \frac{1}{\frac{K^2}{k^2}\cot^2(KR+\xi)+1}\right] \qquad ...(5.7.32)$$

If $k \to 0$ $\sin kR \to 0$ and equation 5.7.32 gives

$$\sigma_{sc} = 4\pi R^2\left[1 - \frac{\tan(KR+\xi)}{KR}\right]^2 \qquad ...(5.7.33)$$

The scattering cross section has been derived in equation (5.7.20). However, the cross section derived in the above equation (5.7.31) is a better approximation.

The resonance in the absorption of the $l = 0$ neutron requires $f_0 = 0$ so that from equation (5.7.28)

$$\cot(KR + \xi) = 0 \qquad ...(5.7.34)$$

In that case equation (5.7.32) gives the scattering cross section at resonance as

$$\begin{aligned}\sigma_{sc} &= 4\pi^2 \lambda^2 (1 - \sin^2 kR)\\ &\simeq 4\pi \lambda^2 (1 - k^2R^2)\\ &= 4\pi(\lambda^2 - R^2)\\ &\cong 4\pi\lambda^2\end{aligned} \qquad ...(5.7.35)$$

At energies away from resonance energy, the scattering cross section is given by equation (5.7.20).

At the resonance energy E_R the logarithmic derivative $f_0 = 0$. In the vicinity of the resonance at channel energy E we have

$$f_0(E) = f_0(E_R) + (E - E_R)\frac{\partial f_0}{\partial E}\Big|_{E_R} + ... \qquad ...(5.7.36)$$

Now we define

$$\left.\frac{\partial f_0}{\partial E}\right)_{E_R} = \frac{-2kR}{\Gamma} \qquad ...(5.7.37)$$

so that

$$f_0(E) = \frac{-2kR}{\Gamma}(E - E_R) \qquad \text{...(5.7.38)}$$

which gives

$$a = \frac{-2}{\Gamma}(E - E_R) \quad \text{and } b = 0 \qquad \text{...(5.7.39)}$$

Substituting the value of a in equation (5.7.32) and neglecting terms k^2R^2 and $k^2R^2(E - E_R)$ we get

$$\sigma_{sc} = 4\pi\lambda\!\!\!^{-2}\frac{\Gamma^2/4}{(E - E_R)^2 + \Gamma^2/4} \qquad \text{...(5.7.40)}$$

5.8 RESONANCE SCATTERING WITH REACTION

In the above discussion we have assumed that the outgoing wave is not present ($A = 0$ in equation 5.7.22). This is pure resonance scattering as $b = 0$. In actual practice we have potential scattering, resonance scattering and reaction, so that both a and b of equation (5.7.8) have to be finite. Let us assume that a wave of finite though small amplitude goes out after the neutron interacts inside the nucleus. Thus in equation 5.7.22 we have $B > A > 0$. We can put

$$A = -Be^{2i\xi}\, e^{-2q} \qquad \text{...(5.8.1)}$$

when, $2q > 0$

The wave function for the particle inside the nucleus can be written as

$$\psi_i(r) = -\frac{B}{r}(e^{2i\xi} - e^{-2q} \cdot e^{iKr} - e^{-iKR})$$

$$= -\frac{B}{r}e^{i\xi}e^{-q} \cdot 2i\sin(Kr + \xi + iq) \qquad \text{...(5.8.2)}$$

The logarithmic derivative is obtained as

$$f_0 = KR\cot(KR + \xi + iq) \qquad \text{...(5.8.3)}$$

If $q << 1$ we have

$$f = KR\frac{\cos(KR + \xi + iq)}{\sin(KR + \xi + iq)}$$

$$= KR\frac{\cos(KR + \xi)\cos(iq) - \sin(KR + \xi)\sin(iq)}{\sin(KR + \xi)\cos(iq) + \cos(KR + \xi)\sin(iq)}$$

$$= KR\frac{\cos(KR + \xi) - (iq)\sin(KR + \xi)}{\sin(KR + \xi) + (iq)\cos(KR + \xi)}$$

$$= KR\frac{\cot(KR + \xi) - iq}{1 + iq\cot(KR + \xi)}$$

$$f_0 = KR\frac{(1-q^2)\cot(KR+\xi) - iq\{1+\cot^2(KR+\xi)\}}{1+q^2\cot^2(KR+\xi)} \qquad \ldots(5.8.4)$$

For resonance to occur we again have the condition,

$$\cot(KR+\xi) = 0$$

so that
$$f_0 = -iqKR \qquad \ldots(5.8.5)$$

In the vicinity of the resonance we now have

$$f(E) = (E - E_R)\left(\frac{\partial f}{\partial E}\right)_{E_R, q\to 0} + \left(\frac{\partial f}{dq}\right)_{E_R, q\to 0} \cdot q \qquad \ldots(5.8.6)$$

As before we define

$$\left.\frac{\partial f}{\partial E}\right)_{E_R} = \frac{-2kR}{\Gamma_{sc}}$$

$$\left.\frac{\partial f}{\partial q}\right)_{E_R} = \frac{k}{K}\frac{\Gamma}{\Gamma_{sc}} \qquad \ldots(5.8.7)$$

Substituting in (5.8.7) we get

$$f(E) = -\frac{2kR}{\Gamma_{sc}}(E - E_R) - ikR\frac{\Gamma_r}{\Gamma_{sc}} \qquad \ldots(5.8.8)$$

Equation 5.7.8 gives

$$a = -\frac{2}{\Gamma_r}(E - E_R), \qquad b = -\frac{\Gamma_r}{\Gamma_{sc}} \qquad \ldots(5.8.9)$$

If we put $\Gamma = \Gamma_r + \Gamma_{sc}$ we get the scattering and the reaction cross section as (neglecting k^2R^2 and $kR(E - E_R)$)

$$\sigma_{sc} = 4\pi\lambda\!\!\!^{-2}\left(\frac{\Gamma_{sc}}{\Gamma}\right)^2 \cdot \frac{1}{\left[\frac{2}{\Gamma}(E-E_R)\right]^2 + 1}$$

$$= \pi\lambda\!\!\!^{-2}\frac{\Gamma_{sc}^2}{(E-E_R)^2 + \Gamma^2/4} \qquad \ldots(5.8.10)$$

and
$$\sigma_r = \frac{\Gamma_r}{\Gamma_s} \cdot \sigma_{sc}$$

$$= \pi\lambda\!\!\!^{-2}\frac{\Gamma_r\Gamma_s}{(E-E_R)^2 + \Gamma^2/4} \qquad \ldots(5.8.11)$$

The width of the resonance is $\pm\Gamma/2 = \Gamma$.

The maximum cross section at resonance $E = E_R$ is given as

$$\sigma_{sc} = 4\pi\lambda\!\!\!^{-2}\left(\frac{\Gamma_s}{\Gamma}\right)^2$$

$$\sigma_r = 4\pi\lambda\!\!\!^{-2}\left(\frac{\Gamma_r\Gamma_s}{\Gamma^2}\right) \quad \text{...(5.8.12)}$$

If the spin of the incident neutron, the target nucleus and the compound nucleus are taken into account then the cross section have to be multiplied by the statistical factor

$$\frac{2I_C + 1}{(2I_a + 1)(2I_A + 1)} \quad \text{...(5.8.13)}$$

The derivation of the above resonance reaction and scattering cross section has been on the assumption that there is a single excited state of width Γ giving rise to the resonace.

The strong interaction model of Bohr discussed above assumes a compound nucleus having sharp energy levels with long life times (10^{-16} sec). The cross section for scattering and absorption of a neutron in the neighbourhood of a compound state is given by the Breit-Wigner one level formula. In many nuclei at low and medium incident energies gaint resonances have been observed. These resonances have half width which may be as large as 1 MeV and vary with the atomic mass of the target nucleus. The dependence of cross section with mass number and bombarding energies can not be explained on the basis of compound nucleus formation.

5.9 OPTICAL MODEL OF NUCLEAR REACTIONS

The explanation of the existence of the gaint resonance is usually based on the optical model. The optical model explains the behaviour of the reaction cross section at both low energies and high energies. At low energies the model deals with the energy average of the cross section. If the incident neutron energy is not sharply defined, a number of resonances may be covered in the energy spread. One can then average the cross section and look for its energy and mass number dependence.

In optical model the potential generated by a nucleus is taken as complex. The imaginary part of the potential describes the inelastic processes. Both real and imaginary parts of the potential are energy dependent. At low energies the real part of the potential is similar to the shell model potential. The imaginary part of the potential affects the absorption of the incident particle.

The simplest form of optical potential is the square well potential.

$$V(r) = -V_0 - i\omega_0 \quad \text{for } r < r_0$$
$$= 0 \text{ for } r > r_0 \quad \text{...(5.9.1)}$$

The smoothed out square well potential given by Woods and Saxon explains the experimental results to a better extent. The Woods Saxon potential is

$$V = -(V_0 + i\omega_0)\left\{1 + \exp\left(\frac{r-R}{r_0}\right)\right\}^{-1}, \quad ...(5.9.2)$$

where V_0, ω_0, R and r_0 are the optical model parameters. Other workers have used more sofisticated forms of potential to take into account the deformed nuclei.

In optical model the cross sections are averaged over energy range Δ which is larger than the separation D between the energy levels of the compound nucleus, the actual cross section may fluctuate considerably with energy due to resonances. However, the average cross section may vary smoothly with the incident energy.

The average of a function $f(E)$ is defined as

$$\langle f(E)\rangle = \int_{-\infty}^{+\infty} \rho(E - E')f(E')dE' \quad ...(5.9.3)$$

where ρ is the weight factor. One can use the following square function for $\rho(E - E')$

$$\begin{aligned} \rho(x) &= 0 && \text{if} \quad x < -\Delta/2 \\ &= \Delta^{-1} && \text{if} \quad -\Delta/2 < x < \Delta/2 \\ &0 && \text{if} \quad x > \Delta/2 \end{aligned}$$

In that case

$$\langle f(E)\rangle = \frac{1}{\Delta}\int_{E-\Delta/2}^{E+\Delta/2} f(E')dE' \quad ...(5.9.4)$$

The energy spread Δ is taken large in comparison to D, the level spacing, so that the averaging is over a large number of resonances. If σ_T is total cross section σ_{el} and σ_r the elastic and reaction cross sections respectively then

$$\sigma_T = \sigma_{el} + \sigma_r$$

We have seen that the solution of the Schrodinger equation gives the wave function $\psi(r)$ in equation (5.5.4), where η is the factor modifying the outgoing wave. The elastic scattering and the reaction cross sections are given as in equation (5.5.8) and (5.5.9) on summation for different l values as

$$\sigma_{el} = \pi\lambda^2\sum_l (2l+1)\,|1 - \eta_l|^2 = \sum_l \sigma_{el}^l$$

$$\sigma_r = \pi\lambda^2\sum_l (2l+1)[1 - |\eta_l|^2] = \sum_l \sigma_r^l$$

$$\sigma_T = \pi\lambda^2\sum_l (2l+1)[2R_l(1-\eta_l)] = \Sigma\sigma_T^l \quad ...(5.9.5)$$

If we take the average of each term in the above equation then

$$\langle\sigma_{el}^l\rangle = \pi\lambda^2(2l+1)\left[1 - \langle\eta_l\rangle - \langle\eta_l^*\rangle + \langle|\eta_l|^2\rangle\right]$$

$$\langle\sigma_{el}^{l}\rangle = \pi\lambda^2(2l+1)\left[|1-\langle\eta_l\rangle|^2 + \left(\langle|\eta_l|^2\rangle - |\langle\eta_l\rangle|^2\right)\right]$$

and

$$\langle\sigma_{r}^{l}\rangle = \pi\lambda^2(2l+1)\left[1-|\langle\eta_l\rangle|^2 - \left(\langle|\eta_l|^2\rangle - |\langle\eta_l\rangle|^2\right)\right]$$

$$\langle\sigma_{T}^{l}\rangle = \pi\lambda^2(2l+1)\left[|1-\langle\eta_l\rangle|^2 + 1 - |\langle\eta_l\rangle|^2\right] \qquad ...(5.9.6)$$

The shape elastic or potential scattering σ_{SE}^{l} for angular momentum l, which takes place without the formation of a compound nucleus is defined as

$$\sigma_{SE}^{l} = \pi\lambda^2(2l+1)\,|1-\langle\eta_l\rangle|^2 \qquad ...(5.9.7)$$

The compound elastic scattering cross section is

$$\sigma_{CE}^{l} = \pi\lambda^2(2l+1)\left[\langle|\eta_l|^2\rangle - |\langle\eta_l\rangle|^2\right] \qquad ...(5.9.8)$$

The cross section for the formation of a compound nucleus is defined as

$$\sigma_{c}^{l} = \pi\lambda^2(2l+1)\left[1-|\langle\eta_l\rangle|^2\right] \qquad ...(5.9.9)$$

We then have

$$<\sigma_{el}> = \sigma_{SE} - \sigma_{CE}$$

$$<\sigma_{r}> = \sigma_{C} + \sigma_{CE}$$

$$<\sigma_{T}> = \sigma_{SE} + \sigma_{C} \qquad ...(5.9.10)$$

The interaction of the incident wave with the real part of the potential gives rise to the shape elastic or potential scattering. The imaginary part of potential gives rise to the formation of compound nucleus which when decays, may yield some compound elastic scattering. In medium and heavy nuclei at incident nuclear energy of 10 MeV or more, the energy width Γ of the compound nucleus is much larger than the level spacing D. In such a case η_l varies slowly with momentum and this implies $\sigma_{CE} \to 0$. At lower bombarding energies σ_{CE} should be taken into consideration.

When the incident particle is inside the nucleus it experiences a potential V and the Schrodinger equation can be written as

$$\frac{\hbar^2}{2m}\frac{d^2\psi}{dr^2} + [E - V(r)]\psi(r) = 0 \qquad ...(5.9.11)$$

$$\frac{\hbar^2}{2m}\frac{d^2\psi}{dr^2} + [(E - i\omega) - (V - i\omega)]\psi = 0 \qquad ...(5.9.12)$$

$(V - i\omega)$ may be interpreted as the complex potential as used in the optical model. The energy eigen values are $(E - i\omega)$ while the eigen function ψ remains same as that for the square well potential. The term $i\omega$ adds upto the imaginary part of the energy in equation (5.9.1). Thus the effect of adding an imaginary part

ω to the potential results in the broadening of the resonance level. The optical model is thus helpful in the interpretation of the giant resonances which arise due to the clustering of the excited states of the compound nucleus.

5.10 DIRECT REACTIONS

Processes like inelastic nuclear collision, stripping and its inverse, the pickup reaction are referred as direct reactions. Such reactions proceed without the formation of a compound nucleus and the time in which the reaction is completed is much shorter than the life time of a compound nucleus. In a direct reaction process, the target nucleus and the incident projectile form a system having a life time of 10^{-22} sec. On the other hand a compound nucleus may have a life time of $10^{-14} - 10^{-17}$ sec. It is difficult to say at what energy a given reaction will proceed according to either one or the other reaction mechanism.

Qualitatively it may be said that at higher projectile energy, direct reaction probability is more.

Direct reaction is a one step process. It has been observed that (d, p) reaction is more probable than the (d, n) reaction. If compound nucleus were formed, due to Coulomb barrier, (d, p) reaction should be less probable. Oppenheimer and Phillips explained the reaction on the basis that deuteron being a loosely bound system (binding energy only 2.2 MeV) gets polarised as it approaches a nucleus. The neutron is captured and the proton moves on. At high deuteron energies the (d, p) and (d, n) stripping reactions are equally probable. In all direct reactions the angular distribution of all the reaction products is peaked in the forward direction with a very small intensity in the backward direction. The forward peaking of the reaction products has also been observed in reactions such as ^{31}P, (α, p) ^{34}S, ^{23}Na (d, p) ^{24}Na, ^{7}Li $(p, d)$$^{6}Li$, ^{13}C $(He^3, \alpha)$$^{12}C$. The reactions of the type (p, d) and (n, d) are known as pickup reactions in which the nucleon passing in the vicinity of the target nucleus, picks up a nucleon from it forming a deuteron. The (α, p) reaction is explained as the stripping of a triton from the α-particle. The (p, α) is the inverse pickup reaction.

Reaction of the type (p, p'), (n, n') , (α, α') and (d, d') can also be analysed acording to direct reaction mechanism. The incident particle interacts with a nucleus at the surface of the target nucleus and looses some of its energy and escapes from it.

The analysis of the angular distribution of the inelastically scattered particles gives information about the quantum numbers of the initial and final states of the traget nucleus. Consider the (p, p') reaction. Let $\vec{L}_i$, $\vec{S}_i$, $\vec{L}_f$, $\vec{S}_f$ be the orbital and spin angular momenta of the incoming and the outgoing proton respectively. I_i and I_f are the spin of target nucleus before and after the reaction respectively, then we have

$$\vec{L}_i + \vec{I}_i + \vec{S}_i = \vec{I}_f + \vec{L}_f + \vec{S}_f \qquad ...(5.10.1)$$

From the above relation taking components along z direction, we have

$$I_i + I_f + 1 \geq | l_f - l_i | \geq | I_f - I_i - s_i + s_f |_{\min}$$

The change orbital angular momentum $| l_f - l_i |$ is odd, if the two states I_i and I_f have opposite parity and even, if the two states have same parity. The change in orbital angular momentum Δl of incident particle can be written as

$$\Delta l = | k_f - k_i | R_0 \qquad ...(5.10.2)$$

where $\hbar k_i$ and $\hbar k_f$ are the initial and final momenta of the proton. If we have for momentum transfer

$$\hbar|\vec{K}_f - \vec{K}_i| = \hbar K$$

Now

$$K^2 = K_i^2 + K_f^2 - 2K_iK_f \cos\theta$$

As Δl has only some fixed value, the scattering angle θ also has corresponding fixed values. For these values of θ, the differential scattering cross section should show a maxima.

If we neglect the spin of the incoming particle as in (α, α') reaction one can write the probability of transition as

$$P = \frac{2\pi}{\hbar} | H_{if} |^2 \cdot \frac{dn}{dE}$$

where $$\frac{dn}{dE} d\Omega = \frac{Vp^2 dp}{(2\pi\hbar)^3 dE} d\Omega$$

$$= \frac{Vp^2}{(2\pi\hbar)^3 v_f} d\Omega$$

The differential cross section is

$$\frac{1}{V} v \frac{d\sigma}{d\Omega} d\Omega = \frac{2\pi}{\hbar} | H_{if} |^2 \frac{Vp^2}{(2\pi\hbar)^3 \cdot v_f} d\Omega$$

$$\therefore \qquad \frac{d\sigma}{d\Omega} = \frac{m^2}{4\pi^2\hbar^4} \frac{k_f}{k_i} | V \cdot H_{if} |^2 \qquad ...(5.10.3)$$

We assume that the target nucleus consists of an α particle N and the remaining nucleus C. The incident α particle interacts with the remaining nucleus C as well as the α particle N. Let both the interactions be spherically symmetric. Let the two interactions be $U_{\alpha N}$ $(r_{\alpha N})$ and $U_{\alpha C}$ $(r_{\alpha C})$ where $r_{\alpha N}$ and $r_{\alpha C}$ refer to the radius vector of the incident particle with respect to centres of mass of N and C respectively. Let the initial state of the target nucleus be $\psi^\circ(r_{NC})$ and its orbital angular momentum $l = 0$ and total spin I. The final nucleus has an angular momentum l' and total angular momentum I'. In the final state the wave function of the nucleus would be $\psi_I(r_{NC})$. The function $\psi_I^\circ(r_{NC})$ having $l = 0$ is spherically symmetric. The function $\psi_I^l(r_{NC})$ has an angular dependence given by combination of wave function $Y_m^l(\theta, \phi)$ with $m = l, l - 1, \ldots l$. It is assumed that the incident α-

particle interacts with N only so that the matrix element corresponding to $U_{\alpha N}$ $(r_{\alpha N})$ can be written as

$$|H_{if}| = \int \psi^*_{I'}(r_{NC}) e^{-i\vec{K}_f \vec{r}_{\alpha I}} U_{\alpha N}(\vec{r}_{\alpha N}) \cdot e^{i\vec{K}_i \vec{r}_{\alpha I}} \cdot \psi_0(r_{NC}) \cdot d\vec{r}_{\alpha I} d\vec{r}_{NC} \qquad ...(5.10.4)$$

We replace $r_{\alpha I}$ as

$$\vec{r}_{\alpha I} = \vec{r}_{\alpha N} + \frac{M_C}{M_N + M_C} \vec{r}_{NC} \qquad ...(5.10.5)$$

where M_N is the mass of remaining nucleus C and M_N mass of the α particle N. This gives

$$H_{if}(U_{\alpha N}) = \int dr_{\alpha N} \exp(-i\vec{K} \cdot \vec{r}_{\alpha N}) U_{\alpha N}(r_{\alpha_N}) \cdot \int d\vec{r}_{NC} \exp(-i\vec{K}' \vec{r}_{NC}) \cdot \psi_0 \psi^*_{c'} r_{NC} \qquad ...(5.10.6)$$

where $\vec{K} = \vec{K}_f - \vec{k}_i$

and $$\vec{K}' = \vec{K}' \frac{M_C}{M_N + M_C} \qquad ...(5.10.7)$$

Assuming the z-axis along k, we expand the plane wave as

$$e^{ikz} = \sum_L 4\pi(2L+1) i^L Y^o_L(\cos\theta) \cdot j_L(Kr) \qquad ...(5.10.8)$$

Taking into account the orthogonality of the spherical harmonics, the integral over r_{NC} is proportional to

$$\int j'_l (K'r) \psi_0(r) \psi^*_I(r) r^2 dr$$

Here $\psi^*_I(r)$ is only the radial part of the wave function. The wave function ψ_0 and ψ^*_I vary rapidly beyond the nuclear surface where $r = R$. The integrand is then largest near the nuclear surface, so that the integral is proportional to $j_l(kR_l)$.

Due to the short range of nuclear forces we can assume $U_{\alpha N} = U^\circ_{\alpha N}\, \delta(r_{\alpha N})$. This reduces the integral over $r_{\alpha N}$ as $U^\circ_{\alpha N}$.

The matrix element due to the Hamiltonian is small due to the orthogonality of ψ_0 and ψ_I and can be neglected.

It is thus observed that the matrix element $|H_{if}|$ depends upon the scattering angle θ through the Bessel function $j_i(k, r)$ where

$$K^2 = K_i^2 + K_f^2 - 2K_i K_f \cos\theta = \frac{M_N + M_C}{M_C} K'^2$$

so that $$K' = \frac{M_C}{M_N + M_C} K = \frac{M_C}{M_I} K \qquad ...(5.10.9)$$

If the target nucleus has l different from zero, the matrix element H_{if} would have summation over all values of L where $|l - l'| \le L \le l + l'$. As the Bessel function resembles damped oscillations, the differential cross section also shows a similar behaviour.

Putting the value of $|H_{if}|^2$ it is clear that the differential cross section becomes

$$\frac{d\sigma}{d\Omega} \propto |j_L(K' R_l)|^2$$

This type of angular distribution is observed not only in inelastic scattering but also in stripping and pickup reactions.

EXERCISES

1. Show that in elastic collision between particles of equal masses, the kinetic energy in the centre of mass system is half of that in laboratory system.
2. A neutron of mass M_N collides with an atom of mass M_A with K.E. E_0. Show that the minimum energy of the scattered neutron is $E_0 = \frac{M_A}{M_A + M_N}$.
3. Consider the reaction, $d + {}^{12}_{6}C \rightarrow {}^{13}N + n - 0.28$ MeV. What is the threshold energy in laboratory system for the above reaction.
4. A ^{12}C beam hits a deuterium target and gives rise to the above reaction. What is the threshold energy for the reaction.
5. A proton beam of kinetic energy 5 MeV hits a ^{6}Li target giving rise to the following reaction, $p + {}^{6}Li \rightarrow {}^{3}He + {}^{4}He + 4.02$ MeV. What are the energies of the reaction products in the laboratory system.
6. What would be the size of deuteron if its binding energy were only 0.5 MeV.
7. One gram of copper containing gold impurity is irradiated to a thermal neutron flux of 10^{13} n/sec. for 10 hours. The cross section σ for ^{197}Au is 96 barns and half life of ^{198}Au is 2.7 days. Each disintegration of ^{198}Au may be assumed to emit one photon of energy 412 KeV. If the irradiated sample emits 10^4 photons per second what is the percentage impurity of gold in copper.
8. Isotopic abundance of ^{202}Hg is 30%. A one gram sample of soil is irradiated to a neutron flux of 10^{13} n/sec. for 10 hours. The sample is found to emit 10 photons of 280 KeV energy per seconds for the decay of ^{203}Hg. What is the percentage of mercury in the soil sample.

   σ for ^{202}Hg = 5 barns. $T_{1/2}$ of ^{203}Hg = 47 d.
9. Consider the following reactions and decay

   $$^{23}Na + p \rightarrow n + {}^{23}Mg + Q$$

   $$^{23}Mg \rightarrow {}^{23}Na + \beta^+ \qquad E_{\beta^+} = 3 \text{ MeV}$$

   What is the threshold energy for the (p, n) reaction in laboratory system.
10. The thermonuclear reaction which can be used in power production is

    $$D + D \rightarrow {}^{3}He + n + 3.27 \text{ MeV}$$

    If the deuteron nuclei interact at rest, calculate the final kinetic energies of the reaction products. If the deuterium nuclei must come within a distance of 10^{-11} cm. of each other, what energy must be supplied to overcome the Coulomb repulsion. At what order of magnitude of temperature, the reaction can take place.
11. Show that for maximum value of the slow neutron reaction cross section, the scattering cross section is equal to reaction cross section. Hence show that

    $$\sigma_{r\,max} = \frac{0.5 \times 10^6}{E_r(eV)}$$ barns, where E_r is the resonance energy.

6

INTERACTION OF RADIATION WITH MATTER

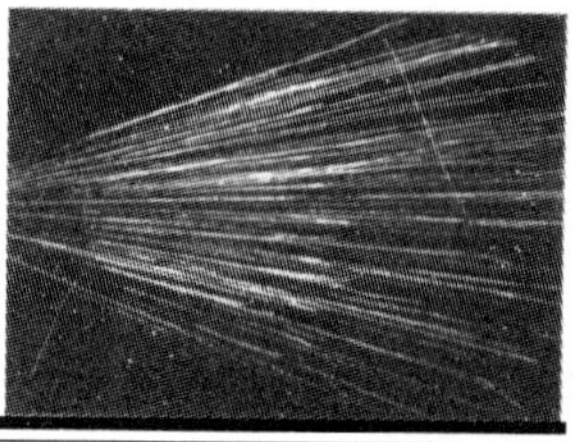

6.1 INTERACTION OF RADIATIONS WITH ATOMS

When radiations *e.g.* α-particles, β-particles, γ-rays, neutrons or charged particles enter any matter, they interact with its atoms. The type of interaction depends upon the nature of radiations and their energy. The interactions that can take place are (*a*) nuclear reactions (*b*) elastic and inelastic scattering by the atomic nuclei or (*c*) electromagnetic interaction with the atomic electrons.

(*a*) Nuclear Reactions

A charged particle can induce nuclear reaction in an atomic nucleus of the medium if its kinetic energy is more than the Coulomb barrier of the nucleus. Neutrons being electrically neutral do not experience any Coulomb barrier and can be absorbed by nuclei even when their kinetic energy is very low. Gamma rays induce nuclear reaction only if their energy is much greater than the binding energy of a nucleon in the nucleus. β-particles *i.e.* electrons and positrons do not experience the nuclear potential and do not give rise to nuclear reactions.

(*b*) Elastic and Inelastic Scattering

The nuclear potential and the Coulomb potential due to the atomic nucleus can scatter an incoming charged particle without any change in its kinetic energy. This process is known as elastic scattering. For inelastic scattering to take place, the kinetic energy of the incoming particle should be greater than the energy of the first excited state of the scattering nucleus.

(*c*) Electromagnetic Interaction

Charged particles moving in matter interact with the atomic electrons through its Coulomb potential. The gamma-rays interact through its electromagnetic field with the atomic electrons.

6.1.1 Interaction of Heavy Charged Particles with Matter

A charged particle passing in the vicinity of an atom produces a strong electrostatic force on the electrons. There is a probability that the incoming particle may transfer enough energy to an electron so as to excite it to higher enery states and sometimes to pull it out of the atom, producing ionization. The energy imparted to an atomic electron reduces the kinetic energy of the incoming particle. Usually the energy imparted to the electron is much larger than its binding energy in the atom (at least in the outer shells) and the electron may be considered as an unbound free particle. As the incoming particle looses some of its kinetic energy in each such collision, it comes to rest after a number of collisions.

Let a particle with charge ze, velocity V and mass M pass at a distance b from a free electron with mass m and charge e (Fig. 6.1). It is assumed that the incoming particle is so heavy ($M >> m$) that after a collision with an electron it does not deviate from its path. Naturally this assumption will not hold good when the incoming particle is an electron or a positron. One also assumes that $V << c$ and that during the time of collision with the incoming particle, the electron does not move.

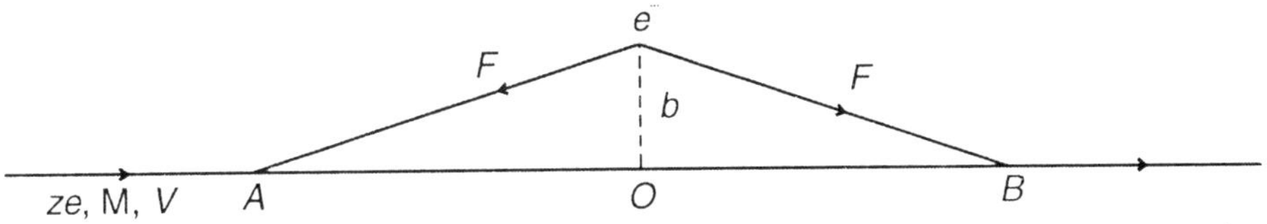

Fig. 6.1: Particle with charge ze, mass M and velocity V passes at a distance b from an electron

As the charged particle passes by the electron with impact parameter b (distance of closest approach) it imparts an impulse to the electron. During the passage of the particle the resultant of the forces at the electron, when the particle is at the point A and at point B ($AO = OB$) is only in the vertical direction. For every point A on one side of O there is an equivalent point B on the other side. The resultant impulse given to the electron is thus in a direction perpendicular to the trajectory of the incoming charged particle. This impulse is equal to the momentum $p_\perp$ given to the electron, so that,

$$p_\perp = \int_{-\infty}^{+\infty} F_\perp dt$$

$$= \int_{-\infty}^{+\infty} e \in_\perp dx \cdot \frac{dt}{dx}$$

$$= \frac{e}{V} \int_{-\infty}^{\infty} \in_\perp dx \qquad \ldots(6.1.1)$$

where $\in_\perp$ is the component of the electric field due to the charged particle perpendicular to its trajectory. To calculate $\in_\perp$ one assumes a hollow cylindrical surface of radius b around the trajectory of the charged particle. The electric flux perpendicular to the cylindrical surface of length dx is $2\pi b dx \epsilon_\perp$.

The electric field intensity at the whole length of the cylindrical surface is, by Gauss's law, proportional to the charge enclosed, so that,

$$\int_{-\infty}^{+\infty} \epsilon_\perp \, dx = K\frac{4\pi ze}{2\pi b} = \frac{K2ze}{b}, \qquad K = \frac{1}{4\pi \epsilon_0} \quad \ldots(6.1.2)$$

Using eqn. (6.1.2) in equation 6.1.1 gives the momentum imparted to the electron as,

$$p_\perp = K \cdot \frac{2ze^2}{bV} \quad \ldots(6.1.3)$$

The kinetic energy imparted to the electron due to the impulse is,

$$\Delta E = \frac{p_\perp^2}{2m} = \frac{2K^2z^2e^4}{mV^2} \cdot \frac{1}{b^2} \quad \ldots(6.1.4)$$

ΔE is the kinetic energy acquired by each of the electrons situated at a distance b from the trajectory of the charged particle. Let n be the density of electrons in the medium through which the charged particle is moving. In a cylindrical layer of unit length, radius b and thickness db with the particle trajectory as the axis, (Fig. 6.2) the number of electrons is $2\pi bdbn$. The energy lost by the incoming particle to all such electrons in unit length is,

$$dE_b = 2\pi nbdb\Delta E$$

$$= \frac{K^2 \cdot 4\pi z^2e^4n}{mV^2} \cdot \frac{db}{b} \quad \ldots(6.1.5)$$

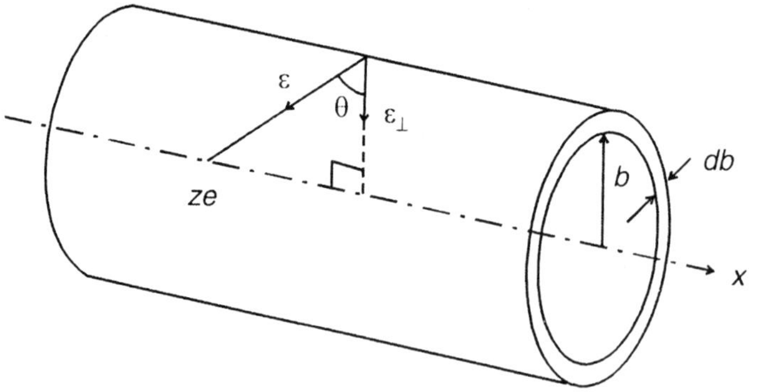

Fig. 6.2: Annular cylinder with radius b and thickness db with particle trajectory as axis

The total energy loss by the particle in unit length, to the electrons in a cylindrical shell with radii b_{min} and b_{max} is

$$-\frac{dE}{dx} = K^2 \cdot \frac{4\pi z^2e^4n}{mV^2} \int_{b_{min}}^{b_{max}} \frac{db}{b}$$

$$-\frac{dE}{dx} = K^2 \cdot \frac{4\pi z^2 e^4 n}{mV^2} \ln \frac{b_{max}}{b_{min}} \qquad ....(6.1.6)$$

Nonrelativistically, the duration of the impulse to the electron is $\tau = b/V$ approximately. The electron, which is bound in the atom, has a vibration frequency ν such and that it's excitation energy is $h\nu$. If the time of impact τ is much greater than the time of vibration ($\tau >> 1/\nu$), there is no transfer of energy to the electron. Relativistically, however, the electric field of the incident particle is contracted in the direction of motion and $\in_\perp$ is increased by a factor $1/\sqrt{(1-\beta^2)}$ where $\beta = V/c$, where c is velocity of light. The impulse gets sharpened and the time of impact is now $(b/V)\sqrt{(1-\beta^2)}$. The total impulse remains unchanged as it is the product of $\in_\perp$ and τ. The maximum value of b is such that

$$\frac{b}{V}\sqrt{(1-\beta^2)} < \frac{1}{\nu}$$

or $$b_{max} = V / \nu\sqrt{(1-\beta^2)}$$

As the electrons in different shells of the atom have different energies of excitation, one has to take an appropriate average $\bar{\nu}$ for the frequency of the electron. Thus,

$$b_{max} = V / \bar{\nu}\sqrt{(1-\beta^2)} \qquad ....(6.1.7)$$

To obtain the value of b_{min} consider the frame of reference in which the incoming particle is at rest and the electron is moving with velocity V. The momentum of the electron would be $mV(1-\beta^2)^{-1/2}$. The de Broglie wavelength of this electron is,

$$\lambda = \frac{h}{p} = \frac{h\sqrt{(1-\beta^2)}}{mV}$$

The value of the impact parameteı $b < \lambda$ has no meaning. Hence the minimum value of b is b_{min} given as,

$$b_{min} = \frac{h\sqrt{(1-\beta^2)}}{mV} \qquad ...(6.1.8)$$

Substituting the values of b_{max} and b_{min} in equation (6.1.6) gives the energy loss per unit length by the incoming charged particle as,

$$-\frac{dE}{dx} = \frac{K^2 4\pi z^2 e^4 n}{mV^2} \ln \frac{mV^2}{h\bar{\nu}(1-\beta^2)} \qquad ...(6.1.9)$$

In this expression $h\bar{\nu}$ is the average excitation potential of the atom of the medium or absorber through which the charged particle is passing, $-dE/dx$ is known as the specific energy loss by the particle. Exact calculations for the specific energy loss due to excitation and ionization by heavy charged particles of mass M with kinetic energy $E < Mc^2$ is given as,

$$-\frac{dE}{dx} = \frac{K^2 4\pi z^2 e^4 n}{mV^2}\left\{\ln\frac{2mV^2}{\bar{I}} - \ln(1-\beta^2) - \beta^2 - \delta - u\right\} \quad ...(6.1.10)$$

where $\bar{I} = h\bar{\nu}$ is the mean excitation potential of the absorbing atom. The term δ depends upon the density of the electrons in the absorber material. The atoms of the medium near the trajectory of the incoming particle get polarized, leading to a decrease in the electric field acting on remote electrons, reducing thereby the energy loss.

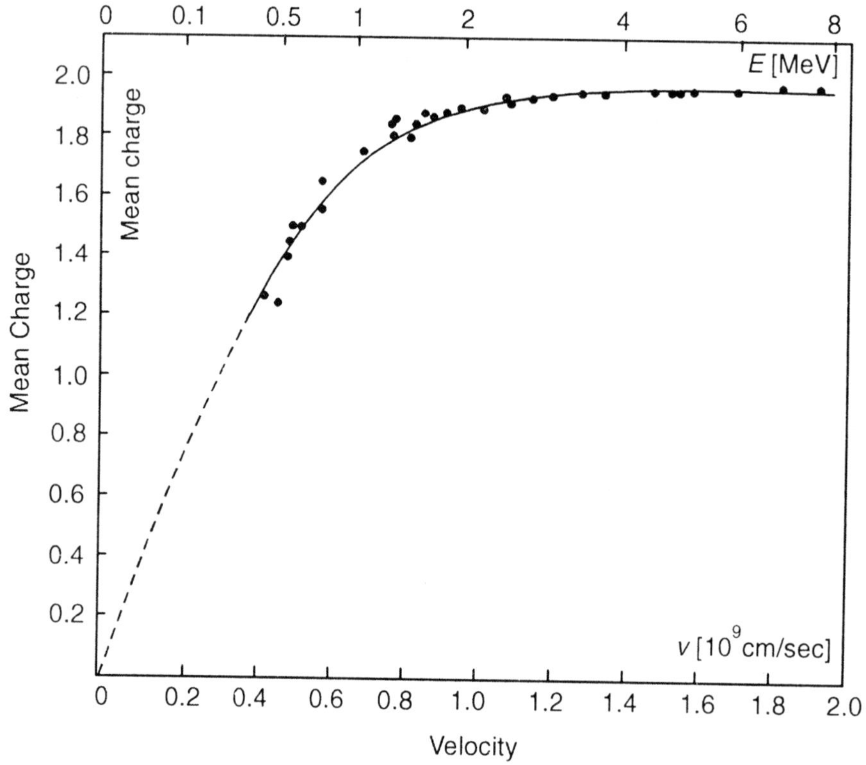

Fig. 6.3: Mean effective charge of an α-particle in a medium as a function of its velocity [T. Hall. (1950). *Phys. Rev.*, **79**: 504]

The last term u is the correction term for relatively low energies of the ionizing particle. When the particle velocity is comparable with the velocity of K or I shell electrons, the later stop participating in the collisions and the specific energy loss decreases. The energy of the particle at which this correction must be introduced increases with the atomic number Z of the medium. For $Z = 30$ the correction for K-shell electron should be introduced, for protons below 50 MeV energy. For L-shell electrons this correction term is appreciable below 10 MeV proton energy. At low energies a positively charged incident particle may capture an electron from the medium for short intervals of time. This reduces the effective charge of the incident particle, thus reducing the specific energy loss. Figure 6.3 shows the mean effective charge of an α-particle as a function of velocity. The curve is almost independent of the medium atleast for $Z < 50$. It is found that the maximum specific ionization for α-particles is at about 0.5 MeV energy.

The specific energy loss as a function of the particle energy is shown in Fig. 6.4. The part *AB* of the curve represents the effect of the term u. The curve *BCD* represents the $1/V^2$ dependence. At relativistic energies, the velocity V changes little and *CD* is asymptotic at $V = c$. However, the term under logarithm changes as V approaches c giving rise to an increase in the specific energy loss as represented by part *CE* of the curve.

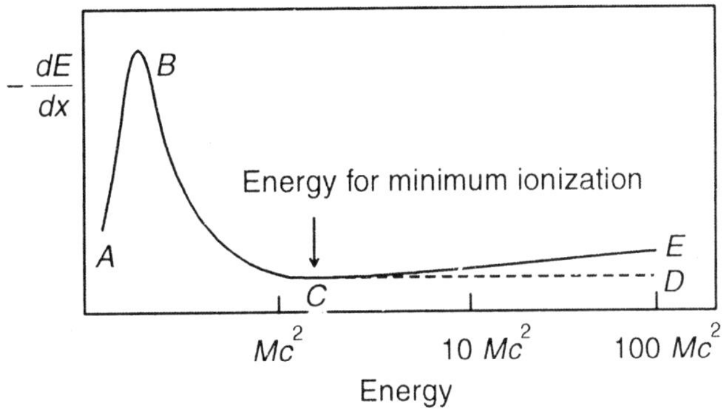

Fig. 6.4: Specific energy loss of a charged particle as a function of its kinetic energy in units of Mc^2. Part *AB* of the curve occupies energy range of approximately $10^{-3}\ Mc^2$

At medium energy both the terms δ and u are insignificant. The quantity under logarithm changes very little with the energy and mass of the ionizing particle. The main variation of the specific energy loss is then given by the quantity outside the bracket, so that,

$$-\frac{dE}{dx} \approx \left(\frac{K^2 4\pi e^4}{m}\right) \cdot \frac{z^2 n}{V^2} \qquad \ldots(6.1.11a)$$

$$\approx \left(\frac{K^2 4\pi e^4}{m}\right) \cdot \frac{z^2 n M}{2E} \qquad \ldots(6.1.11b)$$

As the incoming charged particle moves in the absorber, its energy decreases and correspondingly the specific energy loss increases. This is clearly shown by the Bragg curve for alpha-particles shown in Fig. 6.5 where the specific energy loss of an alpha-particle along its track, is shown. As the alpha-particle moves along its track, its energy decreases and specific energy loss increases. Before the end of the range the specific energy loss is maximum. Near the end of the track the energy of the alpha-particle is very low and the term u in expression (6.1.10) becomes significant, reducing the specific energy loss.

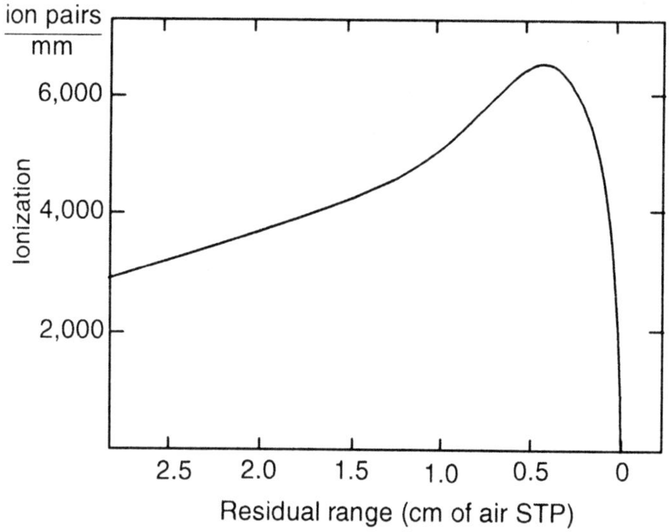

Fig. 6.5: Specific ionization by an α-particle along its trajectory (Bragg Curve). The maximum specific ionization is at the residual α-particle energy of about 0.5 MeV. [W. Riezler and H. schepers (1961). *Ann. Phys.*, **8**: 270]

The effect of the absorber manifests in the term n, the electron density and the average excitation potential $\bar{I}$. For an absorber with atomic number Z the electron density is $n = Zn_a$ where n_a is the atomic density, which is given as,

$$n_a = \rho N_o / A$$

hence electron density

$$n = Z\rho N_o / A \qquad \text{...(6.1.12)}$$

where N_0 is the Avogadro number, and ρ is absorber density and A the mass number of the absorber atom. Substituting the value of n in equation (6.1.11) one gets,

$$-\frac{dE}{dx}\frac{1}{\rho} \approx \left(\frac{K^2 2\pi e^4 N_0}{m}\right) \cdot \frac{z^2 M}{E}\frac{Z}{A} \qquad \text{...(6.1.13)}$$

$-dE/\rho dx = dE/dt$ is the mass stopping power and its unit is energy per (mass/unit area) or Jm^2/kg in mks units. Generally the unit of mass stopping power used is MeV cm^2/gm. The quantity Z/A in expression (6.1.13) is nearly 1/2 for all elements and its variation with Z is small. The mass stopping power is thus almost independent of the atomic number Z of the absorber and one can write,

$$-\frac{dE}{dt} \propto \frac{z^2 M}{E} \qquad \text{...(6.1.14)}$$

The mass stopping power depends mainly on the charge, mass and kinetic energy of the incoming particle. If the quantity $-dE/dt$ is plotted as a function of E in silicon, it gives curves shown in Fig. 6.6. The curves for different charged particles *e.g.* α-particles, deuteron, protons, ^3H (triton) and ^3He are all well separated. Thus measuring $-dE/dx$ (or dE/dt) and energy E of an incoming particle one can distinguish between various charged particles emitted in a nuclear reaction.

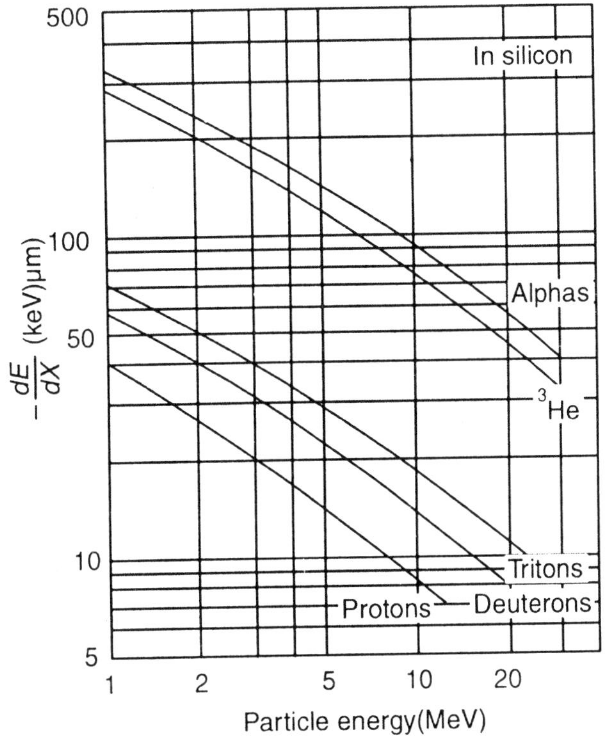

Fig. 6.6: Specific energy loss of α-particles, $^3He^{++}$, tritons, deuterons and protons in silicon as a function of its kinetic energy. [D. J. Skyrme (1967). *Nucl. Instr. Meth.*, **57**: 61]

The effective ionization potential $\bar{I}$ was assumed by Block to be proportional to Z and approximately expressed as,

$$\bar{I} = 11.5Z \text{ eV} \qquad \ldots(6.1.15)$$

The excitation potential for various elements has been measured by number of workers. Table 6.1 gives the excitation potential for a number of elements and gases. It is found that the excitation potential is same whether the ionizing particle is electron or α-particle. By a curve fitting method Sternheimer has given the following relation for the excitation potential $\bar{I}$.

$$\frac{\bar{I}}{Z} = 12 + \frac{7}{Z} \qquad \text{for } Z < 13$$

$$\frac{\bar{I}}{Z} = 9.76 + 58.8\,Z^{-1.19} \quad \text{for } Z > 13 \qquad \ldots(6.1.16)$$

The above relations for excitation potential are all approximate. It is perhaps best to use the experimental values of the excitation potential.

TABLE 6.1: MEAN EXCITATION POTENTIAL FOR SOME ELEMENTS AND GASES

Medium	*Z*	$\bar{I}$	$\frac{\bar{I}}{Z}$
H_2	1	15.6	15.6
Li	3	34.0	11.3
Be	4	60.4	15.1
C	6	76.4	12.7
Al	13	150	11.5
Fe	26	243	9.4
Cu	29	279	9.6
Ag	47	422	9.0
Sn	50	472	9.4
W	74	680	9.2
Pb	82	737	9.0
U	92	853	9.3
He	2	36.0	18.0
Ne	10	125	12.5
A	18	198	11.0
Kr	36	396	11.0
Xe	54	529	9.8
Air	7.22	80.1	11.1
N_2	7	81.2	11.6
O_2	8	91.2	11.4
CO	7	81.2	11.6
CO_2	7.35	118	16.0
CH_4	2	41.6	20.8
C_2H_6	2.25	45.9	20.6
C_3H_8	2.36	480	20.3
C_4H_{10}	2.43	45.3	18.7

6.1.2 Energy Loss by Electrons

The expression (6.1.10) for the specific energy loss has to be modified in the case of electrons. Electrons being light particles travel with relativistic velocities and are indistinguishable from the electrons with which they collide. Electrons come close to the atomic nucleus and suffer Coulomb scattering by large angles. While scattering, the electrons experience a radial acceleration and emit electromagnetic radiations known as Bremsstrahlung radiations. The energy of the Bremsstrahlung radiations is derived from the kinetic energy of the incident particle. The electrons while traversing an absorber thus loose energy by two processes—ionization of medium and Bremsstrahlung radiations.

An electron after collision with an electron of the medium acquires a transverse component of momentum equal to the momentum imparted to the atomic electron. Unlike a heavy particle, the transverse velocity of the electron is not negligible and should be taken into consideration. For collisions between identical particles quantum mechanical calculations require the exchange phenomenon to be taken into account. Bethe has given the specific energy loss of electrons by ionization of the medium as,

$$\left(-\frac{dE}{dx}\right)_{ion} = \frac{K^2 2\pi e^4 n}{mV^2}\left[\ln\frac{mV^2 T}{2\bar{I}^2(1-\beta^2)} - \ln 2\left(2\sqrt{1-\beta^2}-1+\beta^2\right)\right.$$

$$\left. +1-\beta^2+\frac{1}{8}\left(1-\sqrt{1-\beta^2}\right)^2-\delta\right] \quad \ldots(6.1.17)$$

where T is the relativistic kinetic energy of the electron.

The specific energy loss by the process of Bremsstrahlung radiation has also been calculated by Fermi. It is found that,

$$\frac{-\left(\frac{dE}{dx}\right)_{rad}}{-\left(\frac{dE}{dx}\right)_{ion}} = \frac{ZE(\text{MeV})}{700} \quad \ldots(6.1.18)$$

The radiation loss of energy of an electron becomes significant in heavy elements and at high particle energies (above 2 MeV).

6.1.3 Absorption Curve and Range of Charged Particles

A charged particle, during its passage in matter looses energy till it comes to rest. The linear distance traversed by the particle is known as the range of the particle in that medium. The energy loss dE in traversing a distance dX is given as

$$-dE = -\frac{dE}{dx}\cdot dx$$

The distance R over which the particle looses whole of its kinetic energy E_0 is given as

$$R = \int_{E_0}^{E=0}\frac{dE}{dE/dx}$$

It is not convenient to calculate the range of a particle using the above equation. The range of α- particles, β- particles and protons of different energies have been measured by observing their absorption curves in different absorbers.

α-particles have a small range. Generally the absorption curve for α-particles is measured in air. A suitable arrangement to study the absorption curve of α-particles is to mount an α-source and a surface barrier detector in a vacuum tight vessel in which air pressure can be varied Fig. 6.7. The pressure of air in the chamber can be measured with a suitable manometer. The distance between the souce S and the detector D is measured with a travelling microscope. The thickness of the air column between the source and detector can be determined in gm/cm^2 from the air pressure and the temperature in the chamber.

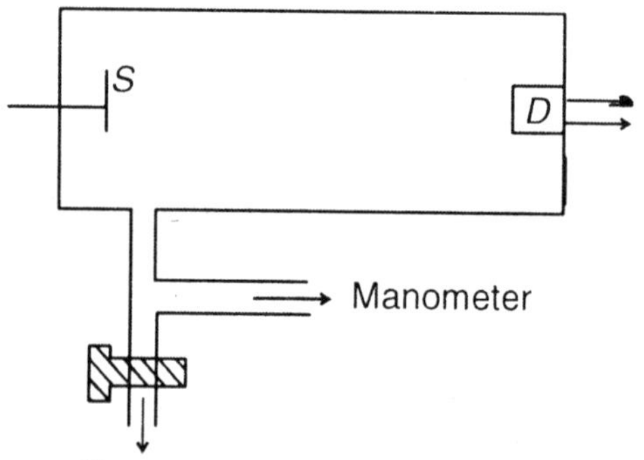

Fig. 6.7: Schematic diagram of the vaccum chamber to measure range of α-particle. The α-source S and the detector D are mounted at a fixed distance

When the counting rate in the detector is plotted as a function of thickness of absorber (air plus detector window), it is found to be constant upto a certain point. The counting rate then decreases suddenly. The end part of the absorption curve on an extended scale is shown in Fig. 6.8. Part BA of the curve extends to zero absorber thickness. Part BCD of the absorption curve shows a fall in the counting rate with finite slope. It implies that different α-particles all having the same energy are stopped in different thicknesses of the absorber. This is known as straggling of α-particles and arises due to the statistical nature of the energy loss. The absorber thickness R_0 in which half of the α-particle are stopped is known as the range of the α-particles. The intercept of the straight line part of the absorption curve with the x-axis is known as the extrapolated range denoted by R_e.

An emperical relation for the range R of α-particles of kinetic energy E(MeV) in dry air at 15°C and 760 mm of Hg pressure is given as

$$R(\text{cm}) \approx 0.32\ E^{3/2} \quad \text{...(6.1.19)}$$

The range of protons upto an energy of 200 MeV in dry air at 15°C and 760 mm of Hg pressure is emperically given as

$$R(\text{cm}) \approx 100(E/9.3)^{1.8} \quad \text{...(6.1.20)}$$

where the proton energy is in MeV units.

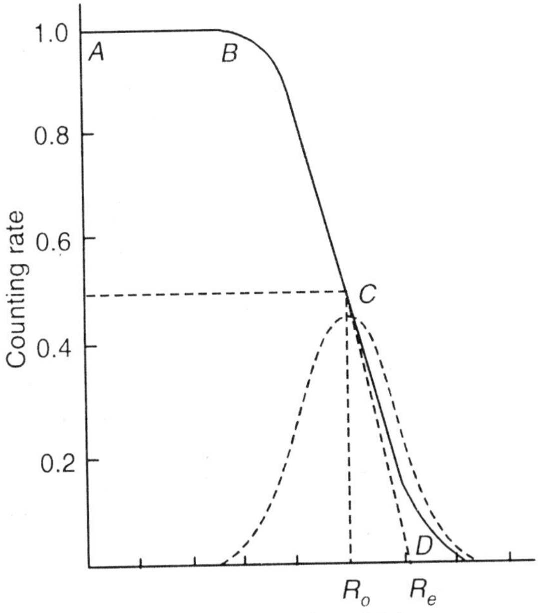

Fig. 6.8: Absorption curve for α-particles on an extended scale. The differential curve is Gaussian with maxima at R_0. The width of the curve gives the straggling of the particles

Unlike heavy charged particles, electrons travelling in a medium do not have a straight line trajectory. Due to frequent scattering, the actual path length an electron traverses, is much larger then the linear distance. The path length is different for different electrons of the same kinetic energy. Consequently there is a large variation in the range of monoenergetic electrons.

The range of electrons is usually measured in aluminium absorber. An arrangement to study the absorption curve for electrons is shown in Fig. 6.9. A beam of electrons from a source *S* passes through aluminium absorbers *A* and detected in the detector *D*. The detector could be an end window G.M. counter with known window thickness.

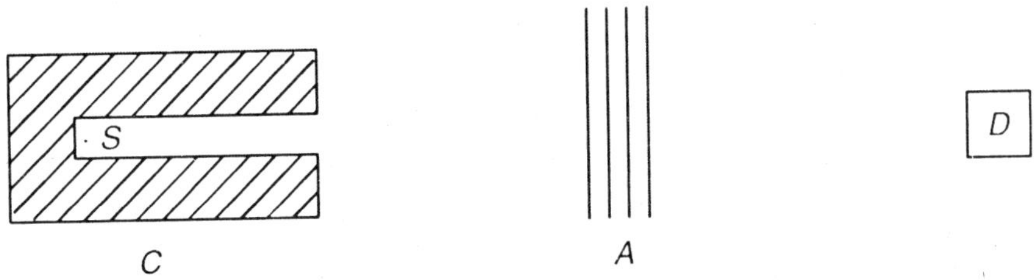

Fig. 6.9: Experimental arrangement to study the absorption of electrons and β-particles. The detector, absorber and source distances are adjusted according to requirement of the experiment

There are very few radioactive sources which emit monoenergetic electrons (internal conversion electrons). Monoenergetic electrons from electron accelerators are used for the study of their absorption curve. The absorption curves for some energies of electrons are shown in Fig. 6.10. The absorption curves show an almost linear fall in transmitted intensity. There is a long tail in the absorption curve making it difficult to determine the range of the particles. The linear part of the absorption curve is extrapolated and its intercept with the x-axis is called the "practical range" R_p of the electrons. The practical range R_p in aluminium for monoenergetic electrons of energy $E(E > 0.6$ MeV) is found to be given as

$$R_p\ (\text{gm/cm}^2) = 0.526E\ (\text{MeV}) - 0.094 \quad \ldots(6.1.21)$$

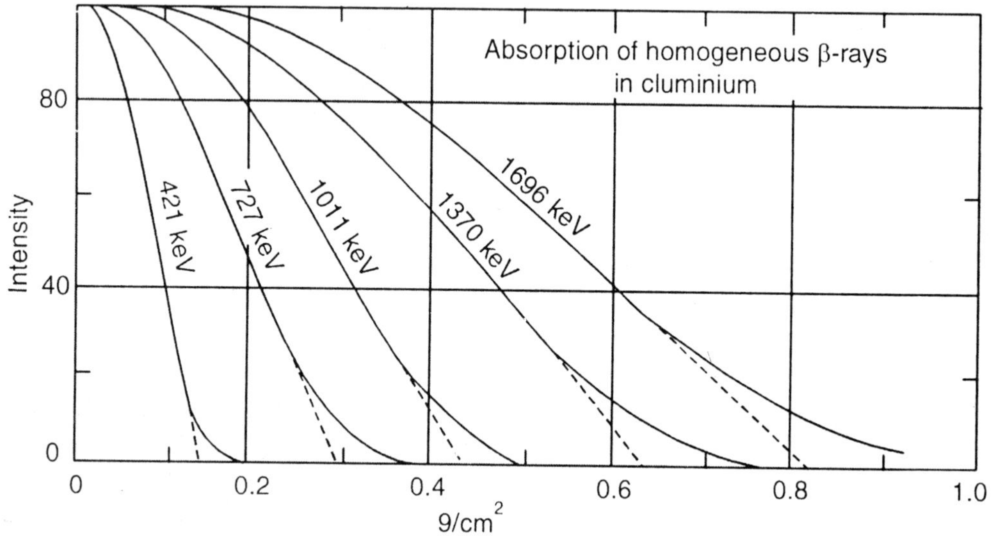

Fig. 6.10: Absorption curve for mono-energetic electrons of different energies. The intercept of the dashed line gives the practical range R_0 [(J. Marshall and A.G. Ward. *Can. J. Research* (1937), **A15**: 39]

The β-particles emitted from a radioactive isotope have a continuous energy spectrum with a definite end point. The maximum energy of the β-spectrum is specific to the radioactive isotope. The absorption curve of *RaE* β-particles in aluminium is shown in Fig. 6.11. The curve *A* is the absorption curve as observed. When the background due to cosmic rays and other sources is subtracted, curve *B* is obtained. At the end point, the curve (on semi-log scale) makes an asymptotically vertical intercept with the x-axis. This corresponds to the range R_0 of the β-particles. The initial part of the absorption curve-for absorber thickness upto about 25mg/cm^2 depends upon the geometry of the apparatus. After the initial portion, the curve shows an exponential absorption. In this region the number N of β-particles transmitted through an absorber of thickness x and density ρ can be represented as

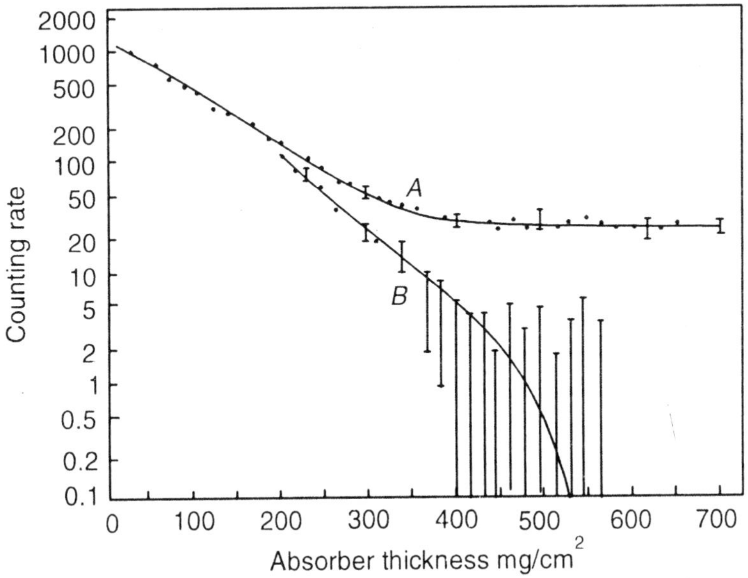

Fig. 6.11: Absorption curve of β-particles from a *RaE* source, curve *A* is composit for β-particles and background. Curve *B* is after subtracting the background

$$N = N_o e^{-\frac{\mu}{\rho}x} \qquad \ldots(6.1.22a)$$

where N_o is the number of incident β-particles, and μ is the mass absorption coefficient. The maximum range R_o and the mass absorption coefficient are functions of the maximum energy of the β-particles. The linerar absorption coefficient μ/ρ in aluminium and the maximum energy E_o (MeV) of the β-particles are emperically given as

$$\frac{\mu}{\rho} = 17.0 E_o^{-1.43} \qquad \ldots(6.1.22b)$$

The exponential absorption of β-particles is purely accidental. It is useful in calculating the dose received from β-particles in different depths of the tissue. In cancer therapy one can thus control the dose given to a particular tissue of the patient.

Emperically the range of β-particles in aluminium and its maximum energy E_o (MeV) is found to be given as

$$R_o\,(\text{gm/cm}^2) = 0.542\,E_o - 0.133 \qquad E_o > 0.8 \text{ MeV}$$

$$R_o\,(\text{gm/cm}^2) = 0.407\,E^{1.38} \qquad \text{for } E_o < 0.8 \text{ MeV}$$

$$\ldots(6.1.23)$$

The range energy curve for β-particles in aluminium is shown in Fig. 6.12.

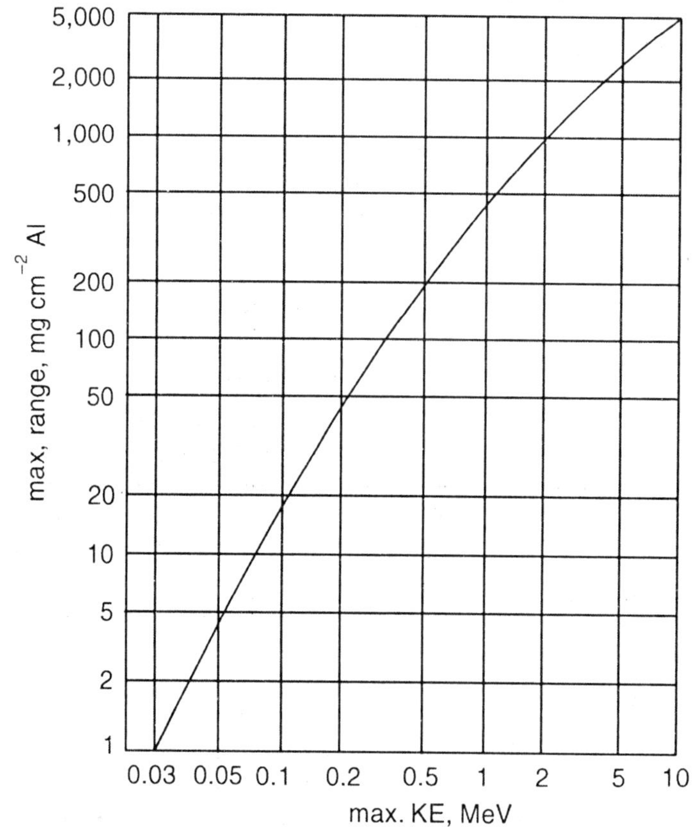

Fig. 6.12: Range energy plot for β-particle. [Kai Seigbahn. "*Beta- and Gamma-ray Spectroscopy*".] New Holland Publishing Co. Amsterdam, (1955)]

6.1.4 Straggling of α-particles

The specific energy loss calculated by equation (6.1.10) is an average for a large number of α-particles. For individual α-particle the actual value fluctuates around the mean value with the result that for a given path length the ionization produced fluctuates and the range of the particle fluctuates correspondingly. The differential absorption curve ABCD of Fig. 6.8 represents the variation of the range of the α-particles. This variation in the range is known as the straggling of α-particles. The range of all charged particles shows the straggling effect. The fluctuation in the specific energy loss is explained on the statistical nature of the number of collisions made by a charged particle in traversing a fixed distance in a medium.

The average number of collisions N_{bo} with impact parameter between b and $b + db$ made by a charged particle in traversing a distance x in a medium is

$$N_{bo} = 2\pi\, b\, db\, .\, n\, .\, x$$

For any particular particle if the number of collisions is N_b, then the average of $N_b = < N_b > = N_{bo}$.

As the number of collisions varies statistically, the variance in the number of collisions is

$$\left\langle (N_b - N_{bo})^2 \right\rangle = \left\langle N_b{}^2 \right\rangle - N_{bo}^2 = N_{bo}$$

The corresponding variance in the energy loss E_b in traversing a distance x is

$$\left\langle E_b^2 \right\rangle - E_{bo}^2 = \left(\left\langle N_b \cdot E_r \right\rangle - N_{bo} E_r \right)^2$$

$$= N_{bo} E_r^2 \qquad \text{.... (6.1.24)}$$

where E_{bo} is the average energy loss, E_r is the energy loss in a collision with impact parameter b and is given as (eqn. 6.1.4)

$$E_r = \frac{2K^2 z^2 e^4}{mV^2 b^2}$$

substituting the value of N_{bo} and E_r in eqn. (6.1.24) gives

$$\left\langle E_b^2 \right\rangle - E_{bo}^2 = \frac{K^4 8\pi n z^4 e^8}{m^2 V^4} \cdot \frac{db}{b^3} \cdot x \qquad \text{...(6.1.25)}$$

The variance in $\sigma_x^2(E)$ energy loss in traversing distance x for all impact parameters between b_{min} and b_{max} is given as

$$\sigma_x^2(E) = \frac{K^4 8\pi n z^4 e^8 x}{mV^2} \int_{b_{min}}^{b_{max}} \frac{db}{b^3}$$

$$= \frac{K^4 4\pi n z^4 e^8 x}{mV^2} \left[\frac{1}{b_{min}^2} - \frac{1}{b_{max}^2} \right] \qquad \text{...(6.1.26)}$$

As $b_{max} >> b_{min}$

$$\sigma^2(E) = \frac{K^4 4\pi n z^4 e^8}{mV^4} \frac{1}{b_{min}^2} \qquad \text{...(6.1.26}a\text{)}$$

When a heavy particle with velocity V collides with a light particle, the maximum velocity imparted to it is $2V$. Correspondingly the maximum energy transfer to the light particle is $1/2$ m $(2V)^2 = 2mV^2$. Applying this classical argument in the case of α-particle-electron collision, one gets from equation (6.1.4).

$$2mV^2 = \frac{2K^2z^2e^4}{mV^2b_{\min}^2}$$

or
$$\frac{1}{b_{\min}^2} = \frac{m^2V^4}{K^2z^2e^4}$$

Substituting in eqn. (6.1.26*a*) gives

$$\sigma_x^2(E) = K^2 4\pi nz^2e^4x$$

This is the variance in the energy loss in traversing a distance x. The variance in specific energy loss in the medium of atomic density N and atomic number Z is

$$\sigma^2(E) = K^2 4\pi nz^2e^4 NZ \qquad \ldots (6.1.27)$$

In the case of α-particles the average energy loss in traversing the range R is E_o, the energy of the α-particle. One can assume that the loss of energy by α-particles traversing a distance equal to the range R has a Gaussian distribution with the variance given by expression (6.1.27). This variance in energy loss manifests as straggling in the range or the path length in which an α-particle looses exactly the same amount of energy, E_o. The standard deviation in energy loss $\sigma(E)$ and that in range R_o denoted by $\sigma(R)$ are related as

$$\frac{\sigma(E)}{\sigma(R)} = \left|\frac{dE}{dx}\right| \qquad \ldots. (6.1.28)$$

It is clear from the above expression that the range of α-particles of a definite energy E_o would have a Gaussian distribution around R_o. This is shown by the differential curve in Fig. 6.9.

In the above discussion, only those collisions between α-particles and electrons are considered in which the energy transfer is a very small fraction of the total energy of the incoming particle, as also the straggling in its energy loss. There may be some collisions with atomic nuclei in which an incoming α-particle may loose an appreciable fraction of its energy. Such collisions are rare and do not effect the average energy loss, but they do influence the fluctuations appreciably. This gives rise to a tail in the Gaussian distribution on the side of higher energy loss. This effect is easily seen in the spectrum of electrons passing through thin absorbers as shown in Fig. 6.13.

It has been shown above that the range of a heavy particle with a given energy E_o has a Gaussian distribution around the average range R_o. The probability $P(R)$ for observing a range between R and $R + dR$ for an α-particle is given as

$$P(R)dR = \frac{1}{2s} \exp - \left[\frac{\pi}{4s^2} (R - R_o)^2 \right] \qquad \text{.... (6.1.29)}$$

where the parameter's' is defined as

$$s = \left[\frac{\pi}{2} \left\langle (R_e - R_o)^2 \right\rangle \right]^{\frac{1}{2}} \qquad \text{.... (6.1.30)}$$

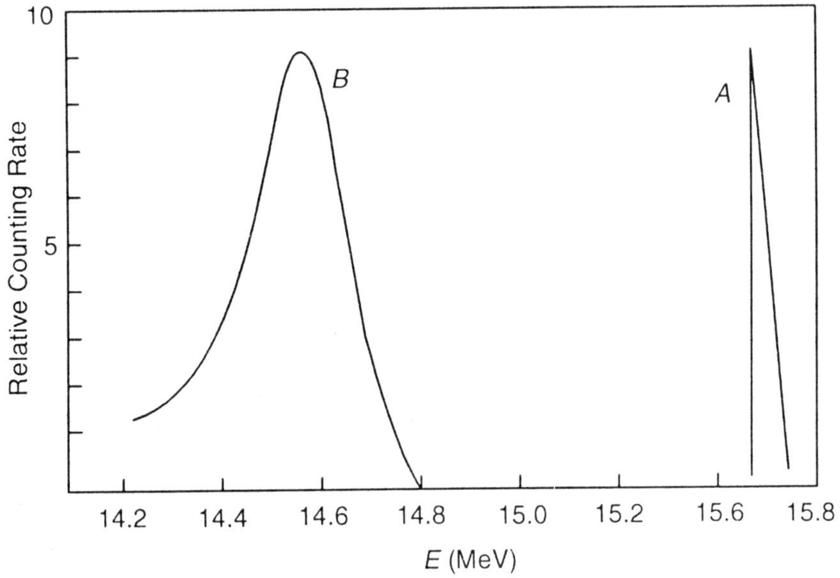

Fig. 6.13: Energy spread of 15.7 MeV electrons after passing through 0.86 gm/cm^2 thick aluminium absorber. Curve *A* is spectrum of incident beam and curve *B* is that of the transmitted beam. [E. L. Gold-wasser, F. E. Mills and A. O. Hanson, (1952), *Phys. Rev.*, **88**: 1137].

and it is called the straggling of range and is the difference between the average range R_o and the extrapolated range R_e (Fig. 6.8). The straggling s for α-particles is about 1-2 per cent while for low energy protons it is about 2-5 per cent.

6.2 INTERACTION OF GAMMA - RAYS WITH MATTER

Gamma-rays are electromagnetic radiations of very small wavelength. The gamma rays, generally have energies between 10 keV and 10 MeV. Gamma rays are emitted when a nucleus in its excited state makes transition to the ground or other excited states. If similar energy radiations are emitted in the electronic transitions, they are called *X*-rays. Gamma-rays and *X*-rays of the same energy (and wavelength) behave identically.

When gamma-rays pass through a medium, they are scattered or absorbed by one of the following processes.

(1) Elastic scattering
(2) Compton scattering
(3) Photoelectric effect
(4) Pair production
(5) Nuclear reaction.

6.2.1 Elastic Scattering

In the classical model, an electromagnetic wave sets a charged particle in oscillating motion in the direction of its E (electric) vector. The frequency of oscillations is same as that of the incident wave. The oscillating charged particle radiates electromagnetic waves. If the incident waves are unpolarized, the angular distribution of the scattered radiations is proportional to $(1 + \cos^2\theta)$, where θ is the angle of scattering of the wave. If the electromagnetic wave of large wave length (low energy) impinges on a free electron, the electron may absorb only negligible recoil momentum from the scattered radiations. The scattered waves have same frequency as the incident wave and are coherent. This is known as Thomson scattering. The total cross section σ for scattering is

$$\sigma = \frac{8\pi}{3} r_o^2 \qquad \text{... (6.2.1)}$$

$$r_o = \frac{e^2}{m_o c^2} = 2.82 \times 10^{-15}\,\text{m}$$

As the energy of the incident radiation increases, its momentum $\vec{p}_i$ increases. If $\vec{p}_\theta$ is the momentum of the scattered photon, the change in momentum $\vec{p}_i - \vec{p}_\theta$ is appreciable. This difference in the two momenta is taken up by the electron, giving it some kinetic energy. This kinetic energy has to come from the energy of the incident radiation. Thus the scattered radiation no longer has the same energy as the incident radiation. The scattering is no longer elastic or coherent. This becomes a case of Compton scattering. In the region of gamma ray energies, there is no Thomson scattering by free electrons.

6.2.2 Nuclear Thomson Scattering

The atomic nucleus of the medium is a free, massive charged particle and can give rise to Thomson scattering. Due to its large mass, the recoil momentum produces only a very small kinetic energy of the nucleus. The energy of the scattered radiation does not differ much from the energy of the incident radiation and it is thus a case of elastic and coherent scattering. The differential cross section per unit solid angle for Thomson scattering by a nucleus of charge Z and mass M is given as

$$\frac{d\sigma}{d\Omega} = \frac{Z^4 e^4}{M^2 c^4} \cdot \frac{1}{2}(1 + \cos^2\theta) \qquad \text{...(6.2.2)}$$

The total scattering cross section σ_o is obtained by integrating the above equation over the 4π solid angle, so that

$$\sigma_o = \frac{8\pi}{3} \frac{Z^4 e^4}{M^2 c^4} \qquad \text{...(6.2.3)}$$

6.2.3 Rayleigh Scattering

Classically one assumes that an electron bound in an atom makes oscillatory motion with frequency ν such that its binding energy in the atom is $h\nu$. When an electromagnetic wave of frequency ν_o falls on an atom it sets almost all the electrons in forced oscillations. These electrons radiate waves of the frequency ν_o coherently. The amplitudes of the waves scattered by all the electrons are added. The superimposition of the various amplitudes results in the maxima of intensity of scattered waves in the forward direction. In this process the recoil momentum is taken up by the whole atom. The phenomenon is known as Rayleigh scattering.

The approximate theory for coherent scattering (Rayleigh scattering) of gamma-rays gives the differential cross-section as a product of two factors. The first factor is the probability that the incident radiation is scattered by a certain angle θ and transfers the recoil momentum $(\vec{p}_i - \vec{p}_0)$ to an atomic electron as though it were free. This factor may be taken as the Thomson scattering cross section. The second factor is the probability that the Z electrons of the atom take up the recoil momentum without absorbing any energy. This factor is the square of an atomic structure factor F, which is the ratio of the amplitude of the radiation, scattered by the atom to the amplitude which a single electron would scatter. It has been shown that for coherent scattering, more than three quarter of the scattering takes place at angles less than a characteristic angle θ_c such that

$$\sin\frac{\theta_c}{2} = 0.026 Z^{\frac{1}{2}} \left(\frac{m_o c^2}{h\nu_o}\right) \qquad \text{.... (6.2.4)}$$

where $h\nu_o$ is the energy of the incident gamma-ray photon. For angles $\theta < \theta_c$ the scattering cross section is proportional to Z^2 and is independent of the photon energy. When $\theta > \theta_c$ the different scattering cross-section in solid angle $d\Omega$ is given as

$$\frac{d\sigma}{d\Omega} = \frac{8.73 \times 10^{-33}}{\sin^3\theta/2} \left(\frac{Z m_o c^2}{h\nu_o}\right)^3 \cdot \frac{1}{2}\left(1 + \cos^2\theta\right) \qquad \text{...(6.2.5)}$$

The calculations of the coherent Rayleigh scattering cross section is not straight forward and depends upon the model of the atomic structure.

When the energy of the incident gamma radiations is high, the frequency of the forced oscillations of an electron is higher than its natural frequency and there is a probability that the electron may leave the atom. This represents photo electric effect. It is thus seen that the photoelectric effect and the Rayleigh scattering are related phenomenon.

6.2.4 Delbruk Scattering

In the strong electromagnetic field of the nucleus, a gamma ray photon may produce a virtual electron positron pair, which may combine to give rise to a elastically scattered photon. This is known as Delbruk scattering or nuclear potential scattering and is purely a quantum phenomenon. When the energy of the incident gamma-ray photon is more than the rest mass of the e^+ - e^- pair ($E_r > 2mc^2$) the pair may come out. Both pair production and Delbruk scattering have an appreciable magnitude for gamma-ray energy $E_r >> 2mc^2$. Experimental studies at gamma-ray energy from 1 to 3 MeV have failed to prove conclusively the presence of Delbruk scattering. It has however been confirmed in scattering of gamma-rays of energies 9MeV, 17 MeV and 87 MeV. It may be seen that Delbruk scattering and pair production are related phenomena.

6.2.5 Nuclear Resonance Scattering

If the energy of the gamma-ray photon is just right to excite the nucleus to one of its excited states, it is absorbed resonantly. The photon is re-emitted in different directions when the nucleus makes transition to its ground state. The cross section for resonant scattering by the atomic nucleus is greater than the cross section for Compton scattering of gamma-rays. Nuclear resonant scattering is a very rare phenomenon below 5 MeV photon energy and can be safely neglected.

Nuclei show two very broad excited states in the energy region 12 to 25 MeV. The first, known as magnetic dipole giant resonance, occurs at an energy between 12-20 MeV. The second excited state, known as electric quadrupole giant resonance, has an excitation energy between 15-25 MeV. If the energy of an incident gamma-ray is above 15 MeV, the giant resonances could contribute to nuclear resonance scattering.

All the processes giving rise to elastic scattering of gamma-rays are coherent which means that the final amplitude of scattered wave 'a' is the sum of the amplitudes of all the coherent processes. Thus

$$a = a_{Th} + a_R + a_D + a_N$$

where a_{Th}, a_R, a_D, a_N are the amplitudes for nuclear Thomson, Rayleigh, Delbruk and nuclear resonance scattering respectively. The intensity of the scattered radiation is given by a^2.

6.2.6 Compton Scattering

In the very low energy region (visible region) a free electron can scatter electromagnetic radiations without change of energy (Thomson scattering). If the energy of the electromagnetic quanta increases, there is a change in the energy of the scattered quantum and one has a case of Compton scattering. Compton scattering can be understood on the basis of particle property of the electromagnetic radiations. The photon on being scattered by an electron imparts to the electron the recoil momentum and hence kinetic energy. If the energy imparted to the electron is large enough, the bond of the electron in the atom breaks and it behaves as a free charged particle.

Consider a photon of energy $h\nu_0$ and momentum $h\nu_0/c$ interacting with a free electron of mass m_o at rest. The photon is scattered at an angle θ and electron at angle ϕ with respect to the incident direction as shown in Fig. 6.14. The energy of the scattered photon is $h\nu'$. One can obtain the energy of the photon scattered at an angle θ by applying the laws of conservation of energy and momentum along the incidence direction and perpendicular to it, giving (v is velocity of electron and $\beta = v/c$)

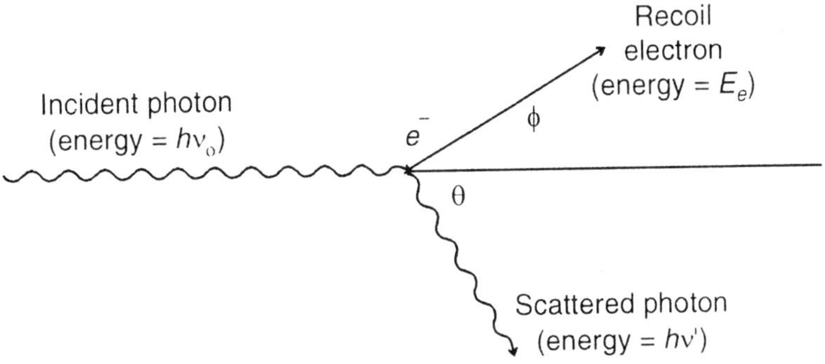

Fig. 6.14: Compton scattering of a photon of energy $h\nu_o$ by a stationary electron

$$h\nu_0 = h\nu' + m_o c^2 \left\{ \frac{1}{(1-\beta^2)^{1/2}} - 1 \right\} \qquad \text{....(6.2.6)}$$

$$\frac{h\nu_0}{c} = \frac{h\nu'}{c}\cos\theta + \frac{m_o\beta c}{(1-\beta^2)^{1/2}}\cos\phi \qquad \text{....(6.2.7}a\text{)}$$

$$0 = \frac{h\nu'}{c}\sin\theta + \frac{m_o\beta c}{(1-\beta^2)^{1/2}}\sin\phi \qquad \text{....(6.2.7}b\text{)}$$

From the above three equations the three unknowns $h\nu'$, ϕ and E_e can be determined in terms of θ. Equations (6.2.7a) and (6.2.7b) give respectively

$$h\nu_o - h\nu'\cos\theta = \frac{m_o\beta c^2}{\sqrt{(1-\beta^2)}}\cos\phi \quad ...(6.2.8a)$$

$$-h\nu'\sin\theta = \frac{m_o\beta c^2}{\sqrt{(1-\beta^2)}}\sin\phi \quad ...(6.2.8b)$$

On squaring and adding both the above equations one gets

$$(h\nu_o)^2 + (h\nu')^2(\sin^2\theta + \cos^2\theta) - 2h\nu_o h\nu'\cos\theta$$

$$= \frac{m_o^2\beta^2c^4}{1-\beta^2}(\cos^2\phi + \sin^2\phi)$$

$$= \frac{m_o^2c^4}{1-\beta^2} - m_o^2c^4 \quad ...(6.2.9)$$

Equation 6.2.6 gives

$$(h\nu_o)^2 + (h\nu')^2 - 2h\nu_o h\nu' + m_o^2c^4 + 2m_oc^2(h\nu_o - h\nu') = \frac{m_o^2c^4}{1-\beta^2} \quad ...(6.2.10)$$

Subtracting (6.2.10) from (6.2.9) gives

$$2h\nu_o\, h\nu'\,(1 - \cos\theta) - 2m_oc^2\,(h\nu_o - h\nu') = 0 \quad ...(6.2.11)$$

or $2h\nu_o\, h\nu'\,(1 - \cos\theta) = 2m_oc^2\,(h\nu_o - h\nu')$

or $h\nu'\,\{h\nu_o\,(1 - \cos\theta) + m_oc^2\} = m_oc^2\, h\nu_o$

or
$$h\nu' = \frac{h\nu_o}{1 + \frac{h\nu_o}{m_oc^2}(1-\cos\theta)}$$

$$= \frac{h\nu_o}{1 + 2\alpha\sin^2\theta/2} \quad ...(6.2.12)$$

where

$$\alpha = \frac{h\nu_o}{m_oc^2}$$

The equation (6.2.12) gives the energy of the photon scattered by an angle θ. In terms of the wave length λ of the incoming radiation and λ' of the scattered radiation, equation (6.2.12) gives

$$\Delta\lambda = \lambda' - \lambda = \frac{h}{m_o c}(1 - \cos\theta)$$

$$= \lambda_c (1 - \cos\theta) \qquad ...(6.2.13)$$

where $\lambda_c = h/m_o c$ is called the Compton wavelength of the electron and its value is equal to 0.0243 Å, so that

$$\Delta\lambda = 0.0243\,(1 - \cos\theta)\ \text{Å}$$

$$= 0.0486 \sin^2 \frac{\theta}{2}\ \text{Å} \qquad ...(6.2.14)$$

It may be observed that the change in wave length $\Delta\lambda$ is independent of the wave length of the incoming radiation.

The kinetic energy E_e of the recoiling electron is the difference in the energies of the incident and the scattered photon so that

$$E_e = h\nu_o \frac{a(1 - \cos\theta)}{1 + \alpha(1 - \cos\theta)} \qquad ...(6.2.15)$$

In terms of the angle ϕ, the recoil energy of the electron is given by

$$E_e = h\nu_o \frac{2\alpha\cos^2\phi}{(1+\alpha)^2 - \alpha^2\cos^2\phi} \qquad ...(6.2.16)$$

and the angle ϕ is given as

$$\tan\phi = \frac{\cot\frac{\theta}{2}}{1+\alpha} \qquad ...(6.2.17)$$

It may be observed from the above equations that
when $\theta = 0$, $\phi = 90°$ and when $\theta = 180°$, $\phi = 0°$

This shows that the scattered electron always moves in the forward quadrants. The minimum energy of the scattered electron is zero when $\phi = 90°$. The maximum energy of the electron when $\phi = 0$ and $\theta = 180°$ is $h\nu_o = \frac{2\alpha}{1+2\alpha}$. For very high energy gamma-rays, $\alpha >> 1$ or $h\nu_o >> m_o c^2$, $E_{e_{max}} = h\nu_o - \frac{m_o c^2}{2}$, correspondingly the energy of the back scattered photon is $m_o c^2/2$. The energy of a high energy photon scattered by $\theta = 90°$ according to equation (6.2.12) is $h\nu' = h\nu_o/1 + \alpha \approx m_o c^2$.

The differential cross-section or the probability of Compton scattering of a photon of energy $h\nu_o$ by an angle θ in a solid angle $d\Omega = 2\pi \sin\theta\, d\theta$ has been given by Klein and Nishina as

$$\frac{d\sigma}{d\Omega}=\frac{r_o^2}{2}\left(\frac{\nu'}{\nu_o}\right)^2\left(\frac{\nu_o}{\nu'}+\frac{\nu'}{\nu_o}-2+4\cos^2\Theta\right) \quad ...(6.2.18)$$

where r_o is the classical radius of the electron, and Θ the angle between the electric vectors of the incident and scattered waves. The incoming and the outgoing waves both are polarized. Generally all polarizations of the scattered gamma-rays are detected. Integrating over all polarizations of the outgoing wave one gets.

$$\frac{d\sigma}{d\Omega}=\frac{r_o^2}{2}\left(\frac{\nu'}{\nu_o}\right)^2\left(\frac{\nu_o}{\nu'}+\frac{\nu'}{\nu_o}-2\sin^2\theta\cos^2\eta\right) \quad ...(6.2.19)$$

where θ is the angle of scattering and η is the angle between the plane of polarization of the incident gamma-rays and the plane of scattering. If the incoming gamma-rays are unpolarized one can integrate over the angles η giving.

$$\frac{d\sigma}{d\Omega}=\frac{r_o^2}{4}\left(\frac{\nu'}{\nu_o}\right)^2\left(\frac{\nu_o}{\nu'}+\frac{\nu'}{\nu_o}-\sin^2\theta\right) \quad ...(6.2.20)$$

The Klein-Nishina relation for differential scattering is found to be correct to about 15 per cent. The total scattering cross section σ obtained by integrating ever values of θ between 0 and π is correct to about 3 per cent and is given as

$$\sigma=\int_0^\pi 2\pi\sigma(\theta)\sin\theta d\theta$$

$$=2\pi r_o^2\left[\frac{1+\alpha}{\alpha^2}\left\{\frac{2(1+\alpha)}{1+2\alpha}-\frac{1}{\alpha}\ln(1+2\alpha)\right\}+\frac{1}{2\alpha}\ln(1+2\alpha)-\frac{1+3\alpha}{(1+2\alpha)^2}\right] \quad ...(6.2.21)$$

where $\alpha=\frac{h\nu_o}{m_oc^2}$. For very low energy gamma-rays such that $h\nu << m_oc^2$ the Klein Nishina cross section becomes

$$\sigma=\sigma_{Th}\left[1-2\alpha+5.2\alpha^2-13.3\alpha^3\right] \quad ...(6.2.22)$$

where $\sigma_{Th}=\frac{8\pi}{3}r_o^2$ is the Thomson scattering cross-section.

The Compton scattering cross section has been calculated per electron. In a material with atomic number Z the cross section for scattering per atom is $Z.\sigma$.

The angular distribution of the Compton scattered gamma-rays is given by eqn. (6.2.20) and depends upon the energy of the radiations. Fig. 6.15 shows the angular dependence of the differential scattering cross section for different energy gamma rays ($\alpha = h\nu_o/m_oc^2$). It may be observed from the figure that at high gamma-ray energy the scattered radiations go predominantly in the forward direction. At very low energy $\alpha \approx 0$ the angular distribution is same as that for Thomson scattering.

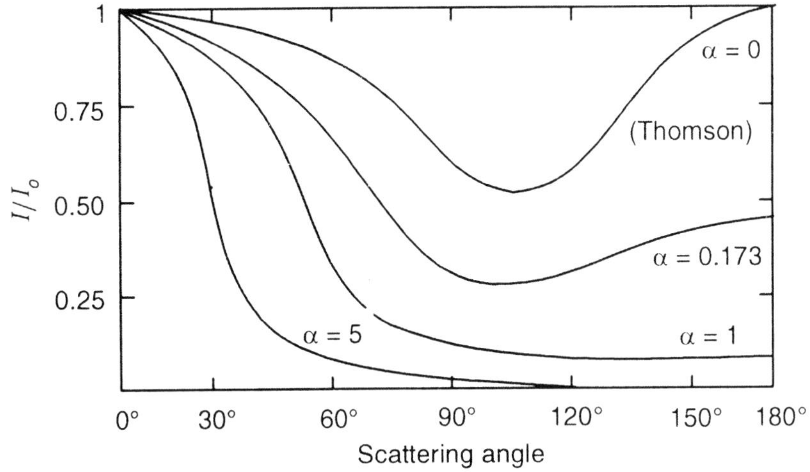

Fig. 6.15: Normalised intensity of different energy Compton scattered γ-rays as a function of scattering angle. ($\alpha = h\nu_o/m_oc^2$)

The Klein Nishina differential cross-section for giving an electron a recoil energy in the interval T and $T + dT$ is given as

$$\frac{d\sigma}{dT} = \frac{\pi r_o^2}{\alpha m_o c^2}\left[2 + \left(\frac{T}{h\nu_o - T}\right)^2 \left\{\frac{1}{\alpha^2} + \frac{h\nu_o - T}{h\nu_o} - \frac{2}{\alpha}\left(\frac{h\nu_o - T}{T}\right)\right\}\right] \quad \ldots(6.2.23)$$

This cross-section eqn. (6.2.23) when plotted as a function of the electron energy gives the energy spectrum of the Compton electrons produced in a medium by gamma-rays of given energy $h\nu_o$. Fig. 6.16 shows the cross-section $d\sigma/dT$ hence the Compton electron energy spectrum. The spectrum shows an abrupt fall at an energy $h\nu_o$. $(2\alpha/1 + 2\alpha)$. This is known as the Compton edge.

6.2.7 Measurement of Polarization of Gamma-rays

In many experiments it is desirable to measure the polarization of certain gamma-rays. The polarization of gamma-rays is defined as

$$P = \frac{N_{\parallel} - N_{\perp}}{N_{\parallel} + N_{\perp}} \quad \ldots(6.2.24)$$

where $N_{\parallel}$ and $N_{\perp}$ are the number of photons with their electric vector electric parallel and perpendicular to a certain reference plane, respectively. In experiments using cascade gamma-rays, emitted from a radioactive nucleus, the plane in which the two cascade photons are detected is the reference plane. In nuclear reactions the incident beam and the emitted gamma-ray define the reference plane.

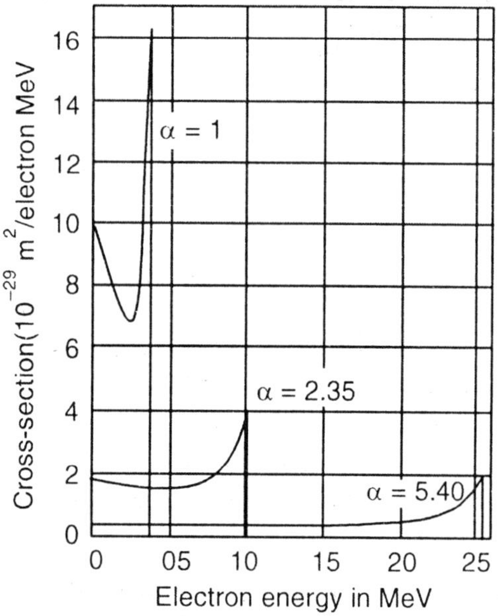

Fig. 6.16: Energy spectra of Compton scattered electrons for different energy γ-rays ($\alpha = h\nu_0/m_0c^2$)

The cross-section for Compton scattering according to equation (6.2.19) is minimum for $\eta = 0°$ (plane of the incident electric vector) and maximum for $\eta = 90°$. The difference in the cross-section at $\eta = 0°$ and at $\eta = 90°$ depends upon the incident photon energy $h\nu_0$ and the scattering angle θ. For a given incident photon energy, suitable value of the scattering angle θ can be calculated, where the difference in cross-section at $\eta = 0$ and $\eta = 90°$ is maximum. For an arbitrary reference plane the polarization of gamma-rays is defined as in equation (6.2.24).

The gamma-rays whose polarization is to be measured are Compton scattered by the plastic scintillator of a scintillation counter, at a fixed angle θ as shown in Fig. 6.17. The scattered photons are detected by a NaI(Tl) scintillation counter. The pulses from the two scintillation counters are fed to a time coincidence circuit of short resolving time. A coincidence is recorded when a photon scattered from plastic scintillator is detected in the NaI(Tl) counter. A short resolving time of the coincidence circuit ensures that unrelated events triggering the two scintillation counters and giving chance coincidences are minimised. The observed number of coincidences at a position of the counter must be corrected for chance coincidences and back ground coincidences due to cosmic rays and room background. The true coincidence counting rates are recorded when the NaI(Tl) counter is in the reference plane and also when it is in a plane perpendicular to the reference plane. If the two counting rates are C_o and $C_\perp$ the polarization is given by

$$P = \frac{C_\perp - C_o}{C_\perp + C_o}$$

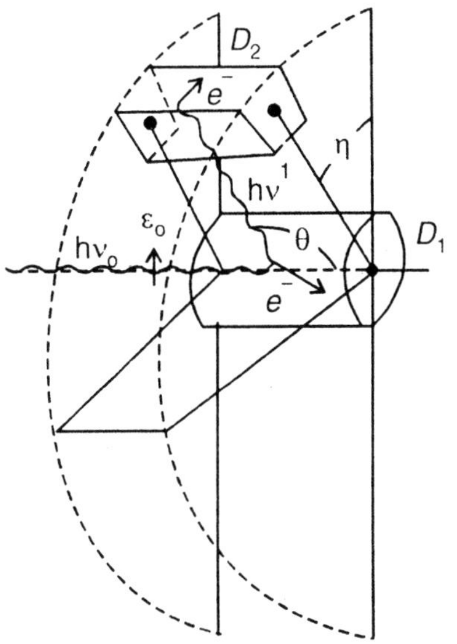

Fig. 6.17: Schematic diagram of an apparatus used for measuring polarization of γ-rays. Two scintillation counters are used in coincidence. The photon $h\nu_0$ scattered at angle θ by plastic scintillator D_1 is counted in a NaI (*Tl*) scintillation counter D_2. Detector D_2 can be rotated with respect to axis of detector D_1. Coincidences between D_1 and D_2 are recorded for η = o and $\eta = 90°$ with respect to the reference plane.

The measurement of polarization of gamma rays is difficult because of a large background and chance counting rate and small difference between $C_\perp$ and C_o. It is essential to shield both the scintillation counters with thick lead shields. Wherever possible the coincidence circuit should be triggered by the accelerator beam or pulses from a third detector detecting cascade radiations. This reduces the chance counting rates considerably.

Measurement of polarization of gamma rays helps in establishing the multipolarity of the gamma ray transition in the nucleus.

6.2.8 Photoelectric Effect

When a gamma-ray photon interacts by photoelectric effect, it imparts all its energy to an electron bound in an atom. Consequently the electron is ejected from the atom with kinetic energy $E = h\nu_o - \phi_I$, where ϕ_I is the ionization energy or binding energy of the electron in its shell. The recoil momentum is taken up by the whole atom. Photoelectric effect in a particular shell of an atom can take place only if the photon energy is greater than the binding energy of the electron in the shell. The probability of photoelectric effect taking place is greatest for the *K*-shell of an atom.

The energy dependence of the cross section for photo electric effect can be considered for three energy regions as follows:

(*a*) The photon energy is in the vicinity of the binding energy of the electrons in an atomic shell.
(*b*) Photon energy is greater than the binding energy of the electrons.
(*c*) At very high photon energies such that $h\nu_o >> m_o c^2$

In the first region the cross-section shows sharp discontinuities at the binding energies of electrons in various shells as shown in Fig. 6.18. As the photon energy is reduced from well above the K-shell binding energy, the photoelectric cross section increases. Just below the K-shell binding energy, I_K the cross section reduces sharply and the first spike appears.

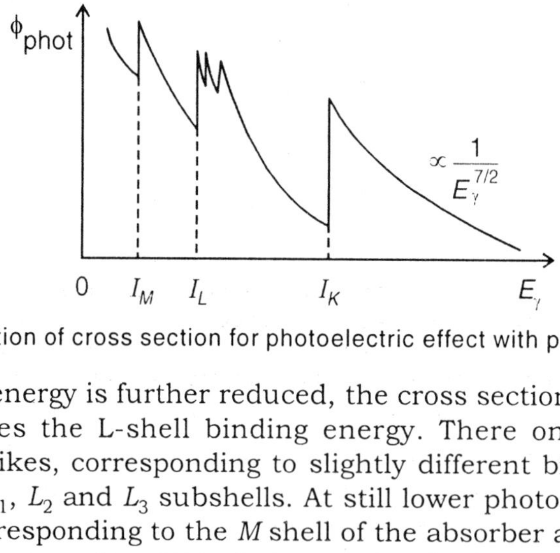

Fig. 6.18: Variation of cross section for photoelectric effect with photon energy $E\gamma$

As the photon energy is further reduced, the cross section increases till the energy approaches the L-shell binding energy. There one observes three closely spaced spikes, corresponding to slightly different binding energies of electrons in the L_1, L_2 and L_3 subshells. At still lower photon energy, one can observe spike corresponding to the *M* shell of the absorber atom.

In the second case, when the photon energy is greater than the K-shell binding energy, the cross section σ_{Ph} for photoelectric effect is given as

$$\sigma_{Ph} = \left(\frac{32}{\alpha^7}\right)^{\frac{1}{2}} \alpha_o^4 \cdot Z^5 \sigma_{Th} \quad \text{...(6.2.25)}$$

where σ_{Th} is the Thomson scattering cross-section, $\alpha = h\nu_o / m_o c^2$, $\alpha_o = e^2/hc = 1/137$ is fine structure constant and Z is the atomic number of the absorbing material. It may be observed that the cross section increases as Z^5 and reduces as $(h\nu_o)^{-\frac{7}{2}}$.

At relativistic energy where $\alpha >> 1$ the photoelectric cross section is given as

$$\sigma_{Ph} = \frac{1.5}{\alpha}\alpha_o^4 \cdot Z^5 \sigma_{Th} \qquad \ldots(6.2.26)$$

The energy dependence is now only as $(h\nu_o)^{-1}$.

The angular distribution of the ejected photo electrons is pronounced in the forward direction. As the energy of the gamma-ray photon increases, the photoelectron emission becomes more and more peaked in the forward direction. This behaviour of the differential photoelectric cross section $\sigma(\theta)$ is shown in Fig. 6.19.

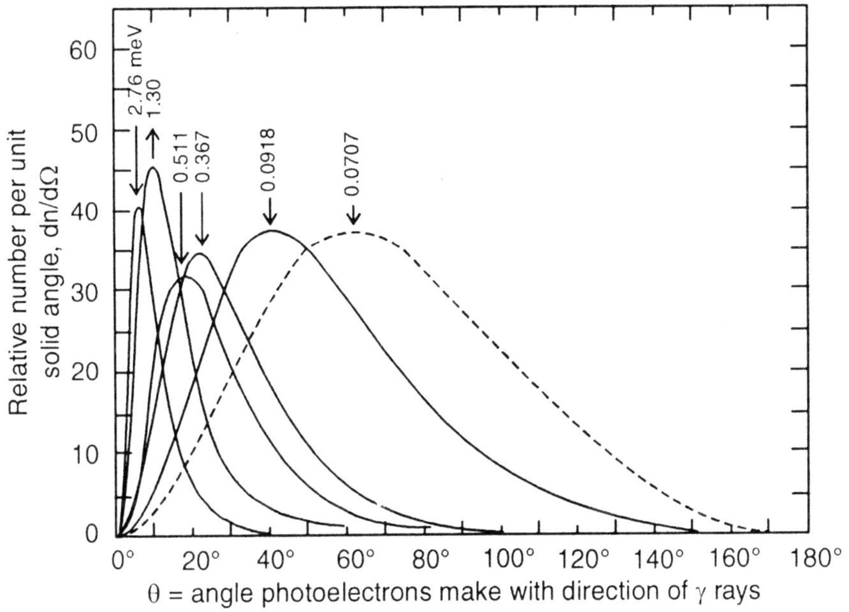

Fig. 6.19: Calculated directional distribution of photo-electrons per unit solid angle [C. M. Davisson and R. D. Evans (1952), *Rev. Mod. Phys.*, **24**: 79].

6.2.9 Differential Absorption of Gamma-rays

The phenomenon of the absorption edges in the photoelectric cross-section is made use of in the higher absorption of slightly higher energy gamma-rays in comparison to that of lower energy gamma-rays. If a beam of two gamma-rays differing slightly in energy is passed through an absorber, the lower energy gamma-ray is normally absorbed more than the higher energy radiations. If however, the two gamma-rays are passed through an absorber whose *K*-shell binding energy is just below the higher energy gamma-ray and above the energy of second gamma-ray radiation, the first radiation will be absorbed more (due to larger photoelectric cross section) than the lower energy gamma-ray. Similar differential absorption of *X*-rays or gamma-rays can be obtained using the *L* absorption edge. One can achieve almost four times more absorption of the higher energy gamma-rays. One has to select a suitable element or its salt whose *K*, or *L* shell binding energy is just below the higher energy gamma rays.

6.2.10 Auger Effect

When photoelectric effect takes place, a vacancy is created in the shell of an atom. The electrons in the outer shells fall in the vacancy emitting fluorescent radiations. If the vacancy is created in the K-shell, the whole K-series of *X*-rays are emitted. It is possible that at times, instead of the emission of fluorescent radiations, an electron from the next outer shell is ejected. A vacancy in the K-shell may be filled up by an *L* electron and instead of *X*-rays another *L* electron is emitted with energy I_K-$2I_L$, where I_K and I_L are the ionization energies of the *K* and the *L*-shell respectively. It is also possible that the vacancy in the K-shell is filled by an *L*-electron and a *M* electron is ejected with energy $I_K - I_L - I_M$. This process of electron emission is known as Auger effect after its discoverer. The emitted electrons are called Auger electrons.

The emission of fluorescent radiation and Auger electrons are competing processes. The number of *X*-ray photons emitted per vacancy in a given shell is called the fluorescent yield of the shell and is denoted by β. A more detailed study of the phenomenon may distinguish the fluorescent yields β for different upper and lower shells, so that

$$\beta_k = \beta_{KL} + \beta_{KM} + \beta_{KN} \quad ...(6.2.27)$$

where β_{KM}, β_{KM} indicate the partial fluorescent yields for filling the vacancy in the K-shell by transition from *L*, *M*, *N* shells. The fluorescent yield increases with the atomic number of the absorber as shown in Fig. 6.20.

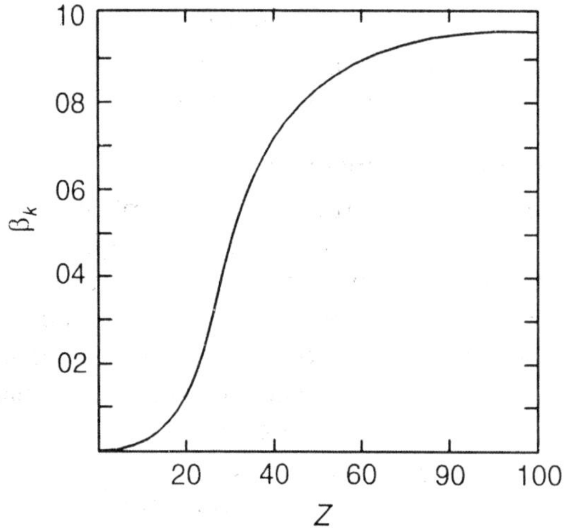

Fig. 6.20: Fluorescent yield for *K*-shell as a function of atomic number *Z* of the atom

6.2.11 Pair Production

An energetic photon, in the strong electromagnetic field of a nucleus, can transform into an electron-positron pair. The recoil momentum is taken up by the nucleus so as to conserve energy and momentum in the transformation. The threshold energy for the process of pair production is the rest mass energy of the pair *i.e.* $2m_oc^2 = 1.022$ MeV in the centre of mass system. The kinetic energy of the electron positron pair is $h\nu_o - 2m_oc^2$. The cross section σ_{pair} for the formation of electron – positron pair at relativistic energy $h\nu_o >> 2m_oc^2$ is

$$\sigma_{pair} = \frac{r_o^2 Z^2}{137} \quad ...(6.2.28)$$

where Z is the atomic number of the nucleus. The process becomes appreciable above the gamma-ray energy of 3MeV.

6.2.12 Positron Annihilation

When an electron–positron pair is created in a medium, both the particles move in forward direction. Both the particles loose their energy by ionization in the medium and by radiation and come to rest. The electron gets attached to some atom to form a negative ion. The positron may capture an electron and form a hydrogen atom like system with the two particles revolving around each other. This system is known as positronium. In a small time interval, the pair annihilates emitting two photons each of energy $m_oc^2 = 0.51$ MeV moving in opposite directions so as to conserve momentum.

For the two annihilation photons moving in opposite directions, the $e^+ - e^-$ can annihilate only if their spins are in opposite direction and their angular momentum is zero. This correspond to the $^1S_0$ state of the positronium. If the relative velocity v of the positron with respect to the electron is high, the positronium can be said to be formed in an unbound state. For such an unbound state the cross-section for $e^+ - e^-$ annihilation resulting in two photons is given as

$$\sigma_{2\gamma} = 4\pi r_o^2 c / v \quad ...(6.2.29)$$

where r_o is the classical electron radius and c the velocity of light. If N is the density of electrons in the medium, the probability $\lambda_{2\gamma}$ for two photon annihilation is given as

$$\lambda_{2\gamma} = N\, v\, \sigma_{2\gamma}$$

$$= 4\pi r_o^2 Nc \quad ...(6.2.30)$$

If the velocity v of the positron is low, the positronium is formed in a bound state and the electron density N in equation (6.2.30) is replaced by the probability of the overlap of the electron positron clouds. The probability of two photon annihilation is than given as

$$\lambda_{2\gamma} = \sigma_{2\gamma}\, v\, |\psi(0)|^2$$

$$= \frac{1}{2}\left(\frac{e^2}{hc}\right)^5 \frac{mc^2}{h}\frac{1}{n^3}$$

$$= \frac{1}{2}\left(\frac{e^2}{hc}\right)^5 \frac{mc^2}{h}\frac{1}{n^3}$$

$$= \frac{1}{125\times10^{-10}}\frac{1}{n^3}\,\text{sec}^{-1}, \qquad \text{... (6.2.34)}$$

where a_o is the Bohr radius and n the total quantum number of the orbit.

The probability of forming positronium in the triplet state is three times that of the singlet state. Annihilation of electron positron in the triplet state results in the emission of three photons. The probability of three photon annihilation is only 1/372 of that of two photon annihilation. Thus the life time of three photon annihilation is 372 times the life time for two photon annihilation. However, if the formation of positronium takes place in the presence of molecules having magnetic moment, the triplet state changes to singlet state by collisions, which decays with a life time of 1.25×10^{-10}sec. The measured mean life of the triplet state is thus the mean life for conversion of the state into singlet state. One can observe the true life time of the triplet state when $e^+ - e^-$ annihilation takes place in gases like freon (CCl_2F_2) whose molecules do not have any magnetic moment.

6.2.13 Photo Nuclear Reactions

If the energy of the incoming photon is greater than the binding energy of a nucleon in the nucleus, it may be absorbed with the emission of a nucleon. The probability of photonuclear reaction also known as nuclear Compton effect is appreciable at photon energies above 10 MeV.

6.2.14 Absorption of Gamma-rays

When a beam of gamma rays passes through an absorber, the photons may be absorbed (photoelectric effect, pair production and photonuclear reaction) or may be scattered (Compton and elastic scattering) away from the beam. The transmitted beam intensity is attenuated. The total cross section σ for the attenuation of the beam is the sum of cross sections for all the processes, so that

$$\sigma = \sigma_{Ph} + \sigma_{Pair} + \sigma_{Nuc} + Z\sigma_c + \sigma_{el} + \sigma_{Rays} \qquad \text{...(6.2.32)}$$

where σ_{Ph}, σ_{Pair}, σ_{Nuc}, $Z\sigma_c$, and σ_{el} are the cross sections per atom for photoelectric, pair production, nuclear reaction, Compton scattering and elastic scattering, respectively. Unless the beam is very narrow, the Rayleigh scattering is not observable. The elastic scattering cross section is generally very small (except at resonance energy) compared to that for other process.

If N is the number of atoms per unit volume of an absorber the linear absorption is given by the relation

$$I = I_o e^{-N\sigma x} = I_o e^{-\mu x} \qquad \text{... (6.2.33)}$$

where I_o and I are the intensities of the incident and the transmitted beam of gamma rays through an absorber of thickness x. The quantity μ is known as the linear absorption coefficient of the gamma ray. In cgs system μ is cm^{-1} and thickness x is measured in centimetres.

The mass absorption coefficient μ_m is defined as

$$\mu_m = \mu/\rho \qquad \text{...(6.2.34)}$$

where ρ is the density of the absorber. The thickness of the absorber t is measured in gm/cm^2 and one has $t = x.\ \rho$.

Equation (6.2.33) gives

$$I = I_o e^{-\mu x} = I_o e^{-\frac{\mu}{\rho} \cdot x\rho} = I_o e^{-\mu_m \cdot t} \qquad \text{...(6.2.35)}$$

The mass absorption coefficient $\mu_m = N\sigma/\rho$ depends upon the energy of the gamma-ray and the atomic number of the absorber. At gamma-ray energies below 100 keV and absorbers of high Z, the photo electric absorption of the beam is predominant. The Compton effect is the dominant mode upto about 2 MeV. At gamma-ray energies above 10 MeV the pair production is the dominant mode of absorption of the gamma rays. Fig. 6.21 shows the variation of the cross section for the three major processes as a function of energy.

6.3 INTERACTION OF NEUTRONS WITH MATTER

Neutrons are neutral particles, a little heavier than the protons. Neutrons are an integral constituent of a nucleus in which they are bound by strong interaction. A free neutron decays by β^--emission to a proton with half life of 10.25 minutes. Neutrons are produced in nuclear reactions induced in nuclei by high energy protons, deuterons, α-particles and other heavy particles. A neutron does not experience any Coulomb barrier and is easily absorbed by a nucleus. The compound nucleus formed by the absorption of a neutron is in a highly excited state and may decay with the emission of a neutron, a proton, an α-particle or high energy gamma-rays. When gamma-rays are emitted, the ground state of the nucleus so formed is often radioactive and decays with the emission of β-rays. At times the neutron, rather than being absorbed by the nucleus, is scattered by its nuclear potential, contributing to the elastic scattering. Thus when neutrons pass through a medium, they interact with the atomic nuclei of the medium giving rise to

(1) Nuclear reactions
(2) Inelastic scattering
(3) Elastic scattering

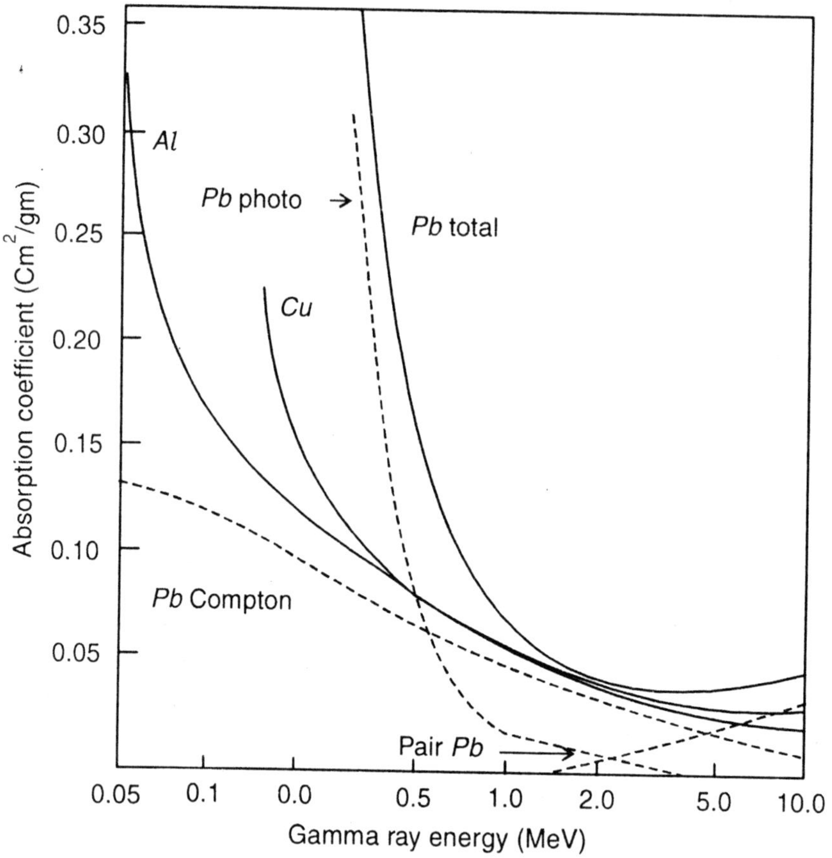

Fig. 6.21: Variation of mass absorption coefficient for γ-rays in lead, copper and aluminium as a function of energy. Dashed curves show the contributions of Compton effect, photo electric effect and pair production in lead. [G. White Grodstein, (1957), *Natl. Bur. Standard Circular* 583].

6.3.1 Nuclear Reactions

The compound nucleus formed by the capture of a neutron is always in an excited state of energy equivalent to the binding energy of the neutron. If the absorbed neutron has a kinetic energy, the excitation energy is still higher. The following nuclear reactions are generally observed.

(*a*) (n, p) reaction in which the excited compound nucleus emits a proton. The Q value of this reaction is generally positive. The reaction has an appreciable cross section for high energy neutrons because in that case the outgoing proton has high enough energy to overcome the Coulomb barrier. In very light nuclei the Coulomb barrier is low and even thermal neutrons (0.025 eV) can induce the (n, p) reaction *e.g.*

$$^{14}_{7}N + n \rightarrow ^{14}_{6}C + p + 0.6 \text{ MeV}$$

This is the reaction by which the neutrons present in cosmic rays produce carbon 14 in the atmosphere.

(*b*) (n, α) reaction in which an α-particle is emitted. The reaction with positive Q value generally requires neutrons of energy greater than 0.5 MeV so that emergent α-particle may overcome the Coulomb barrier. In the case of light nuclei, where the Coulomb barrier is low and the Q value is positive, the reaction has appreciable cross section even for thermal neutrons. Two important reactions, which are employed in the detection of neutrons are

$$^{6}_{3}Li + ^{1}_{0}n \rightarrow ^{3}_{1}H + ^{4}_{2}He + 4.5 \text{ MeV}$$

$$^{10}_{5}B + ^{1}_{0}n \rightarrow ^{7}_{3}Li + ^{4}_{2}He + 2.8 \text{ MeV}$$

(*c*) (n, 2n) and (n, np) reactions in which two particles are emitted. The Q values for these reactions are negative and about 8 MeV. Due to Coulomb barrier for proton emission, the (n, 2n) reaction is generally more probable of the two.

(*d*) (n, f) or fission reaction in which a heavy nucleus like ^{235}U or ^{233}U absorbs a neutron and fissions in two nearly equal parts with release of a large amount of energy (about 180 MeV). In the fission process a number of neutrons and gamma rays are also emitted. The fission fragments are always radioactive.

(*e*) (n, γ) reaction in which the compound nucleus, being unable to emit a particle (due to energy considerations) emits gamma-rays. The excited compound nucleus makes transitions to lower energy excited states which decay to still lower energy states, till finally the ground state of the compound nucleus is reached. The nucleus so formed is usually radioactive decaying by β-emission.

The cross section for the absorption of neutrons is high at low energies, in fact below about 1 keV energy the cross section varies as $1/v$ where v is the velocity of the neutron. In most of the nuclei, the neutron absorption shows resonance phenomenon at thermal energies and the cross-section at times increases to a few thousand barns (10^{-28} m^2). Some significant neutron capture reactions are with ^{10}B in which the capture curve follows the $1/v$ law upto 1000 eV neutron energy Fig. 6.22. Indium shows a resonance capture cross section of 26400 barns at 1.46 eV Fig. 6.23. For the nucleus ^{113}Cd, the resonance cross section of 6000 barns is observed at neutron energy of 0.176 eV. Natural cadmium has a neutron capture cross section of 2500 barns at thermal energy. Fig. 6.24.

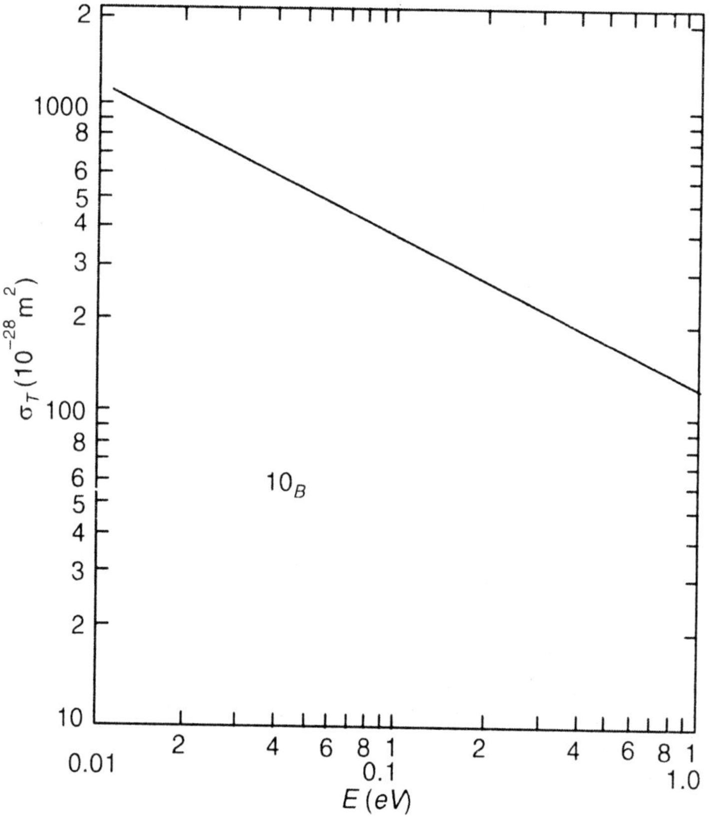

Fig. 6.22: Total absorption cross-section for neutrons in boron as a function of energy. The corss-section follows the 1/*V* law. [D. J. Hughes "*Neutron Cross Sections*" *AE CU-2046*, 1952]

(*f*) (n n') reaction, where a high energy neutron is absorbed by a nucleus and re-emitted with a lower energy, leaving the nucleus in an excited state. The inelastic scattering is possible whenever there is an excited state in the absorbing nucleus, having energy less than the kinetic energy of the neutron.

(*g*) Elastic scattering in which the incoming neutron is scattered by a nucleus without any change of energy. The elastic scattering can take place either by the nuclear potential or by the re-emission of the neutron after absorption. In either case the neutron is scattered in a certain direction and the nucleus recoils to conserve energy and momentum. The kinetic energy of the recoiling nucleus is derived from the energy of the neutron.

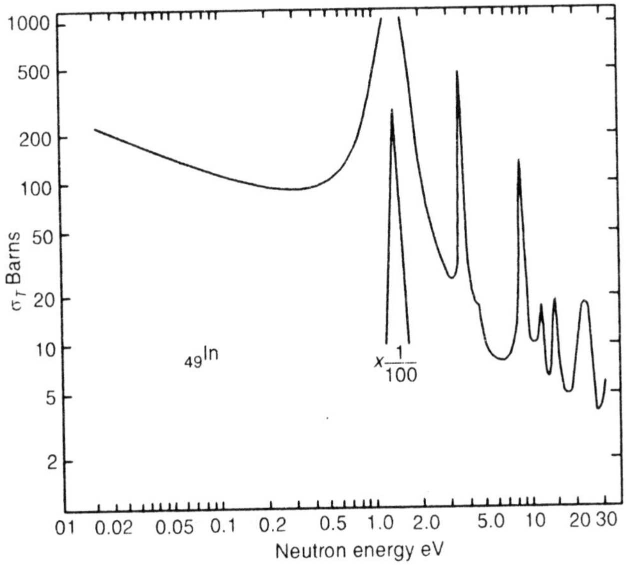

Fig. 6.23: Total absorption cross-section for neutrons in natural indium as function of energy. Strong resonance at neutron energy of 1.46 eV is evident. [D. J. Hughes *AECU*, (1952), 2046]

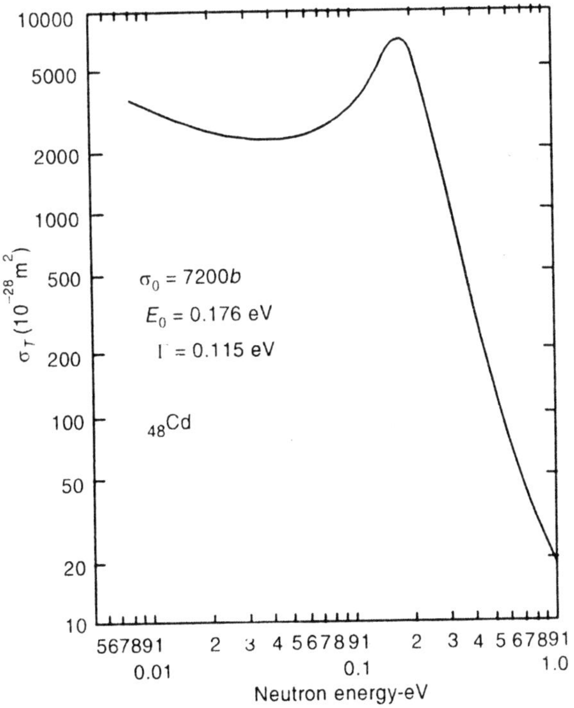

Fig. 6.24: Total absorption cross-section for neutrons in natural cadmium as function of energy. [D. J. Hughes, *AECU*, (1952), 2046]

6.3.2 Slowing Down of Neutrons

The slowing down process for the neutrons gained importance because of large capture cross-section (fission cross section) of thermal neutrons in uranium. In a fission reaction neutrons of about 2 MeV energy are produced and their slowing down, therefore, becomes essential so that a sustained chain reaction in a reactor can be achieved.

A fast neutron (energy around 2 MeV) when collides elastically with a nucleus of mass number. A, loses nearly 1/A of its initial energy, taking the mass of a neutron as unity. It is because of this, in a nuclear reactor, elements used to slow down fission neutrons are of low mass numbers or low atomic weights. Such elements are known as moderators. The materials which are commonly used as moderators are water (H_2O), heavy water (D_2O), Graphite (C), Beryllium (Be) and Beryllium oxide (BeO). The first excited state or the first resonance level, for these light material mostly lie around 0.1 MeV or so. Therefore, the collisions of high energy neutron with the moderator nuclei for most part of its slowing down are elastic. Inelastic collisions have negligible contribution to the slowing down process. As a matter of fact, inelastic collisions are significant only when neutrons collide with certain nuclei of high mass number like iron and uranium and even for these material the neutron must have at least 0.1 MeV energy. When the neutron energy is above 1 eV then during slowing down it is possibile to neglect the thermal motion and the binding energy of the atoms of the material. The atoms can be regarded as free and at rest. The collisions are then essentially elastic, like the collisions between billiard balls.

The neutrons travelling in a medium (or a moderator) make elastic collisions with atomic nuclei and are gradually slowed down. After making a large number of successive collisions, the lowest energy which the neutrons attain is the thermal energy kT, where T is the absolute temperature of the medium. At room temperature 300K (27°C), the neutrons are in thermal equilibrium with their surrounding. They have a Maxwellian distribution, with a maximum probable velocity of 2.2×10^3 m/s corresponding to the kinetic energy of 0.025 ev. Such neutrons are known as Thermal Neutrons. By applying principles of conservation of momentum and energy (in elastic collisions) it is possible to obtain the energy loss of neutrons during slowing down.

We have described in detail (Chapter 2) the elastic collision of two masses, and obtained expressions for energy after collision and the scattering angle. The actual event takes place in a frame of reference called Laboratory frame, while it is convenient to treat the same in a centre of mass frame of reference. We have discussed the collisions in both these frames and have obtained the relationship of the quantities of interest, like energy, scattering angle in the two frames. (transformation from one frame into another).

If E_1 is energy of a neutron, before collision in laboratory system and E_1' after collision then from equation (2.11.21) we have

$$E'_1 = E_1 \cdot \frac{(1 + A^2 + 2A\cos\theta_c)}{(1 + A)^2} \qquad ...(6.3.1)$$

where $A = \frac{m_2}{m_1}$, m_1 being the mass of the neutron and m_2 the mass of the target nucleus (Scattering nucleus) θ_c is angle of scattering of the neutron in the centre of mass system.

If $$E_1 = \frac{1}{2} m_1 u^2$$

and $$E'_1 = \frac{1}{2} m_1 u_{1'}^2$$

and $$\frac{E'_1}{E_1} = \frac{\frac{1}{2} m_1 u_1^2}{\frac{1}{2} m_1 u^2} = \frac{u_1^2}{u^2} = \frac{1 + A^2 + 2A\cos\theta_c}{(1 + A)^2} \qquad ...(6.3.1)$$

Let us put

$$\alpha = \left(\frac{A - 1}{A + 1}\right)^2 \qquad ...(6.3.2)$$

so that

$$\frac{1}{2}(1 - \alpha) = \frac{2A}{(1 + A)^2}, \quad \frac{1}{2}(1 + \alpha) = \frac{1 + A^2}{(1 + A)^2}$$

and then equation (6.3.1) becomes

$$\frac{E'_1}{E_1} = \frac{1}{2}\left[(1 + \alpha) + (1 - \alpha)\cos\theta_c\right] \qquad ...(6.3.3)$$

The ratio $\frac{E'_1}{E_1}$ has a maximum value *i.e.* the loss in energy in a collision is minimum, when $\theta_c = 0$ *i.e.* for glancing collision; $\cos\theta_c = 1$ and equation (6.3.3) gives

$$\frac{E'_{1\max}}{E_1} = 1 \text{ or } E_{1\max} = E_1 \qquad ...(6.3.4)$$

The energy of neutron before collision and after collision are equal, it suffers no loss of energy in a collision.

The minimum value of the ratio E'_1/E_1 *i.e.* the maximum loss of energy would occur when $\theta_c = \pi$ *i.e.* in a head on collision. The value of $\cos\theta_c = -1$ and then equation (6.3.3) gives

$$\frac{E'_{1\min}}{E_1} = \alpha$$

or
$$E'_{1\min} = E_1\alpha \qquad ...(6.3.5)$$

Equation (6.3.5) gives the minimum value of the energy to which a neutron energy can be reduced in an elastic scattering. The maximum fractional loss of energy in a collision is given by

$$\frac{E'_1 - E'_{1\min}}{E_1} = 1 - \alpha \qquad ...(6.3.6)$$

The maximum energy loss in a collision is, therefore,

$$E'_1 - E'_{1\min} = (\Delta E)_{\max} = E_1(1 - \alpha) \qquad ...(6.3.7)$$

Since α is related to the mass number A (eqn. 6.3.2) of the moderator nucleus, as such the loss in neutron energy in a collision will also depend on A. For hydrogen or a proton, $A = 1$ and hence $\alpha = 0$. It follows, therefore, that a neutron may possibly lose all its kinetic energy in a single collision with a hydrogen nucleus (proton). This is for the fact that masses of the neutron and the proton are almost equal. For $A > 1$ we can expand α in inverse power of A as

$$\alpha = \left(1 - \frac{1}{A}\right)^2 \left(1 + \frac{1}{A}\right)^{-2}$$

$$= 1 - \frac{4}{A} + \frac{8}{A^2} - \frac{12}{A^3} \qquad ...(6.3.8)$$

Thus in a single collision, the neutron will suffer a maximum fractional loss in energy of about 4% for $A = 100$ and about 33% for $A = 12$ (carbon). Lighter is the target nucleus, larger is the loss in energy of a neutron in a single collision with the target nucleus.

In order to calculate the average energy loss in a collision, we must know the probability for a neutron to be scattered through an angle θ_c and $\theta_c + d\theta_c$ in the centre of mass system. For the energies less than a few MeV but above thermal energies, it is reasonable to assume that the neutron scattering is isotropic in centre of mass system. At low thermal energies the vibrational energy of a nucleus become significant in comparison to the kinetic energy of the neutron and this assumption therefore breaks down.

Assuming that the scattering in C.M. system is spherically symmetric, the probability that a neutron is scattered in an angle θ_c and $\theta_c + d\theta_c$ is

$$p(\theta_c)d\theta_c = \frac{d\Omega}{4\pi} = \frac{2\pi \sin\theta_c d\theta_c}{4\pi} \qquad ...(6.3.9)$$

Since the neutron energy E_1' after a single collision is uniquely related to the scattering angle θ_c in c-system eqn. (6.3.3), the neutrons with the energies lying between E_1' and $E_1' + dE_1'$ are the same which are scattered within an angle θ_c and $\theta_c + d\theta_c$ after collision. This means that the probability $p(\theta_c)$ (6.3.9) and the probability $p(E_1')\, dE_1'$ for the neutron energy to lie between E_1' and $E_1' + dE_1'$ are equal. Thus, we can write

$$p(E'_1)dE'_1 = \frac{1}{2}\sin\theta_c dE'_1 \cdot \frac{2}{E_1(1-\alpha)\sin\theta_c}$$

$$= \frac{dE'_1}{E_1}(1-\alpha) \qquad \text{...(6.3.11)}$$

It may be noted that the probability $p(E'_1)$ per unit energy interval about E_1 for the neutron of energy E_1 when scattered after one collision has a value only within the range $\alpha, E_1 \le E_1' \le E_1$ (eqn. 6.3.7), and outside this range $p(E_1')$ is zero. The eqn. (6.3.11) also indicates that after a single collision neutron energies, between E_1 and $E_1\alpha$ are equally probable.

(A) Average Energy Loss Per Collision

Let $\overline{E_1'}$ be the average energy of the neutron after a collision, then

$$\overline{E_1'} = \frac{\int_{\alpha E_1}^{E_1} E'_1\, p(E'_1)dE'_1}{\int_{\alpha E_1}^{E_1} p(dE'_1)dE'_1}$$

Using equation (6.3.11) and taking the denominator to be unity since the total probability being one. αE_1 and E_1 being the minimum and maximum energy of the neutron after a single collision.

$$\overline{E'_1} = \int_{\alpha E_1}^{E_1} E'_1 \frac{1}{(1-\alpha)E_1} dE'_1$$

$$= \frac{1}{(1-\alpha)} \int_{\alpha E_1}^{E_1} E_1' dE_1 = \left(\frac{1+\alpha}{2}\right)E_1 \qquad \text{...(6.3.12)}$$

The average energy loss per collision is

$$\overline{\Delta E} = E_1 - \overline{E'_1} = E_1 - \left(\frac{1+\alpha}{2}\right)E_1$$

or

$$\overline{\Delta E} = \left(\frac{1-\alpha}{2}\right)E_1 \qquad \text{...(6.3.13)}$$

The equation (6.3.13) shows that the average energy lost by a neutron in a collision is directly proportional to its incident energy (for a particular moderator). However, the average fractional loss in energy per collision is

$$\frac{\overline{\Delta E}}{E_1} = \frac{(1-\alpha)}{2} \qquad ...(6.3.14)$$

It is independent of the incident energy. This shows that in a moderator of given A, the average fractional loss in energy per collision is same for all successive collisions.

(b) Average Cosine of the Scattering Angle

For isotropic scattering the average cosine of the scatering angle in the centre of mass system, is given by

$$\overline{\cos\theta_c} = \frac{\int_o^\pi \cos\theta_c \cdot p(\theta) d\theta_c}{\int_o^\pi p(\theta_c) d\theta_c} = 0 \qquad ...(6.3.15)$$

The total probability $\int_o^\pi p(\theta_c)d\theta_c$ being equal to 1.

Assuming isotropic scattering in C.M. system—the average cosine of the scattering angle in Laboratory system (L-system) is defined as

$$\cos\theta_L = \frac{\int \cos\theta_L \cdot p(\theta_c) d\theta_c}{\int p(\theta_c) d\theta_c} \qquad ...(6.3.16)$$

or $$\overline{\cos\theta_L} = \frac{1}{2}\int_o^\pi \cos\theta_L \sin\theta_c d\theta_c$$

From equation (2.11.11)

$$\cos\theta_L = \frac{(1 + A\cos\theta_C)}{(1 + A^2 + 2A\cos\theta_c)^{\frac{1}{2}}} \qquad ...(6.3.17)$$

$$\therefore \quad \overline{\cos\theta_L} = \frac{1}{2}\int_o^\pi \frac{(1 + A\cos\theta_c)\sin\theta_c d\theta_c}{(1 + A^2 + 2A\cos\theta_c)^{\frac{1}{2}}} \qquad ...(6.3.18)$$

Evaluating the integral, we get

$$\overline{\cos\theta_L} = \frac{2}{3A} \qquad ...(6.3.19)$$

The equation (6.3.19), therefore, indicates that the mean value of the cosine of the scattering angle in *L*-system is inversely proportional to the mass number A, of the target nucleus. It also implies that isotropic scattering in *C*-system leads to peaking in the forward direction in *L*-system. This preferential forward scattering in the *L*-system influences the mean distance travelled by a neutron during slowing down. This means that the anisotropy in the scattering in *L*-system causes an increase in the mean free path of the neutron, and the effective mean free path, then, is known as transport mean free path λ_{tr} and is related to scattering mean free path λ_s as

$$\lambda_{tr} = \frac{\lambda_s}{1 - \overline{\cos\theta_L}} = \frac{\lambda_s}{1 - \dfrac{2}{3A}} \qquad \ldots(6.3.20)$$

This relation holds for low absorption of neutrons. Thus the factor $\dfrac{1}{1 - \dfrac{2}{3A}}$ may be regarded as a correction for anisotropic scattering. For large *A*, $\overline{\cos\theta_L}$ is small and $\lambda_{tr} \approx \lambda_s$, *i.e.* the tendency for the forward scattering in *L*-system tends to disappear in the case of scattering from heavy nuclei. We give in Table (6.2) values of the $\overline{\cos\theta_L}$ in the *L*-system for some materials commonly used in nuclear reactors.

TABLE 6.2: VALUES OF THE AVERAGE COSINE OF THE SCATTERING ANGLE IN L-SYSTEM FOR SOME MATERIALS

Material	*A*	$\overline{\cos\theta_L}$
Hydrogen	1	0.667
Deuterium	2	0.333
Beryllium	9	0.074
Boron	10	0.061
Carbon	12	0.056
Uranium	238	0.003

(C) Average Logarithmic Energy Decrement (ξ)

Another useful quantity in the study of slowing down of neutrons is the average logarithmic energy decrement. It is the average value of the decrease in the natural logarithm of the neutron energy per collision. It is sometimes referred as average change in lithargy of a neutron per collision. It is denoted by a symbol ξ. Thus

$$\xi = \overline{\ln \frac{E_1}{E_1'}} \qquad \ldots(6.3.21)$$

where E_1 is the energy of the neutron before collision and E_1'; is its energy after collision.

or
$$\xi = \overline{\ln \frac{E_1}{E_1'}} = \frac{\int_{E_1}^{\alpha E_1} \ln \frac{E_1}{E_1'} \cdot p(E_1')dE_1}{\int_{\alpha E_1}^{E_1} p(E_1')dE_1'} \quad \text{...(6.3.22)}$$

As shown earlier, the denominator in the eqn. (6.3.22) is equal to unity using eqn. (6.3.11) for $p(E_1)\ dE_1'$, we have

$$\xi = \overline{\ln \frac{E_1}{E_1'}} = \int_{E_1}^{\alpha E_1} \ln \frac{E_1}{E_1'} \cdot \frac{dE_1'}{E_1(1-\alpha)} \quad \text{...(6.3.23)}$$

Evaluating the integral, we get

$$\xi = 1 + \frac{\alpha}{1-\alpha} \ln \alpha \quad \text{...(6.3.24)}$$

Using eqn. (6.3.2) for α

$$\xi = 1 + \frac{(A-1)^2}{2A} \ln \frac{(A-1)}{(A+1)} \quad \text{...(6.3.25)}$$

For $A > 10$, we can write

$$\xi \approx \frac{2}{A + \frac{2}{3}} \quad \text{...(6.3.26)}$$

For $A = 2$, use of equation (6.3.26) is in error by only 3% and its use becomes more and more valid as A increases. However, one finds that as A increases ξ decreases and for large A the value of ξ tends to zero. For some materials of interest the values of ξ are given in Table (6.3).

TABLE 6.3: SLOWING DOWN PARAMETERS FOR SOME MATERIALS

Material	*A*	ξ	*Average no. of collisions* $\langle N \rangle$	*Slowing down power* $\xi\Sigma_s$	*Moderating ratio* $\frac{\xi\Sigma_s}{\Sigma_a}$
Hydrogen	1	1.0	15	*	*
Deuterium	2	0.726	20	*	*
Beryllium	9	0.207	70	0.158	143
Boron	10	0.187	77	0.065	62×10^{-5}
Carbon	12	0.158	92	0.060	192
Uranium	238	0.0084	1730	0.003	9×10^{-3}
H_2O	*	0.92 †	16	1.35	71
D_2O	*	0.509 †	28	0.176	5670

*Not defined.
†Effective values.

The knowledge of ξ helps in calculating the average number of collisions needed to slow down a neutron from an initial energy to a certain final energy. For isotropic scattering in C.M. System, if the neutrons have initial energy E_1 and they are to be slowed down to an energy E_2, then the total logarithmic energy decrement is $\ln \frac{E_1}{E_2}$. Then the average number of collisions is given by

$$< N > = \frac{\ln(E_1 / E_2)}{\xi} \quad \ldots(6.3.27)$$

From some material the value of $< N >$ calculated from the Eqn. (6.3.27) are given in the Table 6.3 for E_1 = 2 MeV and E_2 = 1eV.

(D) Slowing Down Power and Moderating Ratio

For an efficient slowing down (good modertor) we need to have large ξ. In addition, for a good modertor, one also requires that the mean free path for collision in the medium be small, which means a large value for the scattering corss-section, Σ_s $\left(\Sigma_s = \frac{1}{\lambda_s}\right)$. This implies that the neutron may suffer large number of collisions in a given time. Therefore, for a good slowing down medium (moderator), the product $\xi\Sigma_s$, called the slowing down power should be large. Another consideration for a material to be a good moderator, is that it should be a very poor absorber of neutrons. Its macroscopic absorption coefficient, Σ_a, should have small value. The ratio of slowing down power $(\xi\Sigma_s)$ to the macroscopic absorption cross-section Σ_a is called as moderating ratio *i.e.*

$$\text{moderating ratio} = \frac{\xi\Sigma_s}{\Sigma_a} \quad \ldots(6.3.28)$$

We have given the values of moderating ratio for some materials in the Table 6.3.

It may be noted that macroscopic scattering cross-section Σ_s is equal to $N\sigma_s$ where σ_s is scattering cross-secton and N is the number of scattering nuclei per unit volume.

If the target or the moderator is a mixture of different materials, the average logarithmic energy decrement is given as

$$\xi = \frac{\xi_1\Sigma_{s1} + \xi_2\Sigma_{s2} + \xi_3\Sigma_{s3} + \ldots}{\Sigma_{s1} + \Sigma_{s2} + \Sigma_{s3} + \ldots} \quad \ldots(6.3.29)$$

The slowing down power (S.D.P.) in such case is

$$S.D.P. = \xi_1\Sigma_{s1} + \xi_2\Sigma_{s2} + \xi_3\Sigma_{s3} + \ldots \quad \ldots(6.3.30)$$

and the moderation ratio (M.R) is

$$M.R = \frac{\xi_1 \Sigma_{s1} + \xi_2 \Sigma_{s2} + \xi_3 \Sigma_{s3} + \ldots}{\Sigma_{a1} + \Sigma_{a2} + \Sigma_{a3} + \ldots} \qquad \ldots(6.3.31)$$

In the equations (6.3.29), (6.3.30) and (6.3.31) ξ_i is the average logarithmic energy decrement, Σ_{s1} and Σ_{ai} respectively are the macroscopic scattering and absorption cross-section for nuclei of type i.

(E) Spatial Distribution of Slowed Down Neutrons

Fermi age (Continuous slowing down)

As pointed out earlier, the process of slowing down of neutrons acquires a very important dimension in the studies of physics of nuclear reactors. For example, the average (net vector) distance between the point at which a neutron is produced and at which it becomes thermal due to the slowing down process determines the leakage of neutrons in a finite medium. If this distance is large, the reactor size will be large, so that the leakage could be reduced before the neutrons reach thermal energies. Therefore, it is important to know the spatial distribution of neutrons of different energies during the course of their slowing down.

Consider a neutron produced during fission (or from a certain source) at an energy E_0 which starts slowing down in a material (moderator). It will travel for a certain time with this energy before it makes a scattering collision with a nucleus. If $\lambda(E)$ is the mean free path of collision and $v(E)$ is the velocity, then the time between two successive collisions is λ/v. After a collision, the energy of the neutron will be decreased and it will then diffuse at this constant lower energy until it makes a collision with another nucleus of the material. Since at each collision the energy is decreased with which the neutron moves, the time between collisions, therefore, on the average increases with decreasing energy. This process of alternate diffusion at constant energy with a collision in between when the energy is decreased (neutron is slowed down) continues till the neutron attains energy equal to the energy of its surroundings, *i.e.* it gets thermalised. We know that in any single collision, the average logarithmic energy drecrement, ξ, is constant and is independent of the energy. If we plot $\ln E$ of a neutron as a function of time, we would obtain a stepped function as shown in the Fig. 6.25. The steps are approximately of constant height equal to ξ. The width or the horizontal line corresponds to the time between two collisions and which increases with time *i.e.* with decreasing energy. These horizontal lines also represent the constant energy at which the neutron diffuses between collisions.

If ξ is sufficiently small *i.e.* the height of the steps in the Fig. 6.25 is small, we may replace the stepped function by a smooth curve as shown in the figure. This implies that a neutron slows down continuously in time. This assumption of continuous slowing down model (Fermi age) can be valid for material or moderators of moderate or high mass number like graphite, beryllium but would not apply for hydrogen or deuterium or even in H_2O or D_2O since it is possible for a neutron to lose all its energy in a single collision, with a hydrogen nucleus.

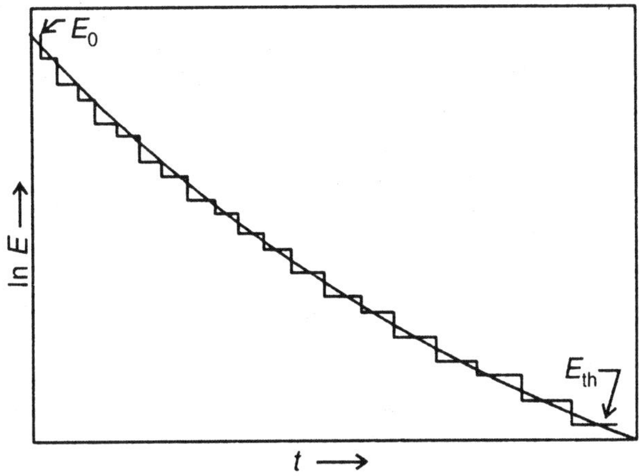

Fig. 6.25: Continuous slowing down approximation for neutrons

6.3.3 Fermi Age Equation

Let us assume that the medium through which slowing down takes place is a non-absorbing medium. The number of scattering collisions in time dt is $v/\lambda_s dt$ since ξ is the average logarithmic energy decrement per collision, the change in the logarithm of neutron energy is given by

$$-d(\ln E) = \xi \frac{v}{\lambda_s} \cdot dt \qquad \text{...(6.3.32)}$$

The logarithmic energy decrement is termed as lethargy, and is represented by u *i.e.*

$$u = \ln \frac{E_o}{E} \qquad \text{...(6.3.33)}$$

where E_o is an arbitrary energy corresponding to zero lethargy and as the energy of the neutron decreases its lethargy increases. From the eqn. (6.3.33)

$du = -d(\ln E)$. Therefore in terms of lethargy

$$du = \xi \frac{v}{\lambda_s} dt \qquad \text{...(6.3.34)}$$

Now, let the number of neutrons per unit volume or the neutron density at a point r and at time t be $n(r, t)$. Then the number of neutrons leaving a unit volume in unit time will be given by $\frac{dn}{dt}$.

We have seen that the neutron diffuses at a constant energy in between scattering collisions. A typical neutron trajectory is a zig zag path as shown in the Fig. 6.26, the path elements being of varying lengths, pointing in different directions with different speeds. These constitute the scattering free path of which the average is the scattering mean free path. After a collision, the direction of the neutron is not known exactly, but it can be given in terms of probability distribution, slowing down process is being statistical in nature. However, when a large number of neutrons are considered, there is always a net movement of neutrons from regions of high neutrons density to a region of low neutron density. The rate at which this motion takes place is known a diffusion.

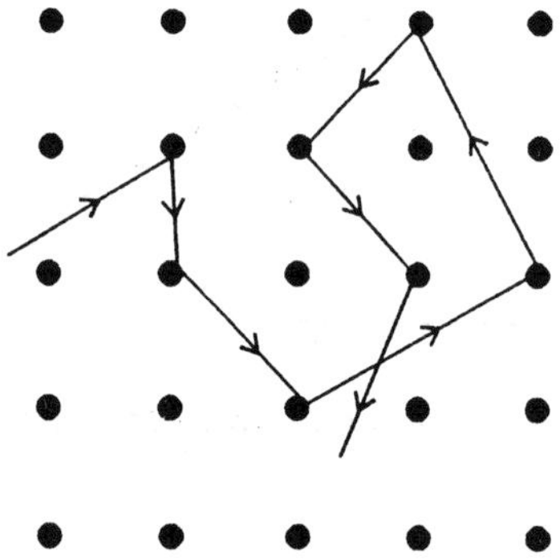

Fig. 6.26: Elastic scattering of neutrons by nuclei in a medium

The diffusion of neutrons in a non-absorbing medium, can be described in terms of the well known "Fick's law of diffusion", which in the present case states that there is a net number of neutrons flowing in unit time through a unit area (held) normal to the direction of flow of the neutrons.

Representing this by $\vec{J}$ then

$$\vec{J} = -D_o \text{ grad } n\,(\vec{r}, t) \qquad \ldots(6.3.35)$$

where n is the neutron density and D_o is a constant called diffusion co-efficient. D_o has the dimensions of $cm^2 sec^{-1}$.

Let us now consider the flow of neutrons through a rectangular volume elementd dV with dimensions $dx\, dy\, dz$ located at a point whose co-ordinates are x, y, z. Fig. 6.27. The number of neutrons moving along z-direction and entering the lower face $dx\, dy$ per second is $J_z\, dx\, dy$, where J_z is the net current along z-direction. The number leaving the upper face per second is $J_{z+dz}\, dx dy$. Now from equation (6.3.35) the net rate of flow of neutrons out of the given volume element dV, through the faces parallel to the x, y plane will be

$$(J_{z+dz} - J_z)\, dx\, dy = -D_o\left[\left(\frac{\partial n}{\partial z}\right)_{z+dz} - \left(\frac{\partial n}{\partial z}\right)_z\right] dx\, dy$$

$$= -D_o \frac{\partial^2 n}{\partial z^2} \cdot dz \cdot dx\, dy$$

$$= -D_o \frac{\partial^2 n}{\partial z^2} dV \qquad \ldots(6.3.36)$$

Similarly the net rate of loss of neutrons from other two faces parallel to (y, z) and (x, z) planes will be $-D_o^2 \frac{\partial^2 n}{\partial x^2} dV$ and $-D_o^2 \frac{\partial^2 n}{\partial y^2} dV$.

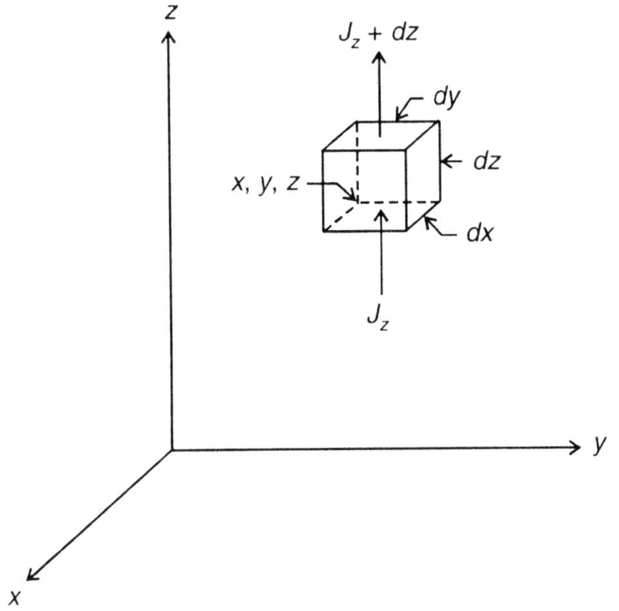

Fig. 6.27: Calculation of neutron leakage

Therefore, the total number of neutrons leaving a unit volume per sec is

$$-D_o\left\{\frac{\partial^2 n}{\partial x^2} + \frac{\partial^2 n}{\partial y^2} + \frac{\partial^2 n}{\partial z^2}\right\} = -D_o \nabla^2 n(\vec{n}, t) \qquad \ldots(6.3.37)$$

where ∇^2 is the Laplacian operator. By definition, eqn. (6.3.37) is equal to $-\frac{dn}{dt}$.

or
$$\frac{dn}{dt} = D_o \nabla^2 n(\vec{r}, t) \qquad \ldots(6.3.38)$$

The eqn. (6.3.38) is a neutron balance equation in a non-absorbing medium far away from the neutron source. It is also the diffusion equation. Once the neutrons are produced then after leaving the source, the number of neutrons which diffuse between the times interval t and $t + dt$ at $\vec{r}$ in a unit volume is, by definition, $n(\vec{r}, t)\,dt$. If u is lithargy (eqn. 6.3.33), then it is possible to transform the variable 't' to the variable 'u' such that the number or neutrons that are within the lithargy interval u and $u + du$ is given as

$$n(\vec{r}, u)\,du = n(\vec{r}, t)\,dt \qquad \ldots(6.3.39)$$

using equation (6.3.34)

$$n(\vec{r},u) = n(\vec{r},t)\frac{\lambda_s}{\xi v} \qquad \ldots(6.3.40)$$

Expressing the equation (6.3.38) in terms of the variable u, we get

$$D_o\nabla^2 n(\vec{r},u)\frac{\xi_v}{\lambda_s} = \frac{dn}{du}\cdot\frac{du}{dt}$$

$$= \frac{du}{dt}\frac{\partial}{\partial u}\left\{\frac{\xi_v}{\lambda_s}\cdot n(\vec{r},u)\right\} \qquad \ldots(6.3.41)$$

But $n(\vec{r}, u)\, v = \phi(\vec{r}, u)$, where $\phi(\vec{r}, u)$ is the neutron flux at r per unit lithargy interval, Now

$$\frac{\xi\cdot\phi(\vec{r},u)}{\lambda_s} = \xi\Sigma_s\phi(\vec{r},u) = q(\vec{r},u) \qquad \ldots(6.3.42)$$

where $q(\vec{r}, u)$ is the slowing down density and $\frac{1}{\lambda_s} = \Sigma_s$ is the macroscopic scattering cross-section. Using $q(\vec{r}, u)$ in equation (6.3.41) we get

$$D_o\nabla^2 q = \frac{\xi v}{\lambda_s}\frac{\partial q}{\partial u} \qquad \ldots(6.3.43)$$

We use a new variable τ which is known as Fermi age and is defined as

$$\tau = \int_o^t D_o dt = \int_o^u \frac{D_o\lambda_s}{\xi v}du \qquad \ldots(6.3.44)$$

where eqn. (6.3.34) is used for dt.

Now $$\frac{\partial q}{\partial u} = \frac{\partial q}{\partial \tau}\cdot\frac{\partial \tau}{\partial u} = \frac{\partial q}{\partial \tau}\frac{D_o\lambda_s}{\xi v} \qquad \ldots(6.3.45)$$

where eqn. (6.3.43) reduces to

$$\nabla^2 q = \frac{\partial q}{\partial \tau} \qquad \ldots(6.3.46)$$

This is called as the *Fermi Age Equation.* The dimensions of Fermi Age τ is $(\text{meter})^2$. It is a symbolic age. Though the Fermi Age τ does not in any way represent the elapsed time, it, however, gives a measure of the time for which neutrons have diffused after being produced at the source (fission) at time $t = 0$. The age of the source neutron say at energy E_o, is zero *i.e.* $t = 0$ at $E = E_o$. It increases with increasing time or decreasing neutron energy. In other words, as the neutron slows down, its Fermi age τ. increases.

The Fermi age equation provides a complete description of neutron distribution in energy (lithargy) and space (r). It has a form similar to heat conduction equation in a continuous solid medium, under non-steady-state condition. The age equation, however, is for steady state condition. The solution of heat conduction equation for different forms are available in literature. It is possible to make use of these and write the solution of age equation for a point source of neutrons emitting S fast neutrons per second of energy E_0 in an infinite homogenous non-absorbing medium. These neutrons undergo slowing down in the medium. the solution of the age equation is then

$$q(\vec{r}, \tau) = \cdot \frac{Se^{-r^2/4\tau}}{(4\pi\tau)^{3/2}} \qquad \ldots(6.3.47)$$

where $q(\vec{r}, \tau)$ is slowing down density for neutrons of age τ at a distance r from the point source. The distribution of slowing down density is thus Gaussian. The shape of the curve for $q(\vec{r}, \tau)$ as a function of r is shown in the Fig. 6.28 for two different values of age τ. We find that for small τ, the curve is narrow and high. Small τ means, the neutrons are near the source hence of large energy. On the other hand large τ, means lower energy, and the curve is, therefore, broad, flat and lower. We expect such a behaviour on physical grounds also. A small value of age τ means that the neutrons have suffered less collisions, resulting little slowing down and so they have not diffused far from the source. Most of the neutrons, therefore, will have high energies and staying near the source ($r = 0$). The curve is high and narrow. For large τ, the neutrons will have undergone considerable slowing down (decrease in energy) and consequently diffused at large distances from the source. The curve thus should be low and broad.

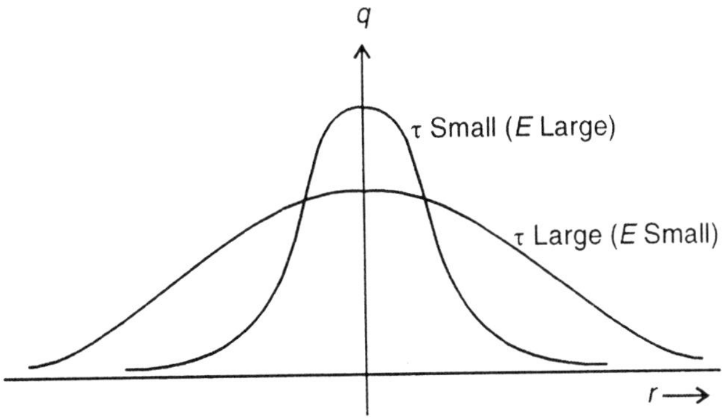

Fig. 6.28: Neutron slowing down distribution about a point source

The width of the Gaussian distribution curve is defined as the value of r at which the slowing down density $q(\vec{r}, \tau)$, becomes $\frac{1}{e}$ of its maximum value. It is given by

$$r_o = 2\sqrt{\tau} \qquad \text{...(6.3.48)}$$

The mean distance from the source of the neutrons at which neutrons get thermalised can be calculated from the equation (6.3.47) and this distance will be given by equation (6.3.48), when $r_o = r_{th}$ and $\tau = \tau_{th}$. Beyond this distance r_{th} from the source, the neutrons diffuse without further losing any energy.

Another important quantity is the mean square distance travelled by the neutrons to attain a given value of τ *i.e.* the mean square (net vector) slowing down distance, denoted by $\overline{r^2}$ about a point source.

$$\overline{r^2} = \frac{\int_o^\infty r^2 q(\vec{r},\tau)4\pi r^2 dr}{\int_o^\infty q(\vec{r},\tau)4\pi r^2 dr} \qquad \text{...(6.3.49)}$$

substituting for $q(\vec{r}, \tau)$ from the eqn. (6.3.47)

$$\overline{r^2} = \frac{\int_o^\infty r^4 S e^{-r^2/4\tau} dr}{\int_o^\infty r^2 S e^{-r^2/4\tau} dr}$$

$$\overline{r^2} = \frac{\int r^4 e^{-r^2/4\tau} dr}{\int_o^\infty r^2 e^{-r^2/4\tau} dr}$$

noting that

$$x^{2n} e^{-ax^2} dx = \frac{1.3.5 \ldots (2n-1)}{2^{n+1} a^n}\sqrt{\frac{\pi}{a}}$$

We obtain

$$\overline{r^2} = 6\tau$$

or

$$\tau = \frac{1}{6}\overline{r^2} \qquad \text{...(6.3.50)}$$

The age τ for a given neutron energy is one-sixth of the mean square distance travelled by the neutrons to attain that energy *i.e.* distance between the source of neutrons where $\tau = 0$ to the point at which its age is τ.

(F) Experimental Determination of Ferimi Age (τ)

The Fermi age of neutrons can be measured by activation method. The detector by which the age is to be measured must have high absorption for neutrons of that particular energy. For example the slowing density can be measured at Indium resonance *i.e.* at energy 1.44 eV. If Indium foils of known thickness are placed at different distances from the neutron source, the saturated activities induced in the foils can represent the slowing down density at 1.44 eV energy as a function of distance. The Indium foils are covered with cadmium to prevent thermal neutrons reaching the Indium foils. The activity induced in the Indium foils will be proportional to the slowing down density q. Therefore, measuring the saturated activities in the foil at different places one knows the values of q as a function of r. Now the equation (6.3.47) gives

$$\ln q = \text{constant} - r^2/4\tau$$

Plotting $\ln q$ as a function of r^2 will give a straight line from which the age τ (E_r) can be calculated at the Indium resonance energy 1.44 eV. Since the Indium resonance energy (1.44 eV) is above thermal energy (0.025 eV). necessary correction is to be applied to $\tau(E_r)$ to get $\tau(E_{th})$. This correction, *i.e.* estimation of τ from 1.44 eV to 0.025 eV has to be done theoretically. The values of τ for thermal neutrons in various materials (moderators) is given in the Table 6.4. We have seen that the Fermi age is defined in terms of the mean square distance travelled by the neutron while slowing down. Therefore, the square root of the age is called the slowing down length. Its value in certain materials is given in the Table 6.4.

TABLE 6.4

Material (Moderator)	*Age* τ *(cm²)*	*Slowing down length* $\sqrt{\tau}$ *cm*
Water (H_2O)	33	5.7
Heavy water (D_2O)	120	11.0
Beryllium (*Be*)	98	9.9
graphite (*c*)	350	18.7

It may be remembered that though one cannot strictly define age for neutrons in moderator like hydrgoen or deuterium, still in these substances, the age may be defined by the equation (6.3.50).

Once the neutron is produced by the source (say fission neutrons), it starts slowing down in a medium. It, therefore, takes certain average time known as slowing down time to slow down to thermal energies and on attaining thermal equilibrium with the material the thermal neutrons diffuse before being absorbed. The average time for which the thermal neutrons diffuse before absorption/leakage is called diffusion time.

(i) Slowing Down Time

Let us consider that the neutrons are slowing down in a non-absorbing material and after a time t let their velocity be v. The average number of collisions in a time interval dt is $\frac{v\,dt}{\lambda_s}$. The average decrease in the logarithm of the energy $-d\ln E$, in time dt is given by equation (6.3.32) *i.e.*

$$-d\ln E = -\frac{dE}{E} = \frac{v\xi}{\lambda_s}dt$$

or

$$dt = -\frac{\lambda_s}{\xi}\sqrt{\frac{m}{2}}\cdot\frac{dE}{E^{3/2}} \qquad \text{...(6.3.51)}$$

and the slowing down time t_s will be

$$t_s = \int_o^{t_s} dt = -\frac{\lambda_s}{\xi}\sqrt{\frac{m}{2}}\int_{E_o}^{E_{th}}\frac{dE}{E^{3/2}}$$

where E_o is the average energy of the neutrons at the source and E_{th} is the average energy of thermal neutrons

$$t_s = \frac{\lambda_s\sqrt{2m}}{\xi}\left(\frac{1}{\sqrt{E_{th}}} - \frac{1}{\sqrt{E_o}}\right) \qquad \text{...(6.3.52)}$$

The calculated slowing down times for neutrons with $E_o = 2$ MeV to $E_{th} = 0.25$ eV for certain materials (moderators) are given in the Table 6.5.

The average diffusion time for thermal neutrons is

$$t_d = \frac{\lambda_a}{v_{th}} = \frac{1}{\Sigma_a v}$$

where λ_a is the absorption mean free path. The mean velocity v_{th} at ordinary temperatures is 2.2×10^5 cm/sec. The values of diffusion time (thermal life time) are given in the Table 6.5.

TABLE 6.5

Material (Moderator)	*s (cm)*	*Slowing down time t_s (sec)*	*Diffusion time t_d(sec)*
Water (H_2O)	1.1	10^{-5}	2.1×10^{-4}
Heavy Water (D_2O)	2.6	4.6×10^{-5}	15×10^{-2}
Beryllium (*Be*)	1.6	6.7×10^{-5}	4.3×10^{-3}
Graphite (*c*)	2.6	1.5×10^{-4}	1.2×10^{-2}

From these values of slowing down time t_s and diffusion time t_d we find that the average time a neutron spends in slowing down from fission energy to thermal energy is negligible compared to the time it spends in diffusing as a thermal neutron.

EXERCISES

1. When a charged particle passes through matter, how the major part of its kinetic energy is lost due to
   (a) Ionization of atoms
   (b) Excitation of atoms
   (c) Knocking out the k-shell electron
   (d) Excitation of the atomic nuclei.
   (e) Scattering by the atomic nuclei.
2. What are δ-rays?
3. What factors control the specific ionization by α-particles at low energies?
4. What is straggling of α-particles?
5. Why mono-energetic electrons do not have a well defined range in an absorber?
6. Why γ-rays do not have a well defined range in an absorber?
7. How different charged particles emited in a nuclear reaction identified.
8. What is Compton edge for Compton scattered electrons.
9. How Compton effect can be used to measure polarization of nuclear γ-rays.
10. How larger absorption of higher energy γ-rays than that of slightly lower energy achieved?
11. What are Auger electrons?
12. What is fluroscence yield of an atom?
13. What is positronium? Which factors effect its life time?
14. How is ^{14}C produced in atmosphere?
15. Why high energy neutron produce greater biological damage than thermal neutrons?
16. What is moderating power of a moderator?
17. What properties are essential for a material to be used as a moderator for neutrons?
18. What is neutron age?
19. Why slowing down of neutrons is necessary in a reactor?

7

NUCLEAR FISSION AND FUSION

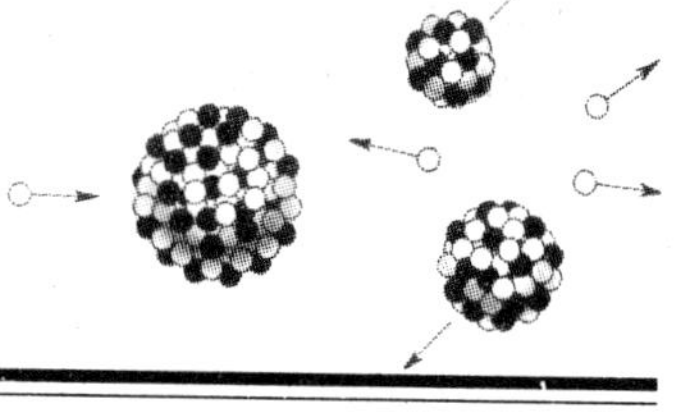

7.1 FISSION AND MECHANISM OF FISSION

After the discovery of the neutron by Chadwick (1932) a systematic study of nuclear reactions induced by neutrons was initiated by E. Fermi and his collaborators. During such studies Otto Hahn and Fritz Strassmann found that when uranium was bombarded by neutrons, the uranium nucleus splits into two lighter nuclei. Lise Meitner and Otto Frisch almost immediately explained the results of these experiments on the basis of liquid drop model of nucleus proposed by N. Bohr. They named this process of splitting of a nucleus as 'Nuclear Fission' drawing analogy from the biological cell division. From the mass difference of reactants (projectile and target nucleus) and the fission products they estimated that about 200 MeV of energy is released in a single fission event. Besides the release of this enormous amount of energy, it was also observed that on an average two to three neutrons were emitted per fission. These fission neutrons can cause further fission thereby inducing a chain reaction. Further investigations revealed that fission occurs within a short time of the order of 10^{-17} sec. of the neutron capture by the uranium nucleus and most of the fission neutrons are ejected almost instantaneously, within a time of the order of 10^{-14} sec. of the fission. These neutrons are called as Prompt neutrons. The product nuclei in which the uranium gets split after neutron capture are termed as fission fragments. After about one minute of the fission event, small fraction of neutrons are ejected from the decay of fission fragments. Such neutrons are called delayed neutrons. The figure (7.1) shows such a fission event of ^{235}U by a neutron. It may be remembered that the fission fragments are radio active because of the excess number of neutrons in comparison with stable nuclei of same Z value. They, therefore, decay emitting β-particles, and γ-rays also, along with delayed neutrons.

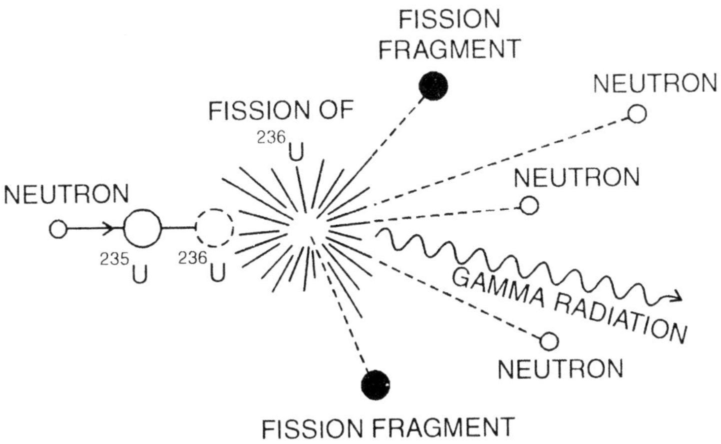

Fig. 7.1: Fission event in U-235

Besides neutrons causing fission, it is now found that high energy particles like deuteron, proton, α-particle, even γ-ray photons and light ions like ^{12}C can cause fission in heavy elements ($Z > 70$). The first explanation of the fission process was given by *L.* Meitner and *O.* Frisch on the basis of liquid drop model of nucleus. Later on the idea of liquid drop of the nucleus was used by Bohr and Wheeler giving a detailed theory of fission. To understand qualitatively the mechanism of fission we consider a nucleus of heavy target material (Uranium) as a spherical drop of incompressible liquid. The stability of such a nucleus against fission depends chiefly on the relative effectiveness of two main forces: (*i*) the long-range Coulomb repulsive forces between protons (the large nucleus contains many nucleons) and (*ii*) the short range nuclear attractive forces between the nucleons. When a nucleus captures a neutron (or any other projectile) a compound nucleus is formed in an excited state. The excitation energy is given by the kinetic energy of the projectile (neutron) and the energy released by way of re-pairing of nucleons inside the necleus. In the Fig. (7.2 *A*) shows the equilibrium shape of the spherical charged compound nucleus. This excited nucleus will undergo surface oscillations which tend to deform it from its spherical shape as shown in Fig. (7.2 *B*). If the excitation energy is small compared to the Coulomb repulsive forces, the surface tension forces will dominate and the nucleus will return to its original spherical shape (ground state), emitting the excitation energy as γ-rays. When the excitation enrgy is large and the oscillations deform the nucleus and the deformation is furthered by the Coulomb forces, such that the surface tension forces become ineffective. At this stage the nucleus acquires a dumb-bell shape [Fig. 7.2 (*C*)]. The process of deformation continues till the separation between the two charge centres reaches a critical value, the Coulomb repulsion between the charge centres pushes them apart. The nucleus splits into two separate nuclei which is a process called as Fission. The stages *D*, *E* and *F* in Fig. (7.2) show the sequence in which the dumb-bell splits into two light nuclei known as fission fragments. The position E at which the two nuclei are just in contact is known as Scission state. The electrostatic potential energy at E is transformed into kinetic energy of fission fragments with which they fly at tremendous speed. The difference in the energy at A and at E is, therefore, the critical energy (E_f) or the fission threshold energy.

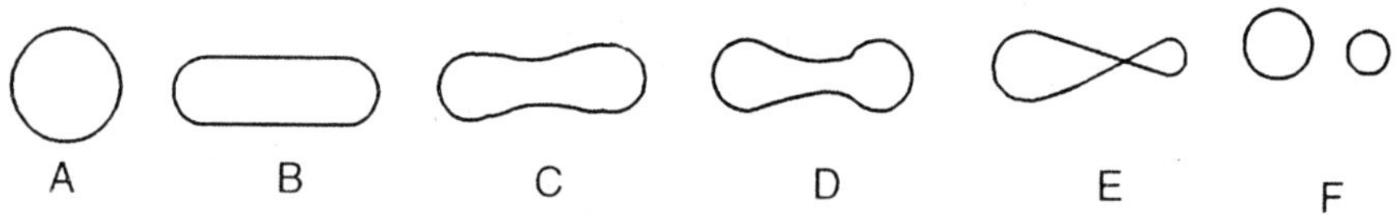

Fig. 7.2: Sequence during the splitting of a nucleus

The potential energy of a nucleus as a function of deformation is shown in the Fig. (7.3). For very heavy nuclei $A > 300$, the Coulomb forces are more effective and the nucleus breaks up into fragments instantaneously as soon as it is formed. This is indicated by the curve (a) in the figure, which shows the potential energy of the two fragments as a function of their separation. In the case of nuclei $A \simeq 236$, the potential energy is shown by curve (b) in the figure. We find that for deformations smaller than a certain critical value marked by D in the figure, the strong attractive nuclear forces dominate the repulsive Coulomb forces and the potential energy follows the curge BC. This is the region of stable oscillations and the nucleus can come to its original spherical shape. For deformations larger than D, the Coulomb forces are no longer compensated and their occurs a short of potential barrier at C and with the increase in deformation, the separation between the charge centres becomes more rapid, as a result the potential energy decreases. This is shown by the curve CK. Thus we find that at the critical deformation D, the potential energy is termed as a fission barrier or critical energy for fission (fission threshold). This corresponds to the breaking of the parent nucleus into two fragments, termed as fission fragments. The stage E in Fig. (7.2) pertains to this stage D. The critical energy is the difference in the energy of the nucleus at the point C and the initial energy of the nucleus at B. It is shown as E_f in the figure. When A is still low, that is for nuclei having $A \approx 100$, the fission barrier is much higher, curve (C) and a projectile (neutron) when captured by such nuclei, does not impart sufficient excitation energy to surmount the barrier. It is interesting to note that the barrier hinders (classically) the nucleus to split if the deformation lies in the region BC. But we know that quantum mechanically, there is always a finite probability for a particle to tunnel through a potential barrier in such cases. This means that even if the energy of deformation is below E_f value, breaking of the nucleus can still take place, though it will take place very slowly. This is the situation of spontaneous fission of a nucleus, which otherwise is stable against fission. (In natural uranium, ^{238}U nucleus undergoes spontaneous fission with a half life of around 10^{16} years). It may be noted that in each case of the Fig. (7.3), the two fission fragments once formed suffer Coulomb repulsion and at infinite separation between them will have kinetic energy determined by the change in the mass (mass difference between the initial nucleus and the mass of the two fragments).

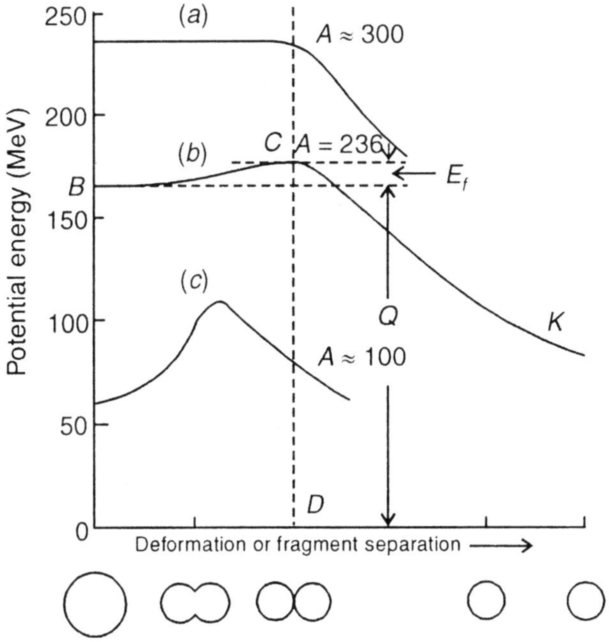

Fig. 7.3: Potential energy of a nucleus as a function of deformation

7.2 CALCULATION OF THE CRITICAL ENERGY BASED ON THE LIQUID DROP MODEL OF THE NUCLEUS

In order to calculate the critical energy of fission E_f, we make certain assumptions:

(*i*) The original nucleus is spherical in its ground state;
(*ii*) The total volume of the system stays constant during the various stages; and
(*iii*) The surface oscillations of the compound nucleus in the excited state introduce deformation in the surface.

For simplicity we regard the fission as symmetiric fission meaning thereby that the two fission fragments are alike in charge and mass (in actual practice, this, however, does not happen). Thus, if the charge and mass of the initial compound nucleus is Z and A respectively, then each fragment will have charge $Z/2$ and mass $A/2$. (We assume both Z and A to be even).

(*i*) The surface energy of the initial spherical drop will be given by

$$E_{si} = S.T = 4\pi R^2 T \qquad ...(7.2.1)$$

where S, T and R are the surface area, Surface Tension and the nuclear radius respectively.

(*ii*) The electrostatic energy of the undeformed nucleus is

$$E_{ei} = \frac{3}{5}\frac{(Ze)^2}{R} \qquad \text{...(7.2.2)}$$

The radius R and mass number A are related through the relation

$$R = r_o A^{\frac{1}{3}} \qquad \text{...(7.2.3)}$$

where the constant $r_o = 1.3 \times 10^{-15}$ m.

The total energy (E_i) of the initial nucleus in its ground state is given by adding E_{si} and E_{ei}. Thus

$$E_i = E_{si} + E_{ei}$$

$$= 4\pi R^2 T + \frac{3}{5}\frac{(Ze)^2}{R}$$

or

$$E_i = 4\pi T r_0^2 A^{\frac{2}{3}} + \frac{3}{5}\frac{(Ze)^2}{r_0 A^{\frac{1}{3}}} \qquad \text{...(7.2.4)}$$

The surface energy of the two fragments when they are just in contact (position D in Fig. (7.2) is,

$$E_{sf} = 2.S.T = 2.4\pi\left(\frac{R}{2^{\frac{1}{3}}}\right)^2 T$$

or

$$E_{sf} = \frac{2.4\pi}{2^{\frac{2}{3}}}\left(r_o A^{\frac{1}{3}}\right)^2 T = 8\pi r_o^2\left(\frac{A}{2}\right)^{\frac{2}{3}} T \qquad \text{...(7.2.5)}$$

where $\left(\frac{R}{2^{\frac{1}{3}}}\right)$ is the radius of each fragment. (This is a consequence of our second assumption). The electrostatic energy of the fragments is the sum of the electrostatic energy of the two undeformed nuclei and the energy arising out of repulsion between them at contact

$$E_{ef} = 2\frac{3}{5}\frac{\left(\frac{Ze}{2}\right)^2}{\frac{R}{2^{\frac{1}{3}}}} + \frac{\left(\frac{Ze}{2}\right)^2}{2.\frac{R}{2^{\frac{1}{3}}}}$$

using, $$R = r_o A^{\frac{1}{3}}$$

$$E_{ef} = \frac{3}{5} \cdot \frac{Z^2e^2}{2.r_o\left(\frac{A}{2}\right)^{\frac{1}{3}}} + \frac{Z^2e^2}{8r_o\left(\frac{A}{2}\right)^{\frac{1}{3}}} \qquad ...(7.2.6)$$

The total energy of fission fragments is given by

$$E_{ff} = E_{sf} + E_{ef}$$

$$= 8\pi T r_o^2\left(\frac{A}{2}\right)^{\frac{2}{3}} + \frac{3}{5}\frac{Z^2e^2}{2r_o\left(\frac{A}{2}\right)^{\frac{1}{3}}} + \frac{Z^2e^2}{8r_o\left(\frac{A}{2}\right)^{\frac{1}{3}}} \qquad ...(7.2.7)$$

The fission energy or the critical energy for fission is the difference of E_{ff} and E_i *i.e.*

$$E_f = E_{ff} - E_i$$

$$= 8\pi T\, r_o^2\left(\frac{A}{2}\right)^{\frac{2}{3}} + \frac{3}{5}\frac{Z^2e^2}{2r_o\left(\frac{A}{2}\right)^{\frac{1}{3}}} + \frac{Z^2e^2}{8r_o\left(\frac{A}{2}\right)^{\frac{1}{3}}}$$

$$-4\pi T\, r_o^2 A^{\frac{2}{3}} - \frac{3}{5}\frac{Z^2e^2}{r_o A^{\frac{1}{3}}}$$

or $$E_f = 4\pi T r_o^2 A^{\frac{2}{3}}\left(2^{\frac{1}{3}} - 1\right) + \frac{3}{5}\frac{Z^2e^2}{r_o A^{\frac{1}{3}}}\left\{\frac{1}{2^{\frac{2}{3}}} + \frac{5}{12 \times 2^{\frac{2}{3}}} - 1\right\}$$

$$= 4\pi T r_o^2 A^{\frac{2}{3}}\left(2^{\frac{1}{3}} - 1\right) - \frac{3Z^2e^2}{10 r_o A^{\frac{1}{3}}} \times \left\{2 - 2^{\frac{1}{3}} - \frac{5 \times 2^{\frac{1}{3}}}{12}\right\} \qquad ...(7.2.8)$$

Putting $a = \left(2^{\frac{1}{3}} - 1\right)$ and $b = \left(2 - 2^{\frac{1}{3}} - \frac{5 \times 2^{\frac{1}{3}}}{12}\right)$

$$E_f = 4\pi T r_o^2 A^{\frac{2}{3}} a - \frac{3}{10}\frac{Z^2 e^2 b}{r_o A^{\frac{1}{3}}} \quad \text{...(7.2.9)}$$

For spontaneous fission, to occur E_f must be zero. Equation (7.2.9) then gives

$$E_f = 0,$$

or $$4\pi T r_o^2 A^{\frac{2}{3}}.a - \frac{3Z^2 e^2 b}{10 r_o A^{\frac{1}{3}}} = 0$$

or $$4\pi T r_o^2 A^{\frac{2}{3}} a = \frac{3Z^2 e^2 b}{10 r_o A^{\frac{1}{3}}}.$$

Now $a = \left(2^{\frac{1}{3}} - 1\right) = 0.26$ and $b = \left(2 - 2^{\frac{1}{3}} - \frac{5 \times 2^{\frac{1}{3}}}{12}\right) = 0.22$

or $$a \approx b\,.$$

$\therefore$ $$4\pi T r_o^2 A^{\frac{2}{3}} = \frac{3Z^2 e^2}{10 r_o A^{\frac{1}{3}}}$$

or $$\left(\frac{Z^2}{A}\right) = \frac{2 \times 4\pi T r_o^2}{\left(\frac{3}{5}.\frac{e^2}{r_o}\right)} \quad \text{...(7.2.10)}$$

The magnitude of the right hand side of the above equation is known from the Weiszacker's mass formula for binding energy of a nucleus. It gives $4\pi T r_o^2 = 17.8$ and $\frac{3}{5}\frac{e^2}{r_o} = 0.71$.

Using these values in equation (7.2.10), we get

$$\frac{Z^2}{A} \simeq \frac{2 \times 17.8}{0.71} \simeq 50 \quad \text{...(7.2.11)}$$

However, a more general acceptable value for this parameter $\left(\frac{Z^2}{A}\right)$ for spontaneous fission is 47.8. In other words, a nucleus would undergo fission instantaneously if $\frac{Z^2}{A}$ is close to this value. If $\frac{Z^2}{A}$ is smaller than this limiting value, the nucleus would be stable against spontaneous fission. Figure (7.4) shows the variation of the life time for spontaneous fission as a function of $\frac{Z^2}{A}$. For ^{235}U, $\frac{Z^2}{A}$ is 36.02 and for ^{239} Pu it is 36.97. These values are much below the limiting value for spontaneous fission and hence need energy to be supplied for fission to occur when a projectile is captured. As an example, for ^{236}U (^{235}U + n), the critical energy for fission is around 5.3 MeV, while the absorption of a neutron of almost zero energy releases an energy of 6.4 MeV (neutron separation energy in the ground state of the nucleus). This acts as the excitation energy for ^{235}U to undergo fission by slow neutrons. While ^{238}U is not fissionable by thermal neutrons. The excitation energy E_X = B.E. of compound nucleus minus the B.E. of target nucleus in their ground states. The neutron separation energy or the excitation energy in ^{239}U (^{238}U + n) is around 4.9 MeV. Whereas the critical energy for fission E_f is 5.8 MeV. Thus ^{238}U has about 1 MeV as fission threshold energy. The increased neutron separation energy in ^{236}U is on account of the fact that $^{235}_{92}U_{143}$ is an even odd nucleus and on the capture of a neutron an even-even (even Z, even N) nucleus is formed, on the other hand $^{238}_{92}U_{146}$ is an even-even nucleus, more tightly bound compared to even-odd nucleus and on the capture of a slow neutron, an even-odd nucleus $^{239}_{92}U_{147}$ is formed. This difference arises, therefore, on account of the pairing term in the Weiszacker's formula for B.E. Thus whenever E_x is greater than the critical energy for fission E_f, the thermal neutron can cause fission. Whenever E_x is less than E_f (E_x–E_f < 0), there will be a threshold on neutron energy in terms of its kinetic energy. We give in Table (7.1), the calculated values of critical energy for fission E_f, its observed values and the corresponding excitation energy E_x, for some of the nuclei. We have also given $\frac{Z^2}{A}$ values for these nuclei.

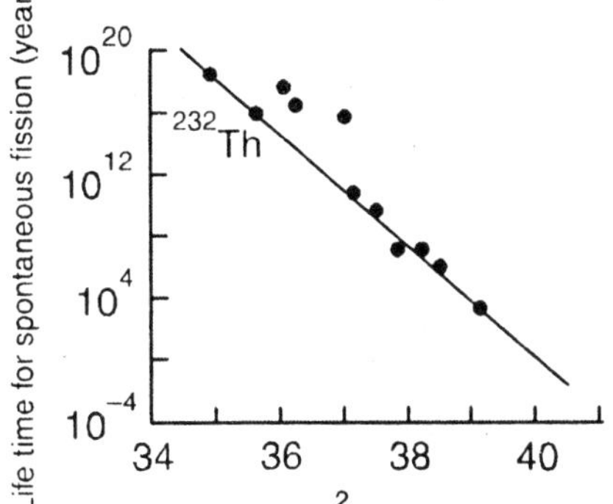

Fig. 7.4: Life time for spontaneous fission

TABLE 7.1: EXCITATION ENERGY AND CRITICAL ENERGY FOR FISSION FOR CERTAIN HEAVY NUCLEI

Target nucleus	*Compound nucleus formed on the capture of a thermal neutron*	*Z^2/A*	*E_f(MeV)*		*E_x (MeV)*
			Calculated-ted (liquid drop)	*Observed*	
${}^{232}_{90}Th_{142}$	$\left[{}^{233}_{90}Th_{143}\right]$	34.91	6.5	5.95	5.1
${}^{233}_{92}U_{141}$	$\left[{}^{234}_{92}U_{142}\right]$	36.33	4.6	5.49	6.6
${}^{235}_{92}U_{143}$	$\left[{}^{236}_{92}U_{144}\right]$	36.02	5.3	5.75	6.4
${}^{238}_{92}U_{146}$	$\left[{}^{239}_{92}U_{147}\right]$	35.56	5.5	5.8	4.9
${}^{239}_{94}Pu_{145}$	$\left[{}^{240}_{92}Pu_{146}\right]$	36.97	4.0	5.48	6.4

From the table we find that nuclei like like^{233}U, ^{235}U, ^{239}Pu which can easily be fissioned by thermal neutrons are known as 'fissile' material and ^{232}Th and ^{238}U which fission only by fast neutrons are referred as 'fissionable' material.

7.3 FISSION PRODUCTS AND ENERGY RELEASED IN FISSION

Let us consider the fision of ^{235}U by thermal neutrons (Fig. 7.1)

$$ {}^{235}_{92}U + {}^{1}_{0}n \rightarrow \left[{}^{236}_{92}U\right]^* \rightarrow 2 \text{ Fission fragments} $$

$$ + x \text{ neutrons} + \gamma + Q. \qquad \text{...(7.3.1)} $$

This primary fission reaction shows that the product contains two middle weight nuclei called fission fragments (shown as X and Y in Fig. 7.1), x denotes the number of neutrons (usually 2 or 3, depending on the fission fragments)a few γ-rays and Q is the energy released (nearly 200 MeV).

The fission fragments are of unequal mass, one is heavier than the other. Such a fission is said to be asymmetric. In the case of ^{235}U fission, it is found that it can fission in 40 different ways producing 80 different nuclei. One or two examples are given here;

$$ {}^{235}U + n \rightarrow \left[{}^{236}U\right]^* \rightarrow {}^{141}_{56}Ba + {}^{92}_{36}Kr + 3n + \gamma + Q $$

$$ \rightarrow {}^{140}_{54}Xe + {}^{94}_{38}Sr + 2n + \gamma + Q . $$

The heavier fission fragments normally lie in the mass range of 125-150 and lighter one in the range of 80-110. The mass distribution of fission fragments is mostly determined in the form of yield. Such a yield curve for fission of ^{235}U by thermal neutron is shown in the Fig. (7.5). The peaks in the curve gives the maximum yield of a particular mass.

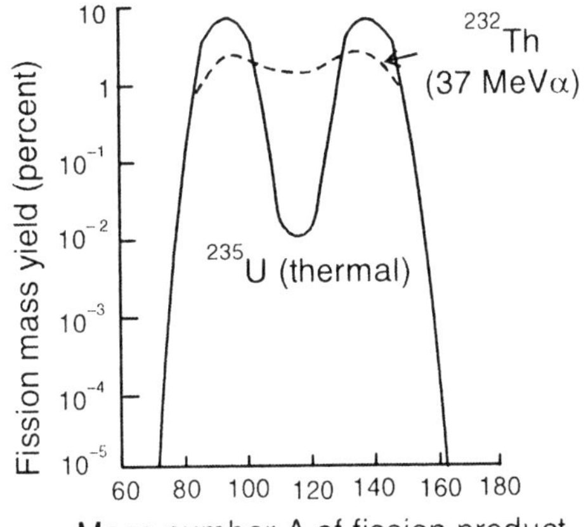

Fig. 7.5: Fission yield

In the lighter group, nuclide of mass (A) around 95 has maximum yield (6.3%) and mass (A) around 140 corresponds to maximum yield (6.3%). These peaks around A = 95 and A = 140 are associated with neutron number around N = 50 and N = 82. Thus a possible tendency of a heavy nucleus to break into fragments having closed shells around the magic numbers 50 and 82 respectively. The total percentage yield of the fission fragment is 200%, because the yield of lighter and heavier groups separately is 100%. Since linear momentum is conserved the fission fragments move exactly in opposite direction carrying most of the momentum ($M_1 V_1 = M_2 V_2$). Therefore, the energy of the fission fragments is inversely proportional to their masses,

$$\frac{E_2}{E_1} = \frac{\frac{1}{2} M_2 V_2^2}{\frac{1}{2} m_1 V_1^2} = \frac{M_1}{M_2}$$

where M_1, M_2, V_1, V_2 are the masses and velocities of two fission fragments.

The energy distribution, therefore, shows two peaks as shown in Fig. (7.6). The higher energy peak, around 95 MeV corresponds to the lighter group and lower energy peak of 65 MeV for the heavier group. The measured energy distribution of fission fragments can be used to obtain their mass distribution and vice versa.

It has been found that symmetric fission due to thermal neutrons is very rare, but becomes probable as the energy of neutrons is increased. For example, the probability of symmetric fission with 14 MeV neutrons increases, the valley in the curve (Fig. 7.5) tends to disappear and almost disappears with neutrons of around 100 MeV energy. Similar behaviour is shown by $^{232}_{20}Th$ for its fission by α-particles of 37 MeV. The other fissile nuclei ^{233}U and ^{239}Pu also show similar behaviour.

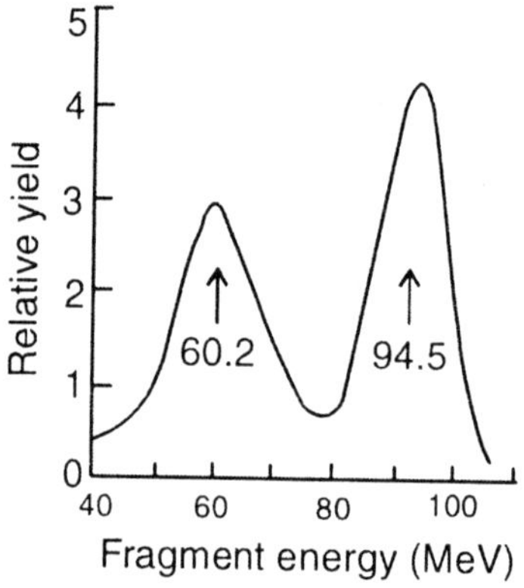

Fig. 7.6: Energy distribution of fission fragments

The mass distribution of excited fission fragments indicates that they carry excess number of neutrons than required for the corresponding stable nuclei of their charge. Some of these neutrons are instantaneously emitted as prompt neutrons leaving the fission fragments still with some excess neutrons. This leaves them β^--radioactive. Thus the fission fragments, once formed decay to a stable nucleus after following a β^--emission decay chain. A decay chain of interest is

$$^{140}_{54}\text{Xe} \xrightarrow[16s]{} {}^{140}_{55}\text{Cs} \xrightarrow[66s]{} {}^{140}_{56}\text{Ba} \xrightarrow[12.8d]{} {}^{140}_{57}\text{La} \xrightarrow[10h]{} {}^{140}_{58}\text{Ce (stable)}$$

This decay chain was first studied in part by Hahn and Strassman, who identified two nuclides ^{140}Ba and ^{140}La which led to the discovery of fission. In some of such decay chains, a β-process may leave a product nucleus in a highly excited state such that neutron emission may become predominant. These neutrons are referred as delayed neutrons. An example of emission of delayed neutrons is shown in Fig. (7.7).

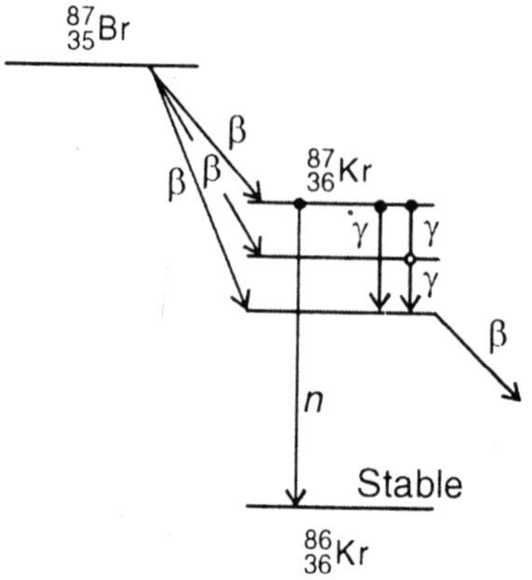

Fig. 7.7: Delayed neutron from the decay of a fission fragment

It is clear that during the β-decay chains γ-rays are also emitted, in addition to the γ-rays (prompt) being emitted during the fission process. These γ-rays are referred as delayed γ-rays.

Energy released during fission is thus distributed as kinetic energy of fission fragments, prompt neutrons, γ-rays both prompt and delayed, β^--particles and neutrinos. On an average the total energy released in one event of fission varies from 190-210 MeV. But for all practical purposes we take the effective value of the energy released per fission to be 200 MeV. The energy released in fission is dissipated into the surrounding environment as heat. The distribution of energy released in the fission of U-235 amongst various products is shown in th table 7.2.

TABLE 7.2: ENERGY RELEASED IN THERMAL FISSION OF ^{235}U

1.	Kinetic energy of fission fragments:		
	Lighter fragment	..	67 MeV
	Heavier fragment	..	100 MeV
2.	Kinetic energy of prompt neutrons	..	5 MeV
3.	Energy of γ-rays (prompt) emitted at the time of fission	..	7 MeV
4.	Decay products:		
	(a) β^--particles	..	5 MeV
	(b) γ-rays	..	5 MeV
	(c) Anti Neutrinos	..	11 MeV
	Total energy	..	200 MeV

Fission Cross-Section

When a neutron is captured by a fissile nucleus (^{233}U or ^{235}U or ^{239}Pu), it does not necessarily lead to fission, even if the energy is available. Because sometimes it so happens that the compound nucleus so formed before it splits, decays to its ground state emitting a γ-ray. Therefore, when a neutron is captured, predominantly two processes compete: (*i*) rediative capture (n, γ) reaction forming ^{236}U nucleus in its ground state; and (*ii*) fission reaction (n, f). Each of the two processes is characterised by its own cross-section. These cross-sections depend on the energy of neutrons used for causing fission. In Table (7.3) we give the values of the cross-section $\sigma\ (n, \gamma)$ and $\sigma\,(n, f)$ for radiative capture and fission for thermal neutrons (energy 0.025 eV or velocity $\simeq$ 2200 m/s). In the table we have also given the average number of neutrons released per fission (denoted by ν) or the fast neutron yield per thermal fission processes.

TABLE 7.3: THERMAL NEUTRON DATA FOR SOME FISSIONABLE MATERIALS

Nucleus	σ *(n, γ) barns*	σ *(n, f) barns*	ν
^{233}U	53	531	2.49
^{235}U	99	582	2.42
Natural Uranium	3.4	4.2	2.42
^{239}Pu	287	743	2.88
^{238}U	2.7	–	–

It should be remembered that the actual number of fission neutrons emitted in any one fission is an integer but fission nucleus can be fissioned in 40 different ways, hence the average number in one fission is not an integer. Nearly 99% percent of the total neutrons emitted are prompt neutrons.

7.4 FISSION CHAIN REACTION

While describing the fission process it was indicated (7.1) that the neutrons produced in fission can induce fission in other fissile nuclei, thereby inducing a chain reaction (Fig. 7.8). Since on the average two neutrons (more correctly 2.5) are produced per fission of a nucleus, all of them may not be available to cause further fission. Some may be absorbed in non-fission processes and some may escape (leak) from the geometry of the system before encountering

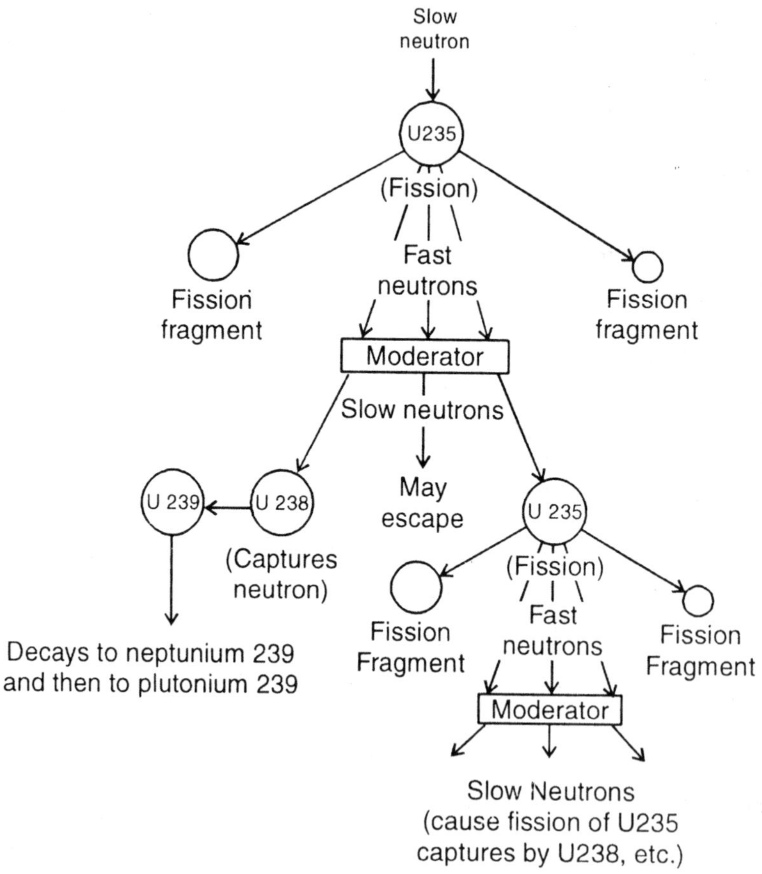

Fig. 7.8: Nuclear fission chain reaction

a fissionable nucleus. Nevertheless, the situation is so arranged that after each fission at least one of the neutrons (average number 2.5) born per fission is always available to produce another fission. We, then, say that a self supporting or a self sustained chain reaction has taken place. In other words, when the rate of production of neutrons equals the rate of loss of neutrons, the reaction is said to be self-supporting or self-sustained. Chain reactions are very common in Chemistry. Combustion is a chain reaction. We know that around 200 MeV of energy is produced per fission, therefore, in a self-sustained chain reaction, an enormous amount of energy is produced. This release of energy takes place in an uncontrolled fashion. Atom bomb is an example of this process. It takes hardly a fraction of a second, for such an explosion to take place. The device where a self-sustained but a controlled chain reaction is maintained is called a "Nuclear Reactor" or just a "Reactor".

When it was thought to use natural uranium which contains mostly ^{238}U isotope, to produce fission chain reaction certain difficulties appeared. The ^{238}U nucleus mostly absorbs neutrons without producing fission (except when neutrons have energy above 1 MeV) and as such chain reaction is not possible in a sample containing natural uranium, because once the fission starts by an external source, the reaction is likely to stop as soon as the external neutron source is removed. A way out of this difficulty was suggested by E. Fermi. The fission cross-section for ^{235}U nucleus is very large for low energy neutrons. In natural uranium about 0.7% is ^{235}U isotope. Fermi, therefore, suggested that an effective use of ^{235}U in natural uranium could be possible if neutrons once produced in fission reaction are slowed down due to collisions with the nuclei of a slowing down medium or a moderating medium, before the neutron is able to interact with other uranium nuclei. Graphite was found to be a suitable material, but a homogeneous mixture of uranium with graphite could not produce a sustained chain reaction. Fermi, therefore, suggested to use a heterogeneous assembly. Rods of uranium were uniformly distributed in a graphite (moderator) medium. A successful self sustained chain reaction in such a heterogeneous assembly was achieved first by Fermi and his co-workers on December 2, 1942. This made a starting point of obtaining energy from fission chain reaction in a controlled manner.

For a self-sustained fission chain reaction the rate of production of neutrons must be equal to the rate of loss of neutrons. In a reactor the loss of neutrons is either through leakage from the assembly or absorption by the medium. The rate of leakage is proportional to the surface area and the production is proportional to the volume of the assemply. Hence for a chain reaction to be sustained there should be an optimum size and mass of the fissionable material. The mass then is known as critical mass and the size (volume) as the critical size.

The condition for a self-sustined chain reaction is usually expressed in terms of a constant k termed as a neutron multiplication factor or reproduction factor. It is defined as the effective number of neutrons born per generation. More correctly it is expressed as the ratio of the number of neutrons resulting from fission in any one generation to the total number lost by both absorption and leakage in the immediately preceding generation.

If $k > 1$, the number of neutrons will increase with time such that the chain reaction becomes divergent and uncontrollable. If $k < 1$, the number of neutrons decreases with time and ultimately the chain reaction in this situation stops. When $k = 1$, the number of neutrons in any two successive generations remains same and the chain reaction continues at a steady rate, independent of time. The reactor under these conditions is termed as super critical ($k > 1$), subcritical ($k < 1$) and critical ($k = 1$). Because of this reason the constant k is also referred as criticality factor.

The growth of neutrons in a fission reaction is expressed by an approximate equation as

$$N_t \simeq N_o \exp. (k - 1)\, t/t_o \qquad \ldots(7.4.1)$$

where N_t is the number of neutrons produced during successive generations in time t starting with N_o neutrons at initial time $t = 0$, t_o is the average time between successive generations (average time between the creation of a neutron and its removal by leakage or absorption) t/t_o is thus the number of successive generations in which the number of neutrons grows to N_t.

The Figure (7.9) shows the variation of neutron population with time in terms of the criticality constant k. The knowledge of multiplication factor is very important in determining whether one is interested in having the release of energy in fission chain reaction in a controlled OR an uncontrolled manner.

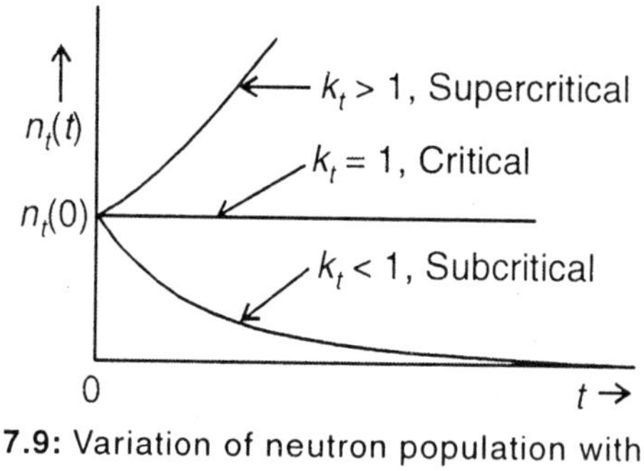

Fig. 7.9: Variation of neutron population with time

An atom bomb is an example of a sustained chain reaction where energy is released in an uncontrolled way in a time of the order of micro seconds. In principle an atomic bomb consists of two subcritical ($k < 1$) assemblies of fissile material (uranium/plutonium) separated by a suitable distance so that each being in a subcritical stage, are unable to produce required explosion. When they are suddenly brought together so that the total mass and size of the one piece so formed exceeds the critical size *i.e.* the assembly becomes supercritical ($k > 1$). A sustained fission chain reaction in an uncontrolled way takes place within less than a micro second. As an example, the total mass of fissile material used in early atom bomb was about 1 kg and $k = 1.1$ (sometimes in atomic bomb the conditions are so arranged that $k = 2$). Since 1 kg of the fuel (uranium or plutonium) would contain about 10^{24} atoms (nuclei) one would need 10^{24} neutrons to cause fission in 1 kg of the fissile material. If N_o = 1 and $t_o \approx 10^{-8}$ sec. *i.e.* 0.001 micro second than it follows from the equation (7.3.1), that the value of $N = 10^{24}$ would be reached in less than a micro second. This means that once the supercritical assembly is formed, then in a time less than a micro-second enough neutrons would be generated in a fission chain starting with a single neutron. 1 kg of the fuel then would be fissioned causing a powerful explosion and releasing enormous energy, about 10^{14} Joules which is equivalent to about 20 kilo tons of TNT.

7.5 NEUTRON LIFE CYCLE AND FOUR FACTOR FORMULA

(*i*) Infinite Assembly

We have seen that thermal neutrons can cause fission in ^{235}U, as such we examine the life history of a thermal neutron once it is produced in an assembly containing natural uranium having 99.3% of ^{238}U isotope and 0.7% of ^{235}U isotope. Once a thermal neutron is born, it may have the following possibilities:

(*i*) It may escape from the assembly (leakage).

(*ii*) It may be absorbed by (*a*) fissile material to cause either a non-fission reaction or a radiative capture reaction; (*b*) non fissile material present in the assembly (like moderator, control material etc.)

It may be noted that for an assembly of very large size, the leakage of neutrons from its surface will be almost zero. As such, we first consider an assembly of infinite size with zero leakage from the surface and start with one thermal neutron. Our assembly contains mostly fissile material (natural uranium) and moderator (low atomic mass material like graphite). This thermal neutron is captured by uranium nucleus. Let η fast neutrons each of energy about 2 MeV are produced. This is average number of fast neutrons produced directly by fission for every thermal neutron absorbed in the fissile material. One should be able to distinguish between η and ν. By definition ν is the total average number of neutrons emitted for each thermal neutron captured in any one fission process. While η corresponds to the average number of neutrons liberated directly (only) by fission for every thermal neutron absorbed in the fuel. That is

$$\eta = \nu \frac{\text{Number of neutrons absorbed in a fission reaction}}{\text{Total number of neutrons absorbed in the fuel}} \qquad ...(7.5.1)$$

These fast neutrons, having energy greater than fission threshold for ^{238}U, may be absorbed by it causing fast fission. Thus the number of fast neutrons will actually be more than η. We denote them by $\eta\epsilon$. It is these neutrons which would be slowing down by colliding with the moderator atoms. ϵ is known is *fast-fission factor* and is defined as the ratio of the total number of (fast) neutrons slowing down past the fission threshold of Uranium-238 (about 1 MeV) to the number produced by thermal neutron fission. It can also be defined as the number of neutrons slowing down past the fission threshold of ^{238}U per primary fission neutron. Consequently, the capture of one thermal neutron in the fuel will result in $\eta\epsilon$ fast neutrons slowing down past the fission threshold of ^{238}U.

When the neutron energy falls below the fission threshold of ^{238}U, the neutrons continue to slow down and during the course of further slowing down, some of the neutrons are captured in non-fission processes in ^{238}U particularly what is known as the resonance abosrption without producing new neutrons. If we denote by p, the fraction escaping this absorption, then the neutrons which would reach thermal energies will be $\eta\epsilon p$. The factor p is called as resonance escape probability. It is defined as the ratio of the thermal neutrons absorbed in the assembly to the total number of neutrons absorbed at all energies below the fast fission threshold.

When these $\eta\epsilon p$ neutrons reach thermal energies they will diffuse for some time in the infinite system and some of them are captured by moderator and other (non fissionable) materials and a fraction 'f' of these (thermal neutrons) is again absorbed by the fissile (fuel) material inducing further fission 'f' is called as thermal utilisation factor. It is defined as the ratio of thermal neutrons absorbed in the fuel (uranium) to the total number or thermal neutrons in the system. Thus out of $\eta\epsilon p$ thermal neutrons, the number absorbed by the fissile material inducing further fission is $\eta\epsilon pf$. In other words, we started with one thermal neutron for producing fission at the next generation we have $\eta p\epsilon f$ thermal neutrons capable of causing further fission. Therefore, at each generation, the number of thermal neutrons absorbed is $\eta\epsilon pf$. We denote this product by k_∞ which is usually called as infinite (medium) multiplication factor or reproduction factor for the infinite system. Consequently

$$k_\infty = \eta\epsilon pf \qquad \text{...(7.5.2)}$$

This formula is known as the "four factor formula". Such a division of the infinite multiplication factor into four parts helps in evaluating the reproduction factor for a multiplying system. As an example, the numerical values of these constants in the case of a reactor at Calder Hall, England are

$$\eta = 1.31,\ \epsilon = 1.02,\ p = 0.82,\ f = 0.88 \text{ which gives } k_\infty = 1.05.$$

For an infinite system $k_\infty = 1$ represents the criticality condition. when an assembly contains mostly ^{235}U, both ϵ and p are close to unity. It is interesting to note that the definition of k_∞ is not unique. One may start from any point in the cycle for the discussion of the neutron cycle. As such k_∞ can also be defined as

$$k_\infty = \frac{\text{Number of primary fission neutrons in one generation}}{\text{Number of primary fission neutrons in the preceding generation}}$$

$$k_\infty = \frac{\text{Number of thermal neutrons absorbed by fuel in one generation}}{\text{Number of thermal neutrons absorbed by fuel in the preceding generation}}$$

(ii) Finite Assembly

When the assembly containing fissile material and moderator material, is of finite size the leakage of neutrons, once they are produced in fission reaction, is possible. The leakage of fast neutrons can take place while they are slowing down and that of the thermal neutrons when they are diffusing in the system. If P_f and P_t denote the non-leakage probabilities for the fast neutrons and thermal neutrons respectively, then for a finite system the multiplication factor k will be given by

$$k = \eta \epsilon p f P_f P_t \quad \text{...(7.5.3)}$$

$$= k_\infty P_f P_t = k_\infty P \quad \text{...(7.5.4)}$$

where $P = P_f P_t$.

The criticality condition for a finite system would be

$$k = 1, \text{ or } k = k_\infty P = 1. \quad \text{...(7.5.5)}$$

Since the non-leakage probability ($P = P_f P_t$) would always be less than unity, it becomes desirable to have k greater than unity for a finite system.

A complete neutron cycle (thermal neutron cycle) for a *finite assembly* is schematically shown in the Fig. (7.10).

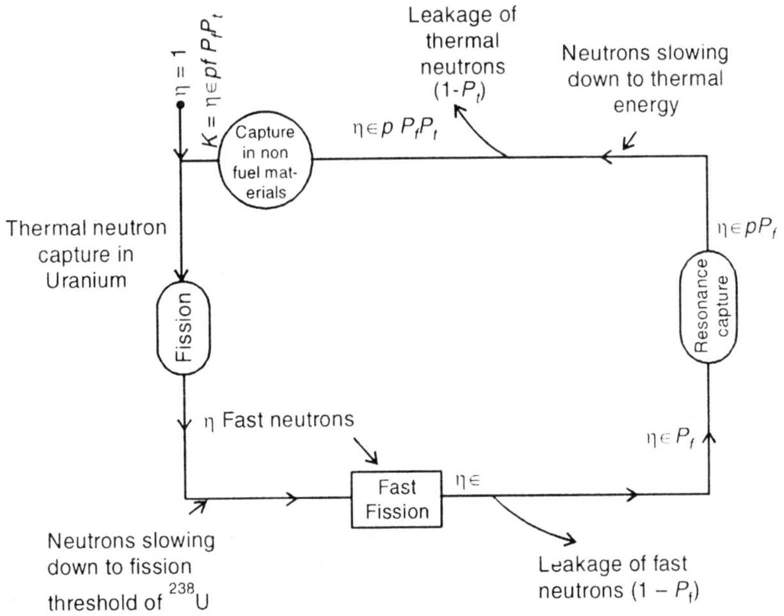

Fig. 7.10: A complete neutron cycle for a finite assembly

Note: For an infinite system, both P_f and P_t will be zero in the schematic diagram for a complete neutron cycle.

It may be mentioned here that if one starts with n thermal neutrons, then at the end of one cycle (one generation), the number of thermal neutrons in an infinite assembly will be $n.\ \eta\epsilon pf$ and by definition

$$k_\infty = \frac{n.\eta \in pf}{n} = \eta \in pf$$

Likewise for a finite assembly

$$k = \frac{n.\eta \in pfP_f P_t}{n} = \eta \in pfP_f P_t \qquad \text{...(7.5.6)}$$

7.6 CALCULATION OF THE CRITICALITY CONSTANT

Having defined the four factor formula (eqn. 7.5.2) for an infinite assembly and the subsequent formula (eqn. 7.5.4) for a finite system, we now consider how the various factors that constitute k_∞ (*i.e.* $\eta\epsilon pf$) and P (*i.e.* $P_f P_t$) are determined. We shall, for this purpose, consider a heterogeneous thermal reactor using natural uranium and graphite.

(*i*) Eta (η)

We have defined (7.5) eta (η) as the average number of neutrons emitted by fission caused by thermal neutrons for every thermal neutron absorbed in the fuel. That is for natural uranium

$$\eta = \nu \frac{\Sigma_f}{\Sigma_a} = \nu \frac{\Sigma_f^{235}}{\Sigma_a^{235} + \Sigma_a^{238}} \qquad \text{...(7.6.1)}$$

$$= \nu \frac{N^{235}\sigma_f^{235}}{N^{235}\sigma_f^{235} + N^{235}\sigma_r^{235} + N^{238}\sigma_a^{238}} \qquad \text{...(7.6.2)}$$

where σ_f^{235}, σ_r^{235} and σ_a^{238} are the fission cross-section of U-235, absorption (radiative capture) cross-section of U-235 and absorption cross section (non-fission) of U-238 respectively. N^{235} and N^{238} stand for the respective number of nuclei per unit volume. For natural Uranium $\frac{N^{238}}{N^{235}} = 140$ and the experimental values for the microscopic cross-section are $\sigma_f^{235} = 582$ barns $\sigma_r^{235} = 99$ barns, $\sigma_a^{238} = 2.8$ barns. Using $\nu = 2.5$ we get $\eta = 1.34$.

For simplicity, we assume that the reactor is cylindrical with uranium rods arranged in a square lattice formed by the moderator, graphite. To make calculations convenient, the lattice is divided into a number of identical cells and suppose that a square cross-section can be treated equivalent to a circular cross-section of the same area. The Fig. (7.11) shows unit cells in a heterogeneous square lattice.

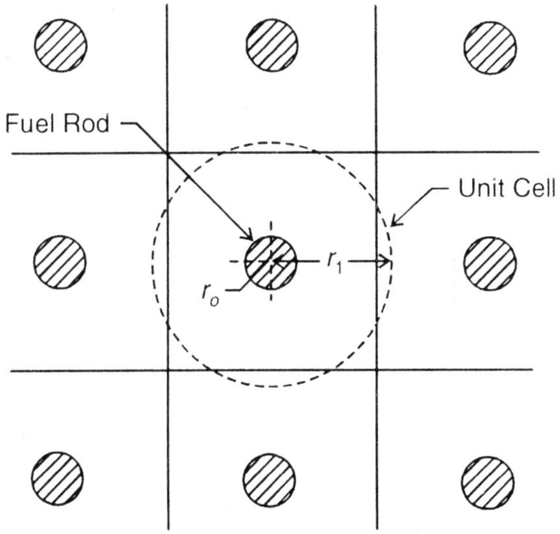

Fig. 7.11: Unit cell in a heterogeneous square lattice

(ii) Fast Fission Factor ($\in$)

The calculation of $\in$ requires the knowledge of fission (σ_f), non-fision capture (σ_a) and elastic (σ_e) and inelastic (σ_i) scattering cross-sections for fast neutrons. Let P be the probability that a primary fission neutron will undergo a collision in the fuel rod where it is produced before actually leaving it. The probability that the fission neutron will be absorbed by the fuel and induce fission is $\frac{\sigma_f}{\sigma}$, where σ is total cross-section *i.e.*

$$\sigma = \sigma_f + \sigma_a + \sigma_i + \sigma_e \qquad \text{...(7.6.3)}$$

Therefore the number of fission neutrons produced inside the fuel rod in the first collision of a primary neutron is $P\left(\nu\frac{\sigma_f}{\sigma}\right)$. The value of ν used here is given by those neutrons only which have energies above the fast fission threshold and as such ν may be slightly less than the actual number per fission. It may be taken as the average number of fission neutrons produced by the thermal neutron in U-235. The number of neutrons making elastic collisions is $p, \frac{\sigma_e}{\sigma}$ Thus the number of neutrons available for further fast fission is

$$N_1 = P\frac{\nu\sigma_f}{\sigma} + P\frac{\sigma_e}{\sigma} \qquad \text{...(7.6.4)}$$

The number of neutrons escaping from the rod without collision is (1-P) and the number of neutrons slowed down below the fission threshold for U-238 due to inelastic collisions is a $P\frac{\sigma_i}{\sigma}$. Thus the number of neutrons which are not capable of inducing fission is

$$N_2 = (1 - P) + P\frac{\sigma_i}{\sigma} \qquad ...(7.6.5)$$

Let P_1 be the probability that a neutron of second generation makes a collision before leaving the fuel rod. The total number of neutrons available for fast fissions will be $PP_1\left(\frac{\nu\sigma_f + \sigma_e}{\sigma}\right)^2$ and the number of neutrons whose energy falls below the fast fission threshold will be

$$N_3 = P\left(\frac{\nu\sigma_f + \sigma_e}{\sigma}\right)\left[(1 - P_1) + \frac{P_1\sigma_i}{\sigma}\right] \qquad ...(7.6.6)$$

Likewise it is possible to get the number of neutrons slowing down past the fast fission threshold by elastic and inelastic collisions in the fuel. Taking that the probability of second and higher generations fission neutrons to make a collision with in the fuel rod will be equal, and if this probability is represented by P'. Then the total number of neutrons slowing down below the fission threshold per neutron produced in thermal fission is by definition, the fast fission factor $\in$. It is then given by the following expression

$$\in = 1 - P + \frac{P\sigma_i}{\sigma} + P.\frac{\nu\sigma_f + \sigma_e}{\sigma}\left(1 - P' + \frac{P'\sigma_i}{\sigma}\right) + ...$$

$$= 1 + \frac{\left[(\nu - 1) - \frac{\sigma_c}{\sigma_f}\right]\frac{\sigma_f}{\sigma}P}{1 - P'\left(\frac{\nu\sigma_f + \sigma_e}{\sigma}\right)} \qquad ...(7.6.7)$$

where $\sigma_c = \sigma - (\sigma_e + \sigma_{in} - \sigma_f)$ is the capture cross-section. P and P' depend on the dimensions of the rods; as an example for a fuel rod of radius r = 1.15 cm, the value of $\in$ is 1.025. For an infinitely large block, say of Uranium $P = P' = 1$ and the maximum value of $\in$ is 1.15.

(*iii*) Resonance Escape Probability (*p*)

The resonance escape probability is defined as the probability that a neutron produced in fission while slowing down will escape capture in resonances at energies above the thermal energy region.

Let $q(E)$ be the slowing down density and ξ is the logarithmic energy decrement, then assuming Fermi's theory of continuous slowing down, the decrease in the logarithm of neutron energy per sec is given by

$$-\frac{d}{dt}\ln E = \xi\frac{v}{\lambda_s} \qquad ...(7.6.8)$$

where $\frac{v}{\lambda_s}$ gives the number of collisions a neutron of velocity v, makes in unit time and the scattering mean free path is given by

$$\lambda_s = \frac{1}{N\sigma_s} = \frac{1}{\Sigma_s} \qquad ...(7.6.9)$$

where N is the number of moderator nuclei per unit volume and σ_s is the scattering cross-section.

Now by definition of the slowing down density

$$n(E)\, dE = q(E)\, dt \qquad ...(7.6.10)$$

where $n(E)\, dE$ is the average number of neutrons that cross the energy interval dE in time dt. Using equation (7.6.8) for dE/dt $(= \xi\, vE/\lambda_s)$ in equation (7.6.10) we get

$$n(E).v = \frac{q\lambda_s}{\xi E} \qquad ...(7.6.11)$$

But the flux of neutrons of energy E is

$$\phi(E) = n(E)v$$

$$\therefore \quad \phi(E) = \frac{q\lambda_s}{\xi E} = \frac{q}{\xi E \Sigma_s} \qquad ...(7.6.12)$$

Assuming that the resonance absorption in uranium at energy E is small so that the neutron flux at energy E does not change, then the decrease in the slowing down density is

$$-dq = \sigma_{ao} N_o \phi(E) V_o / V_1 .dE . \qquad ...(7.6.13)$$

where N_o is the number of uranium (fuel) atoms per unit volume, σ_{ao} is the absorption cross-section section of Uranium. V_o and V_1 are the volumes of Uranium and graphite (moderator) in the lattice. Using eqn. (7.6.12) for $\phi(E)$ in eqn. (7.6.13) we get

$$-dq = \frac{\sigma_{ao} N_o V_o q}{\sigma_{s1} N_1 V_1 \xi E} dE \qquad ...(7.6.14)$$

where $\lambda_s = \frac{1}{\sigma_{s1} N_1}$, σ_{s1} being the scattering cross-section for graphite and N_1 being the number of graphite atoms per unit volume. If Q is the source strength *i.e.* the initial density of neutrons at energy E_o, then the resonance escape probability is given by

$$p(E) = \frac{q(E)}{Q} = \exp\left\{-\int_E^{E_o} \frac{V_o N_o \sigma_{ao}}{V_1 N_1 \sigma_{s1} \xi} \frac{dE'}{E'}\right\} \qquad ...(7.6.15)$$

The above expression for resonance escape probability depends on the fact that absorption in the fuel rod is almost negligible which does not hold and a suitable correction for the depletion in the flux inside the fuel (Uranium) rod has to be used. The numerical value of p for most of the cases is of the order of 0.9. If neutron absorption cross-section is considered, then equation (7.6.12) is modified. It is now expressed as

$$\phi(E) = \frac{q}{\xi E(\Sigma_s + \Sigma_a)}$$

and the resonance escape probability (eqn. 7.6.15) is then given by

$$p(E) = \exp\left\{-\int_E^{E_o} \frac{V_o N_o \sigma_{ao}}{\xi V_1(\Sigma_a + \Sigma_s)} \cdot \frac{dE'}{E'}\right\}$$

$$= \exp\left\{-\int_E^{E_o} \frac{V_o N_o \sigma_{ao}}{\xi V_1(N_o \sigma_{ao} + N_1 \sigma_{s1})} \frac{dE'}{E'}\right\} \quad \ldots(7.6.15i)$$

In writing the above equation, we assumed that the flux does not change. However, in reality the flux inside the moderator and fuel are not the same and this fact must be accounted for. The resonane escape probability now becomes

$$p(E) = \exp\left\{-\int_E^{E_o} \frac{V_o N_o \phi_{ao} \sigma_{ao}}{\xi V_1 \phi_{s1}(N_o \sigma_{ao} + N_1 \sigma_{s1})} \cdot \frac{dE'}{E'}\right\} \quad \ldots(7.6.15ii)$$

(iv) Thermal Utilization Factor (f)

By definition, the thermal utilization factor is

$$f = \frac{\text{Thermal neutrons absorbed in fuel}}{\text{Total thermal neutrons absorbed}}$$

Then in the present case we have (neglecting absorption of neutrons by impurities)

$$f = \frac{N_o V_o \sigma_{ao}}{N_o V_o \sigma_{ao} + V_1 N_1 \sigma_a \bar{\phi}_1 / \bar{\phi}_o} \quad \ldots(7.6.16)$$

or

$$\frac{1}{f} - 1 = \frac{V_1 N_1 \sigma_{a1} \bar{\phi}_1}{V_o N_o \sigma_{ao} \bar{\phi}_o} \quad \ldots(7.6.17)$$

where $\bar{\phi}_1$ and $\bar{\phi}_o$ are the average values of the thermal neutron flux in graphite (moderator) and Uranium (fuel). Since the values of absorption cros-sections for thermal neutrons are usually available, the calculation of the ratio $\bar{\phi}_1 / \bar{\phi}_o$ can be done by finding the thermal neutron flux distribution in the cell using classical diffusion theroy. The value of f for a Uranium-graphite system is around 0.9.

Thus the multiplicity factor k_∞ can be obtained from these values of the four factors. That is

$$k_\infty = \eta \in pf \qquad \text{...(7.5.2)}$$

The maximum value of k_∞ that can be obtained for a graphite moderated natural Uranium reactor is about 1.08. Now to have a self sustained chain reaction the size of the reactor should be such that the non-leakage probability P in it is at least equatl to $1/k_\infty$.

We know that the non-leakage probability P is given by the product of two probabilities *i.e.* $P = P_f P_t$. One is the probability of non-leakage of fast neutrons (P_f) during slowing down and the other is the probability of non-leakage of thermal neutrons (P_t) while diffusing.

Let us first calculate the non-leakage probability (P_f) during the slowing of neutrons from fission to thermal energies. Assuming the continuous slowing down theory of Fermi, the diffusion equation is solved to obtain the slowing down density. Without going into the mathematical details of solving the diffusion equation we find that the slowing down density for thermal neutrons at the point r is given by

$$q(r, \tau_{th}) = \frac{k_\infty}{p} \Sigma_a \phi(r) e^{-B^2 \tau_{th}} \qquad \text{...(7.6.18)}$$

where τ_{th} is the Fermi age for thermal neutrons and B is a constant depending on the properties of the multiplying media. The τ_{th} also gives the slowing down length in the moderator.

The slowing down density of fission neutrons is $\frac{k_\infty}{p}\Sigma_a \phi(r)$. Hence $e^{-B^2\tau_{th}}$ is the fraction of source neutrons that remain inside the reactor and do not escape. In other words $e^{-B^2\tau_{th}}$ is the non-leakage probability (P_f) for the fission neutrons during slowing down,

or $$P_f = e^{-B^2 \tau_{th}} \qquad \text{...(7.6.19)}$$

The non-leakage probability of thermal neutrons (P_f) while diffusing is obtained employing the diffusion theory. The neutron flux in any multiplying system satisfies the equation

$$\nabla^2 \phi + B^2 \phi = 0 \qquad \text{...(7.6.20)}$$

The number of thermal neutrons moving out (leaking) of a unit volume per sec is $-D\nabla^2\phi$ and this can be put equal to $DB^2\phi$ using equation (7.6.20) *i.e.* $-D\nabla^2\phi = DB^2\phi$. The number of thermal neutrons absorbed per unit volume per sec is $\Sigma_a \phi$. Thus the ratio of thermal leakage to thermal absorption is given by

$$\frac{\text{Thermal neutron leakage}}{\text{Thermal neutron absorption}} = \frac{DB^2}{\Sigma_a} = L^2 B^2 \qquad \text{...(7.6.21)}$$

where $\frac{D}{\Sigma_a} = L^2$ is the square of the thermal diffusion length. From the above we find that

$$\frac{\text{Thermal neutron absorption}}{\text{Thermal neutron absorption + Thermal neutron leakage}} = \frac{1}{1 + L^2 B^2} \quad \text{...(7.6.22)}$$

Since the denominator on the left hand side of equation (7.6.22) represents the total number of neutrons that are thermalised, it therefore, follows that $\frac{1}{1+L^2B^2}$ is the probability (P_t) for the non-leakage of thermal neutrons while diffusing *i.e.*

$$P_t = \frac{1}{1 + L^2 B^2} \quad \text{...(7.6.23)}$$

Thus

$$P = P_f P_t = e^{-B^2 \tau_{th}} \frac{1}{1 + L^2 B^2}. \quad \text{...(7.6.24)}$$

A solution of equation (7.6.20) shows that the constant B^2 is inversely related to the size of a reactor. For a large reactor, $B^2\tau_{th}$ is small, as such

$$e^{-B^2 \tau_{th}} \simeq 1 - B^2 \tau_{th} \approx \frac{1}{1 + B^2 \tau_{th}}$$

and thus

$$P = \frac{1}{(1 + B^2 \tau_{th})(1 + L^2 B^2)} \quad \text{...(7.6.25)}$$

If terms of the order of B^4 (being small) are neglected, then the total non-leakage probability is

$$P = \frac{1}{1 + B^2 (L^2 + \tau_{th})} = \frac{1}{1 + B^2 M^2} \quad \text{...(7.6.26)}$$

where $L^2 + \tau_{th} = L^2 + L_s^2 = M^2$ is known as migration area for thermal neutrons.

Now the criticality condition is

$$k = k_\infty P = 1$$

and using equation (7.6.26) for P we get

$$\frac{K_\infty}{1 + B^2 M^2} = 1$$

or

$$\frac{K_\infty - 1}{M^2} = B^2 \quad \text{...(7.6.27)}$$

Thus the knowledge of the four factors ($k_\infty = \epsilon\eta pf$) and the slowing down and diffusion lengths for M^2 will give the size of the reactor at criticality. B^2 is also known as buckling and the table (7.4) gives its expression for different geometry of the reactor.

TABLE 7.4

Geometry	*Buckling* (B^2)
Sphere	$\left(\frac{\pi}{R}\right)^2$
Rectangular parallelopiped	$\left(\frac{\pi}{a}\right)^2 + \left(\frac{\pi}{b}\right)^2 + \left(\frac{\pi}{c}\right)^2$
Cylinder	$\left(\frac{2.405}{R}\right)^2 + \left(\frac{\pi}{H}\right)^2$

The above discussion about the calculation of k, M and B is for a bare reactor *i.e.* a reactor without reflector. In actual practice, reactors have reflectors and as such the calculations for the criticality constant is modified. A detailed study shows that the reactor core diameter can safely be reduced by a value equal to twice the diffusion length, L, of thermal neutrons in the reflector material by using a reflector of thickness of about $3/2\ L$.

7.6.1 Homogeneous Reactor System

When the fuel and moderator are uniformly mixed the fuel and the moderator are in the same physical state, in other words, the fuel is uniformly dispersed throught the moderator, the reactor is termed as homogeneous reactor. In evaluating the reproduction factor k_∞ for an infinite, assembly, we must examine the various factors. For a homogeneous system (no impurities), the volume fraction for the fuel and moderator are indentical and also the neutron flux is same in each constituent.

Thus in equation (7.6.16), $V_o = V_1$ and $\bar{\phi}_1 = \bar{\phi}_o$, hence f can be expressed as

$$f = \frac{N_o \sigma_{ao}}{N_o \sigma_{ao} + N_1 \sigma_{a_1}} = \frac{\sigma_{ao}}{\sigma_{a_o} + \sigma_{a_1} \dfrac{N_1}{N_o}} \qquad \ldots(7.6.28)$$

where σ_{a_o} and σ_{a_1} are the absorbtion cross-section for thermal neutrons.

$\frac{N_1}{N_o}$ gives the ratio of moderator to fuel concentration.

The resonance escape probability for a homogeneous assembly is

$$p(E) = \exp\left\{-\frac{N_{a_o}}{\xi}\int_E^{E_o}\frac{\sigma_{a_o}}{N_{a_o}\sigma_{a_o} + N_{s_1}\sigma_{s_1}}\cdot\frac{dE'}{E'}\right\}$$

$$= \exp\left\{-\frac{1}{\xi}\int_E^{E_o}\frac{\sigma_{a_o}}{\sigma_{a_o} + \frac{N_{s_1}}{N_{a_o}}\sigma_{s_1}}\cdot\frac{dE'}{E'}\right\} \qquad \text{...(7.6.29)}$$

The infinite multiplication factor is given by $k_\infty = \eta f$, because the absorption and production of neutrons take place at a single energy and as such p and $\in$ are equal to 1. Hence for a homogeneous system

$$k_\infty = \eta f = \eta\frac{\Sigma_{a_o}}{\Sigma_{a_o} + \Sigma_{a_1}} = \frac{Z}{Z+1} \qquad \text{...(7.6.30)}$$

where $Z = \frac{\Sigma_{a_o}}{\Sigma_{a_1}}$ (Σ_{a_o} is marcroscopic cross-setion for fuel and Σ_{a_1} is the macroscopic cross-section for the moderator. η is the average number of neutrons produced per neutron absorbed in the fuel).

7.6.2 Reactor Control

Normally a reactor is always built larger than what is required for attaining criticality. The rate at which the neutron density builds up in such a reactor at any time (t) is given by

$$n(t) = n_o \exp(k - t)t / t_o \qquad \text{...(7.6.31)}$$

where t_o is the average life time of neutrons in the reactor and n_o is the initial number of neutrons. If the effect of delayed neutrons that are always produced in the fission of U-235 is neglected, the life time of prompt neutrons comes out to be around a milli-second (0.001 sec). According to the equation (7.6.31), we find that even for $k - 1 = 0.01$, the number of neutrons would increase by a factor of 2×10^4 every second. Since $\frac{n(t)}{n_o} = \exp\frac{0.01}{0.001} = e^{10} \simeq 2.2 \times 10^4$. Such a reactor would not be controlled easily. It is the delayed neutrons which come to the rescue. Though the delayed neutrons are only 0.75 percent of the total number of fission neutrons, they increase the average life time of fission neutrons from 10^{-3} sec to 0.1 sec. Thus the delayed neutrons would increase the neutron number by a factor of 2.7 for every 10 seconds even with the same excess reactivity $(k - 1) = 0.01$. Thus the presence of delayed neutrons produced during fission makes it possible to control the reactor. In a reactor the chain reaction is controlled by moving Cadmium or Boron rods. These elements have very large values for the absorption cross-section of thermal neutrons, as such they can effectively control the thermal neutron flux and also the reactivity. If the chain reaction is to be stopped, the rods are pushed sufficiently inside the reactor so that the rate of absorption becomes greater than the rate of production of neutrons. The neutron flux, as a result, then goes down almost to zero.

7.6.3 Fission Rate and Reactor Power

If the macroscopic fission cross-section is Σ_f an ϕ is the neutron flux of monoenergetic neutrons, the rate of fission per unit volume is given by $\Sigma_f\phi$ where $\Sigma_f = N\sigma_f$ and $\phi = nv$, N being the number of fissile nuclei per unit volume n is neutron density, v is velocity of netutrons and σ_f is the fission cross-section.

We know that 200 MeV of energy is produced (on an average) per fission. This is equal to 3.2×10^{-11} watt-sec per fission so that 3×10^{10} fissions are required to produce 1 watt-sec of energy. If V is volume of a reactor then the number of fissions occuring per second will be $V\Sigma_f\phi$. If the reactor has been operating for some time, then the power P for the reactor is given by

$$P = \frac{V\Sigma_f\phi}{3 \times 10^{10}} \text{ Watts.} \qquad \text{...(7.6.32)}$$

The mass of Uranium-235 used in a reactor of volume V is

$$M = \frac{NV235}{6 \times 10^{23}}$$

Then

$$P = \frac{V.N\sigma_f\phi}{3 \times 10^{10}} = \frac{6 \times 10^{23}.M\sigma_f\phi}{3 \times 10^{10} \times 235}$$

$$= 8 \times 10^{10}\, M\sigma_f\phi \text{ Watts}$$

where M is in grammes and σ_f in barns. (1 barn = 10^{-24}cm^2)

Using $\sigma_f = 580 \times 10^{-24}\text{cm}^2$

$$P = 8 \times 10^{10} \times 580 \times 10^{-24} \text{ M}\phi \text{ Watt.}$$

$$\doteqdot 4.6 \times 10^{-11} \text{ M } \phi \text{ watt.}$$

7.7 NUCLEAR REACTORS

The first reactor built by E. Fermi and his coworkers aroused wide interest in the construction of reactors, not only for academic and research purposes but also to meet our energy requirement. Nuclear reactor is just like a furnace where fuel like Uranium or Plutonium burns (controlled chain reaction) giving neutrons, radio-isotopes and above all, enormous energy in the form of heat which can be utilised to produce steam to run turbines for production of electricity.

Nuclear reactors are highly complex installations and therefore physics of reactors is a vast subject. However, we shall describe some of the general features of nuclear reactors and the basis for their classifications.

7.8 GENERAL FEATURES OF A NUCLEAR REACTOR

All nuclear reactors consists of the following main components:

(*i*) **Fuel:** The fissile material used in a reactor is called the reactor fuel. The commonly used fuels are Uranium isotopes ($^{233}_{92}U$, $^{235}_{92}U$ and $^{238}_{92}U$), Thorium isotope ($^{232}_{92}Th$) and the Plutonium isotope ($^{239}_{94}Pu$). Plutonium is a much better fuel in comparison to Uranium and Thorium, because fission chain reaction can be produced in plutonium both by fast and slow neutrons. It can be obtained from Uranium-238 through the absorption of neutrons according to the decay scheme

$$^{238}_{92}U + {}_0n^1 \rightarrow {}^{239}_{92}U \rightarrow {}^{239}_{93}Np \rightarrow {}^{239}_{94}Pu$$

The central region of the nuclear reactor where the fission reaction takes place is known as reactor core.

(*ii*) **Moderator:** The fast neutrons produced in fission are slowed down to thermal energies by the use of materials called moderators. The moderators are particularly used for thermal reactors. The material of a moderator is of low mass number so that neutron may lose significant amount of energy on colliding with the moderator nuclei till they come to thermal equilibrium with the material. Besides, the moderators should have small neutron absorption cross-sector (σ_a) and large scattering cross-section (σ_s). Most commonly used moderators are graphite, heavy water (D_2O), light water (H_2O), beryllium, beryllium oxide, hydrides of materials like Zirconium hydride (ZrH_2) and some organic liquids like polyphenyls.

(*iii*) **Control Materials or Control Rods:** In order to control the chain reaction and to maintain the multiplication factor so as to ensure the safe operation of a reactor control rods (or plates) are used. The control rods are mostly made of materials such as Cadmium, Boron, Hafnium, Gadolinium. These materials have a large neutron absorption cross-section. The shape and form of the control materials varies from reactor to reactor. For example in a boiling water thermal reactor, Boron is used in the form of Boron Carbide (B_4C) powder stuffed in stainless steel control rods having a cross-section in the form of a plus (+) sign. In some reactors, Ag-In-Cd alloy is used. The control rods are moved in or out of the reactor to control the growth of neutron population (neutron flux).

(*iv*) **Coolants:** To remove the heat generated due to fission in the core (also heat from any other part of the reactor) cooling materials are circulated. These materials are termed as coolants. They can be either liquids or a gas. A reactor coolant should have low neutron absorption cross-section, high thermal capacity, good thermal and radiation stability and should not react chemically with the materials with which they come in contact in the reactor. In thermal reactors ordinary water, steam, heavy water, carbon-dioxide, dry air, nitrogen, helium, are frequently used. In fast reators, liquid sodium or liquid helium is generally used as coolants.

(*v*) **Reactor Shield:** Most of the nuclear reactors are sources of neutrons and γ-rays. To protect the personnel and scientists working around the reactor and also the equipment placed around it, from these hazardous radiations, the reactor vessel is surrounded by a radiation shield. It is generally thick concrete wall about (2m thick) capable of absorbing both neutrons and gamma-radiations. In some cases alternate layers of heavy and light elements are used. Such an arrangement is known as the biological shield or the radiation shield.

(*vi*) **Reflector:** In order to check the escape of neutrons from the core, a region of non-fissionable material is put next to the core. Such a material is known as reflector. In thermal reactors, a moderating material is mostly used as a reflector. In fast reactors, Nickel (Ni) Copper (Cu) are used as reflectors.

These general features of a nuclear reactor are shown as a schematic diagram in the figure (7.12). The heat exchanger in the figure just removes heat from the coolant, which circulates in a closed loop. The water in the heat exchanger boils and steam is produced which is supplied to a turbine to produce electricity.

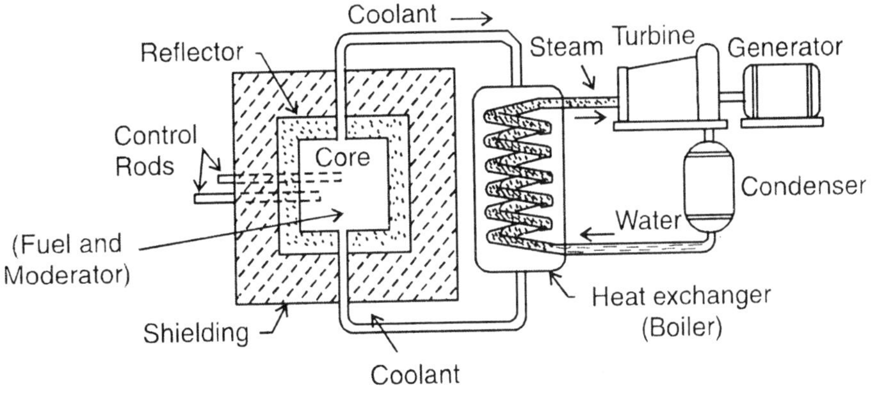

Fig. 7.12: Nuclear power generator

7.9 CLASSIFICATION OF NUCLEAR REACTORS

Reactors have been built for a wide range of uses from power generation to research and testing new components to be used in reactors. Therefore the reactors are classified in different ways. The criteria which is generally applied for classification of reactors are as under.

(*i*) **According to the materials used for fuel, moderator coolant and their physical state:** The reactors using different combination of fuel, moderator, coolant are being referred accordingly. For example, a thermal reactor which uses natural uranium as fuel, graphite as moderator and carbon-di-oxide (gas) as coolant is termed as the graphite moderated gas cooled natural uranium fuelled reactor. When light water is used both as a moderator and as a coolant with enriched uranium (U-235 is more than 0.7%) is referred as light water moderated and cooled enriched uranium reactors. Or simply light water reactors. (These reactors can be pressurized and boiling water types). Likewise we can have 'heavy water reactors', 'gas cooled heavy water moderated reactors', 'light water cooled heavy water moderated reactors'. The fast reactors are classified according to the coolant used. Helium in the gas form or sodium metal in molten form are used as coolants. These reactors are termed as 'gas cooled fast reactors' or 'liquid metal cooled fast reactors'. Some of these types of early reactors are listed in Table (7.5).

(*ii*) **According to the average energy of neutrons which can cause fission:** The majority of nuclear reactors are classified as 'thermal' 'intermediate' and 'fast' according to the neutron energy with which fission occur. In thermal reactors fission is induced by neutrons having energy around 0.025 eV or thermal neutrons. Such reactors are extensively built for research purposes and power generation. In the medium or intermediate energy reactors, the average energy of the neutrons is more than thermal and less than few hundred electron volts (in the range 10-100 eV). Such reactors are sometimes referred as resonance reactors, because the neutron energies are preferably above the resonance region. Intermediate reactors have not been much useful except that they offer a possibility of breeding plutonium-239, but not as efficiently as a fast reactor. Only a few (one or two) reactors of these types have been constructed. In the fast reactors most of the fissions is caused by (fast) neutrons having energies of the order of few hundred kilo electron volts. The most interesting feature of fast reactors is their use in producing (breeding) fissile material like Plutonium-239, Uranium-233.

(*iii*) **According to the geometrical arrangement of fuel and moderator:** From the point of view of their structure, thermal reactors can be classified as homogeneous or heterogeneous nuclear reactors. In homogeneous

TABLE 7.5: SOME OF EARLY RESEARCH REACTORS

Name and Location		*Year*	*Fuel*	*Moderator*	*Coolant*	*Reflector*	*Remarks*
CP-1	-U.S.A.	1942	Natural Uranium	Graphite	Air	Grapbite	World's first reactor
LoPO	-U.S.A.	1944	Enriched Uranium (~15%)	H_2O	H_2O	BeO	First enriched Uranium reactor
CP-3	-U.S.A.	1944	Natural-Uranium	D_2O	D_2O	Graphite	First Heavy water reactor
Clementine	-U.S.A.	1946	Plutonium-239	-	Hg	Graphite	First fast reactor
GLEEP	-U.K.	1947	Natural Uranium	Graphite	Air	Graphite	-
NRX	-Canada	1947	Natural Uranium	D_2O	D_2O	Graphite	High Flux Reactor
Apsara	-India	1956	2% enriched Uranium	H_2O	H_2O	H_2O	First Indian reactor
Cirus	-Inida	1960	Natural Uranium	D_2O	H_2O	Graphite	First Indian D_2O reactor
Purnima	-India	1972	Pu-239	-	Air	Molybdenum and copper	First Indian fast reactor

nuclear reactors the fuel and the moderator are in the same phase (same physical state) both in solid or both in liquid phase. The two are intimately mixed. In the case of heterogeneous reactors, the fuel and the moderator are in different phase. The fuel may be in the form of rods or plates which may be uniformly dispersed in the moderator which may be a liquid (H_2O or D_2O) or a solid (graphite or BeO). The fuel and the moderator are thus geometrically separated. It may be mentioned that most of the present day reactors are heterogeneous type of nuclear reactors.

(*iv*) **According to the main purpose of the reactor:** The classification of reactors according to the purpose is based on the following categories.

(*a*) **Basic research:** These reactors provide facilities for research in various branches of science, for testing new reactor designs, for producing radio-isotopes. Apsara, Cirus, Purnima, Zerlina, Dharuva are examples of research reactors. It may be interesting to know that a single research reactor simultaneously serves many of purposes.

(*b*) **Breeder reactors:** Reactors designed to convert a fertile isotope, ^{232}Th or ^{238}U, into a fissile isotope, of ^{233}U or ^{239}Pu, are called 'converters'. If the amount of so produced fissile isotope is more than what is actually burnt in maintaning the fission chain reaction, the reactors are called breeder reactors.

(*c*) **Power reactors:** Reactors installed to produce power are called as Power reactors. These are both thermal and fast reactors. In India five power reactors are in operation.

Note: For actual description of different types of reactors, one can look into literature, *e.g.*, *Nuclear Reactors* by Kothari *et al.*, Tata McGraw Hill; *Nuclear Reactor Engineering*, Glasstone, Sesonska; Van Nostrand.

7.10 NUCLEAR FUSION

We have seen that in nuclear fission reaction a nucleus of heavy element uranium when captures a neutron breaks into two lighter nuclei and a large amount of energy is released. The explanation of this reaction was found in the binding energies of the nucleus that splits and that of the resulting nuclei. The curve for the binding energy per particle (Fig 1.4 Chapter 1) reveals that BE/A for heavy elements is smaller than the light nuclei of middle mass number elements (fission fragments). A similar situation also apppears towards the beginning of the binding energy curve where the BE/A for lighter nuclei like hydrogen, helium is smaller than the nuclei of middle mass number ($A \sim 60$). This led to believe that release of energy, as in fission, should be possible, if lighter nuclei combine to form a heavier nucleus. Such a process is just the reverse of fission reaction. Thus a nuclear reaction or a process where nuclei of lighter elements are made to combine or fuse together to produce a new heavier nucleus is termed as FUSION. In fusion reaction energy is also released. (It is an exothermic process). This is evident from the fact that the sum of the masses of the fusing lighter nuclei is greater than the fused (product) nucleus and the missing mass is converted into energy. However, in a fusion reaction, the two lighter nuclei, being positively charged, while colliding experience a Coulomb repulsion. This repulsive force acts as a potential barrier and in order to combine, the nuclei must have sufficient kinetic energy so that they may come close enough, where strong nuclear forces produce the necessary fusion effect. Even at low kinetic energies the nuclei may tunnel through the barrier (a quantum mechanical effect). This problem of Coulomb repulsion does not arise in fission, because the neutron which causes fission is electrically neutral. The Coulomb repulsion increases with atomic number (Z), the nuclear fusion is therefore, expected to occur at reasonable kinetic energies for low atomic number (Z) nuclei.

Let us estimate the kinetic energy required to over come the potential barrier. Consider that the two nuclei having charges Z_1e and Z_2e are separated by a distance R, between their centres, equal to the sum of their nuclear radii. the height of the Coulomb barrier will then be given by the electro static potential energy.

$$V_c = \frac{1}{4\pi \epsilon_o} \frac{(Z_1e)(Z_2e)}{R} \quad \ldots(7.9.1)$$

$$= 1.5 \frac{Z_1 Z_2}{R(fm)} \text{MeV} \quad \ldots(7.9.2)$$

If $R \approx 1\ fm$, then

$$V_c = 1.5\ Z_1 Z_2 \text{ MeV} \quad \ldots(7.9.3)$$

This gives the order of magnitude of the kinetic energy of the two nuclei. For example if $Z_1 = Z_2 = 1$, that is if two protons tend to combine, then their relative kinetic energies should be of the order of 1 MeV which is fairly a large amount of energy.

We know that the average kinetic energy of a system of particles having a temperature T K is around kT where k is Boltzmann constant, whose value is 8.6×10^{-11} MeV/K.

Thus an energy of 1 MeV would correspond to a temperature of about 10^{10} K which is much higher than the temperature of the interior of the sun. The temperature of the centre of the sun is about 10^7 K. It is now well established that nuclear fusion has been the source of energy in the sun and the stars.

We, thus, find that for a nuclear fusion reaction to take place, the colliding particles must have sufficient kinetic energy. Such energies ($\approx$1 MeV or so) can be imparted to charged particles by accelerators, but these energetic particles are found to be unable to sustain a fusion reaction. The accelerated particles lose energy in scattering when they strike a target and thus become unable to cause fusion. However, as seen above, in principle it is possible to have such kinetic energies by increasing the temperature and to have kinetic energies of the order of 1 MeV temperatures of around 10^9–10^{10} kelvin would be required. Also at these temperatures the atoms and molecules of the substance lose their extra-nuclear electrons (due to collisions) and there exists just a neutral mixture of positively charged ions and negative electrons. Such a completely ionised system is termed as PLASMA. It is also referred as Fourth state of matter. The energy distribution of plasma particles is similar to the Maxwellian distribution corresponding to the plasma temperature. Some particles in the plasma, will, therefore, have energies much higher (towards the tail of the Maxwellian) than the average energy kT for the temperature T. Such highly energetic particles (nuclei) are able to over come the Coulomb barrier and produce fusion reaction. Since the energy necessary for fusion is provided through high temperature, the reaction is known as *thermo nuclear reaction.*

Since there is a finite probability or corss-section, σ, for a reaction to occur at a certain energy, the acutal reaction rate is therefore, given by the product of the effective cross-section σ at a certain energy and the number of nuclei available at that energy. In Fig. (7.13), Maxwellian distribution of the particles at some absolute temperature T and the cross-section as a function of energy are shown. In the figure is also given the the product of σ and the number of nuclei. It is evident that most of the thermo-nuclear reactions are due to nuclei having energies much having energies much above the avarage energy. Also the reaction rate reaches a maximum at a certain energy.

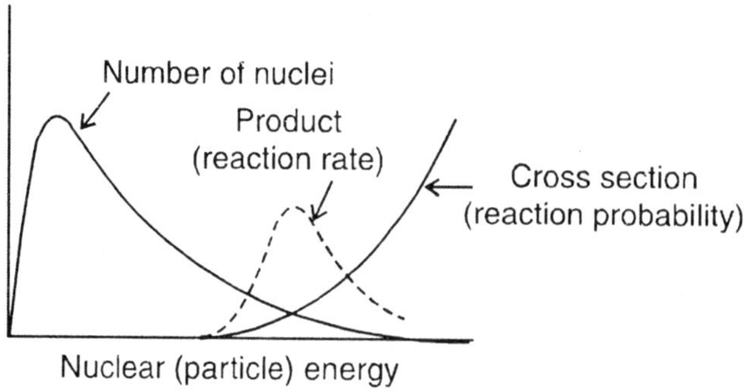

Fig. 7.13: Dependence of rate of nuclear reaction on particle energy at a given temperature

The interior of the sun and other similar stars have temperatures of the order of 10^7 K and as such thermo-nuclear reactions can take place. It is now well established that the thermo-nuclear reactions are main source of energy in the sun and stars. One of the important fusion process occuring in sun is known as Proton-Proton cycle (Bethe and Critchfield, 1938) and Carbon-Nitrogen cycle or just carbon cycle (Bethe, 1939, also Weizsacker, 1939).

7.11 PROTON-PROTON CYCLE

The cycle occurs in the following different steps:

(*i*) To begin with two protons combine to form a deuterium nucleus, through a weak interaction and a positron and a neutrino are emitted

$$ {}_1^1H + {}_1^1H \rightarrow {}_1^2H + \beta^+ + \nu + 0.42 \text{ MeV} $$

(*ii*) The deuteron then captures a proton and forms a ${}_2^3He$ nucleus, emitting a γ-ray

$$ {}_1^2H + {}_1^1H \rightarrow {}_2^3He + \gamma + 5.5 \text{ MeV} $$

The γ-ray takes away the excess energy from the ${}_2^3He$ nucleus and also it conserves the energy and momentum in the reaction.

(*iii*) The ${}_2^3He$ nuclei, once formed in abundance, collide amongst themselves producing a Helium nucleus (α-particles) and two protons are also emitted

$$ {}_2^3He + {}_2^3He \rightarrow {}_2^4He + {}_1^1H + {}_1^1H + 12.8 \text{ MeV} $$

Each ${}_2^3He$ nucleus will need, in turn, three protons for its production *i.e.* reactions (*i*) and (*ii*) have to occur twice. But in the end (reaction (*iii*)) two protons are produced as such effectively only four protons are transformed into one helium nucleus and two positrons, two neutrinos and two gamma rays are emitted. The final reaction, therefore, will be

$$ 4\,{}_1^1H \rightarrow {}_2^4He + 2\beta^+ + 2\nu + 2\gamma + 24.64 \text{ MeV} $$

It may be noted that the γ-rays take away the excess energy from the ${}_2^3He$ nucleus (reaction (*ii*)) and also conserve enrgy and momentum in the reaction. The emission of two positrons conserve charge and two neutrinos conserve lepton number in the fusion reaction. The period of the proton-prtoton cycle in the sun is about 3×10^9 years.

7.12 THE CARBON-CYCLE (THE CARBON-NITROGEN CYCLE)

This cycle takes place in the following steps:

(*i*) ${}_1^1H + {}_6^{12}C \rightarrow {}_7^{13}N + 1.95 \text{ MeV}$

(*ii*) $^{13}_{7}N \rightarrow ^{13}_{6}C + \beta^+ + \nu + 2.22$ MeV

(*iii*) $^{1}_{1}H \rightarrow ^{13}_{6}C \rightarrow ^{14}_{7}N + \gamma + 7.54$ MeV

(*iv*) $^{1}_{1}H \rightarrow ^{14}_{7}N \rightarrow ^{15}_{8}O + \gamma + 7.35$ MeV

(*v*) $^{15}_{8}O \rightarrow ^{15}_{7}N + \beta^+ + \nu + 2.7$ MeV

(*vi*) $^{1}_{1}H + ^{15}_{7}N \rightarrow ^{4}_{2}He + ^{12}_{6}C + 4.96$ MeV

On adding these six equations and cancelling common terms, we find that the net result is again like in *p-p* cycle, 'burning' of four protons producing a Helium nucleus plus two positrons to conserve charge, two neutrinos to conserve lepton numbers and emitting energy of nearly the same amount.

$$4\,^{1}_{1}H \rightarrow ^{4}_{2}He + 2\beta^+ + 2\nu + 26.7 \text{ MeV}$$

Another interesting features of the Carbon cycle is that the reaction starts with $^{12}_{6}C$ which is not used and regenerated at the end. $^{12}_{6}C$ thus acts as a catalyst in the transformation of four protons into a Helium nucleus. It takes around 6×10^9 years for a Carbon atom to go through this cycle in the sun.

It may be noted that the neutrinos emitted in both the cycles carry away some of the energy of reaction, but effectively in both the cycles the available energy of fusion is of the same order of magnitude (~24 MeV)

The relative probabilities for proton-proton cycle and the carbon cycle depend on the temperature. It has been established from the available data that at the interior of the sun the temperature is around 10^7K where the proton-proton cycle is the main contributor of solar energy. In the case of still hotter stars of main-sequence, the carbon cycle predominates in the energy production.

The physical conditions of high temperature and large density of particles inside the sun favour the occurence of thermo nuclear fusion reactions. The large density ensures frequent collision with each other due to their random motion and high temperature ensures sufficient kinetic energy to the particles for overcoming the coulomb potential barrier. Nevertheless, quantum mechanical tunnelling through the potential barrier of relatively low energy nuclei also takes place resulting into the fusion reaction.

Once the thermonuclear fusion takes place in the sun, large amount of energy is liberated and four protons are transformed into a helium nucleus. The reaction is self-sustained and the plasma is contained under gravitational forces of massive bodies. Any slow down of the fusion process (due to cooling of the plasma) is counter balanced by the heating produced through gravitational contraction. The sun and similar stars, thus, behave as self-controlled systems for fusion processes.

The sun has been emitting energy at the rate of 4×10^{26} J per sec, for the last about 10^9 years. It is estimated that with the total mass of hydrogen in the sun (10^{30} Kg), the earth would continue to receive energy as at the present rate for another about 100 billion years or so.

7.13 THERMO-NUCLEAR REACTION ON THE EARTH

The liberation of energy in the fusion of light nuclei has led to explore the possibility of harnessing the energy and using the fusion reactions as a future source of energy. Though the energy released in a fusion reaction is much less than that from a fission reaction, but per kg, the energy released in fusion is greater (10^{14} J/kg) than that in fission $\left(10^3 \frac{\text{J}}{\text{kg}}\right)$. This has further enhanced the importance of fusion reactions as a future source of energy.

We have seen that a self- sustaining fusion chain reaction has been possible in the sun because of high temperature and large density of nuclei existing there. However, the proton-proton-chain and carbon-chain responsible for the steller energy production are extremely slow. One cycle in each chain takes around 10^9 years. The most promising fusion reactions on the earth, therefore, involve deuterium $\left({}_1^2\text{H}\right)$ and tritium $\left({}_1^2\text{H}\right)$ both being isotopes of hydrogen. The fusion reactions are:

(*i*) ${}_1^2\text{H} + {}_1^2\text{H} \rightarrow {}_1^3\text{H} + {}_1^1\text{H} + 4.2\text{ MeV}$ (*D-D* reaction)

(*ii*) ${}_1^2\text{H} + {}_1^2\text{H} \rightarrow {}_2^3\text{He} + {}_0^1\text{n} + 3.3\text{ MeV}$ (*D-D* reaction)

Both these fusion reactions have same probability.

The other pssible fusion reaction is between deuterium and tritium.

(*iii*) ${}_1^2\text{H} + {}_1^3\text{H} \rightarrow {}_2^4\text{He} + {}_0^1\text{n} + 17.6\text{ MeV}$ (*D-T* reaction)

This reaction has large cross-section and it liberates relatively large amount of energy.

(*iv*) ${}_1^2\text{H} + {}_2^3\text{He} \rightarrow {}_2^4\text{He} + {}_1^1\text{H} + 18.3\text{ MeV}$

Since ${}_2^3\text{He}$ is not readily available, the fusion reaction (*iv*) is not found very feasible, even though large amount of energy is liberated in the reaction. The oceans do contain tritium, but in very small amount. Its extraction cannot be economical.

The cross-sections σ for the fusion reactions (*i*), (*ii*) and (*iii*) as a function of deuteron energy are shown in the Fig. (7.14). The fusion cross-section σ is very slow for low energies, but rises rapidly, reaches certain maximum and then slows down again for higher energies.

The energy realeased in deuterium-deuterium reaction is about 2×10^{14} J per kilogram, as such deuterium can be used as a promising fuel for the energy in the future. Deuterium occurs in the nature with a relative abundance of about one atom for about 7000 atoms of hydrogen. The extraction of deuterium from water is also not very expensive. The vast amount of ocean waters, therefore, would be able to provide the world's power requirement for several million years as and when controlled fusion reactions become practical.

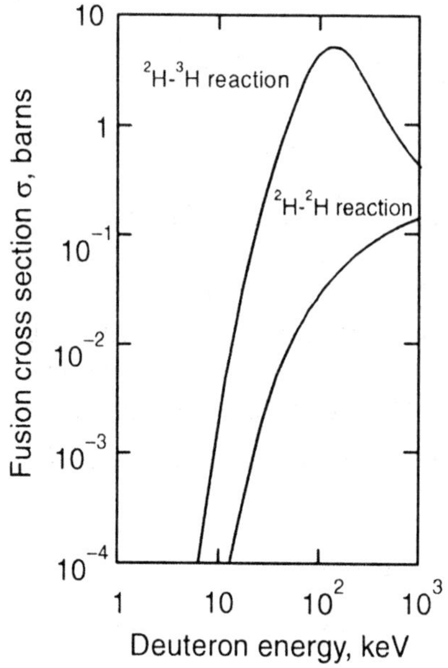

Fig. 7.14: Variation of fusion cross-section with deuterium energy

7.14 CONDITIONS FOR A FUSION REACTOR

If energy liberated in a fusion reaction is to be utilised for useful purposes, then like nuclear reactors where fission takes place in a controlled manner, we must have fusion reaction taking place in a controlled manner. Once fusion reactor becomes feasible then only the energy can be put to use. We have seen that three basic conditions must be satisfied before a fusion reactor can be made successfully operative. The first is a high temperature. The cross-section or the probability for fusion is large for deuteron or tritium energies around 100 to 1000 KeV (Fig. 7.14) which corresponds to temperature around 10^7-10^8 kelvin. The second is high density of nuclei, so that in a given time interval, the number of collisions between nuclei may be large and third condition is that the fusion reaction should be self sustained so that the system of reacting nuclei may be able to stand the time till the nuclear fusion takes palce and also is able to give out more energy than what is spent on the system.

7.15 CRITICAL IGNITION TEMPERATURE

As pointed out earlier, one of the requirements for the controlled release of useful energy is that the fusion reaction must be self-sustaining. The rate of loss of energy from the reacting system must be less than the rate of generation of energy. In other words, once the temperature of the deuterium or its mixture with tritium reaches the point where fusion occurs at sufficient rate than the energy released in fusion must be high enough to maintain the temperature. The loss of energy from the plasma takes place mainly through radiations *i.e.* bremsstrahlung which is chiefly short wave length radiations (X-rays). Therefore, there must strike a balance between the rate of energy generation in fusion reaction and rate of energy lost by radiations which escapes the system. Both these rates increase with temperature, but the energy release increases more rapidly with temerature than rate of energy loss. The Fig. (7.15) shows a variation of these quantities with temperature. There is, therefore, a critical ignitioin temperature t_c which determines the self-sustaining stage of the plasma, such that beyond this temperature the fusion proceeds once it is produced. From the figure we find that the critifal ignition temperature above which the fusion reaction is self sustaining is 5 KeV for a *D-T* fusion reaction and 50 KeV *D-D* reaction (1 KeV $\simeq$ 10^4K).

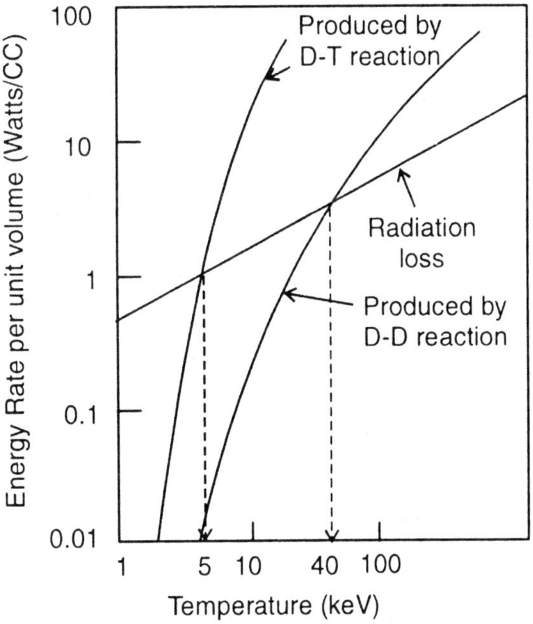

Fig. 7.15: Critical ignition temperature for D.D. and D.T. reactions

7.16 LAWSON CRITERION

At the temperature required for the fusion reaction, the reacting system is in the form of plasma which is fully ionised gas. This plasma must be confined for a suitable length of time so that the reaction may yield sufficient energy to sustain the reaction. However, to obtain a high enough particle density n for a long enough time t in a hot plasma, such that the fusion reaction may be self sustained, is not easy. A criterion was proposed by J.D. Lawson which is referred as Lawson criterion. It is the product of the density of reacting nuclei n and the time t in sec. during which the thermo nuclear reaction takes place (t is the minimum time for which the plalsma must be confined, confinement time). The calculated values for $(n.t.)$are about 10^{20} s/m^3 for D-T reaction and about 10^{22} for D-D reaction.

The estimates of the critical ignition temperature and that of Lawson criterion, suggest that D-T reaction is more promising for thermonuclear self-sustained reaction, than D-D reaction. However, there is a drawback, in that tritium is not easily available. But this can be overcome by producing Tritium through nuclear reaction of Lithium with neutron.

$$^6_3\text{Li} + ^1_0\text{n} \rightarrow ^3_1\text{H} + ^4_2\text{He}$$

In fact, the required tritium can be made by the fusion reaction itself. Once the fusion reaction starts with tritium the neutrons produced in the reaction can be absorbed by Lithium-6 nuclei, present in the ordinary Lithium, placed as a blanket for the reacting assembly. This will produce tritium for the subsequent use in the production of energy through thermonuclear fusion reactions.

7.17 PLASMA CONFINEMENT

At the high temperature (10^7-10^8 K) the gas or the gas mixture is in the form of hot Plasma. But no material object (vessel) can keep such a hot plasma and retain it to ensure a fusion reaction. The Plasma particles, on striking the walls of the vessel will impart their kinetic energy to the walls, which will tend to melt or even vaporised, contaminating the plasma. Secondly the plasma gets cooled and becomes unfit for fusion reaction. As such it becomes essential to devise methods so that the hot plasma, once produced at the required temperature could be kept away from the walls. As pointed out earlier this plasma confinement in sun and stars is taken care of by the enormous gravitational forces of massive bodies.

(*i*) Pinch Confinement or Pinch Effect

Once the principle of producing energy through thermo nuclear reactions was understood and established, attempts were made in most parts of the globe to have plasma at the required kinetic temperature and also to confine the plasma away from the walls of the vessel. The earliest attempts were to make use of an effect known as Pinch effect. In order to achieve the Pinch effect, a strong electric current, of tne order of 10^6 amperes is passed through a low density (low pressure) gas deuterium or a mixture of deuterium and tritium. The gas is ionised due to the heating produced by the current, producing plasma. The plasma (a mixture of ions and electrons) effectively behaves as current carrying conductors. Also the electric currents produce magnetic fields. The magnetic field exerts pressure on the plasma perpendicular to the plasma axis, resulting into the compression (pinching) of the plasma away from the walls of the vessel (may be a toroid or a straight tube). This compression of the plasma is known as Pinch effect. The pinch effect not only keeps the plasma away from the walls, but also heat the plasma to a required temperature. Much success could not be achieved by this method of confinement. The plasma was found to be unstable and other methods had to be devised for the confinement of the plasma.

Magnetic Confinement

The earliest and successful method of confining plasma has been through the application of strong magnetic fields, as proposed by I.E. Tamm in Russia (1950) and L. Spitzer in America (1951).

Two classes of magnetic field geometry are being designed and found suitable for plasma confinement.

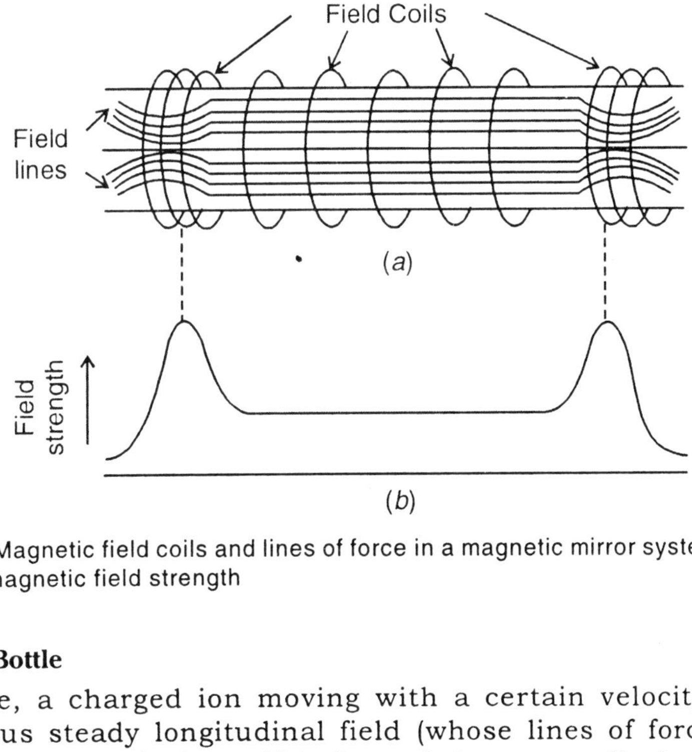

Fig. 7.16: (*a*) Magnetic field coils and lines of force in a magnetic mirror system (*b*) variation of magnetic field strength

(*ii*) Magnetic Bottle

In principle, a charged ion moving with a certain velocity in a non-homogeneous steady longitudinal field (whose lines of force converge) experiences a magnetic force. This force, being perpendicular to both the direction of motion and the direction of field, has a backward component in such a field and also an inward component which causes the particle to move in a helical path. If the backward component is strong enough, then at the converging point of the magnetic field, the ion reverses its direction, and if there is another similar magnetic filed along its reversed path, the ions get confined in the region between these two magnetic fields. The converging magnetic fields act as "magnetic mirrors" and the plasma is effectively confined or trapped in a 'magnetic bottle', for sufficiently long time as required for the fusion reaction to take place. In its simplest form, the device is a straight tube with magnetic coils wound around it in such a way that the magnetic field provided by the coils is considerably stronger at the ends than in the middle. The Fig. (7.16*a*) shows a schematic arrangement of such a device. The variation of the magnetic field is also shown (Fig. 7.16*b*). The stronger magnetic field at the ends constitutes the mirror effect of the magnetic fields.

(*iii*) Toroidal Confinement (The Stellarator)

The other magnetic confinement was first proposed by L. Spitzer in USA. The geometry of the device is in the form of a toroid and named as Stellarator. Here the plasma is confined to a toroidal tube by means of magnetic field such that the lines of force are parallel to the circumference of the tube as shown in the Fig. (7.17). Such a field is known as axial field and is produced by passing electric current in the coil wound round the torus. The field is stronger at the inner perimeter than at the outer perimeter. This non-uniformity of the magnetic field causes the plasma to move towards the walls resulting into non-confinement of the plasma.

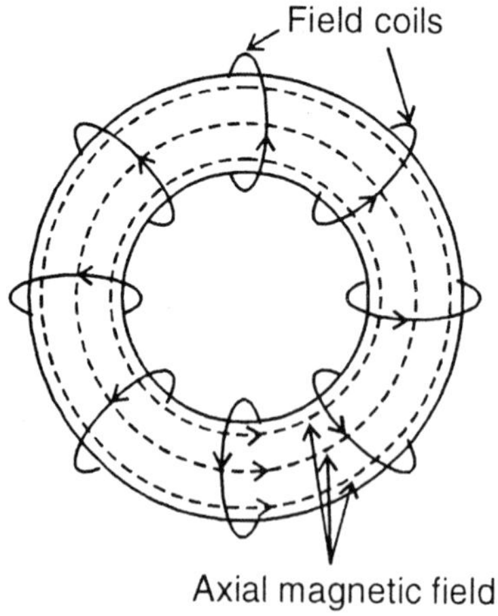

Fig. 7.17: Axial magnetic field in a toroidal tube

Spitzer suggested a modification and the geometry of the stellarator was changed to a shape of the figure of 8 as shown (Fig. 7.18) . The effect of the non-uniformity of the magnetic field in one arm of the toroid is counter balanced by the field in other arm, thereby the confinement of the plasma becomes easier.

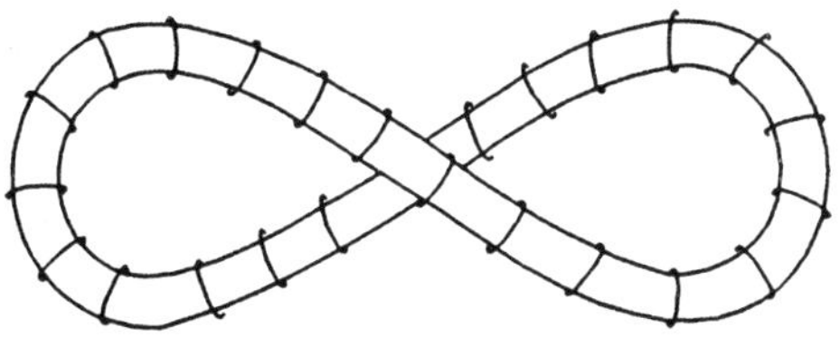

Fig. 7.18: Schematic representation of a stellarator

The Russian version of toroidal confinement of plasma is termed as tokamak (toka means electric, and mak means magnetic). It uses a modified toroidal (doughnut shaped) magnetic field. To prevent the escape of plasma, a poloidal (weaker) magnetic field is generated whose lines of force circles around the toroidal axis. Such a weak field can be produced by a current set up in the plasma itself which is just an induction effect as occurs in a transformer. The Fig. (7.19) shows the doughnut shaped toroid where the poloidal and toroidal magnetic fields are indicated. Those two fields acting together make the path of the plasma ions helical as shown in the figure.

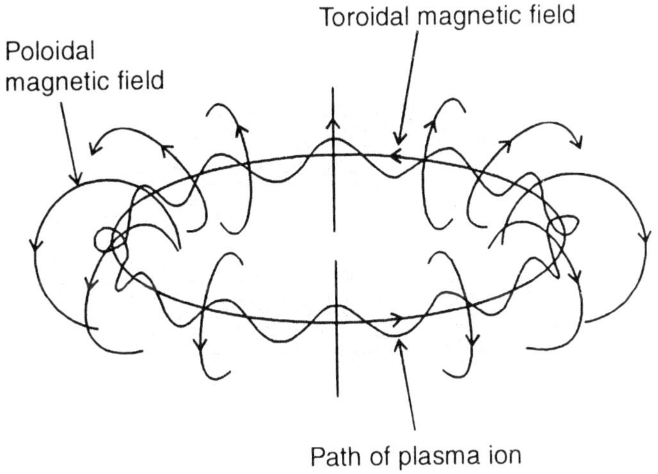

Fig. 7.19: Combined polaroidal and toroidal magnetic fields which confine a plasma in a tokamak

(*iv*) Inertial Confinement

An entirely different procedure which does not employ magnetic fields is called inertial confinement. In this procedure deuterium tritium pellets are heated and compressed using energetic beams. Once the pellets are heated and compressed, implosion takes place raising the temperature to a required value and fusion takes place in a short time of the order of 10^{-8}-10^{-9} sec. If a number of such pellets could be ignited in succession then a steady stream of energy becomes available. These days, high power Laser beams are used to heat and compress tiny pellets (0.1 mm radius) and also beams of charged particles like electrons, protons from particle accelerators are used for this purpose.

7.18 HYDROGEN BOMB AS AN UNCONTROLLED THERMO-NUCLEAR DEVICE

The release of fusion energy is possible only when high temperatures of the order of 10^7K - 10^8K are available. Such high temperatures cannot be produced in Laboratories through conventional methods. However, it was thought that tremendous amount of heat at extremely high temperatures (10^6-10^7 K) produced in the explosion of fission bomb or atomic bomb, could be utilised to trigger a fusion reaction. Such an idea was first suggested by Robert Oppenheimer. Thus, if a chian of fusion reactions could be produced among the deuterium or deuterium-tritium mixture then a very powerful explosion can take place. The hydrogen bomb is a nuclear device where a chain of thermonuclear fusion reactions take place in an uncontrolled manner. It is effectively a fission-fusion bomb, yielding explosive energy equivalent to megaton of TNT. (megaton bomb). Since neutrons are produced in fusion reactions, they can be utilised to cause fission in fissile materials if put around the fusion device. This is the basis of what is referred as fission-fusion-fission bomb.

Thus maintaining hot plasma at ultra high temperatures for long enough time (~ 1 mill-second) to produce fusion reactions in a sustained manner had been a problem. Efforts are going on to solve such problems through out the world including India. Only partial success could be achieved in confining hot and dense plasma for a micro-second.

Though at present there seems to be no hope of a fusion reactor, like a nuclear fission reactor, but the efforts are on and it is hoped that in times to come scientists will be able to produce a miniature sun on the earth 'burning the oceans'. This will then be a major break through in the field of energy sources.

EXERCISES

1. What is the physical significance of fission cross-section for thermal neutrons.
2. Calculate the rate of fission for ^{235}U in order to produce 2 watt of power.
   [6.25×10^{10} fission/sec]
3. How much energy is producd in complete fission of 0.235 Kg of ^{235}U. Energy released per fission being 200 MeV.
   [12×10^{25} MeV]
4. Write the four factor formula for the multiplication for a steady state chain reaction in an infinite assembly and explain the significance of each factor.
5. For a large bare reactor $k = 1.03$ and the neutron migration length is 0.5 m. Determine the critical volume of spherical assembly.
6. Calculate the number of ^{238}U nuclei undergoing spontaneous fission per hour per gm of Uranium.

7. Calculate the thermal utilisation factor, f, for a homogenerous mixture of graphite and enriched uranium-235 having enrichment ratio to be 1. Assume the graphite uranium mass ratio to be 15 and $\sigma_a^c = 4.5$ mb.
8. Describe briefly the principle which leads to pinch effect in a gas plasma.
9. What is a magnetic bottle? How does it work.
10. What is a fusion reactor? In what ways it is different from a fission reactor.

8

NUCLEAR RADIATION DETECTORS

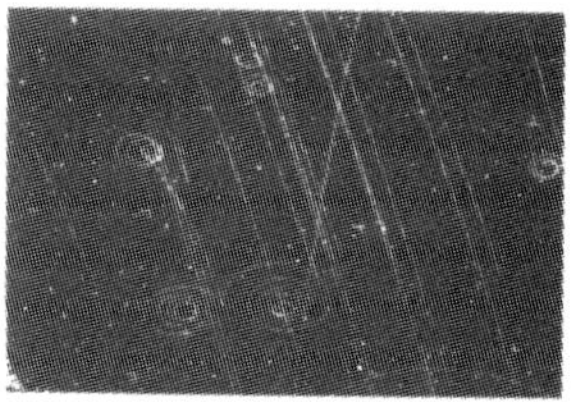

Nuclear radiation detectors play a dominant role in any nuclear physics experiment. Each new development in nuclear radiation detector technology has opened new vistas in nuclear physics. Most radiation detectors are based on the production of excited or ionized atoms/molecules by a charged particle traversing the medium. In gas filled and solid state detectors, the charged particles produced in the medium, due to ionization are, collected to produce an electrical signal. Ionization of atoms in a crystal excites the lattice and under suitable conditions visible or ultraviolet light is emitted. The light is detected by photosensitive devices *e.g.* a photomultiplier tube. In some organic substances the ionization may excite the molecules which emit visible/ ultraviolet light on de-excitation.

8.1 GAS FILLED IONIZATION DETECTORS

The passage of a charged particle in a gas produces excited atoms and ion-electron pairs. If an electric field is present, the ions and the electrons move towards cathode and anode respectively. These moving charges produce an electrical impulse, which is detected with suitable electrical set-up. Gas filled detectors may work in the region of ionization chamber, proportional counter or the Geiger Mueller Counter.

When a charged particle passes through a gas filled detector, a large number of atoms/molecules are excited which de-excite with the emission of visible or ultraviolet photons. A number of atoms are ionized also, giving rise to "primary ionization". The ions and electrons move randomly due to thermal motion. Due to close proximity, the ions and electrons may recombine to form neutral atoms and emit ultraviolet photons. Some times an electron may collide with neutral atoms. If the neutral atom is electronegative *e.g.* oxygen, the electron may attach to it forming a negative ion. The negative ions move with about the same velocity as the positive ions. Usually the presence of electronegative atoms in an ionization counter is not desirable. The randomly moving positive ions collide with neutral atoms. If an atom in which the binding energy of an electron is less than that in the atom of the ion, neutral atom takes up the positive charge of the ion with the emission of excess energy as electromagnetic radiations. The newly created positive ion drifts, till it reaches the cathode.

The recombination of ions and electrons, produced in primary ionization, takes place mostly along the track of the incident charged particle. If there is a general ionization in the counter due to whole volume irradiation, there can be volume recombination. The rate of recombination R is proportional to the densities n_i and n_e of the ions and the electrons respectively so that

$$R = \alpha\, n_e n_i \qquad ...(8.1.1)$$

where α is the recombination coefficient. The recombination of ions and electrons depletes the number of charged particles in the counter volume with an increase in the ultraviolet photons. The ultraviolet photons may eject photoelectrons from the cathode.

When a potential difference is applied between the cathode and the anode, it produces an electric field in the counter volume. The electrons and the ions produced in primary ionization move towards the anode and the cathode respectively. Due to physical separation the recombination rate becomes less. The ions drift towards the cathode with a velocity v such that

$$v = \mu E/P \qquad ...(8.1.2)$$

$$\approx \mu (E/P)^{0.5} \text{ for large values of } E/P$$

where, E is the intensity of the electric field, P the gas pressure in the counter and μ is the mobility of the ions. The mobility of an ion is nearly independent of the gas pressure and the electric field. The value of mobility in different counter gases lies between 1.0 and 1.5×10^{-4} m^2 atoms/volt sec. In the time interval between two collisions with gas molecules, the electron due to their small mass get accelerated much more than the ions. Consequently the electrons get collected at the anode in a very small time period (few microseconds) while the ions take much longer time. Electron mobility is about thousand times that of ion.

Consider a parallel plate ionization chamber filled with pure argon gas. If a potential difference V is applied between the plates separated by distance d, the intensity of the electric field is uniform and equal to V/d (Fig. 8.1).

Consider the primary ionization to be produced at a distance x from the grounded cathode. Let n_o electron-ion pairs be formed in the primary ionization. As the electrons move towards the anode, they induce a negative charge at the anode. The total change induced is proportional to the potential through which the electrons fall while reaching the plate. If V_x is the potential at the point where primary ionisation is formed, the charge induced by the electrons is proportional to $V - V_x$. The positive ions moving towards the cathode also induce a negative charge at the anode proportional to the potential through which they fall *i.e.* V_x. Thus the total charge induced is same as induced by the number n_o of the electrons reaching the anode. If C is the distributed capacitance of the anode, the anode potential changes by a value ΔV where

$$\Delta V = - n_o e/C \qquad \ldots(8.1.3)$$

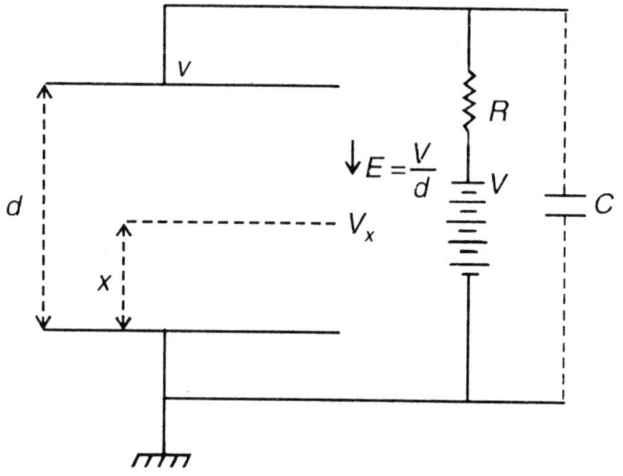

Fig. 8.1: Parallel plate ionization chamber. C is the distributed capacitance of the anode

The potential induced by the electrons is in a much shorter time than that by the ions. The induced charge leaks through the anode load/leak resistance R with a time constant CR. If the time constant CR is less than the collection time of the ions t, one observes only the signal induced by the electrons. Full potential signal at the anode can be observed only if the leakage time constant CR is much greater than the collection time of the ions t'. Fig. 8.2 shows the shape of the signal at the anode of an ionization chamber.

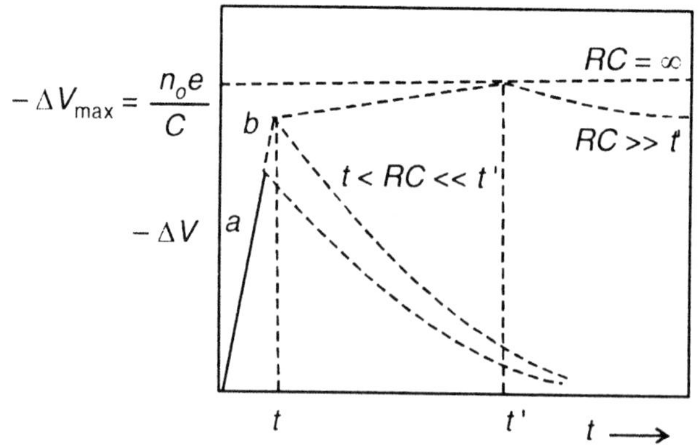

Fig. 8.2: Anode pulse shape. t is the collection time of electrons and t' for the positive ions

Depending upon the trajectory of the ionizing particle, the primary ionization may take place at any distance x from the cathode ($0 \le x \le d$). The actual shape of the signal at the anode is the integrated effect of all the particles produced at different places. The maximum height of the signal remains the same *i.e.* $n_0 e/C$.

When an incident charged particle collides with atoms/molecules of the chamber gas, some of its energy is lost in exciting the atoms to higher energy states. In some collisions an energy greater than the ionization energy of an electron is imparted to the gas atom producing ionization. The electrons produced in the primary ionization may have large enough energy to produce further ionization in the gas (δ-rays). Thus the average energy W lost by the incident particle in producing an electron ion pair is different from the average excitation potential I used in the relation for specific ionization (Chapter 6). As an example the average excitation potential for argon has been measured as $I = 198$ eV, while the average energy lost for producing an electron-ion pair is only 26.4 eV. The ionization energy (binding energy of the outermost electron) of an argon atom is 17.3 eV. As discussed in chapter 6, the average energy required to produce an electron-ion pair in a mixture of argon and 0.12 per cent C_2H_2 is only 21 eV.

The signal produced at the anode is proportional to the energy dissipated by the incident particle in the ion-chamber. If the particle has its full trajectory in the ion chamber, the signal height is proportional to the kinetic energy of the incident particle. The ion chamber can thus be used for the measurement of energy of charged particles. The energy resolution obtainable depends upon the number of electron-ion pairs formed - which follows the normal or Gaussian distribution. The energy resolution R is given as

$$\frac{\Delta E}{E} = R \propto \frac{\sigma(n_0)}{n_0} = \frac{\sqrt{n_0}}{n_0} = \frac{1}{\sqrt{n_0}} \propto \frac{1}{\sqrt{E}} \qquad \ldots(8.1.4)$$

where $\sigma_{n_0} = \sqrt{n_0}$ is the standard deviation in the number n_o of electron ion pairs. A charged particle dissipating 1 MeV energy in pure argon gas produces about 38000 electron ion pairs. The energy resolution available should be about 0.5 per cent. This energy resolution as will be seen later is much better than that obtainable from scintillation detectors. It has been experimentally observed that the energy resolution obtained from ionization detectors is much better than that predicted by equation (8.1.4) This is explained by introducing an empirical factor known as Fano factor which should multiply the variance $\sigma^2_{n_o}$ to give the observed energy resolution. This gives the energy resolution R as

$$R = \sqrt{F\left(\frac{\sigma(n_o)}{n_o}\right)^2} = \sqrt{\frac{F}{n_o}} \qquad \ldots(8.1.5)$$

The Fano factor for pure argon is estimated as 0.17. For a mixture of Argon and 0.5 per cent C_2H_2 the Fano factor is found to be 0.075. The effect of Fano factor is that energy resolution of ionization detector is much better than expected. The pulse height distribution for all events in which same energy is deposited in the ion chamber is Gaussian in shape.

8.2 IONIZATION CHAMBER

Parallel plate ionization chambers are used for special purposes. Usually ionization counters are made with a metallic cylindrical cathode and an axially stretched tungsten wire of diameter about 0.1 mm which acts as the anode (Fig. 8.3).

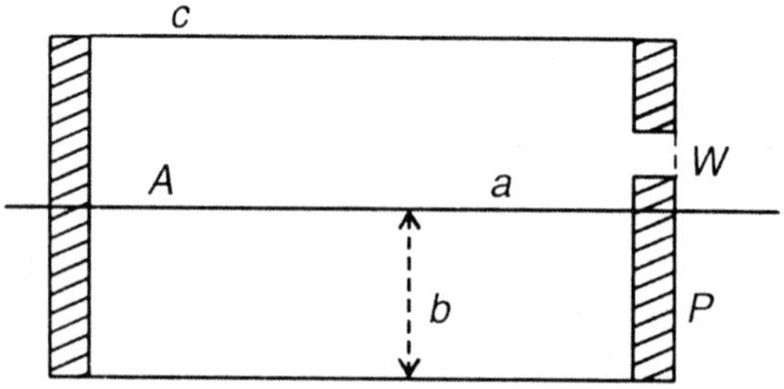

Fig. 8.3: Cyclindrical ionization chamber. *b* is the internal radius of cathode and a the radius of the anode wire. *W* is a thin window for radiation to enter the chamber

In the cylindrical geometry the electric field near the wire is very strong. The electric field at a distance *r* from the axis of the anode wire is given as

$$E(r) = \frac{V}{r \ln b/a} \quad ...(8.2.1)$$

where *V* is the voltage applied to the anode, *a* and *b* are the radii of the anode wire and the inner surface of the cathode respectively. It can be seen from equation (8.2.1) that the electric field reduces by a factor of 10 at a distance of 0.5 mm from the anode wire.

A window is some times provided for the radiations to enter the ionization chamber. It is desirable that the range of the incident radiation is less than the size of the ionization chamber. This determines the size of the chamber and the gas pressure in it. The gases generally used in the ionization chamber are argon, carbondioxide, methane or a mixture of argon and methane. The gas pressure in the chamber is usually below atmospheric pressure. Specially designed ionization chambers have been used much above the atmospheric pressure as well. For the study of low energy α-particles or β-particles, it some times becomes necessary to mount the source inside the ionization chamber itself.

As discussed above, at very low anode voltages, the ion-pairs formed in the primary ionization caused by the dissipation of any energy *E* may recombine to some extent, reducing the charge collected at the anode. As the anode voltage is increased, the electron and ions get separated and full charge produced in the primary ionization is collected. Further increase of the anode voltages does not change the amount of charge collected. The electrical signal generated at the anode is then independent of the anode voltage. If there is another group of particles, which dissipate twice the energy, $2E$ creating twice the number of ion pairs, the signal generated at the anode becomes double as shown in Fig. 8.4.

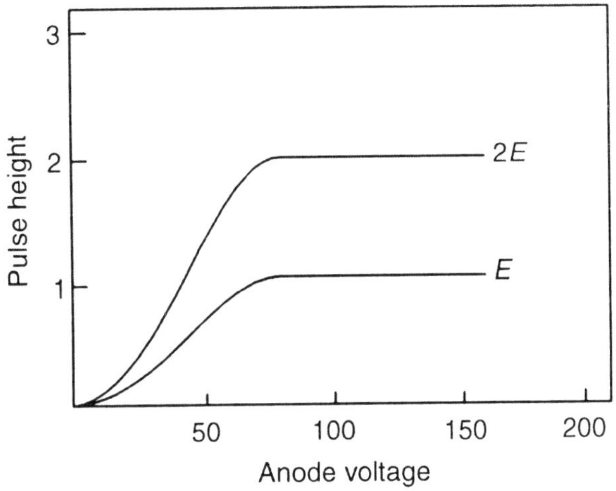

Fig. 8.4: Anode pulse height in an ionization chamber when energy dissipated by incident particle is E and $2E$

If the anode voltage is increased above a certain value, some of the electrons of the primary ionization start to ionize the atoms they collide with. This increases the total charge in the chamber and the anode pulse height increases. The ionization chamber enters the region of proportional counter which shall be discussed later.

Ionization chambers have been extensively used in studying the energy spectra of charged particles. For energy measurement, it is necessary that the whole trajectory of the charged particle lies in the sensitive volume of the ionization chamber. This puts a limit on the energy and the nature of the particles that can be studied. As has been discussed above, the energy resolution obtained with an ionization chamber is proportional to $(F/E)^{1/2}$ where the Fano factor is approximately 0.1 and E is the energy of the particle (whole of which is dissipated in the chamber). The full energy resolution can be obtained if the collection of charge from the whole volume of the ionization chamber is equally efficient. Near the ends of the anode wire the electric field is relatively weak and the charge collection may not be fully efficient. To overcome this problem guard sleeves are provided at the ends of the anode wire. These sleeves are connected to the anode voltage supply directly. The sleeves help define the sensitive volume of the chamber.

In a cylindrical ionization chamber the change of potential near the wire is very large. If the primary ionization is produced at some distance from the anode wire, the electrons fall through a larger potential than the positive ions. The signal generated at the anode is mostly due to the electrons. The positive ions move slowly towards the cathode. To observe the full pulse height generated at the anode, the decay time *CR* of the signal should be much larger than the time of collection of the ions (about 100 μsec).

Very often ionization chambers are employed to monitor the flux of radiation *e.g.* γ-rays and neutrons. In such an application, the ionization is produced in the whole volume of the chamber. The chamber is usually used in the *d.c.* mode by increasing the load resistance such that $CR \sim 0.1$ sec. The continuous current that flows in the ionization chamber is measured using sensitive current meters.

8.3 PROPORTIONAL COUNTERS

As pointed out earlier, the electrons and ions formed in the primary ionization, in the presence of electric field move towards anode and cathode respectively, making collisions with the atoms of the chamber gas. In the time interval between two collisions an electron gains kinetic energy, which is lost in the next collision. If the electric field is intense, the kinetic energy acquired by the electron may become large enough to ionize the atom it collides with. At still higher electric field, the electrons produced in the secondary ionization process also ionize the atoms they collide with. Thus a single electron of primary ionization during its movement towards the anode produces an avalanche of ionized atoms. The size of the avalanche depends upon the electric field intensity in the chamber. Due to the secondary ionization, the total number of ion pairs produced in the chamber is many times those in the primary ionization. The ratio of the total number of ion pairs to that in the primary ionization is known as the gas multiplication factor. This factor increases almost exponentially with the anode voltage. As the electric field is most intense in the vicinity of the anode, the gas multiplication takes place close to the anode wire.

The electrons colliding with neutral atoms ionize some atoms. A large number of atoms are only excited to higher energy states. While de-exciting the atoms emit ultraviolet and visible photons. These photons may eject photoelectrons from the cathode. The photoelectrons moving towards the anode may produce their own avalanche. The positive ions move towards the cathode colliding with other atoms. If by chance the ion is neutralised it may emit ultraviolet photon which may again interact with the cathode. When the positive ion reaches the cathode, it pulls out an electron from the metal surface after giving an energy equal to the work function ϕ of the surface. The ion is then in an excited state with an energy $E_i - \phi$, where E_i is the ionization energy of the atom of the ion. If the energy $E_i - \phi$ is greater than the work function, an electron may be ejected from the surface with an energy $E_i - 2\phi$. This electron moves towards the anode producing its own avalanche. The production of avalanches by photoelectrons and secondary emission electrons (due to ions) tend to change the gas multiplication factor non-linearly with the primary ionization. In counters filled with pure argon gas, the gas multiplication factor is constant and does not depend upon the primary ionization only as long as it is less than 100. At higher gas multiplication, its value varies with the intensity of primary ionization.

A proportional counter can be used for measurement of particle energy only if the gas multiplication factor is constant. A multiplication factor of 100 is rather low. It is possible to increase this factor to a value of 10^4 or 10^5 and still keeping it independent of the primary ionization. To achieve this, the cathode surface is suitably treated to form a coating of oxide of high work function. Increasing the work function reduces the probability of emission of electrons by the ions. Secondly a polyatomic gas *e.g.* methane, carbondioxide or ethyl alcohol is mixed with the counter gas-say argon to an extent of about 10 per cent. This is known as the quenching gas.

The ionization energies of argon atom and alcohol molecule are 15.7 eV and 11.4 eV respectively. On neutralisation the argon ion emits an ultraviolet photon of energy 15.7 eV which is strongly absorbed by the alcohol gas. The ionized alcohol molecule thus formed moves towards the cathode. On neutralisation by capture of an electron either in the counter gas of at the cathode surface, the energy of excitation of the alcohol molecules is distributed to its vibrational states of its component atoms. The atomic bonds break and the alcohol molecule is dissociated into its constituents. The argon ions moving towards cathode collide with the alcohol molecules and charges exchange takes place. The charged alcohol molecule on capture of an electron either in the counter gas or at the cathode surface breaks down into its constituents as discussed above. The presence of the quenching gas thus reduces to a great extent the effect of the ultraviolet photons and the positive ions. It is possible to achieve a gas multiplication factor of $10^4 - 10^5$ which is independent of the primary ionization. The multiplication factor is increased by increasing the anode voltage which increases the electric field in the counter.

If the voltage is increased further such that the multiplication factor is above 10^5, the number of ultraviolet photons and the ions is so large that the quenching gas is not able to nullify their effect. The multiplication factor becomes dependent upon the primary ionization. The signal at the anode is no longer proportional to the energy dissipated by the incident particle in the counter volume. The counter then enters into the region of limited proportionality. In this region due to large charge multiplication, the anode signal is fairly large. A counter in this region is not suitable for studying the energy spectrum of incident particles, however it can be fruitfully used to distinguish between say electrons and α-particles. The only advantage is the large pulse height in comparison to that from a proportional counter. The BF_3 counter used for detection of thermal neutrons are often used in the region of limited proportionality.

With a constant gas multiplication factor, it is expected that the energy resolution obtainable with a proportional counter would be same as that available with an ionization chamber. The individual electron produced in the primary ionization generates its own avalanche and there could be a distribution in the size of avalanches, produced by different electrons. It has been shown that the Fano factor for the proportional counters is greater than that for an ionization chamber by about 0.4. Thus the effective Fano factor for proportional counter is approximately 0.5. The energy resolution is still given by equation 8.5.

Usually proportional counters are used with a gas multiplication factor of about 10^4. This implies that the signal from a proportional counter is greater than that from an ionization chamber by a factor of 10^4. An ionization chamber signal may need an amplification of 10^6, the signal from a proportional counter needs an amplification of only a few hundreds. This simplifies the electronics considerably. The noise problem is also absent in a proportional counter. In an ionization chamber the collection of electrons take a time of 50-100 μsec while in a proportional counter the electrons are collected in a time interval of 1-10 μsec. The collection time depends upon the gas pressure and the anode voltage.

Proportional counters are extensively used in the study of energy spectra of low energy electrons. For studying electron spectra below 10 keV, the source is mounted inside the counter itself. The gas filling is usually helium mixed with 10 per cent methane at a pressure a little below atmospheric pressure. For the study of energy spectrum of X-rays and γ-rays below 20 keV energy, the filling gas is usually krypton or xenon mixed with 10 per cent methane or 5 per cent carbon dioxide, at a pressure of about 50 cm of Hg. The γ-rays enter the proportional counter through a window covered with beryllium sheet. The γ-ray photons produce photoelectrons in the gas which loose all their energy in the counter gas.

The multiplication factor in a proportional counter is very sensitive to the impurities present in the gas. If there is any leakage of air in the counter volume or if occluded gases are emitted from the cathode, the multiplication factor changes. To overcome this problem continuous flow of counter gas was maintained in the counter so that all impurities are washed out. The gas pressure in the gas flow proportional counter is a little above the atmospheric pressure. Such proportional counters are cumbersome to use. It has now become possible to obtain sealed proportional counters which maintain their characteristics for a few years.

8.4 GEIGER MUELLER COUNTER

We have discussed that in a counter working in the proportional counter or limited proportionality region, each electron produced in the primary ionization produces its own avalanche whose size determines the multiplication factor. If the anode voltage is increased beyond the limited proportionality region, the size of the avalanche increases hundred fold or more. All the secondary ionization is in a very small volume around the anode wire. There is a significant amount of recombination of ions and electrons emitting UV photons. The excited atoms of the gas also emit ultraviolet photons. These photons produce photoelectrons in the gas and the avalanche spreads along the length of the wire. The whole length of the anode wire is surrounded with ionization. This is known as the Geiger discharge. In a Geiger discharge the linear density of the ions along the wire is almost 10^8 per centimetre. This density increases sharply with the anode voltage. In a counter operated in the Geiger counter region, a single electron in the counter volume generates the Geiger discharge along the whole length of the anode. The amount of ionization produced is thus independent of the primary ionization. In a Geiger counter an electron and an α-particle produces the same ionization.

The whole Geiger discharge is within a radius of about 0.5 mm around the anode wire. The electrons are quickly collected at the anode in a time of 0.1–0.5 μsec. The positive ion sheath around the wire acts as the anode and the electric field in the counter volume becomes very low. Any further primary ionization caused by a subsequent incident particle can not produce secondary ionization. The counter at this stage becomes insensitive to any incident radiation. Gradually the positive ions move towards the cathode and the electric field in the counter regenerates. When the field is almost fully established, the counter becomes ready to detect another ionizing particle. During the regeneration of the field, the pulses produced by the subsequent primary ionization have an increasing amplitude with time as shown in Fig. 8.5. The time during which the amplitude of the pulses following the main pulse keeps increasing is known as the recovery time of the counter. The time during which the electric field in the counter is too small to produce secondary ionization is known as the dead time of the GM counter (50-100 μs). Usually the time interval in which the GM counter set up can not record two incident particles is known as the resolving time of the GM counter.

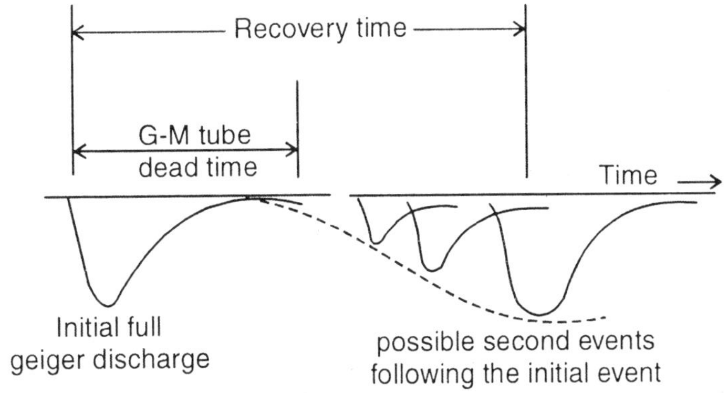

Fig. 8.5: Variation of pulse height when G.M. counter is triggered during its recovery time after a full geiger discharge

If during the period the electric field is regenerating after an avalanche, an electron is produced in the counter volume either by the ultraviolet photons or the positive ions striking the cathode, a second avalanche would start. The process continues and the counter produces a number of spurious pulses. The quenching of the GM counter is achieved by mixing a quenching gas as in the case of a proportional counter discussed above. The quenching gas absorbs the UV photons as well as neutralises the argon ions with large probability (GM counters are usually filled with argon gas at a pressure of 5-10 cm of Hg and quenching gas methane, butane, or ethyl alcohol at a pressure of 2-4 cm of Hg). If the anode voltage on the GM counter is further increased the size of the avalanche becomes very large and the quenching gas is not able to absorb all ultraviolet photons. Some argon ions are also able to reach the cathode and produce secondary electrons. The result is that the counter goes in almost a continuous discharge.

The working of an ionization counter from ionization chamber region to the discharge region can be studied by measuring the height of the anode signal as a function of the anode voltage. Fig. 8.6 shows the various regions of the working of an ionization counter. The two curves in the figure correspond to two different amount of energies (1 MeV and 2 MeV) dissipated by incident particle in the counter volume.

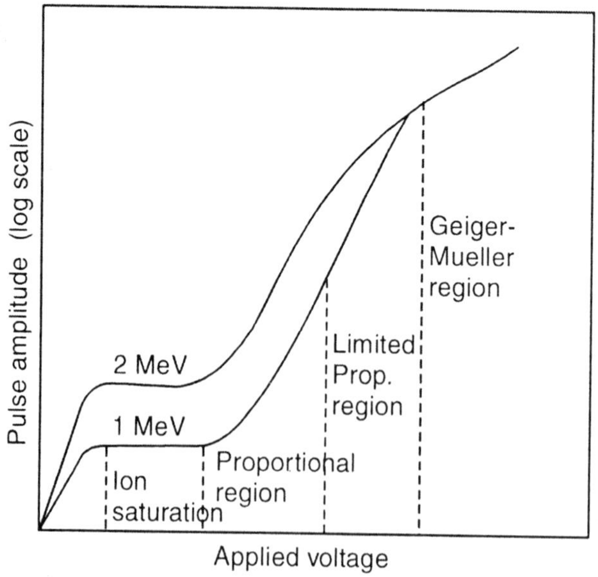

Fig. 8.6: Variation of charge collected at the anode of an ionization counter with the anode voltage

In every Geiger discharge about 10^8-10^9 ion pairs are formed. The number of quenching gas molecules dissociated in each Geiger discharge is about 10^9. In the discharge region the number of molecules of the quenching gas dissociated becomes very large and the gas gets depleted. It is advisable not to use a GM counter in the discharge region for any length of time.

GM counters are fabricated in different sizes, depending upon their use. The size may vary from 40 cm long × 6 cm dia. to 5 cm long × 2 cm dia. The cathode of a GM counter could be copper, brass or stainless steel. The anode wire is invariably of tungsten with diameter about 0.1 mm. For the detection of β-particles, GM counters are fabricated with an end window. One end of the counter tube is closed with thin mica sheet (1.5 to 3.5 mg/cm^2 thickness). The tungsten wire is fixed at the other end of the tube only. Before filling a counter with the counter gas, it is essential to clean thoroughly all the internal surfaces. The counter is given heat treatment under high vacuum for long hours so that all occluded gases from the cathode are released. An end window GM counter is shown in Fig. 8.7.

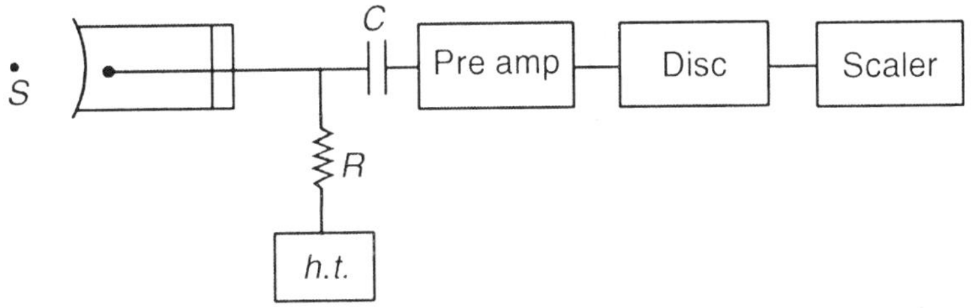

Fig. 8.7: End window G.M. counter and associated electronic set up

8.4.1 Plateau of a GM Counter

The electronic set up for studying the plateau characteristics of an end window GM counter is shown in Fig. 8.7. The voltage pulse produced at the anode with the leak resistance R of about 10^6 ohms passes through a preamplifier. In the GM counter region, the height of the anode pulse is more than a few volts and it needs no amplification. To observe the pulse, in the proportional counter region or the ionization chamber region, suitable amplification of the signal is essential. From the preamplifier the signal passes through a discriminator which is adjusted to bias out the noise pulses. The output of the discriminator is counted in a scaler circuit.

Radiations emitted from the radioactive source S enter the GM counter though the window. Every particle producing ionization in the counter is recorded in the scalar. The β-particles entering the counter produce ionization in the counter gas. The γ-rays entering the counter interact with the cathode mostly by Compton effect. The Compton electron produces ionization in the counter and is recorded. The efficiency of detection of the β-particles is almost 100 per cent and for γ-rays it is 1-3 per cent only. If the half life of the radioactive source S is very long, the average number of radiations entering the counter remains constant during the time of the experiment. The counter thus detects on an average, a fixed number of radiations per unit time. A GM counter detects cosmic rays also. The average number of cosmic ray particles entering the counter is also independent of time. As the anode voltage of the counter is increased from some low value, the counter passes through the proportional counter region and the limited proportionality region. In these regions the anode pulse is too small to be counted. When the anode voltage reaches the GM discharge region, the anode pulse is large enough to be recorded in the scaler. As the anode voltage is increased further, the signal pulse height increases (due to the increase the avalanche size) but as all the radiations producing ionization .in the counter are already counted, the counting rate in the scaler does not increase. Thus beyond the threshold of the Geiger discharge, the counting rate due to the radiations should remain independent of the anode voltage. However as discussed above, there are some spurious Geiger discharges due to inadequate absorption of UV photons and neutralisation of argon ions. These spurious discharges increase with the anode voltage and subsequent increase in the avalanche size. As a result of this, the recorded counting rate increases a little with the anode voltage. The variation of the recorded counting rate as a function of anode voltage of a GM counter is shown in Fig. 8.8. Part *AB* of the curve corresponds to the region where the pulse height is close to the discriminator bias. In the region *BC* where the counting rate is nearly constant, is known as the plateau of the counter. The region *CD* corresponds to the discharge region where the number of spurious pulses increases sharply. The slope of the plateau is mostly due to the unwanted spurious pulses and determines the quality of the detector. The slope of the GM counter plateau is measured as per cent change in counting rate per 100 volt change in the anode voltage. The length *BC* of the plateau and its slope determine the quality of the GM counter. A GM counter with a plateau slope greater than 10 per cent per 100 volts or a plateau length of less than 100 volts is not satisfactory.

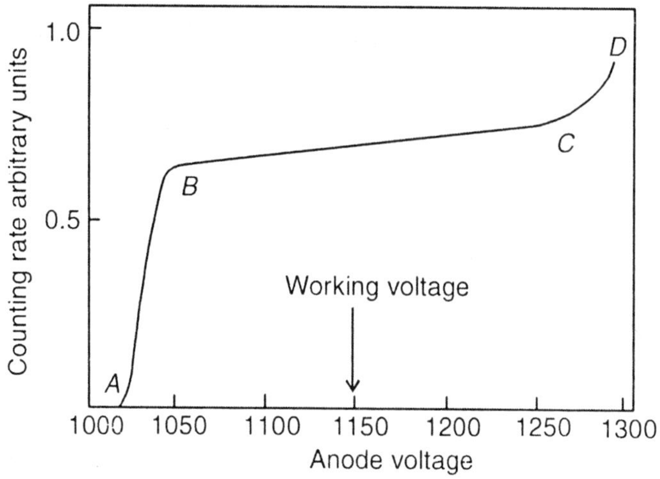

Fig. 8.8: Plateau characteristic of a G.M. counter

The shape of the anode signal is essentially similar to that shown in Fig. 8.2. As the anode voltage in a GM counter is high, the time of collection of electrons is less than a microsecond while that for the ions is 50-100 microseconds. Thus the dead time of a GM counter is approximately 100 microseconds.

Geiger Mueller counters are inexpensive and easy to handle. They however have a low detection efficiency for γ-rays. These detectors can not be used either for energy measurements or for timing experiments where time interval of 10^{-7} seconds or less are measured. A GM counter can be fruitfully used as a random pulse generator with constant pulse height.

8.5 SCINTILLATION DETECTOR

When ionization is produced in certain materials, light is emitted. This light is detected with some suitable device. Rutherford and Co-workers in their experiments on scattering of α-particles used a silver activated zinc sulphide screen as a detector. The α-particles striking the ZnS screen produced a flesh of light which was observed by the eye, through a microscope. In 1944 Curran used a plastic button loaded with activated zinc sulphide as a scintillator. The light produced in the plastic button was detected with a photomultiplier tube. Since then great progress has been made in the development of both scintillators and photomultiplier tubes. The electronics used in scintillation counters has also become more sophisticated. A number of organic and inorganic scintillators and a variety of photomultiplier tubes have been developed. One can choose a scintillator and a photomultiplier tube according to the requirement of the experiment.

A scintillator to be useful in nuclear physics experiment should preferably fulfil certain requirements.

(1) For a given amount of ionization in the scintillator, its light output should be as high as possible. The scintillator should have a high light efficiency.

(2) The scintillator should be transparent to the light it emits.

(3) The wave lengths of the light emitted by a scintillator should lie in the region in which the available photomultiplier tube has maximum detection efficiency.

(4) The intensity of the scintillation light decays exponentially with a definite decay time. If more than one mode of light emission is possible, the scintillator may emit light with more than one decay time. It is desirable that there is a single decay time which should be as short as possible. A short decay time is essential in experiments involving measurements of time intervals in the range of nanoseconds.

(5) When a scintillator is used for the detection of γ-rays, it should have a high density and should be composed of high atomic number elements.

(6) It should be possible to have large single crystals of the scintillator without any cracks or flaws.

(7) As the scintillator has to be coupled to the glass window of the photomultiplier tube, it is desirable that the refractive index of the scintillator material is close to that of glass.

(8) The scintillator should be easy to handle. It should not be too fragile.

(9) It should be possible to cut the scintillator in proper size and shape and polish the same.

(10) The material of the scintillator should not be too toxic.

(11) It is desirable that the light output of the scintillator is proportional to the ionization produced in it.

There is no scintillator which satisfies all the above conditions. The requirements of an experiment determine the choice of the scintillator. The properties of some of the commonly used scintillators are given in Table 8.1. The scintillators are divided in two broad categories-organic and inorganic. The two types of scintillators differ in their constituents as well as in the mechanism of emission of light.

TABLE 8.1: PROPERTIES OF SOME COMMONLY USED SCINTILLATORS

Phosphor	*Density gm/cm³*	*Refractive index*	*Wave length of maximum emission nm*	*Decay time of principal mode μs*	*Relative γ-ray pulse height with Bi-alkali Photomultiplier*
NaI (TI)	3.67	1.85	415	0.23	1.00
CsI (TI)	4.51	1.80	540	1.10	0.45
CsI(Na)	4.51	1.80	420	0.63	0.85
BGO ($Bi_4(GeO_4)_3$	7.13	2.15	505	0.30	0.13
BaF_2	4.89	1.49	310	0.62	0.13
	4.89	–	220	0.0006	0.03
Anthracene	1.25	1.62	447	0.30	0.50
Trans-stiibene	1.16	1.62	410	0.006	0.3
Polystyrene with 16 g/lit of p-terphenyl (plastic)	1.04	1.58	450	0.005	0.2
Toluene + 3 g/lt 2-5 diphenyle oxazole	0.88	1.50	382	0.004	0.2

8.5.1 Organic Scintillators

Organic scintillators are ring compounds composed of carbon, hydrogen and oxygen. When a charged particle produces ionization in an organic scintillator, the electronic states along with their vibrational states of the molecules are excited. The excitation energy of electronic states is about 4 eV while the vibrational energies are about 0.1 eV. The excited vibrational states decay in a very short time, to their associated lowest electronic state. The electronic state decays to the ground state of the molecule and its associated vibrational states, emitting characteristic wavelength light. The coupling between electronic and vibrational states provides a channel for dissipation of the excitation energy of a molecule. This reduces the amount of energy emitted as light. This effect is minimum in molecules where electronic excitation is not accompanied with large configurational changes. Benzene derivatives which fulfil this requirement are often used as scintillators. The wavelength of the emission band of these scintillators increases with the number of conjugated rings in the molecule. Pure anthracene and stilbene crystals have their emission bands in the blue region and are commonly used as scintillators. The emission of light in an organic scintillator is a molecular phenomenon and the scintillation decay time is about 10^{-8} seconds. In all the three phases-solid, liquid and gas, the wavelengths of light emitted by both anthracene as well as stilbene are same.

The molecular excitation in bulk organic scintillator is transmitted to its different regions by some resonance phenomenon. As a result, foreign molecules situated in the matrix of the phosphor and having their characteristic absorption energy lower than the excitation energy of the phosphor, can absorb energy and re-emit. If the transition probability for de-excitation in the foreign molecule is larger than that for the phosphor molecule, most of the light is emitted in the wavelength band characteristic of the foreign molecule. The decay time of light emission remains the same as that of the phosphor molecule. Naphthalene, for example, emits light in the near ultraviolet region. An addition of only 1 per cent anthracene to naphthalene results in the emission of almost the entire excitation energy in the blue region, characteristic of anthracene. Thus anthracene acts as a wavelength shifter. Such binary scintillators using wavelength shifters are used commonly. Plastic scintillator can be prepared by adding 1-4 per cent tetraphenyl butadiene in polystyrene and polymerising it. Some times two wavelength shifters are used in a scintillator. Adding 4 grams of *p*-terphenyl and 8 mg of diphenyl hexatriene in a litre of pure toulene or phenyl cyclohexane gives a liquid scintillator. In this scintillator the solvent emits light of very short wavelength, which is transformed by *p*-terphenyl to a wavelength band at 3500 *A*. This light is shifted by diphenyl hexatriene molecules to the blue region. There is an optimum amount of wavelength shifter that should be added to a phosphor. A larger amount reduces the light output by transferring the excitation energy to its vibrational states, appearing ultimately as heat. Minute quantities of other impurities in the phosphor quench the light emission by transferring the excitation energy to the vibrational states. The organic compounds used in a organic scintillators should be of very high chemical purity.

For the same amount of ionization produced in an organic scintillator, the amount of light emitted depends upon the specific ionization. For the same total ionization, the light output diminishes with the increase in specific ionization. This effect is explained as due to quenching of light emission by the molecules damaged due to ionization. The amount of light emitted by an organic scintillator per unit energy deposited in it by an ionizing particle (dL/dE) is found to be given as

$$\frac{dL}{dE} = \frac{L_o}{1 + a\frac{dE}{dX}} \quad \ldots(8.5.1.1)$$

where L_o and a are constants and dE/dX is the specific energy loss by the particle. The specific energy loss by electrons is very small and its effect can be observed at kinetic energies below 100 keV. Relative to electrons, the light output as a function of energy for different particles stopping in a plastic scintillator Ne123 is shown in Fig. 8.9.

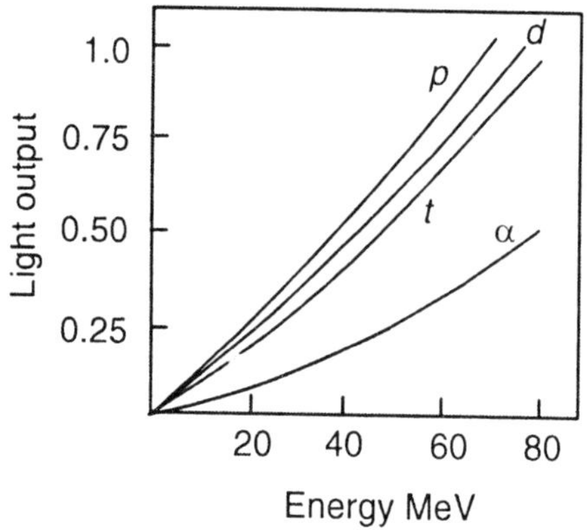

Fig. 8.9: Variation of light output of an organic scintillator NE123 with energy of the incident charged particles

Among pure materials, anthracene and stilbene are the commonly used scintillators. Anthracene has the largest scintillation efficiency (light output for a given dissipated energy). Stilbene has a decay time of 4.5 nsec. Some part of its light is emitted with a decay time of a few hundred nanoseconds. The intensity of the long decay time component of the light increases with the specific ionization by the incident charged particle. The decay of scintillation pulses with time from a stilbene scintillator for different ionizing particles is shown in Fig. 8.10. The shape of the light pulse is used to discriminate electronically between electrons, protons and α-particles. Both anthracene and stilbene crystals are very fragile and difficult to obtain in large sizes.

The light efficiency of liquid scintillators is only about half of that of an anthracene crystal. Liquid scintillators often have two wavelength shifters as discussed earlier. The great advantage of liquid scintillator is that it can be obtained in large quantities. The experiment on the detection of antineutrinos employed 200 litres of a liquid scintillator filled in a glass tank. For specific purposes, liquid scintillators are loaded with heavy elements. Lead loaded liquid scintillators are used for the detection of γ-rays. Thermal neutrons are detected with gadolinium loaded liquid scintillators. Gadolinium captures thermal neutrons with a large cross section and gives rise to capture γ-rays and subsequent β and γ-activity which is detected by the scintillator.

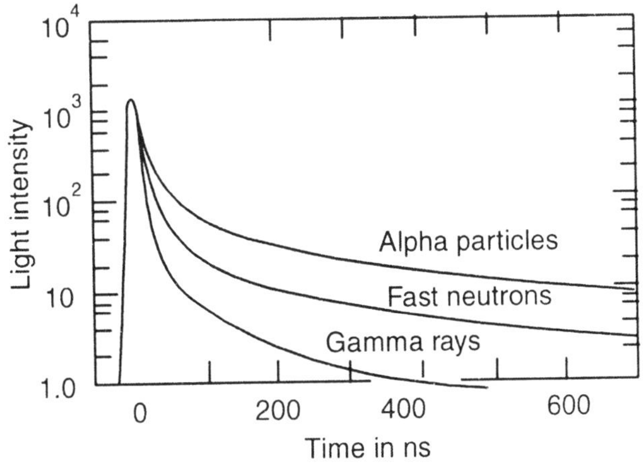

Fig. 8.10: Luminiscence decay of stilbene for different charged particles (From L. M. Bollinger and G. E. Thomas (1961), *Rev. Sci. Instr.*, **32**: 1044)

Plastic scintillators have a light output only as good as the liquid scintillators. These can be obtained in large size blocks and can be easily machined and polished. Boron and gadolinium loaded plastic scintillators are often employed for the detection of thermal neutrons. The scintillation decay time of plastic scintillators is 3-5 nanoseconds. They are often employed in fast timing experiments. Prolonged exposure to ionizing radiations, light and oxygen causes polymer degradation in plastic scintillators.

8.5.2 Inorganic Scintillators

Inorganic scintillators are either activated halide crystals *e.g.* thallium activated sodium iodide or caesium iodide crystals-or pure crystals *e.g.* bismuth germinate (BGO) or barium fluoride (BaF_2) crystals. Thallium activated sodium iodide crystal is the most commonly used inorganic scintillator.

In inorganic scintillators, the emission of light is a crystalline phenomenon. When a charged particle produces ionization in the crystal, the electrons in the valence band are excited to the conduction band. The energy difference between the two bands is generally 6-8 eV. The transition probability for transition of electron from conduction band to valence band is relatively small. The activator which is an impurity in the crystal lattice has its energy levels between the conduction band and the valence band as shown in Fig. 8.11.

The electron which has been excited to the conduction band is free to migrate throughout the lattice. The hole created in the valence band also migrates in the lattice till it meets an impurity atom. The hole ionizes the impurity atom. The electron migrating in the conduction band meets the ionized impurity atom and makes transition to one of its excited states. This excited state on transition to the impurity ground state emits visible light. The scintillation life time is determined by the transition probability of the impurity excited state and is of the order of 10^{-7} seconds. If the free electron makes transition to some metastable state of the impurity atom, the scintillation life time for such a state is longer and gives rise to an after glow in the scintillator. There is some probability that the electron is captured in an excited state which is coupled to the ground state with radiation less transitions. Process through which the conduction band electron returns to the valence band without emitting radiations is known as quenching.

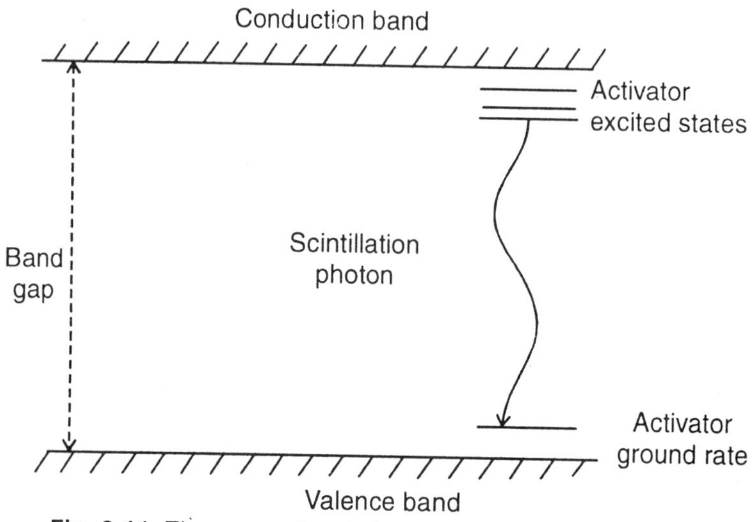

Fig. 8.11: The energy levels in an activated halide crystal

The impurity atom for a scintillator has to be chosen properly so that the scintillation light lies in the desired wavelength band. The concentration of the impurity atoms is also critical. In the case of thallium activated sodium iodide NaI(Tl), it is found that a thallium concentration of about 10^{-3} mole fraction gives the best scintillation efficiency. In the activated scintillators, the energy of the photon emitted is much less than the band gap. The scintillator is therefore totally transparent to the light it emits. In the case of NaI(Tl) scintillator, it has been estimated that one visible photon is emitted for every ion pair formed in the crystal by the incident ionizing particle.

In the case of scintillators of pure materials *e.g.* barium fluoride or bismuth germinate, the process of emission of light is more complex. The luminescence of a BGO crystal is associated with an optical transition of the Bi^{3+} ion which is the major constituent of the crystal. There is a large shift between the optical absorption and emission spectra of the Bi^3 states. This makes the BGO crystal transparent to the light it emits.

The NaI(Tl) scintillator has the largest scintillation efficiency. Its short scintillation decay time of 230 nanoseconds coupled with the large light output makes NaI(Tl) scintillator suitable for studying time interval as short as 10^{-9} seconds. It is possible to grow large size crystals of NaI(Tl). Crystals of NaI(Tl) as large as 30 cm dia × 10cm have been successfully fabricated and used in experiments. Though highly fragile, the NaI(Tl) crystals can be cut and polished. NaI(Tl) scintillation crystals are available with thicknesses from 1 mm to 10 cm. A NaI(Tl) crystal is highly hygroscopic and has to be hermetically seated in an air tight container in dry atmosphere. A scintillator is usually used with a right circular cylindrical shape. The NaI(Tl) crystal is cut to desired size and polished coarsely on the sides and the circular faces. The crystal is mounted in an aluminium can as shown in Fig. 8.12.

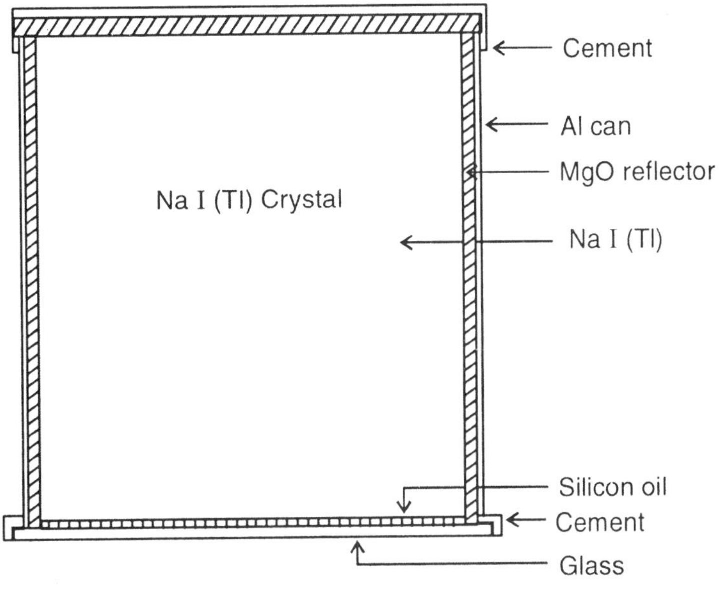

Fig. 8.12: Mounting of a NaI (Tl) crystal

The crystal is coupled to the glass window with the help of silicon grease or silicon oil of very high viscosity. The refractive index of silicon oil is close to that of glass and the scintillator. There is therefore minimum reflection of light going from the scintillator towards the glass window. Inside the aluminium can the scintillation crystal is surrounded with aluminium oxide or magnesium oxide powder which acts as a diffused reflector for light emerging out from the sides of the crystal. Aluminium oxide and magnesium oxide reflect almost 96 per cent of the light falling on it. For a thin scintillator even polished aluminium foil acts as a good reflector. Though NaI(Tl) crystals are very transparent and flawless, the polishing of the surfaces and surrounding them with, aluminium oxide reflector helps in a uniform transmission of light, from every part of the crystal to the glass window and thence to the photocathode of the photomultiplier tube.

Due to the high density, high atomic number of one of its constituents-iodine-and good scintillation efficiency, NaI(Tl) scintillators are used extensively in γ-ray spectroscopy. The scintillation decay time of NaI(Tl) scintillator, at room temperature, is 230 nanoseconds. At 0°C the decay time becomes 400 nanoseconds while at 100°C it is only about 150 ns. At room temperature it is found that NaI(Tl) emits about 9 per cent of its scintillation light with a decay time of 0.15 seconds. This gives rise to a continuous after glow of the NaI(Tl) scintillator which becomes significant at very high counting rates.

The energy resolution of a spectrometer for 662 keV γ-rays from a ^{137}Cs source is generally taken as a measure of the quality of the NaI(Tl) scintillator and the photomultiplier tube employed. With selected photomultiplier tube and NaI(Tl) scintillator, it has been possible to obtain an energy resolution of 7.5 per cent for the ^{137}Cs gamma-rays. The energy resolution deteriorates if there are air bubbles in the coupling silicon oil in the sealed crystal container or between the photocathode and the crystal. If moisture leaks in the housing of the NaI(Tl) crystal, the crystal turns yellow and the light output goes down.

Pure bismuth germinate ($Bi_4Ge_3O_{12}$) crystals having a relative density of 7.3 and large atomic number (Z = 83 for bismuth) have high detection efficiency for γ-rays. The light output of BGO scintillator is reported to be 10-20 per cent of that of NaI(Tl) scintillator. About 10 per cent of the scintillation light has a decay time of 60 nanoseconds while the major component has a decay time of 300 ns. Unlike NaI(Tl) there is no after glow or large decay time in the BGO scintillator. A BGO crystal is very rugged and can be cut and shaped easily. This scintillator is primarily used where high detection efficiency for γ-rays is required. Due to the poor scintillation efficiency the energy resolution obtained with a BGO scintillator is only about 20 per cent for the 662 keV γ-rays of ^{137}Cs.

A pure barium fluoride BaF_2 crystal emits light in the ultraviolet region. About 20 per cent of the scintillation light is emitted with a decay time of 0.6 nanoseconds and the rest with a decay time of 630 nanoseconds. The light output of a BaF_2 scintillator is only about 20 per cent of that of a NaI(Tl) scintillator. The large detection efficiency for γ-rays due to barium and the very small decay time of 0.6 ns make the BaF_2 scintillator ideal for timing experiments using γ-rays. As the light emitted by a BaF_2 crystal is in the ultraviolet region, it is detected with a special photomultiplier tubes with quartz window. White teflon tape is reported to be a good reflector for the ultraviolet light of the scintillator.

Thallium activated cesium iodide is another scintillator which is used in some experiments. Because of the high atomic number of both the constituents of the crystal, it has a very high detection efficiency for γ-rays. The light output of CsI(Tl) is a little better than that of NaI(Tl), but it is at much longer wavelength which does not match with the response of the photomultiplier tube. The effective light yield as detected by the photomultiplier tube is thus much less than that of NaI(Tl) scintillator. A cesium iodide crystal can be cut in very thin slices. The utility of this scintillator lies in the fact that the scintillation decay time is different for different types of ionizing particles *e.g.* electrons, protons and α-particles. The scintillator can be fruitfully used for differentiating between different particles emitted in a nuclear reaction.

There are other inorganic scintillators *e.g.* CsI(Na), CsI(Eu) and Silicate glasses which are some times used for special purposes.

8.5.3 Light Guides

A scintillator has to be coupled to the window of the photomultiplier tube such that there is no loss of light. If the scintillator does not have to be sealed in a container, its flat and polished surface can be coupled directly to the photomultiplier tube window with the help of silicon oil. The canned NaI(Tl) scintillator is also mounted on the photomultiplier tube with silicon oil. When the refractive index of the coupling oil is close to that of glass there is no loss of light due to reflections between the surfaces.

Magnetic field and high temperatures adversely affect the working of a photomultiplier tube. If the scintillator is kept under such conditions, it is necessary to keep a distance between the scintillator and the photomultiplier tube. The coupling between the scintillator and the photomultiplier tube is done through a light guide. A simple light guide is a cylinder of transparent plastic perspex-whose one end is shaped to fit the window of the photomultiplier tube and the other end to the scintillation crystal. Light guides are necessary when the size and shape of the scintillator and the photomultiplier tube window do not match. In a light guide, the light is transmitted directly as also through total internal reflection at the cylindrical surface. A light guide helps in spreading the light coming from the scintillator over the whole area of the photocathode of the photomultiplier tube. This helps in averaging out the uneven response-if any- of the photocathode.

Light guides always absorb some fraction of light coming from the scintillator. The longer the light path, the greater is the absorption. The light guides are therefore used only when necessary.

8.5.4 Photomultiplier Tube

Photomultiplier tube is an important component of a scintillation counter. It converts the very weak light output of a scintillation pulse, that typically consists of a few hundred photons, into a corresponding electrical current pulse. There is a great variety of photomultiplier tubes sensitive to radiant energy in ultraviolet, visible and near infrared regions of electromagnetic spectrum.

The light produced in a scintillator is detected in a photomultiplier tube. A photomultiplier is essential a highly evacuated glass tube whose one end is closed with a flat glass. The electrical connections are taken out from the other end. The flat end is coated inside with a photosensitive material and is called the photocathode. The photoelectrons emitted from the photocathode are guided to an electron multiplier structure called dynodes. The sensitivity or quantum efficiency of a photocathode defined as the number of photoelectrons emitted for every hundred incident photons, depends upon the photosensitive material employed. Multi-alkali material Na_2KSb activated with cesium is used as a photosensitive material in some photomultipliers. Using this compound, it is possible to achieve a quantum efficiency of 30 per cent for light in the blue region. A bialkali compound K_2CsSb activated with oxygen and cesium is found to give still higher quantum efficiency. The coating of the photosensitve material should be uniform over the whole area of the photocathode so that it is uniformly sensitive.

The electrons emitted from the photocathode have very low energy. They are accelerated and focused at the first dynode by the focusing electrodes. The electrons strike the first dynode with kinetic energy equal to the potential difference between the cathode and the dynode (which is generally 100-300 volts). Due to the impact of these electrons, secondary electrons are emitted from the first dynode. The secondary electrons having very low energy are focused on the second dynode by the electrostatic field between the two. The electrons strike the second dynode with a energy of 50-100 eV-accelerated by the potential difference between the two dynodes. The potential on subsequent dynodes is increased by 50-100 volts each and the secondary electrons emitted by one dynode are focused on the next dynode. The number of secondary electrons ejected from a dynode depends upon the interdynode potential difference. This number is usually three to six. The number of electrons is thus amplified at each dynode by a factor of three to six. The electrons ejected from the last dynode are collected at the anode which is at 200-300 volts higher potential than the last dynode. A photomultiplier tube with ten dynodes can easily give a charge gain of 10^4~10^6. A fourteen dynode photomultiplier can give a gain of 10^8-10^9. The inner structure of a photomultiplier tube is shown in Fig. 8.13.

The electrons ejected from the photocathode or the dynodes have very low velocity and their trajectory can be greatly influenced by external electrostatic and magnetic fields. Electrostatic shielding of the photomultiplier assembly is achieved by the metallic coating on the interior wall of the tube and connecting it to the ground. For magnetic shielding the photomultiplier tube is surrounded by a magnetic shield cylinder made generally of Mu-metal. The photomultiplier tube can detect extremely faint light and it is essential that the tube along with the scintillator is enclosed in a light tight container. As an alternative black rubber tape can be wrapped round the tube and the scintillator to make it light tight.

The successively increasing potentials between the photocathode to dynodes and anode are provided by a resistance chain called the bleeder. Figure 8.14 shows the bleeder chain of resistances used to supply voltages to the dynodes and the anode of a twelve dynode photomultiplier tube RCA8575.

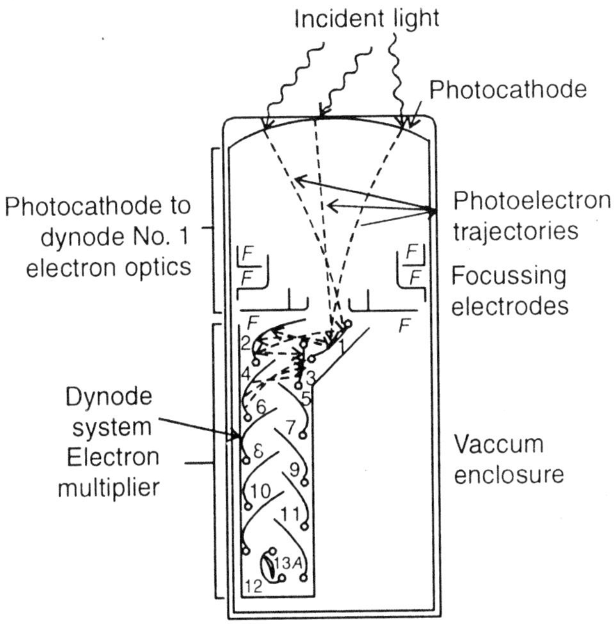

1-12: Dynodes 13*A*: Anode

Fig. 8.13: Internal construction of a photomultiplier tube. [RCS Photomultiplier Manual Series PT-61] RCA Solid State Division. Electro Optics and Devices. Lancaster, 1970

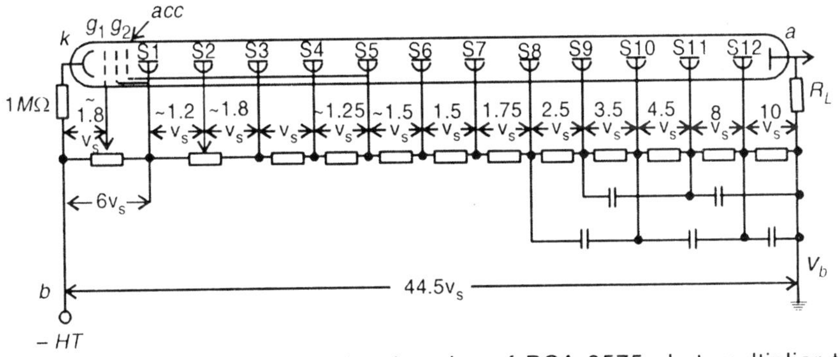

Fig. 8.14: Distribution of voltages for the dynodes of RCA 8575 photomultiplier tube. The values of the bleeder resistances can be obtained for given bleeder current

The charge gain of a photomultiplier tube remains constant unless the interdynode potential changes. As the charge emitted from the last few dynodes, in a high gain photomultiplier, could be quite large, the dynode voltages may change momentarily. To reduce this effect capacitors are connected from the dynodes to the high voltage supply which acts as an *AC* ground.

The electrons leaving the last dynode are collected at the anode where they deposit a charge Q. If C is the distributed capacitance of the anode and the associated wiring, the negative signal produced at the anode is ΔV where,

$$\Delta V = Q/C$$

If the anode leak resistance is R, the signal decays with a time constant CR. The anode signal is fed to a preamplifier which could be only an emitter follower or a source follower. To keep the anode capacitance minimum, the bleeder chain and the preamplifier are mounted at the base of the photomultiplier tube. The output of the preamplifier is fed to a large band width (about 10 MHz) linear amplifier whose output is connected to pulse height analyser. The pulse height analyser is used to study the pulse height spectrum of the anode pulses.

As the gain of the photomultiplier tube is constant, the number of electrons producing the anode signal ΔV is proportional to the number N of the photoelectrons emitted from the photocathode which in turn is proportional to the number of photons emitted by the scintillator. The number of photons L is proportional to the ionization produced in the scintillator which in turn is proportional to the energy dissipated by the incident charged particle. Thus

$$\Delta V \propto Q \propto N \propto L \propto E$$

Strict proportionality between the anode pulse height V and the energy E dissipated in the scintillator is possible only when the following conditions are fulfilled:

(1) The scintillation efficiency of the scintillation crystal should be uniform over the whole volume. This implies a uniform doping of the scintillator with the activator.

(2) All the light produced in any part of the scintillator should reach the photocathode. This condition is fulfilled if the scintillation crystal is perfectly transparent to the light it emits. The crystal should be free of any cracks or other defects. The surfaces of the crystal should be cleaned and polished. The crystal should be surrounded with a good diffuse reflector.

(3) The photo-cathode should have a uniform sensitivity over the whole area illuminated by the scintillator light.

8.5.5 Scintillation Spectrometer

A typical scintillation spectrometer is shown in Fig. 8.15. As discussed earlier the bleeder resistances and the preamplifier are mounted near the base of the photomultiplier tube. After amplification in a linear amplifier the output of the preamplifier is fed to a single channel or a multichannel pulse height analyser. A single channel pulse height analyser gives an output only when the height of the input signal lies between the predetermined voltages V and $V + \Delta V$. The voltage V, adjustable by a 10 turn potentiometer is known as the baseline voltage and ΔV, also adjustable by a potentiometer, is known as the window width. The output of the single channel analyser (SCA) is counted in an electronic scaler.

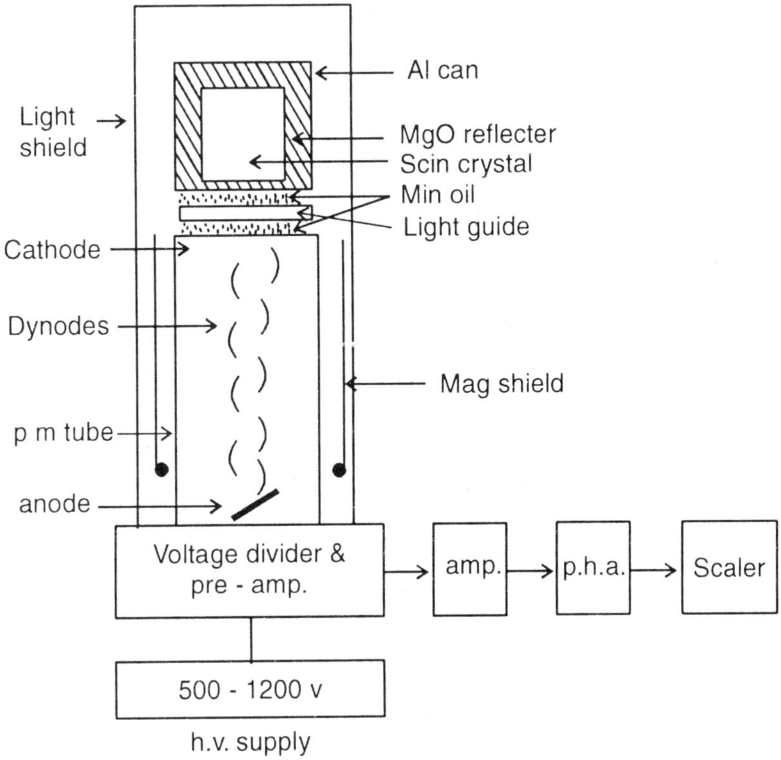

Fig. 8.15: A scintillation counter arrangement

The pulse height spectrum is studied by varying the baseline voltage in small steps of 0.1 volt generally while the window width is kept fixed at 0.05 or 0.1 volt. For each base line voltage the number of output pulses are recorded in the scaler for a fixed interval of time-say one minute. The maximum base line voltage scanned should be less than the pulse height at which the amplifier starts to saturate. The plot of the counting rate recorded at different base line voltages gives the pulse height spectrum. Each pulse height corresponds to a definite energy dissipated in the scintillator by the ionizing particle.

In a multichannel analyser, the pulse height spectrum of all the input pulses is analysed simultaneously. In a few minutes the whole pulse height spectrum can be observed on an oscilloscope screen and recorded if desired. Presently computer based multichannel analysers are most common.

8.5.6 Energy Resolution of a Scintillation Spectrometer

Ideally for a number of particles each depositing exactly the same amount of energy in the scintillator, the height of the output pulses should be same. The pulse height spectrum due to all such particles should be a point at some base line voltage. However the phenomena of production of light photons in the scintillator, emission of photoelectrons from the photocathode and secondary electron emission at the dynodes are all statistical in nature, hence liable to fluctuate statistically. The largest fractional statistical fluctuation occurs where the number of particles is minimum. The photoefficiency of the photocathode is only about 10 per cent, hence the smallest number of particles responsible for formation of the output pulse are the photoelectrons. The fluctuation in the number of electrons arriving at the anode of the photomultiplier tube is governed by the fluctuation in number of photoelectrons. There is some contribution to this fluctuation due to the statistical emission of secondary electrons from the first dynode. The contribution by other dynodes is negligible as the number of electrons emitted from them is quite large. The number of photons emitted by the scintillator is quite large-about ten times the number of photoelectrons-and their contribution to fluctuation in the final pulse height is small.

The fluctuation in the charge collected at the anode of the photomultiplier tube causes a corresponding fluctuation in the height of the output pulses. The amplification of the signal in the preamplifier or the amplifier does not effect the distribution in the pulse height. For the same amount of ionization in a scintillator by a number of charged particles, the pulse height has a Gaussian distribution. Thus a single point spectrum becomes a distribution of Gaussian shape with its maxima corresponding to the energy dissipated. The effect pf resolution on the energy spectrum in a scintillation counter is shown in Fig. 8.16.

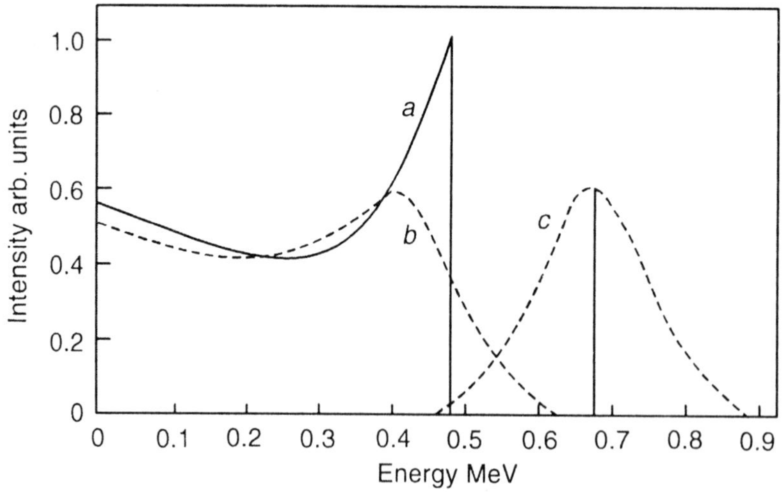

Fig. 8.16: The statistical resolution of the scintillation counter modifies the photo-electron and Compton electron spectra. The Gaussian peak '*c*' is due to photoelectrons. The Compton electron spectrum '*a*' is modified to give spectrum '*b*'

The width of the distribution is determined by the average number of photoelectrons emitted for the particular degree of ionization. The energy resolution of the scintillation spectrometer may be defined as the ratio of the energy spread (due to spread in pulse height) as measured by the spectrometer to the energy dissipated by the ionizing particle in the scintillator. For the Gaussian distribution the energy resolution is given as

$$R = \frac{\Delta E}{E} \propto \frac{\sigma_N}{N} = \frac{1}{\sqrt{N}} = \frac{1}{\sqrt{E}}$$

where N is the average number of photoelectrons emitted from the photocathode when energy E is dissipated in the scintillator. Even if the statistical distribution of secondary electrons emitted from dynodes is taken into consideration the energy resolution R remains proportional to $E^{-1/2}$.

As the output pulse height spectrum for a definite energy dissipation in scintillator is Gaussian, the energy resolution is usually defined as ratio of full width at half maximum (FWHM) of the peak and the peak position on the base line voltage. The relationship between the base line voltage and the energy dissipated (calibration curve) is established by observing the pulse height spectra due to monoenergetic radiations of different energies.

8.5.7 γ-ray Spectroscopy Employing Scintillation Detector

Usually a NaI(Tl) scintillation crystal is used for γ-ray spectroscopy. For γ-rays of energies below 100 keV a centimetre thick crystal can be used. For higher energy γ-rays, a crystal 3.5 cm to 5 cm thick have a fairly good detection efficiency. The photomultiplier tubes generally used have the photocathode diameter of 4.8-5.0 cm. The scintillation crystal could have a diameter between 3.5 to 4.8 cm. Scintillators of diameter greater than 4.8 cm are used in some special experiments only. They are coupled to the photomultiplier tube through a light guide.

When a γ-ray photon strikes a scintillator there is some probability that it may pass through it without interacting. The photon may interact with the scintillator by photoelectric absorption or by Compton scattering. If the energy of the photon is greater than 2 m_0c^2, it may be absorbed by pair production process also. In either of these processes an energetic electron is produced which looses its kinetic energy in the scintillator, producing ionization. The scintillator emits a number of photons which is proportional to the energy deposited by the electron. The pulse height spectrum observed in a scintillation spectrometer represents the spectrum of energies dissipated by the electrons in the scintillator-modified by the statistical fluctuation. When a beam of monoenergetic photons strikes a NaI(Tl) scintillation crystal, the following events contribute to the pulse height spectrum.

(*i*) The photon with energy $h\nu_0$ may produce photoelectric effect in the high atomic number atoms of the scintillator [(iodine in the case of NaI(Tl)]. The photoelectron is ejected with energy $h\nu_0 - I$ where I is the binding energy of the electron. The probability of the photoelectric effect to take place is largest in the *K*-shell of the iodine atom. The ejected electron losses all its energy in the scintillator. The ionized iodine atom captures an electron from the lattice and attains its ground state by emitting *K*-*X*-rays series or Auger electrons. The Auger electrons also deposit their energy in the scintillator. The iodine *K*-*X*-rays with a maximum energy of 31.4 keV are absorbed in the scintillator by photoelectric effect in other orbits or atoms. Thus in steps, the whole energy $h\nu_0$ of the incident photon is dissipated in the scintillation crystal as depicted in Fig. 8.17*a*. Such events give rise to the "full energy peak" or "photopeak" in the pulse height spectrum.

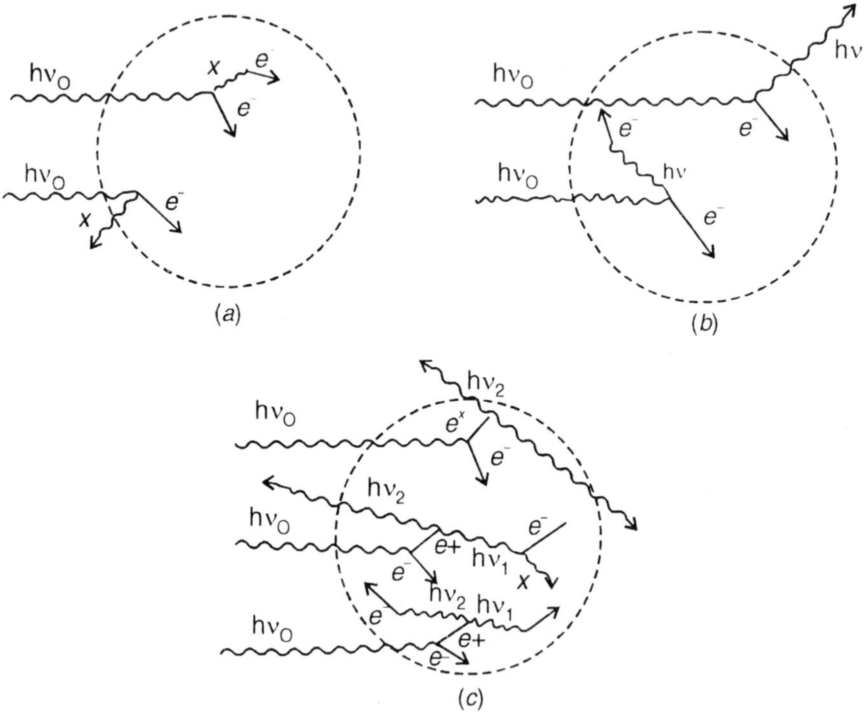

Fig. 8.17: Primary and secondary processes during absorption of γ-rays in a scintillation crystal by (*a*) photoelectric effect (*b*) Compton effect and (*c*) pair production

When the energy of the incident photons is less than about 100 keV, most of them are absorbed in a small thickness of the front surface of the scintillator. There is a small probability that the iodine *K*-*X*-rays emitted subsequently go out of the crystal. In such events the energy lost by the incident photon is $h\nu_0 - h\nu_k$, where $h\nu_k$ is the energy of the iodine *K*-*X*-rays. This results in a peak in the pulse height spectrum at an energy about 31 keV below the full energy peak". Such a peak is called the "iodine X-ray escape peak" such a peak is observed only for γ-rays of energy below 100 keV. (Fig. 8.17*a*) (*ii*) The incident photon may be Compton scattered in the scintillator. The Compton electron losses its energy in the scintillator while the scattered photon has some probability of escaping. The energy of the Compton electron depends upon the angle of scattering of the photon. The maximum energy of the electrons corresponds to the Compton edge as discussed in chapter 6. The spectrum of the Compton electrons is shown in Fig. 6.16.

(*iii*) If the energy of the incident photon is greater than the energy of a positron-electron pair ($h\nu_0 > 2m_0c^2$) pair formation takes place in the scintillator. Both the particles loose their kinetic energy ($h\nu_0 > 2m_0c^2$) in the scintillator. The positron coming to rest, captures an electron from the lattice and annihilates to give two photons each of energy m_0c^2 (0.51 MeV) moving in opposite directions. When both the photons escape from the scintillator, the energy dissipated by the incident γ-ray photon is only $h\nu_0 - 2m_0c^2$. This gives rise to a peak in the pulse height spectrum known as "two annihilation radiation escape peak" and correspond to the energy deposition of $h\nu_0 - 2m_0c^2$.

If the size of the NaI(Tl) scintillator is small-say about one centimetre, the above discussed phenomena only determine the pulse height spectrum. For a thin scintillator the areas under the full energy peak, Compton distribution and the two annihilation radiation escape peak are nearly proportional to the cross-section for the three processes. The pulse height spectrum as discussed earlier is modified due to the energy resolution of the spectrometer.

When the size of the scintillator is large-say 3.5 cm thick or more and diameter greater than 3.5 cm, the Compton scattered photons and the annihilation radiations may interact with the scintillator and modify the pulse height spectrum. The following secondary processes are significant in a large size NaI(Tl) scintillator.

(*a*) The Compton scattered photons-being of low energy may be absorbed in the scintillator by photo electric effect as depicted in Fig. 8.17*b*. In this process the whole energy $h\nu_0$ of the incident photon is dissipated in the scintillator, enhancing the full energy peak and reducing the area of the Compton distribution. This effect is pronounced when the scattered photon energy is low (corresponding to Compton edge). The pulse height spectrum corresponding to the Compton edge is thus suppressed. The photon Compton scattered in the forward direction has fairly large energy (only a little less than $h\nu_0$) and may interact with the scintillator by photoelectric effect or Compton scattering. If the photon is absorbed it contributes to the full energy peak. If it is Compton scattered again, it contributes to the Compton distribution only (Fig. 8.17*b*).

(*b*) The two annihilation photons created in the annihilation of the positron formed in pair production, may interact with the scintillator. If both the photons are absorbed by photoelectric effect, the total energy $h\nu_0$ of the incident photon is deposited in the scintillator. This enhances the full energy peak and reduces the annihilation radiation escape peak (Fig. 8.17*c*). If one of the annihilation radiations escapes the scintillator and the other absorbed by photoelectric effect, the energy deposited in the scintillator is $h\nu - m_0c^2$ which gives rise to a peak at that energy and is known as "one annihilation radiation escape peak". If the annihilation radiations are Compton scattered such that the scattered photon escapes the scintillator, the energy deposited is between $h\nu - m_0c^2$ and $h\nu_0$ and the Compton distribution gets enhanced. The pulse height spectrum due to the 662 keV γ-ray from 137Cs ($h\nu_0 < 2m_0c^2$) and that due to γ-rays from ^{24}Na ($h\nu_0 > 2m_0c^2$) are shown in Fig. 8.18 and Fig. 8.19 respectively.

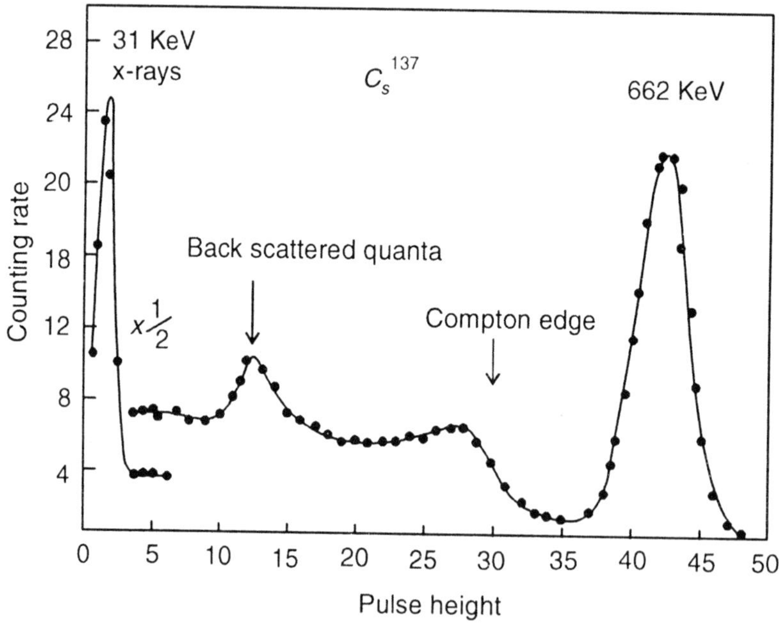

Fig. 8.18: Pulse height spectrum due to 662 KeV γ-rays from ^{137}Cs in a NaI (Tl) scintillation counter. Scintillator size 3.5 cm dia × 3.5 cm. The back scattered peak is due to γ-rays scattered from photomultiplier tube and the surroundings. (Jagdish Varma, Ph.D. Thesis, University of Delhi, 1955)

The calibration of the scintillation spectrometer base line is done by observing the pulse height of full energy peaks of a number of γ-rays. A plot of the energy of the full energy peak and the position of the peak on the base line gives the calibration curve of the scintillation spectrometer.

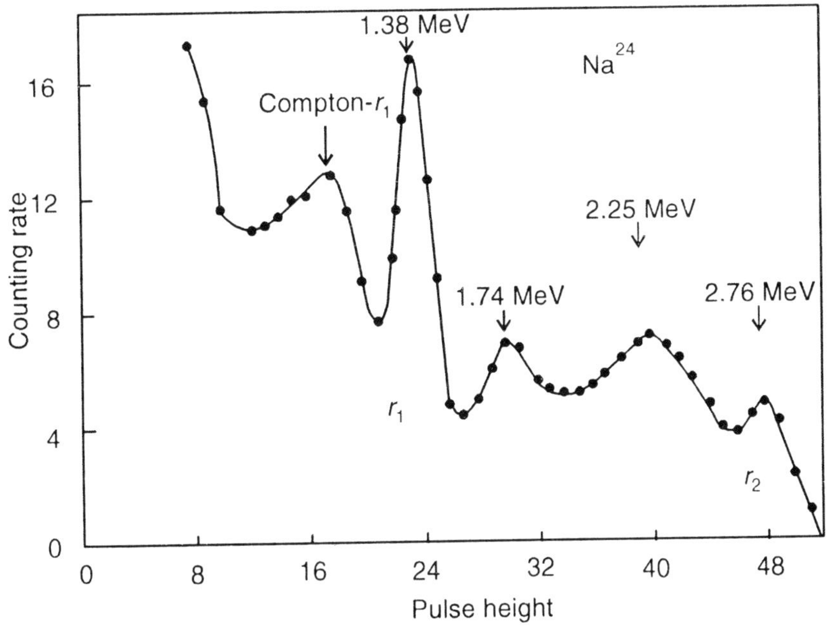

Fig. 8.19: Pulse height spectrum due to 2.76 MeV and 1.38 MeV γ-rays from ^{24}Na. The peaks at 2.25 MeV and 1.74 MeV are due to one and two annihilation radiation escape respectively. (Jagdish Varma, Ph.D. Thesis, University of Delhi, 1955)

8.6 SEMICONDUCTOR DETECTORS

Semiconductor diode detector is a reverse biased P-N junction. The ionization produced by an ionizing radiation in the depletion layer of the diode gives rise to a diode current. The semiconductor detector performs all functions of a gas filled ionization chamber with an advantage that it has high density, small size and better energy resolution. These detectors can be mounted in a vacuum chamber. As has been discussed earlier, the energy resolution arises due to the statistical fluctuation of the number of particles responsible for forming the signal pulse. In a scintillation counter dissipation of about 1-2 KeV energy by the incident radiation gives rise to one photo-electron from the photocathode. In an argon filled ionization chamber an energy of about 37 eV is required to produce one ion pair. In a semiconductor detector only 3.1 eV energy is required to produce an electron hole pair. Thus the number of particles which determine the final energy resolution of the detector is ten to thirty times larger in a semiconductor detector. After taking into consideration the Fano-factor the energy resolution obtainable from a semiconductor detector could be as much as 1 per cent.

8.6.1 Semi-Conductors

Semi-conductors are materials in which the energy gap between the conduction band and the completely filled valence band is 1 to 2 eV. If the band gap is more than 5 eV, the material is a non-conductor. Among the elements only silicon and germanium both group IV elements show semi-conductor behaviour. Both have filled valence bands. The energy gap between conduction and valence band is 1.1 eV for silicon and 0.67 eV for germanium.

At room temperature, in any material, the electrons in thermal equilibrium have Maxwellian distribution of velocity with a mean kinetic energy kT, where k is Boltzman constant and T absolute temperature. At any temperature T, there is a certain probability $p(T)$ that an electron may have an energy greater than the band gap energy Eg of a semiconductor. Such an electron jumps into the conduction band leaving a hole in the valence band. Both the electron and the hole are free to move in the semiconductor and can conduct electricity. The probability $p(T)$ for formation of such an electron hole (e-h) pair per second is given as

$$p(T) = AT^{3/2} \exp\left(-\frac{E_g}{kT}\right) \qquad \text{...(8.6.1)}$$

where A is a constant, characteristic of the material. The e-h pairs so formed drift in the material and recombine, so that at any given time there is an equilibrium density. At 300K the density of such thermally produced e-h pairs is 1.5×10^{10} per cm^3 in silicon and 2.4×10^{13} per cm^3 in germanium. These e-h pairs conduct electricity giving a finite conductivity to the semiconductors. When an electric field is applied across a semi-conductor crystal, the electron and holes acquire a drift velocity (superimposed over their random thermal velocity). The holes move along the electric field and the electrons opposite to it. The drift velocities of the holes and the electrons are given as

$$\vec{v}_h = \mu_h \vec{E}$$

$$\vec{v}_e = \mu_e \vec{E}$$

μ_h and μ_e are mobilities of holes and electrons in the material. The hole mobility μ_h is about half of μ_e. At the liquid nitrogen temperature of 77K, the two mobilities are approximately equal. As the electric field increases, the electron and hole carriers reach a saturation drift velocity of about 10^7 cm/sec.

Some important properties of silicon and germanium semi-conductors are given in Table 8.2.

When a pentavalent impurity like phosphorous is mixed with a semiconductor to an extent of about two parts per million (2 ppm), the impurity atoms occupy the lattice sites. Four of the five valence electrons of the impurity atom get bound in the lattice while the fifth electron being extra, remains as a free electron and moves freely in the lattice. Thus the pentavalent impurity donates a free electron to the lattice and is known as donor. If the impurity atom is trivalent - like aluminium, there is a shortage of one electron in binding the atom in the lattice. This creates a hole in the lattice which is again free to move. Such impurity atoms are called acceptors. Semiconductors which have incorporated pentavalent or trivalent impurities are known as doped semiconductors. The donor impurities make the semiconductor n-type while acceptor impurities make it p-type.

TABLE 8.2: PROPERTIES OF INTRINSIC SILICON AND GERMANIUM

	Si	*Ge*
Atomic number	14	32
Atomic weight	28.09	72.60
Stable isotope mass numbers	28-29-30	70-72-73-74-76
Density (300 K)g/cm^3	2.33	5.32
Atoms/cm^3	4.96×10^{22}	4.41×10^{22}
Dielectric constant	12	16
Forbidden energy gap (300 K); eV	1.115	0.665
Forbidden energy gap (0K); eV	1.165	0.746
Intrinsic carrier density (300 K); cm^{-3}	1.5×10^{10}	2.4×10^{13}
Intrinsic resistivity (300 K); Ω cm	2.3×10^5	47
Electron mobility (300 K); cm^2/V.s	1350	3900
Hole mobility (300 K) cm^2/V.s	480	1900
Electron mobility (77 K); cm^2/V.s	2.1×10^4	3.6×10^4
Hole mobility (77 K); cm^2/V.s	1.1×10^4	4.2×10^4
Energy per electron-hole pair (300 K); eV	3.62	
Energy per electron-hole pair (77 K); eV	3.76	2.96
Fano factor (77 K)	0.09	0.08

The conduction of electricity in the n-type semiconductor is by electrons while in the p-type, it is by holes. The charge carriers contributed by the n-type or p-type doping are known as majority carriers. The charge carriers created by thermal ionization are known as minority carriers. In a n-type semiconductor there is a large number of free electrons and the holes of the minority carriers are completely neutralised leaving only majority and minority electron carriers. Similarly in a p-type semiconductor both majority and minority carriers are holes. If a doped semi-conductor is cooled to say liquid nitrogen temperature of 77K, the number of majority carriers remains the same while the number of minority carriers decreases very much. Depending upon the amount of impurity a semi-conductor could be heavily doped or lightly doped. In a normally doped semiconductor there are about 10^{17} carriers/cm^3.

It is not possible to have an absolutely pure semiconductor, There are always some metallic impurities present which can capture electrons and holes. Even interstitial defects and vacancies in a semi-conductor crystal can cause loss of electrons and holes. At the present time it is possible to have a semiconductor in which the impurity concentration is as low as 10^{-10}.

8.6.2 Diode Detector

Semi-conductor detector can be made by thermally diffusing a donor element at the surface of a *p*-type, or an acceptor element at the surface of a *n*-type semiconductor crystal. The depth and the concentration of the diffused element is suitably controlled. Suppose a donor element *e.g.* phosphorous is diffused on the surface of a *p*-type semiconductor crystal with uniform acceptor impurity concentration N_A. The donor impurity density profile could be as shown by curved N_D in Fig. 8.20. At the surface $N_D >> N_A$ and the semi-conductor there behaves as *n*-type. At some depth the donor density becomes less than that of the acceptor density and the material is again *p*-type. The point at which *n*-type changes into *p*-type is the *N-P* junction and represents a discontinuity in the conduction electron density.

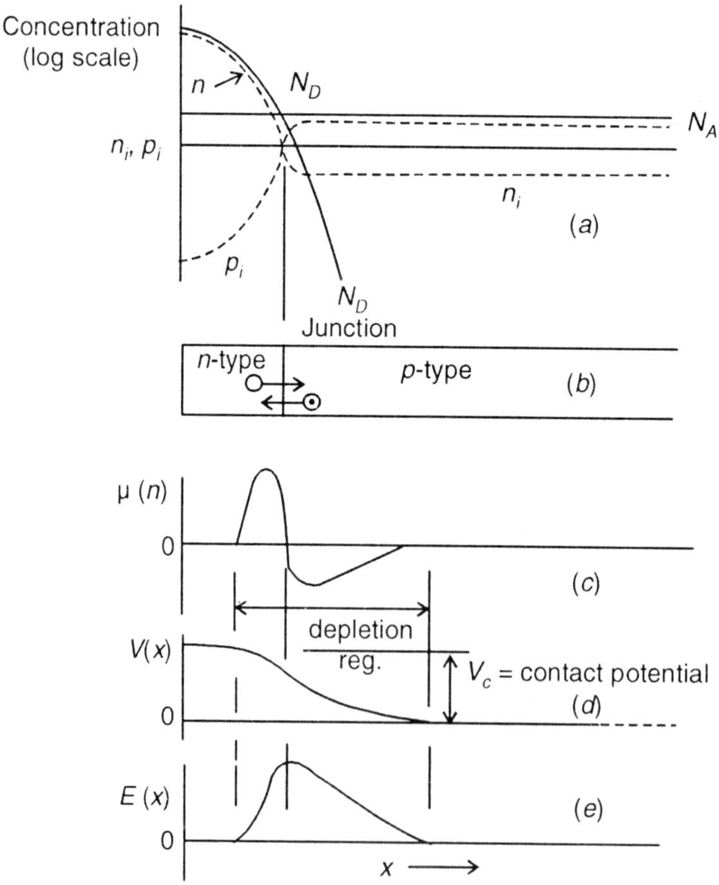

Fig. 8.20: Concentration of donor impurity N_D inside a *p*-type semiconductor near the PN junction is shown in Figure (*a*). n_i and p_i are concentrations of donors and acceptors respectively. The left part to the junction, Figure (*b*), is *n*-type and the right part is *p*-type. Figure (*c*) shows the space charge density near the junction. (*d*) shows the variation of electrical potential in the depletion depth and (*e*) the variation of electric field in the depletion depth

At the junction, due to higher concentration on the n-side the electrons diffuse into the p-region leaving fixed positively charged atoms in the lattice. The electrons reaching the p-region neutralise the holes. Similarly the holes diffuse from the p-region across the junction leaving fixed negative charge in the lattice and neutralising the electrons in the n-region. Thus static positive space charge is produced in the n-region and static negative charge in the p-region (Fig. 8.20*c*). This accumulated space charge creates an electric field which at equilibrium is just adequate to prevent further diffusion of electrons and holes across the junction. A steady state charge distribution is thus established (Fig. 8.20*e*).

The region close to the *P-N* junction, free of all electrons and holes (majority carriers) is called the depletion region (Fig. 8.20*a,c*). This region extends into both the p-type and the n-type side of the junction. If the concentration of the donors and the acceptors on the two sides of the junction are equal, the depletion region extends equal distance in both the regions. Usually there is a marked difference in the doping levels on one side of the junction compared to the other. If the donor density in the n-type region is higher than that of acceptor atoms in the p-type, the depletion region has to extend to a larger distance in the p-region, because the number of electrons and holes neutralising has to be equal. The space charge profile at equilibrium is shown in Fig. (8.20*c*). The potential difference of about 1 volt generated by the space charge across the junction is called the contact potential and is nearly equal to the band gap energy of the semiconductor material. The electric field due to the space charge across the *P-N* junction is shown in Fig. (8.20*e*). The contact potential prevents the movement of electrons and holes across the junction. If however e-h pairs are formed in the depletion region by an ionizing particle, the holes are swept towards the p-side and electrons towards the n-side. The only charges remaining in the depletion region are the immobile ionized donor and filled acceptor sites, which do not contribute to electric conduction. The depletion region exhibits a very high resistivity compared to the n-type or p-type material on either side of the junction.

If a forward bias is applied by making the n-side negative with respect to the p-side, the reverse potential difference is reduced. If the forward bias is greater than the contact potential, the electrons and holes flow across the junction freely and a current flows. This is exactly as in the case of a forward biased diode. If on the other hand, a reverse bias is applied by making the n-side positive with respect to the p-side no current can flow across the junction. The depth of the depletion layer increases with the applied reverse bias. The increase in the depletion layer is larger in the region with smaller density of the majority carriers. If the donor density is very much greater than the acceptor density the increase in the depletion depth is almost totally in the p-region and this depth d is given as

$$d = \left(\frac{2 \in V}{eN}\right)^{\frac{1}{2}}$$

where N is the acceptor concentration ($N_D >> N_A$)and $\in$ the dielectric constant of the semiconductor. The maximum electric field occurs at the point of transition from n-type to p-type material and is given as

$$E_{\max} = \left(\frac{2VNe}{\sigma}\right)^{\frac{1}{2}}$$

where σ is the conductivity of the semiconductor. This maximum electric field could be as high as $10^6 - 10^7$ volts/m.

If the acceptor density in the p-type material is very small, it is possible to increase the depletion layer over the whole thickness of the p-type crystal wafer. The smallest density of the acceptor is of course determined by the purity of the semiconductor. The higher the resistivity of the intrinsic semiconductor (before doping) the smaller can be the density of the doping impurity.

Under the condition of reverse bias, if an ionizing radiation enters the depletion layer, it produces e-h pairs which are immediately collected due to the strong electric field in the region. The resulting pulse produced has a height proportional to the number of e-h pairs formed-exactly as in the case of ionization chamber. The pulse height is thus a measure of the energy lost by the incoming radiation in the depletion layer. The thickness of the n-type layer deposited on a p-type crystal wafer (or p-type layer deposited on a n-type wafer) is only a few micron thick and serves as a window for the incoming ionizing particle to enter the depletion layer. On the other side of the p-type wafer a highly doped p-type layer is deposited. The ohmic contact with this layer is made by depositing a thin layer of gold on it. The highly doped p-type layer on the lightly doped p-type wafer acts as a blocking contact. There is no semiconductor junction between the highly doped and lightly doped regions. However, the highly doped layer provides the non-injecting conditions necessary to suppress leakage current due to minority carrier motion across the junction. The maximum reverse bias for any diode detector should of course not exceed the breakdown voltage of the diode. Semiconductor detectors are made in different configurations some of which are discussed below:

(i) Diffused Junction Detector

In this type of detector one surface of a homogenous p-type crystal is exposed to vapours of donor impurity phosphorous. This converts the exposed surface of the crystal to n-type. Thus a junction is formed close to the surface. The depth of the n-type layer is typically about 0.1 to 2.0 microns (μm). As the surface layer gets heavily doped, the depletion layer extends only in the p-type material. The n-type layer acts as the window for the ionizing particle to enter. It does not contribute to the formation of the e-h pairs and acts only as a dead layer. The thickness of this dead layer can be measured experimentally.

(ii) Surface Barrier Detector

In these detectors a very thin layer of aluminium is deposited on a *p*-type crystal surface. The aluminium surface acts as the *n*-type contact. The aluminium film acts as a thin window for the incoming radiations. Surface barrier detector can also be made by etching the surface of a *n*-type crystal and depositing a thin layer of gold on the same. The process of depositing gold is so manipulated that a very thin layer of oxide is formed on the surface. The etching of the crystal surface creates a high density of electron traps thus forming a depletion layer. The formation of the oxide layer is found to play an important role on the performance of the detector. The disadvantage of the surface barrier detectors is that they are sensitive to visible light. The aluminium or the gold film are so thin that it is transparent to visible light. The detector surface gets easily deteriorated by dust and vapours.

(iii) Ion Implanted Layer Detectors

In these detectors the doping impurities are implanted on a *p*-type or *n*-type crystal surface by exposing it to suitable ion beams accelerated to an energy of about 10 keV. The beams of phosphorus or boron are usually used for the purpose. The ion beams have a definite range in the crystal which can be increased by increasing the accelerating voltage. By ion implantation, the density of the added impurity can be closely controlled. After implantation the crystal is annealed at a temperature of about 500°C. Compared to surface barrier detectors the ion implanted detectors are more stable and less subject to ambient temperatures. The detectors can be formed with extremely thin entrance window.

(iv) Fully Depleted Detectors

As discussed above, when a reverse bias is applied to a *P-N* junction, the depletion layer extends in the region with smaller number of carriers. In fully depleted detectors the depletion depth extends over the whole thickness of the semi-conductor crystal wafer. In such detectors a *N-P* junction is made by depositing a heavy concentration of say *n*-type impurity on the surface of mildly *p*-type very high purity semiconductor. To maximise the depletion depth at a given reverse bias, the concentration of the doping impurity in the high purity side of the junction is minimised. The heavily doped layer at the surface of the wafer can be very thin, providing entrance window for the incoming radiations.

In a fully depleted detector, the whole reverse bias voltage drop is across the depletion region. If the reverse bias is sufficiently high, the electric field over the whole depletion depth becomes uniform resulting in a uniform collection of *e-h* pairs formed by the incoming radiations. The ohmic connection at the back of the wafer, away from the *P-N* junction is made through a blocking (heavily doped) *p* or *n* type contact.

A fully depleted detector can be used as a transmission detector in which the incoming radiation looses only part of its kinetic energy in the fixed thickness of the detector and then goes out to enter another thick detector. The pulse height at the output of the detector is a measure of the specific energy loss ($-dE/dx$) of the incoming radiation in the detector material. Such detectors are commercially available in the thickness range of about 50 to 2000 microns. After passing through a fully depleted dE/dx detector the incident particle is stopped in a thick detector, where it looses whole of its kinetic energy. This detector pulse height measures the total energy E of the incident particle. Generally a dE/dx detector and the E detector are coupled together. The measurement of dE/dx and E of a particle is employed to identify its nature.

When high voltage of about 1000 volt is applied as reverse bias to a junction detector, leakage currents can be very high. Some part of the leakage current arises due to the presence of minority carriers in the depletion layer. This bulk leakage current is proportional to the area of the P-N junction and is generally quite small. Some fraction of leakage current arises due to the spontaneous formation of e-h pairs in the depletion layer due to thermal energy. This leakage current is minimised by cooling the detector to liquid nitrogen temperature (77K). As the band gap in silicon is much greater (1.1 eV) than that in germanium (0.67 eV), silicon detectors can some times be used even at room temperature. To get the best results, it is desirable to cool the silicon detector as well. Leakage current can also arise due to surface leakage of charge due to humidity, dust, oil etc. Surface leakage is minimised by the use of guard rings. In the present day fabrication techniques, where detectors are manufactured by ion implantation, the junction edges are buried in the silicon wafer which reduces the surface leakage current significantly. It is a regular practice to monitor regularly the leakage current in a detector. Any sudden increase in the leakage current points to an abnormal behaviour of the detector.

Fluctuations in the leakage current effect the energy resolution of the detector. Johnson noise associated with series resistance or poor electrical contacts also give rise to a noise which makes the energy resolution poor. The broadening ΔE_{noise} of the energy peak of the incident radiation, due to the noise is given as

$$(\Delta E_{\text{noise}})^2 = (\Delta E_{\text{bulk}})^2 + (\Delta E_{\text{surface}})^2 + (\Delta E_{\text{Johnson}})^2$$

The noise widths combine in quadrature with the broadening due to the statistical fluctuation in the generation of electron-hole pairs to give the over all width which is characteristic of a detector.

The rise time of the electrical signal generated by the collection of electron-hole pairs in a junction detector depends upon the transit time of the charge carriers. Generally the rise time of the output signal in junction detectors is 2 to 5n sec.

8.6.3 Ge(Li) and HP Ge Detectors

The usual junction detectors can not have a depletion layer thicker than a millimetre or two. As such these detectors are not suitable for the detection of gamma rays. It has been shown that the depletion depth in a junction detector is given as

$$d = \left(\frac{2 \in V}{eN}\right)^{\frac{1}{2}}$$

where V is the reverse bias and N the net impurity concentration in the bulk semiconductor. The maximum reverse bias is limited by the break down voltage. The depletion depth can thus be increased by reducing net impurity concentration. Techniques have been developed to purify germanium to a point where impurity level is only 1 part in 10^{12} which corresponds to $N = 10^{10}$ impurity atoms/cm^3. These impurities could render the germanium either n type or p type. Detectors made with such germanium are known as high purity germanium (HP Ge) detectors and can have a depletion depth of a few centimetres or more. It is not yet possible to purify silicon to the same degree.

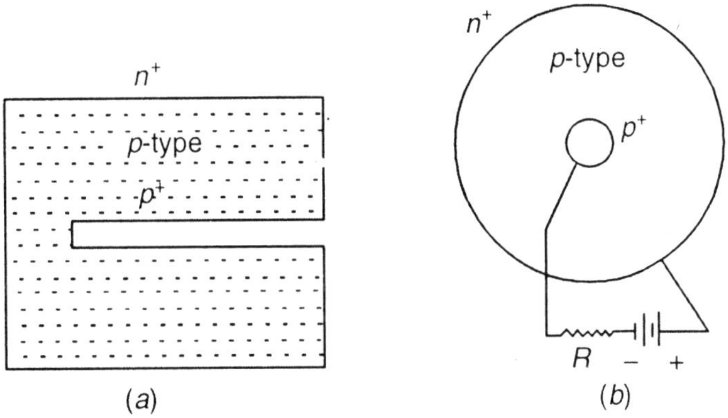

Fig. 8.21: A Coaxial HP Ge detector. n^+ and p^+ denote the highly doped regions

A second approach to reducing the net impurity concentration is to create a compensated semiconductor-germanium or silicon in which the residual impurities are balanced by an equal concentration of dopant atoms. If the impurity is n-type, it is neutralised by p-type dopant and vice-versa. Silicon and germanium crystals are grown with p-type impurities. Lithium atoms are allowed to drift in the crystal body and the acceptor impurities are exactly balanced over a thickness of upto 2 cm. The resulting compensated material has many of the properties of an intrinsic semiconductor. Even if the compensation is not exact, the resulting impurity level may be low enough such that the drifted region can be depleted over its whole length. Such detectors are known as Ge(Li) or Si (Li) detectors. The diffusion of lithium in the lattice of the semiconductor crystal depends upon the temperature and while fabricating a detector the same has to be closely monitored. The lithium atoms can easily diffuse in the germanium lattice even at room temperature. The Ge(Li) detectors have therefore to be stored at liquid nitrogen (L.N.) temperature. Si(Li) detectors can be stored at room temperature. HP Ge detectors can also be stored at room temperature. However to use them, both Si(Li) and HP Ge detectors have to be cooled to *LN* temperature. With the easy availability of HP Ge detectors the Ge(Li) detectors have now gone out of use.

Generally coaxial HPGe and Si(Li) detectors are used. Fig. 8.22 shows the construction of such a detector. The high purity germanium could be slightly *p*-type or *n*-type. For an *n*-type bulk, the outer surfaces are made heavily *p*-type and connected to negative potential. For making a blocking contact the co-axial inner surface is made heavily *n*-type and connected to positive potential. At the applied reverse bias, generally 3000 to 5000 volts, the depletion depth extends from surface to the centre of the cylinder. The detector is placed in a properly evacuated housing and the preamplifier is mounted close to it. The detector and the preamplifier are both cooled to *LN* temperature to reduce the thermal noise. Keeping the detector in vacuum saves it from deposition of moisture.

The HPGe detector being of higher atomic number material, is used for the detection of gamma rays while Si (Li) detectors are used mainly for the detection of *X*-rays. The outer electrode of the detector is generally grounded. The electrical signal generated at the central electrode is amplified. The pulse height analysis of the signal is done on a multichannel pulse height analyser. The pulse height spectrum observed with the gamma-rays from ^{133}Ba in an HPGe detector is shown in Fig. 8.22.

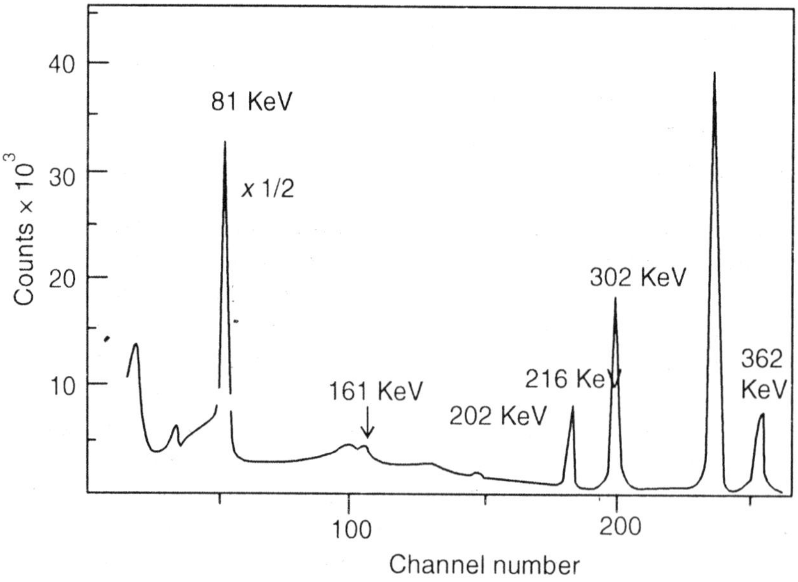

Fig. 8.22: Pulse height spectrum due to γ-rays from ^{133}Ba in a HPGe detector

The gamma-rays interact with the germanium detector in the same way as they do with a NaI (Tl) scintillator. Infact, the detection efficiency of a NaI (Tl) scintillator is more than that of a germanium detector. The great advantage of the HPGe detector is its very high resolution. The full energy peaks in the pulse height spectrum come out very high in comparison to the Compton distribution. The measurement of the energy of the γ-rays becomes very accurate.

The pulse rise time obtainable from a HPGe detector is about 100 ns which is comparable with that obtainable with scintillation detectors. The great advantage of these detectors is the very high energy resolution.

8.7 CERENKOV DETECTORS

When a charged particle moves in a medium with velocity greater than the velocity of light in the medium, it emits electromagnetic radiations. If βc is the velocity of the charged particle and n the refractive index of the medium, the condition for emission of electromagnetic radiation is $\beta > 1/n$. A particle moving with velocity nearly c in glass of refractive index 1.5 emits about 2.50 photons per centimetre of its path. This light is about 100 times weaker than the light emitted by an organic scintillator. The angle θ with respect to the particle trajectory, at which the radiations are emitted is given by the relation

$$\cos\theta = \frac{1}{n\beta}$$

Thus selecting the angle at which light is emitted, one can select the velocity βc with which the particle is moving. The emitted light is usually detected by a photomultiplier tube.

Cerenkov detectors are extensively used in high energy physics, to select particles moving with a certain velocity. Cerenkov radiations are independent of the mass of the emitting charged particle. Semiclassically it has been shown that the number of photons dn emitted in the frequency range ν and $\nu + d\nu$ per unit length of the path of the moving charged particle is given as

$$dn = \frac{4\pi e^2}{\hbar c^2}\sin^2\theta d\nu \qquad ...(8.7.1)$$

In the spectral region between 3000Å and 6000Å the above equation gives the number of photons emitted per centimetre length $dn = 750\sin^2\theta$.

8.8 PHOTOGRAPHIC EMULSION

The blackening of a photographic plate by the radiation from a piece of ore was responsible for the discovery of radioactivity by Becquerel. It has been found that the blackening of a photographic plate is roughly proportional to the product of the intensity of X-rays and the time of exposure.

Special photographic plates with silver bromide rich emulsion are prepared for the detection and study of charged particles emitted in a high energy nuclear reaction. A charged particle traversing in the emulsion produces about 30 grains per 100 micron of its path in a developed plate. The density of grains along the trajectory of the particle is proportional to the specific energy loss dE/dx of the particle. If the particle is moving with non-relativistic velocity, grain density gives information about its velocity. The grain density is measured by measuring the distance between two developed grains under a powerful microscope. If the particle comes to rest in the emulsion, its range can also be measured.

In high energy nuclear physics experiments a stack of nuclear emulsion plates is irradiated to the incoming particles in such a way that their trajectory is parallel to the plane of the plates. In this way a particle traverses much of its path in a single emulsion plate. The plates in a stack are properly numbered and so arranged that if a particle pesses out of one plate, its trajectory can be followed in subsequent plates. After irradiation to very high energy accelerates, the plates are taken away, developed suitably and studied under a high power microscope with facility to measure the position of the track in all the three x, y and z directions. This allows a three dimensional study of the particle track.

The advantage of photographic emulsion detector is that data can be recorded and then studied far away from the accelerator. The whole data is preserved in the plates for all time to come. The disadvantage of nuclear emulsion plate detector is that it takes a very long time to analyse the data and collect necessary statistics. One requires the help of a large number of trained scanners to scan the photographic plates and make measurements. The nuclear emulsion plates have now given way to bubble chambers.

8.9 CLOUD CHAMBER

CTR Wilson observed that super saturated water vapours tend to condense on ions and form droplets. These droplets when illuminated by strong light can be seen as well as photographed. A cloud chamber is an enclosure filled with air and vapour of either water or alcohol. The chamber is connected to a piston so that the air inside the chamber can be expanded adiabatically. The adiabatic expansion cools the air, making the vapours supersaturated. At this moment, if an ionizing particle is traversing the chamber, it produces a track of ionized particles. The supersaturated vapours condense on the ions forming droplets. The droplets grow rapidly in size. At this moment a strong flash of light is switched on and the track of the droplets is photographed.

The Wilson cloud chamber can be placed in a magnetic field which produces a curvature in the path of each charged particle. From the curvature of the track in the cloud chamber, the charge and the momentum of the incident particle can be determined. The number density of the droplets along the track gives the specific energy loss of the particle.

Wilson cloud chamber has been responsible for the discovery of a number of particles *e.g.* e^+, Λ, μ^+, $\pi^\pm$ present in the cosmic rays.

8.10 BUBBLE CHAMBER

In 1952 Glaser developed the bubble chamber in which a liquid, super heated above its boiling point, forms bubbles around the ions present in the liquid. To start with, the chamber is filled with a liquid at the pressure and temperature such that it is below the boiling points. If the pressure is suddenly reduced, the boiling point also gets lowered and the liquid forms bubbles around the ions present in it. These tiny bubbles are visible under strong light. The bubble chamber is in operation only for the time the liquid remains superheated. This time is about 10 milliseconds.

Bubble chambers have been operated using various liquids *e.g.* hydrogen, deuterium (heavy hydrogen), helium, propane and xenon. The advantage of bubble chamber is that the detecting medium is much heavier than air or gas in a cloud chamber. The tracks in a bubble chamber are photographed by two or more cameras so that a three dimensional picture of the tracks can be obtained. The data can be analysed using computers. Recently bubble chamber which is 72 inches long and contains 500 litres of liquid hydrogen has been fabricated. Bubble chamber can be placed in a magnetic field which permits the measurement of charge and momentum of the particles producing tracks.

8.11 SPARK CHAMBER

A spark chamber is a stack of parallel plates with a gap of about 8 mm. Alternate plates are grounded while the other plates are applied a voltage of about 10000 volts for a short duration of time-about 0.2 μsec. The chamber is filled with a noble gas *e.g.* argon at atmospheric pressure. If a passing charged particle has produced ionization in the chamber in a time period of 0.5 μsec preceding the application of the high voltage, a spark is produced in the chamber. These sparks are visible as well as audible. The photograph of the chamber at the time of the formation of the spark gives the track of the particle between different plates. After the chamber is photographed, a cleaning electric field is applied which clears the chamber of all charges. The chamber is again ready to detect another charged particle.

The spark chamber as well as bubble chamber and Wilson chamber can be triggered at a suitable moment by other detectors detecting the nuclear reaction event. Spark chambers are also used in high energy experiments where the particle can traverse a number of plates of the chamber.

EXERCISES

1. Why the energy required to produce an ion pair in an ionization chamber is much less than the ionization potential of the gas?
2. What is Fano Factor ? How does it effect the energy resolution of (*a*) scintillation counter (*b*) ionization chamber and (*c*) solid state detector.
3. How does a poisonous gas affect a proportional counter?
4. How does the quenching gas help in increasing the usable ion multiplication factor in a proportional counter?

5. Why a parallel plate ion chamber can not be used as a proportional counter?
6. What factors effect the rise time and decay time of the anode signal in a proportional counter?
7. Discuss the evolution of a Geiger discharge in a G.M. counter.
8. What factors determine the slope of the plateau of a G.M. Counter?
9. What factors determine the shortest resolving time that can be achieved using scintillation detectors?
10. How does the iodine X-ray escape peak arise in the pulse height spectrum due to γ-rays in a NaI(TI) scintillation counter?
11. How is the non-linearity in the light output of organic scintillators explained?
12. What is an intrinsic semiconductor?
13. How doping can be used to reduce the number of carriers in an intrinsic semiconductor?
14. What factors effect the energy resolution available from a semiconductor detector?
15. Why a bubble chamber is superior to a Wilson cloud chamber?

9

ACCELERATORS

9.1 INTRODUCTION

Accelerators are machines which are used to produce charged and highly energetic nuclear and sub nuclear particles. These high energy charged particles can be for variety of applications like producing artificial transmutation of elements, for studying the nuclear structure and for altering the properties of materials etc. The existing accelerators are classified, according to the energy of the particles they produce, namely into low ($\cong$ 100 KeV), medium ($\cong$ 10 MeV to 1 GeV) and high energy (1GeV) accelerators. The Van de Graaff, Cockroft Walton, the Dynamitrons and the Linac etc. come under the low energy accelerators, the Cyclotron, the Tandem accelerators Synchrocyclotron, Linac etc. are the examples of medium energy accelerators and the Synchrotron, Storage rings are the so called high energy accelerators. In addition to these ion accelerators there are electron accelerators called Betatrons, electron Linacs and electron Dynamitrons.

9.2 BASIC COMPONENTS OF ACCELERATORS

The accelerators make use of electromagnetic fields for increasing the kinetic energy of the charged particles. Electric fields are used for accelerating the ions and magnetic fields are used to bring back repeatedly the accelerated ions into the electric field again to gain more energy *e.g.* Cyclic accelerators. All the low energy accelerators which accelerate the particles in linear paths use strong electric fields. These accelerators are also called potential drop or potential difference accelerators and they consist of a high voltage terminal supported by an insulating column, a high voltage generator and an acceleration tube called acceleration column. To have a convenient size these accelerators use high pressure gas (SF_6) for electrical insulation so that high voltage points can be kept at convenient distances. The electrostatic potential difference between the high voltage terminal and the ground potential accelerates charged particles to an energy equal to the charge times the terminal voltage ($E = qV$). As the DC potential is increased to obtain higher energy particles, the insulation problems are encountered inspite of using high pressure gas and hence it ultimately led *to* the development of successive reapplication (RF) of the same moderate electric field to the particle beam in a phase compatible way and this led to the development of LINACS. An ion source produces the ion species to be accelerated and a variety of ion sources are available commercially. The details of each of the above components of the accelerators are given below.

9.2.1 The Ion Sources

All the ion sources use the principle of electron collisions for producing ions for acceleration. Some of the popular ion sources use RF discharge to produce the ions and these have a long life as there is no limitation of the wearing of the Cathode filaments. The modern ion sources use heated tungsten cathode for producing large ion densities and a modified version of the same is the Duoplasmatron source where an axial magnetic field is used to concentrate the ions into a thick beam. These sources can give large amounts of currents. A recent addition to the ion source list and is mostly used now is the PIG ion source where electron oscillations are employed to enhance the ionisation density in comparison with the electron density and when operated in the pulse mode can deliver large currents without heating the cathode, thus increasing the life of the source. The ion source produces only positive ions through ionisation and for such accelerators which need negative ions *e.g.* Tandem accelerators, the electrons produced in the ionisation are made use of to attach themselves to the neutral gas atoms. For the high energy machines, the lower energy machines are used as the injectors after the initial acceleration.

9.2.2 High Voltage Generators

Low and Medium accelerators used either electronic voltage doubler units (*e.g.* Cockcroft-Walton) or charge transfer processes (like in Van de Graaff and Tandem accelerator). In Linacs the power supply is a radio frequency generator and is applied to successive cylinders in such a way that the beam enters each cylinders in the same phase. In cyclotrons the RF power is applied to the half of the circular structure (the so called D's) in such a way that the ion beam is in phase with the RF power when it enters the D's.

9.2.3 Acceleration Column

In the potential drop accelerators, the acceleration column is made up of many tubes made of insulating material generally of ceramic cylinder of several inches long and several inches in diameter and are connected to each other by Vaccum seals and the whole structure is evacuated such that the ion beam is not scattered while passing through the acceleration column.

9.2.4 Bending and Analysing Magnet

The initial ion energies are not all same when they enter the acceleration column. Thus the accelerated ions will have a range of energies and hence the accelerated beam has to be analysed. For this purpose a suitable magnet is used and the magnetic field of this magnet can be varied so that different energy ions can be delivered to the beam tubes which carry the beams to the target. These analysing magnets can also be used to bend the beam into the desired direction and such magnets are called Bending and Analysing Magnets.

9.3 BASIC PRINCIPLES OF ACCELERATORS

9.3.1 Cockcroft-Walton Generator

This was the first accelerator constructed. This is an electrostatic potential drop accelerator which employed the voltage multiplier principle. This accelerator utilized two stacks of series connected capacitors. One capacitor stack is fixed in voltage, except for voltage ripple, with one terminal connected to the ground and the other to the accelerating column which became the load. One terminal of the second capacitor stack is connected to a transformer giving peak voltage of ±Volts and tne two capacitor stacks are connected in series through rectifiers. The Fig. 9.1. illustrates the Cockcroft Walton accelerator. The voltage at the transformer point connected to the first capacitor in the first capacitor stack varies between –V to +V and the voltages at every point in the second stack Oscillates over a voltage of $2V$ volts. For the first negative half cycle B becomes negative with respect to A and the rectifier R_1 conducts and C_1 is charged to V volts. In the second half cycle B becomes positive w.r.t. A and hence R_1 does not conduct but the charge at point C with respect to ground is increased to $2V$ volts, through capacitor C_1. In this seco d cycle C is positive and hence R_2 conducts making C_2 to share the charge at the point C. After a few half cycles the point D will attain the voltage $2V$ volts. As point D becomes reference point for the rectifier R_3, the R_3 and R_4 repeat the charging process and thus making all points along the second stack oscillate over a voltage range of $2V$. Thus over a number of half cycles the charge is transferred stepwise from ground to the high voltage terminal. The high voltage terminal attains a voltage equal to 2VN where N is the number of stages. However above 500 KV, the voltage falls significantly from this expression due to stray capacitances and other leaks. In order to avoid this, one may use two transformers and two capacitor and this arrangement is called symmetrical cascade rectifier. The high voltage terminal is connected to the accelerating column which acts like the load. If the load current is I and f is the frequency of the power supply and for a stack of capacitors of all equal value C, the terminal voltage will have a ripple R.

$$R = IN(N+2)/16fC \qquad ...(9.3.1)$$

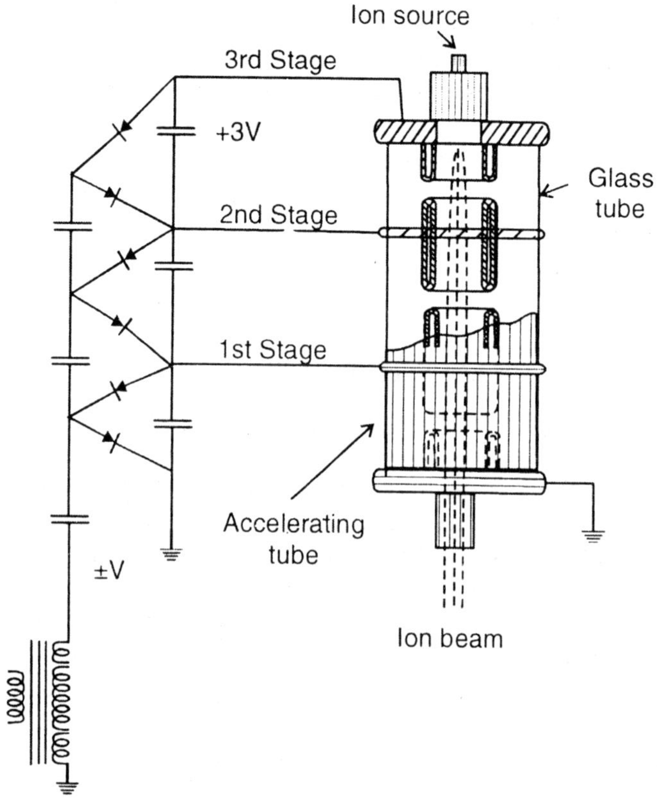

Fig. 9.1: Schematic diagram of a three stage cascade accelerator

The ripple can be decreased either by increasing the value of C or f and increasing f to RF values is more convenient than having big capacitors which will bring the associated insulation problems. Cockcroft-Walton (CW) accelerators are made for voltage ranging from few hundred KeV to few MeV. A *CW* accelerator operating at 1 million volts in air is so large that it must be housed in a very large room to avoid sparking. As voltage on the terminal is increased, the size and the space requirements increase rapidly and hence the upper practical limit to operate these accelerators in air is about 1.5 MV. These accelerators though can give only relatively low energies, are fairly stable and can deliver very high currents. They are, now a days mostly used for neutron studies and low energy experimental studies in material science and atomic physics.

9.3.2 Van de Graaff Accelerator

Robert Van de Graaff built the first electrostatic accelerator in 1929 based on the principle of pumping the charge from ground to the high voltage terminal through a running insulated belt. The Fig. 9.2 illustrates the schematic of this accelerator. It has a voltage dome in the form of a sphere, insulating support cylinder, and a motor driven insulting belt. The ion source is kept in the high voltage dome and the acceleration column running parallel to the belt accelerates the ions down to the ground. The target is kept at the ground potential. The dome is pressurised with the insulating gas, generally SF_6. In order to built the high voltage potential into the dome the electrical charge is sprayed on to the insulating motor driven belt and is carried into the smooth and well rounded metal dome, where the charge is removed, which raises the voltage of the dome. Spraying of the charge on to the belt and its removal in the dome are carried by a series of corona needles at both the places. Belt charging electrodes are to be well shielded from the voltage terminal and charge must be carried well within the sphere before it is removed. Charging current is completely independent of Voltage on the terminal and the voltage will rise till it is discharged by a load current through the acceleration tube.

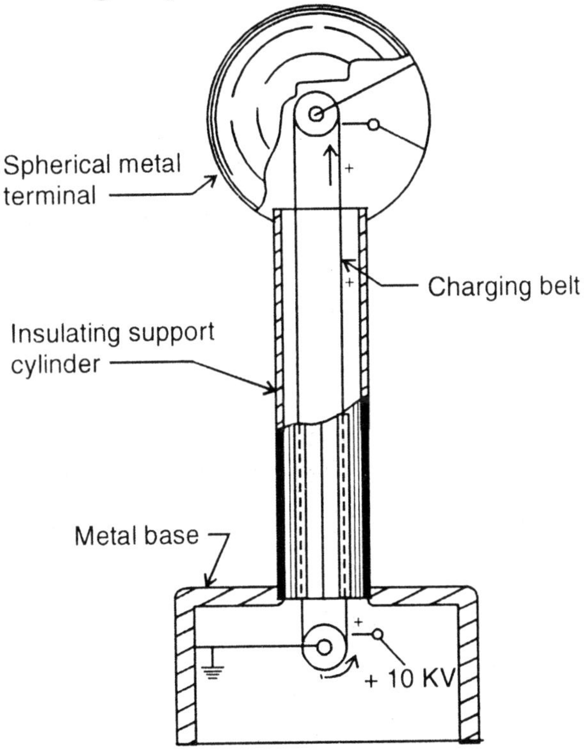

Fig. 9.2: Essential components of a Van de Graaff accelerator

9.3.3 Two Stage Tandem Van de Graaff Accelerators

On a novel idea put forward by W. H. Bennett and L. W. Alverez the two stage Tandem accelerators were developed in 1958 using the same single high voltage dome but accelerating the ions once from ground to high voltage and then from high voltage to ground in series, thus increasing the energy of the ions and hence the name tandem accelerator. Figure 9.3 shows the schematic of a tandem electrostatic accelerator. In these machines since the first acceleration is done from ground potential to high voltage terminal, one needs to produce negative ions at the ion source. Atoms of a large proportion of the elements form stable negative ions. A singly charged negative Oxygen ion gains an energy of 10 MeV as it goes from ground to the terminal of an accelerator operating at 10 MV. The Oxygen atoms at this potential will be stripped off their electrons to make them positive ion and are made to accelerate from high voltage to the ground potential. If all the eight electrons of the Oxygen atom are stripped off they can be accelerated to gain 80 MeV energy and when they finally emerge the ion energy will be a total of 90 MeV. These machines cannot accelerate atoms which cannot be made into negative ions.

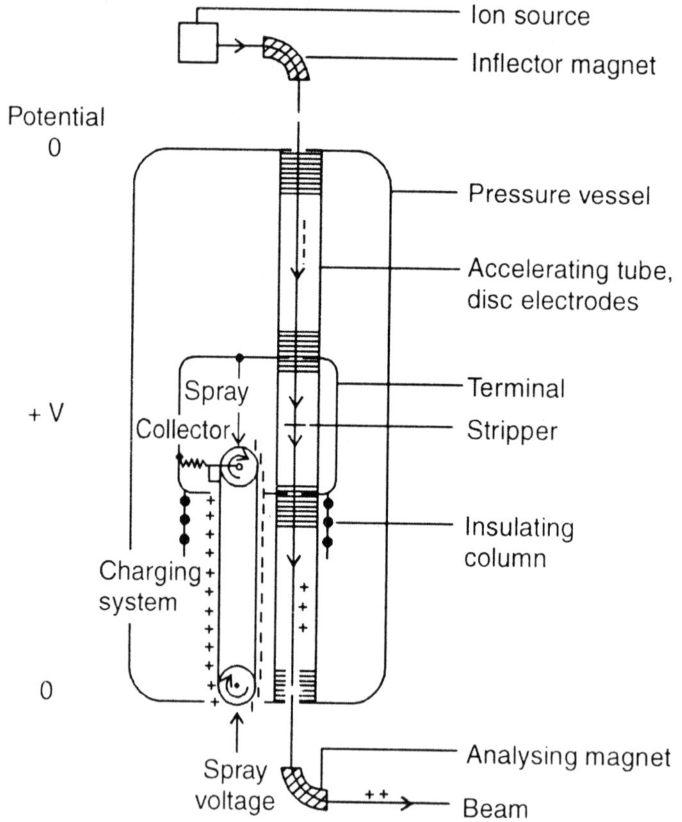

Fig. 9.3: Tandem Van de Graaff accelerator with pressure tank

9.3.4 Pelletron Accelerators

If the charging belt in the tandem accelerator is replaced by a charging chain of pellets then the accelerator is called pelletron accelerator. The chain consists of steel cylinders in the form of pellets joined by links of solid insulating material such as nylon. The chain is intrisically spark protected.

In the modern tandem accelerators the charging chain of pellets are used instead of the usual charging belts, because the pellet charging current is adjustable and highly uniform so that the terminal voltage are more precise and stable with very small ripple voltage. Hence these machines are more stable and have better energy resolution. The pellets are charged inductively so that there is no contact with charging electrodes. For a positive terminal voltage, positive charge is induced on the pellets at a motor driven pulley at ground potential. This charge is removed at a pulley in the terminal and the pellets are charged negatively, with the result the upgoing and downgoing pellets are charged equally. Pelletron accelerators accelerate most of the heavy ions except those that cannot be produced into negative ions. Many low and medium energy ion implantation machines work on the principle of pelletron tandem accelerators. In India there are two Pelletrons—one at TIFR which has a dome voltage of 14 MV and another at Nuclear Science Centre, Delhi whose dome voltage is 15 MV.

9.3.5 Folded Tandem Accelerators

In order to reduce the size of the accelerator and also to economise on the insulating gas and the building costs these folded version of the tandem accelerators are developed. Here both acceleration columns are put in same single insulating column with a 180° magnet in the terminal to steer the beam from one tube into the other. The folded design not only gives a compact system but it also locates both the ion source and sample chamber near the system control panel.

9.4 LINEAR ACCELERATORS (LINACS)

Linear Accelerators are the accelerators which accelerate the nuclear particles in straight line paths by using a series of small impulses repeatedly. They need powerful microwave power sources for this purpose. An oscillating electric field applied to radio-frequency cavities imparts the individual impulses for accelerating the particle. The phase of the radio-frequency electric field is matched with the time of entering of the particle into individual cavity so that the particle energy continuously increases. Fig. 9.4 shows resonant cavities excited in the TM_{010} electromagnetic mode.

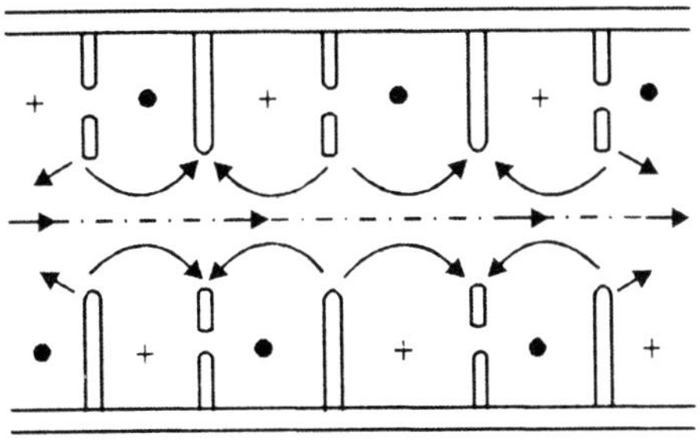

Fig. 9.4: Resonance cavities of a linear accelerator excited in the TM mode

9.4.1 Drift Tube Linacs

Phase stability and phase matching are the principles used in the Linacs for the acceleration of the particles. A series of field free drift tubes of increasing length, separated by small accelerating gaps are connected together alternately and these alternate sets of drift tubes are connected to the r.f. power supply. Thus as long as the particle is inside a drift tube it is in a field free space and in between the tubes the particle gets accelerated. The length of the drift tube is adjusted in such a way that when the particle enters the gap, the field is always in phase to accelerate it. If the voltage across the gap is V and the initial energy of the particles is eV then, the energy of the particle when it enters the nth tube is neV. We can calculate the length of the nth drift tube by taking the velocity of the particle $v_n = (2neV/M)^{1/2}$ where M is the mass of the particle and the frequency of the oscillator $\nu = c/\lambda$. The length of the nth drift tube will be $L_n = \frac{1}{2}\, v_n \lambda/c$, if the flight time for half a cycle is used for the particle. Hence one can see that $L_n \propto n^{1/2}$. Figure 9.5 shows the schematic diagram of the principle of linear accelerator.

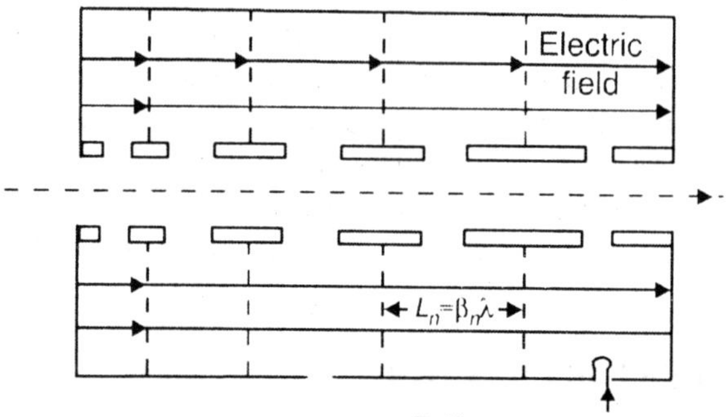

Fig. 9.5: Drift tubes of a linear accelerator

The principle of phase stability allows for the design of the drift tubes in such a way that the exact resonance condition need not be adhered to, because it is enough if a particle crossing a gap at a phase angle arrives at the next gap with the same phase angle. However, this phase stability is associated with a radial de-focussing of the accelerated particles by the radio-frequency electric fields. The present day Linacs use the quadrupole magnets or solenoidal magnetic fields for focussing the radial beams.

Two versions of Linacs are available named as the travelling wave type or the standing wave type, depending on the coupling of the cavity and the r. f. power phase shift per cavity in operation. In the travelling wave Linacs the coupling between adjacent cavities results in a phase shift less than 180° and in the standing wave type the energy is strongly reflected from the end of each cavity chain and the cavity oscillations are 180° out of phase. The early electron Linacs are of travelling wave type and all proton Linacs are of standing wave type.

9.5 CYCLOTRON

In 1929 Ernest O. Lawrence developed the first circular accelerating machine called cyclotron. He started with the idea of producing beams of high energy ions without using very high electrostatic voltages. For this he used the basic principle of bringing the ions repeatedly into the same electrostatic field to accelerate them to higher and higher energies. To bring the ions back into the same field he used magnetic field of suitable strength. The cyclotron designs have developed over the years and we have three successive generations in them-the classic design, the frequency modulated and the sector focused. The classic cyclotron is the prototype not only of the later types but also of a wider class of magnetic resonance accelerators which include Synchrotrons.

9.5.1 Basic Design of Cyclotron

The Cyclotron consists of a dc electromagnet, two hollow metal boxes which are flat D shaped copper electrodes called 'dees', a radio-frequency electric field driven by r-f oscillator, a vacuum tank, an ion source, internal targets, and a deflection system to bring the ion beams out of the accelerator. The electromagnet having circular pole pieces with a gap small compared to their diameter, produces an uniform axially symmetric magnetic field. The dees are placed in the magnet gap and radio-frequency oscillating electric field is produced along the dee gap by the r.f. oscillator. The ion source is kept between the dees and is supplied with a gas flow of neutral atoms. All these are housed in a vacuum tank (Fig. 9.6).

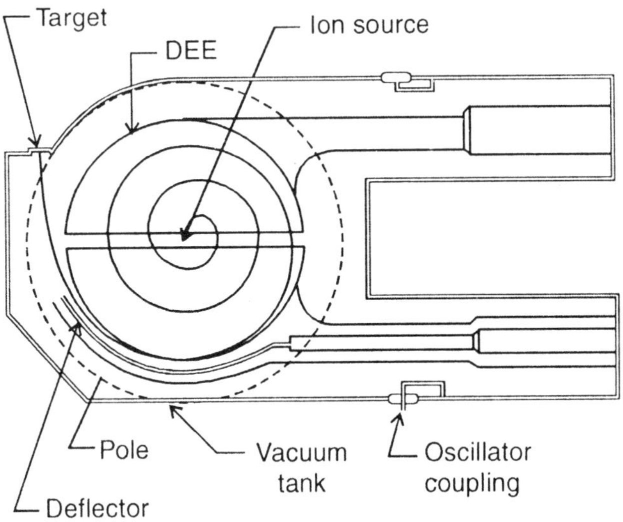

Fig. 9.6: Schematic diagram of a conventional cyclotron

9.5.2 Cyclotron Resonance Condition

A charged particle of mass m and charge q moving with a velocity v enters into a magnetic field induction B at right angles experiences a centripetal Lorentz force $F = qvB$ and as a result the particle moves in a circular orbit of radius r. The centripetal acceleration is given by $a = v^2/r$. From the Newton's second law

$$qvB = mv^2/r \text{ or } r = mv/qB \quad ...(9.5.1)$$

The angular velocity ω is given by $\omega = v/r = qB/m$ and the frequency ν measured in rev./sec. is given by

$$\nu = \omega/2\pi = qB/2\pi m \quad ...(9.5.2)$$

It can be noted that the frequency of revolution ν is independent of velocity v of particle. Fast moving particles move in large circles and the slow moving particle move in small circles but the time taken to complete one revolution is the same for all the particles. The frequency ν is the characteristic frequency for a charged particle in a field and is called *Cyclotron Frequency*.

It is apparent in a cyclotron with a fixed radio frequency-equal to the cyclotron frequency-particles with nearly the same value of q/m *e.g.* deuterons and alpha particles, can be accelerated. By fine tuning the radio frequency, it is possible to accelerate ions of carbon and nitrogen also. The cyclotron frequency depends upon the relativistic mass of the particle being accelerated. With velocity, the relativistic mass increases and the cyclotron frequency becomes less than the radio frequency. At a certain velocity, the particles cross the dee gap out of phase with the radio frequency and the particles instead of gaining, loose energy. This puts a limit on the energy to which particles can be accelerated in a cyclotron. Cyclotron is usually used to accelerate alpha particles to energies up to about 100 MeV.

The kinetic energy of the particles accelerated in the cyclotron depends on the radius of the dees 'R'. From (eq. 9.5.1) the velocity of the particle circulating at radius R is given by

$$v = qBR/m \quad ...(9.5.3)$$

and the kinetic energy

$$T = \frac{1}{2}mv^2 = \frac{1}{2}m \times q^2B^2R^2/m^2 = (qBR)^2/2m \quad ...(9.5.4)$$

An ion moving in a circle around the ion source gains energy on crossing the dee gap each time if the r. f. frequency is equal to the cyclotron frequency. Thus the electric field will reverse every time the ion comes out of a dee and enters the other dee. Between crossings the ions are shielded inside the dees from the electric field. Because of the resonance condition the ions will stay in phase and their radii and energy will grow in every turn until the ion reaches the periphery of the dee. The beam is then taken out of the cyclotron by means of a deflecting plate, charged to a high negative potential.

It may be observed that the energy to which a charged particles can be accelerated in a cyclotron, depends upon the radius of the dee and the magnetic field B of the electromagnet. Made with especially purified iron the electromagnet can produce a field of about 20000 Gauss. The weight of the electromagnet varies almost as cube of the dee radius. These factors place constraints on the size of a cyclotron hence on the maximum energy available. With the development of superconducting magnets, it is possible to obtain high energy with a small size machine.

As the ions are accelerated they tend to drift away from magnet mid-plane and strike the inside surface of a dee, unless a vertical focussing force is provided. This is achieved by adjusting the magnetic field B in such a way that it decreases slightly with increasing radius. This causes the magnetic field lines to curve as shown in Fig. 9.7 The Lorentz force which is perpendicular to these curved lines of force has a small component directed toward median plane. This causes the ions to oscillate slowly up and down executing harmonic motion about the median plane. Their frequency is determined by the difference between the centripetal acceleration and the inward radial Lorentz force. These vertical and radial motions are called betatron oscillations since such oscillations occur in betatrons where the focussing is similar. The vertical and radial oscillation frequencies can be written as $\omega_v = \omega n^{1/2}$ and $\omega_r = \omega(1 - n)^{1/2}$ respectively. The radial variation of the magnetic field B can be represented as $B = B_o (r_o/r)^n$ where B_o is the field at a fixed radius r_0 and the field index n can be written as $n = (- r/B)\, dB/dr$. For $0 < n < 1$ both frequencies are real, giving stable motions. At resonance $(\omega_v/\omega_r = 2)$ a transfer of energy from radial oscillations to vertical oscillations can take place resulting in a beam loss. This happens at an index of $n = 0.2$.

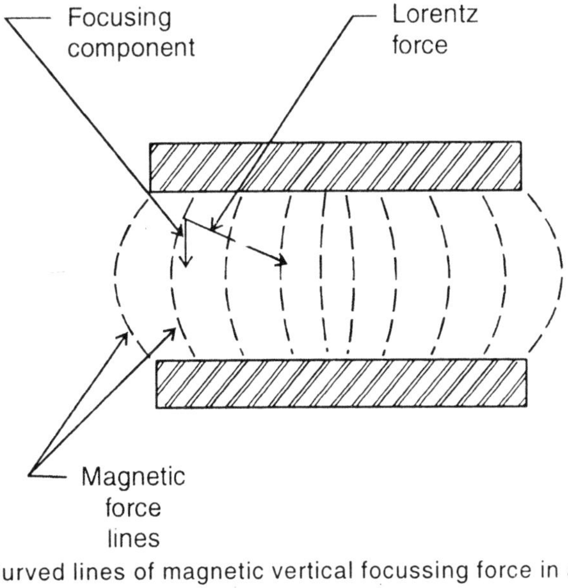

Fig. 9.7: Curved lines of magnetic vertical focussing force in a cyclotron

As has been pointed out, the weight of the electromagnet and the relativistic increase in the mass of the accelerated particle place constrains on the highest particle energy available from a cyclotron. With the increase in the mass of the accelerated particle, its revolution frequency reduces as $\omega = eB/m$. To overcome the problem, the frequency of the radio-frequency oscillator is changed so as to keep in phase with the cyclotron frequency of the particle. Modulation of the radio-frequency is done 0.1 to 1000 times a second. At the maximum energy attained by the particle, it is extracted out of the dees. Such machines are known as synchrocyclotrons. Another method to overcome the limitation of energy due to relativistic increase in mass is to use the so-called azimuthally varying field (AVF) cyclotrons. These machines use increasing magnetic field with increasing radius to compensate for the increasing mass of the particles.

9.6 BETATRON

Betatron accelerates electrons to high energies. The electrons move in a circular, evacuated, doughnut shaped vacuum chamber, placed between the pole pieces of a specially designed electromagnet. They are accelerated by the electromagnetic induction due to the changing magnetic field of the electromagnet. The magnetic field is perpendicular to the plane of the electron orbit. The electromagnet is energized by passing alternating current from the normal mains supply with a frequency of 50-60 Hz. through the coils *CC* and *C' C'* as shown in Fig. (9.8*a*). The electrons are injected in the chamber from an electron gun mounted in the doughnut itself, just outside the equilibrium orbit Fig. (9.8*b*). The electron gun consists of a heated filament, a focussing grid and a positive electrode to accelerate the electrons to energy of about 50 Kev. The injected electrons after executing a few damped oscillations are caught in the equilibrium orbit.

As the velocity of the electrons, circulating in the vacuum chamber is very high, the magnetic field varies very slowly compared to the frequency of revolution of the electrons in the equilibrium orbit. An electron with momentum p_t at any instant of time t will be constrained to move in the equilibrium orbit of radius R, provided that the magnetic field B_t at the time is such that

$$B_t\, e.R = p_t \qquad ...(9.6.1)$$

The tangential force F_t acting on the electron at time t is

$$F_t = dp_t/dt = eR(dB_t/dt) \qquad ...(9.6.2)$$

The work done on the electron in one revolution is $2\pi R \cdot F = 2\pi R^2 e(dB_t/dt)$. The inducted electromotive force E_t generated due to the changing magnetic field at the equilibrium orbit is

$$E_t = -(d\varphi_t/dt) \qquad ...(9.6.3)$$

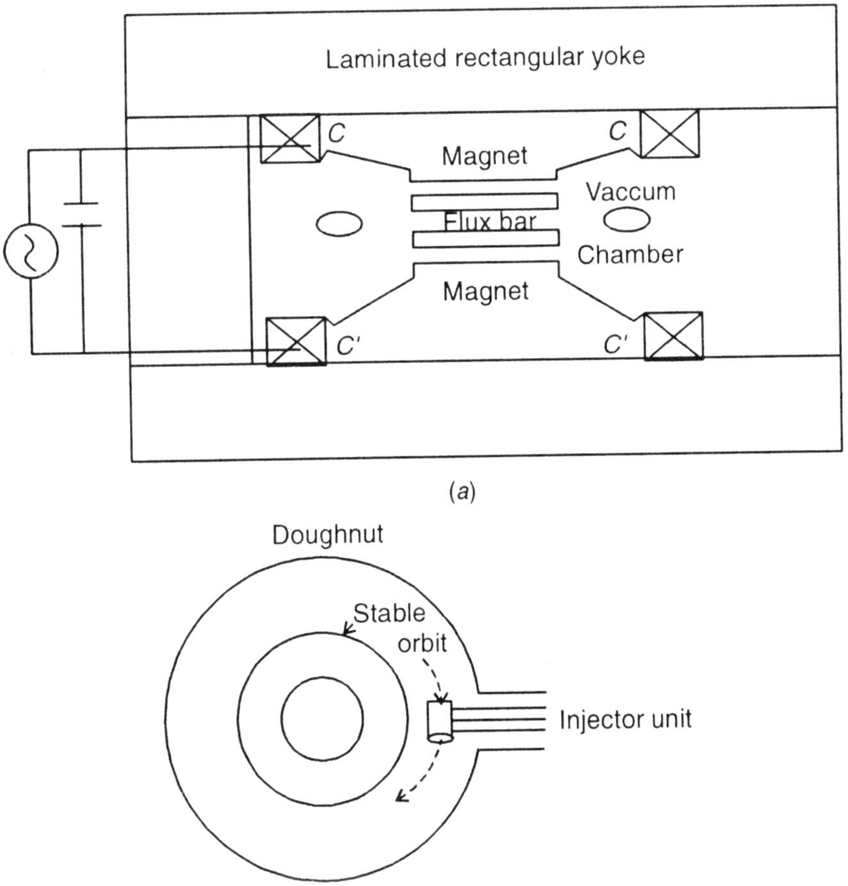

(a) (b)

Fig. 9.8: (*a*) Schematic diagram of a betatron. (*b*) Position of electron injector and equilibrium orbit in the doughnut of a betatron

where φ_t is the magnetic flux at the time *t*. The work done on the electron in one revolution is equal to $e\,(d\varphi_t/dt)$. Equating the two expressions for the work on the electron in one revolution we get

$$2\pi R^2(dB_t/dt) = (d\varphi_t/dt) \qquad ...(9.6.4)$$

If at time $t = 0$ $B_0 = 0$ and $\varphi_0 = 0$ then integrating the above expression gives

$$2\pi R^2 B_t = \varphi_t \qquad ...(9.6.5)$$

This equation shows that the total flux linked with the equilibrium orbit should at all times be equal to twice that, which would be obtained if the magnetic field B_t were uniform throughout the area of the orbit. This condition should be maintained at all times, if the electron is not to move away from the equilibrium orbit. The above condition requires special design of the magnetic pole pieces, so that the field in the central region is stronger than that at the electron orbit This is achieved by inserting high permeability flux bars between the pole pieces The magnet is fabricated from laminations to reduce eddy current losses. The doughnut shaped vacuum chamber is made of glass, ceramic or polymer resin and coated inside with a thin layer of silver. This avoids the accumulation of charges. The capacitor placed across the alternating current supply provides the correction for power factor. This is essential as there is a huge power loss in the resistance of the magnet coils.

To check the electrons form drifting away from the equilibrium orbit, due to small perturbations, it is necessary to provide both median plane and radial focusing. This is done by shaping the magnetic pole pieces, to give a variation in magnetic field, as is done in the case of a cyclotron.

The magnetic field varies sinusoidally with the mains frequency. The electrons are injected at a time, shortly after the field has crossed the value zero. The electrons are accelerated by the increasing magnetic flux linked with the electrons orbit until the maximum of the field is reached. Beyond this point the magnetic field starts reducing and if the electrons are not taken out, they would start to decelerate. At the maximum energy point, the electron beam is extracted from the equilibrium orbit by suddenly sending a large pulse of current through the auxiliary coils wound round the primary magnet coils. This increases the magnetic field temporarily and the electron orbit expands. The electron beam is either taken out of the chamber or allowed to hit a tungsten target fixed near the electron gun. The striking beam of electrons produces X-rays moving in the forward direction.

The betatron of the General Electric Research Laboratories U.S.A. has a pole face of 76 inches and the equilibrium orbit of 66 inches. The electromagnet has a weight of 135 tons and consumes a power of 200 kilowatts. It produces a beam of 100 MeV electrons.

To estimate the energy which can be imparted to the electrons in a betatron, we assume that the magnetic flux φ_t varies sinusoidally with the mains supply of frequency f so that

$$\varphi_t = \varphi \sin \omega_t$$

where φ is the maximum amplitude of the flux, and $\omega = 2\pi f$. The energy imparted to the electron in one revolution due to the induced electromotive force is

$$e(d\varphi_t/dt) = \omega e \varphi \cos \omega_t \qquad \ldots(9.6.6)$$

The period for which the electron is accelerated in time of a quarter cycle is given by $\pi/2\omega$. The average energy imparted to the electron in one revolution is

$$(2\omega / \pi) \int_{o}^{\frac{\pi}{2\omega}} (\omega\, e \cos \omega t)\, dt = (2\omega / \pi) e\varphi \qquad \ldots(9.6.7)$$

The speed of 1 MeV electron is about $0.94c$ hence one can assume that the electron travels throughout its motion with a speed almost same as the speed of light, c. The total distance traversed by the electron during the accelerating quarter cycle is $c(\pi/2\omega)$. The number of revolutions N, the electron makes during this time interval is given by

$$N = (c\pi/2\omega)/2\pi R = (c/4\omega R) \quad \ldots(9.6.8)$$

Multiplying the number of revolutions with the average energy gained in each revolution, the total energy to which the electron is accelerated is W where

$$\begin{aligned} W &= (c/4\omega R) \cdot (2\omega e\varphi/\pi) \\ &= (ec\varphi/2\pi R) \\ &= (ec\pi R^2 B)/2\pi R \\ &= ecRB/2 \quad \ldots(9.6.9) \end{aligned}$$

It is apparent that the energy to which electrons can be accelerated varies as the radius of the equilibrium orbit and the intensity of the magnetic field at the orbit.

As the electrons are accelerated in a circular orbit they radiate energy. An upper limit to the electron energy available from a betatron is reached when the radiation losses from the circulating electrons become equal to the energy gained by them in a revolution. The energy loss by radiation can be reduced by increasing the radius of the equilibrium orbit- there by increasing the size of the magnet, or by increasing the frequency of the current supplied to the electromagnet. Both the alternatives are difficult to achieve. To obtain higher energies one uses a linear accelerator or a synchrotron.

9.7 ELECTRON SYNCHROTRON

In an electron synchrotron, the electrons initially accelerated to about 50 keV energy (when their speed is about $0.5c$) are injected into a doughnut shaped vacuum chamber and are guided into an equilibrium orbit of radius R by a magnetic field. The electrons are then accelerated to about 2 MeV energy by the betatron action, by changing the magnetic field of the electromagnet. The pole pieces of the electromagnet are cylindrical as shown in Fig. 9.9. The pole pieces do not extend to the central region of the doughnut. In the centre of the vacuum chamber there are flux bars to increase the magnetic flux. This helps to fulfil eqn (9.6.5), but only to a limited extent, since by the time the electrons acquire an energy of 2 Mev , the flux bars get saturated, putting an end to the betatron action. At this energy, the electrons moving with speed of $0.98c$ circulate with almost fixed frequency $f = c/2\pi R$ in the equilibrium orbit. Any further increase in energy of the electrons will manifest as increase in their mass. The enhanced mass will reduce the frequency of circulation of the electrons. This means that the electron orbit will tend to expand. The electrons can be kept in the equilibrium orbit by increasing the magnetic field B, while small increments of energy beyond 2 MeV are being imparted to the electrons in such a way that B/m (m is relativistic mass) remains constant,. In this way the electrons continue to move in the equilibrium orbit even when their energy increases. The electrons are provided energy by a radio-frequency accelerating voltage from an osciiiator having a fixed frequency equal to $c/2\pi R$.

As the electromagnet is ring shaped, the amount of iron in the magnet of a synchrotron is much less than in a betatron, with the same output energy. The magnet is made of laminations to minimize the eddy current losses. The energy losses are further minimized by resonating the magnet coil with a shunt capacitor to the frequency of the alternating mains supply. The magnetic field is suitably shaped, as in the case of a betatron, to achieve axial and radial focussing of the electron beam. Thus in an electron synchrotron, the betatron action is used to accelerate the electrons only upto 2 MeV energy. Beyond this, the energy is supplied to the electrons by the radio-frequency voltage. The accelerating radio-frequency voltage is applied across two cylindrical electrodes placed inside the vacuum chamber, such that the equilibrium orbit passes along their axis. By choosing a proper phase of the accelerating voltage with the circulating electrons, they can be made to gain the same amount of energy in each revolution, thus exhibiting phase stability.

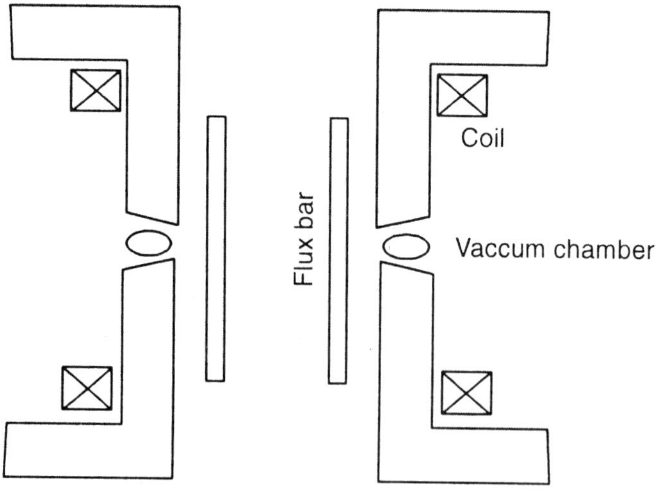

Fig. 9.9: Position of flux bars and vaccum chamber in an electron synchrotron

In an electron synchrotron the accelerating peak voltage of the radio-fraquency oscillator can be much higher than in a betatron. There is thus large energy increment in a revolution and the electrons gain the final energy in fewer turns than in a betatron designed for the same energy . This reduces the loss of energy due to radiation to a great extent. The 70 MeV electron synchrotron of General Electric Company U.S.A. has an electromagnet whose weight is only 8 tons. The equilibrium orbit radius is only 29.6cm, with the peak magnetic field of 8100 Gauss and the 163 MHz radio-frequency voltage of 1000 Volts. The electromagnet is operated at the mains frequency of 60 Hz.

In the "race track" design of a synchrotron, the doughnut consists of four quadrants connected by short straight sections. These straight sections are in field free space, as the electromagnet is made in four sections corresponding to the four quadrants as shown in Fig. 9.10. The field free sections are used to place the electron injector the radio-frequency accelerating gap and the target at convenient places. The functioning of this synchrotron is similar to the earlier type. Initially the electrons are accelerated by the betatron action and then by synchrotron action. Just before the magnetic field reaches its maximum value, the accelerating radio-frequency oscillator is switched off. As the magnetic field keeps on rising without the electrons gaining any energy, the radius of the electron orbit starts decreasing until they strike the target projecting from the side of the inner edge of the vacuum chamber producing X-rays.

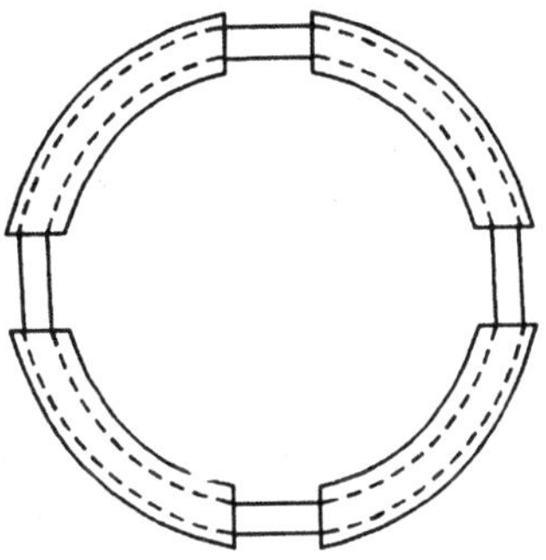

Fig. 9.10: Straight section electron synchrotron

Acceleration of electrons much above 100 MeV is done in linear accelerators. These accelerators have been used upto an energy of 600 MeV. Synchrotron machines are used to accelerate protons also. In these machines the protons are injected after accelerating them to energy of about 20 MeV.

9.7.1 Phase Stable Synchrotron

These machines are now used to produce the synchrotron radiation. The particle beam accelerated by it are used in medical application and treatment. These are ring shaped accelerators and use the principle of synchronous nature of acceleration process called phase stability as detailed above. The very high energy versions of synchrotrons are called storage rings. In a typical synchrotron or in a storage ring the particles travel millions of kilometers in an evacuated pipe of few centimeters in diameter. The largest synchrotrons in operation are 1 TeV (10^{12}eV) proton synchrotron at Fermi National Laboratory in Illinois, which is 6.3 km in circumference and the 3 TeV proton synchrotron which is 21 Km in circumference in Russia and a 100 GeV electron-positron synchrotron 27 km in circumference at CERN, Geneva. Fig. 9.11 gives a schematic of the proton synchrotron.

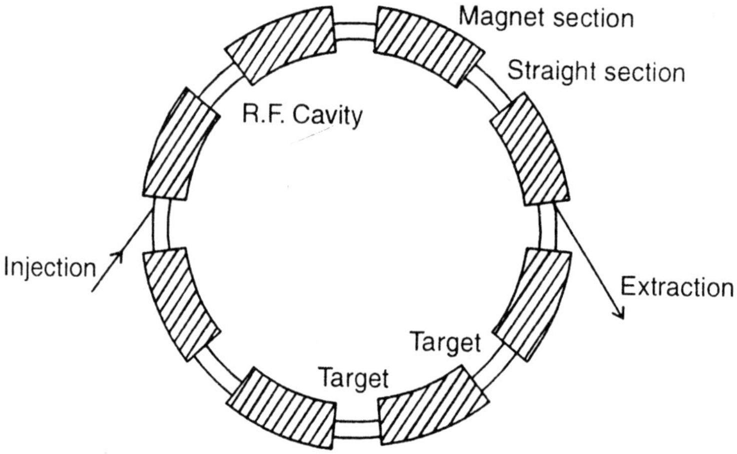

Fig. 9.11: Schematic diagram of a proton synchrotron

A particle in the beam, moving with momentum p in a magnetic field B, will have a radius of curvature $R = pc/qB$ where c is the velocity of light and q is the net electric charge of the particle. The kinetic energy T can be related to the relativistic momentum as

$$(pc)^2 = T^2 + 2Tm_0c^2$$

m_0 is the rest mass of the particle; for protons $m_0c^2 = 938$ MeV and for electrons and positrons it is 0.511 MeV. In the non-relativistic limit we can write the above expression as $(pc)^2 = 2Tm_0c^2$ as the kinetic energy is smaller than rest mass and in the relativistic limit where the kinetic energy is much larger than the rest mass, we have $pc = T$. In the phase stable mode the revolution frequency of the ion f_0 which is equal to oscillator frequency can be written in the non relativistic limit is

$$f_0 = v/2\pi R = pqB/2\pi mpc = qB/2\pi mc \qquad \text{...(9.7.2)}$$

In the total relativistic limit $m = m_0 + T/c^2$, where m is the total relativistic mass which includes both the rest mass and the mass due to kinetic energy.

9.7.2 Storage Rings and Collider Rings

Synchrotrons which can store a particle beam for long periods, typically several hours or days are called storage rings. For this purpose a very good high quality vacuum in the acceleration chamber and a very small orbit perturbing electromagnetic fields are very essential. These storage rings are used for producing synchrotron radiation and nuclear reactions.

The synchrotrons which are used to collide the beam of particles are called collider rings. In these collider rings two counter rotating beams are brought into collision at one or more points in the ring. The two counter-rotating beams can be antiparticles of each other for *e.g.* Protons and antiprotons or electrons and positrons. If both the beams of particles have the same energy then one synchrotron ring can accelerate and store both beams in the same vacuum chamber. CERN, Geneva has a 400 GeV proton-antiproton Collider storage ring and Fermi Lab., Chicago, USA also has a 1 TeV proton-antiproton Collider. Several proton synchrotrons are built to make proton-proton collision to study nuclear reactions and 30 GeV interacting storage ring at CERN is one such example.

Colliding beams can deliver much more energy in a collision than a beam of same energy delivering to a stationary target. Because the mass of a very high energy proton is much more than a stationary proton and in such a collision a very little energy transfer is expected where as two very high energy protons having equal masses can deliver the entire energy of both the particles to the collision and can cause nuclear reactions. The relative advantage increases as the square root of beam energy.

9.7.3 Synchrotron Radiation

The electromagnetic radiation emitted when charged particles moving at very high speed (almost at the speed of light) are accelerated, is called synchrotron radiation. Electrons moving at relativistic speeds when passed through a uniform magnetic field move in circular orbits and emit the electromagnetic radiation due to acceleration caused by the Lorentz force. This radiation is also called Bremstrahlung if the electron radiates in the field of a nucleus. Such radiations are seen to be emitted from astronomical objects like pulsars in the crab nebula. These radiations can be produced in high energy synchrotrons and storage rings. The radiated power in such an emission varies as fourth power of the energy of the particle and inverse fourth power of the mass of the particle and can be written as

$$p \propto (E / m)^4 \frac{1}{R^2}$$

where R is the radius of the particle motion. It is evident that significant radiated power can be obtained from light particle like electrons and positrons when accelerated to high energies typically greater than 500 MeV. The synchrotron radiation extends from the far infrared into the *X*-ray region in a smooth continuum. The high brilliance, natural collimation and linearly polarised nature of the synchrotron radiation are very useful properties for many unique experimental research studies in diverse field as Physics, Chemistry and Biology.

10

ELEMENTARY PARTICLES

10.1 ELEMENTARY PARTICLES

The concept of an elementary particle has originated from man's age old search for an ultimate and indivisible unit as the constituent of matter. The beginning was made through the discovery of atoms. Towards the end of the nineteenth century, electron was discovered to exist inside the atoms. This followed by the discovery of nucleus at the center of the atom. The nucleus was found to contain positively charged protons and electrically neutral neutrons. These three, elementary particles, to begin with, were regarded for quite sometime, as the basic constituents of matter. The interaction between charged particles, known as electromagnetic interaction, takes place through the emission and absorption of light quanta, called as photons. The explanation of beta-decay of radioactive nuclei postulated by W. Pauli, brought the existence of an electrically neutral particle of almost zero rest mass. It was named as neutrino P.A.M. Dirac predicted from theoretical considerations, the positive electron called as positron, which was observed experimentally through the studies of cosmic rays. The positron was regarded as anti particle to electron. This led to believe that particles known so far must have their associated antiparticles. Photons were found to be their own antiparticles. The strong interaction (so called nuclear force) between nucleons (neutrons and protons) was possible only through the exchange of pions. Pions were observed in cosmic rays. Another elementary particle called as muon was also observed in cosmic rays. During the last 40 years or so, many particles have been added to the list of elementary particles some are more elementary than others in the sense that few of the particles are stable while many others are unstable. Besides, some of the particles are relatively stable having mean life times shorter than 10^{-10} sec. Some 100 or so very short lived particles with mean lives shorter than 10^{-21} have been identified. Such particles are known as resonances.

The discoveries of all such elementary particles have been through experiments, involved mathematical calculations or on cosmic rays or through the use of particle accelerators.

10.2 CLASSIFICATION OF PARTICLES

All the so far known elementary particles have been grouped under two main classes. The particles which can interact mainly by strong (nuclear) interaction are known as Hadrons and the particles which respond only to weak interaction are known as Leptons. The hadrons are further subdivided into two groups Baryons and Mesons. Baryons include nucleons (protons and neutrons) and hyperons. There are four main hyperons. Lambda, ($\mathring{\Lambda}$), Sigma ($\Sigma^{\pm}$, Σ°), χ (Ξ°, Ξ^{-}) and omega (Ω^{-}). All the nucleons and hyperons have their respective anti particles.

There are three main members of meson group. Pions ($\pi^{\pm}$, π^{0}), Kaons ($K^{\pm}$, K^{0}) and eta (η^{0}). All these mesons have their respective anti particles Leptons include electron (e^{-}) and its anti particle positron (e^{+}), muon (μ^{-}) and its anti particle (μ^{+}) and neutrino (ν), and anti neutrino (ν^{+}).

Besides these, there are two particles, photons and gravitons, which have been included in the list of elementary particles. These particles have been listed in Table 10.1.

10.3 PROPERTIES OF THE PARTICLES

10.3.1 Baryons

Among the baryons, the proton is relatively stable. However, the current theories of elementary particles indicate that proton should be unstable. It should decay into leptons. The expected mean life, 10^{30} years, is very long and as such the proton decay has not been confirmed.

Neutrons remain stable so long as they stay inside a nucleus. But in free space a neutron undergoes β-decay in about 12 minutes. ($n \rightarrow p + \bar{e} + \nu$). The hyperons are heavier than nucleons and all are strange particles. The baryons are fermions.

(i) $\mathring{\Lambda}$ **Hyperon:** It was the first hyperon discovered in the studies of cosmic rays. The rest mass of $\mathring{\Lambda}$ hyperon as estimated from its most probable decay is around 2183 m_e and the corresponding rest mass energy being around 1116 Mev. The decay (through strong interactions) reactions are

$$\mathring{\Lambda} \rightarrow p + \pi^{-} \qquad (66\%)$$
$$\mathring{\Lambda} \rightarrow n + \pi^{o} \qquad (34\%)$$

The mean life time of $\mathring{\Lambda}$-hyperon is 2.6×10^{-10} sec. Since $\mathring{\Lambda}$ appears only in one charge state (zero charge) it is an iso spin singlet $T = 0$ with $T_3 = 0$ state. Its spin is ½ and has even parity. The strangeness quantum number or the strangeness number for $\mathring{\Lambda}$ particle is $S = -1$ (Table 10.1).

TABLE 10.1: ELEMENTARY PARTICLE CHARACTERISTICS

Class	*Name*	*Particle*	*Anti particles*	*Rest Mass in MeV*	*Mean life (Sec.)*	*Spin ($\hbar$)*	L_e	L_μ	*B*	*S*	*Y*	*T*	T_z
							(antiparticles have opposite signs)						
Mass less Boson	Photon	γ	(γ)	0	Stable	1							
	Graviton	g	g	0	Stable	2							
Lepton (Fermions)	Electron	e^-	e^+	0.51	Stable	½	+1	0	0				
	Muon	μ^-	μ^+	106	2×10^{-6}	½	0	+1	0				
	e-neutrino	ν_e	$\bar{\nu}_e$	0	Stable	½	+1	0	0				
	μ-neutrino	ν_μ	$\bar{\nu}_\mu$	0	Stable	½	0	+1	0				
Hadrons Mesons (Boson)	Pion	π^+	π^-	140	2.6×10^{-8}	0	0	0	0	0	0	1	+1
		π^0	(π^0)	135	0.8×10^{-16}	0	0	0	0	0	0	1	0
		π^-	π^+	140	2.6×10^{-8}	0	0	0	0	0	0	1	−1
	Kaon	K^+	K^-	494	1.24×10^{-8}	0	0	0	0	+1	+1	½	+½
		K^0	$\overline{K^0}$	494	1.24×10^{-8}	0	0	0	0	+1	+1	½	−½
	Eta	η^0	(η^0)	549	7×10^{-19}	0	0	0	0	0	0	0	0
Hardrons Baryons (Fermion)	Nucleon												
	Proton	p	$\bar{p}$	938.3	Stable	½	0	0	+1	0	+1	½	+½
	Neutron	n	$\bar{n}$	938.6	930	½	0	0	+1	0	+1	½	−½
	Lambda Hyperon	$\overset{o}{\Lambda}$	$\overline{\overset{o}{\Lambda}}$	1116	26×10^{-10}	½	0	0	+1	−1	0	0	0
	Sigma hyperon	Σ^+	$\overline{\Sigma^-}$	1189	0.8×10^{-20}	½	0	0	+1	−1	0	1	+1
		Σ^0	$\overline{\Sigma^0}$	1192	5.8×10^{-20}	½	0	0	+1	−1	0	1	0
		Σ^-	$\overline{\Sigma^+}$	1197	1.5×10^{-10}	½	0	0	+1	−1	0	1	−1
	xi hyperon	Ξ^0	$\overline{\Xi^0}$	1315	3×10^{-10}	½	0	0	+1	−2	−1	½	+½
		Ξ^-	$\overline{\Xi^+}$	1321	1.7×10^{-10}	½	0	0	+1	−2	−1	½	−½
	omega hyperon	Ω^-	$\overline{\Omega^+}$	1672	1.3×10^{-10}	3/2	0	0	+1	−3	−2	0	0

(*ii*) The sigma hyperons: There are three types of sigma hyperons Σ^+, Σ^- and Σ^0. The known decay modes (strong interaction) for the charged sigma hyperons are (Table 10.2).

$$\Sigma^+ \rightarrow p + \pi^0 \qquad (53\%)$$
$$\Sigma^+ \rightarrow n + \pi^+ \qquad (47\%)$$

Only one decay mode is observed for Σ^-.

$$\Sigma^- \rightarrow n + \pi^- \qquad (100\%)$$

The rest masses of Σ^+ and Σ^- as estimated from the Q-values of most probable decay reactions, are $m(\Sigma^+) = 2327\ m_e$ and the rest mass energy 1189 MeV

and $m(\Sigma^-) = 2342\ m_e$ and $m(\Sigma^-)c^2 = 1197$ MeV

The neutral sigma hyperon decays by electro magnetic interaction.

$$\Sigma^0 \rightarrow \overset{\circ}{\Lambda} + \gamma$$

The rest mass of Σ^o has been estimated to be

$m\,(\Sigma^0) = 2333$ MeV and the corresponding rest mass energy to be $m(\Sigma^0)c^2 = 1192$ MeV

The mean life times of sigma hyperons respectively are

$$\tau(\Sigma^+) = 0.8 \times 10^{-10} \text{ sec.}$$
$$\tau(\Sigma^-) = 1.5 \times 10^{-10} \text{ sec.}$$
$$\tau(\Sigma^0) = 5.8 \times 10^{-20} \text{ sec.}$$

The sigma hyperons, each, are spin ½ particles with even parity. As there are three charge states, the hyperons belong to an isospin $T = 1$ state with $T_3 = +1$, 0 and -1 respectively. The strangeness number $S = -1$ for all the three sigma hyperons.

***(iii)* Xi (ksi) Hyperon:** There are two types of Xi hyperons Ξ^0 and Ξ^- these hyperons decay into cascade decay and as such they are also known as cascade particles. The decay modes of the Xi-hyperons are

$$\begin{array}{l} \Xi^0 \rightarrow \overset{\circ}{\Lambda} + \pi^0 \\ \qquad\ \downarrow \\ \qquad\ p + \pi^- \end{array}$$

and

$$\begin{array}{l} \Xi^- \rightarrow \overset{\circ}{\Lambda} + \pi^- \\ \qquad\ \downarrow \\ \qquad\ p + \pi^- \end{array}$$

The rest mass and the rest mass energy of Xi-hyperons have been estimated as

$$m\,(\Xi^0) = 2573\ m_e \text{ and } m\,(\Xi^0)c^2 = 1315 \text{ MeV}$$
$$m\,(\Xi^-) = 2585\ m_e \text{ and } m\,(\Xi^-)c^2 = 1321 \text{ MeV}$$

The mean life times are

$$\tau(\Xi^0) = 3 \times 10^{-10} \text{ sec.} \qquad \tau(\Xi^-) = 1.7 \times 10^{-10} \text{ sec}$$

The spin of both the Xi hyperon is ½$\hbar$ and each has even parity. [The spin parity is expressed as $½^+$]. The strangeness number for each hyperon is $S = -2$. The iso-spin for Xi hyperon is $T = ½$ with $T_3 = +½$ for $= \Xi^0$ and $T_3 = -½$ for $= \Xi^-$ hyperon.

***(iv)* Ω^- hyperon:** It is the heaviest hyperon. It undergoes into cascade decay through weak interactions. The sequence of Ω^- decay is

$$\Omega^- \rightarrow \Xi^0 + \pi^-$$

$$\downarrow$$

and $$\Xi^- \rightarrow \mathring{\Lambda} + \pi^0$$

$$\downarrow$$

$$p + \pi^-$$

The rest mass has been estimated to be equal to $m(\Omega^-) = 3272\ m_e$ and the rest mass energy $m(\Omega^-)c^2 = 1672$ MeV. It is the only hyperon whose spin is 3/2 and has even parity (spin-parity $3/2^+$). The mean life $\tau(\Omega^-)$ is 1.3×10^{-10} sec. it is an iso-spin singlet state with $T = 0$ and $T_3 = 0$. The strangeness number is $S = -3$.

10.3.2 Mesons

These are particles having rest mass between electron mass and proton mass. All the mesons have spin zero and as such they are bosons.

Pi-mesons or Pions

Production and properties.

As predicted by H. Yukawa, pions are carriers of strong nuclear interaction (Chap. 2). They were first discoverea by C. F. Powell in 1947 in the studies of cosmic rays. The result of cosmic ray investigations revealed that pions are produced by nucleon-nucleon collision and also by the interaction of high-energy photons with nucleons. This led to believe that pions can be produced artificially using high energy particle accelerators. The pions were first produced in 1948 when a strong beam of 380 MeV alpha particles obtained from the synchrocyclotron at the Berkeley laboratory in U.S.A, were made to strike a carbon target. The expected pions were detected in photographic emulsion. The first pion so detected was negatively charged. The most suitable targets for the production of pions through nucleon-nucleon collision have been hydrogen, and deuterium. The π^+-mesons are produced through the following reactions.

$$p + p \rightarrow D + \pi^+$$

$$p + p \rightarrow p + n + \pi^+$$

Also a collision between a high energy proton with a neutron in deuteron can result into the production of π^+ mesons. The possible reactions are

$$p + n \text{ (in d)} \rightarrow n + n + \pi^+$$

$$p + n \text{ (in d)} \rightarrow p + p + \pi^-$$

High energy γ-ray photons (about 500 MeV and above) are able to produce charged pions.

$$p + \gamma \rightarrow n + \pi^+$$

$$p + \gamma \rightarrow p + \pi^-$$

The neutral pions π^0 can be produced through nucleon (N) nucleon (N) collision or through the collision of photons with a nucleon.

When protons with energy around 175 MeV from the Berkeley accelerator were made to impinge on a target of carbon or beryllium, then π^0-mesons were produced, which were detected through their decay products. The reaction can be expressed as

$$p + p \rightarrow p + p + \pi^0$$

Like wise energetic photons (around 330 MeV) when collide with hydrogen target, π^0 mesons are produced. The reactions is

$$p + \gamma \rightarrow p + \pi^0$$

As the high energy neutrons are available now, they can be used to produce pions. The possible reactions are expressed by the following:

$$n + n \rightarrow n + p + \pi^-$$
$$n + p \rightarrow n + n + \pi^+$$
$$n + p \rightarrow p + p + \pi^-$$

10.4 PROPERTIES OF PIONS

10.4.1 Mass of Pions

The mass of charged pions as estimated from the muon mass which are produced on their decay

$$\pi^\pm \rightarrow \mu^\pm + \nu_\mu$$

By determining the kinetic energy of the muons, the mass of pions is estimated.

The pion mass is also determined from their production in accelerator experiments. The rest mass of positive and negative pion is same. It is

$m(\pi^\pm) = 273.2\ m_e$ and the rest mass energy

$$m(\pi^\pm)c^2 = 140 \text{ MeV}$$

The mass of neutral pion π^0 is derived from the energy of two γ-ray photons into which π^0 decays.

$$\pi^0 \rightarrow \gamma + \gamma$$

Another method for the mass of neutral pion is through the interaction of a negatively charged pion with protons at rest.

$$\pi^- + p \rightarrow n + \pi^0$$

The energy of the neutron released is measured. The experiment thus gives the difference in mass of π^- and that of π^0. Knowing the mass of π^-, the mass of π^0 is obtained. The rest mass of π^0 comes out to be

$m(\pi^0) = 264 m_e$ and $m(\pi^0)c^2 = 135$ MeV

The mass of π^0 is less than the mass of $\pi^\pm$. This difference is associated with electromagnetic effects.

10.4.2 The Decay of Pions

The pions are unstable particles. If the energy of charged pion is less than about 150 MeV, positive pions do not interact with nucleons and they decay according to the scheme

$$\pi^+ \rightarrow \mu^+ + \nu_\mu$$

Since μ^+ is an antiparticle to μ^- the ν_μ must be a neutrino to conserve muonic lepton number. The decay occurs due to weak interaction. The parity is not conserved. The pion may also decay, to a very small extent, in the following way.

$$\pi^+ \rightarrow e^+ + \nu_e$$

The negative pion π^- decays as

$$\pi^- \rightarrow \mu^- + \bar{\nu}_\mu$$

and $$\pi^- \rightarrow e^- + \bar{\nu}_e$$

In the decay of $\pi^\pm$, the ejected muons $\mu^\pm$ always have the same kinetic energy, about 4.2 MeV. This shows that pion decay is a two particle process. Unlike the decay of π^+ pions, the neutral pion π^0 decays by electromagnetic interaction into two photons.

$$\pi^0 \rightarrow \gamma + \gamma \qquad (98.8\%)$$

Another, possible decay mode, though to very small extent, is

$$\pi^0 \rightarrow \gamma + e^+ + e^- \qquad (1.2\ \%)$$

In this mode, it appears as if one of the photon has converted into an electron-positron pair.

10.4.3 Mean Life Time (τ)

The average lifetime of positively charged pion has been determined measuring the time elapsed between the arrival π^+ pion and its decay. The mean life of π^+ pion has been found to be equal ιo 2.61×10^{-8} sec. The negative pion π^- generally reacts with nucleus before it decays, as such its lifetime in free state can not be measured. However, the meanlife time of π^- meson is measured by observing its decay as done in the case of π^+ pions. It is found that the average time is same for π^- pions. As such it has been accepted that the mean life time of charged pions $\pi^\pm$ is equal to 2.61×10^{-8} sec. since neutral pion π^0 decays under electro magnetic interaction its life time, as expected, is much shorter. The mean life time of π^0 pion has been determined by observing the decay of K^+ meson in nuclear emulsion ($K^- \rightarrow \pi^+ + \pi^0$). The life time of a π^0 meson is of the order of 10^{-16} sec. The accepted value $\tau(\pi^0) = 0.8 \times 10^{-16}$ sec.

10.4.4 Spin and Parity

The intrinsic spin of charged pions $\pi^\pm$ has been found to be zero. The spin of π^0 can not be determined directly, but it is known that π^0 decays into two photons, as such its spin should also be zero. The parity of the three pions is odd (–).

Pion is a triplet charge state as such its isospin is $T = 1$ with $T_3 = \pm 1$ for $\pi^{\pm}$ pions and $T_3 = 0$ for π^0 pion.

10.4.5 The Intrinsic Parity of the Pion

Consider that a beam of π^- mesons moves through liquid deuterium. Many pions, after being slowed down are captured by deuterons. The most probable reaction to occur is

$$\pi^- + d \rightarrow n + n$$

In the ground state, the orbital angular momentum of the π^- is zero and since the spin of pion is zero, the total angular momentum of the pionic (mesic) atom is that of the deuteron, which is $J = 1$. The angular momentum is to be conserved, therefore, the two neutrons must also have total angular momentum of 1. The possible states for the two neutrons, as permissible by Pauli's exclusion principle are:

Triplet states $S = 1$ (spins parallel) ${}^3P_0$, ${}^3P_1$ ${}^3P_2$, ${}^3F_2$, ${}^3F_3$, ${}^3F_4$ ...etc.

with $l = 1, 3, 5, \ldots$, etc.

Singlet states, $S = 0$ (spins anti parallel) : ${}^1S_0$, ${}^1D_2$, ${}^1G_4$ ...

With $l = 0, 2, 4, \ldots$, etc.

The only state with $J = 1$ is ${}^3P_1$ which must be the state of the two neutrons corresponding to an orbital angular momentum $l_n = 1$.

Let P_π, P_p and P_n be the parities of pion and proton and neutron in the deuteron and the orbital augural momentum of deuteron be l_d and the orbital angular momentum of the two neutrons be l_n. Then the conservation of parity in strong interaction demands that

$$p_\pi p_p p_n (-1)^{l_d} = p_n^2 (-1)^{l_n}$$

But the ground state of deuteron has $l_d = 0$ and $l_n = 1$, as shown above. Then $P_\pi\ P_p = -\ P_n$. Since the parity of a proton and a neutron is assumed to be same. (they are different charge states of a nucleon) Hence $P_\pi = -1$ and thus parity of π^- pion is odd (–). It is, therefore, assumed that π^+ pion which is the antiparticle of π^- pion must also have odd (–) parity. π^0 pion has odd parity, which is based on the fact that $\pi^- + d \rightarrow \pi^- + 2n^0$ process does not occur.

10.5 K-MESONS OR KAONS

Unlike pions Kaons or K-mesons are not predicted but they are observed in cosmic rays or produced in laboratories using high energy particle accelerators. In one of the cloud chamber photographs of cosmic rays, it was found from the decay of particles, that there are two particles having different decay modes but each having nearly same mass of around 1000 m_e. The particle decaying into three pions was termed as τ-meson and the other decaying into two pions was called as θ meson.

$$\tau \rightarrow 3\pi$$
$$\theta \rightarrow 2\pi$$

This behavior of the particle led to a puzzle which, for quite sometime, was known as τ-θ puzzle. It was found that both τ and θ have almost same mass and same average life time. Though both τ and θ are produced in strong interaction their mean lives of decay is around 10^{-8} sec which is fairly long on nuclear time scale ($\sim 10^{-23}$ sec). The puzzle could be resolved only on theoretical considerations. The decay products of each τ and θ are pions which have odd parity, and spin zero. Since parity is multiplicative, the spin parity of τ is 0^- and that of θ is 0^+. It was suggested by T. D. Lee and C. N. Yang that τ and θ were the same particle (which was actually named as Kaon). Having two modes of decay in which parity is not conserved. This idea of non-conservation of parity in weak interaction was later confirmed experimentally.

The τ and θ mesons are charged K-mesons. The Kaons are found to exist in two charged states $K^{\pm}$ and two neutral forms.

Kaons or K mesons are produced in strong interaction in nucleon-nucleon collision or in the pion-nucleon collision. When high energy protons or pions collide with nuclei kaons are produced. The possible reactions for their production are

$$p + p \rightarrow p + p + K^+ + K^-$$
$$p + p \rightarrow p + n + K^- + \overline{K}_0$$
$$p + \pi^- \rightarrow \mathring{\Lambda} + K^0$$

In these reactions, $\overline{K}_0$ is antiparticle to K_0

The K^+ and K^- mesons decay in at least six different modes (including the decay modes of τ and θ). As a matter of fact, the decay modes of K^- meson are regarded same as those of K^+ meson. Since K^- mesons are easily captured in matter (as π^- pions). The mass of K^+ meson is estimated from the Q-values of the decay modes. The rest mass of K^+ meson has been found to be

$$m\,(K^+) = 966 \text{ me and rest mass energy}$$

$$m\,(K^+)c^2 = 494 \text{ MeV.}$$

The rest mass and rest mass energy of K^- meson are taken to be same as those of K^+ meson.

The mean life time for decay of K^+ and K^- mesons is

$$\tau(K^{\pm}) = 1.24 \times 10^{-8}\text{sec.}$$

The spin parity of $K^{\pm}$ is 0^-.

[The negative sign ( ) for odd parity used as a super script over the spin zero (0)]

THE VARIOUS (SIX) DECAY MODES OF K$^{\pm}$ MESON WITH THEIR RELATIVE PROBABILITIES

$K^{\pm} \rightarrow \mu^{\pm} \nu_{\mu}$	~ 63%
$K^{\pm} \rightarrow \pi^{\pm} \pi^{0}$	~ 21%
$K^{\pm} \rightarrow 2\pi^{\pm}$	~ 6%
$K^{\pm} \rightarrow \pi^{0} e^{\pm} \nu_{e}$	~ 5%
$K^{\pm} \rightarrow \pi^{0} \mu^{\pm} \nu_{\mu}$	~ 3%
$K^{\pm} \rightarrow 2\pi^{0} \pi^{\pm}$	~ 2%$^{\pm}$

One finds that both leptonic and non-leptonic decays of $K^{\pm}$ take place.

The K^- meson is antiparticle to K^+ meson. The iso-spin for K^+ is $T = ½$ and $T_3 = ½$ (Table 10.1).

10.5.1 Neutral Kaon or K^o Meson

Curiously enough, like two charged kaons, two neutral forms of kaons have been observed. The two have different decay modes and also different mean life times. One decays into two pions with a life time of around 10^{-10} sec. and the other into three pions with a relatively longer mean life time (~ 10^{-8}sec.). Such dual characteristics led to their explanation with the help of symmetry principles (charge conjugation '*c*' and parity '*p*'). It is then regarded that there are two K neutral kaons K^o and $\overline{K^0}$, such that $\overline{K^0}$ is antiparticle of K_o but the two have different decay modes. The two together form a charge conjugate pair. The conjugate pair forms two mixed states (as explained quantum mechanically) such that one state has longer mean life time with three pions as decay products and the other has shorter life time with two pions as decay products.

Thus

$$K_L^o \rightarrow \pi^+ \pi^- \pi^0, 3\pi^0, \qquad \tau = 5.7 \times 10^{-8} \text{ sec.}$$

$$K_S^o \rightarrow \pi^+ \pi^-, 2\pi^0, \qquad \tau = 0.87 \times 10^{-10} \text{ sec.}$$

Here

$$K_L^o = \frac{1}{\sqrt{2}}(K^o + \overline{K}^0)$$

$$K_S^o = \frac{1}{\sqrt{2}}(K^o + \overline{K}^0)$$

where the symbols include the wave functions *e.g.*

$$|K_L^o\rangle = \frac{1}{\sqrt{2}}\left[|K^o\rangle + |\overline{K}^o\rangle\right]$$

$$|K_S^o\rangle = \frac{1}{\sqrt{2}}\left[|K^o\rangle - |\overline{K}^o\rangle\right]$$

10.6 η^o- MESON (ETA-NEUTRAL ETA MESON)

The η^o-meson was first observed in a bubble chamber where highly energetic π^+ meson beam (~1.2 BeV/c) was allowed to react with deuterons in liquid deuterium. The reaction is

$$\pi^+ + d \rightarrow p^+ + p^+ + \eta^o$$

The main modes of decay of η^o are by electromagnetic interaction into two photons [$\eta^o \rightarrow 2\gamma$, (31.4%)] and two photons plus a neutral pion [$\eta^o \rightarrow 2\gamma + \pi^o$, (20.5%)] other decays are through weak interactions, one decay mode is into three neutral pions [$\eta^o \rightarrow 3\pi^o$, (21%)] and other into two charged pions plus a neutral pion [$\eta^o \rightarrow \pi^+ + \pi^- + \pi^0$, (22.4%)].

The estimated rest mass of η^0 is $m(\eta^0)$ = 1074 m_e and the rest mass energy $m(\eta^0)\, c^2$ = 549 MeV. The spin parity of η^0 is 0^-. Its iso spin is $T = 0$ since only one state exist, $T_3 = 0$. Its antiparticle is η^0 itself. The mean life time τ for η^0 meson is

$$\tau(\eta^0) = 7 \times 10^{-19} \text{ sec.}$$

The characteristics of pions, kaons and η^o meson are given in Table (10.1) and their decay modes in Table (10.2) respectively.

10.7 LEPTONS

As mentioned earlier the main members of Lepton group are electrons ($\bar{e}$) negatively charged muons (μ^-), neutrinos (ν) and their respective anti particles, positron (e^+), positively charged muon (μ^+) and anti neutrino ($\bar{\nu}$). It is significant that each type of lepton is associatea with a neutrino *i.e.* the electron neutrino (ν_e) and the muon neutrino (ν_μ). Thus there are eight known leptons. The electron and positron and the corresponding neutrino and anti neutrino have been described earlier in (Chap.3). Here we describe the production and properties of muons.

10.7.1 Production of Muons

Muons are mostly produced in the laboratory in the weak decay of pions,

$$\pi^\pm \rightarrow \mu^\pm + \nu_\mu$$

In high energy nucleon-nucleon collision, pions are produced which on decay produce muons. Muons are also produced, to a small extent in the weak decay of K-mesons, and some hyperons.

$$K^\pm \rightarrow \mu^\pm + \nu_\mu$$

$$\overset{o}{\Lambda} \rightarrow \mu^- + p + \bar{\nu}_\mu$$

$$\Sigma^\pm \rightarrow \mu^\pm + n + \nu$$

Muons pairs ($\mu^+\mu^-$) can be produced by photons of energy around 211 Mev. Such photo production of pairs is like the production of electron-positron

($\bar{e}\,e^+$) pair by γ-rays. The negative muon is found to be very much similar to an electron,. There is no neutral muon

10.7.2 Muon-Decay

In the free state most of the muons decay according to the following decay mode

$$\mu^- \rightarrow e^- + \nu_\mu + \bar{\nu}_e$$

$$\mu^+ \rightarrow e^+ + \bar{\nu}_\mu + \nu_e$$

ν_μ and ν_e are muon and electron neutrinos and $\bar{\nu}_\mu$ and $\bar{\nu}_e$ are the corresponding muonic antineutrino and electronic anti neutrino. In the decay products the electron and positrons can be observed, but neutrinos can not be observed. As the tracks in the chamber or in emulsion show a continuous distribution of electron energy, the conservation of energy and momentum require the emission of three particles it may be seen that in the above decay scheme the lepton number L is conserved.

$$\mu^- = e^- + \nu_\mu + \bar{\nu}_e \qquad \mu^+ \rightarrow e^+ + \nu_e + \bar{\nu}_\mu$$

$$L = 1 \quad 1 \quad +1 \quad -1 \qquad L = -1 \quad -1 \quad -1 \quad +1$$

If the two neutrinos emitted in the muon decay are related (electron neutrino, ν_e and its anti neutrino $\bar{\nu}_e$ or muon neutrino ν_μ and its anti neutrino $\bar{\nu}_\mu$) then at some stage the two should annihilate producing γ-rays. Since no such gamma rays are ever observed in muon decay indicates that the two neutrinos are different. Hence one is electron-neutrino, ν_e and the other muon neutrino, ν_μ. The $\bar{\nu}_e$ and $\bar{\nu}_\mu$ are corresponding anti neutrinos.

10.7.3 Muon Life Time (τ)

The average or mean life time of free muons was initially determined from cosmic ray studies. Later on accelerators based experiments were designed where pions were produced and from the decay of these pions the time interval between the arrival of muons μ^+ and its subsequent decay into positrons was determined.

The negative muon μ^- has the same life time. Hence the mean life time of muon is

$$\tau(\mu\pm) = 2.2 \times 10^{-6} \text{ sec}$$

10.7.4 Muon-Mass

Like the life times, the early determination of muon mass was done from the kinematics of the tracks in the cloud chamber or photographic emulsion. The rest mass of both positive and negative muons has been found to be 206.8 m_e and the corresponding rest mass energy is 105.7 MeV.

TABLE 10.2: PARTICLE–DECAY MODES

	Particle	*Decay mode*	*Relative probability %*	*Half-life, s*
	Photon Graviton	Stable Stable		
Leptons	Neutrino Electron Muon	Stable Stable $\mu^- \to e^- + \nu + \overline{\nu}$		1.52×10^{-6}
	Pion	$\pi^+ \to \mu^+ + \nu$ $\to e^+ + \nu$ $\pi^0 \to \gamma + \gamma$ $\to \gamma + e^+ + e^-$	100 $\sim 10^{-4}$ 99 1	1.80×10^{-8} 6×10^{-17}
Mesons	Kaon	$K^+ \to \mu^+ + \nu$ $\to \pi^+ + \pi^0$ $\to 2\pi^+ + \pi^-$ $\to \pi^0 + e^+ + \nu$ $\to \pi^0 + \mu^+ + \nu$ $\to \pi^+ + 2\pi^0$	63 21 5.6 4.8 3.4 1.7	8.56×10^{-9}
		$K^0 \to \pi^+ + \pi^-$ $\to 2\pi^0$	35 15	6.0×10^{-11}
	Eta	$\eta^0 \to \gamma + \gamma$ $\to \pi^0 + \gamma + \gamma$ $\to 3\pi^0$ $\to \pi^+ + \pi^- + \pi^0$ $\to \pi^+ + \pi^- + \gamma$	33 20 20 5	$< 10^{-16}$
Baryons	Proton	Stable		
	Neutron	$\eta^0 \to p^+ + e^- + \overline{\nu}$		$\sim 7.0 \times 10^2$
	Lambda	$\mathring{\Lambda} \to p^+ + \pi^-$ $\to n^0 + \pi^0$	66 34	1.76×10^{-10}
	Sigma	$\Sigma^+ \to p^+ + \pi^0$ $\to n^0 + \pi^+$	53 47	5.6×10^{-11}
		$\Sigma^0 \to \mathring{\Lambda} + \gamma$		$< 7 \times 10^{-15}$
		$\Sigma^- \to n^0 + \pi^-$		1.1×10^{-10}
	Xi	$\Xi^0 \to \mathring{\Lambda} + \pi^0$		2.0×10^{-10}
		$\Xi^0 \to \mathring{\Lambda} + \pi^-$		1.2×10^{-10}
	Omega	$\Omega^- \to \mathring{\Lambda} + K^-$ $\to \Sigma^0 + \pi^-$	50 50	10^{-10}

*To obtain the decay of antiparticles, change all particles into antiparticles on both sides of the equations.

As mentioned earlier that muons are weakly interactiing particles and behave like heavy electrons. This fact has been examined through interaction of muons with matter and other particles.

Muonic-Atom (μ-Atom)

The muons have relatively longer life times as such it is possible that during its journey through matter, a negative muon (μ^-) may replace an electron temporarily forming a muonic atom. The radius of the first Bohr Orbit in such a μ-atom (muonic atom) will be smaller by a factor of ~1/207, the ratio of the masses of electron and muon. (m_e/m_μ). The muon in its bound state in a muon atom can come very close to the nucleus, specially in high Z substances resulting into its capture through electromagnetic interaction—a behaviour similar to the orbital electron capture. The capture of μ^- by a nucleus usually takes place from the first orbit of μ-atom (K-shell). The muon (μ^-) interacts with the proton in the nucleus producing a neutron and a μ-neutrino.

$$\mu^- + p \rightarrow n + \nu_\mu$$

The positive muon μ^+ mainly under goes decay in the matter, since it can not be captured inside the nucleus, because of Coulomb repulsion of the nucleus. Like positronium, a temporary combination of positron with electron moving round it, a transient system of (μ^+ e^-) known as muonuim has been observed. It is not stable. Its life time is around 2×10^{-6} sec. Unlike (e^+ e^-) annihilation in positronium, into photons no such annihilation of (μ^+ e^-) in muonium is observed. Such annihilation of (μ^+ e^-) into photons is prohibited because of the violation of leptonic number conservation.

Like ordinary electron, μ^- has magnetic moment and spin magnetic moment $\mu_\mu^- = 1$ Bohr magneton and spin $S_\mu = \frac{1}{2}\hbar$

10.8 FUNDAMENTAL INTERACTIONS

We have seen that a large number of elementary particles (~200) have been observed having such a complex behavior in their production and decay. It becomes necessary to formulate the various types of interactions (forces) involved between these particles. Such interactions have been found to fall under four general categories which in turn govern the behavior of all observable physical systems comprising the particles. These interactions in order of decreasing strength are (1) strong nuclear interactions (2) Electromagnetic interactions (3) Weak interactions and (4) gravitational interactions.

(*i*) The strong interaction was first manifested as the strong nuclear binding force between neutrons and protons inside the nuclei of atoms,. They operate upto a short range of the order of $\frac{\hbar}{m_\pi c} \approx 10^{-15}$ m. Other strong interaction processes are the scattering of mesons and nucleons by nucleons. Strong interactions involve mesons and baryons. The carriers of strong interactions between nucleons are Yukawa particles, the pions ($\pi^{\pm}$, π^0).

Now it is well established that hadrons are composed of more fundamental particles, known as Quarks. In the analysis of strong interaction between hadrons it is ultimately found to be the strong interaction between quarks and the field particle or the carrier of the interaction between quarks is through the exchange of gluons; a mass less particle.

(*ii*) The electromagnetic interaction operates between all electrically charged particles, and the carrier of the interaction being light quanta; photons. The range of electro magnetic interaction is infinite. Its strength is less than the strong interactions and is determined by the fine structure constant $\alpha = 1/137$. Creation of electron positron pair, from gamma rays and annihilation of the e^+ e^- pair into gamma rays are examples of electro magnetic interaction.

(*iii*) The weak interaction differs from the first two in the sense that it does not produce attractive force. It is a weak nuclear interaction which is responsible for decay of elementary particles like muons, pions, kaons and certain hyperons. The beta decay of radioactive nuclei takes place under weak interaction. The beta decay of a free neutron is an example of weak interaction. The range of weak interaction is short ($\sim 10^{-17}$m) and as such the carriers of the weak interaction are heavy particles, termed as intermediate bosons. There are two intermediate bosons, $W^{\pm}$ and Z^o. Their mass is nearly 30 times the proton mass, the rest mass energy of $W^{\pm}$ bosons is around 81 GeV and that of Z^o is 93 GeV. The $W^{\pm}$ boson has charge equal to $\pm e$ and their spin is 1. While the Z-boson is electrically neutral and it also has spin 1.

A peculiar characteristic of weak interaction is that, parity is not conserved in weak interactions.

(*iv*) The fourth and the weakest of the fundamental interactions is the gravitational interaction. It operates between mass particles and massive bodies and has infinite range. It is presumed that graviton, a mass less spin 2 particle is a carrier of the gravitational force. It travels with speed of light but it has not been confirmed so far.

The characteristics of these four fundamental interactions are given in Table (10.3)

TABLE 10.3: FOUR FUNDAMENTAL INTERACTIONS

Interactions	*Particles Affected*	*Relative Strength*	*Range*	*Particles Exchanged*
Strong	Quarks	1	10^{-15} m	Gluons
	Hardrons			Mesons (Pions, Kaons)
Electromagnetic	Charged Particles	10^{-2}	Very long almost infinite	Photons
Weak	Quarks and Leptons	10^{-13}	10^{-17} m	Intermediate Bosons (*W*, *Z*)
Gravitational	All elementry to Massive bodies	10^{-39}	Very long, almost infinite	Gravitions, (have not been experimentally detected so far)

10.9 CONSERVATION LAWS AND QUANTUM NUMBERS FOR ELEMENTARY PARTICLES

All physical processes are subjected to certain conservation laws, which permit a certain process to occur and forbid the occurrence of other. The elementary particles, their production, decay and their reaction with other particles, are also governed by conservation laws or invariance principles. The Kinematic conservation laws are (1) conservation of momentum (2) conservation of angular momentum (3) conservation of energy. When particles move with high energy, moving with relativistic velocities, the rest mass energy should be considered for the conservation of energy. These three laws hold good for all the four types of interactions discussed above. In order to bring certain order in the complex picture of the elementary particle world, few other conservation laws were introduced. They are (4) conservation of Baryons (5) conservation of Leptons (6) conservation of charge (7) conservation of strangeness (8) conservation of iso-spin.

At this stage, one question comes up. Why so many conservation laws ? The answer lies in the complex behaviour of elementary particles. For example: consider the processes $\gamma \to e^- + p^+$, $\mathring{\Lambda} \to \overline{p^-} + \pi^+$ or $n \to \pi^0 + \gamma$ or $\pi^+ + p^+ \to \Sigma^+ + \pi^+$ which obey the first three conservation laws, even they comply with the conservation of conventional charge, but in actual practice these processes are not observed. What forbids them is explained by the other conservation laws (4) to (8). We shall now discuss these laws.

(*i*) **Conservation of Baryons:** Let us introduce a baryon quantum number *B*. Such that $B = +1$ for all baryon particles and $B = -1$ for anti baryon particles and $B = 0$ for all non baryon particles. The conservation demands that in any process, the total baryon number must remain unchanged (remains invariant). Thus neutrons, protons and hyperons have $B = +1$ and their corresponding anti particles have $B = -1$. Mesons which are non-baryon, have $B = 0$.

Consider the process of neutron decay

$$n \rightarrow p + e^- + \bar{\nu}$$
$$B = 1 \quad 1 \quad 0 \quad 0$$

We find that baryon number is conserved

Like wise in the process

$$\overset{\circ}{\Lambda} = \overline{p^-} + \pi^+$$
$$B = 1 \quad -1 \quad 0$$

Here the baryon number is not conserved, hence the process does not occur.

Like wise, consider the proton-proton collision resulting into an anti proton and three protons

$$p^+ + p^+ \rightarrow \overline{p^-} + p^+ + p^+ + p^+$$
$$B = 1 \quad 1 \quad -1 \quad 1 \quad 1 \quad 1$$

We find that baryon number is conserved. Here the charge is also conserved. This process has actually been observed experimentally for the production of anti protons employing protons with energy greater that 5.6 GeV to collide with protons at rest.

(ii) Conservation of Leptons: Like baryons, a Leptonic quantum number L is alloted to Leptons, such that $L = 1$ for leptons and $L = -1$ for anti leptons. For non-lepton particles, $L = 0$. As we have seen that amongst the group of Leptons we have electrons, muons and corresponding neutrino.

A lepton number $L_e = 1$ is assigned to electron e^- and to electron neutrino ν_e and $L_e = -1$ to their anti particles; positron e^+ and electron anti neutrino $\bar{\nu}_e$. Likewise $L_\mu = +1$ for muon μ and muon neutrino ν_μ and $L_\mu = -1$ for anti muon and muon anti neutrino $\bar{\nu}_\mu$. For all other non-Leptonic particles $L = 0$. Thus for any process the total number of Leptons must remain constant (invariant or unchanged) L_e and L_μ, however conserve separately.

Example: β-decay of a neutron

$$n \rightarrow p^+ + e^- + \bar{\nu}$$
$$L_e = 0 \quad 0 \quad +1 \quad -1$$

Lepton number is conserved. Another examples of Lepton conservation is pion decay, muon decay and pair production.

$$\pi^- \rightarrow \mu^- + \bar{\nu}_\mu$$
$$L_\mu = 0 \quad +1 \quad -1$$

$$\mu^- \rightarrow e^- + \nu_\mu + \nu_e$$
$$L_\mu = +1 \quad 0 \quad 1 \quad 0$$
$$L_e = \quad 0 \quad 1 \quad 0 \quad 1$$

$$\gamma \rightarrow e^+ + e^-$$
$$L_e = 0 \quad -1 \quad +1$$

However, the decay process $\mu^- \rightarrow e^- + \gamma$, though theoretically possible, the conservation of Lepton number is violated. Hence does not occur, as is clear from the following, the electron Lepton number being different from muon lepton number

$$\begin{array}{llll} & \mu^- = & e^- + & \gamma \\ L_e = & 0 & 1 & 0 \\ L_\mu = & 1 & 0 & 0 \end{array}$$

(iii) Conservation of strangeness: In order to characterize hyperons and kaons Gell-Mann and Nishijima attached (independently) a quantum number, S, called as strangeness quantum number to the particles. The particles and anti particles have opposite strangeness. The conservation of strangeness required that in any process the total strangeness must remain the same. The assignment of S to different particles is given in table (10.1). In pion-nucleon collision, for example, two strange particles (a hyperon and a kaon) are produced according to the reaction (strong interaction).

$$\begin{array}{lllll} & \pi^- + & p \rightarrow & \Sigma^- + & K^+ \\ S = & 0 & 0 & -1 & 1 \end{array}$$

Here strangeness is conserved.

Also in the collision of high energy protons that involve strong interaction, strangeness is conserved.

$$\begin{array}{lllllll} & \pi^- + & p \rightarrow & \mathring{\Lambda} + & K^o + & p^+ + & \pi^+ \\ S = & 0 & 0 & -1 & +1 & 0 & 0 \end{array}$$

It may be noted that hyperon or kaon is never produced alone, because, then strangeness is violated in strong interaction. For example.

$\pi^- + p$ never produce $n + K^o$. Since strangeness of pions and nucleons is zero where as for Kaon (K^o) it is +1.

When hyperons decay, strangeness is not conserved. The decay being through weak interaction, hence strangeness is not conserved in weak interaction. It may, however change by 0, ± 1 ($\Delta S = \pm 1$) as is clear from the following examples.

$$\begin{array}{llll} & \mathring{\Lambda} \rightarrow & p^+ + & \pi^- \\ S = & -1 & 0 & 0 \end{array}$$

$$\begin{array}{llll} & \Sigma^+ \rightarrow & p^+ + & \pi^0 \\ S = & -1 & 0 & 0 \end{array}$$

$$\begin{array}{llll} & \Xi^+ \rightarrow & \mathring{\Lambda} + & \pi^- \\ S = & -2 & -1 & 0 \end{array}$$

The decay of hyperons and kaons proceed through weak interaction and accordingly the process is extremely slow. On a time scale it is around 10^{-10} sec, compared to the time of the order of 10^{-23} sec, involved in their production.

(iv) Conservation of Hypercharge: A new parameter, termed as hyper charge Y is introduced alongwith strangeness. It is defined as $Y = B + S$ where B is baryon number and S is strangeness. Hypercharge is conserved in strong interaction. The hypercharge assigned to different particles is given in table (10.1). Wherever B and S are conserved, the hypercharge is also conserved. Hypercharge is thus conserved or is invariant under strong and electro magnetic interaction and not conserved in weak interactions.

(v) Conservation of Iso-Spin: From the list of particles-anti particles. we find that certain members can be grouped having same mass and interaction, but different charges. Such groups are called "multiplets" and it is possible to think that the members of the multiplets as representing different charge states of a certain single entity. Thus a quantum number T is associated with each multiplet, known as isotopic spin quantum number or iso-spin quantum number. And $(2T + 1)$ gives the number of members in the multiplet. Thus $T = 0, \frac{1}{2}, 1$ gives singlet, doublet, triplet states having one, two or three charge states in the respective multiplets. The $\mathring{\Lambda}$, Ω^- and η^o each has $T = 0$ since each occur in only a singlet state on the other hand the nucleons has $T = \frac{1}{2}$, it is a doublet, the two charge states are proton and neutron. The pion multiplet is $T = 1$ state. It is a triplet charge state, since $(2x1 + 1) = 3$. (π^+, π^0 and π^-) are the three members of this triplet state. As described earlier the isotopic spin T bears some resemblance with angular momentum vector, such that its components along say Z-direction correspond, to the quantum number T_3. The possible values of T_3 are given by $T_3 = T\ T - 1 \ldots 0 \ldots -T$ so that T_3 is half integral if T is half integral and integral or zero. if T is integral. T is same for particle and anti particle multiplets but opposite value for T_3. For example, the iso spin for nucleon is $T = \frac{1}{2}$ which means that T_3 can be $+\frac{1}{2}$ or $-\frac{1}{2}$. Customorily $T_3 = \frac{1}{2}$ represents proton and $T_3 = -\frac{1}{2}$ a neutron. In the case of pions, $T = 1$ and $T_3 = 1, 0$ and -1 such that $T_3 = 1$ for π^+, $T_3 = 0$ represents π^0 and $T_3 = -1$ represents π^-. In this way iso-spin quantum number can be assigned to other particles.

When certain particles are involved in an interaction then the T and T_3 both together may or may not be conserved. It is found that the total iso-spin T and T_3 both are conserved in strong interactions. In the electro magnetic interactions or weak interaction T is not conserved.

The charge Q of a particle can now be defined in terms of hyper charge and T_3 as

$$Q = e(T_3 + Y/2) = e\left(T_3 + \frac{B+S}{2}\right)$$

This shows that each allowed orientation of the iso-spin vector T is connected to the charge of the particle. For nucleon multiplet $T_3 = ½$ for proton $B = 1$ and $S = 0$ so that $Q = e$ while for neutron $T_3 = -½$ and $B = 1$ and $S = 0$ such that $Q = 0$.

The charge Q and baryon number B are conserved in all interactions. Also whenever S is conserved T_3 must be conserved.

The conservation of conventional electric charge is a manifestation of the invariance of the laws of electro magnetic field (Maxwell's equations) with respect to a "gauge" transformation.

10.10 CONSERVATION LAWS AND SYMMETRY

The conservation laws are related to the invariance of physical entities under certain types of symmetry properties of the physical systems.

(i) **Conservations of Linear Momentum and Space Translation:** It holds good for all types of interactions since the total momentum of an isolated system is constant, it remains invariant under translation of space *i.e.* translation symmetry is exhibited. The conservation of linear momentum is a result of translation invariance of the laws describing the system.

(ii) **Conservation of Angular Momentum and Space Rotation:** The physical laws describing any system remains invariant under rotation in space-that is space is isotropic. It does not depend on the orientation of the system in space. Since the total angular momentum of a system is constant, it follows therefore, that the conservation of angular momentum follows from the rotational symmetry of the space in which the system is described.

The conservation of angular momentum holds for all types of interactions.

(iii) **Conservation of Energy and Time Translation:** The invariance under time translation means that a physical system, once it is formed, does not depend on time. It remains in the state at any other time. The laws which describe the physical system will remain invariant under translation of time. Thus with respect to any chosen origin of time conservation of energy of a system is a consequence of the invariance of the laws under time translation. It holds good for all types of interactions.

(iv) **Conservation of Parity:** It is related to the invariance of physical laws under inversion of the space coordinates such that x, y, z are replaced by $-x$, $-y$, $-z$ (this is equivalent to combined reflection and rotation through 180°). The conservation of parity, thus, states that the parity is conserved in a process if the mirror image (parity operation) of the process is also a process occurring in nature.

Parity is conserved in all processes where strong and electro magnetic interactions are involved.

Parity operator P is assigned values +1 and −1 since two successive parity operation on a system brings back the system to its initial state such that $P^2 = 1$, so that $P = \pm 1$ If $\psi(\vec{r})$ is the state function of a system, then

$$P\psi(\vec{r}) = \pm\psi(-\vec{r})$$

when $P\psi(\vec{r}) = +\psi(-\vec{r})$, the parity of the state is said to be even and for $P\psi(\vec{r}) = -\psi(-\vec{r})$, the parity is said to be odd. Thus the operator P has two eigen values +1 (even parity), −1 (odd parity). The conservation of parity demands that the parity operator must commute with the Hamiltonian of a system. If H is Hamiltonian operator, then $\vec{P}\cdot\vec{H} = \vec{H}\cdot\vec{P}$. For strong and electromagnetic interaction, parity is conserved, as such P is said to be a good quantum number.

Every elementary particle has been assigned an intrinsic parity. The proton, neutron, electron, Λ, Σ, Ξ, Ω are assigned positive or even parity. The pions, K-mesons and η^0 meson, have odd or negative parity as they are involved in strong interactions with nucleons.

The non-conservation of parity in weak interaction has been described in β-decay (Chap. 3). It was further explained by Lee and Yang in the case of decay of Kaons. The problem of parity conservation in the case of kaons which at that time was known as $\tau - \theta$ puzzle was solved by Lee and Yang with the suggestion that parity is not conserved in weak interactions. They suggested an experiment which was performed by C.S. *Wu* confirming the violation of parity in weak interactions (Chap. 3).

(v) Conservation of Charge Parity (C-Parity): The symmetry operation which corresponds to the conservation of charge parity is called as charge conjugation. It, therefore, means that the total charge Q comprising all types of charges, electronic (e) baryonic (B) hypercharge Y and leptonic (L) is replaced by $-Q$, that is e, B, $Y$and L are changed to their opposite charges under charge conjugation. It also amounts that under charge conjugation, all particles are changed to their respective anti particles. Charge conjugation, when applied to neutral particles and photons, does not change them, but the symmetry operation leaves the particle to itself. Such particles are said to be self-conjugate γ, π^0, η^0 are examples of self-conjugate particles. Positronuim (e^+ e^-) is a self adjoint system. Charge conjugation or C-parity is conserved in strong and electromagnetic interaction. Like parity P it is not conserved in weak interactions. Also like P operator, the C-operator has values (+1) for even C-parity and (-1) for odd C-parity. For π^0 pion C is even, since charge conjugation of π^0 pion transforms it to π^0. But the decay of pions is not C-parity invariant. A π^+ pion becomes π^- pion on charge conjugation as such π^+ decay becomes π^- decay on charge conjugation and the neutrino in the two cases are different. The antineutrino in π^- decay has different helicity than the helicity of neutrino in π^+ decay. (Chap. 3)

(vi) Time Parity *T* or Time Reversal: It is related with the change of time (t) with ($-t$) in the wave function. If a physical process remains same under time reversal, it is said to be invariant and time parity is said to be conserved. Under time reversal, physical quantities like velocity, momentum and angular momentum are reversed but the rate of change of momentum or the force is invariant under time reversal because both p and t go over to ($-p$) and ($-t$) on time reversal. The equations of motion remain invariant under time reversal. Strong, electromagnetic and weak interaction are conserved under time reversal. Time reversal transforms wave function to its complex conjugate *i.e.*

$$T\psi(x, t) = \psi^*(x, -t)$$

The electric field E is invariant but the magnetic field B is not invariant under time reversal, because velocities and hence currents reverse on time reversal. The Lorentz force $\vec{F} = e(\vec{E} + \vec{\upsilon} \times \vec{B})$ remains invariant under time reversal.

10.10.1 Combined CP-Invariance

We have seen that weak interactions violate both parity inversion and charge conjugation, as has been shown by the example of decay of π^+ pion. However if one performs, say space inversion to the π^+ decay and then charge conjugation to this transformed system then a final state is obtained which actually occurs in nature. This is shown in the following sequence of decay.

$\pi^+ \rightarrow \mu_L^+ + \nu_R$	$\pi^+ \rightarrow \mu_R^+ + \nu_L$	$\pi^+ \rightarrow \mu_L^- + \nu_L$
	P-inversion	*C*-inversion
(*i*)	(*ii*)	(*iii*)

The sequence indicates that (*iii*) is the result of combined *CP* operation and the decay so obtained does actually occur. The suffix (*L*) and (*R*) indicates the left-handedness and right handedness. Neutrino and antineutrinos have opposite helicities.

The order of operation *CP* or *PC* is immaterial in examining the combined effect.

It has been shown that parity is not conserved in the weak interaction (β-decay) of cobalt-60 nucleus (Chap. 3). But the same β-decay, when examined under *CP* operation, then it is found that even weak interaction is invariant under *CP* operation, as shown by the sequence (*i*), (*ii*) and (*iii*) for π^+ decay. However, there is an example in the decay of neutral K-meson (K^0) under weak interaction where even *CP* invariance is violated though the violation is of very small extent. Let us examine this decay of K^0 meson. The state of K^0

and $\overline{K^0}$ in weak interaction, will behave as if they constitute two states K_L^0 and K_S^0 represented as

$$|K_L^0\rangle = \frac{1}{\sqrt{2}}\left[|K^0\rangle + |\overline{K^0}\rangle\right]$$

$$|K_S^0\rangle = \frac{1}{\sqrt{2}}\left[|K^0\rangle - |\overline{K^0}\rangle\right]$$

The wave functions of the respective states (K_L^0, K_S^0, K^0, $\overline{K^0}$) are associated with the corresponding symbols. The parity and charge conjugation operation on K^0 and $\overline{K^0}$ may be represented as $P|K^0\rangle \to -|K^0\rangle$, since kaon has (-ve) odd parity, (the sign of the wave function is changed). Now $-C|K^0\rangle \to -|\overline{K^0}\rangle$, since charge conjugation transforms particle into its anti particle.

Thus $$CP|K_L^0\rangle = CP\frac{1}{\sqrt{2}}\left[|K^0\rangle + |\overline{K^0}\rangle\right] = \frac{1}{\sqrt{2}}\left[CP|K^0\rangle + CP|\overline{K^0}\rangle\right]$$

$$= \frac{1}{\sqrt{2}}\left[|-\overline{K^0}\rangle - |K^0\rangle\right] = -\frac{1}{\sqrt{2}}\left[|K^0\rangle + |\overline{K^0}\rangle\right] = -|K_L^0\rangle$$

Since $CP|K^0\rangle \to -|\overline{K^0}\rangle$, $\left[\text{since } P|\overline{K^0}\rangle = -|\overline{K^0}\rangle \text{and} - C|\overline{K^0}\rangle = -|K^0\rangle\right]$

Likewise

$$CP\,|K_S^0\rangle = |K_S^0\rangle$$

Thus $|K_S^0\rangle$ is even while $|K_L^0\rangle$ is odd under *CP* operation. This invariance or symmetry should be applicable to their decay products also. The decay of $|K_S^0\rangle$ into $\pi^+ + \pi^-$ or $\pi^0 + \pi^0$ is permitted. This is consistent with the application of *C* and *P* to the $\pi^\pm$ system or $2\pi^0$ system. The K_L^0 particle, however can decay through $\pi^+ + \pi^- + \pi^0$ or $3\pi^0$ and application of *CP* to the either mode of the decay, will result into the change of sign of the wave function. Thus *CP* invariance holds whenever K_S^0 decays into two pions and K_L^0 into three pions. But experimentally it has been found that about 0.3% of the K_L^0 decays is through two pions, This, therefore, indicates that even *CP* is violated in weak interaction.

10.10.2 Combined CPT or CPT Theorem

This theorem states that all physical laws (all interactions) are considered to be invariant under the three combined operations *C P T*, charge conjugation (*C*), space inversion or parity (*P*) and time reversal (*T*). The order of operation is not important. The interesting feature of *CPT* invariance is that violation of any one is compensated by the combined effect of remaining two.

We have seen that *CP* invariance does not hold in the case of weak interaction. If time reversal symmetry is also violated in weak interactions then under such condition *CPT* operation will be invariant. Such a possible explanation have been accepted when it was found in experiments on the decay of neutral kaons that in about 0.3% of the cases time reversal may violate in weak interactions.

10.11 RESONANCES

The life of elementary particles is fairly long by particle time scale. (10^{-10} sec or so) and some of them are even stable particles. In addition to these, the laboratory studies for search of particles have led to the collection of information which indicates existence of very short lived particles. They are known as resonance particles or resonances. Their life time is so short (10^{-20} sec or less) they do not leave any track in the bubble or spark chambers. This makes their direct detection more difficult, but during production and decay. These resonances can be classified according to baryon number, hyper charge and isotopic spin and are represented by same symbols as those baryons and mesons.

Resonances are found both amongst hyperons and mesons. They are produced in high energy collision between hadrons and mesons by strong interactions. Also they are observed in scattering or reactions taking place between elementary particles obtained from particle accelerators.

The first resonance was discovered as a result of the analysis of proton antiproton annihilation. From the analysis of the energies of the products (pions) it was concluded that three pions amongst the five pions are from the decay of a short lived particle having a rest mass of about 548 MeV. This was designated as η^0 resonance. The process can be expressed as

$$p + \overline{p} \rightarrow \eta^0 + \pi^+ + \pi^-$$

$$\eta^0 \searrow \pi^+ + \pi^- + \pi^0$$

The life time of η^0 resonance is around 10^{-20} sec. Another example for the identification of resonance particle is, say, the, bombardment of proton by energetic π^+ pions and a certain reaction is studied

$$p + \pi^+ \rightarrow \pi^+ + p + \pi^+ + \pi^- + \pi^0$$

The effect of the interaction of pion with proton is to produce three new pions, when the number of such events observed is plotted against the total energy (Rest energy + Kinetic energy), the curve shows a strong peak at around 785 MeV and a some what weaker peak around 548 MeV. This shows that the reaction proceeds via the creation of an intermediate particle which can have mass either 785 MeV or 548 MeV.

Resonances decay by means of strong interaction which accounts for their very short lives. Some of the decays are

$$\eta_1^0 \to \pi^+ + \pi^- + \pi^0 \qquad \mathring{\Lambda}_1 \to \Sigma^0 + \pi^0$$
$$K_1^+ \to K^+ + \pi^0 \qquad N_1^+ \to \Sigma^0 + \pi^0$$

The scattering cross section for the scattering of high energy pions from protons is measured as a function of the energy of incident pion. A plot shows peaks at certain pion energies. In one of the experiment it was found that scattering of π+ from proton appeared for pion energy of ~200 MeV in Laboratory system and the rest mass energy can be calculated from the energy of the scattered particle (pion) corresponding to the peak (maximum) in the cross section. As the peak is an indication of the resonances, the average life time of the resonance is estimated from the uncertainty principle, $\tau = \Delta t = \frac{\hbar}{\Delta E}$, where ΔE is peak width at half maximum height. The Figure (10.1) shows cross section for scattering of π+ and π^- pions by protons. The peak at ~200 MeV energy when converted to centre of mass of pion-proton, the rest mass of the resonance particle comes around 1236 MeV. The width of the peak at half maximum height is around 120 MeV, which gives mean life of about 10^{-24} sec. Such a short life indicates strong interaction. Large number of resonances have been discovered and Fig. (10.2) shows the masses of the baryon resonances. Similar resonances have observed in mesons also.

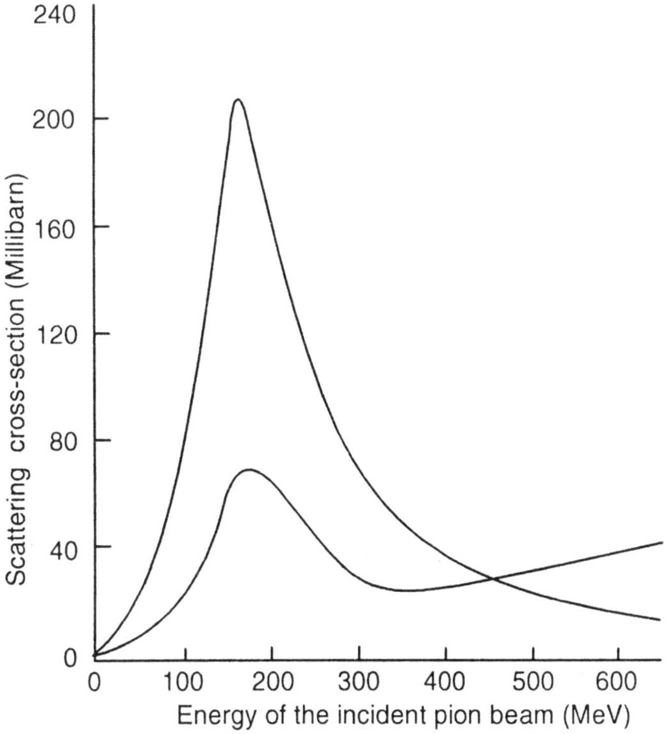

Fig. 10.1: Cross-section for scattering of positive and negative pions by protons (pion-proton resonance)

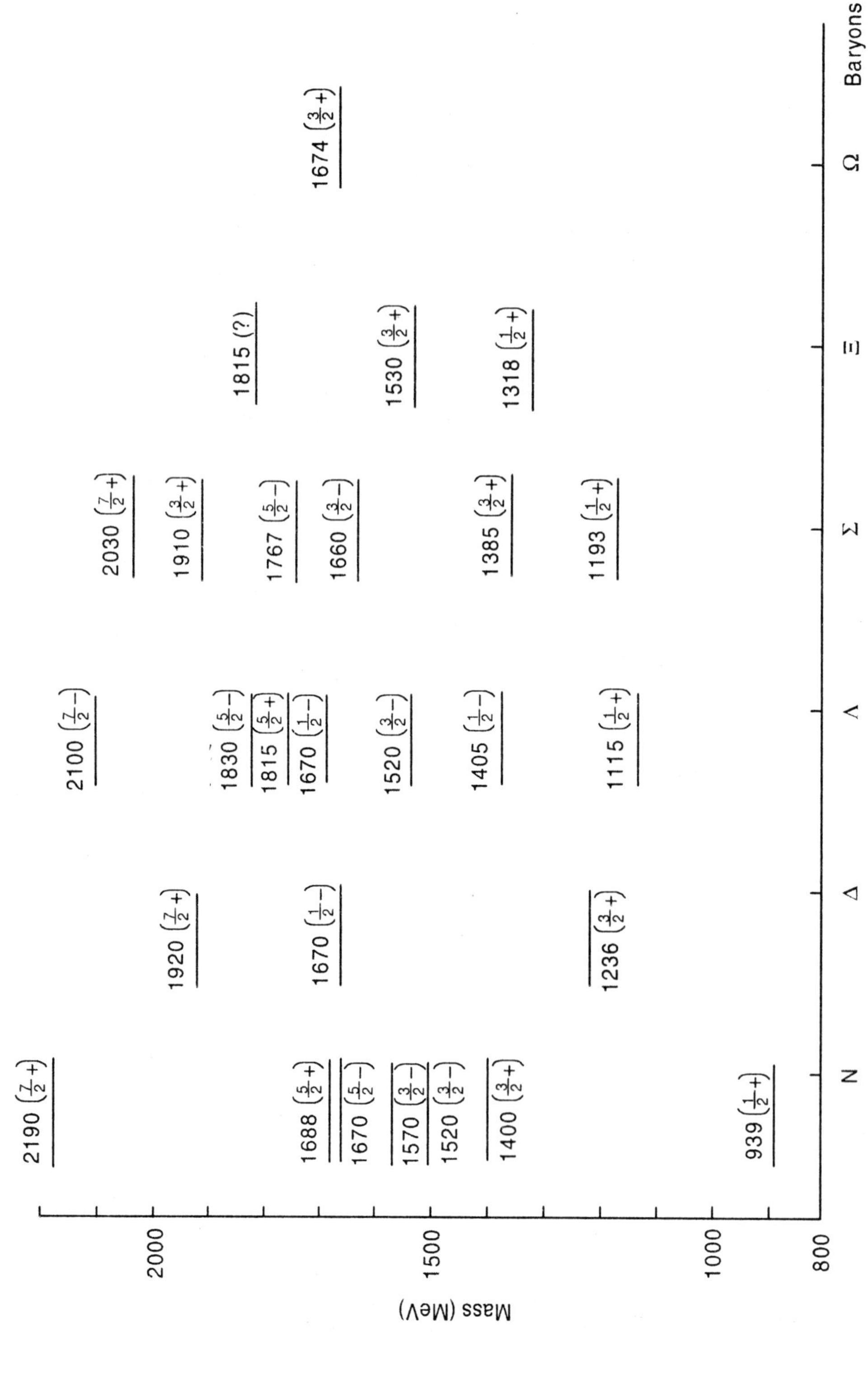

Fig. 10.2: Baryon resonances. The spin and parity of each resonance is given in the brackets

10.12 SYMMETRY CLASSIFICATION OF ELEMENTARY PARTICLES

We have seen that in general every conservation law reveals an underlying symmetry operation (Section 10.10). The set of operators that represents the symmetry constitutes a group. The mathematical theory of groups be regarded as the systematic study of symmetries. The group theory thus, has also been used to arrange the elementary particles into a number of families. The irreducible representation of a group consists of a number of states to which the symmetry operations can be applied. A group operation can transform any one of these states into another in the same representation. The fundamental representation is the one containing the smallest number of states for the particular group and the number oı states in this representation determines the basic dimensions of the group.

10.12.1 Unitary Symmetry

The charge independence of the nuclear force (strong nucleon-nucleon interaction) has led to believe that neutron and proton are two states of the same particle, known as a nucleon. It is, therefore possible to imagine that in the absence of electro magnetic interaction, there exists a group of symmetry operation, which can transform a proton into a neutron or a neutron into a proton. The neutron and proton would then form the fundamental representations of the group. The existence of such symmetry implies that there is a quantity called as iso-spin which remains conserved under strong interactions. The neutron and proton, being a state of fundamental representation, each has isospin (isotopic-spin) $T = ½$ and its third component T_3 (in a hypothetical isospin space) is equal to $+ ½$ for a proton and $T_3 = - ½$ for a neutron. The mathematics of iso-spin is identical to that which describes angular momentum in quantum mechanics. The algebraic structure is defined by the properties of 2×2 matrices which are unitary and which have a determinant equal to +1. This group of matrices is called *SU* (2), it is an abbreviation for "special unitary" (symbol *S* for special and *U* for unitary). The number 2 in the bracket indicates, dimensions of the group. The unitary group is special because a restriction reduces by unity the number of operators in the group. Thus in *SU* (2) group, instead of $2 \times 2 = 4$ operators, there are three operators. The group is said to have three generators.

In *SU* (2), multiplets exist such that they are $(2T + 1)$ in number and all members of the multiplet have the same isospin *T*. The $(2T + 1)$ substates in the multiplet have equally spaced values of T_3 going from $T_3 = + T$ to $T_3 = -T$ with no states missing and these sub-states are identical except for the electric charge (or T_3). For example for pions $T = 1$ and $(2T + 1)$ is equal to 3, so that there are three charge states of pions, with $T_3 = +1$ for π^+, $T_3 = 0$ for π^0 and $T_3 = -T$ for π^-. The iso spin is strictly conserved in strong interactions, the members (sub states) of the multiplet would differ in charge and T_3 and not in mass. The neutral member π^0 in the case of pions and the neutron differ in mass from $\pi^\pm$

pions and proton. This mass difference arises because of electric charge; the *SU* (2) symmetry is violated in the electro magnetic interactions and iso-spin is not conserved in electromagnetic interactions. It is found that symmetry operations group *SU* (2) breaks in the case of weak interactions as well.

10.12.2 The SU(3) Symmetry or the Eight Fold Way

The first successful attempt of classifying the hadrons and resonance states resulted from the extension of the *SU*(2) symmetry to the special unitary group *SU*(3) in three dimensions. Here also multiplets exist that are known as super multiplets. In 1961 M. Gell Mann and Y. Ne'eman independently proposed a scheme where *SU*(3) could be applied into hadron where both iso-spin *T*, and hyper charge Y ($Y = B + S$) are conserved. Since, a three dimensional unitary group has nine ($3 \times 3 = 9$) operators, having determinant equal to +1. but being special the restriction reduces this number by unity leaving eight operators. Gell Mann has therefore named the scheme as eight fold way (after the Great Budha's eight fold path to nirvana through eight right actions).

Gell Mann thus proposed that group of eight baryons should exist in a supermultiplet of the *SU*(3) symmetry, each having spin half and even (+ parity). These baryons are n, p, Λ, Σ^0 Σ^+ Σ^- Ξ^0 Ξ^- like wise there should be group of eight mesons each having spin 0 and odd (–) parity, in the family of supermultiplet of *SU*(3) symmetry. These are K^+, K^0, π^0, π^+, π^-, η^0, $\overline{K}^-$, $\overline{K}^0$. There is a third ten member supermultiplet (decuplet) consisting of baryons each having 3/2 spin and even (+) parity, they are all resonance particles except, Ω^- hyperon. As a matter of fact, Ω^- was predicted by this scheme of classification which was discovered in 1964, confirming the validity of Gell Mann's octet symmetry or the eight fold way.

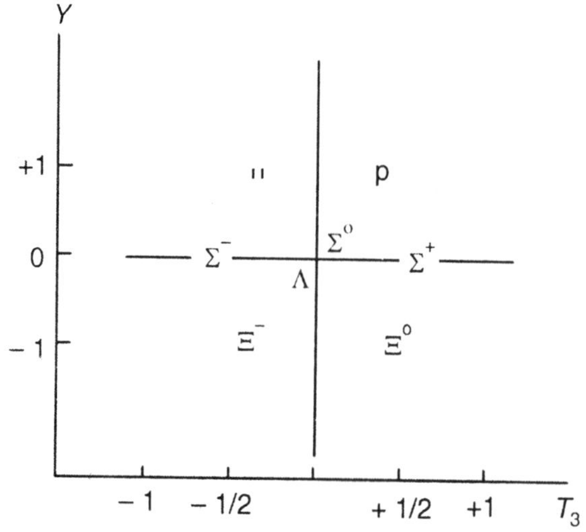

Fig. 10.3: The SU (3) octets of baryons ($J^P = ½^+$)

This $SU(3)$ octet of baryons and mesons and decuplets (10) of baryons are shown in the Figures (10.3), (10.4) and (10.5) respectively. The neutral baryons (Σ^0, $\mathring{\Lambda}$) and the mesons ($\pi^- \eta^0$) lie at the origin of ($T_3 - Y$) axes (of the hypothetical case). in Figs. (10.3) and (10.4) respectively.

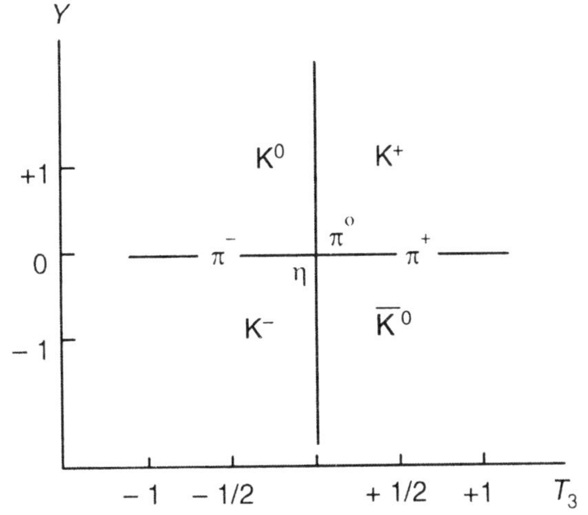

Fig. 10.4: The SU(3) octets of mesons ($J^P = O^-$)

The members of each supermultiplet would all be identical in the absense of any interactions. The octet symmetry is (in mass, charge and strangeness) broken partly when strong interaction operates. And the member of the multiplet each having same charge is further splitted under the action of electromagnetic interactions. As an example let us consider the basic baryon state. The applications of the strong interaction splits it into four components say N (nucleon), Λ, Σ and Ξ, each member having different strangeness. The electromagnetic interactions, further splits each member according to its T_3 value or the charge *i.e.* in the terms of the isospin multiplets. Th isospin of N and Ξ are both half and as such each of them has two components. For Σ baryon $T = 1$, so it has three components, Σ^+ ($T_3 = +1$), $\Sigma^0(T_3 = 0)$ and Σ^- ($T_3 = -1$) and $T = 0$ for $\mathring{\Lambda}$ as such it does not split under electromagnetic interactions. The mass difference between the isospin multiplets is much smaller than the mass difference between supermultiplets obtained from strong interactions, this is so because the strong interaction is more powerfull than the electromagnetic interaction. For example the mass difference between the neutron and the proton is about 1.3 MeV whereas it is around 176MeV between N and $\mathring{\Lambda}$ particles.

10.13 THE QUARK MODEL

All the known hadrons where found to fit into SU(3) families of 1, 8 or 10 members. The success of the eight fold way and its further analytical understanding lead Gell Mann and independently G. Zweig to propose that all hadrons are infact composed of even more elementary constituents. Gell Mann called these constituent particles as quarks. These quarks are of three types (*i*) The *u* (for "up") quark carrying a charge + 2/3 *e* and strangeness zero, (*ii*) the *d* (for "down") quark carries a charge – 1/3 *e* and strangeness zero and (*iii*) the *s* (for "strange") quark having a charge – 1/3*e* and strangeness $S = -1$. To each quark (*q*) there corresponds an antiquark ($\bar{q}$) with the opposite charge and strangeness.

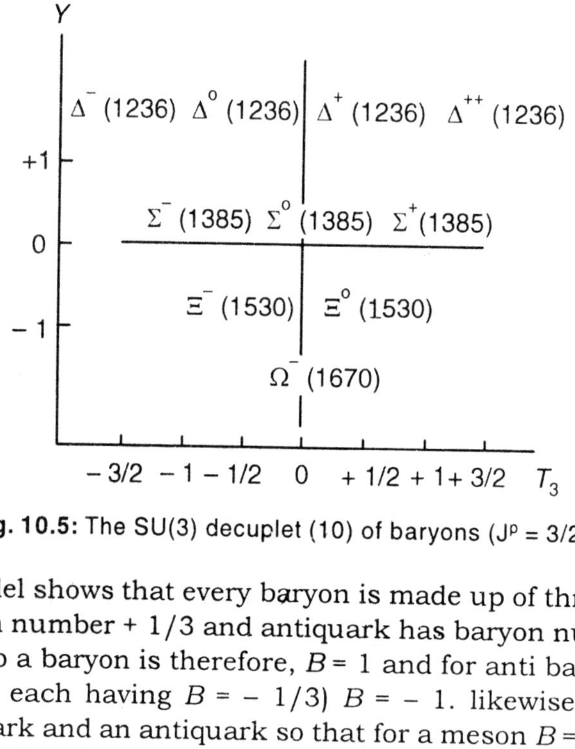

Fig. 10.5: The SU(3) decuplet (10) of baryons ($J^p = 3/2^+$)

The quark model shows that every baryon is made up of three quarks. Each quark has baryon number + 1/3 and antiquark has baryon number – 1/3. the baryon number fo a baryon is therefore, $B = 1$ and for anti baryon (made from three antiquarks each having $B = -1/3$) $B = -1$. likewise every meson is composedof a quark and an antiquark so that for a meson $B = 0$ quarks have a spin half. They are fermions as an example a proton has three quarks (uud), so the baryon number $B = 1/3 + 1/3 + 1/3 = 1$. The electric charge is $(2/3 + 2/3 - 1/3)\,e = 1e$, spin (↑↑↓) = + 1/2 and strangeness $S = 0 + 0 + 0 = 0$. similarly it is possible to tabulate the composition of say π^+ meson which is composed of quark (*u*) and anti quark ($\bar{d}$).

The quark model posed a serious theoretical problem. The presence of two or three quarks of the same kind in a particular particle (for instance two *u* quarks in a proton or three *s* quarks in an Ω^- hyperon) violates the Pauli's exclusion principle because the quarks are fermions and should obey the exclusion principle. In 1964, O. W. Greenberg proposed a way out of this problem. He suggested that quarks are not only of three types (*u*, *d* and *s*), but each of this has an additional property called as 'colour'. Each quark has three colours namely red, green and blue. The anti-quark colours are anti-red, anti-green and anti-blue. The term 'colour' and the red, blue and green should not be confused with traditional colours bearing these names. These are simply labels used to denote three additional properties associated to quarks. Association of the colour hypothesis satisfies the Pauli's exclusion principle, because two or three otherwise identical quarks present in a particle, each will have a different colour and then no two will be identical.

Besides the three quarks (u, d and s), three more quarks are suggested in order to have a significant analysis of the symmetries. These are charm (c), top (t) and bottom (b). Thus in all there are 6 quarks. The various characteristics of these quarks are given in the table (10.4).

The quark exists inside the hadrons, the search to observe quarks has been on for the last 30 or more years. Some experiments were carried out in the late 60's and in early 70's in America at Stanford linear accelerator center and in Europe at CERN. These experiments on inelastic scattering of the high energy electrons (10 GeV) from protons supported the quark model of the protons.

Attempts are being made in the various laboratories of the world to observe quarks in free state. High energy proton beams are being used for the purpose but no convincing results so far been made available. Recently in 1995 claim has been made for the experimental observation of top quark in proton and neutron.

The quarks are massive particles, interacting strongly within hadrons. The strong interaction between quarks is through the exchange of particles called gluons. Gluons are massless particles and travel at the speed of light. The emission and absorption of the gluons by quark changes the quark's colour. The theory to explain the interaction between quarks is known as the quantum chromodynamics.

TABLE 10.4: QUARK CLASSIFICATION (QUANTUM NIMBERS)

Quark	T	T_3	B	S	Y	Q	c	b	T	Mass (Gev.)
u (up)	½	+ ½	1/3	0	1/3	2/3	0	0	0	0.39
d (down)	½	− ½	1/3	0	1/3	− 1/3	0	0	0	0.39
s (strange)	0	0	1/3	− 1	− 2/3	− 1/3	0	0	0	0.51
c (charm)	0	0	1/3	0	1/3	+ 2/3	1	0	0	1.55
b (bottom)	0	0	1/3	0	1/3	− 1/3	0	1	0	≈ 5.4
t (top)	0	0	+ 1/3	0	+ 1/3	+ 2/3	0	0	1	≈ 20

The quantum number T_3, B, S, Y, Q, c, b, t for antiquarks are opposite (reversed) to those of quarks.

EXERCISES

1. Evaluate the quantum numbers T, T_3, B, S and Q for the particles $\mathring{\Lambda}$, $\overline{\Omega}$, $\overline{\Sigma}$.
2. Classify the following processes in terms of the type of interactions.
   (*i*) $\pi^- + p \rightarrow \mathring{\Lambda} + K^0$
   (*ii*) $\pi^0 \rightarrow \gamma + \gamma$
   (*iii*) $\Lambda \rightarrow p + \pi^-$
   (*iv*) $K^0 \rightarrow \pi^+ + \pi^-$
3. Consider the process
   $\pi^- + p \rightarrow \mathring{\Lambda} + K^0$ and
   (*a*) Determine the possible values of T for
   π^-, p and $\mathring{\Lambda}$, K^0 particles
   (*b*) Determine the value of T for the conservation of isotopic spin
   (*c*) Examine the conservation of T_3 and S
4. Examine the conservation laws (conservation of T, T_3, S and Y) in the following processes.
   (*i*) $p^+ + p^+ \rightarrow p^+ + \mathring{\Lambda} + K^+$
   (*ii*) $K^- K^- + p^+ \rightarrow \Omega^- + K^+ + K^0$
   (*iii*) $\pi^+ \rightarrow \mu^+ + \nu + \overline{\nu}$
5. Why must the quarks in a hadron have different colors? Hence write down quark composition of (*i*) π^- (*ii*) Ω^- (*iii*) Σ^0 and (*iv*) K^+

APPENDIX-A

RADIATION HAZARDS AND PROTECTION

Anybody working with radioactive sources or an accelerator should be familiar with the hazards of radiations. The radiations emitted from radioactive sources and accelerators cause damage to the body. The cosmic rays coming from space and radiations emitted from the natural radioactivity present in lime, concrete and sand also affect the body to some extent. While nothing can be done about the radiations from the environment, one should observe necessary precautions while working in the presence of such radiations.

A.1 Radiation Hazards

When a charged particle enters the body tissue, it produces ionization. The electrons thus produced may have enough energy to produce further ionization and cause chemical dissociation of the molecules. Ultimately all the energy of the incoming particle is converted into heat and energy of chemical dissociation. When chemical dissociation takes place in a vital component of a cell, it dies. When a large number of cells of the body are destroyed, it causes weakness and loss of vitality. If by chance the DNA chain in a cell gets distorted the cell may multiply very rapidly, forming a malignant tumour.

The damage to the body due to radiations from external sources is much less than that due to ingested sources. The range of α-rays in tissue is less than a millimeter and entering the body from outside, they are stopped in the outermost layer of the skin, causing little damage. Continuous energy β-rays from radioactive sources loose most of their energy in about two millimeters of tissue and are stopped mostly in the skin. If a radioactive source is ingested, it is distributed in different parts of the body according to its chemical nature. Radioactive Strontium and Radium having properties similar to Calcium, are preferentially absorbed in the bone, while Iodine goes into the thyroid gland.

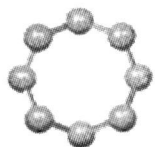

The radiations emitted from the absorbed sources produce severe damage in the neighbouring tissue.

The absorption of gamma-rays takes place by three processes *i.e.* photoelectric effect, Compton effect and pair production. In tissue photoelectric effect predominates for gamma-rays below 100 keV energy. Pair production which is possible for gamma-ray energies above 1.02 MeV is significant only above 10 MeV energy. Compton effect is the most probable process by which the gamma-rays interact with the body. As the body is predominantly composed of light elements, absorption cross-seciton above 100 keV energy is small and the gamma-rays affect the whole body almost uniformly. A good fraction of the gamma-rays pass through the human body without interacting. In all the three processes of interaction, the gamma-rays produce high energy electrons which in turn produce ionization in the tissue resulting in biological damage.

Neutrons are also capable of causing severe biological damage. When neutrons collide with hydrogen atoms, protons with fairly large energy are knocked out. These protons when produced in the tissue, produce intense ionization causing damage. The neutrons could also react with the nitrogen atoms present in the aminoacids of the body and produce energetic protons by the following reactions,

$$^{14}_{7}\text{N} + ^{1}_{0}n \rightarrow ^{14}_{6}\text{C} + ^{1}_{1}\text{H}$$

The proton is emited with high energy and produces ionization in the tissue. The recoiling ^{14}C nucleus breaks the molecular bonds in the cells. The low energy neutrons or thermal neutrons which are encountered mostly near atomic reactors, are easily absorbed by different atoms in the tissue, giving rise to high energy gamma-rays. These captured gamma-rays interact with the tissue and produce damage.

Some of the serious health hazards caused by heavy exposure to radiations are

(1) **Genetic Mutation:** When a gene in the DNA molecule of a cell is distorted or changed, the same distortion is reproduced in the subsequent divisions of the cell. If the affected gene belongs to a cell taking part in reproduction, the progeny may show some mental or physical disability. The harmful genetic mutation gets transmitted to future generations.

(2) **Cancer:** When there is an uncontrolled multiplication of cells whose genes are distorted by ionization, undesirable growth or tumour may be formed.

(3) **Leucopenia:** When the number of white blood cells in the blood reduces, the resistance of the body to infections is lowered.

(4) **Bone Necrosis:** The ionization due to radiations damages the bone marrow which is responsible for producing the red blood cells.

(5) **Sterility:** This is due to the damage in the gonads.

(6) **Epilation:** Heavy exposure to radiations causes the falling of body hairs.

The diagnostic X-ray exposure of different parts of the body also produce tissue damage.

A.2 Radiation Doses

It is necessary to lay down the norms for permissible radiation exposure to which a person may be subjected. For this purpose it is necessary to understand the following terms often used in radiation dosemetry.

Rontgen: Abbreviated as 'R' is "that quantity of X-rays or gamma radiations such that the associated corpuscular emission per 0.001293 gm of air produces ions carrying 1 esu of quantity of electricity of either sign", Production of 1 esu of charge of either sign corresponds to the formation of $1/4.8 \times 10^{-10} = 2.08 \times 10^{9}$ ion pairs. The mass of 0.001293 gm of air corresponds to 1 cm^3 of air at 0°C and 760 mm of Hg pressure. Since the average energy required to produce an ion pair in air is about 32.5 eV, an exposure of $1R$ corresponds to the absorption of 83 ergs energy per gm of air.

Due to different amounts of energy required to produce an ion pair in different substances, exposure of $1R$ corresponds to an absorption of about 96 ergs/gm in water and tissue, 42 ergs/gm in fat and 900 ergs/gm in bone.

Dose: Measured in Rad (Radiation absorbed dose), it is equivalent to the dessipation of energy of 100 ergs/gm in the irradiated sample. It may be seen that dose of 1R is equivalent to 0.83 Rad in air, 0.96 Rad in tissue and 9 Rad in bone.

It may be noticed that while exposure in Rontgen is a measure of the quantity of the incident X-ray or gamma radiations, the dose in Rad is a measure of the energy dissipated by any radiations in the irradiated material.

Relative Biological Effectiveness

For the same dose, the biological damage has been found to depend upon the ionizing power (specific ionization) of the radiations. The relative biological effectiveness-(RBE) is an emperical factor, introduced to take into account the effect of the specific ionization caused by different radiations. It is based upon the effect of 0.2 MeV X-rays as standard. Thus

$$\text{RBE} = \frac{\text{Dose in Rad of 0.2MeV X – rays producing a certain biological effect}}{\text{Dose in Rad of given radiation producing the same boilogical effect}}$$

Table A.1 gives the values of RBE for different radiations. It may be mentioned that the present knowledge of the biological effectiveness of radiations of different specific ionization is rather meagre. For any range of the specific ionization, it is safer to use the higher values in the range of values of RBE.

TABLE A.1: RELATIVE BIOLOGICAL EFFECTIVENESS OF DIFFERENT RADIATIONS

Radiation	*Range of specific ionization (ion pairs/ micron of water)*	*Range of RBE*
γ-rays, *x*-rays, electrons, positrons	100	1
Thermal neutrons	100-650	1-5
Fast neutrons and protons	650-1500	5-10
Deuterons	1500	10-15
α-particles	1500-5000	10-20
Heavy ions	5000	20

Rem (rontgen equivalent men): It is dose in Rad multiplied by the RBE of the radiation. Thus

$$\text{Rem} = \text{Rad} \times \text{RBE}$$

Dose Rate: The units, Rad and Rem express the exposure received by a body over a period of time. For proper control, it is desirable to know the dose received per unit time. If a person is permitted to receive a dose equivalent of 100 (milli rem) evenly distributed over a 40 hour week, he can work in a radiation field where the radiation dose rate is 2.5 mrem/hour.

The permissible dose equivalent that a worker may be exposd in a 40 hour week has been a subject of a number of international conferences. It has been recommended that the whole body dose equivalent should not exceed 100 mrem per week. Different parts of the body are susceptible to radiation demage to different extent. The maximum permissible dose equivalent for different organs of the body is given in Table A.2. The maximum permissible flux of neutrons of different energies that correspond to dose rate of 2.5 mrem/hr is given in Table A.3

TABLE A.2: RECOMMENDED MAXIMUM PERMISSIBLE EXPOSURE TO DIFFERENT ORGANS

Organs	*Maximum permissible exposure*	
	rem/year	*mrem/hour*
Whole body, gonads,	5	2.5
Skin, bone, thyroid	30	15.0
Hands, forearm, feet, ankle	75	37.5
Other single organs	15	7.5

TABLE A.3: MAXIMUM PERMISSIBLE FLUX OF NEUTRONS OF DIFFERENT ENERGIES FOR DOSE RATE OF 2.5 mrem/hr

Neutron Energy (MeV)	*RBE*	*Permissible flux neutrons/sec/cm²*
2.5×10^{-8} (thermal)	3	680
0.1	8	115
1.0	10.5	19
5.0	7.0	15
10.0	6.5	17
50	5.0	10

It is observed that the same exposure, given at small dose rate, causes much less biological damage than that at high dose rate. It is therefore desirable that the permissible yearly dose is distributed evenly in time. This is especially recommended for women in the reproductive age group. In pregnant women the exposure to the embryo in the first two months may be less than one rem but for the next seven months the accumulated dose should not exceed one rem.

A whole body exposure of 25 rem can change the blood count and an exposure of 100 rem can cause radiation sickness *e.g.* nausea, vomiting and blisters on the skin. Half of the persons getting exposure of 400-500 rem may die within one month. An exposure of 600 rem is almost invariably lethal.An exposure of 50000 rem is required to kill batceria and 10^6 rem to inactivate viruses.

To obtain the radiation dose rate due to γ-rays at a point, one should know the energy, flux and the absorption coefficient of the γ-rays. As an example the radiation dose rate at a distance of one meter from a Curie of a radioactive source emitting gamma rays of energy 1 MeV is 0.45 rad/hr.

Traditionally Curie is used as the unit of activity of a radioactive source. A Curie is the activity of 1 g of radium in equilibrium with its daughter products, and is equal to 3.7×10^{10} disintegrations per second. In the S.I. system, the unit of activity is Becquerel after the discoveror of radioactivity Antoine Henry Becquerel. One Becquerel of activity represents one disintegration per second in the radioactive source. This unit is very small and one generally employs mega Becquerel (1 MBq = 10^6 disintegrations/sec) or giga Becquerel (1gBq = 10^9 disintegrations/sec) as the unit. Similarly in the S.I. system the unit of radiation dose is Sievert which is the amount of any radiation which produces the same biological effect as produced by the absorption of an energy of one Joule of X-ray or γ-ray by one kilogram of the body tissue. Thus

1 Sievert = 100 rem.

The weekly dose should normally not exceed one milli Sievert.

A.3 Monitoring of Radiations

Human body cannot sense radiations as it senses heat or light. One has to rely on radiation measuring instruments to measure the dose rate. Radiation monitoring is done employing ionization chambers or G.M. counter. Photographic films are commonly used to monitor the dose of radiations received by a person over a period of time.

A radiation monitor using G.M. counter as detector is inexpensive and simple to construct. This radiation monitor gives only a rough indication of the dose rate. The G.M. counter has a dead time of about 200 μsec and cannot be used where the counting rate is above 10^5 counts per minute. A radiation monitor employing an ionization chamber is most reliable, as it directly measures the charge produced by the ionizing radiation inside the chamber volume. The ionization chamber may have a thin window so that the β-rays may also be detected. The current flowing through the ionization chamber can be calibrated to read the radiation dose rate. It is possible to construct a radiation monitor employing a plastic scintillation counter. Such a monitor could measure the dose rate due to X-rays, γ-rays and neutrons.

Pocket ionization chamber is commonly employed to measure the integrated radiation received by the user. It resembles a fountain pen in appearance and worn in the pocket of the worker. Before using, the ionization chamber is charged to a certain potential using a charger. As the insulation used is extremely good the charge on the wire does not leak even for a few days. When the dosimeter is exposed to radiations, the ionization produced in the air inside the chamber discharges the wire in proportion to the ionization. With the electroscope fitted to the chamber, the residual charge on the wire can be directly read by the worker. The scale in the electroscope is calibrated to read milli rad. It is recommended that pocket dosimeter may be used in addition to other radiation monitoring device *e.g.* filmbadge.

A piece of X-ray film can be used to measure the integrated dose or exposure by the blackening produced by the ionizing radiations. The piece of *X*-ray film approximately 2 cm × 3 cm is covered with black paper. A part of the film is covered with a lead sheet also. After a week, the film is sent to a central processing laboratory where it is developed and the blackening produced is compared with standard films. This gives the exposure that the wearer of the film may have received. Wearing of film badges is mandatory for all persons working near X-ray machines, radioactive sources and accelerating machines. An employer who allows his employees to receive radiation exposures much above the prescribed dose, is liable to pay damages. The records of the exposure as recorded by the film badge is therefore a legal document.

Some crystalline materials when exposed to radiations give rise to either luminiscence or thermoluminiscence. Radiation dose meters can be built out of such materials. Such dosimeters are generally employed when the radiation dose rate is in kilo rads or mega rads.

APPENDIX–B

STATISTICS OF NUCLEAR COUNTING

In quantum mechanics one always calculates the average or expectation value of a physical quantity. The expectation value has significance only for a large ensemble. As an example, the law of decay for a large number N of radioactive particles is

$$-\frac{dN}{dt} = \lambda N$$

where $-dN/dt$ is the rate of decay of the particles, and λ the decay constant is the probabillity of decay of a radioactive atom in unit time. The above law holds good only if N is infinitely large. The life time T of the activity $T = 1/\lambda$ represents the average time for which a radioactive atom lives before decaying. A particular atom can live for any time ranging from zero to infinity, before undergoing decay. All phenomena which are described by a probability function are governed by the laws of statistics.

In practice one can not have an infinitely large ensemble and the number of events recorded in an experiment are finite. The experiment is usually repeated a number of times and a statistical analysis of the data is made to obtain the results. The statistical analysis is useful in calculating the most probable value of the quantity, one is trying to measure. It also helps in estimating systematic errors in the result, as they do not follow the laws of statistics.

In the following treatment, radioactive decay has been taken as the base, but it is equally applicable to all nuclear physics experiments in which a statistically variable quantity is measured.

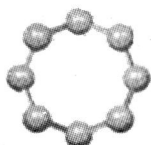

In the radioactive decay, atoms emit some radiations which are detected in a detector with a fixed solid angle. The detection efficiency of a detector is statistically variable to some extent. The number of counts recorded in the detector in a finite interval of time show statistical variations as does the number of decaying atoms. The laws of statistics can be applied to radioactive decay if the following conditions are satisfied.

(*a*) The probability of decay at any time, in a given time interval is same for all atoms of the radioactive source. This implies that all atoms undergoing decay are identical.

(*b*) The decay of an atom does not affect the probability of decay of another atom.

(*c*) The probability of decay of an atom in a given time interval is same for all times.

(*d*) The total number of atoms and the time interval for which observations are made, are reasonably large, so that statistical analysis of the data can be made.

The above conditions are satisfied if the radioactive isotope is pure (not contaminated with radioacctive atoms of other species) and has a half life very much longer than the time in which the experiment is performed. In such a situation the average rate of emission of radiations remains constant.

Let the average counting rate recorded in the detector over an infinitely long interval of time be r. In an interval of time t the true average number of counts expected is $m = rt$. It is observed that the number of counts recorded in time interval t is seldom equal to m. It is usually slightly different. Every time the observation is made, a different number of counts in the same time interval t is recorded. If in the jth observation the number of counts recorded is x_j then

$$\lim_{J \to \infty} \langle x_j \rangle = \bar{x} \approx m \qquad \ldots(B.1)$$

where $\bar{x}$ is the average of the number of counts recorded in j number of observations. In an actual experiment j is always finite. The statistical analysis of the data gives the most probable of value of m and the possible error in its value. If there are certain number of observations which give the recorded number of counts x_j very different from what is expected on the statistical analysis, one has to look for some systematic errors in the experiment.

With r as the true average counting rate, the probability $p_1(dt)$ that one count is recorded in an infinitesimally small time interval dt is given as

$$p_1(dt) = r\,dt \qquad \ldots(B.2)$$

The probability $p_1(dt)$ is extremely small. The probability that two or more would be recorded in time interval dt is vanishingly small, so that $p_1(dt) >> p_2(dt) >> p_3(dt)$. The probability that no counts is recorded in time dt is $p_0(dt)$. The total probability of recording either zero or some counts in time interval dt is one. so that

$$p_0(dt) + p_1(dt) + p_2(dt) + p_3(dt) = 1$$

neglecting $p_2(dt)$, $p_3(dt)$ and so on gives

$$p_0(dt) = 1 - p_1(dt) = 1 - rdt \qquad ...(B.3)$$

The probability that x number of counts are recorded in time interval $(t + dt)$ is equal to the sum of probability of recording x counts in time interval t and no count in interval dt and the probability that $(x - 1)$ counts are recorded in time interval t and one count is recorded in time interval dt. This gives

$$p_x(t + dt) = p_x(t) \cdot p_0(dt) + p_{x-1}(t) \cdot p_1(dt) \qquad ...(B.4)$$

$$= p_x(t)(1 - rdt) + p_{x-1}(t)\, r(dt)$$

$$= p_x(t) - rp_x(t)\, dt + rp_{x-1}(t)(dt) \qquad ...(B.4)$$

The probability $p_x(t + dt)$ can be expanded as

$$p_x(t + dt) = p_x(t) + \frac{d}{dt} p_x(t)dt \qquad ...(B.5)$$

equating equations $(B.4)$ and $(B.5)$ gives

$$\frac{dp_x(t)}{dt} - rp_x(t) = rp_{x-1}(t) \qquad ...(B.6)$$

If $x = 0$ we have

$$\frac{dp_0(t)}{dt} - rp_0(t) = 0, \qquad p_{x-1}(t) \equiv 0 \qquad ...(B.7)$$

This yields

$$p_0(t) = e^{-rt} \qquad ...(B.8)$$

The solution of equation $B.6$ is the Poisson distribution function given as

$$p_x(t) = \frac{(rt)^x}{x!} e^{-rt}$$

$$= \frac{m^x}{x!} e^{-m} \qquad ...(B.9)$$

Thus $p_x(t)$ gives the probability of observing x number of counts when the true average number of counts is m for the same interval of time t. If the number of observations are finite, the true average m can be replaced by the average $\bar{x}$, so that

$$p_x(t) = \frac{\bar{x}^x}{x!} e^{-\bar{x}} \qquad ...(B.10)$$

The probability distribution given by $B.9$ is not symmetric about m. It is stretched towards higher values of x as shown in Fig. $B.1$. When the value of m is 10 or more the probability distribution of $(B.9)$ becomes symmetrical about m and can be replaced by the normal or Gaussian distribution.

$$p(x) = \frac{1}{\sqrt{2\pi D_x}} e^{-\frac{(x-m)^2}{2D_x}} \qquad \ldots(B.11)$$

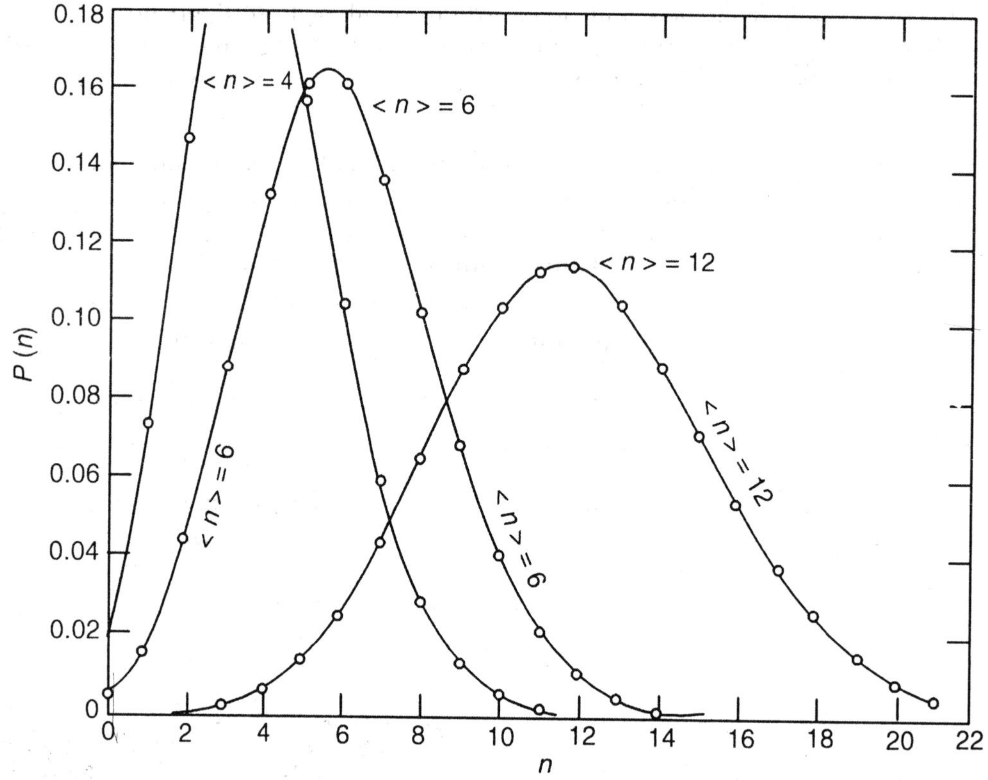

Fig. B.1: Distribution of probability to observe n number of counts when the average is $\bar{n} = 4$, 6 and 10. The distribution for $< n > = 4$ is markedly asymmetric

For a finite number N of observations the dispersion D also known as mean square deviation (denoted by σ^2) is defined as

$$\begin{aligned} D = \sigma^2 &= \frac{1}{N}\sum_{j=1}^{N}(x_j - m)^2 \\ &= <(x_j - m)^2> \\ &= <x^2> - m^2 \end{aligned} \qquad \ldots(B.12)$$

However, for finite number of observations, m is not known exactly and hence D can not be calculated. In such a case a good approximation for dispersion is

$$\sigma^2 = \frac{N}{N-1}\sum_{j=1}^{N}(x_j - \bar{x})^2$$

$$\sigma^2 = \frac{1}{N-1}\sum_{j=1}^{N}(x_j - \bar{x})^2 \qquad ...(B.13)$$

σ is also known as the standard deviation in $\bar{x}$. The above expression for σ^2 differs from equation $B.12$ only in its denominator $N-1$ and in the use of $\bar{x}$ in place of m. The term $N-1$ is correlated with the number of "degrees of freedom" in the dispersion of the data. For N number of observations, this number of degrees of freedom is $N-1$.

Employing equation $B.9$ for the probability distribution, the total probability to observe either zero or any number of counts can be shown to be unity, as expected.

$$\sum_{x=0}^{\infty} p(x) = \sum_{x=0}^{\infty} \frac{m^x}{x!} e^{-m}$$

$$= e^{-m} \sum_{x=0}^{\infty} \frac{m^x}{x!}$$

$$= e^{-m} \cdot e^{+m}$$

since $\sum_{x=0}^{\infty} \frac{m^x}{x!} = e^{+m}$

$$\therefore \sum_{a=0}^{\infty} p(x) = 1 \qquad ...(B.14)$$

It can also be shown that the true average, number of counts for infinite number of observations is $\bar{x} = m$.

$$\bar{x} = \sum_{x=0}^{\infty} x p(x) = \sum_{0}^{\infty} x \frac{m^x}{x!} e^{-m}$$

As $x = 0$ does not contribute to the average, one threfore has

$$\bar{x} = \sum_{x=1}^{\infty} \frac{m^x}{(x-1)!} e^{-m}$$

$$= e^{-m} \cdot m \sum_{x=1}^{\infty} \frac{m^{x-1}}{(x-1)!}$$

$$= e^{-m} \cdot m \sum_{l=0}^{\infty} \frac{m^l}{l!}$$

putting $l = x - 1$

$$= m \cdot e^{-m} \cdot e^{+m}$$

$$= m \qquad ...(B.15)$$

The value of mean square deviation σ^2 or D_x can be calculated as

$$\sigma^2 = D_x = \langle (x-m)^2 \rangle$$

$$\sigma^2 = \sum_{x=0}^{\infty}(x-m)^2 p(x)$$

$$= \sum_{x=0}^{\infty}(x)^2 p(x) - \sum_{0}^{\infty} 2mx\ p(x) + m^2 \sum_{x=0}^{\infty} p(x)$$

$$= \sum_{x=0}^{\infty}(x)\ p(x) = \bar{x} = m$$

and $$\sum_{x=0}^{\infty} p(x) = 1$$

$\therefore$ $$\sigma^2 = \sum_{x=0}^{\infty}(x)^2 p(x) - 2m^2 + m^2$$

$$= \sum_{x=0}^{\infty}(x)^2 p(x) - m^2 \qquad ...(B.16)$$

Now $$= \sum_{x=0}^{\infty}(x)^2 p(x) = \sum_{x=0}^{\infty} x^2 \frac{m^x}{x!} e^{-m}$$

$$= me^{-m}\sum x \frac{m^{x-1}}{(x-1)!}$$

$$= me^{-m}\left[\sum \{(x-1)+1\} \frac{m^{x-1}}{(x-1)!}\right]$$

$$= me^{-m}\left[\sum \left\{(x-1)\frac{m^{x-1}}{(x-1)!} + \sum \frac{m^{x-1}}{(x-1)!}\right\}\right]$$

Putting $l = x - 1$

$$= me^{-m}\left[\sum_{0}^{\infty} l \frac{m^l}{l!} + \sum_{0}^{\infty} \frac{m^l}{l!}\right]$$

$$= m^2 + m \qquad ...(B.17)$$

Substituting in equation (B.16) gives

$$\sigma^2 = m^2 + m - m^2 = m \qquad ...(B.18)$$

The value of the mean square deviation σ^2 is equal to the true average (infinite number of observations) m.

The Gaussian distribution given by equation (B.11) and its comparison with the Poisson distribution of equation (B.10) for $< x > = 20$ is shown in Fig. B.2. For large values of m (>16) the Gaussian and the Poisson distributions overlap.

The standard deviation $\sigma(= \sqrt{m})$ gives the probability that in a particular observations, the number of counts recorded would fall between the values $(m + \sqrt{m})$ and $(m - \sqrt{m})$, is about 68 per cent. The probability that the number of counts recorded in any single observations could lie between y and $y + dy$ is given as

$$p(y)dy = \frac{1}{\sigma\sqrt{2\pi}} e^{-\frac{(y-m)^2}{2\sigma^2}} \cdot dy \qquad \ldots(B.19)$$

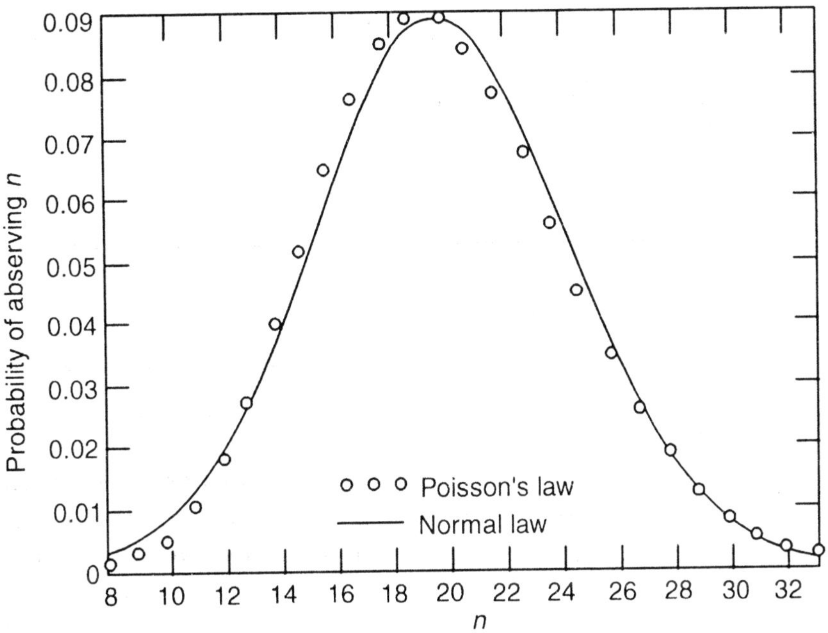

Fig. B2: Comparison of Poisson and Gaussian (normal) distribution for $< n > = 20$

The probability that a measured value of x would lie between $m + y_0$ and $m - y_0$ can be obtained by integrating equation ($B.19$) between the two limits. The probability p_{in} of the observed value of counts would lie within the limits $m + y_0$ and $m - y_0$ and the probability p_{out} of its lying outside the limits is given in table $B.1$ for different values of y/σ. Whereas σ is called the standard deviation, the value $0.6745\,\sigma$ is known as the probable error in m. If the number of recorded counts is 100 the standard deviation is $\sigma = 10$ and the probable error is 6.7.

If in an experiment there is only one observation and the recorded number of counts is x, then one has to take $\bar{x} = x$ and $\sigma = \sqrt{x}$.

Gaussians Distribution from Poisson Distribution

The Poisson distribution is given as

$$p(x) = \frac{m^x}{x!} e^{-m}$$

or $$\log p(x) = x \log m - m - \log (x!) \quad \text{...(B.20)}$$

According to the Stirlings relation

TABLE: B.1

y/σ	0.6745	1	1.5	2	3
P_{in}	0.5	0.683	0.866	0.988	0.99727
P_{out}	0.5	0.317	0.134	0.012	0.00273

$$x! = \sqrt{2\pi x}\, x^x e^{-x}\left(1 + \frac{1}{12x} + \ldots\right)$$

$$\doteq \sqrt{2\pi x}\, x^x e^{-x}, \quad x >> 1$$

This gives

$$\log(x!) = 1/2 \log 2\pi + x \log x - x \quad \text{...(B.21)}$$

Substituting in equation (B.20) gives

$$\log p(x) = x \log m - m + x - x \log x - 1/2 \log 2\pi x$$

on differentiating one gets

$$\frac{d \log p(x)}{dx} = \log m + 1 - 1 - \log x - \frac{1}{2x} \quad \text{...(B.22)}$$

$$= \log m - \log x, \quad \text{since } \frac{1}{2x} << \log x$$

Equating $\frac{d \log p(x)}{dx}$ to zero gives the maxima of the probability distribution at $x = m$. The probability of observing m number of counts in an observation is

$$\log p(m) = -\frac{1}{2} \log 2\pi m$$

or $$p(m) = \frac{1}{\sqrt{2\pi m}} \quad \text{...(B.23)}$$

Differentiating equation B.22 again we get

$$\frac{d^2}{dx^2} \log p(x) = -\frac{1}{x} + \frac{1}{2x^2}$$

since $\frac{1}{x^2} << \frac{1}{x}$

$$\therefore \quad \frac{d^2}{dx^2}\log p(x) = -\frac{1}{x} \qquad \text{...(B.24)}$$

The series expansion of log $p(x)$ can be written as

$$\log p(x) = \log p(m) + \frac{n-m}{1}\left(\frac{\partial}{\partial x}\log p(x)\right)_{n=m}$$

$$+\frac{(n-m)^2}{2!}\left(\frac{\partial^2}{\partial x^2}\log p(x)\right)_{n=m} \qquad \text{...(B.25)}$$

At $n = m$, $\frac{d}{dx}\log p(x) = 0$ sbstituting equation (B.23) and (B.24) in equation (B.25) gives

$$p(x) = \frac{1}{\sqrt{2\pi m}}e^{-\frac{(x-m)^2}{2m}} \qquad \text{...(B.26)}$$

This is the Gaussian or Normal distribution giving the probability of observing x number of counts when the true average number of counts for the same time interval is m.

In the Poisson distribution the variable x takes only integral values, while it can take continuosly varying (including fractional) values in the Gaussian distribution. This difference becomes insignificant when the values of m is very large.

Statistical Error in Functions of Statistically Variable Numbers

It is possible to derive the standard deviation in functions of numbers obeying Gaussion law of probability. The following cases are generally encountered.

(1) If x is a number following normal distribution and a is a constant then

$$z = ax$$

then

$$\sigma(z) = a\,\sigma(x) = a\sqrt{x} \qquad \text{...(B.27)}$$

(2) If x and y are two statistically variable numbers and

$$z = x + y$$

Then

$$\sigma^2(z) = \sigma^2(x) + \sigma^2(y) = x + y$$

$$\sigma(z) = \sqrt{x+y} \qquad \text{...(B.28)}$$

(3) If $z = ax + by$ when a and b are constants
Then

$$\sigma^2(z) = \sigma^2(ax) + \sigma^2(by)$$
$$= a^2x + b^2y$$

$$\sigma(z) = \sqrt{a^2x + b^2y} \qquad \text{...(B.29)}$$

(4) If a measurement is repeated a number of times and the counts recorded for same interval of time, then one does not know the true average m. If x_i is the number of counts recorded in the ith observation then the average number of counts in N observations is

$$\overline{x} = \frac{1}{N}\sum_{i=1}^{N} x_i$$

and

$$\sigma^2 = \sigma^2(\Sigma x_i) = \sum_{1}^{N} x_i = N\overline{x}$$

$$\therefore \qquad \sigma = \sqrt{N \cdot \overline{x}} \qquad \text{...(B.30)}$$

The fractional error δx is defined as

$$\delta(x) = \frac{\sigma(x)}{x} = \frac{\sqrt{x}}{x} = \frac{1}{\sqrt{x}}$$

hence

$$\delta(\Sigma x_i) = \frac{\sqrt{\Sigma x_i}}{\Sigma x_i} = \frac{1}{\sqrt{\Sigma x_i}} = \frac{1}{\sqrt{N} \cdot \sqrt{\overline{x}}}$$

and $$\sigma(\overline{x}) = \frac{\sigma(\Sigma x_i)}{N} = \frac{\sqrt{N\overline{x}}}{N} = \frac{\sqrt{\overline{x}}}{\sqrt{N}} \qquad \text{...(B.31)}$$

As discussed earlier in the absence of the knowledge of true average m, the standard deviation is approximately given as

$$\sigma(\overline{x}) = \frac{\sqrt{\overline{x}}}{\sqrt{N-1}} \qquad \text{...(B.32)}$$

A consequence of the above result is that one observation for five minutes interval for the number of counts would have nearly the same statistical error in the counting rate as five observations of one minute each. There is thus no advantage of taking five observations of one minute each over a single observation of five minutes duration. If, however, five observations of one minute each are taken, one can check if the individual observations fit with the normal distribution. A widely different observation may point to some systematic error in the measurement. It may be observed that the time of performing an experiment increases as square of the factor by which the accuracy of the measurement is increased.

(5) If $$z = x - y$$

$$\sigma^2(z) = \sigma^2(x) + \sigma^2(y)$$

$$= x + y$$

$$\therefore \quad \sigma(z) = \sqrt{x + y}$$

and $$\delta(z) = \frac{\sqrt{x + y}}{x - y} \qquad \ldots(B.33)$$

If the difference between x and y is small the statistical error in the difference is very large.

It may be observed that in both addition and subtraction of statistically varying numbers the dispersion σ^2 in the two quantities is added. To achieve a better accuracy in the experiment the effort should be made to reduce the larger dispersion in the same interval of time for performing the experiment. If x_l and x_b are two counting rates (*e.g.* from a radioactive source and the room background respectively) and the two are counted for time intervals t_l and t_b then for minimum error in the result, the two time intervals should be divided such that

$$\frac{t_b}{T - t_b} = \frac{t_b}{t_l} = \sqrt{\frac{x_b}{x_l}} \qquad \ldots(B.34)$$

where T is the fixed time interval in which the two sets of observations are completed. For small background, it should be counted for smaller interval of time and source counts should be recorded for longer interval of time. From the two observations the counting rates x_l and x_b can be determined. The quantity $x_l - x_b$ gives the counting rate due to the source only.

(6) If $$z = x \cdot y$$

Then $$\delta^2(z) = \delta^2(x) + \delta^2(y)$$

$$\delta(x) = \frac{\sigma(x)}{x} = \frac{1}{\sqrt{x}}; \quad \delta y = \frac{1}{\sqrt{y}}$$

so that $$\delta^2(z) = \frac{1}{x} + \frac{1}{y}$$

$$\therefore \qquad \delta(z) = \frac{\sigma(z)}{z} = \sqrt{\frac{1}{x} + \frac{1}{y}} \qquad \text{...(B.35)}$$

(6) If $z = x/y$

Then $\delta^2(z) = \delta^2(x) + \delta^2(y)$

$$\delta(z) = \frac{\sigma(z)}{z} = \sqrt{\frac{1}{x} + \frac{1}{y}} \qquad \text{...(B.36)}$$

Poisson Distribution in Time

If the true average counting rate in a detector is r, the average time interval between two counts in the detector is $1/r$. Probability that no count is recorded in a time interval t is $p_o(t) = e^{-rt}$. The probability that a count is recorded in time interval dt is $r\,dt$. Thus the probability that no count is recorded in time t and one count recorded in the next time interval dt is

$$p_o(t)dt = r\,e^{-rt}dt \qquad \text{...(B.37)}$$

The mean time interval between the two pulses is given as

$$\bar{t} = \int_0^\infty t \cdot re^{-rt}dt = \frac{1}{r} \qquad \text{...(B.38)}$$

Similarly

$$\bar{t}^2 = \int_0^\infty t^2 p_o(t)dt$$

$$= \int_0^\infty t^2 re^{-rt}dt$$

$$= r\int_0^\infty t^2 e^{-rt}dt$$

$$= 2/r^2$$

The dispersion in time t is D_t where

$$D_t = \bar{t}^2 - \bar{t}^{-2} = \frac{2}{r^2} - \frac{1}{r^2} = \frac{1}{r^2} \qquad \text{...(B.39)}$$

The standard deviation in time interval between two pulses is

$$\sigma_t = \frac{1}{r} \qquad \text{...(B.40)}$$

and fractional variation in time is

$$\frac{\sigma_t}{\bar{t}} = \frac{\frac{1}{r}}{\frac{1}{r}} = 1$$

Thus the time interval between two consequtive pulses may vary by an interval equal to the mean time interval itself.

The probability that after recording one count the next count is recorded after an interval of time T is

$$= \int_T^{\infty} p_0(t)dt = \int_T^{\infty} re^{-rt}dt$$

$$= e^{-rT} \quad \ldots(B.41)$$

This shows that the probability for one count following the previous count in time T increases as T reduces as shown in Fig. B.3.

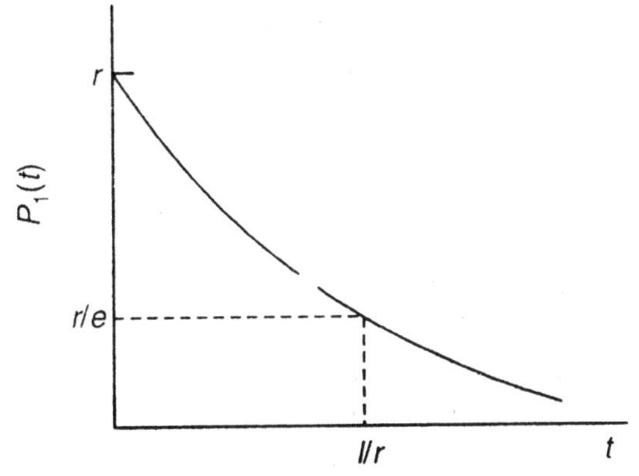

Fig. B.3: Probability of a count following other after a time interval t. r is the average counting rate

Effect of Scaling

Generally the pulses from a radiation detector are counted in a scaling circuit. This electronic circuit gives an output pulse after a predetermined input number of pulses-say s pulses. s is known as the scaling factor. It has been shown that for a counting rate r the average time interval between two pulses from the detector is $\bar{t} = 1/r$ and the fractional fluctuation in this time is $\sigma_t / \bar{t} = 1$. After scaling the average time interval between two out put pulses of the scaling circuit (scaler) is $\bar{t} = s/r$ and the variance in time interval is $\sigma^2(t) = s/r^2$ giving the fractional fluctuation

$$\frac{\sigma(t)}{\bar{t}} = \frac{1}{\sqrt{s}}$$

It is thus seen that the scalar output is more evenly spaced in time in comparison to the input pulses.

Analysis of Experimental Data

Whenever an experiment is performed, there is an effort to obtain some significant information which can be correlated with the physics involved. Physical measurements always have some statistical errors and it is necessary to estimate them. The most common method of analysing experimental data is the method of least squares.

Let an experimentally observed quantity y be a function of an independent variable x (*e.g.* counting rate as a function of time in the radioactive decay experiment). Thus

$$y = f(x)$$

To see whether a relationship between y and x exists, the observed value of y is plotted as a function of x. If a relationship seems to exist, one can find the same, on the assumption that the same relationship exists at the regions of x where measurements have not been made. One can postulate y as the polynominal function of x, so that

$$y = f(x) = a + bx + cx^2 + \ldots \qquad \ldots(B.42)$$

In the least square fit method, one determines the values of the constants a, b, c such that the sum of the squares of the deviation of the observed value y_i in the ith measurement and the calculated value of y_i is minimum. Thus one tries to minimise the expression σ^2.

$$\sigma^2 = \Sigma(y_i - f(x_i))^2 \qquad \ldots(B.43)$$

To make a least square fit one assumes that the variable x for any particular measurement is known without any uncertainty. If the observations of y_i have different degrees of accuracy a weight factor w_i is assigned to each value of y. The weight factor w_i is inversely proportional to the mean square deviation σ_i^2 in the observed value of y_i so that

$$w_i = 1/\sigma_i^2 \qquad \ldots(B.44)$$

For a complicated function $f(x)$ or number of terms in the polynomial of equation (B.42) the minimisation of σ^2 of equation (B.43) is done numerically. There are computer programmes for the same. In many experiments the data can be fitted using one, two or three terms of the polynomial (B.42). We shall discuss here only simple functions $f(x)$.

(a) Suppose the function is

$$y = a \qquad \ldots(B.45)$$

This represents a constant value of y. One minimises

$$\sigma^2 = \sum_{i=1}^{N}(y_i - a)^2$$

with respect to 'a' so that

$$\frac{d}{da}\Sigma(y_i - a)^2 = 0$$

or $$\sum_{i=1}^{N} 2(y_i - a) = 0$$

or $\quad \Sigma y_i - Na = 0$

or
$$a = \frac{\sum\limits_{1}^{N} y_i}{N} = \bar{y} \qquad \text{...(B.46)}$$

The coefficient is the arithmatic mean of all the N observations of y.

(*b*) For the function

$$y = bx \qquad \text{...(B.47)}$$

The function to be minimised is

$$\sigma^2 = \sum_{i=1}^{N} (y_i - bx)^2$$

so that

$$\frac{\partial}{\partial b}\left\{\Sigma (y_i - bx_i)^2\right\} = 0$$

or
$$\frac{\partial}{\partial b}\left\{\Sigma (y_i^2 + b^2x_i^2 - 2bx_iy_i)\right\} = 0$$

or
$$\Sigma (2bx_i^2 - 2x_iy_i) = 0$$

or
$$b = \frac{\sum\limits_{i=1}^{N} x_iy_i}{\Sigma x_i^2} \qquad \text{...(B.48)}$$

deleting the subscript i one can write

$$b = \frac{\Sigma xy}{\Sigma x^2} \qquad \text{...(B.49)}$$

(*c*) For function

$$y = a + bx \qquad \text{...(B.50)}$$

The function is be minimised is

$$\sigma^2 = \sum_{i=1}^{N} (y_i - a - bx_i)^2 \qquad \text{...(B.51)}$$

The expression $B.51$ has to be minimised with respect to both a and b so that one gets

$$\frac{\partial}{\partial a}\left\{\Sigma (y_i - a - bx_i)^2\right\} = 0 \qquad \text{...(B.52)}$$

and
$$\frac{\partial}{\partial b}\left\{\Sigma (y_i - a - bx_i)^2\right\} = 0 \qquad \text{...(B.53)}$$

Equation $B.52$ gives

$$aN + b\,\Sigma x_i - \Sigma y_i = 0 \qquad \ldots(B.54)$$

Equation $B.53$, yields

$$a\sum x_i + b\sum x_i^2 = \sum x_i y_i \qquad \ldots(B.55)$$

Solving equation $(B.54)$ and $(B.55)$ one gets

$$a = \frac{\sum y_i \sum x_i^2 - \sum x_i \sum x_i y_i}{N\sum x_1^2 - (\sum x_i)^2} \qquad \ldots(B.56)$$

$$b = \frac{N\sum x_i y_i - \sum x_i \sum y_i}{N\sum x_i^2 - (\sum x_i)^2} \qquad \ldots(B.57)$$

If w_i is the weight assigned to the ith observation, it effectively means that the ith observation has been repeated w_i times. In such a case it can be shown that

$$D \cdot a = \sum w_i x_i^2 \sum w_i y_i - \sum w_i x_i \sum w_i x_i y_i \qquad \ldots(B.58)$$

$$D \cdot b = \sum w_i \sum w_i x_i y_i - \sum w_i x_i \sum w_i y_i \qquad \ldots(B.59)$$

where,

$$D = \sum w_i \sum w_i x_i^2 - (\sum w_i x_i)^2 \qquad \ldots(B.60)$$

When the experimetnal observations have equal weight ($w_i = 1$) the probable error p_a in a and p_b in b is given by

$$p_a = \sqrt{\frac{\sum x_i^2}{N\sum x_i^2 - (\sum x_i)^2}} \cdot py_o \qquad \ldots(B.61)$$

$$p_b = \sqrt{\frac{N}{N\sum x_i^2 - (\sum x_0)^2}} \cdot py_o \qquad \ldots(B.62)$$

where,

$$py_o = 0.675\sqrt{\frac{\sum (y_i - a - bx_i)^2}{N - 2}} \qquad \ldots(B.63)$$

(d) In the case of a quadratic functions

$$y = a + bx + cx^2 \qquad \ldots(B.64)$$

The sum of the squares of deviation σ^2 is minimised with respect to the three unknowns a, b and c. One thus gets the equations

$$\frac{\partial}{\partial a}\left[\Sigma\left(y_i - a - bx_i - cx_i^2\right)^2\right] = 0 \qquad \text{...(B.65)}$$

$$\frac{\partial}{\partial b}\left[\Sigma\,(y_i - a - bx_i - cx_i^2\right] = 0 \qquad \text{...(B.66)}$$

$$\frac{\partial}{\partial c}\left[\Sigma\,(y_i - a - bx_i - cx_i^2\right] = 0 \qquad \text{...(B.67)}$$

The above three equations yield the following-

$$aN + b\sum x_i + c\sum x_i^2 = \sum y_i \qquad \text{...(B.68)}$$

$$a\sum x_i + b\sum x_i^2 + c\sum x_i^3 = \sum x_i y_i \qquad \text{...(B.69)}$$

$$a\sum x_i^2 + b\sum x_i^3 + c\sum x_i^4 = \sum x_i^2 y_i \qquad \text{...(B.70)}$$

Solving the above equations gives

$$a \cdot D = \begin{vmatrix} \sum y_i & \sum x_i & \sum x_i^2 \\ \sum x_i y_i & \sum x_i^2 & \sum x_i^3 \\ \sum x_i^2 y_i & \sum x_i^3 & \sum x_i^4 \end{vmatrix} \qquad \text{...(B.71)}$$

$$b \cdot D = \begin{vmatrix} N & \sum y_i & \sum x_i^2 \\ \sum x_i & \sum x_i y_i & \sum x_i^3 \\ \sum x_i^2 & \sum x_i^2 y_i & \sum x_i^4 \end{vmatrix} \qquad \text{...(B.72)}$$

$$c \cdot D = \begin{vmatrix} N & \sum x_i & \sum y_i \\ \sum x_i & \sum x_i^2 & \sum x_i y_i \\ \sum x_i^2 & \sum x_i^3 & \sum x_i^2 y_i \end{vmatrix} \qquad \text{...(B.73)}$$

and

$$D = \begin{vmatrix} N & \sum x_i & \sum x_i^2 \\ \sum x_i & \sum x_i^2 & \sum x_i^3 \\ \sum x_i^2 & \sum x_i^3 & \sum x_i^4 \end{vmatrix} \qquad \text{...(B.71)}$$

The calculation of the probable errors in *a*, *b* and *c* are rather time consuming and are best done by a computer programme.

APPENDIX-C

MULTIPOLE EXPANSION OF RADIATION FIELD

The Maxwell's equations for vacuum are written as

$$\vec{\nabla} \times \vec{H} = \varepsilon_o \frac{\partial \vec{E}}{\partial t} \qquad \ldots(C.1)$$

$$\vec{\nabla} \times \vec{E} = -\mu_o \frac{\partial \vec{H}}{\partial t} \qquad \ldots(C.2)$$

$$\vec{\nabla} \cdot \vec{H} = 0 \qquad \ldots(C.3)$$

$$\vec{\nabla} \cdot \vec{E} = 0 \qquad \ldots(C.4)$$

The symbols have their usual meaning. The solution of these equations describes electromagnetic waves with superposition of incoming and outgoing waves of different frequencies. The electric and the magnetic field of the radiation field is given as

$$\vec{E} = \int_0^\infty \left\{ \vec{E}_r(r,w)e^{iwt} + \vec{E}_r^*(r,w)e^{-iwt} \right\} dw \qquad \ldots(C.5)$$

$$\vec{H} = \int_0^\infty \left\{ \vec{H}_r(r,w)e^{iwt} + \vec{H}_r^*(r,w)e^{-iwt} \right\} dw \qquad \ldots(C.6)$$

Substituting the above values of $\vec{E}$ and $\vec{H}$ in equations ($C.1$) and (2) gives

$$\vec{\nabla} \times \vec{H} = -iw\varepsilon_0 \vec{E} \qquad \ldots(C.7)$$

$$\vec{\nabla} \times \vec{E} = iw\mu_0 \vec{H} \qquad \ldots(C.8)$$

Taking curl of the above equation and putting $k^2 = w^2\varepsilon_0\mu_0$ gives

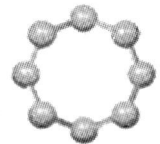

$$(\vec{\nabla} \times \vec{\nabla} \times - k^2)H = 0 \qquad ...(C.9)$$

$$(\vec{\nabla} \times \vec{\nabla} \times - k^2)\vec{E} = 0 \qquad ...(C.10)$$

As the radiation field carries away angular momentum, the solution of equations ($C.9$) and ($C.10$) should be eigen function of angular momentum operator.

The orbital angular momentum operator is defined as

$$\vec{L} = -i\hbar\vec{r} \times \vec{\nabla} \qquad ...(C.11)$$

with its z component as

$$L_z = -i\hbar\left(x\frac{\partial}{\partial y} - y\frac{\partial}{\partial x}\right) \qquad ...(C.12)$$

If solution of equations ($C.9$) and ($C.10$) is an eigen function of L_z then L_z must commute with the operator $\vec{\nabla} \times \vec{\nabla} \times$. The commutation relation gives

$$(\vec{\nabla}\times, L_z)\vec{C} = \frac{\hbar}{i}\left[\vec{\nabla} \times \left(x\frac{\partial}{\partial y} - y\frac{\partial}{\partial x}\right)(\hat{i}C_x + \hat{j}C_y + \hat{k}C_z)\right.$$

$$\left. - \left(x\frac{\partial}{\partial y} - y\frac{\partial}{\partial x}\right)\vec{\nabla} \times (\hat{i}C_x + \hat{j}C_y + \hat{k}C_z)\right] \qquad ...(C.13)$$

where $\hat{i}$, $\hat{j}$, $\hat{k}$, are the unit vectors aslong the three axis and $\vec{C}$ is some vector.

Writting $\nabla = \hat{i}\frac{\partial}{\partial x} + \hat{j}\frac{\partial}{\partial y} + \hat{k}\frac{\partial}{\partial z}$, the above expression can be expanded to give

$$\frac{\hbar}{i}(\vec{\nabla}\times, L_z)\vec{C} = \frac{\hbar}{i}\left[i\left\{\frac{\partial}{\partial y}\left(x\frac{\partial C_z}{\partial y} - y\frac{\partial C_z}{\partial x}\right) - \frac{\partial}{\partial z}\left(x\frac{\partial C_y}{\partial y} - y\frac{\partial C_y}{\partial x}\right)\right\}\right.$$

$$+j\left\{\frac{\partial}{\partial z}\left(x\frac{\partial C_x}{\partial y} - y\frac{\partial C_x}{\partial x}\right) - \frac{\partial}{\partial x}\left(x\frac{\partial C_z}{\partial y} - y\frac{\partial C_z}{\partial x}\right)\right\}$$

$$\left.+k\left\{\frac{\partial}{\partial x}\left(x\frac{\partial C_x}{\partial y} - y\frac{\partial C_y}{\partial x}\right) - \frac{\partial}{\partial y}\left(x\frac{\partial C_x}{\partial y} - y\frac{\partial C_x}{\partial x}\right)\right\}\right]$$

$$-\frac{\hbar}{i}\left[i\left\{\left(x\frac{\partial}{\partial y} - y\frac{\partial}{\partial x}\right)\left(\frac{\partial C_z}{\partial y} - \frac{\partial C_y}{\partial z}\right)\right\}\right.$$

$$+j\left\{\left(x\frac{\partial}{\partial y} - y\frac{\partial}{\partial x}\right)\left(\frac{\partial C_x}{\partial x} - \frac{\partial C_z}{\partial x}\right)\right\}$$

$$+k\left\{\left(x\frac{\partial}{\partial y}-y\frac{\partial}{\partial x}\right)\left(\frac{\partial C_x}{\partial x}-\frac{\partial C_z}{\partial x}\right)\right\}\Bigg] \qquad ...(C.14)$$

Taking the x component of the above expression we get

$$[\vec{\nabla}x, L_z]_x\vec{C} = \frac{\hbar}{i}\left[\left\{\left(x\frac{\partial^2 C_z}{\partial y^2}-\frac{\partial C_z}{\partial x}-y\frac{\partial^2 C_z}{\partial x\partial y}-x\frac{\partial^2 C_y}{\partial y\partial z}+y\frac{\partial^2 C_y}{\partial z\partial x}\right)\right\}\right.$$

$$\left.-\left\{\left(x\frac{\partial^2 C_z}{\partial y^2}-x\frac{\partial^2 C_y}{\partial y\partial z}-y\frac{\partial^2 C_z}{\partial x\partial y}+y\frac{\partial^2 C_y}{\partial z\partial x}\right)\right\}\right]$$

$$= -\frac{\hbar}{i}\frac{\partial C_z}{\partial x}$$

or $$\frac{i}{\hbar}[\vec{\nabla}x, L_z]_y\vec{C} = -\frac{\partial C_z}{\partial x} \qquad ...(C.15)$$

Similarity the y and z components of the commutation relation can be obtained as

$$\frac{i}{\hbar}[\vec{\nabla}\times, L_z]_y\vec{C} = -\frac{\partial C_z}{\partial y} \qquad ...(C.16)$$

and $$\frac{i}{\hbar}[\vec{\nabla}\times, L_z]_z\vec{C} = \frac{\partial C_x}{\partial x}+\frac{\partial C_y}{\partial y} \qquad ...(C.17)$$

combining the three components one can write

$$\frac{i}{\hbar}[\vec{\nabla}\times, L_z]\vec{C} = -\hat{i}\frac{\partial C_z}{\partial x}-\hat{j}\frac{\partial C_z}{\partial y}+\hat{k}\left(\frac{\partial C_x}{\partial x}+\frac{\partial C_y}{\partial y}\right) \qquad ...(C.18)$$

The above equation shows that L_z does not commute with the operator $\vec{\nabla}\times$. Consider an operator S_z defined as

$$\vec{S}_z = i\hbar\vec{k}\times \qquad ...(C.19)$$

so that

$$S_z\vec{C} = i\hbar\vec{k}\times\vec{C} = i\hbar(-\hat{i}C_y+\hat{j}C_x) \qquad ...(C.20)$$

Considering the commutation of S_z with $\vec{\nabla}\times$. We have

$$\frac{i}{\hbar}(\vec{\nabla}\times, S_z)\vec{C} = \vec{\nabla}\times\vec{k}\times\vec{C}-\vec{k}\times\vec{\nabla}\times\vec{C}$$

$$= \left\{\left(\hat{i}\frac{\partial}{\partial x}+\hat{j}\frac{\partial}{\partial y}+\hat{k}\frac{\partial}{\partial z}\right)\times(-\hat{i}C_y+\hat{j}C_x)\right.$$

$$-k\left(\hat{i}\frac{\partial}{\partial x}+\hat{j}\frac{\partial}{\partial y}+\hat{k}\frac{\partial}{\partial z}\right)\times(\hat{i}C_x+\hat{j}C_y+\hat{k}C_z)\Big\}$$

$$=\hat{i}\frac{\partial C_z}{\partial x}+\hat{j}\frac{\partial C_z}{\partial y}-\hat{k}\left(\frac{\partial C_x}{\partial x}+\frac{\partial C_y}{\partial y}\right) \quad ...(C.21)$$

Adding expressions (3.3.17) and (3.3.20) one gets

$$\frac{i}{\hbar}[\nabla\times,(L_z+S_z)]\vec{C}=0 \quad ...(C.22)$$

It is thus seen that the operator $L_z + S_z$ commutes with $\vec{\nabla}\times$. Defining an operator J_z as

$$J_z = L_z + S_z \quad ...(C. 23)$$

We have

$$[\vec{\nabla}\times,J_z]\vec{C}=0 \quad ...(C.24)$$

As in expression (C.19) the operators S_X and S_Y can also be defined. The operators J_X and J_Y can be obtained as

$$\frac{i}{\hbar}J_x = y\frac{\partial}{\partial z}-z\frac{\partial}{\partial y}-\hat{i}\times$$

$$\frac{i}{\hbar}J_y = z\frac{\partial}{\partial x}-x\frac{\partial}{\partial z}-\hat{j}\times \quad ...(C.25)$$

The operator J^2 is defined as

$$J^2 = J_x^2+J_y^2+J_Z^2 \quad ...(C.26)$$

It can be shown J^2 commutes with the operator $\vec{\nabla}\times$ *i.e.*

$$[J^2\vec{\nabla}\times-\vec{\nabla}\times J^2]C=0 \quad ...(C.27)$$

It is observed that both the operators J_z and J^2 commute with $\vec{\nabla}x$. One can write

$$J_z\vec{\nabla}\times\vec{\nabla}\times = \vec{\nabla}\times J_z\vec{\nabla}\times-\vec{\nabla}\times\vec{\nabla}\times J_z \quad ...(C.28)$$

and

$$J^2\vec{\nabla}\times\vec{\nabla}\times = \vec{\nabla}\times J_zJ^2 \quad ...(C.29)$$

Thus both J_Z and J^2 commute with $\vec{\nabla}\times\vec{\nabla}\times$. The solutions of equation (C.10) should therefore be eigen functions of the operators J_Z and J^2.

The operators $\vec{L}$, $\vec{S}$ and $\vec{J}$ are associated with orbital, spin and total angular momentum respectively. L_z, S_z and J_z are the components of the corresponding angular momentum in z direction.

It would be of interest to obtain some properties of spin and orbital angular momentum operators asssocoiated with electromagnetic fields $\vec{E}$ and $\vec{H}$.

(*i*) The operators S is such that

$$S^2 = S_x^2 + S_y^2 + S_z^2 \qquad ...(C.30)$$

where S_x, S_y are similar to S_Z in equation $(C.19)$.

If $\vec{C}$ is a vector then

$$S^2 \cdot \vec{C} = \left(S_x^2 + S_y^2 + S_z^2\right)\vec{C}$$

$$= \left(-\frac{\hbar}{i}\vec{i}\times\right)\left(-\frac{\hbar}{i}\vec{i}\times\right)\vec{C} + \left(-\frac{\hbar}{i}\vec{j}\times\right)\left(-\frac{\hbar}{i}\vec{j}\times\right)\vec{C}$$

$$+\left(-\frac{\hbar}{i}\vec{k}\times\right)\left(-\frac{\hbar}{i}\vec{k}\times\right)\vec{C}$$

$$= -\hbar^2\left(\vec{i}\times\vec{i}\times\vec{C}+\vec{j}\times\vec{j}\times\vec{C}+\vec{k}\times\vec{k}\times\vec{c}\right)$$

Now $\quad \vec{i}\times\vec{C} = -\vec{j}C_z + \vec{k}C_y$

$$\vec{i}\times\vec{i}\times\vec{C} = -\vec{j}C_y - \vec{k}C_z$$

Similarly

$$\vec{j}\times\vec{j}\times\vec{C} = -\vec{i}C_x - \vec{k}C_z$$

$$\vec{k}\times\vec{k}\times\vec{C} = -\vec{i}C_x - \vec{j}C_y$$

so that

$$S^2\vec{C} = 2\hbar^2(\vec{i}C_x + \vec{y}C_y + \vec{k}C_z)$$

$$= 2\hbar^2\vec{C}$$

Quantum mechanically this can be written as

$$S^2\vec{C} = \hbar^2 s(s+1)\vec{C} \qquad ...(C.32)$$

where, $s = 1$

This shows that the spin angular momentum quantum number associated with the electromagnetic wave is $s = 1$.

(*ii*) We have

$$S_z\vec{C} = i\hbar\vec{k}\times\vec{C} = i\hbar(-\vec{i}C_y + \vec{j}C_x)$$

and $\quad S_z^2\vec{C} = -\hbar^2\vec{k}\times\vec{k}\times\vec{C} = \hbar^2(\vec{i}\,C_x + \vec{j}C_y)$

so that $\quad S_z(S_z^2\vec{C}) = i\hbar^3\vec{k}\times(\vec{i}\,C_x + \vec{j}C_y)$

$$= i\hbar^3(\vec{j}C_x - \vec{i}\,C_y)$$

$$= \hbar^2(S_z\vec{C})$$

This gives

$$S_z(S_z^2\vec{C}) - \hbar^2 S_z\vec{c} = 0$$

or $\quad S_z(S_z^2 - \hbar^2)\vec{C} = 0 \qquad \ldots(C.33)$

If $\vec{C}$ is not zero then either $S_z = 0$ or $S_z^2 = \hbar^2$ or $S_z = \pm\,\hbar$. This shows that the z component of the spin angular momentum of the radiation field could have value $S_z = 0$ or $\pm\, 1\hbar$.

(*iii*) It can be shown that

$$J_z\vec{L} = \vec{L}\cdot L_z \qquad \ldots(C.34)$$

We have

$$J_z\vec{L} = L_z\vec{L} + S_z\vec{L}$$

Now $\quad L_z\,L_x - L_x\,L_z = i\hbar L_y$

and $\quad (S_z\vec{L})_x = i\hbar(\vec{k}\times\vec{L})_x$

$$= -\hbar L_y$$

$$L_zL_x - L_xL_z + (S_zL)_x = 0$$

or $\quad L_zL_x + (S_zL)_x = L_xL_z$

Thus $\quad (J_zL)_x = L_xL_z$

Similarly

$$(J_zL)_Y = L_YL_z$$

and $\quad (J_zL)_z = L_xL_z$

so that $\quad J_z\vec{L} = \vec{L}\cdot L_z$

(*iv*) We have

$$J^2 = (\vec{L} + \vec{S})^2$$

$$= L^2 + 2\vec{L}\cdot\vec{S} + S^2$$

This gives

$$J^2\vec{L} - \vec{L}L^2 = (L^2 + 2\vec{L}\cdot\vec{S} + S^2)\vec{L} - \vec{L}L^2$$

$$J^2\vec{L} - \vec{L}L^2 = (L^2\vec{L} - \vec{L}L^2) + 2\vec{L}\cdot\vec{S}\vec{L} + S^2\vec{L}$$

Now $\quad S^2L = \vec{S}\vec{L}\vec{S} = \vec{L}S^2 = L\,2\hbar^2L$

and $\quad 2\vec{L}\cdot\vec{S}\vec{L} = 2(L_xS_x + L_YS_Y + L_zS_z)\vec{L}$

$$= 2i\hbar\,(L_x\vec{i}\times + L_Y\vec{j}\times + L_z\vec{k}\times)\vec{L}$$

$$= 2i\hbar\,(-L_xL_z\vec{j} + L_xL_Y\vec{k} + L_YL_Z\vec{i} - L_YL_x\vec{k} - L_zL_y\vec{i} + L_zL_x\vec{j})$$

$$= 2i\hbar[\vec{i}(L_YL_z - L_zL_Y) + \vec{j}(L_zL_x - L_xL_z) + \vec{k}(L_xL_Y - L_YL_x)]$$

$$= 2i\hbar\cdot i\hbar(\vec{i}L_x + \vec{j}L_Y + \vec{k}L_z)$$

$$= -2\hbar^2\vec{L}$$

Thus we have

$$J^2L = \vec{L}L^2 \qquad \text{...(C.35)}$$

The following properties of the operators $\vec{S}$ and $\vec{L}$ can be shown in a similar way:

(v) $\quad \vec{S}\times\vec{S} = \vec{i}\hbar\vec{s}$ ...(C.36)

(vi) $\quad S^2S_z - S_zS^2 = 0$ ...(C.37)

(vii) $\quad L^2S_z - S_z L^2 = 0$ ...(C.38)

(viii) $\quad L_z\vec{S} - \vec{S}L_z = 0$ and $\vec{L}\cdot\vec{S} - \vec{S}\vec{L} = 0$ ...(C.39)

(ix) $\quad L^2L_z - L_z L^2 = 0$ and $L^2\vec{L} - \vec{L}L^2 = 0$ ...(C.40)

(x) $\quad J^2J_z - J_zJ^2 = 0$ ...(C.41)

It has been shown above that both J_z and J^2 commute with $\vec{\nabla}\times\vec{\nabla}\times$, hence an eigen function of these operators will represent the solution of equation 3.10.8. Consider an eigen function B such that

$$\vec{B} = \vec{L}R_L(r)Y_{LM}(\theta,\phi) \qquad \text{...(C.42)}$$

It can be seen from equations (3.10.23) that

$$J_z\vec{B} = J_z\vec{L}R_L(r)Y_{LM}(\theta,\phi)$$

$$= \vec{L}L_zR_L(r)Y_{LM}(\theta,\phi) \qquad \text{from } (iii)$$

$$= M\hbar\vec{L}R_L(r)Y_{LM}(\theta,\phi)$$

$$= M\hbar\vec{B}$$

where, $L_z = M\hbar$

Similarly

$$J^2\vec{B} = J^2\vec{L}R_L(r)\,Y_{LM}(\theta,\phi)$$

$$= \vec{L}L^2R_L(r)\,Y_{LM}(\theta,\phi) \qquad \text{from (C.35)}$$

$$= \hbar^2(L\,(1+l))\,\vec{B}$$

The function $\vec{B}$ is thus an eigen function of both J^2 and J_z, and can represent $\vec{E}$ and $\vec{H}$ of equation ($C.9$) and ($C.10$). In a free space $\vec{\nabla}\cdot\vec{E}$ as well as $\vec{\nabla}\cdot B$ are zero. Putting this condition one gets

$$(\nabla^2 - k^2)B = 0 \qquad \text{...(C.44)}$$

This equation is similar to the Schrodinger equation and writing ∇^2 in polar coordinates one can separate the variables. The resulting equation of the radial wave function $R(r)$ is given as

$$\left(\frac{d^2}{dr^2} - \frac{L(L+1)}{r^2} + k^2\right)R_L(r) = 0 \qquad \text{...(C.45)}$$

The solution of the above equation is a Bessel function. The vector B can thus be wrtten as

$$\vec{B} = C\vec{L}j_L(kr)Y_{LM}(\theta,\phi) \qquad \text{...(C.46)}$$

C is the constant whose value is obtained by normalisation.

Vector $\vec{B}$ satisfies both equations($C.9$) and ($C.10$) and can be chosen to represent either the magnetic field or the electric field. If $\vec{B}$ is chosen to represent the magnetic field vector H^e_{LM} then

$$H^e_{LM} = C^e\vec{L}J_L(kr)Y_{LM}(\theta,\phi) \qquad \text{...(C.47)}$$

The radial component of the field H^e_{LM} is zero. The radiation field in which the radial component of the magnetic field is zero is called electric multipole field and specified by superscript e. The corresponding electric field is obtained as

$$E^e_{LM} = -\frac{1}{iw\varepsilon_0}\vec{\nabla}\times\vec{H}^e_{LM}$$

$$= -\frac{C^e}{iw\varepsilon_0}\vec{\nabla}\times\vec{L}j_L(kr)Y_{LM}(\theta,\phi) \qquad \text{...(C.48)}$$

If vector $\vec{B}$ represents the electric field then

$$E^m_{LM} = C^m\vec{L}j_L(kr)Y_{LM}(\theta,\phi) \qquad \text{...(C.49)}$$

and the corresponding magnetic field is given as

$$H_{LM}^{m} = \frac{C^m}{i w \mu_0} \vec{\nabla} \times \vec{L} j_L(kr) Y_{LM}(\theta, \phi) \quad \text{...(C.50)}$$

It is observed that in equation (C.49) the radial component of the electric field is zero. Such fields are called magnetic multipole fields and denoted by superscript m. The constants C^e and C^m are obtained by normalising the radiation field such that after the emission of a photon, the field energy is $\hbar w$. In an electromegnetic field half the energy is electric and other half is magnetic so that

$$\varepsilon_0 |\vec{E}|^2 = \mu_0 |H|^2 = \frac{\hbar w}{2} = \frac{\hbar c k}{2} \quad \text{...(C.51)}$$

A single frequency of a wave is obtained if it is enclosed in a large sphere of radius R_0 ($R_0 \to \infty$) so that $kR_0 >> 1$. At $r = R_0$ the radiation field is zero so that

$$j_L(kR_0) = 0$$

Putting the value of $\vec{H}$ in equation (C.51) gives

$$\frac{\hbar c k}{2\mu_0} = C^{e^2} \int_0^{R_0} | j_L(kr) |^2 r^2 dr \int (LY_{LM})^* LY_{LM} d\Omega$$

since $\vec{L}$ is Hermitian, we have

$$\int (\vec{L}Y_{LM})^* \vec{L}Y_{LM} d\Omega = \int Y_{LM}^* \vec{L}\vec{L} Y_{LM} d\Omega$$

$$= \int Y_{LM}^* L^2 Y_{LM} d\Omega$$

$$= L(L+1)\hbar^2 \quad \text{...(C. 52)}$$

For large values of r the Bessel Function

$$j_L(kr) \simeq \frac{1}{kr} \cos\left(kr - \frac{\pi}{2}(L+1)\right)$$

so that

$$\int_0^{R_0} | j_L(kr) |^2 r^2 dr \simeq \frac{1}{k^2} \int_0^{R_0} \cos\left(kr - \frac{\pi}{2}(L+1)\right) dr$$

$$\simeq \frac{R_0}{2k^2} \quad \text{...(C. 53)}$$

Substituting the values from equation (C.53) in equation (C.52) gives

$$C^{e^2} = \frac{ck^3}{\mu_0 R_0 \hbar L(L+1)} \quad \text{...(C.54)}$$

In a similar fashion the value of $(C^m)^2$ is calculated as

$$(c^m)^2 = \frac{ck^3}{\in_0 R_0 \hbar L(L+1)} \qquad \ldots(C.55)$$

Putting the value of c^e and c^m the electric and magnetic multipole fields are given as

$$\vec{E}^e_{LM} = i\sqrt{\frac{ck}{\in_0 R_0 \hbar L(L+1)}}\, \vec{\nabla} x \vec{L} j_L(kr) Y_{LM}(\theta, \phi) \qquad \ldots(C.56)$$

$$\vec{H}^e_{LM} = k\sqrt{\frac{ck}{\mu_0 R_0 \hbar L(L+1)}}\, \vec{L} j_L(kr) Y_{LM}(\theta, \phi) \qquad \ldots(C.57)$$

$$\vec{E}^m_{LM} = k\sqrt{\frac{c \cdot k}{\in_0 R_0 \hbar L(L+1)}}\, \vec{L} j_L(kr) Y_{LM}(\theta, \phi) \qquad \ldots(C.58)$$

$$\vec{H}^m_{LM} = -i\sqrt{\frac{c \cdot k}{\mu_0 R_0 \hbar L(L+1)}}\, \overline{\nabla} \times \vec{L}\, j_L(kr) Y_{LM}(\theta, \phi) \qquad \ldots(C.59)$$

The above equations give the electric and magnetic fields in the radiation field of different multipolarities.

BIBLIOGRAPHY

1. W. Heitler. *The Quantum Theory of Radiations.* Oxford University Press, 1935.
2. B. B. Rossi and Hans H. Staub. *Ionization Chambers and Counters.* McGraw-Hill Book Co., 1949.
3. N. F. Mott and H. B. W. Massey. *The Theory of Atomic Collisions.* Oxford University Press, New York, 1949.
4. E. Fermi. *Nuclear Physics.* The University of Chicago Press, 1950.
5. D. Halliday. *Introductory Nuclear Physics.* John Wiley & Sons, 1950.
6. Glasstone, Samuel and M. C. Edlund. *The Elements of Nuclear Reatcor Theory.* Van Nostrand Princeton, 1952.
7. John M. Blatt and V. F. Weisskopf. *Theoretical Nuclear Physics.* John Wiley & Sons. New York, 1952.
8. Donald J. Hughes. *Pile Neutron Research.* Addison Wesly Publishing Co. Cambridge, 1953.
9. J. Sharpe. *Nuclear Radiation Detectors.* John Wiley & Sons. New York, 1955.
10. R. D. Evans. *The Atomic Nucleus.* Tata McGraw Hill Publishing Co, 1955.
11. L. I. Schiff. *Quantum Mechanics.* McGraw Hill Publishing Co., 1955.
12. M. G. Mayer and J. H. D. Jensen. *Elementary Theory of Nuclear Shell Structure.* John Wiley, New York, 1955.
13. A. E. S. Green. *Nuclear Physics.* McGraw Hill Book Co. New York, 1955.
14. H. A. Bethe and Philip Morrison. *Elementary Nuclear Theory.* John Wiley New York, 1956.
15. N. F. Ramsay. *Molecular Beams.* Oxford University Press, New York, 1956.
16. Kopferman, Hans. *Nuclear Moments.* Academic Press, New York, 1958.
17. L. R. B. Elton. *Introductory Nuclear Theory.* Wiley Interscience New York, 1959.
18. A. M. Weinberg and E. Wigner. *The Physical Theory of Neutron Chain Reactors.* University of Chicago Press, Chicago, 1959.
19. A. W. Wapstra, G. J. Nijgh and R. Van Lieshout. *Nuclear Spectroscopy Tables.* North Holland Publishing Co, Amsterdam, 1959.
20. Ajzenberg-Selove, Fay. *Nuclear Spectroscopy,* Part A and B. Academic Press, New York, 1960.

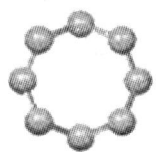

21. W. Tabocman. *Theory of Direct Nuclear Reactions.* Oxford University Press, 1961.
22. L. C. L. Yuan and C. S. Wu. *Methods of Experimental Physics*, Vol. 1-6. Academic Press New York, 1961.
23. E. Segre. *Experimental Nuclear Physics*, Vol. 1-3. Jon Wiley New York, 1962.
24. M. A. Preston. *Physics of Nucleus.* Addison Wesley Reading Mass, 1962.
25. P. M. Endt and P. B. Smith. *Nuclear Reactions.* North Holland Publishing Co, Amsterdam, 1962.
26. M. S. Livingston and J. P. Blewett. *Particle Accelerators.* McGraw Hill, New York, 1962.
27. Ernst Bleuler and George Goudsmith. *Experimental Nucleonics.* Holt Reinhart and Winsten, New York, 1963.
28. A. De Shalit and I. Talmi. *Nuclear Shell Theory.* Academic Press, 1963.
29. J. B. Birks. *The Theory and Practice of Scintillation Counting.* Pergamon Press, Oxford, 1964.
30. S. De Benedetti. *Nuclear Interactions.* John Wiley, New York, 1964.
31. K. Seighbahn. *Alpha Beta and Gamma Ray Spectroscopy.* North Holland Publishing Co, 1965.
32. E. Segre. *Nuclei and Particles.* W. A. Benjamin Inc., 1965.
33. W. J. Price. Nuclear Radiation Detection. McGraw Hill Book Co., New York, 1966.
34. G. Dearnally and D. C. Northrop. *Seminconductor Counters for Nuclear Radiations.* E. F. N. Spon London, 1966.
35. R. R. Roy and B. P. Nigam. *Nuclear Physics.* John Wiley & Sons Inc, New York, 1967.
36. G. E. Brown. *Unified Theory of Nuclear Models and Nuclear Forces.* North Holland Publishing Co. Amsterdam, 1967.
37. K. Kikuchi and M. Kawai. *Nuclear matter and Nuclear Reactions.* North Holland Publishing Co., Amsterdam, 1968.
38. Haro Von Butler. *Nuclear Physics an Introduction.* Academic Press, 1968.
39. P. Marmier and E. Sheldon. *Physics of Nuclei and Particles.* Academic Press, 1969.
40. S. M. Size. *Physics of Semiconductor Devices.* John Wiley & Sons, New York, 1969.
41. D. F. Jackson. *Nuclear Reactions.* Methuen and Co., 1970.
42. *Radiological Health Handbook.* U.S. Dept. of Health, Education and Welfare, 1970.
43. A. Martin and S. Harbison. *An Introduction to Radiation Protection.* Chapman and Hall Ltd., London, 1972.
44. F. S. Goulding and R. H. Pehl. *Nuclear Spectroscopy and Reactions.* Part A Ed. J. Cherney Academic Press, New York, 1974.
45. C. M. Lederer and V. S. Shirley. *Table of Isotopes.* Wiley Interscience New York, 1978.
46. J. Shapiro. *Radiation Protection.* Harvard University Press, Cambridge, Mass. 1981.
47. Yeshwant R. Wagmare. *Introductory Nuclear Physics.* Oxford and I.B.H. Publishing Co., New Delhi, 1981.
48. M.K. Pal. *Theory of Nuclear Structure.* Affiliated East-West Press, New Delhi, 1982.

49. J. E. Coggle. *Biological Effects of Radiations.* Taylor and Francias Ltd. London, 1983.
50. S. S. Kapoor and V. S. Krishnamurthy. *Nuclear Radiation Detection.* Wiley Eastern Ltd. 1986.
51. W. E. Cottingham and D. A. Greenwood. *An Introduction to Nuclear Physics.* Cambridge University Press, 1986.
52. K. N. Mukhin. *Experimental Nuclear Physics.* Mir Publications, Moscow, 1987.
53. Glen F. Knoll. *Radiation Detection and Measurement.* John Wiley, 1989.
54. William R. Leo. *Technique for Nuclear and Particle Physics Experiments.* Narosa Publishing House, 1995.
55. M. G. Mayer and JHD Jensen. *Elementary Theory of Nuclear Shell Structure.* John Wiley, New York, 1948.
56. Bohr. A. and Mottelson B. *Nuclear Structure* Vol. II. Benjamin. 1975.
57. *The Electromagnetic Interaction in Nuclear Spectroscopy.* North-Holland, By Hamilton W. D. (ed.). 1975.
58. *Nuclear Collective Motion*, Methuen, by Rowe D. J. 1970.
59. R.D. Lawson. *Theory of Nuclear Shell Model.* Clarondon Press, Oxford, 1980.
60. M. Stanley Livingston and John P. Blewett. *Particle Accelerators.* McGraw Hill, New York, 1962.
61. Pierre M. Lapostolle and Albert L. Septier (eds). *Linear Accelerators.* North Holland, Amsterdam, 1970
62. D. Allan Browley (ed). *Large Slectrostative Accelerators.* 1974.
63. IEEE Trans. *Nucle. Sci.*, NS-**32**: 2727-29 (1985)
64. Procedins of the XI International conference on cyclotrons and their Applications, Tokyo, Japan, 1986.
65. Hand book of Synchrotron Radiation, Vol. 2, G.V. Marr (ed). New York Holland, Amsterdam, 1987.
66. W. E. Burcham. *Nuclear Physics: An Introduction.* Longmans Green and Co. LTD. 1963.

INDEX

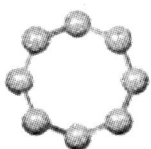

C

D

E